Set Theory

Centre de Recerca Matemàtica
Barcelona, 2003–2004

Joan Bagaria
Stevo Todorcevic
Editors

Birkhäuser Verlag
Basel · Boston · Berlin

Editors:

Joan Bagaria
Departament de Lògica,
Història i Filosofia de la Ciència
Universitat de Barcelona
Baldiri Reixac s/n
08028 Barcelona, Spain
e-mail: bagaria@ub.edu

Stevo Todorcevic
Department of Mathematics
University of Toronto
Toronto, M5S 2E4, Canada
e-mail: stevo@math.toronto.edu

and

Université Paris 7
C.N.R.S., UMR 7056
2, Place Jussieu
Case 7012
75251 Paris Cedex 05, France

2000 Mathematical Subject Classification 03Exx, O3E05, 03E15, 03E30, 03E35, 03E50, 03E55, 03E75

A CIP catalogue record for this book is available from the Library of Congress, Washington D.C., USA

Bibliographic information published by Die Deutsche Bibliothek
Die Deutsche Bibliothek lists this publication in the Deutsche Nationalbibliografie; detailed bibliographic data is available in the Internet at http://dnb.ddb.de

ISBN 3-7643-7691-0 Birkhäuser Verlag, Basel – Boston – Berlin

Part of Springer Science+Business Media
Printed on acid-free paper produced from chlorine-free pulp. TCF ∞
Printed in Germany
ISBN-10: 3-7643-7691-0
ISBN-13: 976-3-7643-7691-8
e-ISBN: 3-7643-7692-9

9 8 7 6 5 4 3 2 1

www.birkhauser.ch

Contents

Foreword

This is a collection of articles on set theory written by some of the participants in the Research Programme on Set Theory and its Applications that took place at the *Centre de Recerca Matemàtica* (CRM) in Bellaterra (Barcelona). The Programme run from September 2003 to July 2004 and included an international conference on set theory in September 2003, an advanced course on Ramsey methods in analysis* in January 2004, and a joint CRM-ICREA workshop on the foundations of set theory in June 2004, the latter held in Barcelona. A total of 33 short and long term visitors from 15 countries participated in the Programme.

This volume consists of two parts, the first containing survey papers on some of the mainstream areas of set theory, and the second containing original research papers. All of them are authored by visitors who took part in the set theory Programme or by participants in the Programme's activities.

The survey papers cover topics as Omega-logic, applications of set theory to lattice theory and Boolean algebras, real-valued measurable cardinals, complexity of sets and relations in continuum theory, weak subsystems of axiomatic set theory, definable versions of large cardinals, and selection theory for open covers of topological spaces.

As for the research papers, they range from topics such as the number of near-coherence classes of ultrafilters, the consistency strength of bounded forcing axioms, $\mathcal{P}_\kappa(\lambda)$ combinatorics, some applications of morasses, subgroups of Abelian Polish groups, adding club subsets of ω_2 with finite conditions, the consistency strength of mutual stationarity, and new axioms of set theory.

We would like to thank all participants in the Programme and its related activities for their effort in making these very successful ventures. We also want to thank the CRM Director and its staff, as well as all the funding institutions, for making the Programme possible.

Joan Bagaria
Stevo Todorcevic
Editors

---

*The volume: Spiros A. Argyros and Stevo Todorcevic, *Ramsey Methods in Analysis*. Advanced Courses in Mathematics CRM Barcelona. Birkhäuser 2005, contains the notes of the course in an expanded form.

Set Theory
Trends in Mathematics, 1–28

An Ω-logic Primer

Joan Bagaria, Neus Castells and Paul Larson

Abstract. In [12], Hugh Woodin introduced Ω-logic, an approach to truth in the universe of sets inspired by recent work in large cardinals. Expository accounts of Ω-logic appear in [13, 14, 1, 15, 16, 17]. In this paper we present proofs of some elementary facts about Ω-logic, relative to the published literature, leading up to the generic invariance of Ω-logic and the Ω-conjecture.

Keywords. Ω-logic, Woodin cardinals, A-closed sets, universally Baire sets, Ω-conjecture.

Introduction

One family of results in modern set theory, called *absoluteness results*, shows that the existence of certain large cardinals implies that the truth values of certain sentences cannot be changed by forcing[1]. Another family of results shows that large cardinals imply that certain definable sets of reals satisfy certain regularity properties, which in turn implies the existence of models satisfying other large cardinal properties. Results of the first type suggest a logic in which statements are said to be valid if they hold in every forcing extension. With some technical modifications, this is Woodin's Ω-logic, which first appeared in [12]. Results of the second type suggest that there should be a sort of internal characterization of validity in Ω-logic. Woodin has proposed such a characterization, and the conjecture that it succeeds is called the Ω-conjecture. Several expository papers on Ω-logic and the Ω-conjecture have been published [1, 13, 14, 15, 16, 17]. Here we briefly discuss

The first author was partially supported by the research projects BFM2002-03236 of the *Ministerio de Ciencia y Tecnología*, and 2002SGR 00126 of the *Generalitat de Catalunya*. The third author was partially supported by NSF Grant DMS-0401603. This paper was written during the third author's stay at the *Centre de Recerca Matemàtica* (CRM), whose support under a Mobility Fellowship of the *Ministerio de Educación, Cultura y Deportes* is gratefully acknowledged. It was finally completed during the first and third authors' stay at the Institute for Mathematical Sciences, National University of Singapore, in July 2005.

[1]Throughout this paper, by "forcing" we mean "set forcing".

the technical background of Ω-logic, and prove some of the basic theorems in this area.

This paper assumes a basic knowledge of Set Theory, including constructibility and forcing. All undefined notions can be found in [4].

1. $\models_\Omega$

1.1. Preliminaries

Given a complete Boolean algebra $\mathbb{B}$ in V, we can define the *Boolean-valued model* $V^{\mathbb{B}}$ by recursion on the class of ordinals On:

$$V_0^{\mathbb{B}} = \emptyset$$
$$V_\lambda^{\mathbb{B}} = \bigcup_{\beta<\lambda} V_\beta^{\mathbb{B}}, \text{ if } \lambda \text{ is a limit ordinal}$$
$$V_{\alpha+1}^{\mathbb{B}} = \{f\colon X \to \mathbb{B} \mid X \subseteq V_\alpha^{\mathbb{B}}\},$$

Then, $V^{\mathbb{B}} = \bigcup_{\alpha\in On} V_\alpha^{\mathbb{B}}$. The elements of $V^{\mathbb{B}}$ are called $\mathbb{B}$-*names.* Every element x of V has a *standard* $\mathbb{B}$-*name* $\check{x}$, defined inductively by: $\check{\emptyset} = \emptyset$, and $\check{x} : \{\check{y} : y \in x\} \to \{\mathbf{1}^{\mathbb{B}}\}$.

For each $x \in V^{\mathbb{B}}$, let $\rho(x) = \min\{\alpha \in On \mid x \in V_{\alpha+1}^{\mathbb{B}}\}$, the *rank* of x in $V^{\mathbb{B}}$.

Given φ, a formula of the language of set theory with parameters in $V^{\mathbb{B}}$, we say that φ is *true* in $V^{\mathbb{B}}$ if its Boolean-value is $1^{\mathbb{B}}$, i.e.,

$$V^{\mathbb{B}} \models \varphi \quad \text{iff} \quad [\![\varphi]\!]^{\mathbb{B}} = 1^{\mathbb{B}},$$

where $[\![\cdot]\!]^{\mathbb{B}}$ is defined by induction on pairs $(\rho(x), \rho(y))$, under the canonical well-ordering of pairs of ordinals, and the complexity of formulas (see [4]).
$V^{\mathbb{B}}$ can be thought of as constructed by iterating the $\mathbb{B}$-valued power-set operation. Modulo the equivalence relation given by $[\![x = y]\!]^{\mathbb{B}} = 1^{\mathbb{B}}$, $V_\alpha^{\mathbb{B}}$ is precisely V_α in the sense of the Boolean-valued model $V^{\mathbb{B}}$ (see [4]):

Proposition 1.1. *For every ordinal α, and every complete Boolean algebra $\mathbb{B}$, $V_\alpha^{\mathbb{B}} \equiv (V_{\check{\alpha}})^{V^{\mathbb{B}}}$, i.e., for every $x \in V^{\mathbb{B}}$,*

$$(\exists y \in V_\alpha^{\mathbb{B}}\, [\![x = y]\!]^{\mathbb{B}} = 1^{\mathbb{B}}) \quad \textit{iff} \quad [\![x \in V_{\check{\alpha}}]\!]^{\mathbb{B}} = 1^{\mathbb{B}}.$$

Corollary 1.2. *For every ordinal α, and every complete Boolean algebra $\mathbb{B}$,*

$$V_\alpha^{\mathbb{B}} \models \varphi \quad \textit{iff} \quad V^{\mathbb{B}} \models \text{“}V_{\check{\alpha}} \models \varphi\text{”}.$$

Notation

i) If $\mathbb{P}$ is a partial ordering, then we write $V^{\mathbb{P}}$ for $V^{\mathbb{B}}$, where $\mathbb{B} = r.o.(\mathbb{P})$ is the *regular open completion* of $\mathbb{P}$ (see [4]).
ii) Given M a model of set theory, we will write M_α for $(V_\alpha)^M$ and $M_\alpha^{\mathbb{B}}$ for $(V_\alpha^{\mathbb{B}})^M = (V_\alpha)^{M^{\mathbb{B}}}$.
iii) *Sent* will denote the set of sentences in the first-order language of set theory.

iv) $T \cup \{\varphi\}$ will always be a set of sentences in the language of set theory, usually extending ZFC.
v) We will write c.t.m. for *countable transitive* $\in$*-model.*
vi) We will write c.B.a. for *complete Boolean algebra.*
vii) For $A \subseteq \mathbb{R}$, we write $L(A, \mathbb{R})$ for $L(\{A\} \cup \mathbb{R})$, the smallest transitive model of ZF that contains all the ordinals, A, and all the reals.

As usual, a *real number* will be an element of the Baire space $\mathcal{N} = (\omega^\omega, \tau)$, where τ is the product topology, with the discrete topology on ω. Thus, the set $\mathbb{R}$ of real numbers is the set of all functions from ω into ω. Throughout this paper, we often talk in terms of generic filters instead of Boolean-valued models. Each way of talking can be routinely reinterpreted in the other.

Let $\mathbb{P}$ be a forcing notion. We say that $\dot{x}$ is a *simple* $\mathbb{P}$*-name* for a real number if:

i) The elements of $\dot{x}$ have the form $((\check{n, m}), p)$ with $p \in \mathbb{P}$ and $n, m \in \omega$, so that $p \Vdash_{\mathbb{P}} \dot{x}(\check{n}) = \check{m}$.
ii) For all $n \in \omega$, $\{p \in \mathbb{P} \mid \exists m$ such that $((\check{n, m}), p) \in \dot{x}\}$ is a maximal antichain of $\mathbb{P}$.

For any forcing notion $\mathbb{P}$ and for all $\mathbb{P}$-names τ for a real, there exists a simple $\mathbb{P}$-name $\dot{x}$ such that $\Vdash_{\mathbb{P}} \tau = \dot{x}$. Hence, any $\mathbb{P}$-generic filter will interpret these two names in the same way.

Let $WF := \{x \in \omega^\omega \mid E_x$ is well founded$\}$, where given $x \in \omega^\omega$, $E_x := \{(n, m) \in \omega \times \omega \mid x(\Gamma(n, m)) = 0\}$, with Γ some fixed recursive bijection between $\omega \times \omega$ and ω. Recall that WF is a complete Π^1_1 set (see [4]).

Let T be a theory whose models naturally contain a submodel N of Peano Arithmetic. A model M of T is an *ω-model* if N^M is standard, i.e., it is isomorphic to ω. In this case, we naturally identify M with its isomorphic copy M' in which $N^{M'}$ is ω.

Stationary Tower Forcing, introduced by Woodin in the 1980's, will be used to prove some important facts about Ω-logic:

Definition 1.3. (cf. [6]) (Stationary Tower Forcing)

i) A set $a \neq \emptyset$ is *stationary* if for any function $F\colon [\cup a]^{<\omega} \to \cup a$, there exists $b \in a$ such that $F''[b]^{<\omega} \subseteq b$.
ii) Given a strongly inaccessible cardinal κ, we define the *Stationary Tower Forcing* notion: its set of conditions is

$$\mathbb{P}_{<\kappa} = \{a \in V_\kappa \mid a \text{ is stationary}\},$$

and the order is defined by:

$$a \leq b \text{ iff } \cup b \subseteq \cup a \text{ and } \{Z \cap (\cup b) \mid Z \in a\} \subseteq b.$$

Fact 1.4. Given $\gamma < \delta$ strongly inaccessible, $a = \mathcal{P}_{\omega_1}(V_\gamma) \in \mathbb{P}_{<\delta}$.

Proof. Given $F\colon [V_\gamma]^{<\omega} \to V_\gamma$, let $x \in [V_\gamma]^{<\omega}$ and let:

$$A_0 = x, \quad A_{n+1} = A_n \cup \{F(y) \mid y \in [A_n]^{<\omega}\}$$

Let $b = \bigcup_{n\in\omega} A_n$. So, $b \in \mathcal{P}_{\omega_1}(V_\gamma)$ and $F''[b]^{<\omega} \subseteq b$. □

Recall the large-cardinal notion of a *Woodin cardinal*:

Definition 1.5. ([10]) A cardinal δ is a *Woodin cardinal* if for every function $f\colon \delta \to \delta$ there exists $\kappa < \delta$ with $f''\kappa \subseteq \kappa$, and an elementary embedding $j\colon V \to M$ with critical point κ such that $V_{j(f)(\kappa)} \subseteq M$.

Theorem 1.6. (*cf.* [6]) *Suppose that δ is a Woodin cardinal and that $G \subseteq \mathbb{P}_{<\delta}$ is a V-generic filter. Then in $V[G]$ there is an elementary embedding $j\colon V \to M$, with M transitive, such that $V[G] \models M^{<\delta} \subseteq M$ and $j(\delta) = \delta$. Moreover, for all $a \in \mathbb{P}_{<\delta}$, $a \in G$ iff $j''\cup a \in j(a)$.*

1.2. Definition of $\models_\Omega$ and invariance under forcing

Definition 1.7. ([17]) For $T \cup \{\varphi\} \subseteq Sent$, let

$$T \models_\Omega \varphi$$

if for all c.B.a. $\mathbb{B}$, and for all ordinals α, if $V_\alpha^{\mathbb{B}} \models T$ then $V_\alpha^{\mathbb{B}} \models \varphi$.

If $T \models_\Omega \varphi$, we say that φ is *Ω_T-valid*, or that φ is *Ω-valid from T*.

Observe that the complexity of the relation $T \models_\Omega \varphi$ is at most Π_2. Indeed, $T \models_\Omega \varphi$ iff

$$\forall \mathbb{B} \forall \alpha (\mathbb{B} \text{ a c.B.a.} \wedge \alpha \in On \to (V_\alpha^{\mathbb{B}} \models T \to V_\alpha^{\mathbb{B}} \models \varphi))$$

The displayed formula is Π_2, since to be a c.B.a. is Π_1 and the class function $\alpha \mapsto V_\alpha^{\mathbb{B}}$ is Δ_2 definable (i.e., both Σ_2 and Π_2 definable) in V with $\mathbb{B}$ as a parameter.

Clearly, if $T \models \varphi$ then $T \models_\Omega \varphi$. Observe, however, that the converse is not true. Indeed, we can easily find Ω_{ZFC}-valid sentences that are undecidable in first-order logic from ZFC, i.e., sentences φ such that $ZFC \not\models \varphi$ and $ZFC \not\models \neg\varphi$. For example, $CON(ZFC)$: For all $\alpha \in On$ and all c.B.a. $\mathbb{B}$, if $V_\alpha^{\mathbb{B}} \models ZFC$, then since $V_\alpha^{\mathbb{B}}$ is a standard model of ZFC, we have $V_\alpha^{\mathbb{B}} \models CON(ZFC)$.

Under large cardinals, the relation $\models_\Omega$ is absolute under forcing extensions:

Theorem 1.8. ([17]) *Suppose that there exists a proper class of Woodin cardinals. If $T \cup \{\varphi\} \subseteq Sent$, then for every forcing notion $\mathbb{P}$,*

$$T \models_\Omega \varphi \quad \text{iff} \quad V^{\mathbb{P}} \models \text{“}T \models_\Omega \varphi\text{”}$$

Proof. $\Rightarrow$) Let $\mathbb{P}$ be a poset. Suppose $\check{\beta}, \dot{\mathbb{Q}} \in V^{\mathbb{P}}$ are such that $V^{\mathbb{P}} \models \text{“}V_{\check{\beta}}^{\dot{\mathbb{Q}}} \models T\text{”}$. By Corollary 1.2, $V^{\mathbb{P}*\dot{\mathbb{Q}}} \models \text{“}V_{\check{\beta}} \models T\text{”}$. By hypothesis, $V^{\mathbb{P}*\dot{\mathbb{Q}}} \models \text{“}V_{\check{\beta}} \models \varphi\text{”}$, and hence $V^{\mathbb{P}} \models \text{“}V_{\check{\beta}}^{\dot{\mathbb{Q}}} \models \varphi\text{”}$.

$\Leftarrow$) Suppose $V^{\mathbb{P}} \models$ "$T \models_\Omega \varphi$". Let $\mathbb{Q}$ be a forcing notion and $\alpha \in On$. Suppose that $V_\alpha^{\mathbb{Q}} \models T$ and G is a V-generic filter for $\mathbb{Q}$. In $V[G]$, let $\kappa = |TC(\mathbb{P})|$, and let $\delta > \kappa, \alpha$ be a Woodin cardinal. Let

$$a = \{X \mid X \prec H_{\kappa^+} \text{ and } X \text{ countable}\}.$$

Notice that $a \in \mathbb{P}_{<\delta}^{V[G]}$. Let $\mathbb{P}_{<\delta}^{V[G]}(a)$ be the forcing $\mathbb{P}_{<\delta}^{V[G]}$ restricted to a.

Let $I \subseteq \mathbb{P}_{<\delta}^{V[G]}(a)$ be a $V[G]$-generic filter. In $V[G][I]$ there is an elementary embedding $j\colon V[G] \to M$ with M transitive such that:

i) $V[G][I] \models M^{<\delta} \subseteq M$,

ii) $(H_{\kappa^+})^V$ is countable in M and $j(\alpha) < \delta$. (See Theorem 1.6.)

$\mathbb{P} \in M$ and the set of dense subsets of $\mathbb{P}$ in V is a countable set in M, so in M there exists a V-generic filter $J \subseteq \mathbb{P}$. Then $V[J] \subseteq V[G][I]$ and for some poset $\mathbb{S} \in V[J]$, there is a $V[J]$-generic $K \subseteq \mathbb{S}$ such that $V[G][I] = V[J][K]$. Since by hypothesis, $V_\alpha^{\mathbb{Q}} \models T$, $V_\alpha^{V[G]} \models T$. Then

$$(V_{j(\alpha)})^M = (V_{j(\alpha)})^{V[G][I]} = (V_{j(\alpha)})^{V[J][K]} \models T.$$

Since $V^{\mathbb{P}} \models$ "$T \models_\Omega \varphi$", $(V_{j(\alpha)})^{V[J][K]} \models \varphi$. So $(V_{j(\alpha)})^M \models \varphi$, and therefore $V_\alpha^{V[G]} \models \varphi$. Thus, $V_\alpha^{\mathbb{Q}} \models \varphi$. □

1.3. Some properties of $\models_\Omega$

Lemma 1.9. *For every recursively enumerable (r.e.) set $T \cup \{\varphi\} \subseteq Sent$, the following are equivalent:*

i) $T \models_\Omega \varphi$.

ii) $\emptyset \models_\Omega$ "$T \models_\Omega \varphi$".

(Note that since T is r.e., "$T \models_\Omega \varphi$" can be written as a sentence in Sent. So, ii) *makes sense.)*

Proof. i) $\Rightarrow$ ii). Let $\alpha \in On$ and $\mathbb{B}$ a c.B.a. Suppose $\beta < \alpha$, and $\dot{\mathbb{Q}}$ is a c.B.a. in $V_\alpha^{\mathbb{B}}$ such that $V_\alpha^{\mathbb{B}} \models$ "$V_{\check\beta}^{\dot{\mathbb{Q}}} \models T$". Then $V_\beta^{\mathbb{B}*\dot{\mathbb{Q}}} \models T$. By i), $V_\beta^{\mathbb{B}*\dot{\mathbb{Q}}} \models \varphi$, and hence $V_\alpha^{\mathbb{B}} \models$ "$V_{\check\beta}^{\dot{\mathbb{Q}}} \models \varphi$".

ii) $\Rightarrow$ i). Suppose $\alpha \in On$, $\mathbb{B}$ is a c.B.a., and $V_\alpha^{\mathbb{B}} \models T$. Fix $\beta > \alpha$, β a limit ordinal. Since T is r.e., if $V_\beta^{\mathbb{B}} \models$ "$\psi \in T$", then $\psi \in T$, and therefore $V_\alpha^{\mathbb{B}} \models \psi$. Thus, $V_\beta^{\mathbb{B}} \models$ "$V_{\check\alpha} \models T$". By *ii)*, $V_\beta^{\mathbb{B}} \models$ "$T \models_\Omega \varphi$". Hence, $V_\beta^{\mathbb{B}} \models$ "$V_{\check\alpha} \models_\Omega \varphi$", and we have $V_\alpha^{\mathbb{B}} \models \varphi$. □

Remarks 1.10. Suppose that ZFC is consistent. For iv) suppose, moreover, that it is consistent with ZFC that $V_\alpha^{\mathbb{B}} \models ZFC$, for some ordinal α and some c.B.a. $\mathbb{B}$. Then,

i) If φ is absolute for transitive sets, then $ZFC \vdash (\varphi \to \emptyset \models_\Omega \varphi)$.

ii) For some $\varphi \in Sent$, $ZFC \nvdash (\varphi \to (\emptyset \models_\Omega \varphi))$.

iii) For some $\varphi \in Sent$, $ZFC \nvdash ((ZFC \models_\Omega \varphi) \to \varphi)$.

iv) For some $\varphi \in Sent$, $ZFC \nvdash ((ZFC \models_\Omega$ "$ZFC \models_\Omega \varphi$"$) \to (ZFC \models_\Omega \varphi))$.

Proof. i) is clear.

ii) holds for every sentence φ that can be forced to be true and false, for example CH.

iii) Let φ="$\exists\beta(V_\beta \models ZFC)$". Let M be a model of ZFC. If for every α and every $\mathbb{B}$ in M, $M_\alpha^\mathbb{B} \not\models ZFC$ (call this Case 1), then $M \models$ "$ZFC \models_\Omega \varphi$" $+ \neg\varphi$. Otherwise, let β be the least such that $M_\beta^\mathbb{B} \models ZFC$, for some $\mathbb{B}$. Then $M_\beta^\mathbb{B}$ is a model of ZFC, call it N, and has the property that for every α and every c.B.a. $\mathbb{C}$, $N_\alpha^\mathbb{C} \not\models ZFC$. So, we are back to Case 1.

iv) Consider the sentence φ="$\exists\beta\exists\gamma(\beta < \gamma \wedge V_\beta \models ZFC \wedge V_\gamma \models ZFC)$". Let M be a model of ZFC such that $M \models \exists\alpha\exists\mathbb{B}(V_\alpha^\mathbb{B} \models ZFC)$. If for every α and every c.B.a. $\mathbb{B}$, $M_\alpha^\mathbb{B} \not\models \varphi$ (call this Case 1), then $M \models (ZFC \models_\Omega$ "$ZFC \models_\Omega \varphi$"$) + \neg(ZFC \models_\Omega \varphi)$.

If for some α and $\mathbb{B}$, $M_\alpha^\mathbb{B} \models \varphi$, then let γ be the least ordinal such that $M_\gamma^\mathbb{B} \models ZFC + \exists\beta(V_\beta^\mathbb{B} \models ZFC)$. Let N be $M_\gamma^\mathbb{B}$. Then N has the property that for every α and every $\mathbb{C}$, $N_\alpha^\mathbb{C} \not\models \varphi$, and so we are back to Case 1. □

Theorem 1.11 (Non-Compactness of $\models_\Omega$). *There is $T \cup \{\varphi\} \subseteq Sent$ such that $T \models_\Omega \varphi$, but for all finite $S \subseteq T$, $S \not\models_\Omega \varphi$.*

Proof. Let φ_0 be the sentence asserting: *There is a largest limit ordinal.*

For each $n \in \omega$, $n > 0$, let φ_n be the sentence asserting: *If α is the largest limit ordinal, then $\alpha + n$ exists.*

Finally, let φ be the sentence that asserts: *Every ordinal has a successor.*

Let $T = \{\varphi_n \mid n \in \omega\}$.

Then, $T \models_\Omega \varphi$. But if $S \subseteq T$ is finite, then $S \not\models_\Omega \varphi$. □

With a bit more work we can show that Compactness of $\models_\Omega$ also fails for $T = ZFC$. Indeed, recall that by Gödel's Diagonalization, for each formula $\psi(x)$, with x the only free variable and ranging over natural numbers, there is a sentence φ such that $ZFC \vdash (\varphi \leftrightarrow \psi(\ulcorner\varphi\urcorner))$, where $\ulcorner\varphi\urcorner$ is the term denoting the Gödel code of φ.

Theorem 1.12. *If ZFC is consistent, then there is a sentence φ such that $ZFC \models_\Omega \varphi$ but for all finite $S \subseteq ZFC$, $S \not\models_\Omega \varphi$.*

Proof. Let $\psi(x)$ be the formula:

$$x \text{ Gödel-codes a sentence } \varphi_x \wedge \forall S(S \textit{ a finite subset of } ZFC \to S \not\models_\Omega \varphi_x)$$

By Gödel's Diagonalization, there is a sentence φ such that $ZFC \vdash (\varphi \leftrightarrow \psi(\ulcorner\varphi\urcorner))$. Let $T \subseteq ZFC$ be finite such that $T \vdash (\varphi \leftrightarrow \psi(\ulcorner\varphi\urcorner))$. Let θ be the conjunction of the set of sentences of T. Then, $\emptyset \vdash \theta \to (\varphi \leftrightarrow \psi(\ulcorner\varphi\urcorner))$.

Claim. $ZFC \models_\Omega \varphi$.

Proof of Claim. Suppose not. Pick α and $\mathbb{B}$ such that $V_\alpha^\mathbb{B} \models ZFC + \neg\varphi$. So, there is $S \in V_\alpha^\mathbb{B}$ a finite set of sentences of ZFC such that $V_\alpha^\mathbb{B} \models$ "$S \models_\Omega \varphi$". Since $V_\alpha^\mathbb{B} \models ZFC$, by reflection, let $\beta < \alpha$ be such that $V_\beta^\mathbb{B} \models S + \neg\varphi$. But since $V_\alpha^\mathbb{B} \models$ "$S \models_\Omega \varphi$", and $V_\beta^\mathbb{B} \models S$, we obtain $V_\beta^\mathbb{B} \models \varphi$, a contradiction. □

Claim. If $S \subseteq ZFC$ is finite then $S \not\models_\Omega \varphi$.

Proof of Claim. Suppose there is $S \subseteq ZFC$ finite such that $S \models_\Omega \varphi$. By Lemma 1.9, $\emptyset \models_\Omega$ "$S \models_\Omega \varphi$". Let $\mathbb{B}$ be a c.B.a.. Since $ZFC \vdash \theta + S$ and $V^{\mathbb{B}} \models ZFC$, by reflection, let α be such that $V_\alpha^{\mathbb{B}} \models \theta + S$. Since $\emptyset \models_\Omega$ "$S \models_\Omega \varphi$", $V_\alpha^{\mathbb{B}} \models$ "$S \models_\Omega \varphi$", i.e., $V_\alpha^{\mathbb{B}} \models (\exists S)(S$ finite and $S \models_\Omega \varphi)$. Hence $V_\alpha^{\mathbb{B}} \models \neg\psi(\ulcorner\varphi\urcorner)$. But since $V_\alpha^{\mathbb{B}} \models \theta$, $V_\alpha^{\mathbb{B}} \models \neg\varphi$, contradicting the assumption that $S \models_\Omega \varphi$. □

2. $\vdash_\Omega$

In order to define the Ω-provability relation $\vdash_\Omega$ (Definition 2.28), the syntactic relation associated to $\models_\Omega$, also introduced by W.H. Woodin, we need to recall some notions that will play an essential part in the definition. Along the way we will also prove some useful facts about these notions.

2.1. Universally Baire sets of reals

The universally Baire sets of reals play the role of Ω-proofs in Ω-logic.

Recall that for an ordinal λ, a *tree* on $\omega \times \lambda$ is a set $T \subseteq \omega^{<\omega} \times \lambda^{<\omega}$ such that for all pairs $(s,t) \in T$, $lh(s) = lh(t)$ and $(s\restriction i, t\restriction i) \in T$ for each $i \in lh(s) \in \omega$. Given a tree on $\omega \times \lambda$, $p[T] = \{x \in \omega^\omega \mid \exists f \in \lambda^\omega (x,f) \in [T]\}$ is the *projection* of T, where $[T] = \{(x,f) \in \omega^\omega \times \lambda^\omega \mid \forall n \in \omega (x\restriction n, f\restriction n) \in T\}$.

Definition 2.1. ([2])

i) For a given cardinal κ, a set of reals A is *κ-universally Baire* (κ-uB) if there exist trees T and S on $\omega^{<\omega} \times \lambda^{<\omega}$, λ some ordinal, such that $A = p[T]$ and $p[T] = \omega^\omega \setminus p[S]$ in any forcing extension by a partial order of cardinality less than κ. We say that the trees T and S witness that A is κ-uB.
ii) $A \subseteq \mathbb{R}$ is *universally Baire* (uB) if it is κ-uB for each cardinal κ.

Proposition 2.2. ([2]). *For $A \subseteq \mathbb{R}$, the following are equivalent:*

i) *A is universally Baire.*
ii) *For every compact Hausdorff space X and every continuous function $f: X \to \mathbb{R}$, the set $f^{-1}(A) = \{x \in X \mid f(x) \in A\}$ has the property of Baire, i.e., there exists an open set $O \subseteq X$ such that the symmetric difference $f^{-1}(A) \triangle O$ is meager.*
iii) *For every notion of forcing $\mathbb{P}$ there exist trees T and S on $\omega \times 2^{|\mathbb{P}|}$ such that $A = p[T] = \omega^\omega \setminus p[S]$ and $V^{\mathbb{P}} \models p[T] = \omega^\omega \setminus p[S]$. We say that the trees T and S witness that A is uB for $\mathbb{P}$.*

The following is a special case of the well-known fact that the well-foundedness of a given tree is absolute to all models of ZFC with the same ordinals.

Proposition 2.3. *Let T and S be trees on $\omega \times \kappa$, for some ordinal κ. Suppose that $p[T] \cap p[S] = \emptyset$. Then, in any forcing extension $V[G]$ we also have that $p[T]^{V[G]} \cap p[S]^{V[G]} = \emptyset$.*

Proof. Towards a contradiction, suppose that $\mathbb{P}$ is a forcing notion, $p \in \mathbb{P}$, τ is a $\mathbb{P}$-name for a real, and $p \Vdash \tau \in p[T] \cap p[S]$.

Let $N \prec H(\lambda)$, λ a large enough regular cardinal, N countable and such that $p, \mathbb{P}, \tau, T, S \in N$. Let M be the transitive collapse of N, and let $\bar{p}, \bar{\mathbb{P}}, \bar{\tau}, \bar{T}$ and $\bar{S}$ be the transitive collapses of $p, \mathbb{P}, \tau, T$ and S, respectively. Thus, in M we have

$$\bar{p} \Vdash_{\bar{\mathbb{P}}} \bar{\tau} \in p[\bar{T}] \cap p[\bar{S}].$$

Let g be $\bar{\mathbb{P}}$-generic over M with $\bar{p} \in g$. So, in $M[g]$, we have

$$\bar{\tau}[g] \in p[\bar{T}] \cap p[\bar{S}].$$

Notice that $p[T \cap N] \subseteq p[T]$ and $p[S \cap N] \subseteq p[S]$. Moreover, $\bar{T} \cong T \cap N$ and $\bar{S} \cong S \cap N$. Hence, since the transitive collapse is the identity on natural numbers, $p[\bar{T}] \subseteq p[T]$ and $p[\bar{S}] \subseteq p[S]$, contradicting the fact that $p[T]$ and $p[S]$ are disjoint. □

Corollary 2.4. *Let T, T' and S be trees on $\omega \times \kappa$, for some ordinal κ. Suppose that $p[T] = p[T']$ and $p[S] = \omega^\omega \setminus p[T]$. If in $V[G]$, $p[S]^{V[G]} = \omega^\omega \setminus p[T]^{V[G]}$, then $p[T']^{V[G]} \subseteq p[T]^{V[G]}$.*

Remark 2.5. In general, under the same assumptions as in the Corollary 2.4, we cannot conclude that $p[T']^{V[G]} = p[T]^{V[G]}$. For instance, one can easily construct trees S and T on $\omega \times \omega$ such that $p[S]$ is the set of reals that take the value 0 infinitely often on the even elements of ω, and $p[T]$ is the set of reals that take the value 0 finitely often on the even elements of ω, and such that S and T will project to the sets with these definitions (and thus to complements) in all forcing extensions. Furthermore, if $\{x_\alpha : \alpha < 2^\omega\}$ is the set of reals (in the ground model) that take the value 0 finitely often on the even elements of ω, and T' is the tree consisting all pairs (a, b) where b is a finite constant sequence with some fixed value $\alpha < 2^\omega$ and a is $x_\alpha \restriction |b|$, then $p[T] = p[T']$ in the ground model, but $p[T] \neq p[T']$ in any forcing extension that adds a real.

By Corollary 2.4, if $A \subseteq \mathbb{R}$ is κ-uB in a model N of ZFC, witnessed by trees T and S, and $N[G]$ is an extension of N by a forcing notion of cardinality less than κ, then $A_G := p[T]^{N[G]}$ is equal to the set of reals in $N[G]$ which are in the projection (in $N[G]$) of some tree in N witnessing that A is κ-uB. Therefore, given $A \subseteq \mathbb{R}$ a uB set, A has a *canonical interpretation* A_G in any set forcing extension $V[G]$ of V, namely:

$$A_G = \bigcup \{p[T]^{V[G]} \mid T \in V \text{ and } A = p[T]^V\}.$$

Thus, if $\mathbb{P}$ is a forcing notion and A is uB for $\mathbb{P}$, witnessed by trees T, S, and also by trees T', S', then in any $\mathbb{P}$-generic extension $V[G]$, $p[T] = p[T'] = A_G$.

Remark 2.6. It is clear from Proposition 2.2 (iii) that a set $A \subseteq \mathbb{R}$ is uB iff for every c.B.a. $\mathbb{B}$, $V^{\mathbb{B}} \models$ "$A_{\dot{G}}$ is uB".

Theorem 2.7. ([2])

i) *Every analytic set, and therefore every coanalytic set, is universally Baire.*

ii) *Every $\boldsymbol{\Sigma}^1_2$ set of reals is uB iff for every set x, $x^\sharp$ exists.*

2.2. *A*-closed models

Let us now define the notion of *A-closed set*, which will be also fundamental for the definition of the Ω-provability relation $\vdash_\Omega$.

Definition 2.8 (**[12]**). Given a uB set $A \subseteq \mathbb{R}$, a transitive $\in$-model M of (a fragment of) ZFC is *A-closed* if for all posets $\mathbb{P} \in M$ and all V-generic filters $G \subseteq \mathbb{P}$,

$$V[G] \models M[G] \cap A_G \in M[G]$$

(i.e., $\Vdash_{\mathbb{P}}$ "$M[\dot{G}] \cap A_{\dot{G}} \in M[\dot{G}]$", where $\dot{G}$ is the standard $\mathbb{P}$-name for the generic filter).

Woodin has given several other definitions of A-closure, but the next proposition shows they are equivalent.

Proposition 2.9. *Given a uB set $A \subseteq \mathbb{R}$ and a transitive model M of ZFC, the following are equivalent:*

a) *M is A-closed.*

b) *For all infinite $\gamma \in M \cap On$, for all $G \subseteq \mathrm{Coll}(\omega, \gamma)$ V-generic,*

$$V[G] \models M[G] \cap A_G \in M[G].$$

c) *For all posets $\mathbb{P} \in M$ and all $\tau \in M^{\mathbb{P}}$, $\{p \in \mathbb{P} \mid p \Vdash^V_{\mathbb{P}} \tau \in A_{\dot{G}}\} \in M$.*

d) *For all infinite $\gamma \in M \cap On$ and all $\tau \in M^{\mathrm{Coll}(\omega,\gamma)}$,*

$$\{p \in \mathrm{Coll}(\omega, \gamma) \mid p \Vdash^V_{\mathrm{Coll}(\omega,\gamma)} \tau \in A_{\dot{G}}\} \in M.$$

e) *For all posets $\mathbb{P} \in M$,*

$$\{(\tau, p) \mid \tau \in M \text{ a simple } \mathbb{P}\text{-name for a real}, p \in \mathbb{P} \text{ and } p \Vdash^V_{\mathbb{P}} \tau \in A_{\dot{G}}\} \in M.$$

f) *For all posets $\mathbb{P}_\gamma = \mathrm{Coll}(\omega, \gamma)$, with $\gamma \in M \cap On$ infinite,*

$$\{(\tau, p) \mid \tau \in M \text{ a simple } \mathbb{P}_\gamma\text{-name for a real}, p \in \mathbb{P}_\gamma \text{ and } p \Vdash^V_{\mathbb{P}_\gamma} \tau \in A_{\dot{G}}\} \in M.$$

Proof. Observe that the implications (a) $\Rightarrow$ (b), (c) $\Rightarrow$ (d) and (e) $\Rightarrow$ (f) are immediate.

(b) $\Rightarrow$ (d). Fix $\gamma \in M \cap On$. Since $M \models ZFC$ and M is transitive, $\mathrm{Coll}(\omega, \gamma) \in M$. Let $\tau \in M^{\mathrm{Coll}(\omega,\gamma)}$. By (b), there exist $p \in \mathrm{Coll}(\omega, \gamma)$ and $\sigma_0 \in M^{\mathrm{Coll}(\omega,\gamma)}$ such that $p \Vdash^V_{\mathrm{Coll}(\omega,\gamma)} M[\dot{G}] \cap A_{\dot{G}} = \sigma_0$. Since $\mathrm{Coll}(\omega, \gamma)$ is homogeneous, we can replace σ_0 with a $\mathrm{Coll}(\omega, \gamma)$-name σ in M such that every condition in $\mathrm{Coll}(\omega, \gamma)$ forces (in V) that $M[\dot{G}] \cap A_{\dot{G}} = \sigma$. Thus, for every $q \in \mathrm{Coll}(\omega, \gamma)$,

$$q \Vdash^V_{\mathrm{Coll}(\omega,\gamma)} \tau \in \sigma \text{ iff } q \Vdash^V_{\mathrm{Coll}(\omega,\gamma)} \tau \in A_{\dot{G}}.$$

Hence, since $\{\mathrm{Coll}(\omega, \gamma), \tau, \sigma\} \subseteq M$ and M is transitive, by absoluteness,

$$\begin{aligned}\{p \in \mathbb{P} \mid p \Vdash^V_{\mathrm{Coll}(\omega,\gamma)} \tau \in A_{\dot{G}}\} &= \{p \in \mathbb{P} \mid p \Vdash^V_{\mathrm{Coll}(\omega,\gamma)} \tau \in \sigma\} \\ &= \{p \in \mathbb{P} \mid p \Vdash^M_{\mathrm{Coll}(\omega,\gamma)} \tau \in \sigma\} \in M.\end{aligned}$$

(d) $\Rightarrow$ (c). Fix a poset $\mathbb{P}$ in M and $\tau \in M^{\mathbb{P}}$. We may assume that τ is a simple $\mathbb{P}$-name for a real. Let $\gamma = |\mathbb{P}|^M$, and let τ^* be the simple $\mathbb{P} \times \mathrm{Coll}(\omega, \gamma)$-name defined by letting $((\check{m,n}),(p,q)) \in \tau^*$ if and only if $((\check{m,n}),p)$ is in τ. Then since $\mathbb{P} \times \mathrm{Coll}(\omega,\gamma)$ has a dense set isomorphic to $\mathrm{Coll}(\omega,\gamma)$, by (d), $\{(p,q) \in \mathbb{P} \times \mathrm{Coll}(\omega,\gamma) \mid (p,q) \Vdash^V_{\mathbb{P}\times\mathrm{Coll}(\omega,\gamma)} \tau^* \in A_{\dot{G}}\} \in M$. Since for all $(p,q) \in \mathbb{P} \times \mathrm{Coll}(\omega,\gamma)$, $(p,q) \Vdash^V_{\mathbb{P}\times\mathrm{Coll}(\omega,\gamma)} \tau^* \in A_{\dot{G}}$ if and only if $p \Vdash^V_{\mathbb{P}} \tau \in A_{\dot{G}}$, the conclusion of (c) follows.

(e) $\Rightarrow$ (a) (similarly for (f) $\Rightarrow$ (b)). Fix a poset $\mathbb{P} \in M$ and suppose $G \subseteq \mathbb{P}$ is V-generic. Let

$$\sigma = \{(\tau,p) \mid \tau \in M \text{ a simple } \mathbb{P}\text{-name for a real}, p \in \mathbb{P} \text{ and } p \Vdash^V_{\mathbb{P}} \tau \in A_{\dot{G}}\}.$$

By (e), $\sigma \in M$. Hence $\sigma \in M^{\mathbb{P}} = V^{\mathbb{P}} \cap M$ and $i_G[\sigma] \in M[G]$.

Claim. $i_G[\sigma] = A_G \cap M[G]$.
Proof of Claim. Suppose $r \in i_G[\sigma]$. Let $p \in G \subseteq \mathbb{P}$ be such that $(\dot{r},p) \in \sigma$ and $i_G[\dot{r}] = r$. Thus $\dot{r}$ is a simple $\mathbb{P}$-name in M for a real and $p \Vdash^V_{\mathbb{P}} \dot{r} \in A_{\dot{G}}$. Hence $r \in A_G \cap M[G]$.

Suppose now $r \in A_G \cap M[G]$. Let $p \in G$ and $\dot{r} \in M^{\mathbb{P}}$ be such that $p \Vdash^V_{\mathbb{P}} \dot{r} \in A_{\dot{G}}$. Let τ be a simple $\mathbb{P}$-name for a real in M such that $p \Vdash^V_{\mathbb{P}} \tau = \dot{r}$. Then $(\tau,p) \in \sigma$ and therefore $r \in i_G[\sigma]$. □

(d) $\Rightarrow$ (f). Fix $\gamma \in M \cap On$. Let $\mathbb{P} = \mathrm{Coll}(\omega,\gamma)$ and $\mathbb{P}' = \mathrm{Coll}(\omega,2^{|\gamma|})$. Let $\langle \tau_\alpha \mid \alpha < 2^{|\gamma|}\rangle \in M$ be an enumeration of all the simple $\mathbb{P}$-names in M for reals. Let $\pi\colon \mathbb{P} \times \mathbb{P}' \to \mathbb{P}'$ be an order-preserving bijection. Define a simple $\mathbb{P} \times \mathbb{P}'$-name σ as follows:

$$\sigma = \{((\check{i,j}),(p,q)) \mid \exists\alpha < 2^{|\gamma|} \text{ such that } q(0) = \alpha \text{ and } ((\check{i,j}),p) \in \tau_\alpha\}$$

Let σ^* be the simple $\mathbb{P}'$-name $\{((\check{i,j}),\pi(p,q)) \mid ((\check{i,j}),(p,q)) \in \sigma\}$.
By (d), $X = \{q \in \mathbb{P}' \mid q \Vdash^V_{\mathbb{P}'} \sigma^* \in A_{\dot{G}}\} \in M$.
Hence,

$$\begin{aligned} Z =& \{(p,q) \in \mathbb{P} \times \mathbb{P}' \mid \pi(p,q) \in X\} = \{(p,q) \in \mathbb{P} \times \mathbb{P}' \mid \pi(p,q) \Vdash^V_{\mathbb{P}'} \sigma^* \in A_{\dot{G}}\} \\ =& \{(p,q) \in \mathbb{P} \times \mathbb{P}' \mid (p,q) \Vdash^V_{\mathbb{P}\times\mathbb{P}'} \sigma \in A_{\dot{G}\times\dot{H}}\} \in M. \end{aligned}$$

Let

$$Y = \{(\tau,p) \mid \exists\alpha < 2^{|\gamma|} \text{ such that } \tau = \tau_\alpha \text{ and } (p,(0,\alpha)) \in Z\}.$$

Since $Z \in M$, $Y \in M$. For $\tau \in M^{\mathbb{P}}$, let $\bar{\tau}$ be the corresponding $\mathbb{P} \times \mathbb{P}'$-name which depends only on the first coordinate. In particular, for each $\alpha < 2^{|\gamma|}$, since $\tau_\alpha \in M^{\mathbb{P}}$, for all $(p,q) \in \mathbb{P} \times \mathbb{P}'$,

$$p \Vdash^V_{\mathbb{P}} (\check{i,j}) \in \tau_\alpha \text{ iff } (p,q) \Vdash^V_{\mathbb{P}\times\mathbb{P}'} (\check{i,j}) \in \bar{\tau}_\alpha.$$

Claim. For each $\alpha < 2^{|\gamma|}$, for all $p \in \mathbb{P}$, $(p,(0,\alpha)) \Vdash^V_{\mathbb{P}\times\mathbb{P}'} \sigma = \bar{\tau}_\alpha$.
Proof of Claim. Let $G = G_1 \times G_2 \subseteq \mathbb{P}\times\mathbb{P}'$ be V-generic such that $(p,(0,\alpha)) \in G$. We check that $i_G[\sigma] = i_G[\bar{\tau}_\alpha]$: If $(i,j) \in i_G[\sigma]$, then for some $(r,s) \in G$, $((\check{i,j}),(r,s)) \in \sigma$, $s(0) = \beta$ for some $\beta < 2^{|\gamma|}$ and $r \Vdash^V_{\mathbb{P}} (\check{i,j}) \in \tau_\beta$. Since $(r,s),(p,(0,\alpha)) \in G$, $\alpha = \beta$ and $(i,j) \in i_G[\bar{\tau}_\alpha]$.

If $(i,j) \in i_G[\bar{\tau}_\alpha]$, let $(r,s) \leq (p,(0,\alpha))$ in G be such that $(r,s) \Vdash^V_{\mathbb{P}\times\mathbb{P}'} (\check{i},\check{j}) \in \bar{\tau}_\alpha$. Then $r \Vdash^V_{\mathbb{P}} (\check{i},\check{j}) \in \tau_\alpha$. Moreover, since $s \leq (0,\alpha)$, $s(0) = \alpha$. Hence, $((\check{i},\check{j}),(r,(0,\alpha))) \in \sigma$ and $(r,(0,\alpha)) \Vdash^V_{\mathbb{P}\times\mathbb{P}'} (\check{i},\check{j}) \in \sigma$. Since $(r,(0,\alpha)) \geq (r,s)$, $(r,(0,\alpha)) \in G$ and $(i,j) \in i_G[\sigma]$. □

Moreover, given $p \in \mathbb{P}$, and τ a simple $\mathbb{P}$-name in M,

$$\begin{aligned}(\tau,p) \in Y &\text{ iff } \exists\alpha < 2^{|\gamma|} \text{ such that } \tau = \tau_\alpha \text{ and } (p,(0,\alpha)) \Vdash^V_{\mathbb{P}\times\mathbb{P}'} \sigma \in A_{\dot{G}\times\dot{H}}\\ &\text{ iff } \exists\alpha < 2^{|\gamma|} \text{ such that } \tau = \tau_\alpha \text{ and } p \Vdash^V_{\mathbb{P}} \tau_\alpha \in A_{\dot{G}}\\ &\text{ iff } p \Vdash^V_{\mathbb{P}} \tau \in A_{\dot{G}}.\end{aligned}$$

Hence,

$$Y = \{(\tau,p) \mid \tau \in M \text{ a simple } \mathbb{P}\text{-name for a real}, p \in \mathbb{P} \text{ and } p \Vdash^V_{\mathbb{P}} \tau \in A_{\dot{G}}\}.$$

(f) ⇒ (e). Fix $\mathbb{P} \in M$. Let $\gamma = |\mathbb{P}|^M$ and $\mathbb{P}_\gamma = \mathrm{Coll}(\omega,\gamma)$. Let $X =$

$$\{(\tau,p) \mid \tau \in M \text{ a simple } \mathbb{P}_\gamma\text{-name for a real}, \ p \in \mathbb{P}_\gamma \text{ and } p \Vdash^V_{\mathbb{P}_\gamma} \tau \in A_{\dot{G}}\}.$$

By f), $X \in M$. In M, let e be a complete embedding of $\mathbb{P}$ into $\mathrm{Coll}(\omega,\gamma)$. As before, e extends naturally to an embedding $e^*\colon M^{\mathbb{P}} \to M^{\mathrm{Coll}(\omega,\gamma)}$ in M. Let

$$Y = \{(\tau,p) \mid \tau \in M \text{ a simple } \mathbb{P}\text{-name for a real}, \ p \in \mathbb{P} \text{ and } p \Vdash^V_{\mathbb{P}} \tau \in A_{\dot{G}}\}.$$

So,

$$Y = \{(\tau,p) \mid \tau \in M \text{ a simple } \mathbb{P}\text{-name for a real}, \ p \in \mathbb{P} \text{ and } (e^*(\tau),e(p)) \in X\}.$$

Thus, $Y \in M$. □

For M countable, the notion of A-closure has a simpler formulation, as shown in Proposition 2.11 below.

Lemma 2.10. *Suppose $A \subseteq \mathbb{R}$ is uB and M is an A-closed c.t.m. of ZFC. Let α be such that M is countable and A-closed in V_α. Suppose $X \prec V_\alpha$ is countable with $\{M,A,S,T\} \in X$, where T and S are trees witnessing that A is ω_1-uB, and N is the transitive collapse of X. Then, for every forcing notion $\mathbb{P} \in M$ and every N-generic filter $g \subseteq \mathbb{P}$, $M[g] \cap A \in M[g]$.*

Proof. Let π be the transitive collapsing function on X. So, $N = \pi"X$. Let $\pi(S) = \bar{S}$ and $\pi(T) = \bar{T}$. Observe that $\pi(M) = M$ and $\pi(A) = A \cap X = A \cap N$. Fix $g \subseteq \mathbb{P} \in M$ N-generic. Since $p[\bar{T}] \subseteq p[T] = A$, writing $(A_g)^{N[g]}$ for $(\pi(A)_g)^{N[g]}$, we have:

$$(A_g)^{N[g]} = (p[\bar{T}])^{N[g]} \subseteq N[g] \cap A$$

and since $p[\bar{S}] \subseteq p[S] = \omega^\omega \setminus A$,

$$N[g] \cap A \subseteq (p[\bar{T}])^{N[g]}.$$

Hence $(A_g)^{N[g]} = N[g] \cap A$. Since M is A-closed in N, $M[g] \cap (A_g)^{N[g]} \in M[g]$. Hence, $M[g] \cap A = M[g] \cap N[g] \cap A = M[g] \cap (A_g)^{N[g]} \in M[g]$. □

If M is a countable transitive model and $\mathbb{P}$ is a partial order in M, we say that a set $\mathcal{G}$ of M-generic filters $g \subset \mathbb{P}$ is *comeager* if there exists a countable set $\mathcal{D}$ of dense subsets of $\mathbb{P}$ (not necessarily in M) such that $\mathcal{G}$ contains the set of M-generic filters that intersect every member of $\mathcal{D}$.

Notice that if $\mathcal{G}$ is comeager, then its complement in the set of all M-generic filters is not comeager. For suppose $\mathcal{D}$ and $\mathcal{D}'$ witness the comeagerness of $\mathcal{G}$ and its complement, respectively. Then, since $\mathcal{D} \cup \mathcal{D}'$ is countable, there is an M-generic filter G that intersects all dense sets in $\mathcal{D} \cup \mathcal{D}'$. But then G would belong to both $\mathcal{G}$ and its complement, which is impossible.

The following provides, in the case of a c.t.m. M, yet another characterization of M being A-closed, in addition to Proposition 2.9.

Proposition 2.11. *Given A a uB set and M a c.t.m. of ZFC, the following are equivalent:*

i) *M is A-closed.*
ii) *for all $\mathbb{P} \in M$, the set of M-generic filters $g \subset \mathbb{P}$ such that*

$$M[g] \cap A \in M[g]$$

is comeager.

Proof. i) $\Rightarrow$ ii). Let $\mathbb{P} \in M$. Let N be as in Lemma 2.10. Since N is countable, there are countably many dense sets of $\mathbb{P}$ in N. Let $\mathcal{D} = \{D_i \mid i \in \omega\}$ be this collection. Let $g \subseteq \mathbb{P}$ be an $(M \cup \mathcal{D})$-generic filter. Since g intersects each dense set in N, g is N-generic and by Lemma 2.10, $M[g] \cap A \in M[g]$.
ii) $\Rightarrow$ i). Let $\mathbb{P} \in M$. Towards a contradiction, let $p \in \mathbb{P}$ be such that $p \Vdash_{\mathbb{P}} M[\dot{G}] \cap A_{\dot{G}} \notin M[\dot{G}]$. By ii), let $\mathcal{D} = \{D_i \mid i \in \omega\}$ be a collection of dense subsets of $\mathbb{P}$ such that for all $(M \cup \mathcal{D})$-generic g, $M[g] \cap A \in M[g]$. Let V_α, α a large-enough uncountable regular cardinal, be such that $M, A, \mathcal{D} \in V_\alpha$. Let T, S be trees witnessing that A is ω_1-uB in V_α. Let $X \prec V_\alpha$ be countable with $\{\mathcal{D}, M, A, T, S\} \in X$ and let N be the transitive collapse of X. Let g be N-generic such that $p \in g$. By elementarity, $p \Vdash^N_{\mathbb{P}} M[\dot{G}] \cap A_{\dot{G}} \notin M[\dot{G}]$. Hence, $M[g] \cap A = M[g] \cap (A_g)^{N[g]} \notin M[g]$. But this contradicts ii), since g is $(M \cup \mathcal{D})$-generic. □

Corollary 2.12. *If M is a c.t.m. of ZFC and A is a uB set, then "M is A-closed" is correctly computed in $L(A, \mathbb{R})$.*

Proof. The next sentence is true in V iff it is true in $L(A, \mathbb{R})$ and says that M is A-closed:

$$\varphi(A, M) := (\forall \mathbb{P} \in M)(\exists \{D_i \mid i \in \omega\} [D_i \subseteq \mathbb{P} \text{ dense} \wedge (\forall g)(g \subseteq \mathbb{P})((g \text{ a filter } M\text{-generic} \wedge (\forall i \in \omega)(g \cap D_i \neq \emptyset)) \rightarrow M[g] \cap A \in M[g])].$$ □

The following alternate form of Proposition 2.11 is sometimes useful.

Lemma 2.13. *Given a uB set $A \subseteq \mathbb{R}$, M a c.t.m. of ZFC, $\mathbb{P} \in M$ a poset, $p \in \mathbb{P}$, and τ a $\mathbb{P}$-name in M for a real, the following are equivalent:*

i) $p \Vdash^V_{\mathbb{P}} \tau \in A_{\dot{G}}$.
ii) *The set of M-generic filters $g \subseteq \mathbb{P}$ such that $p \in g$ and $i_g[\tau] \in A$ is comeager.*

Proof. i) $\Rightarrow$ ii). Let T, S be witnesses for A being ω_1-uB, $A = p[T]$, $\omega^\omega \setminus A = p[S]$. There exists $\dot{z}$ such that for all $i \in \omega$, $p \Vdash^V_{\mathbb{P}} (\tau \restriction i, \dot{z} \restriction i) \in \check{T}$.
Let $\{D_i \mid i < \omega\}$ be such that D_i decides $\dot{z}(i)$, $i \in \omega$, i.e.,

$$D_i = \{q \in \mathbb{P} \mid q \Vdash^V \text{“}\dot{z}(i) = k\text{”, for some } k\}.$$

For all i, D_i is a dense subset of $\mathbb{P}$. Then if $g \subseteq \mathbb{P}$ is M-generic with $p \in g$ and $g \cap D_i \neq \emptyset$ for every $i \in \omega$, g decides $\dot{z}(i)$ and for all $i \in \omega$, $(i_g[\tau] \restriction i, i_g[\dot{z}] \restriction i) \in T$. So $i_g[\tau] \in p[T] = A$.
ii) $\Rightarrow$ i). Let V_α, α a large enough uncountable cardinal, be such that ii) holds in V_α. Let T, S be trees witnessing A is ω_1-uB in V_α. Let $X \prec V_\alpha$ be countable with $\{M, A, T, S\} \in X$ and let N be the transitive collapse of X. Observe that $\pi(A) = A \cap N$ and $\pi(M) = M$, hence $\pi(\mathbb{P}) = \mathbb{P}$ and $\pi(p) = p$. Let $\pi(S) = \bar{S}$ and $\pi(T) = \bar{T}$. By elementarity, there is in N a collection $\{D_i : i \in \omega\}$ of dense subsets of $\mathbb{P}$ such that for all M-generic filters $g \subseteq \mathbb{P}$, if $p \in g$ and $g \cap D_i \neq \emptyset$ for all $i \in \omega$, then $i_g[\tau] \in A \cap N$. Pick any G N-generic with $p \in G$. Since $G \cap D_i \neq \emptyset$ for all i and G is M-generic, by Lemma 2.10, $i_G[\tau] \in A \cap M[G] = (A_G)^{N[G]} \cap M[G]$, so $N[G] \models i_G[\tau] \in A_G$. Since G was an arbitrary N-generic filter containing p, $p \Vdash^N \tau \in A_{\dot{G}}$. By elementarity, $p \Vdash^V_{\mathbb{P}} \tau \in A_{\dot{G}}$. □

For a c.t.m. M, being A-closed is preserved by *most* generic extensions, i.e., by a comeager set of M-generic filters, for any partial order in M.

Proposition 2.14. *For every uB set A, if M is an A-closed c.t.m. and $\mathbb{P}$ is a partial order in M, then the set of M-generic filters $g \subset \mathbb{P}$ such that $M[g]$ is A-closed is comeager.*

Proof. By Proposition 2.11, for each $\mathbb{P}$-name τ in M for a partial order there is a countable set $\mathcal{E}_\tau$ of dense subsets of $\mathbb{P} * \tau$ such that for every $(M \cup \mathcal{E}_\tau)$-generic forcing extension N of M by $\mathbb{P} * \tau$, $N \cap A \in N$. For each $\mathbb{P}$-name σ for a condition in τ and each $E \in \mathcal{E}_\tau$ there is a dense set $D(\tau, E, \sigma)$ of conditions $p \in \mathbb{P}$ for which there is some $\mathbb{P}$-name ρ for a condition in τ such that $(p, \rho) \in E$ and $p \Vdash_{\mathbb{P}} \rho \leq_\tau \sigma$. Let $\mathcal{D}$ be the set of all such sets $D(\tau, E, \sigma)$.

Now suppose that $M[g]$ is a $\mathcal{D}$-generic extension of M by $\mathbb{P}$. Let $\mathbb{Q}$ be a poset in $M[g]$. Then $\mathbb{Q} = i_g[\tau]$ for some $\mathbb{P}$-name $\tau \in M$. Since g is $\mathcal{D}$-generic, for each $E \in \mathcal{E}_\tau$, the set $E^* = \{i_g[\rho] \mid \exists p \in g \text{ such that } (p, \rho) \in E\}$ is dense in $\mathbb{Q}$. Let $\mathcal{E}'$ be the set of these E^*'s, and let $h \subset \mathbb{Q}$ be a $(M[g] \cup \mathcal{E}')$-generic filter. Then

$$g * h = \{(p, \sigma) \in \mathbb{P} * \tau \mid p \in g \text{ and } i_g[\sigma] \in h\}$$

is an $(M \cup \mathcal{E}_\tau)$-generic filter, and so $M[g][h] \cap A \in M[g][h]$. □

Let ZFC^* be a finite fragment of ZFC. Proposition 2.17 below shows that for any uB set A, there is an A-closed c.t.m. M which is a model of ZFC^*. But first let us prove the following:

Lemma 2.15. *If $A \subseteq \mathbb{R}$ is uB and κ is such that $V_\kappa \models ZFC$, then A is uB in V_κ.*

Proof. Let us see that for each poset $\mathbb{P}$ in V_κ there are trees $T, S \in V_\kappa$ such that $p[T] = A$ and $p[S] = \omega^\omega \setminus A$, and for all $\mathbb{P}$-generic filters G over V_κ, $V_\kappa[G] \models p[T] = \omega^\omega \setminus p[S]$. So fix $\mathbb{P} \in V_\kappa$ and suppose S, T witness A is uB for $\mathbb{P}$ in V. Let τ be a $\mathbb{P}$-name in V_κ for the set of reals of the $\mathbb{P}$-extension. Let θ be a large-enough regular cardinal such that $S, T \in H(\theta)$. Take $X \prec H(\theta)$ such that $|X| < \kappa$ and $\{S, T\} \cup \tau \cup A \subseteq X$. Let M be the image of X by the transitive collapse π. Then $\pi(S), \pi(T) \in V_\kappa$ and they witness the universal Baireness of A for $\mathbb{P}$ in V_κ, since $p[T] = p[\pi(T)]$ and $p[S] = p[\pi(S)]$. □

The notion of strong A-closure defined below is not standard. However, as we shall see in Section 2.5 below, the syntactic relation for Ω-logic (Definition 2.28) would not change if strong A-closure is used in place of A-closure.

Definition 2.16. Given $A \subseteq \mathbb{R}$, a transitive $\in$-model M of (a fragment of) ZFC is *strongly A-closed* if for all posets $\mathbb{P} \in M$ and all M-generic $G \subseteq \mathbb{P}$, $M[G] \cap A \in M[G]$.

Notice that by Lemma 2.11, for c.t.m.'s, if A is a uB set, then strong A-closure implies A-closure. Note also that if M is strongly A-closed, $\mathbb{P} \in M$, and $G \subseteq \mathbb{P}$ is M-generic, then $M[G]$ is also strongly A-closed.

Proposition 2.17. *Suppose $A \subseteq \mathbb{R}$ is uB, and κ is such that $V_\kappa \models ZFC$. Then every forcing extension of the transitive collapse of any countable elementary submodel of V_κ containing A is strongly A-closed. In particular, the transitive collapse of any countable elementary submodel of V_κ containing A is A-closed.*

Proof. By Lemma 2.15, A is uB in V_κ. Let $X \prec V_\kappa$ be countable such that $A \in X$. Let M be the image of X by the transitive collapse π. We want to see that any forcing extension of M is strongly A-closed. It suffices to see that M is strongly A-closed. Let $\mathbb{P} \in M$ and let $g \subseteq \mathbb{P}$ be an M-generic filter.

Let S and T be trees in X witnessing the universal Baireness of A for $\pi^{-1}(\mathbb{P})$. Then $\pi(S) = \bar{S}$ and $\pi(T) = \bar{T}$ are trees in M witnessing the universal Baireness of $A \cap M$ for $\mathbb{P}$. If σ is a $\mathbb{P}$-name for a real in M, in $M[g]$, $i_g[\sigma]$ is in $p[\bar{S}]$ or in $p[\bar{T}]$ and not in both, by elementarity of the collapsing map. Thus, since $p[\bar{S}] \subseteq p[S]$ and $p[\bar{T}] \subseteq p[T]$,

$$i_g[\sigma] \in A \text{ iff } i_g[\sigma] \in (p[\bar{T}])^{M[g]}.$$

Hence, $M[g] \cap A = (p[\bar{T}])^{M[g]} \in M[g]$, and M is strongly A-closed. □

Recall the following result of Woodin:

Theorem 2.18 (cf.[7]). *Suppose there is a proper class of Woodin cardinals. Then for every uB set of reals A and every forcing notion $\mathbb{P}$, if $G \subseteq \mathbb{P}$ is a V-generic filter, then in $V[G]$ there is an elementary embedding from $L(A, \mathbb{R}^V)$ into $L(A_G, \mathbb{R}^{V[G]})$ sending A to A_G.*

Corollary 2.19. *Suppose there is a proper class of Woodin cardinals. Then for every uB set of reals A and every forcing notion $\mathbb{P}$, if $G \subseteq \mathbb{P}$ is V-generic, then in $V[G]$, for every formula $\varphi(x,y)$ and every $r \in \mathbb{R}^V$,*

$$L(A, \mathbb{R}^V) \models \varphi(A, r) \text{ iff } L(A_G, \mathbb{R}^{V[G]}) \models \varphi(A_G, r).$$

In particular, if $\varphi(x,y)$ is the formula that defines A-closure, as in Corollary 2.12, it follows that a c.t.m. M is A-closed iff for every (some) generic extension $V[G]$ of V, M is A_G-closed in $V[G]$.

The notion of A-closed model makes sense even for non-well-founded ω-models, i.e., given a uB set $A \subseteq \mathbb{R}$, an ω-model M of (a fragment of) ZFC is *A-closed* if for all posets $\mathbb{P} \in M$, for all $G \subseteq \mathbb{P}$ V-generic,

$$V[G] \models M[G] \cap A_G \in M[G]$$

i.e., $\Vdash_{\mathbb{P}}$ "$M[\dot{G}] \cap A_{\dot{G}} \in M[\dot{G}]$", where $\dot{G}$ is the standard $\mathbb{P}$-name for the generic filter.

However, let us see that the notion of A-closed set is a natural generalization of wellfoundedness.

Lemma 2.20. *Let ZFC^* be ZF minus the Powerset axiom. Suppose N is an ω-model of ZFC^* such that $WF \cap N \in N$. Then for all $x \in \omega^\omega \cap N$, $x \in WF$ iff $x \in WF^N$.*

Proof. $\Rightarrow$) By the downward absoluteness of Π^1_1 formulas for ω-models.

$\Leftarrow$) Suppose $x \in \omega^\omega \cap N$, $x \in WF^N$ and $x \notin WF$. For each n, let $E_x \restriction n = \{m | m E_x n\}$, and let x_n be a real coding $E_x \restriction n$. Since $N \models$ "E_x is well founded" and $WF \cap N \in N$, there is a $n_0 \in \omega$ such that $x_{n_0} \notin WF$ but for all $m E_x n_0$, $x_m \in WF$. Since $E_x \restriction n_0$ is ill founded, there is an $m E_x n_0$ such that $E_x \restriction m$ is ill founded, giving a contradiction. □

Lemma 2.21. *Every ω-model of ZFC which is WF-closed is well founded.*

Proof. Suppose (M, E) is a non well-founded WF-closed ω-model of ZFC. Let γ be an "ordinal" of M which is ill founded in V, let G be M-generic for a partial order in M making γ countable and let x be a real in $M[G]$ coding a wellordering of ω of ordertype γ. Then $x \in WF^{M[G]} \setminus WF$, which by Lemma 2.20 implies that $M[G] \cap WF \notin M[G]$. Since M is WF-closed, by the previous Lemma, $x \notin WF^{M[G]}$. So $E_x \in M[G]$ and is not well founded. Hence $M[G] \not\models$ "Foundation", contradicting the fact that $M \models$ "Foundation" and $M[G]$ is a forcing extension of M. □

Theorem 2.22. *For every ω-model of ZFC, (M, E), the following are equivalent:*

i) *(M, E) is well founded.*
ii) *(M, E) is A-closed for each Π^1_1 set A.*

Proof. i) $\Rightarrow$ ii). Suppose (M, E) is an ω-model of ZFC which is well founded. Fix $A \subseteq \mathbb{R}$ a Π^1_1 set. Let $\mathbb{P} \in M$ and let H be a $\mathbb{P}$-generic over V.

Let $(N, \in)$ be the transitive collapse of (M, E), and let $G = \pi"H$. Since $\pi(\mathbb{P}) \in N$, G is $\pi(\mathbb{P})$-generic over V and N is transitive, G is $\pi(\mathbb{P})$-generic over N. Since Π^1_1 sets are absolute for transitive models of ZFC and A is Π^1_1, in $V[G]$, $A^{N[G]} = N[G] \cap A = N[G] \cap A \cap V[G] = N[G] \cap A^{V[G]}$. And since $A^{V[G]} = A_G$,

$$A^{N[G]} = N[G] \cap A_G \in N[G].$$

Since M is an ω-model, the transitive collapse π is the identity on the reals and therefore,

$$A^{M[H]} = M[H] \cap A_H \in M[H].$$

ii) $\Rightarrow$ i). Suppose (M, E) is A-closed for each Π^1_1 set. Then it is WF-closed, since WF is Π^1_1. So by Lemma 2.21, (M, E) is well founded. □

2.3. AD$^+$

Definition 2.23. (cf.[12]) A set $A \subseteq \mathbb{R}$ is ∞-*Borel* if for some $S \cup \{\alpha\} \subseteq On$ and some formula with two free variables $\varphi(x, y)$,

$$A = \{y \in \mathbb{R} \mid L_\alpha[S, y] \models \varphi(S, y)\}.$$

Assuming $AD + DC$, a set of reals A is ∞-Borel iff $A \in L(S, \mathbb{R})$, for some $S \subseteq \text{Ord}$ (cf. [12]).

Definition 2.24. Θ is the least ordinal α which is not the range of any function $\pi \colon \mathbb{R} \to \alpha$. So, if the reals can be well ordered, then $\Theta = (2^\omega)^+$.

Recall that $DC_{\mathbb{R}}$ is the statement:

$$\forall R(R \subseteq \omega^\omega \times \omega^\omega \wedge \forall x \in \omega^\omega \exists y \in \omega^\omega((x, y) \in R) \to \exists f \in (\omega^\omega)^\omega \forall n \in \omega((f(n), f(n+1)) \in R)).$$

Definition 2.25. (cf.[12]) $(ZF + DC_{\mathbb{R}})$ AD^+ says:

i) Every set of reals is ∞-*Borel,*
ii) If $\lambda < \Theta$ and $\pi \colon \lambda^\omega \to \omega^\omega$ is a continuous function, where λ has been given the discrete topology, then $\pi^{-1}(A)$ is determined for every $A \subseteq \omega^\omega$.

AD^+ trivially implies AD, and it is not known if AD implies AD^+. Woodin has shown that if $L(\mathbb{R}) \models AD$, then $L(\mathbb{R}) \models AD^+$.

AD^+ is absolute for inner models containing all the reals:

Theorem 2.26. (cf.[12])(AD^+) *For any transitive inner model M of ZF with $\mathbb{R} \subseteq M$, $M \models AD^+$.*

Theorem 2.27 ([12]). *If there exists a proper class of Woodin cardinals and $A \subseteq \mathbb{R}$ is uB then:*

1) $L(A, \mathbb{R}) \models AD^+$,
2) *Every set in* $\mathcal{P}(\mathbb{R}) \cap L(A, \mathbb{R})$ *is uB.*

2.4. Definition of $\vdash_\Omega$ and invariance under forcing

Note that the following are equivalent:

i) For all A-closed c.t.m. M of ZFC, all $\alpha \in M \cap On$, and all $\mathbb{B}$ such that $M \models$ "$\mathbb{B}$ is a c.B.a", if $M_\alpha^{\mathbb{B}} \models T$, then $M_\alpha^{\mathbb{B}} \models \varphi$.
ii) For all A-closed c.t.m. M of ZFC, and for all $\alpha \in M \cap On$,
if $M_\alpha \models T$, then $M_\alpha \models \varphi$.

Proof. ii) $\Rightarrow$ i). Let M be an A-closed c.t.m. of ZFC, $\alpha \in M \cap On$, and let $\mathbb{B}$ be such that $M \models$ "$\mathbb{B}$ is a c.B.a". Suppose $M_\alpha^{\mathbb{B}} \models T$ and, towards a contradiction, suppose that, in M, for some $b \in \mathbb{B}$, $b \Vdash$ "$M[\dot{g}]_\alpha \models \neg\varphi$", where $\dot{g}$ is the standard name for the generic filter. By Proposition 2.14, there is g $\mathbb{B}$-generic over M such that $b \in g$ and $M[g]$ is A-closed. We have $M[g]_\alpha \models T$. Hence, by ii) $M[g]_\alpha \models \varphi$, contradicting the assumption that b forced $M[\dot{g}]_\alpha \models \neg\varphi$. □

Definition 2.28 ([17]). For $T \cup \{\varphi\} \subseteq Sent$, we write $T \vdash_\Omega \varphi$ if there exists a uB set $A \subseteq \mathbb{R}$ such that:

1) $L(A, \mathbb{R}) \models AD^+$,
2) Every set in $\mathcal{P}(\mathbb{R}) \cap L(A, \mathbb{R})$ is uB,
3) For all A-closed c.t.m. M of ZFC and for all $\alpha \in M \cap On$, if $M_\alpha \models T$, then $M_\alpha \models \varphi$.

Thus, by Theorem 2.27, if there exists a proper class of Woodin cardinals, $T \vdash_\Omega \varphi$ iff there exists a uB set $A \subseteq \mathbb{R}$ such that 3) above holds.

Notice that, by the equivalence of i) and ii) above, if T is recursive, then point 3) of the last definition can be written as:

3′) For all A-closed c.t.m. M of ZFC, $M \vDash$ "$T \vDash_\Omega \varphi$".

By Theorem 2.27, if there exists a proper class of Woodin cardinals, or if just $L(\mathbb{R}) \models AD$ and every set of reals in $L(\mathbb{R})$ is uB, then for every $T \cup \{\varphi\} \subseteq Sent$, $T \vdash \varphi$ implies $T \vdash_\Omega \varphi$. However, as we would expect, the converse does not hold: Let M be a c.t.m. of ZFC and let $\alpha \in M \cap On$ be such that $M_\alpha \vDash ZFC$. Since M_α is a standard model, $M_\alpha \vDash CON(ZFC)$. This shows $ZFC \vdash_\Omega CON(ZFC)$.

We say that a sentence $\varphi \in Sent$ is *Ω_T-provable* if $T \vdash_\Omega \varphi$. And if A witnesses $T \vdash_\Omega \varphi$, then we say that A is an *Ω_T-proof* of φ, or that A is an Ω-proof of φ from T.

Notice that if A is uB and satisfies 1) and 2) of Definition 2.28, then A *is an Ω_T-proof of φ* iff
$L(A, \mathbb{R}) \vDash \forall M \forall \alpha$ (M is a A-closed c.t.m. of $ZFC \wedge \alpha \in M \cap On \wedge M_\alpha \models T \rightarrow M_\alpha \vDash \varphi$).

It is not very difficult to see that the complexity of the relation $T \vdash_\Omega \varphi$ is at most Σ_3.

Remark 2.29. Arguments in [7] essentially show that if AD^+ holds then there exist A-closed models of ZFC for every set of reals A.

Lemma 2.30. *Given A, B uB sets, the set $C = A \times B$ is uB, and if M is a C-closed c.t.m., then M is both A-closed and B-closed.*

Proof. Given $\gamma \in M \cap On$, let $\mathbb{P} = \mathrm{Coll}(\omega, \gamma)$. For a fixed $\mathbb{P}$-name $\dot{y}$ for an element of $B_{\dot{G}}$,

$$\begin{aligned}&\{(\tau, p) \mid p \in \mathbb{P}, \tau \text{ is a } \mathbb{P}\text{-name for a real number and } p \Vdash^V (\tau, \dot{y}) \in (A \times B)_{\dot{G}}\}\\ &= \{(\tau, p) \mid p \in \mathbb{P}, \tau \text{ is a } \mathbb{P}\text{-name for a real number and } p \Vdash^V_{\mathbb{P}} \tau \in A_{\dot{G}}\}.\end{aligned}$$

Hence if M is C-closed, this set belongs to M and thus M is A-closed. Symmetrically, the same holds for B. □

Corollary 2.31. *Let $T \cup \{\varphi, \psi\} \subseteq Sent$. Suppose that for every uB set A, $L(A, \mathbb{R}) \vdash AD^+$ and every set in $\mathcal{P}(\mathbb{R}) \cap L(A, \mathbb{R})$ is uB. Suppose $T \vdash_\Omega \psi$ and $T \vdash_\Omega \varphi$. If $T \cup \{\psi, \varphi\} \vdash \theta$, then $T \vdash_\Omega \theta$. Hence,*

i) *If $T \vdash_\Omega \varphi$ and $T \vdash_\Omega \psi$, then $T \vdash_\Omega \varphi \wedge \psi$.*
ii) *If $T \vdash_\Omega \varphi$ and $T \vdash_\Omega \varphi \rightarrow \psi$, then $T \vdash_\Omega \psi$.*

Proof. Let A and B be Ω_T-proofs of ψ and φ, respectively. Let us see that $A \times B$ is a Ω_T-proof of θ. Let M be an $(A \times B)$-closed model. Thus, M is both A-closed and B-closed. Suppose $\alpha \in M \cap On$ and $\mathbb{B} \in M$ are such that $M^{\mathbb{B}}_\alpha \models T$. Since M is A-closed, $M^{\mathbb{B}}_\alpha \models \psi$ and since M is B-closed, $M^{\mathbb{B}}_\alpha \models \varphi$. So, $M^{\mathbb{B}}_\alpha \models \theta$. □

The notion of Ω-provability differs from the usual notions of provability, e.g., in first-order logic, in that there is no deductive calculus involved. In Ω-logic, the same uB set may witness the Ω-provability of different sentences. For instance, all tautologies have the same *proof* in Ω-logic, namely, $\emptyset$. In spite of this, it is possible to define a notion of *length of proof* in Ω-logic. This can be accomplished in several ways. For instance: for $A \subseteq \mathbb{R}$, let M_A be the model $L_{\kappa_A}(A, \mathbb{R})$, where κ_A is the least admissible ordinal for $(A, \mathbb{R})$, i.e., the least ordinal $\alpha > \omega$ such that $L_\alpha(A, \mathbb{R})$ is a model of Kripke-Platek set theory. The following result is due to Solovay:

Lemma 2.32. *Assume AD. Then for every $A, B \subseteq \mathbb{R}$, either $A \in M_B$ or $B \in M_A$.*

Proof. Consider the two-player game in which both players play integers so that at the end of the game player I has produced x and player II has produced y. Player I wins the game iff $x \in A \leftrightarrow y \in B$. It τ is a winning strategy for player I, then for every real z, $z \in B$ iff $\tau * z \in A$, and so $B \in M_A$. And if σ is a winning strategy for player II, then for every real z, $z \in A$ iff $z * \sigma \notin B$, and so $A \in M_B$. □

Thus, under *AD*, for $A, B \subseteq \mathbb{R}$, we have $\kappa_A < \kappa_B$ iff $A \in M_B$ and $B \notin M_A$. It follows that $\kappa_A = \kappa_B$ iff $M_A = M_B$.

If A is a uB set of reals that witnesses $T \vdash_\Omega \varphi$, then we can say that κ_A is the *length of the Ω_T-proof* A. Using this notion of length of proof we can find sentences, like the Gödel-Rosser sentences in first-order logic, that are undecidable in Ω-logic. For instance, let $\varphi(A,\theta)$ be the formula:

$$\forall M \forall \alpha((M \text{ is an } A\text{-closed c.t.m. of } ZFC \wedge$$
$$\alpha \in M \cap On \ \wedge \ M_\alpha \models ZFC) \rightarrow M_\alpha \models \theta).$$

Using Gödel's diagonalization, let $\theta \in Sent$ be such that:

$$ZFC \vdash \text{“}\theta \leftrightarrow \forall A(\varphi(A,\theta) \rightarrow \exists B(\varphi(B,\neg\theta) \wedge \kappa_B < \kappa_A))\text{”}$$

Assuming there is a proper class of Woodin cardinals, we have:

$$ZFC \vdash_\Omega \text{“}\theta \leftrightarrow \forall A(\varphi(A,\theta) \rightarrow \exists B(\varphi(B,\neg\theta) \wedge \kappa_B < \kappa_A))\text{”}$$

Suppose $ZFC \vdash_\Omega \theta$ and C witnesses it. Then

$$ZFC \vdash_\Omega \text{“}\forall A(\varphi(A,\theta) \rightarrow \exists B(\varphi(B,\neg\theta) \wedge \kappa_B < \kappa_A))\text{”}$$

is witnessed by some D. Assuming there is an inaccessible limit of Woodin cardinals, we can find a $(C \times D)$-closed c.t.m. M of ZFC with a strongly inaccessible cardinal α, such that M satisfies that for every uB set of reals A, AD^+ holds in $L(A,\mathbb{R})$, and every set of reals in $L(A,\mathbb{R})$ is uB (see 2.27). By reflection, let $\alpha \in M \cap On$ be such that $C \cap M \in M_\alpha$, $M_\alpha \models$ “$C \cap M$ is uB”, and

$$M_\alpha \models ZFC + \forall A(A \text{ is uB} \rightarrow L(A,\mathbb{R}) \models AD).$$

Then, $M_\alpha \models \theta$ and

$$M_\alpha \models \text{“}\forall A(\varphi(A,\theta) \rightarrow \exists B(\varphi(B,\neg\theta) \wedge \kappa_B < \kappa_A)).\text{”}$$

Moreover, $M_\alpha \models \varphi(C \cap M, \theta)$. Hence, in M_α there is B such that $\varphi(B,\neg\theta)$ and $\kappa_B < \kappa_{C\cap M}$. But since $M_\alpha \models$ “$L(B, C\cap M, \mathbb{R}) \models AD$”, by Lemma 2.32, we have $M_\alpha \models B \in M_{C\cap M}$. It follows that:

1. $M_{C\cap M} \models \varphi(C\cap M, \theta)$
2. $M_{C\cap M} \models \varphi(B, \neg\theta)$.

Let $N \in M_{C\cap M}$ be a c.t.m. of ZFC that is both $C \cap M$-closed and B-closed (see Remark 2.29). Then, for any β, if $N_\beta \models ZFC$, we would have $N_\beta \models \theta \wedge \neg\theta$, which is impossible.

An entirely symmetric argument would yield a contradiction under the assumption that $ZFC \vdash_\Omega \neg\theta$, thereby showing that θ is undecidable from ZFC in Ω-logic.

A much finer notion of length of proof in Ω-logic is provided by the Wadge hierarchy of sets of reals (see [9] and [16]).

We shall now see that the relation $\vdash_\Omega$ is also invariant under forcing. In the proof of this, we will use the following result (see [6], Section 3.4).

Theorem 2.33. *Suppose that there exists a proper class of Woodin cardinals, δ is a Woodin cardinal and $j\colon V \rightarrow M[G]$ is an embedding derived from forcing with $\mathbb{P}_{<\delta}$. Then every universally Baire set of reals in $V[G]$ is universally Baire in M.*

Theorem 2.34 ([17]). *Suppose that there exists a proper class of Woodin cardinals. Then for all* $\mathbb{P}$,

$$T \vdash_\Omega \varphi \quad \text{iff} \quad V^{\mathbb{P}} \models \text{“}T \vdash_\Omega \varphi\text{”}$$

Proof. $\Rightarrow$) Let A be an Ω_T-proof of φ.

Then $L(A, \mathbb{R}) \models \forall M \forall \alpha$ (M is a A-closed c.t.m. of $ZFC \wedge \alpha \in M \cap On \wedge M_\alpha \models T \rightarrow M_\alpha \models \varphi$).

Suppose $G \subseteq \mathbb{P}$ is V-generic. By Corollary 2.19, in $V[G]$,

$L(A_G, \mathbb{R}^{V[G]}) \models \forall M \forall \alpha$ (M is a A_G-closed c.t.m. of $ZFC \wedge \alpha \in M \cap On \wedge M_\alpha \models T \rightarrow M_\alpha \models \varphi$).

Since A is uB, by Remark 2.6, A_G is uB in $V[G]$. Hence, A_G is an Ω_T-proof of φ in $V[G]$.

$\Leftarrow$) Assume $V^{\mathbb{P}} \models$ “$T \vdash_\Omega \varphi$”. Let γ be a strongly inaccessible cardinal such that $\mathbb{P} \in V_\gamma$. Pick a Woodin cardinal $\delta > \gamma$. Consider $a = \mathcal{P}_{\omega_1}(V_\gamma) \in \mathbb{P}_{<\delta}$ (see Fact 1.4). Forcing with $\mathbb{P}_{<\delta}$ below a makes V_γ countable, so there is a $\mathbb{P}$-name τ for a partial order such that $\mathbb{P}_{<\delta}(a)$ is forcing-equivalent to $\mathbb{P} * \tau$. Fix $G \subseteq \mathbb{P}_{<\delta}(a)$ V-generic, and let $j \colon V \rightarrow M$ be the induced embedding. Then $j(\delta) = \delta$ and $V[G] \models M^{<\delta} \subseteq M$. We have $V[G] = V[H_0][H_1]$, with $H_0 \subseteq \mathbb{P}$, V-generic. Thus, $V[H_0] \models$ “$T \vdash_\Omega \varphi$”, witnessed by some uB set A. By the other direction of this theorem, $V[G] \models$ “$T \vdash_\Omega \varphi$”, witnessed by A_G. Hence,

$V[G] \models$ “A_G is $uB \wedge \forall N \forall \alpha$ (N is a A_G-closed c.t.m. of $ZFC \wedge \alpha \in N \cup On \wedge N_\alpha \models T \rightarrow N_\alpha \models \varphi$)”.

By Theorem 2.33, A_G is a uB set in M, and since M is closed under countable sequences,

$M \models$ “$\forall N \forall \alpha$ (N is a A-closed c.t.m. of $ZFC \wedge \alpha \in N \cap On \wedge N_\alpha \models T \rightarrow N_\alpha \models \varphi$)”. Thus, $M \models$ “$T \vdash_\Omega \varphi$”. By applying the induced elementary embedding, we have $V \models$ “$T \vdash_\Omega \varphi$”. □

2.5. A-closure vs strong A-closure

Recall (Definition 2.16) that for $A \subseteq \mathbb{R}$, a transitive $\in$-model M of (a fragment of) ZFC is *strongly A-closed* if for all posets $\mathbb{P} \in M$ and all M-generic $G \subseteq \mathbb{P}$, $M[G] \cap A \in M[G]$.

We shall see that the relation $\vdash_\Omega$ would not change if we were to use strong A-closure in place of A-closure in its definition.

Recall the definition of scale on a set of reals (see [9]):

Definition 2.35. If A is a set of reals, then a *scale* on A is a sequence $\langle \leq_i \colon i < \omega \rangle$ of prewellorderings of A satisfying the property that whenever $\langle x_i : i < \omega \rangle$ is a sequence contained in A converging to a real x and $f \colon \omega \rightarrow \omega$ is a function such that

$$\forall i < \omega \ \forall j \in [f(i), \omega) \ (x_{f(i)} \leq_i x_j \wedge x_j \leq_i x_{f(i)}),$$

then x is in A, and for all $i < \omega$ we have $x \leq_i x_{f(i)}$.

If Γ is a pointclass that is closed under continuous preimages, $A \in \Gamma$, and $\langle \leq_i : i < \omega \rangle$ is a scale on A, then $\langle \leq_i : i < \omega \rangle$ is called a Γ-*scale* if there are sets $X, Y \subset \omega \times \omega^\omega \times \omega^\omega$ in Γ (identifying each integer with the corresponding constant function) such that

$$X = \{(i, x, y) \mid x \leq_i y\} = (\omega \times \omega^\omega \times \omega^\omega) \setminus Y \cap (\omega \times \omega^\omega \times A).$$

We say that Γ has the *scale property* if for every $A \in \Gamma$ there is a Γ-scale on A. If there exists a proper class of Woodin cardinals, then the class of uB sets has the scale property (this fact is due to Steel; see, for instance, Section 3.3 of [6]).

If $\langle \leq_i : i < \omega \rangle$ is a scale on a set of reals A, and for each $i \in \omega$ and $x \in A$ we let $\rho_i(x)$ denote the $\leq_i$-rank of x, then the tree

$$S = \{(s, \sigma) \in \omega^{<\omega} \times \mathrm{Ord}^{<\omega} \mid \exists x \in A\ x \restriction |s| = s \wedge \langle \rho_i(x) : i < |s| \rangle = \sigma\}$$

projects to A. We call this *the tree corresponding to the scale.*

The argument below comes from [11].

Theorem 2.36. *Let A be a universally Baire set of reals and suppose that M is an A-closed c.t.m. of ZFC. Let B denote the complement of A. Let $\langle \leq_i^A : i < \omega \rangle$ be a uB scale on A as witnessed by uB sets X and Y, let $\langle \leq_i^B : i < \omega \rangle$ be a uB scale on B as witnessed by uB sets W and Z, and suppose that M is $X \times Y \times W \times Z$-closed. Then M is strongly A-closed.*

Proof. First note that for any well-founded model N, if $\{N \cap X, N \cap Y, N \cap A\} \in N$, then $\langle \leq_i^A \cap N : i < \omega \rangle$ is in N and is a scale for $A \cap N$ in N (and similarly, for W, Z and B). Furthermore, if N is $(X \times Y \times A)$-closed, then for every partial order $\mathbb{P}$ in N there are $\mathbb{P}$-names $\chi_{\mathbb{P}}$, $\upsilon_{\mathbb{P}}$ and $\alpha_{\mathbb{P}}$ such that for comeagerly-many N-generic filters $g \subset \mathbb{P}$, $X \cap N[g] = \chi_g$, $Y \cap N[g] = \upsilon_g$ and $A \cap N[g] = \alpha_g$ (the proof of this is similar to the second parts of the proofs of Lemmas 2.11 and 2.13).

Let γ be an ordinal in M. Since $\mathrm{Coll}(\omega, \gamma)$ is homogeneous and M is $(X \times Y \times A)$-closed, for every pair of conditions p, q in $\mathrm{Coll}(\omega, \gamma)$ there exist M-generic filters g_p and g_q contained in $\mathrm{Coll}(\omega, \gamma)$ such that $p \in g_p$, $q \in g_q$, $M[g_p] = M[g_q]$,

$$i_{g_p}[\chi_{\mathrm{Coll}(\omega,\gamma)}] = i_{g_q}[\chi_{\mathrm{Coll}(\omega,\gamma)}] = M[g_p] \cap X,$$

$$i_{g_p}[\upsilon_{\mathrm{Coll}(\omega,\gamma)}] = i_{g_q}[\upsilon_{\mathrm{Coll}(\omega,\gamma)}] = M[g_p] \cap Y,$$

and

$$i_{g_p}[\alpha_{\mathrm{Coll}(\omega,\gamma)}] = i_{g_q}[\alpha_{\mathrm{Coll}(\omega,\gamma)}] = M[g_p] \cap A.$$

Therefore, for every pair $(a, b) \in \omega^{<\omega} \times \mathrm{Ord}^{<\omega}$, the empty condition in $\mathrm{Coll}(\omega, \gamma)$ decides whether (a, b) is in the tree corresponding to the scale associated to $\chi_{\mathrm{Coll}(\omega,\gamma)}$ and $\upsilon_{\mathrm{Coll}(\omega,\gamma)}$, and therefore the tree T_γ corresponding to this scale in any M-generic extension by $\mathrm{Coll}(\omega, \gamma)$ exists already in M. Since there exists a model N such that $\{N \cap A, N \cap X, N \cap Y\} \in N$ and T_γ is the tree of the scale corresponding to $N \cap X$ and $N \cap Y$ in N, $p[T_\gamma]^V \subset A$ (since X and Y define a scale on A). The remarks above apply to B, W and Z, as well, and so there is a tree S_γ in M which projects in V to a subset of B, and furthermore, T_γ and S_γ project to complements in all forcing extensions of M by $\mathrm{Coll}(\omega, \gamma)$.

Let $\mathbb{P}$ be a partial order in M. Then $\mathbb{P}$ regularly embeds into some partial order of the form $\mathrm{Coll}(\omega, \gamma)$, $\gamma \in On \cap M$. Fixing such a γ, we have that for any $\mathbb{P}$-generic extension N of M, $p[T_\gamma]^N = A \cap N$ and $p[S_\gamma]^N = B \cap N$. □

Let the relation $\vdash_\Omega^-$ be defined as $\vdash_\Omega$ (Definition 2.28) but requiring strong A-closure instead of A-closure. i.e.,

$T \vdash_\Omega^- \varphi$ if there exists a uB set $A \subseteq \mathbb{R}$ such that:

1) $L(A, \mathbb{R}) \models AD^+$,
2) Every set in $\mathcal{P}(\mathbb{R}) \cap L(A, \mathbb{R})$ is uB,
3) For all strongly A-closed c.t.m. M of ZFC and for all $\alpha \in M \cap On$, if $M_\alpha \models T$, then $M_\alpha \models \varphi$.

Since for any uB set A and any c.t.m. M strong A-closure implies A-closure (see Lemma 2.11), clearly $T \vdash_\Omega \varphi$ implies $T \vdash_\Omega^- \varphi$.

Now suppose $T \vdash_\Omega^- \varphi$, witnessed by a uB set A. We would like to see that there is a uB set B such that all B-closed models are strongly A-closed. Theorem 2.36 gives us this, under the assumption that the collection of universally Baire sets has the scale property, which, as we mentioned above, it does when there exist proper class many Woodin cardinals. Even without this assumption one can show that such a B exists, though the proof of this is beyond the scope of this paper. Here is a sketch. Note first that M is a strongly A-closed c.t.m. iff $L(A, \mathbb{R}) \models$ "M is a strongly A-closed c.t.m." So, in $L(A, \mathbb{R})$, A satisfies the following predicate $P(X)$ on sets $X \subseteq \mathbb{R}$:

$$\forall M \forall \alpha (M \text{ a strongly } X\text{-closed c.t.m. of ZFC} \wedge$$
$$\alpha \in M \cap On \ \wedge \ M_\alpha \models T \rightarrow M_\alpha \models \varphi).$$

We now apply Woodin's generalizations of the Martin-Steel theorem on scales in $L(\mathbb{R})$ [8] and the Solovay Basis Theorem (see [3]) to the context of AD$^+$, stated as follows.

Theorem 2.37 (ZF + DC$_\mathbb{R}$). *If AD^+ holds and $V = L(\mathcal{P}(\mathbb{R}))$ then*

- *the pointclass Σ_1^2 has the scale property,*
- *every true Σ_1-sentence is witnessed by a $\utilde{\Delta}_1^2$ set of reals.*

We may then let B be a $\utilde{\Delta}_1^2$ (in $L(A, \mathbb{R})$) solution to $P(X)$. Note that by (2) above, B is uB and, by Theorem 2.26, it is also a witness to $T \vdash_\Omega^- \varphi$. Since $L(A, \mathbb{R}) \models AD^+$, both B and its complement have $\utilde{\Sigma}_1^2$ scales in $L(A, \mathbb{R})$. Those scales are uB (again, by (2) above). So, as in Theorem 2.36, we can find $C \in L(A, \mathbb{R})$ such that if M is a C-closed c.t.m., then M is strongly B-closed. Thus, C witnesses $T \vdash_\Omega \varphi$.

One can formulate a property which roughly captures the difference between A-closure and strong A-closure. We will call this property A-*completeness*, though that term is not standard.

Definition 2.38. Let A be a set of reals. Let us call a c.t.m. M of ZFC A-*complete* if for every forcing notion $\mathbb{P} \in M$, every name for a real $\tau \in M^{\mathbb{P}}$, and every $p \in \mathbb{P}$:

1. If for comeagerly-many M-generic $G \subseteq \mathbb{P}$, $p \in G$ implies $i_G[\tau] \in A$, then for every M-generic $G \subseteq \mathbb{P}$, $p \in G$ implies $i_G[\tau] \in A$.
2. If for comeagerly-many M-generic $G \subseteq \mathbb{P}$, $p \in G$ implies $i_G[\tau] \notin A$, then for every M-generic $G \subseteq \mathbb{P}$, $p \in G$ implies $i_G[\tau] \notin A$.

The conjunction of A-closure and A-completeness implies strong-A-closure.

Lemma 2.39. *Let M be a c.t.m. and A a uB set. If M is both A-closed and A-complete, then it is strongly-A-closed.*

Proof. Fix M and A and suppose M is A-closed and A-complete.

Let

$$\sigma = \{(\tau, p) \mid \tau \in M \text{ a simple } \mathbb{P}\text{-name for a real }, p \in \mathbb{P} \text{ and } p \Vdash^V_{\mathbb{P}} \tau \in A_{\dot{G}}\}.$$

By Proposition 2.9, σ is a $\mathbb{P}$-name that belongs to M.

We claim that for every M-generic $G \subseteq \mathbb{P}$, $i_G[\sigma] = M[G] \cap A$.

So, suppose $G \subseteq \mathbb{P}$ is an M-generic filter. If $\tau \in M$ is a simple $\mathbb{P}$-name for a real and $i_G[\tau] \in A$, then for some $p \in \mathbb{P}$, for a comeager set of M-generic filters g, if $p \in g$, then $i_g[\tau] \in A$. By 2.13, $p \Vdash^V_{\mathbb{P}} \tau \in A_{\dot{G}}$. Hence, $i_G[\tau] \in i_G[\sigma]$.

Now suppose $i_G[\tau] \in i_G[\sigma]$. So, for some $p \in G$, $p \Vdash^V_{\mathbb{P}} \tau \in A_{\dot{G}}$. By 2.13, the set of M-generic filters $g \subseteq \mathbb{P}$ such that $p \in g$ and $i_g[\tau] \in A$ is comeager. But since M is A-complete, for all M-generic $g \subseteq \mathbb{P}$ such that $p \in g$, $i_g[\tau] \in A$. In particular, $i_G[\tau] \in A$. □

Strong A-closure does not imply A-completeness, however. To see this, note that if x is a real and $A = \{x\}$, then every c.t.m. M is strongly-A-closed. But if x is Cohen-generic over M, then M is not A-complete, for if $\mathbb{P}$ is the Cohen forcing, and $\tau \in M^{\mathbb{P}}$ is a name for x, then the set $D = \{p \in \mathbb{P} : \ p \Vdash \tau \neq x\}$ is a dense subset of $\mathbb{P}$ (although $D \notin M$!). So, there is a comeager set of $\mathbb{P}$-generic filters over M such that for each G in the set, $i_G[\tau] \neq x$, i.e., $i_G[\tau] \notin A$. But for some M-generic G, $i_G[\tau] = x \in A$.

Similarly, A-completeness does not imply strong A-closure (and so it does not imply A-closure, either). As an example, let M satisfy ZFC + "$0^\sharp$ does not exist," and let $A = 0^\sharp$ (i.e., $\{n \mid n \in 0^\sharp\}$). Then M is clearly not A-closed, since $M[G] \cap A = A$ for all M-generic $G \subseteq \mathbb{P}$, all $\mathbb{P}$. But M is A-complete. To see this, fix $\mathbb{P}$, p, and τ, and suppose that for comeagerly-many M-generic G, if $p \in G$, then $i_G[\tau] \in A$. It follows then that $X = \{n : \exists p' \leq p \ (p' \Vdash \tau = n)\}$ is contained in A, which in turn implies that $i_G[\tau] \in A$ for all M-generic filters $G \subseteq \mathbb{P}$ that contain p.

3. The Ω-conjecture

Definition 3.1.

i) A sentence φ is *Ω_T-satisfiable* if $T \nvDash_\Omega \neg\varphi$, i.e., there exists α and $\mathbb{B}$ such that $V^{\mathbb{B}}_\alpha \models T + \varphi$.
ii) A set of sentences T is *Ω-satisfiable* if there exists a c.B.a. $\mathbb{B}$ and an ordinal α for which $V^{\mathbb{B}}_\alpha \models T$.

iii) A sentence φ is *Ω_T-consistent* if $T \nvdash_\Omega \neg\varphi$, i.e., for all uB set $A \subseteq \mathbb{R}$ satisfying 1) and 2) of Definition 2.28, there exists a countable transitive A-closed set M such that $M \models ZFC$, and there exists $\alpha \in M \cap On$ such that $M_\alpha \models T + \varphi$.
iv) A set of sentences T is *Ω-consistent* if $T \nvdash_\Omega \bot$, where $\bot$ is any contradiction, i.e., if for all $A \subseteq \mathbb{R}$ uB satisfying 1) and 2) of Definition 2.28, there exists a c.t.m. A-closed $M \models ZFC$ and $\alpha \in M$ such that $M_\alpha \models T$.
v) T is *Ω-inconsistent* if it is not Ω-consistent.

Observe that if AD^+ holds in $L(\mathbb{R})$ and every set of reals in $L(\mathbb{R})$ is uB, then every *Ω_T-consistent* sentence is consistent with T.

Fact 3.2. The following are equivalent for a set of sentences T:

i) T is *Ω-consistent.*
ii) $T \nvdash_\Omega \varphi$ for some φ.
iii) $T \nvdash_\Omega \neg\varphi$ for all $\varphi \in T$, i.e., for all $\varphi \in T$, φ is Ω_T-consistent.

Proof. i) $\Rightarrow$ ii). Trivial.

ii) $\Rightarrow$ iii). Without loss of generality, we may assume that for some uB set A, 1) and 2) of Definition 2.28 hold. Given such an A, by hypothesis there exist an A-closed c.t.m. M and $\alpha \in M \cap On$ such that $M_\alpha \models T + \neg\varphi$. Since $M_\alpha \models \psi$ for all $\psi \in T$, the same M and α witness that $T \nvdash_\Omega \neg\psi$, for all $\psi \in T$.

iii) $\Rightarrow$ i). Without loss of generality, we may assume 1) and 2) of Definition 2.28 hold for some uB set A. Moreover, we may also assume that $T \neq \emptyset$. So, let $\varphi \in T$. By hypothesis there exist an A-closed c.t.m. M and $\alpha \in M \cap On$ such that $M_\alpha \models T + \varphi$. Since $M_\alpha \models T + \neg\bot$, the same M and α witness that $T \nvdash_\Omega \bot$. □

Theorem 3.3 (Soundness, ([12])). *Assume there is a proper class of strongly inaccessible cardinals. For every $T \cup \{\varphi\} \in Sent$, $T \vdash_\Omega \varphi$ implies $T \models_\Omega \varphi$.*

Proof. Let A be a uB set A witnessing $T \vdash_\Omega \varphi$. Fix α and $\mathbb{B}$, and suppose $V_\alpha^{\mathbb{B}} \models T$. Let $\lambda > \alpha$ be a strongly inaccessible cardinal such that $A, \mathbb{B}, T \in V_\lambda$ and $V_\lambda \models$ "$\mathbb{B}$ is a c.B.a.". Take $X \prec V_\lambda$ countable with $A, \mathbb{B}, T \in X$. Let M be the transitive collapse of X, and let $\bar{\mathbb{B}}$ be the transitive collapse of $\mathbb{B}$. By Lemma 2.17 M is A-closed. Hence, if $M_\alpha^{\bar{\mathbb{B}}} \models T$, then $M_\alpha^{\bar{\mathbb{B}}} \models \varphi$. Since $V_\lambda \models$ "$V_\alpha^{\mathbb{B}} \models T$", by elementarity, $M \models$ "$M_\alpha^{\bar{\mathbb{B}}} \models T$". Hence, $M \models$ "$M_\alpha^{\bar{\mathbb{B}}} \models \varphi$". So, again by elementarity, $V_\lambda \models$ "$V_\alpha^{\mathbb{B}} \models \varphi$". Hence, $V_\alpha^{\mathbb{B}} \models \varphi$. □

The assumption of the existence of a proper class of inaccessible cardinals in the Theorem above is not necessary. However, the proof without this assumption is no longer elementary and would take us beyond the scope of this paper.

Thus, if there exists κ such that $V_\kappa \models ZFC + \varphi$, then $ZFC \nvdash_\Omega \neg\varphi$, i.e., φ is Ω_{ZFC}-consistent.

Another consequence of Soundness is that for every finite fragment T of ZFC, an Ω_T-provable sentence cannot be made false by forcing over V.

The following equivalence can be proved without using Theorem 3.3.

Fact 3.4. For every $T \subseteq Sent$, the following are equivalent:

i) For all $\varphi \in Sent$, $T \vdash_\Omega \varphi$ implies $T \models_\Omega \varphi$.
ii) T is Ω-satisfiable implies T is Ω-consistent.

Proof. i) $\Rightarrow$ ii). Suppose T is not Ω-consistent, i.e., $T \vdash_\Omega \perp$. By hypothesis, $T \models_\Omega \perp$ and so for all c.B.a. $\mathbb{B}$ and for all $\alpha \in On$, $V_\alpha^{\mathbb{B}} \not\models T$, and therefore T is not Ω-satisfiable.

ii) $\Rightarrow$ i). Suppose $T \not\models_\Omega \varphi$. Let $\mathbb{B}$ and α be such that $V_\alpha^{\mathbb{B}} \models T$ and $V_\alpha^{\mathbb{B}} \models \neg\varphi$. Then $T \cup \{\neg\varphi\}$ is Ω-satisfiable and therefore Ω-consistent. If $T \vdash_\Omega \varphi$, then $T \cup \{\neg\varphi\} \vdash_\Omega \varphi$. But then $T \cup \{\neg\varphi\} \vdash_\Omega \varphi \wedge \neg\varphi$, a contradiction. □

Thus, by Theorem 3.3 and Fact 3.4, if T is Ω-satisfiable then T is Ω-consistent, i.e., if there exist α and $\mathbb{B}$ such that $V_\alpha^{\mathbb{B}} \models T$, then for every uB set A there exist an A-closed c.t.m. M of ZFC and α in $On \cap M$ such that $M_\alpha \models T$.

Corollary 3.5 (Non-Compactness of $\vdash_\Omega$). *Suppose $L(\mathbb{R}) \models AD$ and every set of reals in $L(\mathbb{R})$ is universally Baire. Then there is a sentence φ such that $ZFC \vdash_\Omega \varphi$ and for all $S \subseteq ZFC$ finite, $S \nvdash_\Omega \varphi$.*

Proof. Take the sentence φ of Theorem 1.12. Suppose $ZFC \nvdash_\Omega \varphi$. Then for each uB set A there is an A-closed c.t.m. M and $\alpha \in M \cap On$ such that $M_\alpha \models ZFC + \neg\varphi$. With the same argument as in the proof of Theorem 1.12 applied to M_α we arrive to a contradiction.

Suppose now there is S finite such that $S \vdash_\Omega \varphi$. Then by Soundness, $S \models_\Omega \varphi$, and this yields a contradiction as in the proof of Theorem 1.12. □

The Ω-*conjecture* says: *If there exists a proper class of Woodin cardinals, then for each sentence of the language of set theory φ,*

$$\emptyset \models_\Omega \varphi \quad \text{iff} \quad \emptyset \vdash_\Omega \varphi.$$

The "if" direction is given by Soundness. So, the Ω-*conjecture* is just Completeness for Ω-logic, i.e., if $\emptyset \models_\Omega \varphi$, then $\emptyset \vdash_\Omega \varphi$, for every $\varphi \in Sent$.

Lemma 3.6. *The following are equivalent:*

i) *For all $\varphi \in Sent$, $\emptyset \models_\Omega \varphi$ implies $\emptyset \vdash_\Omega \varphi$.*
ii) *For every r.e. set $T \cup \{\varphi\} \subseteq Sent$, $T \models_\Omega \varphi$ implies $T \vdash_\Omega \varphi$.*

Proof. i) $\Rightarrow$ ii). Fix T r.e. and φ such that $T \models_\Omega \varphi$. Let $\varphi^* :=$ "$T \models_\Omega \varphi$". By Lemma 1.9, $\emptyset \models_\Omega \varphi^*$, and so by i), $\emptyset \vdash_\Omega \varphi^*$. Hence, there is a uB set A such that for every A-closed c.t.m. $M \models ZFC$, $M \models$ "$\emptyset \models_\Omega \varphi^*$". Then for all $\alpha \in M$, $M_\alpha \models$ "$T \models_\Omega \varphi$". Since $M \models ZFC$, by reflection, $M \models$ "$T \models_\Omega \varphi$". This shows that A witnesses $T \vdash_\Omega \varphi$. □

The Ω-conjecture is absolute under forcing:

Theorem 3.7. *Suppose that there exists a proper class of Woodin cardinals. Then for every c.B.a. $\mathbb{B}$,*

$$V^{\mathbb{B}} \models \Omega\text{-Conjecture} \quad \textit{iff} \quad V \models \Omega\text{-Conjecture}.$$

Proof. By Theorems 1.8 and 2.34, for every c.B.a. $\mathbb{B}$, $\emptyset \models_\Omega \varphi$ if and only if $V^{\mathbb{B}} \models$ "$\emptyset \models_\Omega \varphi$" and $\emptyset \vdash_\Omega \varphi$ if and only if $V^{\mathbb{B}} \models$ "$\emptyset \vdash_\Omega \varphi$". Hence if $V^{\mathbb{B}} \models \Omega$-Conjecture, then $V \models$ "$\emptyset \models_\Omega \varphi$" iff $V^{\mathbb{B}} \models$ "$\emptyset \models_\Omega \varphi$" iff $V^{\mathbb{B}} \models$ "$\emptyset \vdash_\Omega \varphi$" iff $V \models$ "$\emptyset \vdash_\Omega \varphi$". Similarly for the converse. □

Remarks 3.8.

i) Assume $L(\mathbb{R}) \models AD^+$ and every set of reals in $L(\mathbb{R})$ is uB. If T is r.e. and $ZFC \models$ "$T \models_\Omega \varphi$", then $T \vdash_\Omega \varphi$, witnessed by $\emptyset$.

ii) Suppose that ZFC + there exists a strongly inaccessible cardinal is consistent. Let $\varphi =$ "*There is a non-constructible real*". Then,

$$ZFC \nvdash ((ZFC \models_\Omega \varphi) \rightarrow (ZFC \models \text{“}ZFC \models_\Omega \varphi\text{”})).$$

For suppose $V \models ZFC$ + "*There is a non-constructible real*" + $\exists\alpha(V_\alpha \models ZFC)$. Then $ZFC \models_\Omega \varphi$ holds in V. For if γ is an ordinal and $V_\gamma^{\mathbb{B}} \models ZFC$, then $V_\gamma^{\mathbb{B}} \models \varphi$, since $V_\gamma^{\mathbb{B}}$ contains all the reals of V. But, since ZFC plus the existence of a strongly inaccessible cardinal is consistent, there exists in V a model of ZFC + "there exists a strongly inaccessible cardinal" + $V = L$. This model satisfies $ZFC \not\models_\Omega \phi$.

iii) Suppose that ZFC is consistent. Then, for any sentence φ,

$$ZFC \nvdash \neg((ZFC \models_\Omega \varphi) \rightarrow (ZFC \models \text{“}ZFC \models_\Omega \varphi\text{”})).$$

Since there is a model of ZFC + "*There are no models of* ZFC".

Recall that:

i) T is Ω-satisfiable iff there exists a c.B.a. $\mathbb{B}$ and an ordinal α such that $V_\alpha^{\mathbb{B}} \models T$.

ii) T is Ω-consistent iff $T \nvdash_\Omega \perp$.

The following gives a restatement of the Ω-conjecture.

Fact 3.9. The following are equivalent for every $T \subseteq Sent$:

i) For all $\varphi \in Sent$, $T \models_\Omega \varphi$ implies $T \vdash_\Omega \varphi$

ii) T is Ω-consistent implies T is Ω-satisfiable.

Proof. i) $\Rightarrow$ ii). Suppose T is not Ω-satisfiable. Then for all c.B.a. $\mathbb{B}$ and all α, $V_\alpha^{\mathbb{B}} \not\models T$. So, for all $\mathbb{B}$ and all α, if $V_\alpha^{\mathbb{B}} \models T$, then $V_\alpha^{\mathbb{B}} \models \perp$, vacuously. Hence, $T \models_\Omega \perp$. By hypothesis, $T \vdash_\Omega \perp$, and we have that T is Ω-inconsistent.

ii) $\Rightarrow$ i). Suppose $T \nvdash_\Omega \varphi$. Then $T \cup \{\neg\varphi\} \nvdash_\Omega \varphi$, since otherwise $T \vdash_\Omega \neg\varphi \rightarrow \varphi$, and then $T \vdash_\Omega \varphi \vee \varphi$, giving a contradiction. So, $T \cup \{\neg\varphi\}$ is Ω-consistent. Since by hypothesis, $T \cup \{\neg\varphi\}$ is Ω-satisfiable, there are $\mathbb{B}$ and α such that $V_\alpha^{\mathbb{B}} \models T \cup \{\neg\varphi\}$. Therefore $T \not\models_\Omega \varphi$. □

Finally, we note that it is consistent that the Ω-conjecture is true, as Woodin has shown that it holds in fine structural models with a proper class of Woodin cardinals.

References

[1] P. Dehornoy, *Progrès récents sur l'hypothèse du continu (d'après Woodin)*, Séminaire Bourbaki 55ème année, 2002–2003, #915.

[2] Q. Feng, M. Magidor, W.H.Woodin, *Universally Baire Sets of Reals.* Set Theory of the Continuum (H. Judah, W.Just and W.H. Woodin, eds), MSRI Publications, Berkeley, CA, 1989, pp. 203–242, Springer Verlag 1992.

[3] S. Jackson, *Structural consequences of AD*, Handbook of Set Theory, M. Foreman, A. Kanamori and M. Magidor, eds. To appear.

[4] T. Jech, *Set theory*, 3d Edition, Springer, New York, 2003.

[5] A. Kanamori, *The Higher Infinite.Large cardinals in set theory from their beginnings.* Perspectives in Mathematical Logic. Springer-Verlag. Berlin, 1994.

[6] P.B. Larson, *The Stationary Tower. Notes on a course by W. Hugh Woodin.* University Lecture Series, Vol. 32. American Mathematical Society, Providence, RI. 2004.

[7] P.B. Larson, *Forcing over models of determinacy*, Handbook of Set Theory, M. Foreman, A. Kanamori and M. Magidor, eds. To appear.

[8] D.A. Martin, J.R. Steel, *The extent of scales in* $L(\mathbb{R})$, Cabal seminar 79–81, Lecture Notes in Math. 1019, Springer, Berlin, 1983, 86–96.

[9] Y.N. Moschovakis, *Descriptive Set Theory*, Studies in Logic and the Foundations of Mathematics. Vol. 100. North-Holland Publishing Company. Amsterdam, New York, Oxford, 1980.

[10] S. Shelah, W.H. Woodin, *Large cardinals imply that every reasonably definable set of reals is Lebesgue measurable.* Israel J. of Math. vol. 70, n. 3 (1990), 381–394.

[11] J. Steel, *A theorem of Woodin on mouse sets.* Preprint. July 14, 2004.

[12] W.H. Woodin,*The Axiom of Determinacy, Forcing Axioms, and the Nonstationary Ideal.* DeGruyter Series in Logic and Its Applications, vol. 1, 1999.

[13] W.H. Woodin, *The Continuum Hypothesis.* Cori, René (ed.) et al., Logic colloquium 2000. Proceedings of the annual European summer meeting of the Association for Symbolic Logic, Paris, France, July 23–31, 2000. Wellesley, MA: A.K. Peters; Urbana, IL: Association for Symbolic Logic. Lecture Notes in Logic 19, 143–197 (2005).

[14] W.H. Woodin, *The Ω-Conjecture.* Aspects of Complexity (Kaikoura, 2000). DeGruyter Series in Logic and Its Applications, vol. 4, pages 155–169. DeGruyter, Berlin, 2001.

[15] W.H. Woodin, *The Continuum Hypothesis, I.* Notices Amer. Math. Soc., 48(6):567–576, 2001.

[16] W.H. Woodin, *The Continuum Hypothesis, II.* Notices Amer. Math. Soc., 48(7):681–690, 2001; 49(1):46, 2002.

[17] W.H. Woodin, *Set theory after Russell; The journey back to Eden.* In *One Hundred Years of Russell's Paradox*, edited by Godehard Link. DeGruyter Series in Logic and Its Applications, vol. 6, pages 29–48.

Joan Bagaria
Centre de Recerca Matemàtica (CRM)
Apartat 50
E-08193 Bellaterra (Barcelona), Spain

and

ICREA (Institució Catalana de Recerca i Estudis Avançats)
and Departament de Lògica
Història i Filosofia de la Ciència
Universitat de Barcelona
Baldiri Reixac, s/n
E-08028 Barcelona, Spain
e-mail: bagaria@ub.edu

Neus Castells
Departament de Lògica
Història i Filosofia de la Ciència
Universitat de Barcelona
Baldiri Reixac, s/n
E-08028 Barcelona, Spain.
e-mail: n.castells@ub.edu

Paul Larson
Centre de Recerca Matemàtica (CRM)
Apartat 50
E-08193 Bellaterra (Barcelona), Spain

and

Department of Mathematics and Statistics
Miami University
Oxford
Ohio 45056, USA
e-mail: larsonpb@muohio.edu

Set Theory
Trends in Mathematics, 29–54

Upper Semi-lattice Algebras and Combinatorics

M. Bekkali and D. Zhani

To the memory of the first author's mother

Abstract. We characterize upper semi-lattice algebras and study their relationship with other classes of Tail algebras. Using the notion of support of nonzero elements, we derive some finite combinatorics on lengths of elements within this class of algebras.

Mathematics Subject Classification (2000). Primary: 03G, 06E. Secondary: 06A12, 03G10.

Keywords. Tail algebra, semi-group, upper semi-lattice algebra, length of elements.

1. Introduction

Let $(T,<)$ be a partially ordered set and consider the subalgebra $B(T)$ of the power set of T, $\mathcal{P}(T)$ generated by $\{b_t : t \in T\}$, where $b_t := \{x \in T : t \leq x\}$. Then look at the topological closure $\overline{\{\downarrow t : t \in T\}}$ in 2^T, where $\downarrow t := \{u \in T : u \leq t\}$. Analyzing these dual notions, it is natural to ask how the Stone space $\mathcal{U}lt(B(T))$ relates to $\overline{\{\downarrow t : t \in T\}}$ and under what conditions the set of ideals of $(T,<)$, $\mathcal{I}d(T)$ coincides with $\{\downarrow t : t \in T\}$? For the later question, notice that the set of *principal* ideals $\{\downarrow t : t \in T\}$ is, in general, not a closed subspace of 2^T. Now, if $(T,<)$ is an upper semi-lattice then $\{\downarrow t : t \in T\}$ is, indeed, a closed subspace of 2^T and $\mathcal{U}lt(B(T)) \cong \mathcal{I}d(T)(= \{\downarrow t : t \in T\})$. In addition $(\mathcal{I}d(T), \subseteq)$ plays a fundamental role in the Theory of Algebraic Lattices since every algebraic lattice is of this kind, see for instance [6, Theorem 13, p. 106].

On the other hand, by [10, Theorem 7.5, p. 371], every Boolean algebra embeds in some tail algebra $B(T_2)$, where $(T_2,<)$ is a 2-levels poset. Actually, it can be shown that any Stone space X (up to its isolated points) can be obtained by this procedure. Indeed, let B be a Boolean algebra and X be its Stone space and set $T_2 := Clopen(X)\dot{\cup}X$. Then define $<$ on T_2 by: if $a \in X$, $b \in Clopen(X)$, we set $a < b$ whenever $a \in b$. It can be shown that the first derivatives of X and $\mathcal{U}lt(B(T_2))$ are homeomorphic spaces. Notice that $(T_2, <)$ is not an upper semi-lattice: This limitation, which defines the frontier between upper semi-lattices algebras and the general case, is the motivation for our study here.

Throughout this paper 2^T shall be identified with the power set of T denoted by $\mathcal{P}(T)$. 2^T is a Boolean space endowed with Tychonoff's topology for which $\langle \mathcal{U}(A, B) : A, B \in [T]^{<\omega} \rangle$, where $\mathcal{U}(A, B) := \{F \in 2^T : A \subseteq F \text{ and } F \cap B = \emptyset\}$, is a basis for its topology. Next if $(T, \leq)$ is a partial ordering and $t \in T$; define $b_t := \{x \in T : t \leq x\}$ and let $B(T)$, the tail algebra over T, be the subalgebra of the power set algebra of T generated by $\{b_t : t \in T\}$. A partially ordered set $(T, \leq)$ so that $l.u.b.\{x, y\} := x \vee y$ exists in $(T, \leq)$ for all $x, y \in T$ is called an upper semi-lattice and so, $B(T)$ is called the upper semi-lattice algebra over T. For more on tail algebras, interval algebras, trees, pseudo-tree and upper semi-lattices algebras see [1],[2], [8], [10] and [9].

Stone spaces of tail algebras $B(T)$, generated by $\langle b_t = \{x \in T : t \leq x\} : t \in T \rangle$, in the power set algebra of T, can be wild as a family of subsets of the power set of T. Here, we investigate this matter looking at the relationship between Stone space $\mathcal{U}lt(B(T))$ and both $\mathcal{I}d(T)$ and its topological closure, $\overline{\mathcal{I}d(T)}$, in 2^T. Then we give conditions on upper semi-lattices $(T, \leq)$ for which $\mathcal{I}d(T)$ is homeomorphic to the set of initial sections $\mathcal{I}(Q)$ for some poset Q.

Next, when working in upper semi-lattices algebras one notices that nonzero elements in these algebras are built up in a very unique fashion which suggests defining a notion of length on them. Actually, this kind of notion appears in [14] and [15]. Extending the work done in [15], we define the $\triangle$-length μ using the support of elements of a Boolean algebra in a very natural way. Then we describe completely the way it operates on upper semi-lattice algebras. Notice that properties of μ remain valid in any Boolean algebra in which there is a unique decomposition of nonzero elements, using symmetric difference $\triangle$, as is the case of semi-group algebras see, e.g., [5] and [7]. Another notion of length, μ^+, is defined on pseudo treealgebras in a similar way. As it would be noticed, μ^+ is not very interesting on upper semi-lattices algebras but is connected with μ algebraically on interval algebras.

2. Elementary material

2.1. Basic facts from topology

Let $(T, \leq)$ be any non-empty partially ordered set. For $a \in T$ and $A \subseteq T$, set:

$$\uparrow a := \{s \in T : a \leq s\}, \ \ \downarrow a := \{s \in T : s \leq a\}, \ \ b_a = \uparrow a,$$

$$M(A) := \bigcap_{a \in A} \uparrow a, \ \ L(A) := \bigcup_{a \in A} \uparrow a, \ \ M(\emptyset) = T \text{ and } L(\emptyset) = \emptyset.$$

For a poset $(T, \leq)$, $\mathcal{I}(T)$ shall denote the set of initial sections of T (i.e., subsets (possibly empty set) of T that are closed downwards in $(T, \leq)$); $\mathcal{I}d(T)$ shall also denote the set of ideals of T (i.e., all non-empty initial sections $I \in \mathcal{I}(T)$ so that $(\uparrow s) \cap (\uparrow t) \cap I \neq \emptyset$ for all $s, t \in I$). Next, define $\mathcal{I}_{f.i.p}(T)$ to be the set of all $I \in \mathcal{I}(T)$, so that

$$\text{For all } \ A, B \in [T]^{<\omega}, M(A) \nsubseteq L(B), \ \text{ whenever } \ A \subseteq I \ \text{ and } \ B \subseteq T \setminus I \quad (*)$$

A subset $B \subseteq T$ is a finitely generated poset ((f.g.)-poset) in $(T, \leq)$, whenever there is a finite anti-chain $A \subseteq T$ so that $B = L(A)$. Finally, we say that $(T, \leq)$ is a $\vee$-finitely generated poset (($\vee$-f.g.)-poset) whenever for all $s, t \in T$, either $M(\{s,t\}) = \emptyset$ or $M(\{s,t\})$ is an (f.g.)-poset in $(T, \leq)$. Notice that if T is an upper semi-lattice, then $M(\{s,t\}) = \uparrow (s \vee t)$. On the other hand, $\mathcal{I}d(T) \subseteq \mathcal{I}_{f.i.p}(T) \subseteq \mathcal{I}(T)$ always holds.

Examples

1. For a chain $(T, \leq)$, $\mathcal{I}d(T) = \mathcal{I}_{f.i.p}(T) = \mathcal{I}(T)$.
2. For an anti-chain $(T, \leq) = \{t_n : n \in \omega\}$ we have $\mathcal{I}d(T) = \mathcal{I}_{f.i.p}(T) \subsetneqq \mathcal{I}(T)$ (indeed, $I := \{t_0, t_1\} \in \mathcal{I}(T) \setminus \mathcal{I}_{f.i.p}(T)$).
3. Let T be $\{t_n : n \in \omega\} \cup \{s_1, s_2\}$ where $\{t_n : n \in \omega\}, \{s_1, s_2\}$ are anti-chains and $s_1, s_2 \leq t_n$ for all $n \in \omega$. Then, $\mathcal{I}d(T) \subsetneqq \mathcal{I}_{f.i.p}(T) \subsetneqq \mathcal{I}(T)$.

Lemma 2.1. *Let $(T, \leq)$ be a poset. Then*

1. *$\mathcal{I}_{f.i.p}(T) \cup \{\emptyset\}$ is a closed subset of 2^T.*
2. *The following statements are equivalent.*
   i) *$\mathcal{I}_{f.i.p}(T)$ is closed in 2^T.*
   ii) *T is a finitely generated poset.*

Proof. 1. To show that $2^T \setminus (\mathcal{I}_{f.i.p}(T) \cup \{\emptyset\})$ is open, let $I_0 \in 2^T \setminus (\mathcal{I}_{f.i.p}(T) \cup \{\emptyset\})$. Thus, $I_0 \neq \emptyset$ and $I_0 \notin \mathcal{I}_{f.i.p}(T)$.

Case 1. *I_0 is not an initial section.* Thus, there are $p, q \in T$ $(p < q)$ so that $q \in I_0$ and $p \notin I_0$. So, $I_0 \in \mathcal{U}(\{q\}, \{p\}) \subseteq 2^T \setminus (\mathcal{I}_{f.i.p}(T) \cup \{\emptyset\})$.

Case 2. *I_0 does not satisfy* $(*)$. Thus, there are $t_1, \dots, t_n \in I_0, s_1, \dots, s_m \notin I_0$ so that, $\prod_{i=1}^n b_{t_i} \cdot - \sum_{j=1}^m b_{s_j} = \emptyset$. Hence,

$$I_0 \in \mathcal{U}(\{t_1, \dots, t_n\}, \{s_1, \dots, s_m\}) \subseteq 2^T \setminus (\mathcal{I}_{f.i.p}(T) \cup \{\emptyset\}).$$

Therefore $\mathcal{I}_{f.i.p}(T) \cup \{\emptyset\}$ is closed in 2^T.

2. Note that by 1., $\mathcal{I}_{f.i.p}(T) \cup \{\emptyset\}$ is a compact set, but $\mathcal{I}_{f.i.p}(T) \subseteq \mathcal{I}_{f.i.p}(T) \cup \{\emptyset\}$. Thus, i) is equivalent to $\{\emptyset\}$ is open in $\mathcal{I}_{f.i.p}(T) \cup \{\emptyset\}$.

i) implies ii). If $\{\emptyset\}$ is open in $\mathcal{I}_{f.i.p}(T) \cup \{\emptyset\}$, then there is a basic open set V_0 so that $\emptyset \in V_0 \subseteq \{\emptyset\}$, i.e., $V_0 = \{\emptyset\}$. Thus, there are $s_1, \dots, s_m \in T$ so that $V_0 := \{I \in \mathcal{I}_{f.i.p}(T) \cup \{\emptyset\} : s_i \notin I \text{ for } \ i = 1, \dots, m\} = \{\emptyset\}$. Thus, $T = \cup_{i=1}^m b_{s_i}$. Otherwise, pick $t \in T$ so that $s_i \nleq t$ for all $i = 1, \dots, m$. Let $I := \downarrow t$. We have $I \in \mathcal{I}_{f.i.p}(T) \cup \{\emptyset\}$ with $s_i \notin I$ for $i = 1, \dots, m$. Thus, $I \in V_0$ and $I \neq \emptyset$: Contradiction.

ii) implies i). If $T = \cup_{i=1}^n b_{t_i}$ with $n \in \omega \setminus \{0\}, t_i \in T$, then set $V := \{I \in \mathcal{I}_{f.i.p}(T) \cup \{\emptyset\} : t_i \notin I, i = 1, \dots, n\}$. Thus, V is open and $V = \{\emptyset\}$ since every non empty initial section of T contains some t_i for some $i \in \{1, \dots, n\}$. □

Lemma 2.2. *Let $(T, \leq)$ be a poset. Then, $\mathcal{I}_{f.i.p}(T) \cup \{\emptyset\} \cong \mathcal{I}_{f.i.p}(\breve{T})$, where $\breve{T} = T \cup \{\alpha\}, \alpha \notin T$, and α is the least element of $\breve{T}$.*

Proof. Let $\breve{T} = T \cup \{\alpha\}, \alpha \notin T$. Define $\preceq$ on $\breve{T}$ by $x \preceq y$ if and only if $x \leq y$, for all $x, y \in \breve{T} \setminus \{\alpha\}$, and $\alpha \preceq x$, for $x \in T$. Next, define φ from $\mathcal{I}_{f.i.p}(T) \cup \{\emptyset\}$ into $\mathcal{I}_{f.i.p}(\breve{T})$ by $\varphi(I) := I \cup \{\alpha\}$. It is straightforward to check that φ is a homeomorphism. □

Theorem 2.3. *Let $(T, \leq)$ be a poset. If $(T, \leq)$ is a (f.g.)-poset, then $\mathcal{U}lt(B(T))$ and $\mathcal{I}_{f.i.p}(T)$ are homeomorphic spaces. Otherwise, $\mathcal{U}lt(B(T)) \cong_{\text{homeo}} \mathcal{I}_{f.i.p}(T) \cup \{\emptyset\}$.*

Proof. Define θ from $\mathcal{U}lt(B(T))$ into $\mathcal{P}(T)$ by $\theta u := \{t \in T : b_t \in u\}$.

Case 1. *T is an (f.g.)-poset.* Then $T = \cup_{i=1}^{n} b_{t_i}, n \neq 0$, for some $t_1, \ldots, t_n \in T$.

1. For each $u \in \mathcal{U}lt(B(T)), \theta u \neq \emptyset$. For, $b_{t_1} + \cdots + b_{t_n} = T = 1_{B(T)} \in u$. Thus, there is $i \in \{1, \ldots, n\}$ so that $b_{t_i} \in u$. So, $t_i \in \theta u$.

2. $\theta u \in \mathcal{I}_{f.i.p}(T)$. For notice that θu is an initial section which is not empty by 1. Now, let $\{t_1, \ldots, t_n\} \in \theta u$, $\{s_1, \ldots, s_m\} \in T \setminus \theta u$. Thus, $b_{t_i} \subset u$, $b_{s_j} \in u$ Hence, $\prod_{i=1}^{n} b_{t_i} \cdot \prod_{j=1}^{m} (-b_{s_j}) \in u$. Therefore, $\prod_{i=1}^{n} b_{t_i} \cdot -\sum_{j=1}^{m} b_{s_j} \neq 0$, i.e., $M(\{t_1, \ldots, t_n\}) \cap (T \setminus L(\{s_1, \ldots, s_m\}) \neq \emptyset$. Hence, $\theta u \in \mathcal{I}_{f.i.p}(T)$.

3. θ is 1-1 since $(b_t \in u \leftrightarrow b_t \in v) \rightarrow u = v$ for any ultrafilters u, v.

4. $\theta(\mathcal{U}lt(B(T))) = \mathcal{I}_{f.i.p}(T)$. Indeed, by 2., we have $\theta(\mathcal{U}lt(B(T))) \subseteq \mathcal{I}_{f.i.p}(T)$. Now, if $I \in \mathcal{I}_{f.i.p}(T)$ set $\mathcal{F}(I) = \{b_t : t \in I\} \cup \{-b_s : s \notin I\}$. $\mathcal{F}(I)$ has finite intersection property and hence there is u_I, an ultrafilter, so that $\mathcal{F}(I) \subseteq u_I$. We have $\theta u_I = I$. For $t \in I \rightarrow b_t \in \mathcal{F}(I) \subseteq u_I \rightarrow b_t \in u_I \rightarrow t \in \theta u_I$.

Conversely, if $s \in \theta u_I$ then $b_s \in u_I$. Now if $s \notin I$ then $-b_s \in u_I$ which is a contradiction. Thus, $s \in I$.

5. θ is continuous. For let $I = \theta u$ and take $t_1, \ldots, t_n \in I$; $s_1, \ldots, s_m \notin I$. Next, define

$$\mathcal{V}(I) = \{J \in \mathcal{I}_{f.i.p}(T) : t_1, \ldots, t_n \in J \text{ and } s_1, \ldots, s_m \notin J\}.$$

Then $b = \prod_{i=1}^{n} b_{t_i} \cdot \prod_{j=1}^{m} -b_{s_j} \neq 0$. So, $u \in s(b) \subseteq \theta^{-1}(\mathcal{V}(I))$, where s is the Stone representation mapping see [9, p. 99]. So, θ is a continuous bijective mapping between $\mathcal{U}lt(B(T))$ and $\mathcal{I}_{f.i.p}(T)$. Thus, θ is a homeomorphism.

Case 2. *T is not an (f.g.)-poset.*

1. There is $u_0 \in \mathcal{U}lt(B(T))$ so that $\theta u_0 = \emptyset$. Indeed, set $\mathcal{V}_0 = \{-b_s : s \in T\}$. For all $m \neq 0$, for all $s_1, \ldots, s_m \in T$, $\prod_{j=1}^{m} (-b_{s_j}) = -\sum_{j=1}^{m} b_{s_j} \neq 0$ since T is not a (f.g.)-poset. Thus, $\mathcal{V}_0$ has finite intersection property. So, there is u_0 in $\mathcal{U}lt(B(T))$ so that $b_s \notin u_0$ for all $s \in T$. So, $\theta u_0 = \emptyset$.

2. If $\theta u \neq \emptyset$ then $\theta u \in \mathcal{I}_{f.i.p}(T)$. Use a similar argument as in Case 1.(2.).

3. θ is 1-1 by the same argument in Case 1.(3.).

4. The same argument, as before in Case 1, works to show that: $\mathcal{I}_{f.i.p}(T) \subseteq \theta(\mathcal{U}lt(B(T)))$.

Now, by 1. and 2. we have $\theta(\mathcal{U}lt(B(T))) \subseteq \mathcal{I}_{f.i.p}(T) \cup \{\emptyset\}$. Moreover, by 1. we have $\emptyset = \theta u_0 \in \theta(\mathcal{U}lt(B(T)))$. Thus, $\theta(\mathcal{U}lt(B(T))) = \mathcal{I}_{f.i.p}(T) \cup \{\emptyset\}$.

5. Let $u \in \mathcal{U}lt(B(T))$ and set $I = \theta u$.

5.1. $\theta u \neq \emptyset$. By the same argument as in case i) 5. take $t_1, \ldots, t_n \in I,\ s_1, \ldots, s_m \notin I$ and set

$$\mathcal{V}(I) = \{J \in \mathcal{I}_{f.i.p}(T) \cup \{\emptyset\} : t_1, \ldots, t_n \in J \text{ and } s_1, \ldots, s_m \notin J\}.$$

5.2. $\theta u = \emptyset$. Let $s_1, \ldots, s_m \in T$. Set $\mathcal{V}(I) = \{J \in \mathcal{I}_{f.i.p}(T) \cup \{\emptyset\} : s_1, \ldots, s_m \notin J\}$. Thus, $b = \prod_{j=1}^{m} -b_{s_j} \neq 0$ and $u \in s(b) \subseteq \theta^{-1}(\mathcal{V}(I))$, where s is again the Stone representation mapping. Hence, θ is a continuous bijective mapping between $\mathcal{U}lt(B(T))$ and $\mathcal{I}_{f.i.p}(T) \cup \{\emptyset\}$.

Thus, θ is a homeomorphism. □

2.2. Normal form in upper semi-lattice Boolean algebras

Hence, by Lemma 2.2, $(T, \leq)$ shall denote, from now on, any upper semi-lattice with a least element and $B(T)$ the upper semi-lattice algebra over $(T, \leq)$.

Before stating the next lemma set,

$$E := \left\{ b_t \cdot - \sum_{s \in A} b_s : A \text{ is a finite anti-chain in } (T, <) \text{ above } t \right\}.$$

Lemma 2.4.

i) *Each element of E is different from zero;*

ii) *Let p be an elementary product, i.e.,*

$$p = b_{t(1)} \cdots b_{t(n)} \cdot -b_{s(1)} \cdots - b_{s(m)}, \textit{ with } n + m \neq 0.$$

Then,

$$p = 0 \textit{ iff } \begin{cases} n = 0 & \textit{and} \quad \exists i : s(i) = \textit{least element of } T \\ \textit{or} & \\ n \geq 1 & \textit{and} \quad \exists i : s(i) \leq \bigvee_{j=1}^{n} t(j). \end{cases}$$

iii) *Every nonzero elementary product of $\{b_t : t \in T\}$ is an element of E;*

iv) *Every nonzero element b of $B(T)$ is the sum of pairwise nonzero elements of E, i.e., $b = e_1 + \cdots + e_n$, with $e_i \in E$and $e_i \cdot e_j = 0$ for $i \neq j$. Moreover, $e_i = b_{t(i)} \cdot - \sum\{b_s : s \in S(i)\}$, where $S(i)$ is an anti-chain in T and $t(i) < s$ for all $s \in S(i)$.*

Proof. Similar to Lemma 16.3 in [11, Ch. 6; Vol. 1, p. 256]. □

Remark

i) To the contrary of what happens in the case of tree algebras see Lemma 16.6 (b., c., and d.) in [11, Ch. 6; Vol. 1, p. 258], if $e \leq e_1 + \cdots + e_n$, e, $e_i \in E$ and $e_i \cdot e_j = 0$ for $i \neq j$, then it may not be true that $e \leq e_i$, for some i.

ii) Even though each $b \in B(T) \setminus \{0\}$ has a decomposition as $b = e_1 + \cdots + e_n$ with $e_i \in E$ and $e_i \cdot e_j = 0$, this decomposition may not be unique in an upper semi-lattice algebras.

To seek a unique decomposition of nonzero elements of an upper semi-lattice algebra $B(T)$, we start by an example:

Let $(T, \leq)$ be an upper semi-lattice which is not a chain. Hence, pick t_1 incomparable with t_2 in T and put $s = t_1 \vee t_2$. Thus

$$b := \underbrace{(b_{t_1} \cdot -b_s)}_{e_1} + \underbrace{b_{t_2}}_{e_2} = \underbrace{b_{t_1}}_{\varepsilon_1} + \underbrace{(b_{t_2} \cdot -b_s)}_{\varepsilon_2}$$

does not have a unique decomposition using elements of E.

Recall that $\triangle$ denotes the symmetric difference of elements in $B(T)$.

Lemma 2.5. *Let $B(T)$ be an upper semi-lattice algebra. Then:*

1. *For all $s, t \in T$, $b_t \cdot -b_s = b_t \triangle b_{t \vee s}$;*
2. *For every $\{t_i : 1 \leq i \leq m\} \subseteq T$ and $s \in T$, there is $\{s_j : 1 \leq j \leq n\} \subseteq T$ such that $b_{t_1} \triangle \cdots \triangle b_{t_m} \cdot -b_s = b_{s_1} \triangle \cdots \triangle b_{s_n}$;*
3. *For all $b \in E$, $b = \triangle_{i=1}^{k} b_{s_i}$; for suitable k and s_i*
4. *If $a \triangle b = a \triangle b'$, with $a, b, b' \in B(T)$, then $b = b'$;*
5. *Let $\{t_{i_1}, \ldots, t_{i_p}\}$ be the set of minimal elements in $\{t_i \cdot i = 1, \ldots, n\}$. Then: $b_{t_1} \triangle \cdots \triangle b_{t_n} \subseteq b_{t_{i_1}} \cup \cdots \cup b_{t_{i_p}}$, and $t_{i_k} \in b_{t_1} \triangle \cdots \triangle b_{t_n}$ for all $k = 1, \ldots, p$;*
6. *If $b_{t_1} \triangle \cdots \triangle b_{t_n} = b_{s_1} \triangle \cdots \triangle b_{s_m}$, with t_i pairwise different (resp. s_i's). Then there are i, j so that $b_{t_i} = b_{s_j}$.*

Proof. 1. $b_t \triangle b_{t \vee s} = b_t \cdot -b_{t \vee s} + b_{t \vee s} \cdot -b_t = b_t \cdot -b_{t \vee s} = b_t \cdot -b_s$.

2. Let $b := \left(b_{t_1} \triangle \cdots \triangle b_{t_n}\right)$. Then,

$$\begin{aligned} b \cdot -b_s &= \left(b_{t_1} \triangle \cdots \triangle b_{t_n}\right) \cdot -b_s \\ &= \left(b_{t_1} \cdot -b_s\right) \triangle \left(b_{t_2} \cdot -b_s\right) \triangle \cdots \triangle \left(b_{t_n} \cdot -b_s\right) \\ &= \left(b_{t_1} \triangle b_{t_1 \vee s}\right) \triangle \cdots \triangle \left(b_{t_n} \triangle b_{t_n \vee s}\right). \end{aligned}$$

3. Let $b \in E$, $b = b_t \cdot -b_{s_1} \cdot -b_{s_2} \cdots - b_{s_n}$, where $\{s_1, \ldots, s_n\}$ is an anti-chain above t. Next set

$$\triangle(b) := n = |\{s_1, \ldots, s_n\}|.$$

Case 1. $n = 0$ $\quad b = b_t$.

Case 2. $n = 1$ $\quad b = b_t \cdot -b_s = b_t \triangle b_{t \vee s}$ (using 1.)

Suppose that 3. was proved for all b's in E so that $\triangle(b) \leq n - 1$.

Let $b \in E$, $\triangle(b) = n$, $b = b_t \cdot -b_{s_1} \cdot -b_{s_2} \cdots - b_{s_n}$. Let $b' := b_t \cdot -b_{s_1} \cdot -b_{s_2} \cdots - b_{s_{n-1}}$. $b' \in E$ and $\triangle(b') = n - 1$. Hence by induction's hypothesis $b' = b_{\tau_1} \triangle b_{\tau_2} \triangle \cdots \triangle b_{\tau_k}$. So,

$$b = b' \cdot -b_{s_n} = \left(b_{\tau_1} \triangle b_{\tau_2} \triangle \cdots \triangle b_{\tau_k}\right) \cdot -b_{s_n}.$$

Now, by 2., we have: $b = \triangle_{i=1}^{m} b_{t_i}$.

4. It's clear.

5. Obviously $b_{t_1} \triangle b_{t_2} \triangle \cdots \triangle b_{t_n} \subseteq b_{t_1} \cup b_{t_2} \cup \cdots \cup b_{t_n}$. For any $i = 1, \ldots, n$ there is a $j = 1, \ldots, p$ such that $t_{i_j} \leq t_i$ and hence $b_{t_i} \subseteq b_{t_{i_j}}$; so $b_{t_1} \cup b_{t_2} \cup \cdots \cup b_{t_n} \subseteq b_{t_{i_1}} \cup \cdots \cup b_{t_{i_p}}$.

Now suppose that $1 \leq k \leq p$. If $1 \leq l \leq n$ and $l \neq i_k$, then $t_l \nleq t_{i_k}$ and hence $t_{i_k} \notin b_{t_l}$.

Since $\triangle\{b_{t_j} : 1 \leq j \leq n$ and $j \neq i_k\} \subseteq \cup\{b_{t_j} : 1 \leq j \leq n$ and $j \neq i_k\}$, it then follows that $t_{i_k} \in b_{t_1} \triangle \cdots \triangle b_{t_n}$.

6. Let $b = b_{t_1} \triangle \cdots \triangle b_{t_n} = b_{s_1} \triangle \cdots \triangle b_{s_m}$.

Let $\{t_{i_1}, \dots, t_{i_p}\}$ (resp. $\{s_{j_1}, \dots, s_{j_q}\}$) be the set of minimal elements in $\{t_1, \dots, t_n\}$ (resp. $\{s_1, \dots, s_m\}$).

By 5.:

$$b \subseteq b_{t_{i_1}} \cup \cdots \cup b_{t_{i_p}} \qquad (\star)$$

$$b \subseteq b_{s_{j_1}} \cup \cdots \cup b_{s_{j_q}} \qquad (\star\star)$$

$$\forall k = 1, \dots, p \; t_{i_k} \in b; \quad \forall k = 1, \dots, q \; s_{j_k} \in b. \qquad (\star\star\star)$$

Hence, $t_{i_1} \in b$ and thus, by $(\star\star)$, pick k so that $t_{i_1} \in b_{s_{j_k}}$. So, $s_{j_k} \leq t_{i_1}$. Suppose $s_{j_k} < t_{i_1}$, then $s_{j_k} \notin b_{t_{i_1}}$ and thus $s_{j_k} \notin b_{t_{i_d}}$ for all d (otherwise $s_{j_k} \in b_{t_{i_d}}$ implies $t_{i_d} \leq s_{j_k} < t_{i_1}$ and thus $t_{i_d} < t_{i_1}$: contradiction.) So, $s_{j_k} \notin b_{t_{i_d}}$ for each d. Now, by $(\star)$, $s_{j_k} \notin b$ which contradicts $(\star\star\star)$. Therefore, $s_{j_k} = t_{i_1}$ and thus $b_{t_{i_1}} = b_{s_{j_k}}$. □

Theorem 2.6. *Let $B(T)$ be an upper semi-lattice algebra over T. Each nonzero element b of $B(T)$ has a unique decomposition of the form:*

$$b = b_{t_1} \triangle \cdots \triangle b_{t_n}, \text{ for suitable } n, \text{ and } t_i's \text{ so that } t_i \neq t_j \text{ for } i \neq j. \qquad (\triangle)$$

Proof. Existence part. Let $b \in B(T) \setminus \{0\}$. By Lemma 2.4, $b = \sum_{i=1}^{n} e_i$ where $e_i \in E$ and $e_i \cdot e_j = 0$ for $i \neq j$. Hence $b = e_1 \triangle \cdots \triangle e_n$ since $e_i \cdot e_j = 0$. Now, by Lemma 2.5(3.) $e_i = \triangle_{j=1}^{\alpha_i} b_{s_j^i}$, and thus $b = \triangle_{k=1}^{q} b_{t_k}$.

Uniqueness part. We prove the result by induction on $m+n$. Suppose b decomposes as, $b = b_{t_1} \triangle \cdots \triangle b_{t_n} = b_{s_1} \triangle \cdots \triangle b_{s_m}$ with $n \leq m$, s_i and t_j are pairwise distinct. By Lemma 2.5(6.), there are i, j such that $b_{t_i} = b_{s_j}$. Hence,

$$b = b_{t_i} \triangle (\triangle_{l \neq i} b_{t_l}) = b_{t_i} \triangle (\triangle_{k \neq j} b_{s_k}).$$

Now, again by Lemma 2.5(4.), we have $\triangle\{b_{t_k} : 1 \leq k \leq n$ and $k \neq i\} = \triangle\{b_{s_l} : 1 \leq l \leq m$ and $l \neq j\}$.

Case 1. $m = n$ we are done.

Case 2. $m \neq n$ $b_{s_{j_1}} \triangle (b_{s_{j_2}} \triangle \cdots \triangle b_{s_{j_p}}) = 0$

$b_{s_{j_1}} = b_{s_{j_2}} \triangle \cdots \triangle b_{s_{j_p}}$. Hence, by Lemma 2.5(6.) there is k so that $b_{s_{j_1}} = b_{s_{j_k}}$. Hence, $s_{j_1} = s_{j_k}$ with $k \neq 1$. This is a contradiction since all $s_j's$ are distinct. This finishes the proof of the theorem. □

3. Characterization and duality

3.1. Set of ideals of a poset

In this section, we establish necessary and sufficient conditions on a poset T so that the set of ideals $\mathcal{I}d(T)$ is a closed subset of 2^T, and we shall investigate the relationship between $\mathcal{I}d(T)$ and $\mathcal{I}_{f.i.p}(T)$.

Theorem 3.1. *Let $(T, \leq)$ be a poset. Then:*

i) $\mathcal{I}d(T) \subseteq \mathcal{I}_{f.i.p}(T)$,

ii) *If T is a ($\vee$-f.g.)-poset then* $\mathcal{I}d(T) = \mathcal{I}_{f.i.p}(T)$,

iii) *There is a poset $(T_0, \leq)$ so that $\mathcal{I}d(T_0) \subsetneqq \mathcal{I}_{f.i.p}(T_0)$ and $\mathcal{I}d(T_0)$ is not a closed set of 2^{T_0}.*

Proof. i) It suffices to show that for each $I \in \mathcal{I}d(T)$,

$$\prod_{i=1}^{n} b_{t_i} \cdot \prod_{j=1}^{m} -b_{s_j} \neq 0 \text{ for } t_1, \dots, t_n \in I; \quad s_1, \dots, s_m \notin I.$$

ii) Assume that T is a ($\vee$-f.g.)-poset. We need only show that $\mathcal{I}_{f.i.p}(T) \subseteq \mathcal{I}d(T)$. For let $s, t \in I$. We have $b_s \cdot b_t \neq 0$ since $I \in \mathcal{I}_{f.i.p}(T)$. So, $M(\{s,t\}) \neq \emptyset$. But T is a ($\vee$-f.g.)-poset. Hence, there is an anti-chain $\{\alpha_1, \dots, \alpha_n\}$ in $M(\{s,t\})$ so that $b_s \cdot b_t = b_{\alpha_1} + \cdots + b_{\alpha_n}$. So, $b_s \cdot b_t \cdot \prod_{i=1}^{n} -b_{\alpha_i} = 0$. Now, since $I \in \mathcal{I}_{f.i.p}(T)$, it follows that $\alpha_k \in I$ for some k; but, $\alpha_k \in b_s \cdot b_t$. So, $\alpha_k \geq s$ and $\alpha_k \geq t$, i.e., $I \in \mathcal{I}d(T)$.

iii) Set $T_0 = \{t_0, t_1, t_2\} \cup \{\alpha_k : k \in \omega\}$. Now, define $\preceq$ on T_0 by:

1. $t_0 \preceq t_1 \preceq \alpha_k, \quad t_0 \preceq t_2 \preceq \alpha_k, \quad$ for all $k \in \omega$.
2. $t_1 \parallel t_2$ and $\alpha_k \parallel \alpha_l$ for all $k \neq l$, where $\parallel$ is the incomparability sign.

Let $I := \{t_0, t_1, t_2\}$. First, notice that $I \notin \mathcal{I}d(T_0)$ and $I \in \mathcal{I}_{f.i.p}(T_0)$ since

$$b_{t_1} \cdot b_{t_2} \cdot \prod_{i=1}^{n} -b_{\alpha_{k(i)}} \neq 0 \quad \text{for all} \quad n, k(i).$$

So, $\mathcal{I}d(T_0) \subsetneqq \mathcal{I}_{f.i.p}(T_0)$. Now note that $\mathcal{I}d(T_0)$ is not a closed subset of 2^T. For we notice that for each $x \in T_0$, $\mathcal{V}(x) := \{I \in \mathcal{I}d(T_0) : x \in I\}$ is clopen in $\mathcal{I}d(T_0)$. Next, let $I_k :=\downarrow \alpha_k = \{\alpha_k, t_0, t_1, t_2\}$ for each $k \in \omega$, and $V_k := \{I \in \mathcal{I}d(T_0) : \alpha_k \in I\}$. So, $V_k = \{I_k\}$. On the other hand $\{I_k : k \in \omega\} = \{I \in \mathcal{I}d(T_0) : t_1, t_2 \in I\}$. So, $\{I_k : k \in \omega\} = \{I \in \mathcal{I}d(T_0) : t_1 \in I\} \cap \{I \in \mathcal{I}d(T_0) : t_2 \in I\} = \mathcal{V}(t_1) \cap \mathcal{V}(t_2)$. Thus $\{I_k : k \in \omega\}$ is closed in 2^T; but $\{I_k : k \in \omega\} \subseteq \cup_{k \in \omega}\{I_k\}$ where each $\{I_k\}$ is clopen. Now since $\langle \alpha_k : k \in \omega \rangle$ is an infinite anti-chain, $\{I_k : k \in \omega\}$ cannot be compact. This finishes up the proof of Theorem 3.1. □

Theorem 3.2. *Let $(T, \leq)$ be a poset. The following statements are equivalent.*

i) $\mathcal{I}d(T) \cup \{\emptyset\}$ *is a closed set in* 2^T,

ii) $(T, \leq)$ *is a ($\vee$-f.g.)-poset,*

iii) $\mathcal{I}d(T) = \mathcal{I}_{f.i.p}(T)$.

Proof. i) implies ii). Suppose that $\mathcal{I}d(T) \cup \{\emptyset\}$ is compact, let $s, t \in T$ with $M(\{s,t\}) \neq \emptyset$. We need to show that $M(\{s,t\})$ is finitely generated. So, put $M(\{s,t\}) := \{\alpha_k : k \in \sigma\}$. For each $k \in \sigma$ set $I_k :=\downarrow \alpha_k$, $\mathcal{V}_k := \{I \in \mathcal{I}d(T) \cup \{\emptyset\} : \alpha_k \in I\}$, and $F := \{I \in \mathcal{I}d(T) \cup \{\emptyset\} : s, t \in I\}$. F is a closed set and $\mathcal{V}_k$'s are open sets so that $F \subseteq \cup_{k \in \sigma} \mathcal{V}_k$. By compactness of $\mathcal{I}d(T) \cup \{\emptyset\}, F \subseteq \mathcal{V}_{k_1} \cup \cdots \cup \mathcal{V}_{k_n}$. Thus $M(\{s,t\}) = \cup_{i=1}^{n} \uparrow \alpha_{k_i}$. Therefore T is a ($\vee$-f.g.)-poset.

ii) implies iii). By Theorem 3.1(ii).
iii) implies i). If $\mathcal{I}d(T) = \mathcal{I}_{f.i.p}(T)$ then by Lemma 2.1, $\mathcal{I}_{f.i.p}(T) \cup \{\emptyset\}$ is compact and thus so is $\mathcal{I}d(T) \cup \{\emptyset\}$. □

Theorem 3.3. *Let $(T, \leq)$ be a poset. The following statements are equivalent.*

i) *$\mathcal{I}d(T)$ is a closed set in 2^T,*
ii) *$(T, \leq)$ is both a $(\vee$-f.g.) and a (f.g.)-poset,*
iii) *$\mathcal{I}d(T) = \mathcal{I}_{f.i.p}(T)$ and $(T, \leq)$ is an (f.g.)-poset.*

Proof. i) implies ii). If $\mathcal{I}d(T)$ is a closed subset of 2^T, so is $\mathcal{I}d(T) \cup \{\emptyset\}$. Thus, by Theorem 3.2, $(T, \leq)$ is a (v-f.g.)-poset; moreover, $\mathcal{I}_{f.i.p}(T) = \mathcal{I}d(T)$. So, $\mathcal{I}_{f.i.p}(T)$ is closed in 2^T and by Lemma 2.1 we get that T is an (f.g.)-poset.
ii) implies iii). Using Theorem 3.1.
iii) implies i). Using Lemma 2.1. □

Now, the next theorem describes completely the relationship between $\mathcal{I}d(T)$ and $\mathcal{I}_{f.i.p}(T)$ for any poset $(T, \leq)$.

Theorem 3.4. *If $(T, \leq)$ is an (f.g)-poset then $\overline{\mathcal{I}d(T)} = \mathcal{I}_{f.i.p}(T)$; otherwise $\overline{\mathcal{I}d(T)} = \mathcal{I}_{f.i.p}(T) \cup \{\emptyset\}$.*

Proof. We shall show that $\mathcal{I}_{f.i.p}(T) \subseteq \overline{\mathcal{I}d(T)}$. Assume that $\mathcal{I}d(T) \neq \mathcal{I}_{f.i.p}(T)$ and let $I_0 \in \mathcal{I}_{f.i.p}(T) \setminus \mathcal{I}d(T)$. Let $\mathcal{U}(A, B, I_0) := \{I \in \mathcal{I}_{f.i.p}(T) : A \subseteq I, B \cap I = \emptyset\}$ be a basis neighborhood of I_0 in 2^T, $A, B \in [T]^{<\omega}$. Hence, by definition of $\mathcal{I}_{f.i.p}(T)$, $M(A) \nsubseteq L(B)$ since $I_0 \in \mathcal{I}_{f.i.p}(T)$. We shall establish that $I_0 \in \overline{\mathcal{I}d(T)}$ and thus $\mathcal{I}_{f.i.p}(T) \subseteq \overline{\mathcal{I}d(T)}$.

Case 1. $A = \{t_1, \ldots, t_n\}, B = \{s_1, \ldots, s_m\}$.

So, $M(A) \nsubseteq L(B)$ implies $\cap_{i=1}^n b_{t_i} \nsubseteq \cup_{j=1}^m b_{s_j}$. Therefore pick $\alpha \in \cap_{i=1}^n b_{t_i} \setminus \cup_{j=1}^m b_{s_j}$. Now, $\{t_1, \ldots, t_n\} \subseteq \downarrow \alpha$ and $\downarrow \alpha \cap \{s_1, \ldots, s_m\} = \emptyset$. This shows that $\mathcal{U}(A, B, I_0) \cap \mathcal{I}d(T)$ is non empty. Thus $I_0 \in \overline{\mathcal{I}d(T)}$.

Case 2. $A = \{t_1, \ldots, t_n\}, \quad B = \emptyset$.

Note that $M(A) \nsubseteq L(\emptyset) = \emptyset$. Thus $M(A) \neq \emptyset$. So, pick $\alpha \in \cap_{i=1}^n b_{t_i}$. Thus, $\downarrow \alpha \in \mathcal{U}(A, B, I_0) \cap \mathcal{I}d(T)$, so $I_0 \in \overline{\mathcal{I}d(T)}$.

Case 3. $A = \emptyset, B = \{s_1, \ldots, s_m\}$.

Again since $M(A) \nsubseteq L(B)$, we have $T = M(\emptyset) \nsubseteq \cup_{j=1}^m b_{s_j}$. So there is $\alpha \in T \setminus \cup_{j=1}^m b_{s_j}$. Thus $\downarrow \alpha \in \mathcal{U}(A, B, I_0) \cap \mathcal{I}d(T)$ and then $I_0 \in \overline{\mathcal{I}d(T)}$. Assume $(T, \leq)$ is an (f.g)-poset, then, by Lemma 2.1, $\mathcal{I}_{f.i.p}(T)$ is a closed subset of 2^T, and since $\mathcal{I}d(T) \subseteq \mathcal{I}_{f.i.p}(T) \subseteq \overline{\mathcal{I}d(T)}$, we have $\overline{\mathcal{I}d(T)} = \mathcal{I}_{f.i.p}(T)$.

If $(T, \leq)$ is not an (f.g)-poset, then by Lemma 2.1, $\mathcal{I}_{f.i.p}(T) \cup \{\emptyset\}$ is a closed subset of 2^T, and since $\mathcal{I}_{f.i.p}(T) \subseteq \overline{\mathcal{I}d(T)}$ holds always; it follows that $\overline{\mathcal{I}_{f.i.p}(T)} = \overline{\mathcal{I}d(T)}$. Thus, $\overline{\mathcal{I}d(T)} = \mathcal{I}_{f.i.p}(T) \cup \{\emptyset\}$. □

Corollary 3.5. *For any poset $(T, \leq)$ the Stone space $\mathcal{U}lt(B(T))$ of the tail algebra $B(T)$ is $\overline{\mathcal{I}d(T)}$ up to a homeomorphism.*

3.2. Case of upper semi-lattices

Recall that an upper semi-lattice poset $(T, \leq)$ is so that $l.u.b.\{x, y\} := x \vee y$ exists in $(T, \leq)$ for all $x, y \in T$. $(T, \leq)$ is then a ($\vee$-f.g.)-poset and $\mathcal{I}_{f.i.p}(T) = \mathcal{I}d(T)$.

Thus we get, the following corollary.

Corollary 3.6. *Let $(T, \leq)$ be an upper semi-lattice.*

i) *$\mathcal{I}d(T) \cup \{\emptyset\}$ is closed in 2^T,*

ii) *$\mathcal{I}d(T)$ is closed in 2^T if and only if $(T, \leq)$ is an (f.g.)-poset.*

Proof. By Theorem 3.2 and 3.3. □

Next, to study $B(T)$, with $(T, \leq)$ is an upper semi-lattice. We may assume, wlog, that $(T, \leq)$ has a least element by the following proposition.

Proposition 3.7. *Let $(T, \leq)$ be an upper semi-lattice.*

i) *If $(T, \leq)$ is an (f.g.)-poset, then,*

$$\mathcal{U}lt(B(T)) \cong_{\text{homeo}} \mathcal{I}d(T) \cong_{\text{homeo}} \mathcal{I}d(T')$$

where $(T', \preceq)$ is an upper semi-lattice with a least element.

ii) *If $(T, \leq)$ is not an (f.g.)-poset, then,*

$$\mathcal{U}lt(B(T)) \cong_{\text{homeo}} \mathcal{I}d(T) \cup \{\emptyset\} \cong_{\text{homeo}} \mathcal{I}d(\breve{T})$$

where $\breve{T} = T \cup \{\alpha\}$, with $\alpha \notin T$ and α is the least element of $\breve{T}$.

Proof. i) If $(T, \leq)$ is an (f.g.)-poset, then, $T = \cup_{i=1}^{n} b_{t_i}$, with $\{t_1, \ldots, t_n\}$ is an anti-chain in T.

Set, $T' = T$ and for all $x, y \in T'$:

$$x \preceq y \leftrightarrow x \leq y \text{ or } \quad x = t_1.$$

Define the following bijection, φ from $\mathcal{I}d(T)$ into $\mathcal{I}d(T')$ by $\varphi(I) = I \cup \{t_1\}$. To finish up the proof, we show that φ is continuous. Indeed, let $I_0 \in \mathcal{I}d(T)$; $J_0 = \varphi(I_0) = I_0 \cup \{t_1\} \in \mathcal{I}d(T')$.

Let $A, B \in [T]^{<\omega}$ so that $A \subseteq J_0, B \cap J_0 = \emptyset$. Since $t_1 \notin B$, we may assume that $t_1 \notin A$ since each element of $\mathcal{I}d(T')$ contains t_1. Set, $W_0 := \{J \in \mathcal{I}d(T') : A \subseteq J, \quad B \cap J = \emptyset\}$ and $V_0 := \{I \in \mathcal{I}d(T) : A \subseteq I, \quad B \cap I = \emptyset\}$. Then, $I_0 \in V_0 \subseteq \varphi^{-1}(W_0)$. Hence, φ is continuous. By Corollary 3.5 and Corollary 3.6, we have $\mathcal{U}lt(B(T)) \cong_{\text{homeo}} \mathcal{I}d(T)$.

ii) If $(T, \leq)$ is not an (f.g.)-poset, then, by Theorem 2.3 and Lemma 2.2, we have $\mathcal{U}lt(B(T)) \cong_{\text{homeo}} \mathcal{I}_{f.i.p}(T) \cup \{\emptyset\} \cong \mathcal{I}_{f.i.p}(\breve{T})$, since $\mathcal{I}_{f.i.p}(T) = \mathcal{I}d(T)$ and $\mathcal{I}_{f.i.p}(\breve{T}) = \mathcal{I}d(\breve{T})$, we obtain $\mathcal{U}lt(B(T)) \cong_{\text{homeo}} \mathcal{I}d(T) \cup \{\emptyset\} \cong_{\text{homeo}} \mathcal{I}d(\breve{T})$. □

Definition 3.8. *Let B be a Boolean algebra and $H \subseteq B$. H is a disjunctive set whenever:*

i) $0 \notin H$;

ii) *For all $h, h_1, \ldots, h_p \in H$ $[h \leq h_1 + \cdots + h_p \longrightarrow$ There is i so that $h \leq h_i]$*

The next Lemma is Proposition 2.2. in [10].

Lemma 3.9. *For any Boolean algebra B the following statements are equivalent.*

i) $H \subseteq B \setminus \{0\}$ *is disjunctive;*

ii) *For each $M \subseteq H$, there is a homomorphism $f_M : \langle H \rangle \longrightarrow \mathcal{P}(M)$ defined by*

$$f_M h = M \downarrow h;$$

where $M \downarrow h = \{m \in M : m \leq h\}$.

Theorem 3.10. *The following statements are equivalent for any Boolean algebra B.*

i) *B is isomorphic to an upper semi-lattice algebra.*

ii) *B is generated by $H \subseteq B$ so that: $0 \notin H$, H is disjunctive, containing 1 and closed under multiplication.*

Proof. i) implies ii). Let $B = B(T)$ be an upper semi-lattice algebra with T having a least element. Hence $B = \langle H \rangle$, where $H = \{b_t : t \in T\}$. Now, notice that $0 \notin H$ since $b_t > 0$ for all $t \in T$, $1 \in H$ since $b_{0_T} = 1$ and this is true since T has a least element 0_T, and finally H is closed under "$\cdot$" since $b_t \cdot b_{t'} = b_{t \vee t'}$. Moreover, H is a disjunctive set. For let $b_t \leq b_{t_1} + \cdots + b_{t_n}$. So $t \in b_{t_i}$ for some i, and therefore $t \geq t_i$. So $b_t \subseteq b_{t_i}$.

ii) implies i). Let $B = \langle H \rangle$, where $0 \notin H$, $1 \in H$, H disjunctive and closed under "$\cdot$". Put $(T, \preceq) = (H, \geq)$. Note that for all $x, y \in T$, $\sup_{\preceq}(x, y) = \inf_{\leq}(x, y) = x \cdot y \in H = T$. Thus $(T, \preceq)$ is an upper semi-lattice. Again, $\min(T, \preceq) = 1$. Indeed, $x \leq 1$ for all $x \in H$. Hence, $1 \preceq x$ for all $x \in T = H$. So, $(T, \preceq)$ has a least element 1. Next, we claim that $B = \langle H \rangle \cong$ B(T). Indeed, define: f from H into $\mathcal{P}(H)$ by $f(t) = H \downarrow t := \{x \in H : x \leq t\} = \{x \in T : t \preceq x\} = b_t$. Next, since H is disjunctive and by Lemma 3.9, f extends to a homomorphism from $\langle H \rangle$ into $\mathcal{P}(H)$, denoted by f.

First, note that $f(B) = \langle f(H) \rangle = \langle b_t : t \in T \rangle =$ B(T). So, f is onto and we need only show that f is 1-1. For, suppose $\prod_{i=1}^{n} \varepsilon_i f(h_i) = \emptyset$ and show that $\prod_{i=1}^{n} \varepsilon_i h_i = 0$.

Case 1. *For all i, $\varepsilon_i = 1$.*

$\prod_{i=1}^{n} \varepsilon_i f(h_i) = \emptyset \longleftrightarrow \cap_{i=1}^{n} f(h_i) = \emptyset = \cap_{i=1}^{n} H \downarrow h_i \ni h_1 \cdots h_n = h \in H$ since H is closed under "$\cdot$". So this case does not happen.

Case 2. *For all i, $\varepsilon_i = -1$.*

$\prod_{i=1}^{n} -f(h_i) = \emptyset \longleftrightarrow \cup_{i=1}^{n} f(h_i) = H \longleftrightarrow \cup_{i=1}^{n} H \downarrow h_i = H$. Now, since $1 \in H = \cup_{i=1}^{n} H \downarrow h_i$, it follows that $1 \in H \downarrow h_i$ for some i. So, $1 \leq h_i$ and thus $\sum_{i=1}^{n} h_i = 1$, i.e., $\prod_{i=1}^{n} -h_i = 0$.

Case 3. *There are i, j so that $\varepsilon_i = 1$ and $\varepsilon_j = -1$.*

In this case we need to show:

$$\left[(H \downarrow p_1) \cap \cdots \cap (H \downarrow p_k) \subseteq (H \downarrow q_1) \cap \cdots \cap (H \downarrow q_m)\right] \rightarrow \left(p_1 \cdots p_k \leq q_1 + \cdots + q_m\right).$$

Let $p = p_1 \cdots p_k$, $p \in H$ and $p \leq p_i$ for all i. Hence, $p \in \cap_{i=1}^{k} (H \downarrow p_i) \subseteq \cap_{i=1}^{n} (H \downarrow q_i)$. So, there is j so that $p \in H \downarrow q_j$, i.e., $p \leq q_j$. Hence, $p \leq q_1 + \cdots + q_m$. Therefore, $p_1 \cdots p_k \leq q_1 + \cdots + q_m$. This finishes up the proof. □

Recall that $(S,\cdot)$ is called a semi-lattice whenever "$\cdot$" is commutative, associative, and $x^2 = x$ for all $x \in S$. For more details on semi-lattices see [6].

The next theorem characterizes $\mathcal{I}d(T)$, for any upper semi-lattice $(T,\leq)$.

Theorem 3.11. *Let B be a Boolean algebra and set $X = \mathcal{U}lt(B)$ its Stone space. The following statements are equivalent.*

i) *B is isomorphic to $B(T)$, where $(T,\leq)$ is an upper semi-lattice, with a least element.*
ii) *X is homeomorphic to $\mathcal{I}d(T)$, the set of ideals of an upper semi-lattice T with a least element, endowed with Tychonoff's topology inherited from 2^T.*
iii) *X is homeomorphic to $\mathcal{F}(S)$, the set of filters over S, where S is a unitary semi-lattice, endowed with Tychonoff's topology inherited from 2^S.*
iv) *There is a multiplication "$\cdot$" on X so that $(X,\cdot)$ is a unitary semi-lattice and "$\cdot$" is a continuous mapping on $X\times X$ (i.e., $(x,y)\longrightarrow x\cdot y$ is continuous).*

Proof. i) implies ii). Using Proposition 3.7.
ii) implies iii). Set $X = \mathcal{I}d(T)$ as in ii), and let $(T,\leq)$ be an upper semi-lattice with a least element t_0. Put $S := (T,\leq^*)$, where $\leq^*$ is the reversed order of $\leq$. Now, $(S,\wedge)$ is a unitary semi-lattice, where $s\wedge s' = s\vee s'$ in $(T,\leq)$. Notice that t_0 is the unity of $(S,\wedge)$ and an ideal of $(T,\leq)$ is a filter of $(S,\wedge)$. Now, let $\mathcal{F}(S)$ denotes the set of filters of S, endowed with the topology inherited from 2^S. It is clear that $\mathcal{I}d(T)$ and $\mathcal{F}(S)$ are homeomorphic Boolean spaces. Thus iii) follows.
iii) implies iv). Let $X = \mathcal{F}(S)$ as in iii). We shall show that $(\mathcal{F}(S),\cap)$ is a unitary semi-lattice so that φ from $\mathcal{F}(S)\times\mathcal{F}(S)$ into $\mathcal{F}(S)$ by $\varphi(F,G) = F\cap G$ is a continuous mapping. For it's clear that $(\mathcal{F}(S),\cap)$ is a semi-lattice and S is its unity. Next, to show that φ is continuous let $(F_0,G_0)\in\mathcal{F}(S)\times\mathcal{F}(S)$, and pick V a basic open set containing $F_0\cap G_0$. So,

$$V = \{F\in\mathcal{F}(S) : a_i\in F, \text{ and } b_j\notin F \text{ for } i=1,\dots,n;\ \ j=1,\dots,m\},$$

with

$$a_i\in F_0\cap G_0, \text{ and } b_j\notin F_0\cap G_0 \text{ for } i=1,\dots,n;\ \ j=1,\dots,m.$$

Hence,

$$\begin{aligned}\varphi^{-1}(V) &= \{(F,G)\in\mathcal{F}(S)\times\mathcal{F}(S) : (a_i,b_j)\in(F\cap G)\times(\mathcal{F}(S)\setminus(F\cap G))\\ &\qquad \text{for } i=1,\dots,n; j=1,\dots,m\}\\ &= \{(F,G)\in\mathcal{F}(S)\times\mathcal{F}(S) : a_i\in F\cap G \text{ for } i=1,\dots,n\}\\ &\quad \cap\{(F,G)\in\mathcal{F}(S)\times\mathcal{F}(S) : b_j\notin F\cap G \text{ for } j=1,\dots,m\}\\ &= (\{F\in\mathcal{F}(S) : a_i\in F \text{ for } i=1,\dots,n\}\times\\ &\quad \{G\in\mathcal{F}(S) : a_i\in G \text{ for } i=1,\dots,n\})\\ &\quad \cap\cap_{j=1}^m[(\{F\in\mathcal{F}(S) : b_j\notin F\}\times\mathcal{F}(S))\\ &\quad \cup(\mathcal{F}(S)\times\{G\in\mathcal{F}(S) : b_j\notin G\})]\end{aligned}$$

which is clearly an open set containing (F_0, G_0). Thus φ is continuous. The proof of iii) implies iv) is finished.

iv) implies i). Let $X = \mathcal{U}lt(B)$, where $(X, \cdot)$ is a unitary *lower* semi-lattice and $f : X \times X \longrightarrow X$ so that $f(x, y) = x \cdot y$ is continuous. Let $\varphi : X \longrightarrow H$ be a continuous homomorphism, $(H, *)$ be a finite *lower* semi-lattice. If we write $\leq$ for the natural partial ordering on H, then $r(\varphi, h) = \{x \in X : \varphi(x) * h = h\}(= \{x \in X : \varphi(x) \leq h\})$, see [8]. Next let R be the set of $r(\varphi, h)'s$.

1. $R \subseteq Clopen(X)$: Let $\varphi : X \longrightarrow H$ be a continuous homomorphism and $\theta : H \longrightarrow H$ defined by $\theta(y) = y * h$. θ is a continuous function and thus $\psi = \theta \circ \varphi$ defined by $\psi(x) = \varphi(x) * h$ is continuous. Hence, $r(\varphi, h) = \psi^{-1}(\{h\}) \in Clopen(X)$ since $\{h\}$ is clopen in H because H is a finite set.

2. *$Clopen(X)$ is generated by R:* To this end let $x, y \in X, x \neq y$ and, by Numakura's theorem see [12], pick $\varphi : X \longrightarrow H$ a continuous homomorphism with $(H, *)$ a finite semi-lattice so that $\varphi(x) \neq \varphi(y)$. Thus either $\varphi(x) * \varphi(y) \neq \varphi(x)$ or $\varphi(x) * \varphi(y) \neq \varphi(y)$.

Case 1. $\varphi(x) * \varphi(y) \neq \varphi(x)$.

In this case since $\varphi(x) * \varphi(x) = \varphi(x)$ it follows that $x \in r(\varphi, \varphi(x))$ and $y \notin r(\varphi, \varphi(x))$. So, $r(\varphi, \varphi(x))$ separates x and y in X.

Case 2. $\varphi(x) * \varphi(y) \neq \varphi(y)$.

This case is treated similarly.

3. *$R \setminus \emptyset$ is closed under $\cap$:* Indeed, let $r(\varphi, h), r(\psi, k) \in R \setminus \emptyset$, where $h \in H, k \in K$, and $\varphi($ resp. $\psi)$ is a continuous homomorphism from X into H (resp. X into K). Now define $\theta : X \longrightarrow H \times K$ by $\theta(x) = (\varphi(x), \psi(x))$. Denote by $*_H$ (resp. $*_K$) the operation on H (resp. on K) and define $*$ on $H \times K$ by $(a, b) * (c, d) = (a *_H c, b *_K d)$. We claim that $(H \times K, *)$ is a finite semi-lattice. Indeed, $\theta(x \cdot y) = (\varphi(x \cdot y), \psi(x \cdot y)) = (\varphi(x) *_H \varphi(y), \psi(x) *_K \psi(y)) = (\varphi(x), \psi(x)) * (\varphi(y), \psi(y)) = \theta(x) * \theta(y)$. θ is also continuous since φ and ψ are. Thus θ is a continuous homomorphism from X into $H \times K$. Moreover,

$$\begin{aligned} x \in r(\theta, (h, k)) &\leftrightarrow \theta(x) * (h, k) = (h, k) \\ &\leftrightarrow (\varphi(x), \psi(x)) * (h, k) = (h, k) \\ &\leftrightarrow \varphi(x) *_H h = h \text{ and } \psi(x) *_K k = k \\ &\leftrightarrow x \in r(\varphi, h) \cap r(\psi, k). \end{aligned}$$

Thus $r(\varphi, h) \cap r(\psi, k) = r(\theta, (h, k)) \in R$.

Next, we still need to show that $r(\theta, (h, k)) \neq \emptyset$. For we claim that every element of $R \setminus \emptyset$ contains the unity of X. Indeed, let α_0 be this unity of X and let $r(\varphi, h) \in R \setminus \emptyset$ then pick $x_0 \in r(\varphi, h)$. So, $\varphi(x_0) * h = h$ and hence $h = \varphi(x_0) * h = \varphi(\alpha_0 \cdot x_0) * h = \varphi(\alpha_0) * \varphi(x_0) * h = \varphi(\alpha_0) * h$.

So, $\alpha_0 \in r(\varphi, h)$. Thus $\alpha_0 \in r(\varphi, h)$ and $\alpha_0 \in r(\psi, k)$. Hence, $\alpha_0 \in r(\varphi, h) \cap r(\psi, k) = r(\theta, (h, k))$; this shows that $R \setminus \emptyset$ is closed under $\cap$.

4. *$X \in R$ and $\emptyset \in R$:* To see this, note that $\varphi_0 : X \longrightarrow \{0, 1\}$ defined by: $\varphi_0(x) = 0$ is a continuous homomorphism and we have $r(\varphi_0, 0) = X$ and $r(\varphi_0, 1) = \emptyset$.

5. $R \setminus \emptyset$ *is a disjunctive set:* To this end we need to show: *"For each* $n \geq 1$*, for each* $r(\varphi, h), r(\varphi_1, h_1), \ldots, r(\varphi_n, h_n) \in R \setminus \emptyset$ *if* $r(\varphi, h) \subseteq \cup_{i=1}^{n} r(\varphi_i, h_i)$ *then there is* i_0 *so that* $r(\varphi, h) \subseteq r(\varphi_{i_0}, h_{i_0})$*."* Let,

$$r(\varphi, h) \subseteq \bigcup_{i=1}^{n} r(\varphi_i, h_i). \qquad (*)$$

Pick n minimal in $(*)$. If $n = 1$ we are done. Otherwise (i.e., $n \geq 2$), $r(\varphi, h) \nsubseteq \bigcup_{i=1, i \neq j}^{n} r(\varphi_i, h_i)$ and $r(\varphi, h) \cap r(\varphi_j, h_j) \neq \emptyset$, for every $j \in \{1, \ldots, n\}$. Hence, there are $x_1, \ldots, x_n \in r(\varphi, h)$ so that for each $j \in \{1, \ldots, n\}$ $x_j \in r(\varphi, h) \setminus \cup_{i=1, i \neq j}^{n} r(\varphi_i, h_i)$ and since $r(\varphi, h) \subseteq \cup_{i=1}^{n} r(\varphi_i, h_i)$, it follows that $x_j \in r(\varphi, h) \cap r(\varphi_j, h_j)$ and $x_j \notin r(\varphi_i, h_i)$ for $i \neq j$.

Let $x_0 = x_1 \cdots x_n$ be the product of $x_j's$ in X. $x_0 \in r(\varphi, h)$ since $\varphi(x_0) * h = \varphi(x_1) * \cdots * \varphi(x_n) * h = h$ because $x_i \in r(\varphi_i, h_i)$. Now, by $(*)$ there is $k \in \{1, \ldots, n\}$ so that $x_0 \in r(\varphi_k, h_k)$ and thus $\varphi_k(x_0) *_K h_k = h_k$. Let $j \neq k$, hence, $x_j \cdot x_0 = x_j \cdot x_1 \cdots x_n = x_1 \cdots x_n = x_0$ since $x \cdot x = x$. Thus $x_j \cdot x_0 = x_0$. So, $h_k = \varphi_k(x_0) *_K h_k = \varphi_k(x_j \cdot x_0) *_K h_k = \varphi_k(x_j) *_K \varphi_k(x_0) *_K h_k = \varphi_k(x_j) *_K (\varphi_k(x_0) *_K h_k) = \varphi_k(x_j) *_K h_k$. Thus $x_j \in r(\varphi_k, h_k)$ for $j \neq k$ and this contradicts our previous assumption about $x_j's$. Hence, there is i so that $r(\varphi, h) \subseteq r(\varphi_i, h_i)$. Therefore $R \setminus \emptyset$ is closed under $\cap$, disjunctive and generates $Clopen(X)$, hence by Theorem 3.10 $Clopen(X)$ is an upper semi-lattice algebra and the proof of the theorem is completed. □

3.3. Set of initial sections of a poset

Let $(P, \leq)$ be a poset, with a least element. Denote by $\mathcal{I}(P)$ the set of non empty initial sections of P. $\mathcal{I}(P)$ is a Boolean space. In this section, we shall investigate conditions under which $\mathcal{I}d(T)$, where $(T, \leq)$ is a poset, can be represented by $\mathcal{I}(P)$, up to a homeomorphism, for some poset P and conversely. Notice that, when P is a chain, we have $\mathcal{I}(P) = \mathcal{I}d(P)$.

Definition 3.12. *Let* $(T, \leq)$ *be an upper semi-lattice. An element* $a \in T$ *is prime whenever for all* $c, d \in T$ $(a \leq c \vee d \rightarrow a \leq c$ *or* $a \leq d)$. Prim(T) *shall denote the set of prime elements of* T.

Also, we say that Prim(T) *is* $\vee$*-generating set for* T*, whenever for every* $t \in T$ *there are* $t_1, \ldots, t_n \in$ Prim(T) *so that* $t = t_1 \vee \cdots \vee t_n$.

Theorem 3.13. *For any poset* $(P, \leq)$*, with a least element, there exists an upper semi-lattice* T*, with a least element, so that* $\mathcal{I}(P) \cong_{\text{homeo}} \mathcal{I}d(T)$ *where* Prim(T) *is a* $\vee$*-generating set of* T.

Proof. Let $(P, \leq)$ be a poset and set $T := \{\bigcup_{i=1}^{n} (\downarrow t_i) : t_i \in P, n \geq 1\}$. It's clear that $(T, \subseteq)$ is a $\cup$-lattice and if $p_0 = \min(P, \leq)$ then $\downarrow p_0 = \{p_0\} = \min(T, \subseteq)$. Next, define φ from $\mathcal{I}(P)$ into $\mathcal{I}d(T)$ by $\varphi(I) := (\downarrow I) \cap T$, where $\downarrow I := \{J \in \mathcal{I}(P) : J \subseteq I\}$. It's straightforward to see that φ is a homeomorphism.

Let's check, however, that Prim(T) is $\cup$-generating set of T. Indeed, we claim that Prim$(T) = \{\downarrow t : t \in P\}$. For let $a \in T$, $a = \downarrow t \subseteq c \cup d$, with $c = \cup_{i=1}^{p} \downarrow t_i$

and $d = \cup_{j=1}^{q} \downarrow s_j$. So, $t \in (\bigcup_{i=1}^{p} \downarrow t_i) \cup (\bigcup_{j=1}^{q} \downarrow s_j)$. Hence there are i, j so that either $t \in \downarrow t_i$ or $t \in \downarrow s_j$. In the first case, $a = \downarrow t \subseteq \downarrow t_i \subseteq c$; in the second, $a = \downarrow t \subseteq \downarrow s_j \subseteq d$. This shows that $a \in \mathrm{Prim}(T)$. Actually, every element of T, $a = \cup_{i=1}^{n} \downarrow t_i$ where n minimal and $n \geq 2$, is not prime. To see that let $a = \cup_{i=1}^{n} \downarrow t_i$ with n minimal. This means that $t_1, \ldots, t_n$ is an anti-chain in $(P, \leq)$. So, $a \subseteq (\downarrow t_j) \cup (\cup_{i \neq j}^{n} \downarrow t_i)$. But $a \nsubseteq (\downarrow t_j)$ and $a \nsubseteq (\cup_{i \neq j}^{n} \downarrow t_i)$ for each $j \neq i$. This shows that a is not prime in T. Hence, $\mathrm{Prim}(T) = \{\downarrow t : t \in P\}$.

T is obviously $\cup$-generated by $\mathrm{Prim}(T)$. □

Theorem 3.14. *For any upper semi-lattice T, with a least element, so that* $\mathrm{Prim}(T)$ *is $\vee$-generating set of T, there is a poset P so that $\mathcal{I}d(T) \cong_{\text{homeo}} \mathcal{I}(P)$.*

Proof. Let T be an upper semi-lattice, with a least element t_0, so that $\mathrm{Prim}(T)$ is $\vee$-generating set of T, and set $P := \{\downarrow t : t \in \mathrm{Prim}(T)\}$; then $(P, \subseteq)$ is a poset with a least element $\downarrow \{t_0\} = \{t_0\}$. Define φ from $\mathcal{I}d(T)$ into $\mathcal{I}(P)$ by $\varphi(I) = (\downarrow I) \cap P$, where $\downarrow I :=_{def} \{J \in \mathcal{I}d(T) : J \subseteq I\}$. Again it easy to see that $\mathcal{I}d(T) \cong_{\text{homeo}} \mathcal{I}(P)$. □

Let$(T, \leq)$ be a poset with a least element. Theorems 3.13 and 3.14 give indeed, necessary and sufficient conditions on $\mathcal{I}d(T)$ and $\mathcal{I}(P)$ to be homeomorphic spaces. We state these conditions in the following corollary.

Corollary 3.15. *For any ($\vee$-f.g.)-poset $(T, \leq)$, with a least element, the following statements are equivalent.*

i) *There is a poset $(P, \leq)$, with a least element, so that $\mathcal{I}d(T) \cong_{\text{homeo}} \mathcal{I}(P)$.*
ii) *There is an upper semi-lattice T', with a least element, so that $\mathcal{I}d(T) \cong_{\text{homeo}} \mathcal{I}d(T')$, where* $\mathrm{Prim}(T')$ *is a $\vee$-generating set of T'.*

For more details on free poset algebra $F(P)$ ($:= Clopen(\mathcal{I}(P))$, *the set of closed and open sets* of $\mathcal{I}(P)$) see [2].

3.4. Relationship with other classes of Tail algebras

The following theorem gives a concrete construction of semi-group algebras. For let $(M, \wedge)$ be an idempotent semi-lattice with 0 and 1, and let A be the free Boolean algebra with free generators x_p for $p \in M$, and let I be the ideal generated by the set

$$\{(x_p \cdot x_q) \triangle x_{p \wedge q} : p, q \in M\}.$$

Theorem 3.16 (D. Monk).

i) *A/I is a semi-group algebra;*
ii) *Every semi-group algebra is isomorphic to some A/I as above;*
iii) *The Stone space $\mathcal{U}lt(A/I)$ is homomorphic to $\mathcal{F}(S)$, where $(S, \wedge)$ is a unitary meet semi-lattice.*

Proof. i) Take $M = (M, \wedge)$ to be an idempotent semi-lattice with 0 and 1 and form the free Boolean algebra A with free generators x_a for $a \in M$. Then consider, in A, the ideal I generated by the set

$$\{x_0, -x_1\} \cup \{(x_a \cdot x_b)\triangle x_{a\wedge b}\}$$

Where $u\triangle v = (u \cdot -v) + (v \cdot -u)$. Denote elements of A/I by $[u]$ with $u \in A$. Obviously,

$$[x_0] = 0 \text{ and } [x_1] = 1 \tag{1}$$

We claim that A/I is a semi-group algebra on $H := \{[x_a] : a \in M\}$. Clearly 0, $1 \in H$ and H is closed under "$\cdot$". To show that H is disjunctive, suppose that $a, b_1, \dots, b_m \in M \setminus \{0\}$, $m > 0$, and $[x_a] \not\leq [x_{b_i}]$ for all $i = 1, \dots, m$; we want to show that $[x_a] \not\leq [x_{b_1}] + \cdots + [x_{b_m}]$. Suppose to the contrary that this inequality is true. Then $x_a \cdot -x_{b_1} \cdot \cdots \cdot -x_{b_m}$ is in the ideal, so we can write

$$x_a \cdot -x_{b_1} \cdot \cdots \cdot -x_{b_m} \leq x_0 + -x_1 + (x_{c_1} \cdot x_{d_1})\triangle x_{c_1\wedge d_1} + \cdots + (x_{c_n} \cdot x_{d_n})\triangle x_{c_n\wedge d_n}. \quad (*)$$

By freeness, let f be a homomorphism from A into 2 such that $f(x_e) = 1$ if $a \leq e$, and $f(x_e) = 0$ otherwise. If $a \not\leq b_i$ for all $i = 1, \dots, m$, then an application of f to the inequality $(*)$ gives $1 \leq 0$, contradiction. Hence $a \leq b_i$ for some $i = 1, \dots, m$. So $[x_a] \cdot [x_{b_i}] = [x_{a\wedge b_i}] = [x_a]$, as desired.

ii) Let us denote the somewhat concrete Boolean algebra constructed in this way by $S(M)$. Now we show that if B is a semigroup algebra on a subset M, then B is isomorphic to $S(M)$. To see this, let f be a homomorphism from A into B such that $f(x_a) = a$ for every $a \in M$. Thus f maps A onto B. It suffices to show that the kernel of f is I. First, $f(x_0) = 0$ and $f(-x_1) = -f(x_1) = -1 = 0$, so both x_0 and $-x_1$ are in the kernel. Denote the meet operation of B by $\wedge$ to fit into the notation above. If a, $b \in M$, then

$$f(x_a \cdot x_b \triangle x_{a\wedge b}) = (a \wedge b)\triangle(a \wedge b) = 0$$

as desired. So I is a subset of the kernel of f. Now suppose that $u \in A$ is in the kernel of f. Write

$$u = \sum_{i<m} x_{a_{i1}} \cdots x_{a_{ip_i}} \cdot -x_{b_{i1}} \cdots - x_{b_{iq_i}},$$

where we may assume that each p_i and q_i is nonzero; moreover, we may assume that all a_{ij} are nonzero. Take any $i < m$; we show that $x_{a_{i1}} \cdots x_{a_{ip_i}} \cdot -x_{b_{i1}} \cdots -x_{b_{iq_i}}$ is in I. Since f maps this element to 0, by disjunctiveness there is a j such that $a_{i1} \cdots a_{ip_i} \leq b_{ij}$. Hence

$$\begin{aligned}
[x_{a_{i1}} \cdots x_{a_{ip_i}} \cdot -x_{b_{i1}} \cdots - x_{b_{iq_i}}] &\leq [x_{a_{i1}} \cdots x_{a_{ip_i}} \cdot -x_{b_{ij}}] \\
&\leq [x_{a_{i1}\wedge\cdots\wedge a_{ip_i}}] \cdot -[x_{b_{ij}}] \\
&= 0
\end{aligned}$$

as desired.

iii) Let $(S, \wedge)$ be a semi-lattice, let A be the free Boolean algebra with free generators x_p for $p \in S$, and let I be the ideal generated by the set $\{(x_p \cdot x_q)\triangle x_{p\wedge q} : p, q \in S\}$. Then A/I is a semigroup algebra by **i**.

Now recall that a *filter* on $(S, \wedge)$ is a final segment T of S such that for all a, $b \in T$, also $a \wedge b \in T$. We claim that A/I is isomorphic to Clopen($\{T : T$ is a filter of $S\}$). First we note that $\{T : T$ is a filter of $S\}$ is a closed subspace of the Boolean space of all subsets of S, $\mathcal{P}(S)$. Recall that the collection of $\langle \mathcal{U}(C, D) : C, D \in [S]^{<\omega}\rangle$, where $\mathcal{U}(C, D) := \{F \in 2^S : C \subseteq F \text{ and } F \cap D = \emptyset\}$, is a basis for the topology of $\mathcal{P}(S)$.

Next, note that the Stone space of A/I can be taken to be the set of all ultrafilters F on A such that for all p, $q \in S$, the element $-((x_p \cdot x_q)\triangle x_{p\wedge q})$ is in F. For any such ultrafilter F, we define

$$f(F) := \{p \in S : x_p \in F\}.$$

We claim that f is a homeomorphism from the Stone space of A/I onto the space of filters in S. First we check that $f(F)$ is a filter. Suppose that $p \in f(F)$ and $p \leq q$. Thus $x_p \in F$. Hence

$$x_p \cdot -((x_p \cdot x_q)\triangle x_{p\wedge q}) \in F,$$

and

$$x_p \cdot -((x_p \cdot x_q)\triangle x_{p\wedge q}) = x_p \cdot (x_p \cdot x_q \cdot x_p + (-x_p + -x_q) \cdot -x_p) \leq x_q,$$

so $x_q \in F$ and hence $q \in f(F)$. Now suppose that p, $q \in f(F)$. Then

$$x_p \cdot x_q \cdot -((x_p \cdot x_q)\triangle x_{p\wedge q}) \in F,$$

and

$$x_p \cdot x_q \cdot -((x_p \cdot x_q)\triangle x_{p\wedge q}) = x_p \cdot x_q \cdot (x_p \cdot x_q \cdot x_{p\wedge q} + (-x_p + -x_q) \cdot -x_{p\wedge q}) \leq x_{p\wedge q},$$

so $x_{p\wedge q} \in F$ and so $p \wedge q \in f(F)$. Thus $f(F)$ is a filter.

Clearly f is one-one. To show that f is onto, let G be any filter on S. Let F be the ultrafilter on A generated by the set

$$\{x_p : p \in G\} \cup \{-x_p : p \notin G\}.$$

Clearly there is such an ultrafilter on A. We claim that it satisfies the additional condition needed for the Stone space of A/I. For, suppose that p, $q \in S$. To get a contradiction, suppose that $(x_p \cdot x_q)\triangle x_{p\wedge q} \in F$. If p, $q \in G$, then $p \wedge q \in G$, hence $x_p, x_q, x_{p\wedge q} \in F$, so

$$x_p \cdot x_q \cdot x_{p\wedge q} \cdot ((x_p \cdot x_q)\triangle x_{p\wedge q}) \in F;$$

but this element is clearly 0, contradiction. If $p \in G$ and $q \notin G$, then $p \wedge q \notin G$, so $x_p, -x_q, -x_{p\wedge q} \in F$, and a contradiction is easily reached. Similarly if $p \notin G$ but $q \in G$. Finally, if p, $q \notin G$, then $x_{p\wedge q} \notin G$ and a contradiction is easily reached.

So F is in the Stone space of A/I. Clearly $f(F) = G$. So f maps onto.

Finally, f is continuous. For, suppose that $F \in f^{-1}[\mathcal{U}(C,D)]$. Thus $C \subseteq f(F)$ and $D \cap f(F) = \varnothing$. Let $y = \prod_{p\in C} x_p \cdot \prod_{q\in D} -x_q$. Then $F \in s(y) \subseteq f^{-1}[\mathcal{U}(C,D)]$. Here s is used to denote the standard Stone mapping. □

Corollary 3.17. *Every semi-group algebra is isomorphic to an upper semi-lattice algebra as well as any pseudo-tree algebra.*

Proof. By iii) in Theorem 3.16 and 3.11 we have the first part of the statement; now if $(T,<)$ is a pseudo tree then $H := \{b_t : t \in T\}$ generates $B(T)$. Notice that $0,\, 1 \in H$, H is closed under "$\cdot$" and H is disjunctive. Hence $B(T)$ is a semigroup algebra and thus it is an upper semi-lattice. □

4. Finite combinatorics on upper semi-lattices algebras

4.1. △-length μ

Fix $B(T)$ any an upper semi-lattice algebra over T. It follows from Theorem 2.6, that each $b \in B(T) \setminus \{0\}$ has a unique decomposition of the form:

$$b = b_{t_1} \triangle \cdots \triangle b_{t_n}, \text{ for suitable } n, \text{ and } t_i's \text{ so that } t_i \neq t_j \text{ for } i \neq j \qquad (\triangle)$$

Let

$$\operatorname{supp}(b) := \{t_1, \dots, t_n\}$$

and put supp(0)=$\emptyset$.

Definition 4.1.

i) *Let T be an upper semi-lattice. We define the △-length $\mu(x)$ for any $x \in B(T)$ by:*

$$\mu(x) := |\operatorname{supp}(x)|$$

ii) *For any $n \geq 3$ and $\{x_1, \dots, x_n\} \subseteq B(T)$ we set:*

a) $S(x_1,\dots,x_n) := \left|\bigcap_{i=1}^{n} \operatorname{supp}(x_i)\right|$

b) $Z(x_1,\dots,x_n) := \sum_{3\leq k\leq n} (-2)^{k-1} \sum_{1\leq i_1<\cdots<i_k\leq n} S(x_{i_1},\dots,x_{i_k})$.

In particular $Z(x_1,x_2,x_3) = 4S(x_1,x_2,x_3)$.

c) *We say that $\{x_1,\dots,x_n\}$ is a* **3-free set** *whenever for any $u,v,w \in \{x_1,\dots,x_n\}$, $S(u,v,w) = 0$. When $n = 3$ we say that $\{x_1,x_2,x_3\}$ is a* **free set** *instead of 3-free set.*

Lemma 4.2.

i) *If A, B, C, are any sets, then*

a) $|A\triangle B| = |A| + |B| - 2|A\cap B|$.

b) $|C| - 2|A\cap C| - 2|B\cap C| = |A\triangle C| + |B\triangle C| - |A| - |B| - |C|$.

c) $|A\triangle B\triangle C| = |A\triangle B| + |A\triangle C| + |B\triangle C| - (|A|+|B|+|C|) + 4|A\cap B\cap C|$.

ii) *If $\{x,y\} \subseteq B(T)$ then* $\operatorname{supp}(x\triangle y) = \operatorname{supp}(x)\triangle\operatorname{supp}(y)$.

Proof. i) a) We use X' for the complement of X. $|A\triangle B| = |A\cap B'| + |A'\cap B| = |A| - |A\cap B| + |B| - |A\cap B| = |A| + |B| - 2|A\cap B|$.

b) Using a), $|A\triangle C| + |B\triangle C| = |A| + |C| - 2|A\cap C| + |B| + |C| - 2|B\cap C|$, and the desired conclusion follows.

c) Using a) and b),

$$\begin{aligned}|A\triangle B\triangle C| &= |A\triangle B| + |C| - 2|(A\triangle B)\cap C|\\ &= |A\triangle B| + |C| - 2|(A\cap C)\triangle(B\cap C)|\\ &= |A\triangle B| + |C| - 2|A\cap C| - 2|B\cap C| + 4|A\cap B\cap C|\\ &= |A\triangle B| + |A\triangle C| + |B\triangle C| - (|A| + |B| + |C|) + 4|A\cap B\cap C|\end{aligned}$$

as desired.

ii) Let

$$x = b_{t_1}\triangle\cdots\triangle b_{t_n}, \text{ and } y = b_{s_1}\triangle\cdots\triangle b_{s_m}.$$

Put $A = \operatorname{supp}(x) = \{t_1,\dots,t_n\}$ and $B = \operatorname{supp}(y) = \{s_1,\dots,s_m\}$. We have: $x\triangle y = b_{t_1}\triangle\cdots\triangle b_{t_n}\triangle b_{s_1}\triangle\cdots\triangle b_{s_m}$. Note that $b_{t_i}\triangle b_{s_j} = 0$ if $t_i = s_j$, thus $x\triangle y = \triangle\{b_u : u\in(A\cup B)\backslash(A\cap B)\}$. Hence $\operatorname{supp}(x\triangle y) = A\triangle B = \operatorname{supp}(x)\triangle\operatorname{supp}(y)$. □

Corollary 4.3.

i) *For any x, y, z pairwise distinct we have:*

$$\mu(x\triangle y\triangle z) = \mu(x\triangle y) + \mu(y\triangle z) + \mu(z\triangle x) - \Big(\mu(x) + \mu(y) + \mu(z)\Big) + 4S(x,y,z).$$

ii) *For any free set $\{x,y,z\}$ in $B(T)$ we have:*

$$\mu(x\triangle y\triangle z) = \mu(x\triangle y) + \mu(y\triangle z) + \mu(z\triangle x) - \mu(x) - \mu(y) - \mu(z).$$

Lemma 4.4. *If $n\ge 3$ and $\{x_1,\dots,x_n\}\subseteq B(T)$ then:*

a) $$\begin{aligned}S(x_1,\dots,x_{n-2},x_{n-1}\triangle x_n) &= S(x_1,\dots,x_{n-1})\\ &\quad + S(x_1,\dots,x_{n-2},x_n) - 2S(x_1,\dots,x_n).\end{aligned}$$

b) $$Z(x_1,\dots,x_{n-2},x_{n-1}\triangle x_n) = Z(x_1,\dots,x_n) - 4\sum_{i=1}^{n-2} S(x_i,x_{n-1},x_n).$$

Proof. a) Let

$$A = \bigcap_{i=1}^{n-2}\operatorname{supp}(x_i), \quad B = \operatorname{supp}(x_{n-1}) \quad\text{and}\quad C = \operatorname{supp}(x_n).$$

By Lemma 4.2 ii), we have: $\operatorname{supp}(x_{n-1}\triangle x_n) = B\triangle C$, thus:

$$\begin{aligned}S(x_1,\dots,x_{n-2},x_{n-1}\triangle x_n) &= |A\cap(B\triangle C)| = |(A\cap B)\triangle(A\cap C)|\\ &= |A\cap B| + |A\cap C| - 2\,|A\cap B\cap C|\\ &= S(x_1,\dots,x_{n-2},x_{n-1}) + S(x_1,\dots,x_{n-2},x_n)\\ &\quad - 2S(x_1,\dots,x_{n-2},x_{n-1},x_n)\end{aligned}$$

b)
$$\begin{aligned}
&Z(x_1,\ldots,x_{n-2},x_{n-1}\triangle x_n)\\
&= \sum_{k=3}^{n-1}(-2)^{k-1}\sum_{1\le i_1<\cdots<i_k\le n-2} S(x_{i_1},\ldots,x_{i_k})\\
&\quad+\sum_{k=3}^{n-1}(-2)^{k-1}\sum_{1\le i_1<\cdots<i_{k-1}\le n-2} S(x_{i_1},\ldots,x_{i_{k-1}},x_{n-1}\triangle x_n)\\
&\overset{\text{By ii) a)}}{=} \sum_{k=3}^{n-1}(-2)^{k-1}\sum_{1\le i_1<\cdots<i_k\le n-2} S(x_{i_1},\ldots,x_{i_k})\\
&\quad+\sum_{k=3}^{n-1}(-2)^{k-1}\sum_{1\le i_1<\cdots<i_{k-1}\le n-2} S(x_{i_1},\ldots,x_{i_{k-1}},x_{n-1})\\
&\quad+\sum_{k=3}^{n-1}(-2)^{k-1}\sum_{1\le i_1<\cdots<i_{k-1}\le n-2} S(x_{i_1},\ldots,x_{i_{k-1}},x_n)\\
&\quad+\sum_{k=3}^{n-1}(-2)^{k}\sum_{1\le i_1<\cdots<i_{k-1}\le n-2} S(x_{i_1},\ldots,x_{i_{k-1}},x_{n-1},x_n).
\end{aligned}$$

It is not difficult to see that:

$$\begin{aligned}
&\sum_{k=3}^{n-1}(-2)^{k}\sum_{1\le i_1<\cdots<i_{k-1}\le n-2} S(x_{i_1},\ldots,x_{i_{k-1}},x_{n-1},x_n)\\
&\quad=\sum_{k=3}^{n-1}(-2)^{k-1}\sum_{1\le i_1<\cdots<i_{k-2}\le n-2} S(x_{i_1},\ldots,x_{i_{k-2}},x_{n-1},x_n)\\
&\qquad-4\sum_{i=1}^{n-2} S(x_i,x_{n-1},x_n)+(-2)^{n-1}S(x_1,\ldots,x_n).
\end{aligned}$$

Thus,

$$\begin{aligned}
&Z(x_1,\ldots,x_{n-2},x_{n-1}\triangle x_n)\\
&=\sum_{3\le k\le n}(-2)^{k-1}\sum_{1\le j_1<\cdots<j_k\le n} S(x_{j_1},\ldots,x_{j_k})-4\sum_{i=1}^{n-2} S(x_i,x_{n-1},x_n).
\end{aligned}$$

This finishes up the proof of Lemma 4.4. □

Theorem 4.5. *For $n\ge 3$ and $\{x_1,\ldots,x_n\}\subseteq B(T)$ we have:*

$$\mu\left(\underset{1\le i\le n}{\triangle} x_i\right)=\sum_{1\le i<j\le n}\mu(x_i\triangle x_j)-(n-2)\sum_{1\le i\le n}\mu(x_i)+Z(x_1,\ldots,x_n).$$

Note that Theorem 4.5 follows directly from the following theorem that appears in [13].

Theorem. *If $\langle A_i : 1 \le i \le n\rangle$ is a system of sets, then*

$$\left|\underset{1\le i\le n}{\triangle} A_i\right| = \sum_{1\le k\le n} (-2)^{k-1} \sum_{1\le i_1<\cdots<i_k\le n} |A_{i_1}\cap\cdots\cap A_{i_k}|.$$

Actually, Theorem 4.5 implies the above theorem. For let $A_1,\ldots,A_n$ be any finite subsets of a set A. Set $T = \cup_{i=1}^n A_i \cup \{t_0\} \cup \{\infty\}$ with $t_0, \infty \notin \cup_{i=1}^n A_i$. Define $\preceq$ on T so that t_0 and ∞ are respectively the smallest and largest elements of $(T,\preceq)$. Moreover $\cup_{i=1}^n A_i$ is an anti-chain. Now, $(T,\preceq)$ is an upper semi-lattice. So in $B(T)$ we set $x_i = \triangle_{s\in A_i} b_s$ and we have $\mu(x_i) = |A_i|$.

Thus by Theorem 4.5 and Lemma 4.2 i) a) the desired conclusion follows.

Corollary 4.6. *Let $n \ge 3$ and $\{x_1,\ldots,x_n\}$ be a 3-free set in $B(T)$. Then we have:*

$$\mu\left(\underset{1\le i\le n}{\triangle} x_i\right) = \sum_{1\le i<j\le n} \mu(x_i \triangle x_j) - (n-2)\sum_{i=1}^n \mu(x_i).$$

4.2. Length μ^+

Definition 4.7. *Let $E = \{b_t \cdot -\sum_{s\in S} b_s$: S is a finite antichain in T and $t < s$, for all $s \in S\}$ Then, let $e,e' \in E$ and denote by $e\perp e'$ whenever $e.e' = 0$, $\mathrm{supp}^+(e) \nsubseteq \mathrm{supp}^-(e')$ and $\mathrm{supp}^+(e') \nsubseteq \mathrm{supp}^-(e)$. Also, set $e \preceq e'$ whenever $e.e' = 0$ and $\mathrm{supp}^+(e') \subseteq \mathrm{supp}^-(e)$.*

Notice that if e,e' are disjoint elements of E then $\mathrm{supp}^-(e)\cap\mathrm{supp}^-(e') = \emptyset$ and either $e\perp e'$ or $e \preceq e'$ or $e' \preceq e$.

Let $(T,\le)$ be any pseudo-tree, see [10] and denote by $B(T)$ the subalgebra of $\mathcal{P}(T)$ generated by $\langle b_t : t\in T\rangle$ where $b_t := \{u \in T : t \le_T u\}$. See [4] and [11, ch. 6; Vol. 1, p. 255] for treealgebras.

For $t \in T$ and any finite anti-chain $A(t) \subseteq (T,\le)$ above t, $e(t,A(t))$ denotes $b_t \cdot -\sum_{s\in A(t)} b_s$. Let $b \in B(T)\setminus\{0\}$. Recall that b has a unique decomposition of the form (+):

$$b = e(t_1,A(t_1)) + \cdots + e(t_n,A(t_n))$$

where $\{e(t_i,A(t_i)) : 1\le i\le n\}$ are pairwise disjoint and $t_i \notin A(t_j)$ for all i,j distinct; see [1]. Let:

$$\mathrm{supp}^+(b) := \{t_1,\ldots,t_n\}$$
$$\mathrm{supp}^-(b) := A(t_1)\cup\cdots\cup A(t_n).$$

Set

$$\mu^+(b) := |\mathrm{supp}^+(b)| \text{ and } \mu^+(0) = 0.$$

Note that the results of this section, appear in [16], extending the work done in [15]. We shall reproduce, here, the main steps of the proof. Details are left to the reader.

Lemma 4.8. *Let e,e' be disjoint elements of E. Then:*

i) *$e\perp e'$ implies $\mu^+(e+e') = 2$.*
ii) *$e \preceq e'$ implies $\mu^+(e+e') = 1$.*

Proof. Set $b = e + e'$.

Case 1. $e \perp e'$. By [1] this form of b is unique and then $\mu^+(b) = 2$.

Case 2. $e \preceq e'$. Say that

$$e = b_t \cdot - \sum_{i=1}^{n} b_{s_i}, e' = b_{t'} \cdot - \sum_{j=1}^{m} b_{T_j}.$$

Now, since $e \preceq e'$, it follows that there is i_0 so that $t' = s_{i_0}$. So,

$$b = e + e' = b_t \cdot - \left(\sum_{\substack{i=1 \\ i \neq i_0}}^{n} b_{s_i} + \sum_{j=1}^{m} b_{T_j} \right).$$

Hence, $\mu^+(b) = 1$. □

Proposition 4.9. *For any natural number $m \geq 2$, and pairwise disjoint family $\{e_i : i = 1, \ldots, m\} \subseteq E$ we have:*

$$\mu^+ \left(\sum_{i=1}^{m} e_i \right) = \sum_{1 \leq i < j \leq m} \mu^+(e_i + e_j) - m(m-2). \tag{1}$$

The proof is done by induction on m. At the $(m+1)$th step, one may compare e_{m+1} to the previous ones.

Corollary 4.10. *For any x, y, z in $B(T)$ that are pairwise disjoint we have:*

$$\mu^+(x+y+z) = \mu^+(x+y) + \mu^+(x+z) + \mu^+(y+z) - \Big(\mu^+(x) + \mu^+(y) + \mu^+(z)\Big).$$

Proposition 4.11. *For any natural number $m \geq 3$ and $x_1, \ldots, x_m$ in $B(T)$ pairwise disjoint we have:*

$$\mu^+ \left(\sum_{i=1}^{m} x_i \right) = \sum_{1 \leq i < j \leq m} \mu^+(x_i + x_j) - (m-2) \sum_{i=1}^{n} \mu^+(x_i).$$

Proof. By induction. $m = 3$: This is Corollary 4.10. Set $y_i = x_i$ for $1 \leq i \leq m-2, y_{m-1} = x_{m-1} + x_m$. (y_i) are pairwise disjoint elements of $B(T)$. So,

$$\begin{aligned} \mu^+ \left(\sum_{i=1}^{m} x_i \right) &= \mu^+ \left(\sum_{i=1}^{m-2} x_i + x_{m-1} + x_m \right) \\ &= \mu^+ \left(\sum_{i=1} m - 1 y_i \right) = \sum_{1 \leq i < j \leq m-1} \mu^+(y_i + y_j) - (m-3) \sum_{i=1}^{m-1} \mu^+(y_i) \\ &\quad \text{(using induction hypothesis for } y_1, \ldots, y_{n-1}) \end{aligned}$$

$$
\begin{aligned}
&= \sum_{1\le i<j\le m-2} \mu^+(y_i+y_j) + \sum_{i=1}^{m-2} \mu^+(y_i+y_{m-1}) \\
&\quad - (m-3)\sum_{i=1}^{m-2} \mu^+(y_i) - (m-3)\mu^+(y_{m-1}) \\
&= \sum_{1\le i<j\le m-2} \mu^+(x_i+x_j) + \sum_{i=1}^{m-2} \mu^+(x_i+x_{m-1}+x_m) \\
&\quad - (m-3)\sum_{i=1}^{m-2} \mu^+(x_i) - (m-3)\mu^+(x_{m-1}+x_m) \\
&= \sum_{1\le i<j\le m-2} \mu^+(x_i+x_j) + \sum_{1\le i\le m-2} \mu^+(x_i+x_{m-1}) \\
&\quad + \sum_{1\le i\le m-2} \mu^+(x_i+x_m) \\
&\quad + \mu^+(x_{m-1}+x_m) - (m-2)\mu^+(x_{m-1}) \\
&\quad - (m-2)\mu^+(x_m) - (m-2)\sum_{i=1}^{m-2} \mu^+(x_i) \\
&= \sum_{1\le i<j\le m} \mu^+(x_i+x_j) - (m-2)\sum_{i=1}^{m} \mu^+(x_i)
\end{aligned}
$$

which shows that Proposition 4.11 is true for all $m \in \omega$ $(m \geq 3)$. □

Remark. Let $Int(L)$ be an interval algebra and write $b \in Int(L)\setminus\{0\}$ in its normal form as:

$$b = [a_0, a_1) \cup \cdots \cup [a_{2n-2}, a_{2n-1}),$$

where $a_0 < a_1 < \cdots < a_{2n-1} \leq \infty$ ($a_i \in L \cup \{\infty\}$ and $\infty \notin L$). Thus,

$$[a_{2i}, a_{2i+1}) = b_{a_{2i}} \cdot -b_{a_{2i+1}} := e_i.$$

Look at L as pseudotree to write:

$$b = e_0 + \cdots + e_{n-1}.$$

Therefore,

$$\mu^+(b) = |\mathrm{supp}^+(b)| = n.$$

So, Proposition 4.11 applies and hence we have the result of [15].

4.3. μ on interval algebras

Let $Int(L)$ be an interval algebra and write $b \in Int(L)\setminus\{0\}$ in its normal form as:

$$b = [a_0, a_1) \cup \cdots \cup [a_{2n-2}, a_{2n-1}),$$

where $a_0 < a_1 < \cdots < a_{2n-1} \leq \infty$ ($a_i \in L \cup \{\infty\}$ and $\infty \notin L$). Let $\nu(b) = n$. For $t_1 < t_2$ in $L \cup \{\infty\}$, we have $[t_1, t_2) = b_{t_1} \triangle b_{t_2}$. Note that if $t_2 = \infty$, then $b_{t_2} = \infty$, and $[t_1, t_2) = b_{t_1}$. Thus

$$b = \begin{cases} b_{a_0} \triangle b_{a_1} \triangle \cdots \triangle b_{a_{2n-1}} & \text{if } a_{2n-1} \neq \infty \\ b_{a_0} \triangle b_{a_1} \triangle \cdots \triangle b_{a_{2n-2}} & \text{if } a_{2n-1} = \infty. \end{cases}$$

Therefore,

$$\mathrm{supp}(b) = \{a_i : 0 \leq i \leq 2n-1\} \setminus \{\infty\} \subseteq L$$

and,

$$\mu(b) = \begin{cases} 2n & \text{if } a_{2n-1} \neq \infty \\ 2n-1 & \text{if } a_{2n-1} = \infty. \end{cases}$$

Set for $b = 0, \mu(0) = 0$ and $\mathrm{supp}(0) = \emptyset$. Next, call $b \in Int(L) \setminus \{0\}$ a *bounded* element whenever $a_{2n-1} \neq \infty$.

Now, recall that for $x, y, z \in Int(L)$, $S(x, y, z)$ denotes

$$|\mathrm{supp}(x) \cap \mathrm{supp}(y) \cap \mathrm{supp}(z)|.$$

Lemma 4.12.

i) *For any* $b \in Int(L) \setminus \{0\}$*:*

$$\mu(b) = \begin{cases} 2\nu(b) & \text{if } b \text{ is bounded,} \\ 2\nu(b) - 1 & \text{otherwise.} \end{cases}$$

ii) *If* x, y, z *are pairwise disjoint elements of* $Int(L) \setminus \{0\}$*, then we have the following:*

a) *At most one among* x, y, z *is not bounded.*

b) $S(x, y, z) = 0$.

Proof. It's clear. □

We may derive the following Corollary see [15].

Corollary 4.13. *Let* $x, y, z \in Int(L)$ *be a pairwise disjoint family. Then:*

$$\mu^+(x + y + z) = \mu^+(x + y) + \mu^+(x + z) + \mu^+(y + z) - \mu^+(x) - \mu^+(y) - \mu^+(z).$$

Proof. By Lemma 4.12 $S(x, y, z) = 0$ and thus, by Corollary 4.3:

$$\mu(x \triangle y \triangle z) = \mu(x \triangle y) + \mu(x \triangle z) + \mu(y \triangle z) - \mu(x) - \mu(y) - \mu(z).$$

Now, since $x \cdot y = x \cdot z = y \cdot z = 0$ it follows then that $\triangle$ is $+$ (in this case) and we have:

$$\mu(x + y + z) = \mu(x + y) + \mu(x + z) + \mu(y + z) - \mu(x) - \mu(y) - \mu(z).$$

Case 1. *x, y, z are bounded.* In this case, $x + y, x + z, y + z, x + y + z$ are bounded and by Lemma 4.12 we have:

$$2\nu(x + y + z) = 2\nu(x + y) + 2\nu(x + z) + 2\nu(y + z) - 2\nu(x) - 2\nu(y) - 2\nu(z).$$

So,

$$\nu(x + y + z) = \nu(x + y) + \nu(x + z) + \nu(y + z) - \nu(x) - \nu(y) - \nu(z).$$

Thus by Lemma 4.12 the only remaining case is:

Case 2. *There is exactly one element among* x, y, z *which is not bounded.* Say x. Thus $x, x+y, x+z$ and $x+y+z$ are not bounded and $y, z, y+z$ are. Again, by Lemma 4.12:

$$2\nu(x+y+z)-1 = 2\nu(x+y)-1+2\nu(x+z)-1+2\nu(y+z)-2\nu(x)+1-2\nu(y)-2\nu(z).$$

Thus,

$$\nu(x+y+z) = \nu(x+y)+\nu(x+z)+\nu(y+z)-\Big(\nu(x)+\nu(y)+\nu(z)\Big).$$

This finishes up the proof of Corollary 4.13. □

Remark. If x, y, z are pairwise disjoint in an interval algebra $Int(L)$, then $S(x, y, z) = 0$ and not all three of x, y, z are unbounded; but the converse is not true in general. Indeed, let

$$a_0 < b_0 < a_1 < b_1 < c_1 < c_2 \text{ in } L.$$

Then, by setting $x = [a_0, a_1), y = [b_0, b_1), z = [c_1, c_2)$ it follows that x, y and z are bounded and $S(x, y, z) = 0$ but $x \cdot y = [b_0, a_1)$. So, the condition $S(x, y, z) = 0$ and not all three of x, y, z are unbounded is weaker than "pairwise disjointness". Thus one can write down formula using ν and $\triangle$ (instead of $+$) whenever the assumption $S(x, y, z) = 0$ holds and not all three of x, y, z are unbounded.

Proposition 4.14. *Let* $x, y, z \in Int(L)$ *so that* $S(x, y, z) = 0$ *and not all three of* x, y, z *are unbounded. Then,*

$$\nu(x\triangle y\triangle z) = \nu(x\triangle y)+\nu(x\triangle z)+\nu(y\triangle z)-\Big(\nu(x)+\nu(y)+\nu(z)\Big).$$

Proof. Since $S(x, y, z) = 0$, it follows by Corollary 4.3 that:

$$\mu(x\triangle y\triangle z) = \mu(x\triangle y)+\mu(x\triangle z)+\mu(y\triangle z)-\Big(\mu(x)+\mu(y)+\mu(z)\Big).$$

Case 1. x, y, z *are bounded.* In this case $\mu = 2\nu$ and we are done.

Case 2. x *is unbounded and* y, z *are bounded.* Thus, $x, x\triangle y, x\triangle z$ and $x\triangle y\triangle z$ are unbounded; but $y\triangle z, y, z$ are bounded. So,

$$2\nu(x\triangle y\triangle z)-1 = 2\nu(x\triangle y)-1+2\nu(x\triangle z)-1+2\nu(y\triangle z)-2\nu(x)+1-2\nu(y)-2\nu(z)$$

and again we get what we are looking for.

Case 3. x *is bounded;* y, z *are unbounded.* Here again, $x, y\triangle z, x\triangle y\triangle z$ are bounded and $y, z, x\triangle y, x\triangle z$ are not bounded. So,

$$2\nu(x\triangle y\triangle z) = 2\nu(x\triangle y)-1+2\nu(x\triangle z)-1+2\nu(y\triangle z)-2\nu(x)-2\nu(y)+1-2\nu(z)+1$$

which is the relation we want to prove. This ends up the proof of the proposition. □

Acknowledgment

The authors appreciate all suggestions and comments that the referee has made. Next, the first author thanks the Board of Direction of Mathematical Research

Center (Barcelona) for financial support and accommodations that were arranged for him during the workshop on *Advanced Course on Ramsey Methods in Analysis*. Also, he thanks Professors S. Todorcevic and J. Bagaria for their invitation to this course as well as J.D. Monk for all discussions he had with him during the preparation of this work.

References

[1] M. Bekkali. *Pseudo Treealgebras.* Notre Dame Journal of Formal Logic, volume 42, Number 1 (101–108), 2001

[2] M. Bekkali and D. Zhani. *Tail and free poset algebras.* Revista Matematica Complutence(17)Núm.1, (2003), 169–179.

[3] M. Bekkali and D. Zhani. *Upper semi-lattice algebras.* Accepted in New Zealand Journal of Mathematics (2004).

[4] G. Brenner, J.D. Monk. *Tree algebras and chains.* Lecture Notes in Mathematics (1004) (Springer-Verlag), 54–66.

[5] E. Evans. *The Boolean ring universal over a meet semilattice.* J. Austral. Math. Soc.(23) (Series A)(1977), 402–415.

[6] G. Grätzer. *General Lattice Theory.* Second edition (2003), Birkhäuser.

[7] L. Heindorf. *Boolean semigroup rings and exponentials of compact zero-dimensional spaces.* Fundamenta Mathematicae (135) (1990), 37–47.

[8] L. Heindorf. *On subalgebras of Boolean interval algebras.* preprint, 1995.

[9] S. Koppelberg. *Handbook on Boolean Algebras.* Vol. 1, Ed. Monk, J.D., and Bonnet, R., North Holland, 1989.

[10] S. Koppelberg, and J.D. Monk. *Pseudo-Trees and Boolean algebra.* Order 8, 359–374, 1992.

[11] J.D. Monk, R. Bonnet. (editors). *Handbook on Boolean Algebras.* (3 volumes) North-Holland. Amsterdam 1989.

[12] K. Numakura. *Theorems on compact totally disconnected semigroups and lattices.* Proc. Amer. Math. Soc. 8, 623–636, 1957.

[13] D.R. Popescu, I. Tomescu. *Negative cycles in complete signed graphs.* Discrete Applied Mathematics 68 (1996), 145–152.

[14] M. Pouzet, I. Rival. *Every countable lattice is a retract of a direct product of chains.* Algebra universalis (18) (1984), 295–307.

[15] H. SiKaddour. *A note on the length of members of an interval algebra.* Algebra universalis (44) (2000), 195–198.

[16] D. Zhani. *Length of elements in pseudo tree algebras* . To appear in New Zealand Journal of Mathematics

M. Bekkali
P.O. Box 2414, Fez 30000, Morocco
e-mail: `bekka@menara.ma` (corresponding author)

D. Zhani
UFR of Applied and Discrete Mathematics (**MDA**)
Faculty of Sciences and Technology, Fez, Morocco
e-mail: `zhanidriss@hotmail.fr`

Set Theory
Trends in Mathematics, 55–82

Small Definably-large Cardinals

Roger Bosch

To Ramon Bastardes, in memoriam

Abstract. We study the definably-Mahlo, definably-weakly-compact, and the definably-indescribable cardinals, which are the definable versions of, respectively, Mahlo, weakly-compact, and indescribable cardinals. We study their strength as large cardinals and we show that the relationship between them is almost the same as the relationship between Mahlo, weakly-compact and indescribable cardinals.

Mathematics Subject Classification (2000). 03E55, 03E47.

1. Introduction

This paper is a survey on small definably-large cardinals. Recall that, roughly speaking, many large cardinals are regular infinite cardinals that enjoys some property related to its subsets that makes the cardinal large. The definably-large cardinal version of a large cardinal will be an inaccessible cardinal κ which enjoy the same property as the corresponding large cardinal but only for subsets of κ that are first-order definable in V_κ. For instance, a definably-Mahlo cardinal will be an inaccessible cardinal κ such that the set of all inaccessible cardinals below it has nonempty intersection with every club subset of κ which is first-order definable over V_κ. For another example, a definably-Π^1_1-indescribable cardinal will be an inaccessible cardinal κ which is indescribable by means of Π^1_1 sentences using first-order definable subsets of V_κ as predicates. The reason for which we demand inaccessibility instead of regularity is to ensure that we are in the presence of a true large cardinal notion (see for example [10], Exercise IX 1.11.1).

This paper was partially written during a research stay of the author at the *Centre de Recerca Matemàtica* (CRM), at the Universitat Autònoma de Barcelona. The author was partially supported by the research projects: BFM2002-03236 of the Spanish Ministry of Science and Technology, PR-01-GE-10-HUM of the Government of the *Principado de Asturias*, and 2002GR-00126 of the *Generalitat de Catalunya*.

The study of definable versions of large cardinal begins in the early 70's with two papers of Aczel and Richter ([1] and [2]) where they realize that there is a strong analogy between inaccessible cardinals and admissible ordinals closed under inductive definitions. They call such ordinals *recursively-inaccessible cardinals.* Their work was continued until the first years of the 80's by Richter himself ([28] and [29]), Kranakis ([18], [19], [21], [22], [23] and [24]) and Kranakis and Phillips ([25]). All these papers are devoted to study other properties of admissible ordinals which make them analogous to other large cardinals. Thus, they define what we may call *the recursively-Mahlo*, *the recursively-weakly-compact*, *the recursively-indescribable*, *the recursively-Erdös* and *the recursively-Ramsey cardinals.*

In 1981, Kaufmann, studying the elementary end-extensions of models of set theory, introduced a definable version of measurable cardinals ([15], see also [16]). But the study of the definable versions of measurable cardinals was only continued in the restricted context of constructible sets ([17] and [20]) and in the context of *recursively-measurable cardinals* ([4] and [5]), that is, for admissible ordinals that are analogous to measurable cardinals. Thus, almost all work in this field was done in the limited context of admissibility and the constructible universe.

Then, in the mid 80's, the study in the field died out. There is, to my knowledge, only one exception: the paper [3] of Andretta, where a definable version of Woodin cardinals is defined and used. The reason of this lack of interest lies, presumably, in the fact that there were no known applications for recursively-large cardinals.

Notice that definably-large cardinals are very different from recursively-large ones. Definably-large cardinals are true large cardinals, since they are inaccessible. Moreover, definably-large cardinals are defined in the broader context of V and not in the narrow context of L or admissibility theory. But there is one more difference between definably-large and recursively-large cardinals which makes the former a very interesting field of study for set theory: some kinds of definably-large cardinals have been used to prove exact equiconsistency results for statements involving only definable sets. For instance, Leshem ([26]) introduced the definably-indescribable cardinals and then used them to prove the equiconsistency of "There exists a definably-Π^1_1-indescribable cardinal" and "$\aleph_1$ has the definable tree property" (that is, there are no Aronszajn trees which are definable over H_{ω_1}). Other equiconsistency results using definably-large cardinals can be found in [6], [7] and in [8]. Therefore, definably-large cardinals are presumably the kind of large cardinals necessary to provide exact equiconsistency results for sentences involving only definable sets of reals. This makes definably-large cardinals a worthy field of study.

In this paper we study three kinds of small definably-large cardinals: the definably-Mahlo cardinals (Section 2), the definably-weakly-compact cardinals (Section 3) and the definably-indescribable cardinals (Section 4). In all these sections, we focus our work on three main topics. The first one is to determine their place in the consistency-strength hierarchy of large cardinals. We prove that all these cardinals are consistency-wise below a Mahlo cardinal. For the second one, recall that a definably-large cardinal is an inaccessible cardinal κ such that the definable subsets of κ enjoy some given property. Thus, using the Levy hierarchy of formulas, we can divide every definably-large cardinal notion into countably many types of cardinals depending upon the complexity of the definition of the sets which enjoy the property. Our second task will be to study the inner hierarchy of types of a given kind of definably-large cardinal. Finally, we also study the relationship between these three kinds of definably-large cardinals. We show that it reflects almost exactly the relation between Mahlo, weakly-compact and indescribable cardinals.

This paper owes very much to Richter, Aczel, Kranakis, and Baeten's papers on the recursively-large cardinals, since many results in them can be generalized to, and provide insight on, the definably-large cardinals. I also want to express my acknowledgement to Joan Bagaria who first aimed me to study this field and later read preliminary versions of this paper and suggested several improvements of the proofs. Finally, I also would like to thank to an anonymous referee by his useful comments and remarks.

2. Definably-Mahlo cardinals

Definition 2.1. Let κ be a cardinal. $C \subseteq \kappa$ is a $\underset{\sim}{\Sigma}_n$*-closed and unbounded* subset in κ, a $\underset{\sim}{\Sigma}_n$*-club* for short, iff C is a club in κ and there exists a Σ_n formula $\varphi(x; y)$ and $a \in V_\kappa$ such that for every $\alpha < \kappa$,

$$\alpha \in C \text{ iff } V_\kappa \models \varphi(\alpha; a),$$

i.e., C is a club definable over V_κ by means of a Σ_n formula with possibly parameters. Similarly, we define $\underset{\sim}{\Pi}_n$*-club* by substituting Π_n for Σ_n in the above definition. A $\underset{\sim}{\Delta}_n$*-club* in κ is a club in κ that is both $\underset{\sim}{\Sigma}_n$ and $\underset{\sim}{\Pi}_n$. A $\underset{\sim}{\Sigma}_\omega$-club is a club in κ that is definable in V_κ possibly with parameters.

$S \subseteq \kappa$ is a $\underset{\sim}{\Sigma}_n$*-stationary* subset of κ iff for all $\underset{\sim}{\Sigma}_n$-club C in κ, $S \cap C \neq \emptyset$. (Notice that we do not require that S itself be Σ_n-definable.) Similarly, we can define the $\underset{\sim}{\Pi}_n$*-stationary*, $\underset{\sim}{\Delta}_n$*-stationary* and the $\underset{\sim}{\Sigma}_\omega$*-stationary* subsets of κ.

We can also define the lightface forms of these notions, namely, the Σ_n (Π_n, Δ_n, Σ_ω) clubs, and the corresponding Σ_n (Π_n, Δ_n, Σ_ω) stationary sets, by requiring that the clubs are definable without parameters.

We will use the notation $V_\alpha \preccurlyeq_n V_\beta$, α and β any ordinals, to indicate that V_α is a Σ_n-elementary substructure of V_β. The next proposition shows that the above classification of clubs is not trivial.

Proposition 2.2. *Suppose that κ is an inaccessible cardinal and let $C_n = \{\alpha < \kappa : V_\alpha \preccurlyeq_n V_\kappa\}$. Then, for every $n \geq 1$, C_n is a Π_n-club in κ which is not a $\underset{\sim}{\Sigma}_n$-club.*

Proof. It is easy to see that C_n is a closed and unbounded subset of κ. So, we only need to check that C_n is Π_n-definable, but not Σ_n-definable, over V_κ. We first prove that C_n is Π_n-definable.

If $n = 1$, since $V_\alpha \preccurlyeq_1 V_\kappa$ iff α is a strong limit cardinal, C_1 is the club of all strong limit cardinals below κ. Thus for all $\alpha < \kappa$, $\alpha \in C_1$ iff V_κ models

$$(\forall \beta < \alpha)(\exists \gamma < \alpha)\forall f(f \text{ is a function} \wedge \operatorname{dom}(f) = \gamma \wedge \\ \wedge (\forall x \in \operatorname{ran}(f))(x \subseteq \beta) \rightarrow f \text{ is not 1-1})$$

which is a Π_1 formula.

If $n > 1$, let $\sigma_n(v_0, v_1)$ be the Σ_n formula that defines the satisfaction relation for Σ_n formulas in V_κ. Then, for every $\alpha < \kappa$, $\alpha \in C_n$ iff

$$V_\kappa \models (\forall \ulcorner \varphi \urcorner \in \Sigma_n)(\forall a \in V_\alpha)(\sigma_n(\ulcorner \varphi \urcorner, a) \rightarrow Sat(V_\alpha, \ulcorner \varphi \urcorner, a)), \qquad (*)$$

where $\ulcorner \varphi \urcorner$ denotes the Gödel number of φ, Σ_n denotes the set of (Gödel numbers of) Σ_n formulas and $Sat(v_0, v_1, v_2)$ denotes the Δ_1 satisfaction relation for sets and formulas. Since the map $\alpha \longmapsto V_\alpha$ is Π_1 and $n > 1$ the right-hand side sentence of $(*)$ is Π_n.

Now suppose that for some $n \geq 1$ there are $a \in V_\kappa$ and a Σ_n formula $\varphi(x; y)$ such that for all $\alpha < \kappa$,

$$\alpha \in C_n \text{ iff } V_\kappa \models \varphi(\alpha; a).$$

We may assume that for all $\alpha \in C_n$, $a \in V_\alpha$ (otherwise, let $\gamma = \operatorname{rk}(a)$ and work with the $\underset{\sim}{\Sigma}_n$-club $C_n \setminus \gamma + 1$). Let β be the least element of C_n. Since

$$V_\kappa \models \exists \alpha\, \varphi(\alpha; a),$$

$a \in V_\beta \preccurlyeq_n V_\kappa$ and the right-hand formula above is Σ_n,

$$V_\beta \models \exists \alpha\, \varphi(\alpha; a).$$

But then, there is $\alpha < \beta$ such that $\alpha \in C_n$. A contradiction with the minimality of β. □

Note that the proposition above is optimal in the following sense: for $n = 0$, $C_0 = \kappa$ is a Σ_0-club in κ.

Definition 2.3 ([6]). κ is a $\underset{\sim}{\Sigma}_n$*-Mahlo cardinal* ($\underset{\sim}{\Pi}_n$*-Mahlo cardinal*, $\underset{\sim}{\Delta}_n$*-Mahlo cardinal*) iff κ is an inaccessible cardinal and the set of all inaccessible cardinals below κ is $\underset{\sim}{\Sigma}_n$-stationary (respectively, $\underset{\sim}{\Pi}_n$-stationary, $\underset{\sim}{\Delta}_n$-stationary).

κ is $\underset{\sim}{\Sigma}_\omega$*-Mahlo* if it is $\underset{\sim}{\Sigma}_n$-Mahlo for every $n \in \omega$.

Similarly, we may define the lightface Σ_n*-Mahlo*, Π_n*-Mahlo*, Δ_n*-Mahlo*, and Σ_ω*-Mahlo cardinals.*

Remark 2.4. Given a cardinal κ, let us denote the set of all inaccessible cardinals below κ by I. Moreover, for every n, with I_n we denote the set of all inaccessible cardinals $\lambda < \kappa$ such that $V_\lambda \preccurlyeq_n V_\kappa$. Therefore, for every $n \in \omega$, $I_n = I \cap C_n$. Note that for every inaccessible cardinal κ, $I = I_0 = I_1$. Note also that, by the proposition above, for every inaccessible cardinal κ, κ is a Π_n-Mahlo ($\underset{\sim}{\Pi}_n$-Mahlo) iff the set I_n is a Π_n-stationary (respectively, $\underset{\sim}{\Pi}_n$-stationary) subset of κ.

It is clear that every Mahlo cardinal is a $\underset{\sim}{\Sigma}_\omega$-Mahlo cardinal. Moreover, the consistency strength of the existence of a $\underset{\sim}{\Sigma}_\omega$-Mahlo cardinal is much lower than the existence of a Mahlo cardinal:

Theorem 2.5 ([6]). *If κ is a Mahlo cardinal, then the set of $\underset{\sim}{\Sigma}_\omega$-Mahlo cardinals below κ is a stationary subset of κ.*

Proof. Suppose that κ is a Mahlo cardinal. Let C be a club in κ. For every natural number n, the following sentence, call it θ_n, holds it in $\langle V_\kappa, \in, C\rangle$:

> *For all x and for all φ, if φ is a Σ_n formula with parameter x that defines a closed unbounded set D of ordinals, then there exists an inaccessible cardinal $\beta \in D$.*

Let λ be an inaccessible cardinal such that $\langle V_\lambda, \in, C \cap V_\lambda\rangle \preccurlyeq \langle V_\kappa, \in, C\rangle$. So, for every n,

$$\langle V_\lambda, \in, C \cap V_\lambda\rangle \models \theta_n.$$

Let us check that λ is a $\underset{\sim}{\Sigma}_\omega$-Mahlo cardinal: Fix some $n \in \omega$ and let $E \subseteq \lambda$ be a $\underset{\sim}{\Sigma}_n$-club in λ. So, there is some $a \in V_\lambda$ and some Σ_n formula φ with parameter a that defines E in V_λ. By θ_n, there is an inaccessible $\beta \in E$.

Finally, since $C \cap \lambda$ is unbounded in λ, $\lambda \in C$. □

Thus, definably-Mahlo cardinals are consistency-wise below Mahlo cardinals.

We begin the study of definably-Mahlo cardinals by giving other characterizations of them. The equivalence between (1) and (2) below is due essentially to Baeten ([4]). Recall that a function $f : \kappa \to \kappa$ is normal iff it is increasing and continuous.

Theorem 2.6. *Suppose that $n \geq 2$ and κ is a cardinal. Then, the following are equivalent:*

1. *κ is $\underset{\sim}{\Pi}_n$-Mahlo.*
2. *κ is inaccessible and for every Π_{n+2} formula $\varphi(x)$ and every $a \in V_\kappa$, if $V_\kappa \models \varphi(a)$, then there is $\lambda \in I_n$ such that $V_\lambda \models \varphi(a)$.*
3. *κ is inaccessible and every normal function $f : \kappa \to \kappa$ which is Σ_{n+1} definable with parameters over V_κ has an inaccessible fixed point.*

The equivalence also holds for Π_n-Mahlo cardinals, Π_{n+2} sentences and normal functions which are Σ_{n+1} definable without parameters over V_κ.

Proof. (1 ⇒ 2) Suppose that κ is a $\underset{\sim}{\Pi}_n$-Mahlo cardinal and that $\varphi(x)$ is a Π_{n+2} formula such that for some $a \in V_\kappa$, $V_\kappa \models \varphi(a)$. For simplicity, we may assume that $\varphi(x)$ is of the form $\forall y \exists z \psi(x, y, z)$ where $\psi(x, y, z)$ is a Π_n formula.

Define $C \subseteq \kappa$ as follows: for all $\alpha < \kappa$,

$$\alpha \in C \text{ iff } \alpha \in C_n \wedge V_\alpha \models \varphi(a).$$

We claim that C is a $\underset{\sim}{\Pi}_n$-club in κ.

First note that for all $\alpha < \kappa$, $\alpha \in C$ iff

$$V_\kappa \models \alpha \in C_n \wedge Sat(V_\alpha, \ulcorner\varphi\urcorner, a).$$

Since C_n is Π_n definable over V_κ, *Sat* is a Δ_1 relation in V_κ, the map $\alpha \longmapsto V_\alpha$ is Π_1 and $n \geq 2$, the right-hand formula is Π_n. So, C is a Π_n definable with a as parameter.

C is unbounded. Fix $\beta < \kappa$ and define a sequence $\langle \alpha_k : k \in \omega \rangle$ of ordinals below κ as follows:

$\underline{k = 0}$: α_0 is the least ordinal in C_n greater than β such that $a \in V_{\alpha_0}$.
$\underline{k + 1}$: Suppose α_k defined. Since $V_\kappa \models \forall y \exists z \psi(a, y, z)$, for every $b \in V_{\alpha_k}$ fix some $c_b \in V_\kappa$ such that $V_\kappa \models \psi(a, b, c_b)$. Let α_{k+1} be the least ordinal in C_n greater than α_k such that $\{c_b : b \in V_{\alpha_k}\} \subseteq V_{\alpha_{k+1}}$.

Let $\alpha = \sup_{k \in \omega}(\alpha_k)$. Then, $\alpha \in C_n$. Moreover, if $b \in V_\alpha$, then there is $k \in \omega$ such that $b \in V_{\alpha_k}$. But then, since $c_b \in V_\alpha \preccurlyeq_n V_\kappa$, $V_\alpha \models \psi(a, b, c_b)$. Hence, $\alpha \in C$. We can prove that C is a closed subset of κ with a similar argument.

So, since I is a $\underset{\sim}{\Pi}_n$-stationary subset of κ, there is $\lambda \in C \cap I$. Therefore, $\lambda \in I_n$ and $V_\lambda \models \varphi(a)$.

($2 \Rightarrow 3$) Let $f : \kappa \to \kappa$ be a normal function which is Σ_{n+1} definable with parameters over V_κ. Since f is normal, V_κ satisfies

(i) $\forall \alpha \beta (\alpha < \beta \leftrightarrow f(\alpha) < f(\beta))$.
(ii) $\forall \alpha \beta (\alpha \text{ is limit} \wedge f(\alpha) = \beta \to$
$\to (\forall \gamma < \beta)(\exists \delta < \beta)(\exists \eta < \alpha)(\gamma < \delta \wedge f(\eta) = \delta))$.
(iii) $\forall \alpha \exists \beta \, f(\alpha) = \beta$.

Note that, since f is Σ_{n+1} function, (i), (ii) and (iii) are Π_{n+2} formulas with parameters. So, by (2) of the Theorem, there exists $\lambda \in I_n$ such that V_λ satisfies them. We claim that λ is a fixed point of f. It is clear that $\lambda \leq f(\lambda)$. On the other hand, since $V_\lambda \preccurlyeq_n V_\kappa$ and f is a function, $f^{V_\lambda} = f \upharpoonright \lambda$. Thus,

$$f(\lambda) = \bigcup \text{ran}(f \upharpoonright \lambda) = \bigcup \text{ran}(f^{V_\lambda}) \leq \lambda.$$

($3 \Rightarrow 1$) Let C be a $\underset{\sim}{\Pi}_n$-club in κ. Let f be the function enumerating C. That is, for all $\alpha, \beta \in \kappa$, $f(\alpha) = \beta$ iff V_κ satisfies:

(i) $\alpha = 0 \wedge \beta \in C \wedge (\forall \gamma < \beta)(\gamma \notin C)$ or
(ii) $\alpha > 0 \wedge \beta \in C \wedge (\forall \gamma < \beta)(\gamma \in C \to (\exists \delta < \alpha)(\gamma = f(\delta)))$.

It is easy to see that, since C is a $\underset{\sim}{\Pi}_n$-club, (i) and (ii) give a Σ_{n+1} definition of f. Thus, by (3), there exists an inaccessible $\lambda < \kappa$ such that $f(\lambda) = \lambda$. Thus $\lambda \in C$ and κ is a $\underset{\sim}{\Pi}_n$-Mahlo cardinal. □

Corollary 2.7. *Let κ be a cardinal. Then, for every $n \geq 2$, the following are equivalent:*

1. *κ is $\underset{\sim}{\Pi}_n$-Mahlo.*
2. *κ is $\underset{\sim}{\Sigma}_{n+1}$-Mahlo.*
3. *κ is $\underset{\sim}{\Delta}_{n+1}$-Mahlo.*

Moreover, the lightface version also holds.

Proof. We only prove $(1 \Rightarrow 2)$, since the other implications are trivial. So, suppose that κ is a $\underset{\sim}{\Pi}_n$-Mahlo cardinal and let C be a $\underset{\sim}{\Sigma}_{n+1}$-club in κ. Let $\varphi(x; y)$ be the Σ_{n+1} formula that, with $a \in V_\kappa$ as parameter, defines C. Since C is a club,

$$V_\kappa \models \forall\alpha\exists\beta(\alpha < \beta \wedge \varphi(\beta; a)).$$

Since the right-hand formula is Π_{n+2}, by Theorem 2.6, there is $\lambda \in I_n$ such that

$$V_\lambda \models \forall\alpha\exists\beta(\alpha < \beta \wedge \varphi(\beta; a)).$$

Since $V_\lambda \preccurlyeq_n V_\kappa$, $C^{V_\lambda} \subseteq C \cap \lambda$. Therefore $C \cap \lambda$ is unbounded and, hence, $\lambda \in C$. □

In view of the above corollary, henceforth, for $n > 2$, we only work with Π_n and $\underset{\sim}{\Pi}_n$-Mahlo cardinals.

For $n = 0$, it is clear that for every cardinal κ, κ is $\underset{\sim}{\Delta}_0$-Mahlo iff it is $\underset{\sim}{\Sigma}_0$-Mahlo iff it is $\underset{\sim}{\Pi}_0$-Mahlo. And the same holds for the lightface version. Therefore, we only work with Π_0 and $\underset{\sim}{\Pi}_0$-Mahlo cardinals. Moreover, from the definitions follow that κ is Π_0-Mahlo iff κ is inaccessible but not the least, and that κ is $\underset{\sim}{\Pi}_0$-Mahlo iff κ is inaccessible and limit of inaccessible cardinals.

Using the fact that for all inaccessible cardinals $\lambda < \kappa$, the Π_2-formulas are downward absolute for V_λ, V_κ, we have that a proof similar to that of Corollary 2.7 shows that every $\underset{\sim}{\Pi}_0$-Mahlo cardinal is $\underset{\sim}{\Sigma}_1$-Mahlo and that every Π_0-Mahlo cardinal is Σ_1-Mahlo. But note that if κ is a Π_1-Mahlo, then $C = \{\alpha < \kappa : (\exists\lambda \in \alpha)(\lambda \text{ is inaccessible})\}$ is a Π_1-club of κ. Hence, there is $\lambda < \kappa$ such that λ is an inaccessible cardinal greater than the least inaccessible cardinal. Thus, the first Π_0-Mahlo is not Π_1-Mahlo. And, if κ is $\underset{\sim}{\Pi}_1$-Mahlo, then the set of inaccessible cardinals below κ is unbounded and, therefore, the set of all cardinals below κ that are the limit of inaccessible cardinals is a Π_1-club. So, every $\underset{\sim}{\Pi}_1$-Mahlo is the limit of inaccessible cardinals that are themselves limits of inaccessible cardinals. Hence, the first $\underset{\sim}{\Pi}_0$-Mahlo cardinal is not $\underset{\sim}{\Pi}_1$-Mahlo. Therefore, for $n = 1$, we only work with Π_1 and $\underset{\sim}{\Pi}_1$-Mahlo cardinals.

We know from Corollary 2.7 that for every κ, κ is a $\underset{\sim}{\Pi}_2$-Mahlo (Π_2-Mahlo) cardinal iff it is $\underset{\sim}{\Delta}_3$-Mahlo ($\Delta_3$-Mahlo) iff it is $\underset{\sim}{\Sigma}_3$-Mahlo ($\Sigma_3$-Mahlo). Moreover,

Proposition 2.8. *Every $\underset{\sim}{\Delta}_2$-Mahlo cardinal is $\underset{\sim}{\Sigma}_2$-Mahlo. And the same holds for the lightface version.*

Proof. Let κ be a $\underset{\sim}{\Delta}_2$-Mahlo cardinal and let $C \subseteq \kappa$ a $\underset{\sim}{\Sigma}_2$-club. Let $\varphi(x)$ the Σ_2-formula that, with parameter a, defines C in V_κ. Let $C' \subseteq \kappa$ defined by letting, for every $\alpha < \kappa$

$$\alpha \in C' \text{ iff } V_\alpha \models \forall\beta\exists\gamma(\beta < \gamma \wedge \varphi(\gamma)).$$

It is clear that C' is a $\underset{\sim}{\Delta}_2$-club of κ. But then, if $\lambda \in C'$ is inaccessible, since $V_\lambda \preccurlyeq_1 V_\kappa$, $C \cap \lambda$ is unbounded in λ and, hence, $\lambda \in C$. □

Question 1. *Is every $\underset{\sim}{\Pi}_1$-Mahlo (Π_1-Mahlo) cardinal a $\underset{\sim}{\Sigma}_2$-Mahlo ($\Sigma_2$-Mahlo) cardinal? If $V = L$, using the Δ_1 map $\alpha \mapsto L_\alpha$ instead of $\alpha \mapsto V_\alpha$, it is easy to prove that the equivalences of Theorem 2.6 holds for all $n \in \omega$. Hence, if $V = L$, Corollary 2.7 holds for all $n \in \omega$.*

We now show that the definably-Mahlo cardinals also form a proper hierarchy.

Theorem 2.9. *Let κ be a cardinal and $n \in \omega$.*

1. *If κ is a Π_n-Mahlo cardinal with $n > 1$, then the set of all $\lambda < \kappa$ such that λ is a $\mathbf{\Pi}_{n-1}$-Mahlo cardinal is Π_n-stationary. Thus, the least $\mathbf{\Pi}_{n-1}$-Mahlo cardinal is not Π_n-Mahlo.*
2. *If κ is a $\mathbf{\Pi}_n$-Mahlo cardinal, then the set of all $\lambda < \kappa$ such that λ is a Π_n-Mahlo cardinal is $\mathbf{\Pi}_n$-stationary. Thus, the least Π_n-Mahlo cardinal is not $\mathbf{\Pi}_n$-Mahlo.*

Proof. (1) Let κ be a Π_n-Mahlo cardinal. Since I_n is a Π_n-stationary subset of κ, we only need to show that every $\lambda \in I_n$ is a $\mathbf{\Pi}_{n-1}$-Mahlo cardinal. So, let $\lambda \in I_n$ and let $C \subseteq \lambda$ be a $\mathbf{\Pi}_{n-1}$-club in λ. Let $\varphi(x, y)$ be the Π_{n-1} formula that, with parameter $a \in V_\lambda$, defines C in V_λ. Since C is a club in λ,

$$V_\lambda \models \forall\alpha((\forall\beta < \alpha)(\exists\gamma < \alpha)(\beta < \gamma \wedge \varphi(\gamma, a)) \to \varphi(\alpha, a)).$$

But, since φ is a Π_{n-1} formula, the right-hand side formula is Π_n. Since $V_\lambda \preccurlyeq_n V_\kappa$,

$$V_\kappa \models \forall\alpha((\forall\beta < \alpha)(\exists\gamma < \alpha)(\beta < \gamma \wedge \varphi(\gamma, a)) \to \varphi(\alpha, a)).$$

Moreover, since C is a club in λ, $V_\kappa \models (\forall\beta < \lambda)(\exists\gamma < \lambda)(\beta < \gamma \wedge \varphi(\gamma, a))$. Thus, $V_\kappa \models \varphi(\lambda, a)$ and hence $V_\kappa \models \exists\mu(\mu \text{ is inaccessible} \wedge \varphi(\mu, a))$. But then, since φ is a Π_{n-1} formula, "μ is inaccessible" is a Π_1 predicate over μ and $n > 1$,

$$V_\lambda \models \exists\mu(\mu \text{ is inaccessible} \wedge \varphi(\mu, a)).$$

Therefore, there is an inaccessible cardinal $\mu < \lambda$ such that $\mu \in C$ and λ is a $\mathbf{\Pi}_{n-1}$-Mahlo cardinal.

(2) Let κ be a $\mathbf{\Pi}_n$-Mahlo and let I be the $\mathbf{\Pi}_n$-stationary set of all inaccessible cardinals below κ. For each $\lambda \in I$, let $F_\lambda = \langle \ulcorner\varphi_m^\lambda\urcorner : m \in \omega\rangle$ be a sequence of all Π_n formulas that define (without any parameter) a Π_n-club in V_λ.

Note that each F_λ is essentially a ω-sequence of natural numbers. Hence, since κ is inaccessible, there are $< \kappa$ many of them. Let $\langle (F_\lambda)_\alpha : \alpha < \mu\rangle$, $\mu < \kappa$, be an enumeration of them. Let f be the map sending every $\lambda \in I$ to F_λ. We may think of f as a map from I into μ.

Claim 2.10. *There is a $\mathbf{\Pi}_n$-stationary subset S of κ such that f is constant on S.*

Proof. Otherwise, let for every $\alpha < \mu$, $X_\alpha = \{\lambda \in I : f(\lambda) \neq \alpha\}$ and fix for every $\alpha < \mu$ a $\mathbf{\Pi}_n$-club C_α such that $C_\alpha \subseteq (\kappa \setminus X_\alpha)$.

Let $C = \bigcap_{\alpha<\mu} C_\alpha$. Clearly, C is a club. Moreover, for every $\beta < \kappa$,

$$\beta \in C \text{ iff } V_\kappa \models \forall\alpha\neg\sigma_n(\ulcorner\neg\psi_\alpha\urcorner, a_\alpha^\frown\beta)$$

where for every $\alpha < \mu$, ψ_α is the Π_n formula that, with parameter $a_\alpha \in V_\kappa$, defines the $\mathbf{\Pi}_n$-club C_α. So, C is definable by means of a Π_n formula with parameter $\langle\langle\psi_\alpha, a_\alpha\rangle : \alpha < \mu\rangle$ over V_κ. Hence C is a $\mathbf{\Pi}_n$-club in V_κ.

Therefore, $S = I \cap C$ is a $\mathbf{\Pi}_n$-stationary subset of κ. But, if $\lambda \in S$ then for some $\alpha < \mu$, $f(\lambda) = \alpha$ but $\lambda \in C \subseteq C_\alpha$. A contradiction. □

Let S be as in the claim above and let α be such that for all $\lambda \in S$, $f(\lambda) = \alpha$. Note that for each $\lambda \in S$, F_λ is the same. Note also that we may assume that for every inaccessible $\lambda \in S$, $V_\lambda \preccurlyeq_n V_\kappa$ since $C_n = \{\alpha < \kappa : V_\alpha \preccurlyeq_n V_\kappa\}$ is a Π_n-club in κ. So, for every $\lambda \in S$ and $m \in \omega$, φ_m^λ defines a Π_n-club, C_m^λ, in V_λ and, hence, for every $\lambda \in S$, φ_m^λ defines a Π_n-club, C_m, in V_κ. Moreover, for all $\lambda, \lambda' \in S$ if $\lambda < \lambda'$ then, since $V_\lambda \preccurlyeq_n V_{\lambda'}$,

$$C_m^\lambda = C_m^{\lambda'} \cap \lambda.$$

Let $D = \bigcap_{m \in \omega} C_m$. Clearly, D is a club, and as in the claim, it is definable (with parameters) over V_κ by means of a Π_n formula. So, D is a $\underset{\sim}{\Pi}_n$-club in κ. Thus, $S \cap D$ is a $\underset{\sim}{\Pi}_n$-stationary subset of κ.

Now, let $\lambda \in S \cap D$ but not the least such. We claim that λ is a Π_n-Mahlo cardinal. Clearly, λ is inaccessible. So, let C be a Π_n-club in λ. Then, for some $m \in \omega$, $C = C_m \cap \lambda$. Let $\lambda_0 \in S \cap D$, $\lambda_0 < \lambda$. Thus, $\lambda_0 \in C_m \cap \lambda$. Thus, λ is Π_n-Mahlo. □

Note that the proof of clause (1) of Theorem 2.9 also shows that the set of the $\underset{\sim}{\Sigma}_2$-Mahlo cardinals below a Π_2-Mahlo is a Π_2-stationary set. Therefore, the least $\underset{\sim}{\Sigma}_2$-Mahlo cardinal is not Π_2-Mahlo. However, Theorem 2.9 leaves open two questions. The first is about the relation between Σ_ω-Mahlo and $\underset{\sim}{\Sigma}_\omega$-Mahlo cardinals. The second, because of the restriction to $n > 1$, is about the relation between $\underset{\sim}{\Pi}_0$-Mahlo and Π_1-Mahlo cardinals.

The answer to the first question follows from the following theorem:

Theorem 2.11. *Suppose that κ is a Π_n-Mahlo cardinal with $n > 3$. Then, κ is a $\underset{\sim}{\Pi}_{n-2}$-Mahlo cardinal*

Proof. We only need to show that the set I_{n-1} is unbounded in κ. Since then, by downward absoluteness, for every Π_n formula $\varphi(x)$ and every $a \in V_\kappa$, if $V_\kappa \models \varphi(a)$, then there exists $\lambda \in I_{n-2}$ such that $V_\lambda \models \varphi(a)$. Therefore, by Theorem 2.6, κ is a $\underset{\sim}{\Pi}_{n-2}$-Mahlo cardinal.

But suppose that I_{n-1} is bounded in κ. Let $C = \{\alpha < \kappa : I_{n-1} \subseteq \alpha\}$. It is clear that C is a club of κ. Moreover, for every $\alpha < \kappa$, $\alpha \in C$ iff

$$V_\kappa \models \forall \lambda (\lambda \text{ inaccessible} \wedge V_\lambda \preccurlyeq_{n-1} V_\kappa \rightarrow \lambda \in \alpha).$$

Since "λ is inaccessible" is a Π_1 predicate on λ and "$V_\lambda \preccurlyeq_{n-1} V_\kappa$" is a Π_{n-1} predicate on λ, the above formula is Π_n and hence, C is a Π_n-club of κ. But, since I_n is a Π_n-stationary subset of κ, $C \cap I_n \neq \emptyset$. A contradiction. □

Corollary 2.12. *Every Σ_ω-Mahlo cardinal is a $\underset{\sim}{\Sigma}_\omega$-Mahlo cardinal.*

To answer the second question, namely, the relation between $\underset{\sim}{\Pi}_0$-Mahlo and Π_1-Mahlo cardinals, let us first fix some notation.

Notation 2.13. *Let m_0 denote the least Π_0-Mahlo cardinal and $\underset{\sim}{m}_0$ the least $\underset{\sim}{\Pi}_0$-Mahlo cardinal. Similarly, with m_1 and $\underset{\sim}{m}_1$ we denote, respectively, the least Π_1-Mahlo and the least $\underset{\sim}{\Pi}_1$-Mahlo.*

We know, by remarks following Corollary 2.7, that $m_0 < \underset{\sim}{m}_0$, $\underset{\sim}{m}_0 < \underset{\sim}{m}_1$ and $m_0 < m_1$. By Theorem 2.9, we know that $m_1 < \underset{\sim}{m}_1$. Finally, from the following proposition we get that $m_1 < \underset{\sim}{m}_0$:

Proposition 2.14. $m_1 < \underset{\sim}{m}_0$.

Proof. Let $\kappa_{\mathfrak{c}^+}$ denote the least inaccessible cardinal such that $|I| = (2^{\aleph_0})^+$, where, recall, I denotes the set of all inaccessible cardinals below $\kappa_{\mathfrak{c}^+}$. We will prove that $m_1 < \kappa_{\mathfrak{c}^+}$. Since $\underset{\sim}{m}_0$ is the least inaccessible limit of inaccessible cardinals, $\kappa_{\mathfrak{c}^+} < \underset{\sim}{m}_0$ and hence $m_1 < \underset{\sim}{m}_0$.

Notice that, since there are at most countably many sentences of the language of Set Theory, there are at most $2^{\aleph_0}$ consistent and complete theories in that language. Since $|I| = (2^{\aleph_0})^+$, there are $\mu, \lambda \in I$ such that $\mu \neq \lambda$ but $\mathrm{Th}(V_\mu) = \mathrm{Th}(V_\lambda)$. We may assume that $\mu < \lambda$. We claim that λ is a Π_1-Mahlo cardinal. Let C be a Π_1 club on λ. Thus,

$$V_\lambda \models \forall\alpha\exists\beta(\alpha < \beta \wedge \beta \in C).$$

Then, since $\mathrm{Th}(V_\mu) = \mathrm{Th}(V_\lambda)$,

$$V_\mu \models \forall\alpha\exists\beta(\alpha < \beta \wedge \beta \in C).$$

Thus, since $V_\mu \preccurlyeq_1 V_\lambda$, $C^{V_\mu} = C \cap V_\mu = C \cap \mu$, $C \cap \mu$ is unbounded in μ and $\mu \in C$. Therefore, $\lambda < \kappa_{\mathfrak{c}^+}$ is a Π_1-Mahlo cardinal. □

3. Definably-weakly-compact cardinals

Definably-weakly-compact cardinals were introduced by Leshem in [26], where they are called "Π^1_1-reflecting", and in [8]. In both cases, they are defined using a definable version of the Π^1_1-indescribability of weakly-compact cardinals.

Recall that a Π^1_1 formula is a second-order formula which is of the form $\forall X\varphi(X)$ where X is a second-order variable and $\varphi(X)$ is a first-order formula with X as predicate. Also, recall that $\langle V_\kappa, \in, R\rangle \models \forall X\varphi(X)$ iff for all $A \in P(V_\kappa)$, $\langle V_\kappa, \in, R\rangle \models \varphi(A)$.

Definition 3.1 ([8]). Let κ be a cardinal and $n \in \omega$. κ is $\underset{\sim}{\Sigma}_n$*-weakly compact* ($\underset{\sim}{\Sigma}_n$-w.c., for short), iff κ is inaccessible and for every $R \subseteq V_\kappa$ which is definable by a Σ_n formula (with parameters) over V_κ and every Π^1_1 sentence Φ, if

$$\langle V_\kappa, \in, R\rangle \models \Phi$$

then there is $\alpha < \kappa$ (equivalently, unboundedly-many $\alpha < \kappa$) such that

$$\langle V_\alpha, \in, R \cap V_\alpha\rangle \models \Phi.$$

That is, κ *reflects* Π^1_1 *sentences with* $\underset{\sim}{\Sigma}_n$ *predicates.* κ is $\underset{\sim}{\Pi}_n$*-weakly compact*, (Π_n-w.c., for short), is defined analogously by substituting $\underset{\sim}{\Pi}_n$ for Σ_n in the definition above. Thus, an inaccessible cardinal κ is $\underset{\sim}{\Pi}_n$-w.c. iff it reflects Π^1_1 sentences with $\underset{\sim}{\Pi}_n$ predicates. An inaccessible cardinal is $\underset{\sim}{\Delta}_n$*-weakly compact* ($\underset{\sim}{\Delta}_n$-w.c., for short) iff it reflects Π^1_1 sentences with $\underset{\sim}{\Delta}_n$ predicates.

We can also define the lightface forms of these cardinal notions, namely, the Σ_n (Π_n, Δ_n) *weakly-compact*, henceforth Σ_n-w.c. (respectively, Π_n-w.c., Δ_n-w.c.), by requiring that the corresponding predicates are definable without parameters.

Definition 3.2 ([26]). A cardinal κ is $\underset{\sim}{\Sigma}_\omega$-*weakly compact*, $\underset{\sim}{\Sigma}_\omega$-w.c., for short, iff κ is $\underset{\sim}{\Sigma}_n$-w.c. for every $n \in \omega$. Similarly, using the lightface version, we can define the lightface version of this cardinal notion, namely, the Σ_ω-*weakly-compact* (Σ_ω-w.c., for short).

Theorem 3.3. *Let κ be a cardinal. For all $n \in \omega$, the following are equivalent:*

1. *κ is $\underset{\sim}{\Sigma}_n$-w.c.*
2. *κ is $\underset{\sim}{\Pi}_n$-w.c.*
3. *κ is $\underset{\sim}{\Delta}_{n+1}$-w.c.*

Moreover, the lightface version also holds.

Proof. ($3 \Rightarrow 1$) is obvious. So we only prove ($1 \Rightarrow 2$) and ($2 \Rightarrow 3$).
($1 \Rightarrow 2$). It is clear that κ is inaccessible. Thus, we only need to prove that κ reflects Π^1_1 sentences with $\underset{\sim}{\Pi}_n$ predicates.

Suppose that $R \subseteq V_\kappa$. Let us define for every Π^1_1 formula Φ where R appears as a predicate, $\widetilde{\Phi}$ as the formula obtained from Φ by substituting every occurrence of the subformula Rx, where x is a first-order variable, for $\neg Rx$.

It is easy to show by induction on the complexity of formulas that for every Π^1_1 formula $\Phi(x_0, \ldots, x_{m-1})$, and every α and every $a_0, \ldots, a_{m-1} \in V_\alpha$

$$\langle V_\alpha, \in, R\rangle \models \Phi(a_0, \ldots, a_{m-1}) \text{ iff } \langle V_\alpha, \in, (V_\alpha \setminus R)\rangle \models \widetilde{\Phi}(a_0, \ldots, a_{m-1}).$$

Now assume that $R \subseteq V_\kappa$ is definable by means of a Π_n formula over V_κ with parameters and Φ is a Π^1_1 sentence. By the above remark, $\langle V_\kappa, \in, R\rangle \models \Phi$ iff $\langle V_\kappa, \in, V_\kappa \setminus R\rangle \models \widetilde{\Phi}$. Now, since $V_\kappa \setminus R$ is Σ_n definable over V_κ with parameters, there is $\alpha < \kappa$ such that $\langle V_\alpha, \in, (V_\kappa \setminus R) \cap V_\alpha\rangle \models \widetilde{\Phi}$. But then, $\langle V_\alpha, \in, R \cap V_\alpha\rangle \models \Phi$.
($2 \Rightarrow 3$). Since κ is inaccessible, we only need to prove that κ reflects Π^1_1 sentences with $\underset{\sim}{\Delta}_{n+1}$ predicates. Thus, let $R \subseteq V_\kappa$ be a Δ_{n+1} predicate and let Φ be a Π^1_1 sentence with R as predicate such that

$$\langle V_\kappa, \in, R\rangle \models \Phi.$$

Since R is a Δ_{n+1} predicate, there exists two Π_n formulas, $\varphi(x, y, z_0, \ldots, z_m)$ and $\psi(x, y, z_0, \ldots, z_k)$, and $a_0, \ldots, a_m, b_0, \ldots, b_k \in V_\kappa$ such that

$$\langle V_\kappa, \in, R\rangle \models \forall x(Rx \leftrightarrow \exists y \varphi(x, y, a_0, \ldots, a_k) \leftrightarrow \exists y \psi(x, y, b_0, \ldots, b_l)).$$

Let $S_0, S_1 \subseteq V_\kappa \times V_\kappa$ defined as: for all $x, y \in V_\kappa$,

$$\langle x, y\rangle \in S_0 \quad \text{iff} \quad V_\kappa \models \varphi(x, y, a_0, \ldots, a_k),$$
$$\langle x, y\rangle \in S_1 \quad \text{iff} \quad V_\kappa \models \psi(x, y, b_0, \ldots, b_l).$$

Then S_0 and S_1 are Π_n relations in V_κ.

For every Π^1_1 formula Ψ, where R appears as a predicate, let $\widetilde{\Psi}$ be the formula obtained from Ψ by substituting every occurrence of the subformula Rx, where x

is a first-order variable, for $\exists y S_0 xy$ and every occurrence of the subformula $\neg Rx$ for $\exists y S_1 xy$. Note that $\widetilde{\Psi}$ is also a Π^1_1 sentence.

It is clear that

$$\langle V_\kappa, \in, S_0, S_1\rangle \models \widetilde{\Psi} \wedge \forall x(\exists y S_0 xy \leftrightarrow \neg\exists y S_1 xy).$$

Thus, since κ reflects Π^1_1-sentences with $\underset{\sim}{\Pi}_n$ predicates, there is $\alpha < \kappa$ such that

$$\langle V_\alpha, \in, S_0 \cap V_\alpha, S_1 \cap V_\alpha\rangle \models \widetilde{\Psi} \wedge \forall x(\exists y S_0 xy \leftrightarrow \neg\exists y S_1 xy).$$

Fix such an ordinal α. Let $R^V_\alpha = \{x \in V_\alpha : \exists y\ \langle x, y\rangle \in S_0 \cap V_\alpha\}$. We claim that

$$\langle V_\alpha, \in, R \cap V_\alpha\rangle \models \Phi.$$

Note that, since $\langle V_\alpha, \in, S_0 \cap V_\alpha, S_1 \cap V_\alpha\rangle \models \forall x(\exists y S_0 xy \leftrightarrow \neg\exists y S_1 xy)$,

$$\langle V_\alpha, \in, R^{V_\alpha}\rangle \models \Phi.$$

Now, if $x \in R^{V_\alpha}$, then there exists y such that $\langle x, y\rangle \in S_0 \cap V_\alpha \subseteq S_0$, and thus $x \in R$. Otherwise, if $x \notin R^{V_\alpha}$, then there exists y such that $\langle x, y\rangle \in S_1 \cap V_\alpha \subseteq S_1$, and thus $x \notin R$. Therefore, $R^{V_\alpha} = R \cap V_\alpha$ and κ is $\underset{\sim}{\Delta}_{n+1}$-w.c. □

Leshem (see [26], Lemma 3.1) has proved that the consistency strength of $\underset{\sim}{\Sigma}_\omega$-w.c. cardinals, and hence, of all definably-weakly-compact, is below a Mahlo cardinal.

3.1. Some other characterizations

Recall that the weakly-compact cardinals can be characterized in several ways including: κ has the extension property, $\mathcal{L}_{\kappa,\kappa}$ satisfies the Weak-Compactness Theorem, κ has the tree property and κ has the partition property. We will show that some definable versions of these properties also characterize the definably-weakly-compact cardinals.

Definition 3.4 (**[8] and [26]**)**.** Let κ be a cardinal and $n \in \omega$. κ has the $\underset{\sim}{\Sigma}_n$-*extension property* iff κ is inaccessible and for every $R \subseteq V_\kappa$ which is Σ_n definable (with parameters) over V_κ, there is a transitive set M and $R^M \subseteq M$ such that $\kappa \in M$ and $\langle V_\kappa, \in, R\rangle \preccurlyeq_n \langle M, \in, R^M\rangle$. κ has the $\underset{\sim}{\Sigma}_\omega$-*extension property* iff for every $n \in \omega$, κ has the $\underset{\sim}{\Sigma}_n$-extension property.

We also may define the lightface Σ_n-*extension property* and Σ_ω-*extension property* by requiring that the predicates are definable without parameters.

Let κ be an infinite cardinal. Recall that $\mathcal{L}_{\kappa,\kappa}$ denotes any first-order language with infinite disjunctions and conjunctions of length less than κ and infinite quantifications over less than κ individual variables. Recall that a set Γ of $\mathcal{L}_{\kappa,\kappa}$ sentences is *satisfiable* iff it has a model; and it is μ-*satisfiable*, μ a cardinal, iff every subset of Γ of cardinality less than μ has a model. Recall that κ is a weakly-compact cardinal iff $\mathcal{L}_{\kappa,\kappa}$ satisfies the Weak-Compactness Theorem: Every set of at most κ sentences of $\mathcal{L}_{\kappa,\kappa}$ which is κ-satisfiable, is satisfiable. Thus, it is a natural question to ask whether the definably-weakly-compact cardinals also satisfy some form of the Weak-Compactness Theorem.

Note that, if κ is an inaccessible cardinal, we may code all formulas of $\mathcal{L}_{\kappa,\kappa}$ into sets of V_κ and formulate the satisfaction relation for set structures in V_κ as a Δ_1 definable class in V_κ (see [14], I.4). Recall also that for κ inaccessible the following Löwenheim-Skolem Theorem holds for $\mathcal{L}_{\kappa,\kappa}$: if Γ is a satisfiable set of $\mathcal{L}_{\kappa,\kappa}$ sentences of cardinality at most κ, then it has a model of cardinality at most κ. Note also that in the same vein as for the $\mathcal{L}_{\omega,\omega}$ language of set theory, for every $n \in \omega$, we may define the set of Σ_n and the set of Π_n formulas of $\mathcal{L}_{\kappa,\kappa}$. We also may define the satisfaction relation for proper classes and Σ_n sentences of the $\mathcal{L}_{\kappa,\kappa}$ by means of a Σ_n formula of the language $\mathcal{L}_{\omega,\omega}$ of set theory.

Theorem 3.5. *Let κ be a cardinal and $n > 1$. Then the following are equivalent:*

1. *κ is a $\underset{\sim}{\Sigma}_n$-w.c.*
2. *κ is inaccessible and $\mathcal{L}_{\kappa,\kappa}$ satisfies the Weak-Compactness Theorem for $\underset{\sim}{\Delta}_{n+1}$ sets of sentences: For every collection Γ of $\mathcal{L}_{\kappa,\kappa}$ sentences of cardinality κ which is Δ_{n+1} definable (possibly with parameters) over V_κ, if Γ is κ-satisfiable, then it is satisfiable.*
3. *κ has the $\underset{\sim}{\Sigma}_n$-extension property.*
4. *For every Π^1_1 formula $\Phi(x_0, \ldots, x_k)$, if $a_0, \ldots, a_k \in V_\kappa$ are such that $V_\kappa \models \Phi(a_0, \ldots, a_k)$, then there is $\lambda \in I_n$ such that $a_0, \ldots, a_k \in V_\lambda$ and $V_\lambda \models \Phi(a_0, \ldots, a_k)$.*

The equivalence also holds for Σ_n-w.c., Δ_{n+1} sets of sentences of $\mathcal{L}_{\kappa,\kappa}$, the Σ_n-extension property, and Π^1_1 sentences.

Proof. ($1 \Rightarrow 2$) Suppose that κ is a $\underset{\sim}{\Sigma}_n$-w.c. cardinal. Towards a contradiction, suppose that there is a set Γ of sentences of $\mathcal{L}_{\kappa,\kappa}$ which is definable in V_κ by means of a Σ_{n+1} formula and by means of a Π_{n+1} formula (both possibly with parameters) of the language of set theory, and such that Γ is κ-satisfiable but not satisfiable. We may assume that Γ is a set of sequences of length less than κ of elements of V_κ.

Let Ψ be the Π^1_1 sentence saying that κ is inaccessible and let Φ be the following Π^1_1 sentence, in the language for $\langle V_\kappa, \in \Gamma\rangle$, saying that there is no model for Γ:

$$\forall X \exists x (\Gamma x \wedge X \not\models x).$$

Then, since $\langle V_\kappa, \in, \Gamma\rangle \models \Psi \wedge \Phi$ and Γ is a $\underset{\sim}{\Delta}_{n+1}$ predicate, there is $\lambda < \kappa$ such that

$$\langle V_\lambda, \in, \Gamma \cap V_\lambda\rangle \models \Psi \wedge \Phi.$$

So, λ is inaccessible and $\Gamma_\lambda = \Gamma \cap V_\lambda$ is a set of at most $\lambda < \kappa$ sentences of $\mathcal{L}_{\lambda,\lambda}$. Moreover, since $\langle V_\lambda, \in, \Gamma \cap V_\lambda\rangle \models \Phi$, there is no model of Γ_λ included in V_λ. But then, since λ is inaccessible, by the Löwenheim-Skolem Theorem for $\mathcal{L}_{\lambda,\lambda}$, there is no model for Γ_λ. A contradiction.

($2 \Rightarrow 3$) Suppose now that (2) holds and let $R \subseteq V_\kappa$ which is Σ_n definable (possibly with parameters) over V_κ. Let

$$\Gamma = \Delta \cup \{\text{"}c \text{ is an ordinal"}\} \cup \{\alpha < c : \alpha < \kappa\}$$

where Δ is the set of all Σ_n sentences of the language $\mathcal{L}_{\kappa,\kappa}$ true in the model $\langle V_\kappa, \in, R, x\rangle_{x\in V_\kappa}$ and c is a new individual constant.

It is clear that Γ is a set of cardinality κ of $\mathcal{L}_{\kappa,\kappa}$-sentences.

Note that Γ is a Δ_{n+1} set over V_κ. Let $\sigma_n^R(v_0, v_1)$ be the Σ_n formula of the language of set theory defining the satisfaction relation for classes and Σ_n formulas of the language $\mathcal{L}_{\kappa,\kappa}$ with $\in$ and a predicate R as non-logical symbols. That is, for every Σ_n formula $\varphi(x)$ of $\mathcal{L}_{\kappa,\kappa}$ and every $a \in V_\kappa$,

$$\langle V_\kappa, \in, R\rangle \models \varphi(a) \text{ iff } V_\kappa \models \sigma_n^R(\ulcorner\varphi\urcorner, a).$$

Let $\sigma_n(v_0, v_1)$ be the Σ_n formula defining the satisfaction relation for classes and for Σ_n formulas of set theory. Then, for every Σ_n formula φ of $\mathcal{L}_{\kappa,\kappa}$ (possibly with parameters) and every $a \in V_\kappa$, $\ulcorner\varphi(a)\urcorner \in \Delta$ iff

$$V_\kappa \models \sigma_n^R(\ulcorner\varphi(x)\urcorner, a) \wedge (\forall x \in R)\sigma_n(\ulcorner\psi\urcorner, x) \wedge \forall x(\sigma_n(\ulcorner\psi\urcorner, x) \to x \in R)$$

where $\psi(x)$ is the Σ_n formula that (possibly with parameters) defines R in V_κ. Note that since both formulas $\sigma_n(v_0, v_1)$ and $\sigma_n^R(v_0, v_1)$ are Σ_n, Δ and, hence, Γ are Δ_{n+1} sets in V_κ.

Moreover, if $\Gamma_0 \subseteq \Gamma$ is such that $|\Gamma_0| < \kappa$, then Γ_0 has a model (namely, V_κ with c interpreted as some ordinal greater than all α's mentioned in Γ_0).

Thus, Γ has a model $\langle A, E^A, R^A, x^A\rangle_{x\in V_\kappa}$. We may assume that $V_\kappa \subseteq A$, $E^A \cap (V_\kappa \times V_\kappa) = \in$, $R^A \cap V_\kappa = R$ and $x^A = x$ for all $x \in V_\kappa$. Moreover, $\langle V_\kappa, \in, R\rangle \preccurlyeq_n \langle A, E^A, R^A\rangle$, because $\langle A, E^A, R^A, x^A\rangle_{x\in V_\kappa}$ satisfies all Σ_n sentences true in $\langle V_\kappa, \in, R, x\rangle_{x\in V_\kappa}$. Finally, $\langle A, E^A\rangle$ is well founded since $\langle A, E^A, R^A, x^A\rangle \models \Gamma$ and the sentence

$$\neg\exists v_0 \exists v_1 \cdots \exists v_n \cdots \bigwedge_{n\in\omega} (v_{n+1} \in v_n)$$

belongs to Γ. Let $\langle M, \in, R^M\rangle$ be the Mostowski's collapse of $\langle A, E^A, R^A\rangle$. Then it is a transitive Σ_n-elementary extension of $\langle V_\kappa, \in, R\rangle$ such that $\kappa \in M$.

($3 \Rightarrow 4$) Suppose that κ has the $\widetilde{\Sigma}_n$-extension property. Let $\Phi(x_0, \ldots, x_k) = \forall X\varphi(X, x_0, \ldots, x_k)$ be a Π_1^1 formula and assume that $a_0, \ldots, a_k \in V_\kappa$ are such that $V_\kappa \models \Phi(X, a_0, \ldots, a_k)$.

Recall that $C_n = \{\alpha < \kappa : V_\alpha \preccurlyeq_n V_\kappa\}$ is a Π_n-club of κ. Then,

$$\langle V_\kappa, \in, C_n\rangle \models \forall X\varphi(X, a_0, \ldots, a_k) \wedge \forall\alpha\exists\beta(\alpha < \beta \wedge C_n\beta).$$

By the $\widetilde{\Sigma}_n$-extension property for κ, fix some transitive Σ_n-elementary extension $\langle M, \in, C_n^M\rangle$ of $\langle V_\kappa, \in, C_n\rangle$ such that $\kappa \in M$. Note that, since $V_\kappa \subseteq M$, $V_\kappa^M = V_\kappa$, and hence, $V_{\kappa+1}^M \subseteq V_{\kappa+1}$. Thus,

$$M \models (\forall Y \in V_{\kappa+1})(V_\kappa^M \models \varphi(Y, a_0, \ldots, a_k)).$$

Moreover, $C_n^M \cap V_\kappa^M = C_n \cap \kappa = C_n$. Thus,

$$\langle M, \in, C_n^M\rangle \models (\forall\alpha < \kappa)(\exists\beta < \kappa)(\alpha < \beta \wedge C_n\beta).$$

Finally, note that, $M \models$ "κ is inaccessible", since otherwise, it will be accessible in the universe.

Thus, $\langle M, \in, C_n^M \rangle$ models that there exists μ satisfying the conjunction of the following:

1. μ is inaccessible.
2. $(\forall Y \in V_{\mu+1})(V_\mu^M \models \varphi(Y, a_0, \ldots, a_k))$.
3. $(\forall \alpha < \mu)(\exists \beta < \mu)(\alpha < \beta \wedge C_n \beta)$.

But, since the map $\alpha \longmapsto V_{\alpha+1}$ is Π_1, and "μ is inaccessible" is a Π_1 predicate, the sentence saying that there exists μ satisfying the conjunction of (1), (2) and (3) above is a Σ_2 formula with C_n as predicate and $a_0, \ldots, a_k$ as constants. Therefore, since $n > 1$, $\langle V_\kappa, \in, C_n \rangle$ satisfies the same sentence. Fix some witness λ for the sentence. Then, λ is an inaccessible cardinal, $V_\lambda \models \Phi(a_0, \ldots, a_k)$ and, since C_n is unbounded in λ, $\lambda \in C_n$, i.e., $V_\lambda \preccurlyeq_n V_\kappa$.

($4 \Rightarrow 1$) Suppose now that R is a Σ_n definable subset of V_κ and that Φ is a Π_1^1 sentence such that

$$\langle V_\kappa, \in, R \rangle \models \Phi.$$

Let $\varphi(x, y_0, \ldots, y_k)$ be the Σ_n formula that defines R in V_κ with parameters $a_0, \ldots, a_k \in V_\kappa$. Let $\Phi'(y_0, \ldots, y_k) = \Phi(Rx/\varphi(x, y_0, \ldots, y_k))$, i.e., the Π_1^1 formula (with $y_0, \ldots, y_k$ as the only free individual variables) obtained by substituting every occurrence of the formula Rx for the formula $\varphi(x, y_0, \ldots, y_k)$ that defines the predicate. Then, clearly, $V_\kappa \models \Phi'(a_0, \ldots, a_k)$.

Hence, there is $\lambda \in I_n$ such that $V_\lambda \models \Phi'(a_0, \ldots, a_k)$. Now, since $V_\lambda \preccurlyeq_n V_\kappa$ and $a_0, \ldots, a_k \in V_\lambda$, $R \cap V_\lambda = R^{V_\lambda}$. Therefore, $\langle V_\lambda, \in, R \cap V_\lambda \rangle \models \Phi$. □

The equivalence of (1) and (4) above is from [8]. As an immediate consequence we get the following corollary. The equivalence between (1) and (3) was proved by Leshem (see [26], Theorem 3.2).

Corollary 3.6. *Let κ be a cardinal. Then the following are equivalent:*

1. *κ is a $\underset{\sim}{\Sigma}_\omega$-w.c. cardinal.*
2. *κ is inaccessible and $\mathcal{L}_{\kappa,\kappa}$ satisfies the Weak-Compactness Theorem for definable sets of sentences.*
3. *κ has the $\underset{\sim}{\Sigma}_\omega$-extension property.*
4. *For every $\underset{\sim}{\Pi}_1^1$ formula $\Phi(x_0, \ldots, x_k)$ and every $n > 1$, if $a_0, \ldots, a_k \in V_\kappa$ are such that $V_\kappa \models \Phi(a_0, \ldots, a_k)$, then there is $\lambda \in I_n$ such that $a_0, \ldots, a_k \in V_\lambda$ and $V_\lambda \models \Phi(a_0, \ldots, a_k)$.*

Moreover, the lightface version also holds.

Finally, note that in the proof of ($3 \Rightarrow 4$) of Theorem 3.5, since the club C_n is Π_n-club, we really only used that κ has the Σ_n-extension property. Thus, we have:

Corollary 3.7. *Let κ be a cardinal and $n > 1$. Then, κ is Σ_n-w.c. iff it is $\underset{\sim}{\Sigma}_n$-w.c. Therefore, κ is Σ_ω-w.c. iff κ is $\underset{\sim}{\Sigma}_\omega$-w.c.*

In view of the above corollary, henceforth we only work with Σ_n-w.c. and Σ_ω-w.c. cardinals.

3.1.1. The definably tree property

Definition 3.8. Let κ be a cardinal and $n \in \omega$. A tree $T = \langle T, \leq_T \rangle$, $T \subseteq V_\kappa$, is a $\underset{\sim}{\Sigma}_n$*-tree* iff there are Σ_n formulas $\varphi_T(x)$, $\varphi_{\leq_T}(x, y)$ and $\varphi_{\mathrm{ht}_T}(x, y)$ possibly with parameters in V_κ such that for every $t, t' \in V_\kappa$ and every $\alpha < \kappa$,

$$\begin{aligned} t \in T \quad & \text{iff} \quad V_\kappa \models \varphi_T(t), \\ t \leq_T t' \quad & \text{iff} \quad V_\kappa \models \varphi_{\leq_T}(t, t'), \\ t \in T_\alpha \quad & \text{iff} \quad V_\kappa \models \varphi_{\mathrm{ht}_T}(t, \alpha) \end{aligned}$$

where T_α is the αth level of T. Similarly, we define the $\underset{\sim}{\Pi}_n$*-trees* by substituting Π_n for Σ_n in the above definition. T is a $\underset{\sim}{\Delta}_n$*-tree* iff it is both a $\underset{\sim}{\Sigma}_n$-tree and a $\underset{\sim}{\Pi}_n$-tree.

We also may define the lightface versions of these notions, namely, the Σ_n*-tree* (Π_n*-trees*, Δ_n*-tree*) by requiring that the trees are definable without parameters.

Definition 3.9 ([8])**.** Let κ be a cardinal and $n \in \omega$. κ *has the* $\underset{\sim}{\Sigma}_n$*-tree property* iff κ is inaccessible and every κ-tree which is a $\underset{\sim}{\Sigma}_n$ tree has an unbounded branch. κ *has the* $\underset{\sim}{\Pi}_n$*-tree property* and κ *has the* $\underset{\sim}{\Delta}_n$*-tree property* are defined by substituting $\underset{\sim}{\Pi}_n$, respectively $\underset{\sim}{\Delta}_n$, for $\underset{\sim}{\Sigma}_n$ in the above definition. κ *has the* $\underset{\sim}{\Sigma}_\omega$*-tree property* iff for every $n \in \omega$, κ has the $\underset{\sim}{\Sigma}_n$-tree property.

Similarly, we may define the lightface Σ_n*-tree property*, Π_n*-tree property*, Δ_n*-tree property*, and Σ_ω*-tree property*.

Theorem 3.10 ([8])**.** *For every $n \in \omega$, if κ is Σ_n-w.c., then κ has the $\underset{\sim}{\Sigma}_n$-tree property.*

Proof. Suppose that κ is a Σ_n-w.c. cardinal and let T be a κ-tree over V_κ that is Σ_n definable (with parameters) over V_κ. Suppose that T has not an unbounded branch of length κ. So, every branch $B \subseteq T$ has cardinality less than κ and, hence, $B \in V_\kappa$.

Let Ψ be the Π^1_1 sentence expressing that κ is inaccessible.

Let Φ be the following Π^1_1 sentence:

$$\forall B(B \text{ is a branch of } T \rightarrow \exists x\ B = x).$$

Let F be the function with domain κ such that $F(\alpha) = T_\alpha$, the αth level of T. F is $\underset{\sim}{\Delta}_{n+1}$ on V_κ. Let φ be the following first-order sentence:

$$\forall \alpha(\alpha \text{ is an ordinal } \rightarrow \exists x\ F(\alpha) = x).$$

Thus,

$$\langle V_\kappa, \in, T, F \rangle \models \Phi \wedge \Psi \wedge \varphi.$$

Hence, there is $\lambda < \kappa$ such that

$$\langle V_\lambda, \in, T \cap V_\lambda, F \cap V_\lambda \rangle \models \Phi \wedge \Psi \wedge \varphi.$$

Fix some $t \in T_\lambda$. Let $\mathrm{pred}(t) = \{t' \in T : t' <_T t\}$. It is clear that $\mathrm{pred}(t)$ is a branch through $T \cap V_\lambda$. So, $\mathrm{pred}(t) \in V_\lambda$ and, hence, since λ is inaccessible, $|\mathrm{pred}(t)| < \lambda$. A contradiction. □

As an immediate corollary of the above theorem, we get:

Corollary 3.11 ([26]). *If κ is a Σ_ω-w.c. cardinal, then κ has the $\underset{\sim}{\Sigma}_\omega$-tree property.*

3.1.2. The definably partition property. Recall that if κ is a cardinal and $n \in \omega$ is a natural number, $[\kappa]^n$ is the set of all subsets of κ with exactly n elements.

Definition 3.12. Given a cardinal κ, natural numbers n, m, $n > 0$, and a function $F : [\kappa]^n \longrightarrow m$, a set $H \subseteq \kappa$ is *homogeneous* for F iff $F''[H]^n = \{i\}$ for some $i \in m$.

Definition 3.13 ([8]). Let κ be a cardinal. κ *has the $\underset{\sim}{\Sigma}_n$-partition property* iff κ is an inaccessible cardinal and for every function $F : [\kappa]^2 \longrightarrow \{0,1\}$ which is Σ_n definable over V_κ with parameters there exists a homogeneous $H \subseteq \kappa$ of cardinality κ. In this case we write $\kappa \rightarrow_{\underset{\sim}{\Sigma}_n} (\kappa)^2$. κ *has the $\underset{\sim}{\Pi}_n$-partition property* and κ *has the $\underset{\sim}{\Delta}_n$-partition property*, $\kappa \rightarrow_{\underset{\sim}{\Pi}_n} (\kappa)^2$ and $\kappa \rightarrow_{\underset{\sim}{\Delta}_n} (\kappa)^2$ respectively, are defined by substituting Π_n, respectively Δ_n, for Σ_n in the above definition. Finally, κ *has the $\underset{\sim}{\Sigma}_\omega$-partition property*, $\kappa \rightarrow_{\underset{\sim}{\Sigma}_\omega} (\kappa)^2$ for short, iff for all $n \in \omega$, $\kappa \rightarrow_{\underset{\sim}{\Sigma}_n} (\kappa)^2$.

We can also define the lightface versions of these properties, namely, *the Σ_n-partition property, the Π_n-partition property, the Δ_n-partition property,* and *the Σ_ω-partition property* (that we denote $\kappa \rightarrow_{\Sigma_n} (\kappa)^2$, $\kappa \rightarrow_{\Pi_n} (\kappa)^2$, $\kappa \rightarrow_{\Delta_n} (\kappa)^2$, and $\kappa \rightarrow_{\Sigma_\omega} (\kappa)^2$, respectively) by requiring that the function is definable without parameters.

Theorem 3.14 ([8]). *For every $n > 0$, if κ has the $\underset{\sim}{\Sigma}_n$-tree property, then $\kappa \rightarrow_{\underset{\sim}{\Sigma}_n} (\kappa)^2$. Moreover, the lightface version also holds.*

Proof. Clearly κ is inaccessible. Let $F : [\kappa]^2 \longrightarrow \{0,1\}$ be a Σ_n definable (possibly with parameters) partition over V_κ. Note that since $\mathrm{dom}(F) = [\kappa]^2$ is a Δ_0 class in V_κ, F is a Δ_n function (see [13], Lemma 13.10).

For every $\beta < \kappa$, let $f_\beta : \beta \longrightarrow \{0,1\}$ such that for all $\alpha < \beta$, $f_\beta(\alpha) = F(\{\alpha,\beta\})$. Let $T = \{f_\beta \restriction \gamma : \gamma \leq \beta < \kappa\}$ ordered by extension. Note that $f \in T$ iff V_κ models that

$$\exists\beta\gamma(\gamma \leq \beta \wedge \mathrm{dom}(f) = \gamma \wedge (\forall\alpha < \gamma)(\forall i \in \{0,1\})(F(\{\alpha,\beta\}) = i \leftrightarrow f(\alpha) = i).$$

So, since $F(\{\alpha,\beta\}) = i$ is a Δ_n fact in V_κ, T is Σ_n definable (with the same parameters as the function) over V_κ.

It is clear that for every $\beta < \kappa$, f_β belongs to level β of T. So, $\mathrm{ht}(T) = \kappa$. It is also clear that for every $f \in T$, $f \in T_\beta$ iff $\mathrm{dom}(f) = \beta$. Hence, T is a $\underset{\sim}{\Sigma}_n$-tree in V_κ. Moreover, for every $\beta < \kappa$, $T_\beta \subseteq 2^\beta$ and so, since κ is inaccessible, $|T_\beta| < \kappa$. Therefore T is a $\underset{\sim}{\Sigma}_n$-tree that is a κ-tree.

So, since κ has the property of $\underset{\sim}{\Sigma}_n$-tree, there is an unbounded branch B trough T. Let $\{t_\xi : \xi < \kappa\}$ be an increasing enumeration of B. For every $i \in \{0,1\}$, let

$$H_i = \{\xi < \kappa : t_\xi^\frown\langle i\rangle \in B\}.$$

We claim that for every $i \in \{0,1\}$, H_i is an homogeneous subset of κ for F. Fix $\alpha, \beta, \gamma \in H_i$ with $\alpha < \beta < \gamma$. Since $t_\alpha^\frown \langle i \rangle \subseteq t_\beta$ and $t_\beta^\frown \langle i \rangle \subseteq t_\gamma$,

$$F(\{\alpha, \beta\}) = t_\beta(\alpha) = i = t_\gamma(\beta) = F(\{\beta, \gamma\}).$$

So, the H_i are homogeneous, $i \in \{0,1\}$. Since $|B| = \kappa$, either $|H_0| = \kappa$ or $|H_1| = \kappa$. Therefore, $\kappa \to_{\Sigma_n} (\kappa)^2$. □

As an immediate corollary, we get:

Corollary 3.15. *If κ has the $\underset{\sim}{\Sigma}_\omega$-tree property, then $\kappa \to_{\underset{\sim}{\Sigma}_\omega} (\kappa)^2$. Moreover, the lightface version also holds.*

Lemma 3.16 ([23]). *Assume $V = L$. For every $n > 0$, $\kappa \to_{\underset{\sim}{\Sigma}_n} (\kappa)^2$ implies that for every Π^1_1 formula $\Phi(x_0, \ldots, x_k)$ and $a_0, \ldots, a_k \in L_\kappa$ such that $L_\kappa \models \Phi(a_0, \ldots, a_k)$, there is $\lambda < \kappa$ with $L_\lambda \preccurlyeq L_\kappa$ such that $L_\lambda \models \Phi(a_0, \ldots, a_k)$.*

Finally, we have,

Theorem 3.17 ([8]). *($V = L$) Let κ be a cardinal. Then, for every $n > 1$, the following are equivalent:*

1. *κ is a Σ_n-w.c. cardinal.*
2. *κ has the property of $\underset{\sim}{\Sigma}_n$-tree.*
3. *$\kappa \to_{\underset{\sim}{\Sigma}_n} (\kappa)^2$.*

Proof. ($1 \Rightarrow 2$) follows from Theorem 3.10.
($2 \Rightarrow 3$) follows from Theorem 3.14.
($3 \Rightarrow 1$) follows from Lemma 3.16 (this is the only place where $V = L$ is used) and (4) of Theorem 3.5. □

Therefore,

Corollary 3.18 ($V = L$). *Let κ be a cardinal. Then, the following are equivalent:*

1. *κ is a Σ_ω-w.c. cardinal.*
2. *κ has the property of the $\underset{\sim}{\Sigma}_\omega$-tree.*
3. *$\kappa \to_{\underset{\sim}{\Sigma}_\omega} (\kappa)^2$.*

Question 2.

1. *Suppose that κ has the property of the $\underset{\sim}{\Sigma}_n$-tree. Does it follow that κ is a Σ_n-w.c. cardinal? And for some $m < n$, does it imply that κ is Σ_m-w.c.?*
2. *If we remove our demand that κ is inaccessible and work with trees which are definable over H_κ, the set of all sets of cardinality hereditarily less than κ, instead of V_κ, then Leshem ([26]) has proved that the answer to the above question is no, since it is consistent ω_1 has the property of $\underset{\sim}{\Sigma}_\omega$-tree. But note, since we may express that "for every function F from ω to ordinals there exists an ordinal α such that $F(n) \neq \alpha$, for all n" with a Π^1_1 sentence which is true in V_{ω_1}, ω_1 does not reflect Π^1_1 sentences.*
3. *Is $V = L$ a necessary hypothesis to prove Theorem 3.17?*

3.2. Their own hierarchy

We will show that every Σ_n-w.c. cardinal is $\underset{\sim}{\Pi}_n$-Mahlo, but that the least $\underset{\sim}{\Pi}_n$-Mahlo cardinal is not Σ_n-w.c. We also prove that the least Σ_n-w.c. is not $\underset{\sim}{\Pi}_{n+1}$-Mahlo. From above it follows that every Σ_ω-w.c. is $\underset{\sim}{\Sigma}_\omega$-Mahlo. Finally, we prove that the set of $\underset{\sim}{\Sigma}_\omega$-Mahlo cardinals below a Σ_ω-w.c. is a $\underset{\sim}{\Sigma}_\omega$-stationary set.

Theorem 3.19 ([8]). *Every Σ_n-w.c. cardinal κ is a $\underset{\sim}{\Pi}_n$-Mahlo cardinal and the set of $\underset{\sim}{\Pi}_n$-Mahlo cardinals below κ is $\underset{\sim}{\Pi}_n$-stationary.*

Proof. Suppose that κ is Σ_n-w.c. Let C be a $\underset{\sim}{\Pi}_n$-club in κ. Let Φ be the Π^1_1 sentence expressing that κ is inaccessible. Let σ be the first-order sentence expressing that C is unbounded. Then,

$$\langle V_\kappa, \in, C\rangle \models \Phi \wedge \sigma.$$

So, there is $\alpha < \kappa$ such that

$$\langle V_\alpha, \in, C \cap V_\alpha\rangle \models \Phi \wedge \sigma.$$

Therefore α is inaccessible and, since $C \cap V_\alpha = C \cap \alpha$ is unbounded in α, $\alpha \in C$.

Note that "every $\underset{\sim}{\Pi}_n$ club in κ contains an inaccessible cardinal" is expressible by means of a first-order sentence. Therefore, the above argument shows that there is a $\underset{\sim}{\Pi}_n$-stationary set of $\underset{\sim}{\Pi}_n$-Mahlo cardinals below κ. $\square$

Corollary 3.20. *Every Σ_ω-w.c. cardinal is $\underset{\sim}{\Sigma}_\omega$-Mahlo.*

To prove that the set of $\underset{\sim}{\Sigma}_\omega$-Mahlo cardinals below a Σ_ω-w.c. cardinal is a $\underset{\sim}{\Sigma}_\omega$-stationary set we need a couple of lemmas. Given a limit ordinal α, let S be the satisfaction relation class for V_α. That is,

$$S(\ulcorner\varphi\urcorner, a) \text{ iff } V_\alpha \models \varphi(a).$$

Note that $S \subseteq V_\alpha$. Let $sat(X)$ be the (canonically defined) second order formula such that $sat(S)$ iff S is the satisfaction class (see [9]). More precisely $\mathrm{sat}(S)$ holds iff

$$\forall\ulcorner\varphi\urcorner\forall a(S(\ulcorner\varphi\urcorner, a) \leftrightarrow \varphi(a)).$$

Moreover, there exists a first-order formula with S as a predicate saying that S is the satisfaction relation for V_α. That is, there exists a first-order formula $\theta(x, y)$ such that for every limit ordinal α

$$\langle V_\alpha, \in, R\rangle \models \mathrm{sat}(S) \text{ iff } \langle V_\alpha, \in, R, S\rangle \models \forall\ulcorner\varphi\urcorner\forall a\, \theta(\ulcorner\varphi\urcorner, a).$$

Thus, we can prove the following lemma:

Lemma 3.21. *There is a Σ^1_1 formula $\Phi(x, y)$ and a Π^1_1 formula $\Psi(x, y)$ such that for every first-order formula $\varphi(y)$ and every limit ordinal α, every $a \in V_\alpha$ and every $R \subseteq V_\alpha$,*

$$\langle V_\alpha, \in, R\rangle \models \varphi(a) \textit{ iff } \langle V_\alpha, \in, R\rangle \models \Phi(\ulcorner\varphi\urcorner, a) \textit{ iff } \langle V_\alpha, \in, R\rangle \models \Psi(\ulcorner\varphi\urcorner, a).$$

Proof. Since $S \in V_{\alpha+1}$ and since there is only one $S \subseteq V_\alpha$ such that $\text{sat}(S)$, let $\Phi(x,y)$ be the following Σ_1^1 formula:

$$\exists X(\text{sat}(X) \wedge X(x,y)).$$

And let $\Psi(x,y)$ the following Π_1^1 formula:

$$\forall X(\text{sat}(X) \rightarrow X(x,y)).$$

Clearly both formulas work. □

Using Lemma above it is easy to prove the following

Lemma 3.22. *There exists a Π_1^1 sentence Θ such that for every cardinal κ, $V_\kappa \models \Theta$ iff κ is a $\underset{\sim}{\Sigma}_\omega$-Mahlo cardinal.*

Proof. Let Θ be the conjunction of the Π_1^1 sentence expressing that κ is an inaccessible cardinal with the following Π_1^1 sentence:

$$\begin{aligned}&\forall \ulcorner\varphi\urcorner \forall a(\forall\alpha\exists\beta(\alpha < \beta \wedge \Phi(\ulcorner\varphi\urcorner, \beta^\frown a))\wedge \\ &\quad \wedge \forall\alpha((\forall\beta < \alpha)(\exists\gamma < \alpha)(\beta < \gamma \wedge \Psi(\ulcorner\varphi\urcorner, \gamma^\frown a)) \rightarrow \Phi(\ulcorner\varphi\urcorner, \alpha^\frown a)) \rightarrow \\ &\qquad \rightarrow \exists\mu(\mu \text{ inaccessible} \wedge \Psi(\ulcorner\varphi\urcorner, \mu^\frown a))).\end{aligned}$$

Note that, since Φ is Σ_1^1 and Ψ is Π_1^1, the above is a Π_1^1 sentence. It is easy to see that for every κ, $V_\kappa \models \Theta$ iff κ is an inaccessible cardinal and for every first-order formula $\varphi(x,y)$ which defines a club in κ with a as parameter, there exists an inaccessible cardinal μ such that $\varphi(\mu, a)$. Thus, $V_\kappa \models \Theta$ iff κ is a $\underset{\sim}{\Sigma}_\omega$-Mahlo cardinal. □

Theorem 3.23. *Let κ be a Σ_ω-w.c. cardinal. Then there is a $\underset{\sim}{\Sigma}_\omega$-stationary subset of $\underset{\sim}{\Sigma}_\omega$-Mahlo cardinals below κ. Thus, the least $\underset{\sim}{\Sigma}_\omega$-Mahlo cardinal is not Σ_ω-w.c.*

Proof. As in Theorem 3.19, but using the Π_1^1 sentence Θ of Lemma above instead of the Π_1^1 sentence asserting that κ is inaccessible. □

Moreover,

Theorem 3.24. *Let κ be a Π_{n+1}-Mahlo cardinal, $n \geq 1$. Then, the set of all $\lambda < \kappa$ such that λ is Σ_n-w.c. is a Π_{n+1}-stationary subset of κ. Thus, the least Σ_n-w.c. cardinal is not a Π_{n+1}-Mahlo cardinal.*

Proof. Let κ be a Π_{n+1}-Mahlo cardinal. Let $\lambda \in I_{n+1}$. We claim that λ is a Σ_n-w.c. cardinal. Let Φ be a Π_1^1 sentence such that $V_\lambda \models \Phi$. For the sake of simplicity, we may assume that Φ is of the form $\forall X\varphi(X)$ where $\varphi(X)$ is a first-order formula with X as predicate. Since $V_\lambda, V_{\lambda+1} \subseteq V_\kappa$,

$$V_\kappa \models (\forall X \in V_{\lambda+1})(V_\lambda \models \varphi(X)).$$

Thus, since $\lambda \in I_{n+1} \subseteq I_n$,

$$V_\kappa \models \exists\mu(\mu \in I_n \wedge (\forall X \in V_{\mu+1})(V_\mu \models \varphi(X))).$$

But the right-hand formula is Σ_{n+1}. Hence, since $V_\lambda \preccurlyeq_{n+1} V_\kappa$,

$$V_\lambda \models \exists\mu(\mu \in I_n \wedge (\forall X \in V_{\mu+1})(V_\mu \models \varphi(X))).$$

Therefore, there exists $\mu < \lambda$ such that $\mu \in I_n$ and $V_\mu \models \Phi$. Hence, by (4) of Theorem 3.5, λ is a Σ_n-w.c. cardinal. □

Corollary 3.25. *Let $n \geq 1$. If κ is a Σ_{n+1}-w.c. cardinal, then there is a $\underset{\sim}{\Pi}_{n+1}$-stationary subset of κ of Σ_n-w.c. cardinals. Thus, the least Σ_n-w.c. cardinal is not a Σ_{n+1}-w.c. cardinal.*

4. Definably-indescribable cardinals

Recall that a Π^m_n formula is a formula of order $m+1$ which is of the form $\forall X_0 \exists X_1 \dots Q X_{n-1} \varphi(X_0, \dots, X_{n-1})$ where Q is $\exists$, if n is even, or is $\forall$, if n is odd, $X_0, \dots, X_{n-1}$ are $(m+1)$th-order variables and $\varphi(X_0, \dots, X_{n-1})$ is a formula of order m with at least $X_0, \dots, X_{n-1}$ as predicates of order $m+1$. A Σ^m_n formula is the negation of a Π^m_n formula.

Definition 4.1. Let κ be a cardinal and $n, m, k \in \omega$, $n, m, k \geq 1$. κ *reflects Π^m_n sentences with $\underset{\sim}{\Sigma}_k$ predicates* iff for every $R \subseteq V_\kappa$ which is definable by means of a Σ_k formula (possibly with parameters) over V_κ and every Π^m_n sentence Φ, if

$$\langle V_\kappa, \in, R \rangle \models \Phi$$

then there is $\alpha < \kappa$ such that

$$\langle V_\alpha, \in, R \cap V_\alpha \rangle \models \Phi.$$

We define κ *reflects Π^m_n sentences with $\underset{\sim}{\Pi}_k$ predicates* and κ *reflects Π^m_n sentences with $\underset{\sim}{\Delta}_k$ predicates* by substituting Π_k (respectively, Δ_k) for Σ_k in the above definition.

We define κ *reflects Σ^m_n sentences with $\underset{\sim}{\Sigma}_k$ predicates* by substituting Π^m_n for Σ^m_n in the above definition. Similarly for κ *reflects Σ^m_n sentences with $\underset{\sim}{\Pi}_k$ predicates* and κ *reflects Σ^m_n sentences with $\underset{\sim}{\Delta}_k$ predicates.*

We can also define the lightface κ *reflects Π^m_n sentences with Σ_k predicates, κ reflects Π^m_n sentences with Π_k predicates, κ reflects Π^m_n sentences with Δ_k predicates, κ reflects Σ^m_n sentences with Σ_k predicates, κ reflects Σ^m_n sentences with Π_k predicates*, and *κ reflects Σ^m_n sentences with Δ_k predicates* by requiring that the predicate is definable without parameters.

As in the above section, we can prove the following fact

Fact 4.2. *Let κ be a cardinal. For all $n, m, k \in \omega$, the following are equivalent:*

1. *κ reflects Π^m_n (Σ^m_n) sentences with $\underset{\sim}{\Sigma}_k$ predicates.*
2. *κ reflects Π^m_n (Σ^m_n) sentences with $\underset{\sim}{\Pi}_k$ predicates.*
3. *κ reflects Π^m_n (Σ^m_n) sentences with $\underset{\sim}{\Delta}_{k+1}$ predicates.*

Moreover, the lightface version also holds.

Definition 4.3. Let κ be a cardinal and and $n, m, k \in \omega$, $n, m, k \geq 1$. κ is a $\underset{\sim}{\Pi}^m_n$-*k-indescribable cardinal* iff κ is inaccessible and reflects Π^m_n sentences with $\underset{\sim}{\Sigma}_k$ predicates. κ is a $\underset{\sim}{\Pi}^m_n$-*ω-indescribable cardinal* iff κ is $\underset{\sim}{\Pi}^m_n$-k-indescribable for every $k \in \omega$. We define $\underset{\sim}{\Sigma}^m_n$-*$k$-indescribable cardinal* and $\underset{\sim}{\Sigma}^m_n$-*$\omega$-indescribable cardinal* by substituting Π^m_n for Σ^m_n in the above definition. κ is $\underset{\sim}{\Pi}^\omega_\omega$-*$k$-indescribable* iff for all $n, m \geq 1$, κ is $\underset{\sim}{\Pi}^m_n$-k-indescribable. Finally, κ is $\underset{\sim}{\Pi}^\omega_\omega$-*$\omega$-indescribable* iff for all $k \geq 1$, κ is $\underset{\sim}{\Pi}^\omega_\omega$-$k$-indescribable.

Similarly, we may define the lightface Π^m_n-*k-indescribable cardinal*, Π^m_n-*ω-indescribable cardinal*, Σ^m_n-*k-indescribable cardinal*, Σ^m_n-*ω-indescribable cardinal*, Π^ω_ω-*k-indescribable*, and Π^ω_ω-*ω-indescribable* by requiring that the predicate is definable without parameters.

In [26], the $\underset{\sim}{\Pi}^m_n$-ω-indescribable cardinals were defined and it was shown that the set of all $\underset{\sim}{\Pi}^\omega_\omega$-$\omega$-indescribable cardinals below a Mahlo cardinal is stationary. Thus, definably-indescribable cardinals are, consistency-wise, below a Mahlo cardinal. In this section we will give a detailed account of the position of all definably-indescribable cardinals in the hierarchy of large cardinals and in their own hierarchy.

First, note that, since the inaccessible cardinals are Σ^1_1-indescribable without any restriction on the definability of added predicates, we have:

Proposition 4.4. *The following are equivalent:*

1. *κ is a $\underset{\sim}{\Sigma}^1_1$-$n$-indescribable cardinal, for some $n \in \omega$.*
2. *κ is a $\underset{\sim}{\Sigma}^1_1$-$\omega$-indescribable cardinal.*
3. *κ is an inaccessible cardinal.*

Moreover, the lightface version also holds.

Recall that, by Theorem 3.5, κ is a $\underset{\sim}{\Pi}^1_1$-k-indescribable ($\underset{\sim}{\Pi}^1_1$-ω-indescribable) cardinal iff it is Σ_k-w.c. (respectively, Σ_ω-w.c.). And the same holds for the lightface versions of the definably-indescribable cardinals.

Note that if $m \leq m'$, $n \leq n'$ and some of the two inequalities is strict, then κ is $\underset{\sim}{\Pi}^{m'}_{n'}$-$k'$-indescribable ($\underset{\sim}{\Sigma}^{m'}_{n'}$-$k'$-indescribable) implies κ is both $\underset{\sim}{\Pi}^m_n$-k-indescribable and $\underset{\sim}{\Sigma}^m_n$-$k$-indescribable, for all $k \leq k'$. Hence, for every $m, n, k \geq 1$, every $\underset{\sim}{\Pi}^m_n$-k-indescribable, and if $n > 1$ or $m > 1$ also every $\underset{\sim}{\Sigma}^m_n$-$k$-indescribable cardinal, is Σ_k-w.c, and therefore, by Theorem 3.19, $\underset{\sim}{\Pi}_k$-Mahlo. Note also that the same holds for the lightface versions of the definably-indescribable cardinals.

We now prove that the least $\underset{\sim}{\Pi}^m_n$-k-indescribable and the least $\underset{\sim}{\Sigma}^m_n$-$k$-indescribable cardinals are below the least Π_{k+1}-Mahlo cardinal. First note that we can easily generalize the proof of the equivalence between (1) and (4) of Theorem 3.5 to prove:

Theorem 4.5. *Let κ be a cardinal and $m, n, k \geq 1$. Then, the following are equivalent:*

1. *κ is a $\underset{\sim}{\Pi}^m_n$-k-indescribable ($\underset{\sim}{\Sigma}^m_n$-$k$-indescribable) cardinal.*

2. *For every* Π^m_n *(respectively,* Σ^m_n*) formula* $\Phi(x_0, \dots, x_k)$*, if* $a_0, \dots, a_k \in V_\kappa$ *are such that* $V_\kappa \models \Phi(a_0, \dots, a_k)$*, then there is* $\lambda \in I_k$ *such that* $V_\lambda \models \Phi(a_0, \dots, a_k)$*.*

Moreover, the above equivalence also holds for Π^m_n*-k-indescribable (*Σ^m_n*-k-indescribable) and* Π^m_n *(respectively,* Σ^m_n*) sentences.*

Thus, as in Theorem 3.24 we get:

Theorem 4.6. *Let* $k \geq 1$ *and let* κ *be a* Π_{k+1}*-Mahlo cardinal. Then, for all* $m, n \geq 1$*, the set of all* $\lambda < \kappa$ *such that* λ *is* $\underset{\sim}{\Pi}^m_n$*-k-indescribable (*$\underset{\sim}{\Sigma}^m_n$*-k-indescribable) is a* Π_{k+1}*-stationary subset of* κ*. Thus, the least* $\underset{\sim}{\Pi}^m_n$*-k-indescribable (*$\underset{\sim}{\Sigma}^m_n$*-k-indescribable) cardinal is not a* Π_{k+1}*-Mahlo cardinal.*

Proof. The proof is similar to that of Theorem 3.24. We only give the proof for $\underset{\sim}{\Pi}^m_n$-k-indescribable cardinals. The proof for $\underset{\sim}{\Sigma}^m_n$-$k$-indescribability is similar. We only work with the case $m = 1$, since the proof is analogous for the other cases. So, let κ be a Π_{k+1}-Mahlo cardinal and $\lambda \in I_{k+1}$. Let $\Phi(x_0, \dots, x_l)$ be the Π^1_n formula $\forall X_0 \exists X_1 \dots Q X_{n-1} \varphi(x_0, \dots, x_l)$, where Q is $\forall$, if n is odd, or $\exists$, if n is even. Let $a_0, \dots, a_l \in V_\lambda$ be such that

$$V_\lambda \models \Phi(a_0, \dots, a_l).$$

Since $V_\lambda, V_{\lambda+1} \subseteq V_\kappa$,

$$V_\kappa \models (\forall X_0 \in V_{\lambda+1}) \dots (Q X_{n-1} \in V_{\lambda+1})(V_\lambda \models \varphi(a_0, \dots, a_l)).$$

Note that the above right-hand formula is equivalent both to a Σ_2 and to a Π_2 formula with λ and $a_0, \dots, a_l$ as parameters. Since $\lambda \in I_{k+1} \subseteq I_k$,

$$V_\kappa \models \exists \mu (\mu \in I_k \wedge (\forall X_0 \in V_{\mu+1}) \dots (Q X_{n-1} \in V_{\mu+1})(V_\mu \models \varphi(a_0, \dots, a_l))).$$

Since $k \geq 1$, the formula above is Σ_{k+1} and hence, since $V_\lambda \preccurlyeq_{k+1} V_\kappa$,

$$V_\lambda \models \exists \mu (\mu \in I_k \wedge (\forall X_0 \in V_{\mu+1}) \dots (Q X_{n-1} \in V_{\mu+1})(V_\mu \models \varphi(a_0, \dots, a_l))).$$

Therefore, there exists $\mu < \lambda$ such that $\mu \in I_k$ and $V_\mu \models \Phi(a_0, \dots, a_l)$. Hence, by Theorem 4.5, λ is a $\underset{\sim}{\Pi}^1_n$-k-indescribable cardinal. □

The proof above shows that if κ is a Π_{k+1}-Mahlo cardinal, then every $\lambda \in I_{k+1}$ is a $\underset{\sim}{\Pi}^\omega_\omega$-$k$-indescribable cardinal. Thus we have:

Corollary 4.7. *Let* $k \geq 1$*, and let* κ *be a* Π_{k+1}*-Mahlo cardinal. The set of all* $\lambda < \kappa$ *such that* λ *is* $\underset{\sim}{\Pi}^\omega_\omega$*-k-indescribable is a* Π_{k+1}*-stationary subset of* κ*. Thus, the least* $\underset{\sim}{\Pi}^\omega_\omega$*-k-indescribable cardinal is not a* Π_{k+1}*-Mahlo and hence is not* Σ_{k+1}*-w.c.*

We also get the following corollary from Theorem 4.6 together Theorem 3.19

Corollary 4.8. *Let* $k \geq 1$*. If* κ *is a* Π^m_n*-*$(k+1)$*-indescribable (*Σ^m_n*-*$(k+1)$*-indescribable) cardinal, then there is a* Π_k*-stationary subset of* κ *of* $\underset{\sim}{\Pi}^m_n$*-k-indescribable (*$\underset{\sim}{\Sigma}^m_n$*-k-indescribable) cardinals. Thus, the least* $\underset{\sim}{\Pi}^m_n$*-k-indescribable (*$\underset{\sim}{\Sigma}^m_n$*-k-indescribable) cardinal is not a* Π^m_n*-*$(k+1)$*-indescribable (*Σ^m_n*-*$(k+1)$*-indescribable) cardinal.*

Notation 4.9. *Let $1 \leq m, n, k \leq \omega$. Let us denote with $\underset{\sim}{\sigma}{}^m_n$-$k$ and $\underset{\sim}{\pi}{}^m_n$-$k$, respectively, the least $\underset{\sim}{\Sigma}{}^m_n$-$k$-indescribable cardinal and the least $\underset{\sim}{\Pi}{}^m_n$-$k$-indescribable cardinal. We also use σ^m_n-k and π^m_n-k to denote the least Σ^m_n-k-indescribable cardinal and the least Π^m_n-k-indescribable cardinal. Recall that we denote with m_k and $\underset{\sim}{m}_k$, respectively, the least Π_k-Mahlo and the least $\underset{\sim}{\Pi}_k$-Mahlo cardinals.*

From the theorem above and its corollaries, we have:

$$m_k < \underset{\sim}{m}_k < \pi^1_1\text{-}k, \underset{\sim}{\sigma}{}^m_n\text{-}k, \underset{\sim}{\pi}{}^m_n\text{-}k, \underset{\sim}{\pi}{}^\omega_\omega\text{-}k < m_{k+1} < \underset{\sim}{m}_{k+1} < \pi^1_1\text{-}(k+1), \pi^\omega_\omega\text{-}(k+1).$$

To study the inner hierarchy of Σ^m_n-k-indescribable and Π^m_n-k-indescribable cardinals for some fixed $k \geq 1$, we need the following lemmas.

Lemma 4.10. *Let α be a limit ordinal. Then,*

1. *For every $m, n \geq 1$ there exists a Π^m_n formula $\Upsilon^m_n(x, y)$ with the property that for every Π^m_n formula $\Theta(y)$ there exists $r \in \omega$ such that for every $a \in V_\alpha$,*
$$V_\alpha \models \Theta(a) \textit{ iff } V_\alpha \models \Upsilon^m_n(r, a).$$
2. *For every $m, n \geq 1$ there exists a Σ^m_n formula $\Upsilon^m_n(x, y)$ with the property that for every Σ^m_n formula $\Theta(y)$ there exists $r \in \omega$ such that for every $a \in V_\alpha$,*
$$V_\alpha \models \Theta(a) \textit{ iff } V_\alpha \models \Upsilon^m_n(r, a).$$

Proof. We only prove (1). The proof of (2) is similar. We also only prove (1) for $m = 1$. The rest of the cases are the same. Recall the Π^1_1 formula $\Psi(x, y)$ and the Σ^1_1 formula $\Phi(x, y)$ of Lemma 3.21. Let $\Upsilon^1_n(x, y)$ be the formula:

$$\forall X_0 \exists X_1 \ldots \forall X_{n-1} \Psi(x, y),$$

if n is odd, or the formula

$$\forall X_0 \exists X_1 \ldots \exists X_{n-1} \Phi(x, y),$$

if n is even.

Now, let α be a limit ordinal and suppose that $\Theta(y)$ is the Π^1_n formula $\forall X_0 \ldots Q X_{n-1} \varphi(X_0, \ldots, X_{n-1}, y)$. Then, for all $a \in V_\alpha$

$$V_\alpha \models \Theta(a) \text{ iff } V_\alpha \models \Upsilon^1_n(\ulcorner \varphi \urcorner, a).$$ □

Lemma 4.11. *For every $m, n, k \geq 1$, if $m \neq 1$ or $n \neq 1$, then*

1. *There is a Σ^m_n sentence $\underset{\sim}{\Phi}{}^m_{n,k}$ such that $V_\kappa \models \underset{\sim}{\Phi}{}^m_{n,k}$ iff κ is a $\underset{\sim}{\Pi}{}^m_n$-$k$-indescribable cardinal.*
2. *There is a Π^m_n sentence $\underset{\sim}{\Psi}{}^m_{n,k}$ such that $V_\kappa \models \underset{\sim}{\Psi}{}^m_{n,k}$ iff κ is a $\underset{\sim}{\Sigma}{}^m_n$-$k$-indescribable cardinal.*

Moreover, the lightface version also holds. That is, there is a Σ^m_n (Π^m_n) sentence $\Phi^m_{n,k}$ (respectively, $\Psi^m_{n,k}$) such that V_κ models $\Phi^m_{n,k}$ ($\Psi^m_{n,k}$) iff κ is a Π^m_n-k-indescribable (respectively, Σ^m_n-k-indescribable) cardinal.

Proof. (1) Let $\Upsilon^m_n(x,y)$ be the Π^m_n formula of Lemma above. Let $\underset{\sim}{\Phi}^m_{n,k}$ be the conjunction of the Π^1_1 formula saying that κ is inaccessible with the following sentence:

$$\forall \ulcorner\varphi\urcorner \forall a(\Upsilon^m_n(\ulcorner\varphi\urcorner, a) \to \exists\lambda(\lambda \in I_k \wedge V_\lambda \models \Upsilon^m_n(\ulcorner\varphi\urcorner, a)).$$

Note that $V_\lambda \models \Upsilon^m_n(\ulcorner\varphi\urcorner, a)$ is equivalent both to a Σ_2 and to a Π_2 formula. Thus, since $\Upsilon^m_n(x,y)$ is a Π^m_n formula, $\underset{\sim}{\Phi}^m_{n,k}$ is a Σ^m_n sentence. It is easy to see that $V_\kappa \models \underset{\sim}{\Phi}^m_{n,k}$ iff κ is a Π^m_n-k-indescribable cardinal.

(2) The proof is the same, but using the Σ^m_n formula $\Upsilon^m_n(x,y)$ of the Lemma above. □

Theorem 4.12. *For every $m, n, k \geq 1$, if $m \neq 1$ or $n \neq 1$, then*

1. *If κ is $\underset{\sim}{\Pi}^m_{n+1}$-$k$-indescribable, then the set of all $\underset{\sim}{\Pi}^m_n$-k-indescribable cardinals and the set of all $\underset{\sim}{\Sigma}^m_n$-$k$-indescribable cardinals are $\underset{\sim}{\Pi}_k$-stationary subsets of κ.*
2. *If κ is $\underset{\sim}{\Sigma}^m_{n+1}$-$k$-indescribable, then the set of all $\underset{\sim}{\Pi}^m_n$-k-indescribable cardinals and the set of all $\underset{\sim}{\Sigma}^m_n$-$k$-indescribable cardinals are $\underset{\sim}{\Pi}_k$-stationary subsets of κ.*

Moreover the lightface version also holds. Thus, the least $\underset{\sim}{\Pi}^m_n$-k-indescribable ($\underset{\sim}{\Sigma}^m_n$-$k$-indescribable) cardinal is neither a $\underset{\sim}{\Pi}^m_{n+1}$-$k$-indescribable nor a $\underset{\sim}{\Sigma}^m_{n+1}$-$k$-indescribable cardinal. And the least Π^m_n-k-indescribable (Σ^m_n-k-indescribable) cardinal is neither a Π^m_{n+1}-k-indescribable nor a Σ^m_{n+1}-k-indescribable cardinal.

Proof. We only prove (1). The proof of (2) is the same. So, assume that κ is a $\underset{\sim}{\Pi}^m_{n+1}$-$k$-indescribable cardinal. Let C be a $\underset{\sim}{\Pi}_k$ club on κ. Let σ be the first-order sentence with C as predicate saying that C is unbounded. Then,

$$\langle V_\kappa, \in, C\rangle \models \sigma \wedge \underset{\sim}{\Phi}^m_{n,k}$$

where $\underset{\sim}{\Phi}^m_{n,k}$ is the Σ^m_n sentence of the Lemma above expressing that κ is a $\underset{\sim}{\Pi}^m_n$-k-indescribable cardinal. Since $n > 1$, the right-hand side sentence is a Π^m_{n+1} sentence (in fact, it is Σ^m_n) and hence there is $\lambda < \kappa$ such that

$$\langle V_\lambda, \in, C \cap V_\lambda\rangle \models \sigma \wedge \underset{\sim}{\Phi}^m_{n,k}.$$

Thus, λ is a $\underset{\sim}{\Pi}^m_n$-k-indescribable cardinal and, since $C \cap V_\lambda = C \cap \lambda$ is unbounded in λ, $\lambda \in C$. Thus, the set of $\underset{\sim}{\Pi}^m_n$-k-indescribable cardinals below κ is a $\underset{\sim}{\Pi}_k$-stationary set. The same proof, but using the Π^m_n sentence $\underset{\sim}{\Psi}^m_{n,k}$ of the Lemma above, shows that the set of $\underset{\sim}{\Sigma}^m_n$-$k$-indescribable cardinals below κ is a $\underset{\sim}{\Pi}_k$-stationary set. □

Theorem 4.13. *Let $k, m \geq 1$. Then, for all $n > 1$,*

1. *The least $\underset{\sim}{\Pi}^m_n$-k-indescribable cardinal is not Σ^m_n-k-indescribable.*
2. *The least $\underset{\sim}{\Sigma}^m_n$-$k$-indescribable cardinal is not Π^m_n-k-indescribable.*

The same holds for the least Π^m_n-k-indescribable and the least Σ^m_n-k-indescribable.

Proof. We only prove (1). Suppose that κ is the least $\underset{\sim}{\Pi}^m_n$-k-indescribable cardinal. Towards a contradiction, assume that κ is also Σ^m_n-k-indescribable. Since κ is $\underset{\sim}{\Pi}^m_n$-k-indescribable, $V_\kappa \models \underset{\sim}{\Phi}^m_{n,k}$. And, since κ is Σ^m_n-k-indescribable, there is $\lambda < \kappa$ such that $V_\lambda \models \underset{\sim}{\Phi}^m_{n,k}$. Hence, $\lambda < \kappa$ and λ is $\underset{\sim}{\Pi}^m_n$-k-indescribable. A contradiction. □

Summarizing, we have proved that for all $m, n, k \geq 1$,

$$\underset{\sim}{\sigma}{}^m_n\text{-}k, \underset{\sim}{\pi}{}^m_n\text{-}k < \underset{\sim}{\sigma}{}^m_{n+1}\text{-}k, \underset{\sim}{\pi}{}^m_{n+1}\text{-}k$$

and

$$\sigma^m_n\text{-}k, \pi^m_n\text{-}k < \sigma^m_{n+1}\text{-}k, \pi^m_{n+1}\text{-}k$$

and

$$\underset{\sim}{\sigma}{}^m_n\text{-}k \neq \underset{\sim}{\pi}{}^m_n\text{-}k, \pi^m_n\text{-}k$$

and

$$\underset{\sim}{\pi}{}^m_n\text{-}k \neq \underset{\sim}{\sigma}{}^m_n\text{-}k, \sigma^m_n\text{-}k$$

and

$$\sigma^m_n\text{-}k \neq \pi^m_n\text{-}k$$

Question 3.

1. *What is the relationship between σ^m_n-k and π^m_n-k? Is it $\sigma^m_n\text{-}k < \pi^m_n\text{-}k$ or $\pi^m_n\text{-}k < \sigma^m_n\text{-}k$? What about $\underset{\sim}{\sigma}{}^m_n$-$k$ and $\underset{\sim}{\pi}{}^m_n$-$k$? Recall that, if σ^m_n and π^m_n denote, respectively, the least $\underset{\sim}{\Sigma}{}^m_n$-indescribable and the least Π^m_n-indescribable cardinals then, Moschovakis (see [27]) proved that for every $m > 1$ and every $n \in \omega$, $\sigma^m_n < \pi^m_n$ in L. Hauser (see [11] and [12]) proved that it is consistent that for every $m > 1$ and every $n \in \omega$, $\pi^m_n < \sigma^m_n$.*
2. *It is clear that $\pi^m_n\text{-}k \leq \underset{\sim}{\pi}{}^m_n\text{-}k$. But, is it $\pi^m_n\text{-}k < \underset{\sim}{\pi}{}^m_n\text{-}k$ or $\pi^m_n\text{-}k = \underset{\sim}{\pi}{}^m_n\text{-}k$? What about the relationship between σ^m_n-k and $\underset{\sim}{\sigma}{}^m_n$-$k$?*

References

[1] P. Aczel and W. Richter, "Inductive definitions and analogues of large cardinals" in W. Hodges, Ed. *Conference in Mathematical Logic-London'70 (Proceedings of a Conference; Bedford College, London 1970)*, (Springer, Berlin, 1972).

[2] P. Aczel and W. Richter: "Inductive definitions and reflecting properties of admissible ordinals" in J.E. Fenstad and P.G. Hinman, Eds. *Generalized Recursion Theory* (North-Holland, Amsterdam, 1974).

[3] A. Andretta: "Large cardinals and iteration trees of height ω", *Annals of Pure and Applied Logic*, **54** (1991), 1–15.

[4] J. Baeten: "Filters and ultrafilters over definable subsets of admissible ordinals" in G.H. Müller and M.M. Richter, eds. *Models and sets* (*Aachen, 1983*) (Springer-Verlag, Berlin, 1984).

[5] J. Baeten: *Filters and ultrafilters over definable subsets of admissible ordinals*, (Centrum voor Wiskunde en Informatica, CWI, Amsterdam, 1986).

[6] J. Bagaria and R. Bosch: "Solovay models and forcing extensions", *The Journal of Symbolic Logic*, **69** (2004), 742–766.

[7] J. Bagaria and R. Bosch: "Proper forcing extensions and Solovay models", *Archive for Mathematical Logic*, **43** (2004), 739–750

[8] J. Bagaria and R. Bosch: "Generic absoluteness under projective forcing", forthcoming.

[9] K. Devlin: "Indescribability properties and small large cardinals" in G.H. Müller, A. Oberschel and K. Potthoff, eds. *ISILC Logic Conference. Proceedings of the International Summer Institute and Logic Colloquium, Kiel, 1974* (Springer-Verlag, Berlin, 1975).

[10] F.R. Drake: *Set Theory. An introduction to large cardinals*, (North-Holland, Amsterdam, 1974)

[11] K. Hauser: "Indescribable cardinals and elementary embeddings", *The Journal of Symbolic Logic*, **56** (1991), 439–457.

[12] K. Hauser: "The indescribability of the order of the indescribable cardinals", *Annals of Pure and Applied Logic*, **57** (1992), 45–91.

[13] T. Jech: *Set Theory. The Third Millennium Edition, Revised and Expanded*, (Springer, Berlin, 2003)

[14] A. Kanamori: *The Higher Infinite*, (Springer, Berlin, 1997)

[15] M. Kaufmann: "On existence of Σ_n end extensions" in M. Lerman, J.H. Schmerl and R.I. Soare, Eds. *Logic Year 1979–80* (Springer, Berlin, 1981).

[16] M. Kaufmann: "Blunt and topless end extensions of models of set theory", *The Journal of Symbolic Logic*, **48** (1983), 1053–1073.

[17] M. Kaufman and E. Kranakis: "Definable ultrapowers over admissible ordinals", *Zeitschrift für Mathematische Logik und Grundlagen der Mathematik*, **30** (1984), 97–118.

[18] E. Kranakis: "Reflection and partition properties of admissible ordinals", *Annals of Mathematical Logic*, **22** (1982), 213–242.

[19] E. Kranakis: "Invisible ordinals and inductive definitions", *Zeitschrift für Mathematische Logik und Grundlagen der Mathematik*, **28** (1982), 137–148.

[20] E. Kranakis: "Definable ultrafilters and end extensions of constructible sets", *Zeitschrift für Mathematische Logik und Grundlagen der Mathematik*, **28** (1982), 395–412.

[21] E. Kranakis: "Definable Ramsey and definable Erdös cardinals", *Archiv für Mathematische Logik und Grundlagenforschung*, **23** (1983), 115–128.

[22] E. Kranakis: "Stepping up lemmas in definable partitions" *The Journal of Symbolic Logic*, **49** (1984), 22–31.

[23] E. Kranakis: "Definable partitions and reflection properties for regular cardinals", *Notre Dame Journal of Formal Logic*, **26** (1985), 408–412.

[24] E. Kranakis: "Definable partitions and the projectum", *Zeitschrift für Mathematische Logik und Grundlagen der Mathematik*, **31** (1985), 351–355.

[25] E. Kranakis and I. Phillips, "Partitions and homogeneous sets for admissible ordinals" in G.H. Müller and M.M. Richter, eds. *Models and sets (Aachen, 1983)* (Springer-Verlag, Berlin, 1984)

[26] A. Leshem: "On the consistency of the definable tree property on $\aleph_1$", *The Journal of Symbolic Logic*, **65** (2000), 1204–1214.

[27] Y.N. Moschovakis, "Indescribable cardinals in L", *The Journal of Symbolic Logic*, **41** (1976), 554–555

[28] W. Richter: "Recursively Mahlo ordinals and inductive definitions" in R.O. Gandy and C.M.E. Yates, Eds. *Logic Colloquium'69: Proceedings of the Summer School and Colloquium in Mathematical Logic, Manchester, August of 1969* (Amsterdam, North-Holland, 1971)

[29] W. Richter: "The least Σ^1_2 and Π^1_2 reflecting ordinals" in G. H. Müller, A. Oberschel and K. Potthoff, eds. *ISILC Logic Conference. Proceedings of the International Summer Institute and Logic Colloquium, Kiel, 1974* (Springer-Verlag, Berlin, 1975)

Roger Bosch
Dpto. de Filosofía
Universidad de Oviedo
Tte. Alfonso Martínez, s/n
E-33071 Oviedo (Spain)
e-mail: `roger@uniovi.es`

Set Theory
Trends in Mathematics, 83–120

Real-valued Measurable Cardinals and Well-orderings of the Reals

Andrés Eduardo Caicedo

Abstract. We show that the existence of atomlessly measurable cardinals is incompatible with the existence of well-orderings of the reals in $L(\mathbb{R})$, but consistent with the existence of well-orderings of the reals that are third-order definable in the language of arithmetic. Specifically, we provide a general argument that, starting from a measurable cardinal, produces a forcing extension where $\mathfrak{c}$ is real-valued measurable and there is a Δ^2_2-well-ordering of $\mathbb{R}$. A variation of this idea, due to Woodin, gives Σ^2_1-well-orderings when applied to $L[\mu]$ or, more generally, $\Sigma^2_1(\mathrm{Hom}_\infty)$ if applied to nice inner models, provided enough large cardinals exist in V. We announce a recent result of Woodin indicating how to transform this variation into a proof from large cardinals of the Ω-consistency of real-valued measurability of $\mathfrak{c}$ together with the existence of Σ^2_1-definable well-orderings of $\mathbb{R}$. It follows that if the Ω-conjecture is true, and large cardinals are granted, then this statement can always be forced.

However, we introduce a strengthening of real-valued measurability (real-valued hugeness), show its consistency, and prove that it contradicts the existence of third-order definable well-orderings of $\mathbb{R}$.

This work deals with consistency results within the theory of real-valued measurable cardinals and draws from Chapter 3 of the author's dissertation [11], written at the University of California, Berkeley, under the supervision of John R. Steel and W. Hugh Woodin. The author wishes to thank both of them for their guidance and patience. He also wishes to thank the referee for comments that helped to improve the presentation significantly.

1. Basics of random forcing

This section is included in order to make this paper reasonably self-contained, and we do not claim much originality here other than by way of exposition. The main references for the theory of real-valued measurable cardinals are [38] and [19], see also [32] and [22]. For whatever modest contributions in this section are due to us, see after Fact 1.27. Our notation is mostly standard, see [24], [31], and [26] for whatever notions we leave undefined. ZFC^- denotes ZFC without the Power-Set axiom. We start by defining our basic objects:

Definition 1.1. A cardinal κ is *real-valued measurable*, $\mathsf{RVM}(\kappa)$, iff it is uncountable and there is a κ-additive probability measure $\nu : \mathcal{P}(\kappa) \to [0,1]$ that is null on singletons. We call ν a *witnessing* probability.

A real-valued measurable cardinal κ is *atomlessly measurable* iff there is an atomless witnessing probability ν.

That ν is κ-additive means that whenever $\gamma < \kappa$ and $\langle A_\alpha : \alpha < \gamma \rangle$ is a sequence of disjoint subsets of κ, then

$$\nu\left(\bigcup_{\alpha<\gamma} A_\alpha\right) = \sum_{\alpha<\gamma} \nu(A_\alpha) := \sup\left\{\sum_{\alpha\in\mathcal{F}} \nu(A_\alpha) : \mathcal{F} \subset \gamma \text{ is finite}\right\}.$$

Of course, this implies in particular that only countably many of the A_α have positive measure: Otherwise, for some n,

$$B_n = \left\{\alpha < \gamma : \nu(A_\alpha) > \frac{1}{n+1}\right\}$$

would be infinite, contradicting that ν is bounded above by 1. See also Claim 1.30.

That ν is atomless means that whenever $0 < \nu(A)$, there is $B \subset A$ with $0 < \nu(B) < \nu(A)$. We leave it as an easy exercise for the reader to see that in this case, for any ε with $0 < \varepsilon < \nu(A)$, there is $B \subset A$ with $\nu(B) = \varepsilon$ (or see [26, Lemma 2.6] for a hint on how to proceed).

The following is due to Ulam [43], who also introduced the concept:

Theorem 1.2. *If* $\mathsf{RVM}(\kappa)$, *then κ is either measurable or atomlessly measurable, in which case* $\kappa \leq \mathfrak{c}$. □

Definition 1.3. Let ν be a complete measure on some set X. Then

$$\mathcal{N}_\nu := \{\, Y \subseteq X : \nu(Y) = 0 \,\}$$

is the ideal of *ν-null sets.*

Since $\mathrm{add}(\mathcal{N}_\nu)$ is necessarily a regular cardinal, we have the following useful fact:

Fact 1.4. *Suppose* $\mathsf{RVM}(\kappa)$ *and ν is a witnessing probability. Then:*

1. $\kappa = \mathrm{add}(\mathcal{N}_\nu)$ *is regular.*
2. $\mathcal{N}_\nu$ *is an $\aleph_1$-saturated ideal on κ.* □

Remark 1.5. In fact, if $\kappa \leq \mathfrak{c}$ is real-valued measurable, then κ is weakly Mahlo, the κth weakly Mahlo, etc. Recall that κ is *weakly Mahlo* iff it is uncountable and

$$\{\, \rho < \kappa : \rho \text{ is regular} \,\}$$

is stationary in κ. One can see this as a corollary of Theorem 1.6, see Corollary 1.24. That κ is weakly inaccessible follows immediately from Fact 1.4 and the existence of Ulam matrices on successor cardinals, see [31, Theorem II.6.11].

The following basic characterization is due to Solovay, and will be essential for our arguments:

Theorem 1.6. $\mathsf{RVM}(\kappa)$ *iff there is* $\lambda \geq \omega_1$ *such that*

$$V^{\mathtt{Random}_\lambda} \models \exists j : V \overset{\prec}{\longrightarrow} N, \quad \mathrm{cp}(j) = \kappa,$$

where $\mathtt{Random}_\lambda$ *is the forcing for adding* λ *many random reals.*

If $\kappa \leq \mathfrak{c}$ *and* $\mathsf{RVM}(\kappa)$, *then in addition we can require that*

$$V^{\mathtt{Random}_\lambda} \models {}^\omega N \subseteq N.$$

As far as the author can see, the statement of Theorem 1.6 has not appeared explicitly in print. It can certainly be glimpsed in the arguments of [38] (see especially [38, §6]) and it is well known to experts in the area, see for example [22].

Definition 1.7. Specifically, $\mathtt{Random}_\lambda$ is the collection of Borel subsets of 2^λ, modulo null sets, where the measure φ is defined as follows:

- For $J \subset \lambda$, J finite, and $z \in 2^J$, the *cylinder* determined by J, z is
$$C = C_{J,z} := \{\, x \in 2^\lambda : x \restriction J = z \,\}.$$
For such a C, define $\varphi(C) := 2^{-|J|}$.
- The cylinders generate the product topology on 2^λ. Extend φ to a Borel measure by:
$$\varphi(B) := \inf \left\{ \sum_n \varphi(C_n) : B \subseteq \bigcup_n C_n, \ C_n \text{ a cylinder} \right\}$$
for B a Borel subset of 2^λ.

Remark 1.8. In fact, we can extend φ to a complete measure in the standard way. Some presentations of random forcing assume that we are working with this completion and not just with its restriction to Borel sets. For the purposes of forcing, the resulting Boolean algebras are equivalent, and we can ignore the difference.

Definition 1.9. Let $\mathbb{B}$ be a σ-complete Boolean algebra. A *'probability measure'* on $\mathbb{B}$ is a function $\nu : \mathbb{B} \to [0, 1]$ such that

1. $\nu(a) = 0$ iff $a = \mathsf{0}$.
2. $\nu(\mathsf{1}) = 1$.
3. ν is σ-additive: If $\{\, a_n : n \in \omega \,\}$ is an antichain in $\mathbb{B}$, so $a_n \cdot a_m = \mathsf{0}$ whenever $n \neq m$, then
$$\nu\left(\sum_n^{\mathbb{B}} a_n \right) = \sum_n \nu(a_n).$$

We call $(\mathbb{B}, \nu)$ a *measure algebra.*

Fact 1.10.

1. *For all* λ, $\mathtt{Random}_\lambda$ *is ccc and, therefore, a complete Boolean algebra.*

2. *The map* ν : $\texttt{Random}_\lambda \to [0,1]$ *given by* $\nu([X]) = \varphi(X)$, *where* φ *is as described above and* $[X]$ *denotes the equivalence class of the Borel subset* $X \subseteq 2^\lambda$, *is a 'probability measure', so* $(\texttt{Random}_\lambda, \nu)$ *is a measure algebra.*

Proof. That $\texttt{Random}_\lambda$ is ccc follows from Claim 1.30. Since it is σ-complete and ccc, it is a complete Boolean algebra. A proof of 2 can be found in [18], see Remark 1.11 below. □

Remark 1.11. Given any probability space $(X, \mathcal{P}, \mu)$, $\mathcal{P}/\mathcal{N}_\mu$ can be turned into a measure algebra by exactly the same construction as in *2* of Fact 1.10, see [18]. More significantly,

Fact 1.12. *Any measure algebra is isomorphic (as measure algebra) to one of the form* $\mathcal{P}/\mathcal{N}_\mu$ *for some probability space* $(X, \mathcal{P}, \mu)$, *where* $\mathcal{P}/\mathcal{N}_\mu$ *is a measure algebra with the 'probability measure' described in Fact* 1.10.2. □

This is a consequence of the so-called Loomis-Sikorski theorem (due to von Neumann) stating that any σ-complete Boolean algebra is isomorphic (as a Boolean algebra) to $\Sigma/\mathcal{I}$ for some σ-algebra Σ of subsets of some set X, and some σ-complete ideal $\mathcal{I}$ on Σ. See [27] and [18] for details.

Definition 1.13.

1. For $\mathbb{B}$ a complete Boolean algebra, the *generating number* of $\mathbb{B}$ is $\tau(\mathbb{B}) := \min\{\, |X| : X$ generates $\mathbb{B}$ (as a complete algebra) $\}$.
2. $\mathbb{B}$ is τ*-homogeneous* iff[1] $\tau(\mathbb{B}) = \tau(\mathbb{B}\restriction a)$ for any $a \neq \mathsf{0}$.

Fact 1.14.

1. *If* $\mathbb{B}$ *is a complete Boolean algebra which is homogeneous in the forcing sense*[2], *then* $\mathbb{B}$ *is* τ*-homogeneous.*
2. *Let* λ *be a cardinal. Then* $\texttt{Random}_\lambda$ *is homogeneous. Thus, it is* τ*-homogeneous, and* $\tau(\texttt{Random}_\lambda) = \lambda$. □

Theorem 1.15 (Maharam, see [18, Theorem 3.8]**).** *If* $\mathbb{B}$ *is a complete* τ*-homogeneous measure algebra, then it is isomorphic as a measure algebra to exactly one* $\texttt{Random}_\lambda$ *up to the cardinality of* λ. □

Maharam's theorem is actually much more general than we have stated, but this particular case is all we need.

Fact 1.16. *If* $\mathbb{B} \lessdot \texttt{Random}_\lambda$ *(i.e.,* $\mathbb{B}$ *is a complete subalgebra of* $\texttt{Random}_\lambda$*), then there is a condition* $p \in \mathbb{B}$ *(equivalently, there is a dense set of such conditions) such that* $\mathbb{B}\restriction p \cong \texttt{Random}_\gamma$ *for some* γ. □

Notice that, conversely, if $\gamma < \lambda$, then $\texttt{Random}_\gamma \lessdot \texttt{Random}_\lambda$.

[1] For $p \in \mathbb{B} \setminus \{\mathsf{0}\}$, $\mathbb{B}\restriction p$ is the Boolean algebra of elements of $\mathbb{B}$ below p.

[2] I.e., for any $p, q \in \mathbb{B} \setminus \{\mathsf{0}\}$ there are $\mathsf{0} < r \leq p$ and $\mathsf{0} < s \leq q$ such that $\mathbb{B}\restriction r$ and $\mathbb{B}\restriction s$ are isomorphic.

Remark 1.17. The version of Fact 1.16 for Cohen forcing is true for $\lambda \leq \omega_1$ (see [28] or [6]) but false for $\lambda \geq \omega_2$, see [29].

The following is [32, Theorem 3.13].

Fact 1.18. *Let* $\mathbb{B} \lessdot \mathtt{Random}_\lambda$. *Then*

$$1 \Vdash_{\mathbb{B}} (\mathtt{Random}_\lambda)^V/\mathbb{B} \cong \dot{\mathtt{Random}}_\gamma \text{ for some } \gamma. \qquad \square$$

The following is [32, Lemma 3.12].

Fact 1.19. *If* $W \supseteq V$ *is an outer model and* G *(identified as a subset of* λ*) is* $(\mathtt{Random}_\lambda)^W$*-generic over* W*, then* G *is* $(\mathtt{Random}_\lambda)^V$*-generic over* V*. In particular, for any* $\mathbb{P}$*,* $\mathtt{Random}_\lambda$ *completely embeds into* $\mathbb{P} * \dot{\mathbb{Q}}$*, where* $\dot{\mathbb{Q}}$ *is a* $\mathbb{P}$*-name for* $(\mathtt{Random}_\lambda)^{V^{\mathbb{P}}}$. $\square$

Proof of Solovay's Theorem 1.6. ($\Leftarrow$) Suppose

$$V^{\mathtt{Random}_\lambda} \models \exists j : V \xrightarrow{\prec} N, \quad \mathrm{cp}(j) = \kappa.$$

Let $\varphi : \mathtt{Random}_\lambda \to [0,1]$ be the 'probability measure' associated to $\mathtt{Random}_\lambda$. In V, we want to define a probability measure on subsets of κ. Fix names j and N such that

$$[\![N \text{ is a transitive inner model and } j : V \xrightarrow{\prec} N, \mathrm{cp}(j) = \kappa]\!] = 1.$$

For $A \subseteq \kappa$, let $\nu(A) := \varphi[\![\kappa \in j(A)]\!]$, so $\nu : \mathcal{P}(\kappa) \to [0,1]$. It is easy to verify that ν is as wanted[3].

($\Rightarrow$) Suppose $\mathsf{RVM}(\kappa)$. Let ν be a witness, and let $\mathbb{B}_\nu = \mathcal{P}(\kappa)/\mathcal{N}_\nu$. Since $\mathcal{N}_\nu$ is $\aleph_1$-saturated, $\mathbb{B}_\nu$ is complete (by the Smith-Tarski theorem [26, Proposition 16.5]), and we may assume (by reducing to a subset if necessary) that $\mathbb{B}_\nu \cong \mathtt{Random}_\lambda$ for some λ: Necessarily, for some $X \subseteq \kappa$, $X \notin \mathcal{N}_\nu$, we must have that $\mathcal{P}(X)/\mathcal{N}_\nu$ is τ-homogeneous because τ is a decreasing ordinal-valued function and therefore eventually constant. By replacing ν with $\hat{\nu} : Y \mapsto \nu(X \cap Y)$, we may as well assume $X = \kappa$. That $\mathbb{B}_\nu \cong \mathtt{Random}_\lambda$ for some λ now follows from Maharam's Theorem 1.15. If $\kappa \leq \mathfrak{c}$ then $|\mathbb{B}_\nu| \geq \mathfrak{c}$.

Let G be $\mathbb{B}_\nu$-generic over V. Then G is essentially a V-ultrafilter on κ, and we can form $\pi : V \to \mathrm{Ult}(V, G)$ in $V[G]$. But the saturation of $\mathcal{N}_\nu$ ensures that the ultrapower is well founded, and therefore isomorphic to a transitive class N. Let $j : V \xrightarrow{\prec} N$ denote the corresponding embedding, coming from π via the Mostowski collapse. Then $\mathrm{cp}(j) = \kappa$, and since $\mathbb{B}_\nu \cong \mathtt{Random}_\lambda$, we are done, except for the claim that $\lambda \geq \omega_1$. For this, see [21, §2], where it is shown that in fact $\lambda \geq \kappa^+$.

See Fact 1.20 and Remark 1.21 for the proof that ${}^\omega N \subseteq N$. $\square$

Fact 1.20. *Suppose* $\mathsf{RVM}(\kappa)$ *and* $\mathtt{Random}_\lambda$*,* j *and* N *are as in Solovay's theorem. Then* $\mathbb{R}^N = \mathbb{R}^{V^{\mathtt{Random}_\lambda}}$.

[3]Those uncomfortable with our use of proper classes are advised to consult [38] for a first-order treatment.

Proof. This is standard from the theory of saturated ideals: In fact, using the notation from the theorem, if G is $\mathbb{B}_\nu$-generic over V, then $V[G] \models {}^\omega N \subseteq N$. $\square$

Remark 1.21. A strong version of Fact 1.20 is that we can in fact assume that ${}^\kappa N \subset N$:

Use notation as above. We claim first that for every term $\dot{b}$ in $\mathtt{Random}_\lambda$ for an element of the ground model V, there is a function $f \in V$ such that

$$[\![[f]_N = j(\dot{b})]\!].$$

In effect, suppose $[\![\dot{b} \in V]\!] = \mathtt{1}$ and let $A = \{ a_\xi : \xi \text{ is a possible value of } \dot{b} \} \in V$ be a maximal antichain in $\mathcal{P}(\kappa)/\mathcal{N}_\nu$, so for each $a_\xi \in A$, $a_\xi \Vdash \dot{b} = \xi$. Since $\mathcal{N}_\nu$ is ω_1-saturated, A is countable, so we may assume A is a partition of κ, i.e., $a_\xi \subseteq \kappa$ for each $a_\xi \in A$ and $a_\xi \cap a_\zeta = \emptyset$ whenever $\xi \neq \zeta$. In V, define $f : \kappa \to V$ by

$$f(\eta) = \text{the unique } \xi \text{ such that } \eta \in a_\xi.$$

Then $[\![[f]_N = j(\dot{b})]\!] = \mathtt{1}$, since for any ξ, $a_\xi \Vdash [f]_N = [c_\xi]_N$, so $\mathtt{1} \Vdash [f]_N = [c_{\dot{b}}]_N = j(\dot{b})$. This easily leads to a proof that, in $\mathtt{Random}_\lambda$, N is closed under ω-sequences and, in fact, under sequences of length $< \kappa$.

Without loss of generality, the null ideal $\mathcal{N}_\nu$ is normal (see Corollary 1.23), so the identity represents κ in the ultrapower N. Assuming normality of $\mathcal{N}_\nu$, we prove that it is in fact closed under κ-sequences.

Given any term $\langle \tau_\alpha : \alpha < \kappa \rangle$ for a κ-sequence in $V[G]$ of elements of N, let $\langle \rho_\alpha : \alpha < \kappa \rangle$ be a term for a κ-sequence of functions in V such that for each α,

$$[\![\rho_\alpha \in V \cap {}^\kappa V \text{ and } \tau_\alpha = [\rho_\alpha]_N]\!] = \mathtt{1}$$

and then a sequence $\langle f_\alpha : \alpha < \kappa \rangle$ of functions $f_\alpha : \kappa \to V$ can be chosen in V so $[\![[f_\alpha]_N = j(\rho_\alpha)]\!] = \mathtt{1}$. But $[\![[\rho_\alpha]_N = j(\rho_\alpha)(\kappa)]\!] = \mathtt{1}$.

Letting $g : \kappa \to V$ be the function in V given by $g(\beta) = \langle f_\alpha(\beta) : \alpha < \beta \rangle$ for all $\beta < \kappa$ then, in $V[G]$,

$$[g]_N = j(g)(\kappa) = \langle j(f_\alpha)(\kappa) : \alpha < \kappa \rangle = \langle [f_\alpha]_N : \alpha < \kappa \rangle .$$

Hence, ${}^\kappa N \subseteq N$. In particular, $\mathcal{P}^{V[G]}(\kappa) \subseteq N$.

Remark 1.22. Suppose $\mathsf{RVM}(\mathfrak{c})$ and ν is a witness such that $\mathcal{P}(\mathfrak{c})/\mathcal{N}_\nu$ is homogeneous. As mentioned above, it follows that $\mathcal{P}(\mathfrak{c})/\mathcal{N}_\nu \cong \mathtt{Random}_\lambda$ for some $\lambda \geq \omega_1$. It is a result of Gitik and Shelah that in fact $\lambda = 2^\mathfrak{c}$, see [22, Theorem 1.1].

Solovay's characterization allows for easy proofs of several results of the classical theory of real-valued measurability. For example:

Corollary 1.23 (Solovay [38]). *If* $\mathsf{RVM}(\kappa)$ *then there is a witnessing probability* ν *such that* $\mathcal{N}_\nu$ *(see Definition* 1.3*) is a normal ideal.*

Proof. Suppose $\mathsf{RVM}(\kappa)$. Let λ be such that in $V^{\mathtt{Random}_\lambda}$ there is $j : V \to N$ with $\mathrm{cp}(j) = \kappa$, and define ν as in the proof of Theorem 1.6. Then $\mathcal{N}_\nu$ is a normal ideal: Suppose $\langle A_\alpha : \alpha < \kappa \rangle$ is a sequence of subsets of κ such that $[\![\kappa \in j(A_\alpha)]\!] = 0$ for all α. Then certainly $[\![\exists \alpha < \kappa \, (\kappa \in j(A_\alpha))]\!] = 0$, i.e., $[\![\kappa \in j\big(\nabla_{\alpha<\kappa} A_\alpha\big)]\!] = 0$. $\square$

Corollary 1.24 (Solovay [38]). *If* $\mathsf{RVM}(\kappa)$ *then* κ *is weakly Mahlo, the weakly Mahlo cardinals are stationary below* κ*, etc.*

Proof. Suppose $\mathsf{RVM}(\kappa)$. Let λ, j, ν be as in the proof of Theorem 1.6. If κ is not weakly Mahlo, then $A = \{\alpha < \kappa : \text{cf}(\alpha) < \alpha\}$ contains a club in κ and therefore $[\![\kappa \in j(A)]\!] = 1$, i.e., $\text{cf}(\kappa) < \kappa$ in $V^{\mathtt{Random}_\lambda}$. But this is impossible, since $\mathtt{Random}_\lambda$ is ccc. The same argument shows that the non-weakly Mahlo cardinals are in $\mathcal{N}_\nu$, etc. □

Corollary 1.25 (Silver, see [26, Proposition 7.12]). *If* $\mathsf{RVM}(\kappa)$ *then the tree property holds for* κ*.*

Proof. Suppose $\mathsf{RVM}(\kappa)$, and let ν be a witnessing probability. Suppose $\mathcal{T}$ is a κ-tree. Without loss, $\mathcal{T} = (\kappa, <_\mathcal{T})$. As usual, we will identify $\mathcal{T}$ and its levels $\mathcal{T}_\alpha$, $\alpha < \kappa$, with the underlying subsets of κ. Our convention is that trees grow upward, so if 0 is the root of $\mathcal{T}$, $0 <_\mathcal{T} a$ for any other $a \in \mathcal{T}$, etc. Let λ be such that in $V^{\mathtt{Random}_\lambda}$ there is $j : V \to N$ with $\text{cp}(j) = \kappa$. Work in $V^{\mathtt{Random}_\lambda}$.

Then $j(\mathcal{T}) \restriction \kappa = \mathcal{T}$. Let $\mu = j(\nu)$, so μ witnesses $\mathsf{RVM}(j(\kappa))$ inside N. For $\alpha < j(\kappa)$, let $A_\alpha = \{\beta : \alpha <_{j(\mathcal{T})} \beta\}$. Since μ is $j(\kappa)$-complete, $\mu(\mathcal{T}) = 0$ and there is some $\alpha \in j(\mathcal{T})_\kappa$ such that $\mu(A_\alpha) > 0$.

Let $b = \{\beta \in \mathcal{T} : \beta <_{j(\mathcal{T})} \alpha\}$, and let $\langle b_\gamma : \gamma < \kappa \rangle$ be its $<_\mathcal{T}$-increasing enumeration. Then $\mu(A_{b_\gamma}) \geq \mu(A_{b_\rho})$ whenever $\gamma < \rho$. Since $\kappa > \omega$, for some $\rho < \kappa$ we must have $\mu(A_{b_\rho}) = \mu(A_{b_\tau})$ for all $\tau > \rho$.

For $\beta < \kappa$, $b_\rho <_\mathcal{T} \beta$, let $B_\beta = \{\gamma < \kappa : \beta <_\mathcal{T} \gamma\}$. Notice that $\mu(A_\beta) = j(\nu(B_\beta)) = \nu(B_\beta)$ for any such β. Let $\varepsilon = \nu(B_{b_\rho})$. Then $\forall \beta >_\mathcal{T} b_\rho$, either $\nu(B_\beta) = \varepsilon$, or $\nu(B_\beta) = 0$ (If $0 < \nu(B_\beta) < \varepsilon$, and $\beta \in \mathcal{T}_\gamma$, then $\beta \neq b_\gamma$ and $B_\beta \cap B_{b_\gamma} = \emptyset$. But then $\nu(B_{b_\gamma}) \leq \nu(B_{b_\rho} \setminus B_\beta) < \varepsilon$, a contradiction.)

Let $\underline{b} = \{\beta : \beta \leq_\mathcal{T} b_\rho \text{ or } (b_\rho <_\mathcal{T} \beta \text{ and } \nu(B_\beta) = \varepsilon)\}$. Then $b = \underline{b} \in V$ is a κ-branch through $\mathcal{T}$. □

Stripping away the fat from the above argument allows us to weaken the hypothesis of Corollary 1.25 to the existence of a λ-saturated ideal on κ for some $\lambda < \kappa$ (see [26, Proposition 16.4]).

Corollary 1.26 (Kunen, see [19, Theorem 5N]). *If* $\mathsf{RVM}(\mathfrak{c})$ *then* $\diamondsuit_\mathfrak{c}$ *holds.* □

This follows from applying to the context of real-valued measurability the standard proof of $\diamondsuit_\kappa$ for κ measurable, we leave the details to the interested reader.

The main result on preservation of real-valued measurability is the following.

Fact 1.27 (Solovay [38, Theorem 7]). *Suppose* $\mathsf{RVM}(\kappa)$*. Then* κ *stays real-valued measurable after forcing with any* $\mathtt{Random}_\lambda$ *or more generally (by Maharam's theorem), with any measure algebra.* □

Remark 1.28. I do not know if Solovay's characterization allows for an 'elementary embeddings' proof of Fact 1.27: If $\mathsf{RVM}(\kappa)$ and λ, j, N are as in the proof of

Theorem 1.6, so $V^{\mathtt{Random}_\lambda} \models j : V \xrightarrow{\prec} N$, $\mathrm{cp}(j) = \kappa$, then if γ is an ordinal such that $[\![\gamma > j(\kappa)]\!] = \mathbb{1}$, say, it is not clear how to lift j to an embedding

$$\hat{j} : V^{\mathtt{Random}_\gamma} \xrightarrow{\prec} N^{\mathtt{Random}_{j(\gamma)}}$$

in $V^{\mathtt{Random}_\lambda * \mathtt{Random}_\mu}$ for some appropriate μ, which seems to be the natural way using elementary embeddings of arguing about Fact 1.27. Even if this is possible, Solovay's original argument from [38] would not be superseded; for example, Solovay's argument indicates natural ways in which new measures can be produced from the ones witnessing $\mathsf{RVM}(\kappa)$. For more on this approach, consider Kunen's proof that $\mathsf{RVM}(\kappa)$ implies the partition relation $\kappa \to (\kappa, \lambda)^2$ for any $\lambda < \omega_1$. See [18] for this argument.

1.1. Absolutely ccc forcing

We now argue that if $\mathbb{P}$ is ccc and $\mathbb{F} = \mathtt{Random}_\lambda$ for some λ, then $\mathbb{P}$ is still ccc in $V^{\mathbb{F}}$.

Definition 1.29. $\mathbb{Q} \in V$ is *absolutely ccc* iff for all outer models $W \supseteq V$, $W \models \mathbb{Q}$ is ccc.[4]

For example $\mathtt{Coll}(\omega, <\omega_1)$, $\mathtt{Add}(\omega, 1)$ (the forcing for adding one Cohen real), and any σ-centered poset are absolutely ccc. The class of absolutely ccc posets is closed under finite support products and finite support iterations[5]. The following example is slightly more interesting, and we will have several occasions to use it.

Claim 1.30. *All measure algebras, in particular all* $\mathtt{Random}_\lambda$*, are absolutely ccc.*

Proof. Let $\mathbb{P} = (\mathbb{B}, \nu) \in V$ be a measure algebra, and let $W \supseteq V$ be an outer model. Let $\omega_1 = \omega_1^W$.

Suppose in W that $\langle b_\alpha : \alpha < \omega_1 \rangle$ is an ω_1-antichain in $\mathbb{B} \setminus \{\mathbb{0}\}$. Then we can assume that for some $n > 0$, $\nu(b_\alpha) > 1/n$ for all α. This is a contradiction: For any $N \in \mathbb{N}$ the sequence $\langle b_m : m < N \rangle$ is in V and since the b_α form an antichain, we have that $\nu(\sum^{\mathbb{B}}_{m<N} b_m) = \sum_{m<N} \nu(b_m) > \frac{N}{n} > 1$ if N is sufficiently large. □

Claim 1.31. *If* $\mathbb{P}$ *is ccc and* $\mathbb{Q}$ *is absolutely ccc, then* $V^{\mathbb{Q}} \models \mathbb{P}$ is ccc.

Proof. Since $\mathbb{P} \times \mathbb{Q} \cong \mathbb{P} * \check{\mathbb{Q}} \cong \mathbb{Q} * \check{\mathbb{P}}$, it suffices to see that $V^{\mathbb{P}} \models \mathbb{Q}$ is ccc, but this holds by hypothesis. □

Corollary 1.32. *Let* $\mathbb{F} = \mathtt{Random}_\lambda$ *and let* $\mathbb{P}$ *be ccc. Then* $\mathbb{P}$ *is ccc in* $V^{\mathbb{F}}$. □

Corollary 1.33. *The existence of atomlessly measurable cardinals is independent of the existence of Suslin trees.*

---

[4] A possible metatheory in which this definition takes place is Morse-Kelley. For a ZFC rendering, restrict the outer models to those of the form $V^{\mathbb{F}}$ for $\mathbb{F} \in V$ a poset.

[5] For products, this follows from [31, Theorem II.1.9]. Since the finite support iteration of ccc posets is again ccc, the result for iterations follows easily from the definition of absolutely ccc, because if $\mathbb{P} \in V$ is the finite support iteration of a family $\langle \mathbb{P}_\alpha, \dot{\mathbb{Q}}_\alpha : \alpha < \lambda \rangle$, then in any outer model $W \supseteq V$, $\mathbb{P}$ densely embeds into the finite support iteration (in the sense of W) of $\langle \mathbb{P}_\alpha, \dot{\mathbb{Q}}_\alpha : \alpha < \lambda \rangle$.

Proof. Let κ be measurable, and suppose S is a Suslin tree. Then

$$1 \Vdash_{\mathtt{Random}_\kappa} \mathsf{RVM}(\mathfrak{c}) \text{ and } S \text{ is ccc},$$

by Corollary 1.32. Thus, $V^{\mathtt{Random}_\kappa} \models$ There is a Suslin tree.

The other direction is immediate from a result of Laver (see [8, Theorem 3.2.31].) Namely, if $\mathsf{MA}_{\aleph_1}$ holds then for any κ,

$$V^{\mathtt{Random}_\kappa} \models \text{Every Aronszajn tree is special.}$$

In particular, if κ is measurable and MA holds, then $V^{\mathtt{Random}_\kappa}$ is a model of $\mathsf{RVM}(\mathfrak{c})$ where there are no Suslin trees. □

More interesting consequences of the fact that $\mathtt{Random}_\lambda$ is absolutely ccc are explored throughout the paper.

Stronger versions of the following theorem can be obtained, but this suffices for our purposes. Notice the particular case where κ is measurable, so $\mathbb{B}_\nu$ is trivial and $G \in V$.

Theorem 1.34. *Suppose* $\mathsf{RVM}(\kappa)$ *and let* ν *be a witnessing probability such that* $\mathbb{B}_\nu = \mathcal{P}(\kappa)/\mathcal{N}_\nu$ *is homogeneous. Let* G *be* $\mathbb{B}_\nu$*-generic over* V*, and in* $V[G]$ *let* $j : V \to N$ *be the associated generic embedding. Then the forcing* $j(\mathtt{Random}_\kappa)/\mathtt{Random}_\kappa$ *is isomorphic in* $V[G][H]$ *to* $\mathtt{Random}_{j(\kappa)}$*, where* H *is* $\mathtt{Random}_\kappa$*-generic over* $V[G]$.

Proof. Start by noticing that $(\mathtt{Random}_\kappa)^V \in N$. In N,

$$j(\mathtt{Random}_\kappa) = \mathtt{Random}_{j(\kappa)},$$

so $\mathtt{Random}_\kappa \lessdot j(\mathtt{Random}_\kappa)$, and the quotient forcing makes sense. Let $\underline{H}$ be the canonical $\mathtt{Random}_\kappa$ name for the generic filter and recall that, by definition, $j(\mathtt{Random}_\kappa)/\mathtt{Random}_\kappa$ is (a $\mathtt{Random}_\kappa$ name for) the forcing

$$\mathbb{P} = \{\, q \in j(\mathtt{Random}_\kappa) : q \text{ is compatible with every } p \in \underline{H}\,\}.$$

Consequently, fix H a $\mathtt{Random}_\kappa$ generic over $V[G]$ and therefore over N, and work in $V[G][H]$.

- *In* $N[H]$, $\mathbb{P} \cong \mathtt{Random}_{j(\kappa)}$.
  By Fact 1.18.
- *In* $V[G][H]$, $\mathbb{P}$ *is a* σ*-complete homogeneous boolean algebra.*
  Recall that ${}^\omega N \subset N$, and therefore (by the ccc of $\mathtt{Random}_\kappa$) ${}^\omega N[H] \subset N[H]$, from which σ-completeness in $V[G][H]$ follows. Homogeneity is clear, since $\mathbb{P}$ is already homogeneous in $N[H]$.
- *In* $V[G][H]$, $\mathbb{P}$ *is a complete measure algebra.*
  The 'probability measure' witnessing $\mathbb{P}$ is a measure algebra in $N[H]$ is a 'probability measure' in $V[G][H]$, since $N[H]$ is closed under ω-sequences. Hence, $\mathbb{P}$ is a measure algebra. It is ccc, by Claim 1.30. Completeness follows.
- *In* $V[G][H]$, $\mathbb{P}$ *is isomorphic to some* $\mathtt{Random}_\rho$ *and, in fact,*

  $$\mathbb{P} \cong \mathtt{Random}_{|j(\kappa)|}.$$

  This follows now from Maharam's theorem. This completes the proof. □

For a generalization, see the first paragraph of the proof of Claim 3.5.

Theorem 1.34 will prove useful in the following sections, where we obtain the consistency of a third-order definable well-ordering of $\mathbb{R}$ together with real-valued measurability of the continuum. That we cannot improve the complexity of this well-ordering in a straightforward fashion is the content of Theorem 2.5 below.

2. Third-order definability

Recall that $\mathsf{HOD}_{\mathbb{R}}$ denotes the class of sets hereditarily ordinal definable using the elements of $\mathbb{R}$ as parameters. $\mathtt{Add}(\kappa, \lambda)$ is the standard forcing for adding λ many Cohen subsets of κ.

Lemma 2.1. *Let G be $\mathbb{F}$-generic over V, where $\mathbb{F} = \mathtt{Add}(\omega, \lambda)$ or $\mathbb{F} = \mathtt{Random}_\lambda$, $\lambda \geq \omega_1$. Let $\mathbb{R} = \mathbb{R}^{V[G]}$. Then, in $V[G]$, there is $\mathbb{R}_0 \subset \mathbb{R}$ and a nontrivial elementary embedding $j : \mathsf{HOD}_{\mathbb{R}_0} \xrightarrow{\prec} \mathsf{HOD}_{\mathbb{R}}$ such that $j \restriction \mathrm{ORD} = \mathrm{id}$.*

Corollary 2.2. *Let $\mathbb{F} = \mathtt{Add}(\omega, \lambda)$ or $\mathbb{F} = \mathtt{Random}_\lambda$ where $\lambda \geq \omega_1$. Then in $V^{\mathbb{F}}$, $\mathsf{HOD}_{\mathbb{R}} \models \neg\mathsf{AC}$ and therefore no relation in $\mathsf{HOD}_{\mathbb{R}}$ defines a well-ordering of $\mathbb{R}$. In particular, $V^{\mathbb{F}} \models L(\mathbb{R}) \models \neg\mathsf{AC}$.*

Proof. In $V^{\mathbb{F}}$ there is a transitive class $N \neq \mathsf{HOD}_{\mathbb{R}}$ and an elementary embedding $j : N \xrightarrow{\prec} \mathsf{HOD}_{\mathbb{R}}$ that does not move the ordinals. It follows from [26, Proposition 5.1] that the Axiom of Choice fails in N and therefore in $\mathsf{HOD}_{\mathbb{R}}$. Since a well-ordering of $\mathbb{R}$ in $\mathsf{HOD}_{\mathbb{R}}$ would induce a well-ordering of $\mathsf{HOD}_{\mathbb{R}}$ in $\mathsf{HOD}_{\mathbb{R}}$, the result follows. □

Remark 2.3. Corollary 2.2 is known, although the proof presented here seems to be new. See for example [31, Exercises VII.E].

Proof of Lemma 2.1. Let G be $\mathbb{F}$-generic over V, where $\mathbb{F}$ is as in the statement of the lemma. By standard arguments (by Maharam's Theorem 1.15 for the case $\mathbb{F} = \mathtt{Random}_\lambda$), $G \cong G_0 \times G_1$, where G_0 is $\mathbb{F}$-generic over V and G_1 is $\mathbb{F}^{V[G_0]}$-generic over $V[G_0]$. Let $\mathbb{R}_0 = \mathbb{R}^{V[G_0]}$ and $\mathbb{R}_1 = \mathbb{R}^{V[G]}$. In $V[G]$ we define a nontrivial $j : \mathsf{HOD}_{\mathbb{R}_0} \xrightarrow{\prec} \mathsf{HOD}_{\mathbb{R}_1}$ such that $j \restriction \mathrm{ORD} = \mathrm{id}$.

For this, notice that any $x \in \mathsf{HOD}_{\mathbb{R}_i}$, $i = 0, 1$, has the form $\tau(\vec{r}, \vec{\alpha})$ where $\vec{r} \in \mathbb{R}_i$, $\vec{\alpha} \in \mathrm{ORD}$, and τ is some term in the language of $\mathsf{HOD}_{\mathbb{R}}$.[6]

Define j by
$$j\big(\tau(\vec{r}, \vec{\alpha})^{\mathsf{HOD}_{\mathbb{R}_0}}\big) = \tau(\vec{r}, \vec{\alpha})^{\mathsf{HOD}_{\mathbb{R}_1}}.$$
We claim j works.

[6] This language expands the language of set theory by closing under *weak Skolem functions*, i.e., those giving definable terms, so for $\varphi(x, y)$ a formula and z a set, $\tau_\varphi(z)$ is defined iff $\exists! x\, \varphi(x, z)$, and $\tau_\varphi(z) = u$ iff $\varphi(u, z)$. We cannot simply use (definable) Skolem functions, since AC fails in $\mathsf{HOD}_{\mathbb{R}}$. If the reader does not want to bother formalizing this language, it suffices that for every $x \in \mathsf{HOD}_{\mathbb{R}}$ there is a formula $\phi(v_1, \vec{v}_2, \vec{v}_3)$ in the language of set theory, and there are reals $\vec{r}$ and ordinals $\vec{\alpha}$ such that
$$\mathsf{HOD}_{\mathbb{R}} \models x = \{\, y : \phi(y, \vec{r}, \vec{\alpha})\,\}. \qquad (\dagger)$$
The reader should have no problem using (†) to translate our use of terms into standard notation.

Let $\varphi(v_0, \dots, v_n)$ be any formula, let $\tau_0(\vec{v}_0, \vec{v}_1), \dots, \tau_n(\vec{v}_0, \vec{v}_1)$ be terms, and let $x_0, \dots, x_n \in \mathsf{HOD}_{\mathbb{R}_0}$ be given by $x_i = \tau_i(\vec{r}_i, \vec{\alpha}_i)^{\mathsf{HOD}_{\mathbb{R}_0}}$. By composing each τ_i with some projections and some recursive surjections $\pi_0 : \mathbb{R} \to \mathbb{R}^{<\omega}$ and $\pi_1 : \mathrm{ORD} \to \mathrm{ORD}^{<\omega}$, we may assume $\vec{r}_i = r$, $\vec{\alpha}_i = \alpha$ for all i. Let $\psi(v_0, v_1) \equiv \varphi(\tau_0(v_0, v_1), \dots, \tau_n(v_0, v_1))$ and $\mu(v_0, v_1) \equiv \mathsf{HOD}_{\mathbb{R}} \models \psi(v_0, v_1)$.

The whole point of the argument is that there is a set $X \in \mathcal{P}_{\omega_1}(\lambda)$ such that $r \in V[G_0 \restriction X]$, and there are $\mathbb{F}^{V[G_0 \restriction X]}$-generics over $V[G_0 \restriction X]$, $\mathcal{G}_0$ and $\mathcal{G}_1$, such that $V[G_0 \restriction X][\mathcal{G}_0] = V[G_0]$ and $V[G_0 \restriction X][\mathcal{G}_1] = V[G]$.

Then

$$\begin{aligned} \mathsf{HOD}_{\mathbb{R}_0} \models \psi(r, \alpha) &\iff V[G_0] \models \mu(r, \alpha) \\ &\iff \exists p \in \mathcal{G}_0 \big(V[G_0 \restriction X] \models p \Vdash_{\mathbb{F}} \mu(\check{r}, \check{\alpha})\big) \\ &\overset{(*)}{\iff} V[G_0 \restriction X] \models \mathbb{1}_{\mathbb{F}} \Vdash_{\mathbb{F}} \mu(\check{r}, \check{\alpha}) \\ &\overset{(*)}{\iff} \exists q \in \mathcal{G}_1 \big(V[G_0 \restriction X] \models q \Vdash_{\mathbb{F}} \mu(\check{r}, \check{\alpha})\big) \\ &\iff V[G] \models \mu(r, \alpha) \\ &\iff \mathsf{HOD}_{\mathbb{R}_1} \models \psi(r, \alpha), \end{aligned}$$

where $(*)$ holds by the weak homogeneity of $\mathbb{F}$. Recall that a forcing $\mathbb{P}$ is *weakly homogeneous* (see [26, before Proposition 10.19]) iff for all $p, q \in \mathbb{P}$ there is an automorphism π of $\mathbb{P}$ such that $\pi(p)$ is compatible with q. Clearly, $\mathbb{F}$ is weakly homogeneous. It is a basic result in the theory of forcing ([26, Proposition 10.19]) that if $\mathbb{P}$ is weakly homogeneous, $\phi(v_1, \dots, v_n)$ is a formula in the forcing language, all of its free variables displayed, and $x_1, \dots, x_n \in V$, then either $\mathbb{1} \Vdash_{\mathbb{P}} \varphi(\check{x}_1, \dots, \check{x}_n)$ or else $\mathbb{1} \Vdash_{\mathbb{P}} \neg\varphi(\check{x}_1, \dots, \check{x}_n)$.

The chain of equivalences shown above implies immediately that j is well defined and elementary. By definition, $j \restriction \mathrm{ORD} = \mathrm{id}$, and we are done. □

Remark 2.4. Notice that with the same notation as above,

$$j \restriction L(\mathbb{R}_0) : L(\mathbb{R}_0) \xrightarrow{\prec} L(\mathbb{R}_1).$$

Essentially the same argument shows that if $\omega_1 \leq \lambda_1 \leq \lambda_2$, H_1 is $\mathtt{Random}_{\lambda_1}$- (respectively, $\mathtt{Add}(\omega, \lambda_1)$-) generic over V, and H_2 is $\mathtt{Random}_{\lambda_2}$- (respectively, $\mathtt{Add}(\omega, \lambda_2)$-) generic over $V[H_1]$, then in $V[H_1][H_2]$ there is a nontrivial embedding

$$j : L(\mathbb{R}^{V[H_1]}) \xrightarrow{\prec} L(\mathbb{R}^{V[H_1][H_2]})$$

such that $j \restriction \mathrm{ORD} = \mathrm{id}$. To see this, it suffices to argue that if $\varphi(x, y)$ is a formula, r is a real, α is an ordinal, and $\dot{\mathbb{R}}$ is a term (for the appropriate forcing) for the reals of the generic extension, then $\mathbb{1} \Vdash_{\mathtt{Random}_{\lambda_1}} L(\dot{\mathbb{R}}) \models \varphi(r, \alpha)$ iff $\mathbb{1} \Vdash_{\mathtt{Random}_{\lambda_2}} L(\dot{\mathbb{R}}) \models \varphi(r, \alpha)$.

Suppose $|\lambda_1| < |\lambda_2|$, and let $\mathbb{P}$ be the forcing for collapsing λ_2 to λ_1 with countable conditions, so $\mathbb{P}$ does not add any reals. By Fact 1.19, if $\mathcal{G}$ is $\mathtt{Random}_{\lambda_1}$-generic over $V^{\mathbb{P}}$ then $\mathcal{G}$ is $\mathtt{Random}_{\lambda_1}$-generic over V. The same holds for $\mathtt{Random}_{\lambda_2}$-generic filters, and we are done by weak homogeneity: In $V^{\mathbb{P}}$, $\mathtt{Random}_{\lambda_1}$ and

$\texttt{Random}_{\lambda_2}$ are equivalent. Let H be $\mathbb{P}$-generic over V and let G be $\texttt{Random}_{\lambda_1}$-generic over $V[H]$. Then the reals of $V[H][G]$ and the reals of $V[G]$ coincide. But G is also $\texttt{Random}_{\lambda_2}$-generic over $V[H]$.

Let r be a real in $V[G]$. Recall that $\texttt{Random}_\lambda$ and $\texttt{Random}_\omega * \texttt{Random}_\lambda$ are isomorphic for any cardinal λ, by Maharam's Theorem 1.15, so we may write $G \cong G_0 \times G_1$ where $r \in V[G_0]$, G_0 is generic over V for a forcing isomorphic to $\texttt{Random}_\omega$, and G_1 is generic over $V[G_0]$ for a forcing isomorphic to $\texttt{Random}_{\lambda_1}$. Let α be an ordinal and let φ be a formula. It follows that

$$\begin{aligned} V[H][G] \models L(\mathbb{R}) \models \varphi(r,\alpha) &\iff V[G] \models L(\mathbb{R}) \models \varphi(r,\alpha) \\ &\iff \exists p \in G_1 \left(V[G_0] \models p \Vdash_{\texttt{Random}_{\lambda_1}} \varphi(\check{r},\check{\alpha})\right) \\ &\iff V[G_0] \models \mathbb{1} \Vdash_{\texttt{Random}_{\lambda_1}} \varphi(\check{r},\check{\alpha}), \end{aligned}$$

and exactly the same argument with $\texttt{Random}_{\lambda_2}$ instead of $\texttt{Random}_{\lambda_1}$ shows that

$$V[H][G] \models L(\mathbb{R}) \models \varphi(r,\alpha) \iff V[G_0] \models \mathbb{1} \Vdash_{\texttt{Random}_{\lambda_2}} \varphi(\check{r},\check{\alpha})$$

and, therefore,

$$V[G_0] \models \mathbb{1} \Vdash_{\texttt{Random}_{\lambda_1}} \varphi(\check{r},\check{\alpha}) \iff V[G_0] \models \mathbb{1} \Vdash_{\texttt{Random}_{\lambda_2}} \varphi(\check{r},\check{\alpha}),$$

as we needed to show (notice we can ignore G_0 if $r \in V$).

The argument for Cohen forcing is identical.

Theorem 2.5. *If $\kappa \leq \mathfrak{c}$ and $\mathsf{RVM}(\kappa)$ then no well-ordering of $\mathbb{R}$ belongs to $L(\mathbb{R})$.*

Proof. The argument is standard. Assume by contradiction that $\mathsf{RVM}(\kappa)$ and there is $\varphi(x,y,z,w)$ a formula in the language of $L(\mathbb{R})$ such that for some real t and ordinal α, the relation between reals

$$r < s \quad\iff\quad L(\mathbb{R}) \models \varphi(r,s,t,\alpha)$$

is a well-ordering of $\mathbb{R}$. The least such α is definable in $L(\mathbb{R})$, so there is such a formula φ' all of whose parameters are reals. Let λ be as above, so in $V^{\texttt{Random}_\lambda}$ there is an embedding $j : V \stackrel{\prec}{\longrightarrow} N$ such that $\mathrm{cp}(j) = \kappa$ and ${}^\omega N \subseteq N$. Then

$$j{\restriction}L(\mathbb{R})^V : L(\mathbb{R})^V \stackrel{\prec}{\longrightarrow} L(\mathbb{R})^{V^{\texttt{Random}_\lambda}}.$$

In particular, there is $t \in \mathbb{R}^V$ such that $\varphi'(x,y,t)$ still defines a well-ordering of $\mathbb{R}^{V^{\texttt{Random}_\lambda}}$. This is impossible by Lemma 2.1 because $\lambda \geq \omega_1$. □

It follows in particular that no projective (i.e., second-order in the language of arithmetic) formula defines a well-ordering of the reals if there are atomless real-valued measurable cardinals.

Definition 2.6. A Σ^2_n formula is a formula ψ over a three-sorted structure of the form

$$(\mathcal{P}(\mathcal{P}(\mathbb{N})), \mathcal{P}(\mathbb{N}), \mathbb{N}, \in, \dots)$$

such that

$$\psi \equiv \exists X_1 \subseteq \mathcal{P}(\mathbb{N})\, \forall X_2 \subseteq \mathcal{P}(\mathbb{N}) \dots \varphi(X_1, X_2, \dots),$$

where there are n alternations of quantifiers over subsets of $\mathcal{P}(\mathbb{N})$, and φ is a projective statement, i.e., it only involves quantification over $\mathbb{N}$ and $\mathcal{P}(\mathbb{N})$.

It is standard to refer to a Σ^2_n statement as being *third-order* (in the language of arithmetic); similarly, a projective statement is usually called second-order (in the language of arithmetic). An equivalent formulation is mentioned below, in Remark 2.8.

We define Π^2_n, Δ^2_n as usual: A statement is Π^2_n iff its negation is Σ^2_n and it is Δ^2_n iff it is simultaneously equivalent to Σ^2_n and Π^2_n statements. Notice that if a linear ordering of $\mathbb{R}$ is Σ^2_n or Π^2_n, then it is automatically Δ^2_n: Suppose $\phi(x, y)$ is a Σ^2_n formula defining a linear ordering. Then $\neg\phi(r, s)$ iff $s = r$ or $\phi(s, r)$.

We close this section with a fact that (we hope) helps to understand the form taken by the well-orderings obtained in the following sections. The point is that we want to codify definability computations in the language of set theory within the language of third-order arithmetic.

We state the fact in a somewhat informal manner, to emphasize its flexibility. Here, $\mathsf{ZFC}^{-\varepsilon}$ is a sufficiently strong fragment of ZFC. For a specific version, we can take $\mathsf{ZFC}^{-\varepsilon}$ to mean $\mathsf{ZFC}^{-} + \mathcal{P}(\mathbb{R})$ exists (considering a large H_η instead of V_η in the proof below), or $\mathsf{ZFC}{\restriction}\Sigma_{200}$, i.e., ZFC with replacement restricted to Σ_{200} statements.

Fact 2.7. *Let $\varphi(\vec{x})$ be a Σ^2_1-formula. Then there is ψ, and a transitive structure $M \models \mathsf{ZFC}^{-\varepsilon}$ such that $\mathbb{R} \subseteq M$, $|M| = \mathfrak{c}$, or even ${}^{\omega}M \subseteq M$, such that for all reals $\vec{r}$,*

$$\varphi(\vec{r}) \iff M \models \psi(\vec{r}).$$

Proof. The existence of such an M is easily seen to be equivalent to a Σ^2_1-formula. Conversely, given φ, let η be large enough, so for any $\vec{r}$,

$$\varphi(\vec{r}) \iff V_\eta \models (\mathcal{P}(\mathbb{R}), \mathbb{R}, \omega, \dots) \models \varphi(\vec{r}),$$

and we can take as M a suitable substructure of V_η. □

Remark 2.8. In fact, the pointclass Σ^2_n can be identified by this method with the class $\Sigma_n(H_{\mathfrak{c}^+}, \in, H_{\omega_1}, H_\omega)$, where H_{ω_1} and H_ω are seen as parameters and therefore quantification over them is considered bounded.

Fact 2.7 and Remark 2.8 motivate the general structure of the constructions that produce Σ^2_n-well-orderings: A model needs to be produced satisfying certain first-order property ψ (somehow related to properties of the surrounding universe). Since the model can resemble the first-order theory of the surrounding universe as much as necessary, the need to satisfy ψ is in general not the main problem and is expected in practice to be achieved by forcing. The difficulty arises in trying to isolate the model or models that we have in mind from possibly fake ones, which can be thought of as proving a "correctness" theorem. This suggests the need to establish some kind of "thinness" condition, usually in tension with the width the forcing extension provides, this being in practice the main source of complications when implementing this strategy. This general framework will be illustrated with the results of this paper.

3. Σ^2_2-well-orderings

We begin with a construction based on a technique which goes back at least to Woodin's work on generalized Prikry forcing. Starting with a measurable cardinal, this technique produces a model where the cardinal is real-valued measurable, and the generic codes a subset of the reals. This construction is a prototype of several arguments showing the consistency of $\mathsf{RVM}(\mathfrak{c})$ with different kinds of definable well-orderings, and we illustrate some of them. In section 6 we show how, working over $L[\mu]$, a variation due to Woodin of the construction given in this section establishes the consistency of real-valued measurability of the continuum together with a Δ^2_1-well-ordering of $\mathbb{R}$. Here we obtain the consistency of $\mathsf{RVM}(\mathfrak{c})$ together with a Δ^2_2-well-ordering of $\mathbb{R}$ without any restrictions in the large cardinal structure of the universe. The combinatorial tool we use to carry out our coding was first considered in [3], in the presence of MA.

Theorem 3.1. *If κ is measurable in V and $2^\kappa = \kappa^+$, then there is a forcing $\mathbb{F}$ of size κ such that*

$$1 \Vdash_{\mathbb{F}} \mathfrak{c} = \kappa,\ \mathsf{RVM}(\mathfrak{c}),\ \text{and there is a } \Delta^2_2 \text{ well-ordering of } \mathbb{R}.$$

Proof. By a preliminary forcing, if necessary, we may assume GCH holds below κ (see for example [25]).

Let $\mathbb{Q} = \mathtt{Random}_\kappa$. Let $\mathbb{P}$ be the Easton product over the inaccessible cardinals $\lambda < \kappa$ of $\prod_{n\in\omega} \mathtt{Add}(\lambda^{+1+3n}, \lambda^{+3+3n})$, where the product is inverse (i.e., fully supported). Let $\mathbb{S} = \mathbb{P} \times \mathbb{Q}$, and let $G_\mathbb{P} \times G_\mathbb{Q}$ be $\mathbb{S}$-generic over V.

The proof rests on a "lifting" argument, which we isolate as follows:

Claim 3.2. *Let $j : V \to N$ be an ultrapower embedding by a normal measure on κ. Then $j(\mathbb{Q})/\mathbb{Q}$ is isomorphic to an appropriate random forcing in any intermediate model between $V[G_\mathbb{Q}]$ and $V_1 := V[G_\mathbb{Q}][G_\mathbb{P}]$, inclusive, i.e., for any such model M there is a λ such that*

$$M \models j(\mathbb{Q})/\mathbb{Q} \cong \mathtt{Random}_\lambda.$$

There is $G^* \in V$ such that:

- *If H is $j(\mathbb{Q})/\mathbb{Q}$-generic over V_1 then, in $V_1[H]$, j lifts to*

$$j_2 : V_1 \to N[G_\mathbb{P}][G^*][G_\mathbb{Q}][H]$$

and therefore (by Solovay's Theorem 1.6) $\mathsf{RVM}(\mathfrak{c})$ holds in V_1.

- *The restriction of j_2 to $V[G_\mathbb{P}]$ is $j_1 : V[G_\mathbb{P}] \to N[G_\mathbb{P}][G^*]$ (so κ remains measurable in $V[G_\mathbb{P}]$). There is a forcing $\mathbb{P}_{tail} \in N$ such that $j(\mathbb{P}) = \mathbb{P} \times \mathbb{P}_{tail}$, and G^* is $\mathbb{P}_{tail}$-generic over N.*

Similarly, the restriction of j_2 to $V[G_\mathbb{Q}][H]$ is $j_3 : V[G_\mathbb{Q}] \to N[G_\mathbb{Q}][H]$ and witnesses $\mathsf{RVM}(\mathfrak{c})$ in $V[G_\mathbb{Q}]$. Finally, $\mathbb{R}^{V[G_\mathbb{Q}]} = \mathbb{R}^{V[G_\mathbb{Q}][G_\mathbb{P}]}$.

Proof. We begin by showing:

Subclaim 3.3. *$\mathbb{P}$ preserves the measurability of κ. In fact, there is $G^* \in V$ such that whenever G is $\mathbb{P}$-generic over V, $G \times G^*$ is $j(\mathbb{P})$-generic over N, and we can lift j to an embedding $j_1 : V[G] \to N[G \times G^*]$.*

(*Cf.* [23, Lemma 2.2.4] or [13, Fact 3.1].)

Proof. By elementarity, in N, $j(\mathbb{P}) = \mathbb{P} \times \mathbb{P}_{\text{tail}}$ where $\mathbb{P}_{\text{tail}}$ is the Easton product of $\prod_{n\in\omega} \mathtt{Add}(\lambda^{+1+3n}, \lambda^{+3+3n})$, the product being inverse (i.e., fully supported) and λ ranging over the inaccessible cardinals $\lambda \in [\kappa, j(\kappa))$. In N, this set is κ^+-closed. But ${}^{\kappa}N \subset N$, so in fact it is κ^+-closed in V. Now notice that $|\mathcal{P}^N(\mathbb{P}_{\text{tail}})| = |(2^{j(\kappa)})^N| = |j(2^\kappa)| \leq (2^\kappa)^\kappa = 2^\kappa = \kappa^+$, where the last equality holds by hypothesis. Thus, the number of dense subsets of $\mathbb{P}_{\text{tail}}$ which belong to N is at most κ^+, and a straightforward induction lets us build (in V) a decreasing sequence of conditions which meet all of them. The filter G^* they generate is therefore $\mathbb{P}_{\text{tail}}$-generic over N.

It remains to argue that if G is $\mathbb{P}$-generic over V, then $G \times G^*$ is $j(\mathbb{P})$-generic over N, which amounts to showing that G and G^* are mutually generic. If so, j lifts to j_1 in the usual way[7]: For σ a $\mathbb{P}$-name, $j_1(\sigma_G) := j(\sigma)_{G\times G^*}$. The standard argument (see [13, Fact 2.1]) proves that j_1 is well defined and elementary.

But mutual genericity is clear: Since $N[G^*] \subseteq V$, if G is $\mathbb{P}$-generic over V, it is also $\mathbb{P}$-generic over $N[G^*]$.

This completes the proof of Subclaim 3.3. □

Notice that $\mathbb{P}$ is ω_1-closed, so $\mathbb{R}^{V[G_{\mathbb{P}} \times G_{\mathbb{Q}}]} = \mathbb{R}^{V[G_{\mathbb{Q}}]}$.

Subclaim 3.4. *In $V[G_{\mathbb{P}}][G_{\mathbb{Q}}][H]$, j_1 lifts to*

$$j_2 : V[G_{\mathbb{P}}][G_{\mathbb{Q}}] \to N[G_{\mathbb{P}}][G^*][G_{\mathbb{Q}}][H].$$

The restriction of j_2 to $V[G_{\mathbb{Q}}]$ is an embedding

$$j_3 : V[G_{\mathbb{Q}}] \to N[G_{\mathbb{Q}}][H]$$

definable in $V[G_{\mathbb{Q}}][H]$.

Proof. As expected, simply set

$$j_2(\tau_{G_{\mathbb{Q}}}) = j_1(\tau)_{G_{\mathbb{Q}}\frown H},$$

for τ a $\mathbb{Q}$-name in $V[G_{\mathbb{P}}]$. As before, j_2 is well defined and elementary. Since j_1 extends j, $j_3 = j_2 \restriction V[G_{\mathbb{Q}}] : V[G_{\mathbb{Q}}] \to N[G_{\mathbb{Q}}][H]$ is given by $j_3(\tau_{G_{\mathbb{Q}}}) = j(\tau)_{G_{\mathbb{Q}}\frown H}$ for τ a $\mathbb{Q}$-name in V, and is definable in $V[G_{\mathbb{Q}}][H]$ as claimed.

This completes the proof of Subclaim 3.4. □

The proof of Claim 3.2 is complete. □

---

[7]This is the standard way of showing that if ρ is measurable, then it is still real-valued measurable in $V^{\mathtt{Random}_\rho}$.

In $V[G_{\mathbb{Q}}]$, let $A = \langle r_\alpha : \alpha < \kappa \rangle$ be a well-ordering of $\mathbb{R}$. In $V[G_{\mathbb{Q}}][G_{\mathbb{P}}]$, define g as follows:

Let $\langle \delta_\alpha : \alpha < \kappa \rangle$ enumerate in V the inaccessible cardinals below κ. Let G_α be the part of $G_{\mathbb{P}}$ which is generic for $\prod_{n\in\omega} \mathtt{Add}(\delta_\alpha^{+1+3n}, \delta_\alpha^{+3+3n})$. Write $G_\alpha \cong \prod_{n\in\omega} G_\alpha(n)$, where $G_\alpha(n)$ is the part of G_α generic for $\mathtt{Add}(\delta_\alpha^{+1+3n}, \delta_\alpha^{+3+3n})$. Then

$$g = \prod_{\alpha<\kappa} G^*_\alpha,$$

where $G^*_\alpha = \prod_{n\in\omega} G^*_\alpha(n)$ and

$$G^*_\alpha(n) = \begin{cases} G_\alpha(n) & \text{if } n \in r_\alpha, \\ 1_{\mathtt{Add}(\delta_\alpha^{+1+3n}, \delta_\alpha^{+3+3n})} & \text{if } n \notin r_\alpha. \end{cases}$$

Claim 3.5. *$\kappa = \mathfrak{c}$ stays real-valued measurable in $V[G_{\mathbb{Q}}][g]$.*

Proof. By Theorem 1.34, $j(\mathbb{Q})/\mathbb{Q}$ is isomorphic to $\mathtt{Random}_{j(\kappa)}$ in $V[G_{\mathbb{Q}}]$. It follows that in $V[G_{\mathbb{Q}}][G_{\mathbb{P}}]$ as well as in $V[G_{\mathbb{Q}}][g]$, $j(\mathbb{Q})/\mathbb{Q}$ is still a complete measure algebra, since the forcing for which g is generic is a factor of $\mathbb{P}$, which is ω_2-closed in V and therefore ω_2-dense in $V[G_{\mathbb{Q}}]$ by Easton's lemma, see [13, Fact 4.1]. Since $j(\mathbb{Q})/\mathbb{Q}$ was homogeneous in $V[G_{\mathbb{Q}}]$, it is still homogeneous in $V[G_{\mathbb{Q}}][g]$ and in $V[G_{\mathbb{Q}}][G_{\mathbb{P}}]$. We conclude that $j(\mathbb{Q})/\mathbb{Q}$ is still isomorphic to $\mathtt{Random}_{j(\kappa)}$, by Maharam's Theorem 1.15.

Let H be $j(\mathbb{Q})/\mathbb{Q}$-generic over $V[G_{\mathbb{Q}}][G_{\mathbb{P}}]$. We will show that in $V[G_{\mathbb{Q}}][g][H]$, j lifts to

$$j^* : V[G_{\mathbb{Q}}][g] \to N[j^*(G_{\mathbb{Q}})][j^*(g)].$$

This amounts to defining $j^*(G_{\mathbb{Q}})$ and $j^*(g)$, and checking that the induced map j^* is well defined and elementary. Once this is done, Solovay's Theorem 1.6 implies the claim.

Set $j^*(G_{\mathbb{Q}}) = G_{\mathbb{Q}}{}^\frown H$. To define $j^*(g)$, it suffices to define $j^*(g)_{[\kappa,j(\kappa))}$ (so $j^*(g) = g{}^\frown j^*(g)_{[\kappa,j(\kappa))}$). The intention is that the definition of $j^*(g)$ copies that of g, so we must start by defining $j^*(A)$.

Since $A \in V[G_{\mathbb{Q}}]$, $j_3(A) \in V[G_{\mathbb{Q}}][H]$. We set $j^*(A) = j_3(A)$ (with j_3, etc, as in Claim 3.2). The key observation is that we do not really need a whole $j(\mathbb{P})$-generic to define $j^*(g)_{[\kappa,j(\kappa))}$, but a $\mathbb{P}_{\text{tail}}$-generic suffices: Remember that G^*, as built in Subclaim 3.3, is in V. We can now set

$$j^*(g)_{[\kappa,j(\kappa))} := \prod_{\alpha\in[\kappa,j(\kappa))} \prod_{n\in\omega} G^{**}_{\alpha,n},$$

where $G^{**}_{\alpha,n}$ is the $\mathtt{Add}(\delta_\alpha^{+1+3n}, \delta_\alpha^{+3+3n})$-generic added by G^* to N, if $n \in j_3(r)_\alpha$, and the trivial condition otherwise.

Here, $\langle \delta_\alpha : \alpha < j(\kappa) \rangle = j(\langle \delta_\alpha : \alpha < \kappa \rangle)$ is the increasing enumeration of the inaccessible cardinals in N below $j(\kappa)$ and $j_3(A) = \langle j_3(r)_\alpha : \alpha < j(\kappa) \rangle$ is the well-ordering of the reals of N .

Extend j^* to a map

$$j^* : V[G_{\mathbb{Q}}][g] \to N[j^*(G_{\mathbb{Q}})][j^*(g)]$$

in the usual way. Notice that j^* is simply the restriction of j_2, as defined in Subclaim 3.4, to $V[G_{\mathbb{Q}}][g]$. This proves j^* is well defined and elementary. Finally, notice that j^* is definable in $V[G_{\mathbb{Q}}][g][H]$. This concludes the proof of Claim 3.5. □

Remark 3.6. The argument just given is quite general. It works as long as $\mathbb{P}$ is a reasonably definable product of sufficiently closed small forcings. The set we called A can code any subset of the reals in $V[G_{\mathbb{Q}}][G_{\mathbb{P}}]$. By coding A inside a "subproduct" g of $G_{\mathbb{P}}$, we avoid having to set up any sort of book-keeping devices in the ground model in order to define the well-ordering alongside the iteration. As a matter of fact, we do not need to worry about defining in the ground model (as an iteration or otherwise) the forcing whose generic is g.

Notice also that, in spite of this generality, some argument was required, since it is not necessarily true that if W is a forcing extension of V preserving $\mathsf{RVM}(\kappa)$, then any intermediate extension $V \subseteq M \subseteq W$ satisfies $\mathsf{RVM}(\kappa)$ as well. This observation (with κ measurable in V and W) is due to Kunen, see [30]; we present in section 4 a proof of this result, different from the argument in [30]. The proof in section 4 uses the technique illustrated in this section: Starting with an embedding $j : V \to N$, we find in V an N-generic filter for a sufficiently closed forcing notion living in N. A proof dealing specifically with atomless measurability has been produced by Gitik, see [22, Theorem 2]; Gitik's proof can be seen as an elaboration of the argument in section 4, and the reader may find it profitable to read section 4 before consulting [22].

Now we continue with the proof of Theorem 3.1. All what remains is to see that we can "decode" the well-ordering A from g in a Σ^2_2-way in $V[G_{\mathbb{Q}}][g]$. The forcing $\mathbb{F}$ is then the factor of $\mathbb{S}$ for which $G_{\mathbb{Q}} \times g$ is a generic.

The key to our coding is the following notion (see [4]):

Definition 3.7. Let λ be regular. The *club base number* for λ is

$$\min\{\,|X| : X \subseteq \mathcal{P}(\lambda) \text{ and } \forall \text{ club } C \subseteq \lambda\, \exists \text{ club } D \in X\ (D \subseteq C)\,\}.$$

So the club base number for λ is the coinitiality of the club filter at λ, ordered under inclusion. Any collection X of club subsets of λ realizing the minimum above generates the club filter at λ by closing under supersets.

If λ is regular and $2^\lambda = \lambda^+$, then the club base number for λ is λ^+, while if λ^{++} Cohen subsets of λ are added, their closures are club sets containing no club from the ground model, and mutual genericity guarantees that the club base number at λ is λ^{++}.

It follows that in $V[g]$ the inaccessible cardinals below κ are just the δ_α, $\alpha < \kappa$, and the club base number for δ_α^{+1+3n} is either δ_α^{+2+3n} or $\delta_\alpha^{+3(n+1)}$ depending on whether $G^*_\alpha(n)$ is trivial or not, since the base number for $\delta_{\alpha_1}^{+1+3n}$ is not affected by forcing with (a subproduct of) $\prod_{m\in\omega} \mathtt{Add}(\delta_{\alpha_2}^{+1+3m}, \delta_{\alpha_2}^{+3+3m})$ for $\alpha_2 \neq \alpha_1$.

Maybe a more detailed argument is in order: Let $\lambda < \kappa$ be inaccessible, let $n < \omega$, and write $\mathbb{P} \cong \mathbb{P}_{\lambda,n} \times \mathtt{Add}(\lambda^{+1+3n}, \lambda^{+3+3n}) \times \mathbb{P}^{\lambda,n}$, where $\mathbb{P}_{\lambda,n}$ corresponds to the factors of $\mathbb{P}$ that add Cohen subsets to cardinals strictly smaller than λ^{+1+3n}, and $\mathbb{P}^{\lambda,n}$ corresponds to those factors that add Cohen subsets to strictly bigger cardinals. Then $\mathbb{P}^{\lambda,n}$ is sufficiently closed that it cannot ("by accident") add a subset of λ^{+1+3n}, while $\mathbb{P}_{\lambda,n}$ satisfies a sufficiently small chain condition that any club subset of λ^{+1+3n} that it adds contains a club in the ground model. Finally, $G_{\mathbb{Q}}$ is added by ccc forcing, so it does not affect any of the club base numbers that concern us. It follows that in $V[G_{\mathbb{Q}}][G_{\mathbb{P}}]$ the only club base numbers that are affected are those that we have explicitly changed by means of $G_{\mathbb{P}}$, and therefore in $V[G_{\mathbb{Q}}][g]$ we have coded A by means of the club base numbers which have been altered.

Now observe that in $V[G_{\mathbb{Q}}][g]$ we can define A, or rather the corresponding order relation $<_A$ on $\mathbb{R}$ as follows:

Let $\Psi(M)$ denote the conjunction of the following requirements:

$M \models \mathsf{ZFC}^-$, M is transitive, $|M| = \mathfrak{c}$, and $\mathbb{R} \subseteq M$. (Notice that this implies $\mathrm{ORD}^M \geq \mathfrak{c}$.) Moreover,

1. M computes cofinalities correctly, that is, if $\lambda, \mu \in M$, and there is $f : \lambda \to \mu$ cofinal, then there is such an $f \in M$.
2. For all $C \subseteq \lambda < \mathfrak{c}$ club, there is $D \in M$, $D \subseteq \lambda$ club, such that $D \subseteq C$.
3. M computes club base numbers correctly, that is, for all $\lambda < \mathfrak{c}$, $\mathcal{F} \subseteq \mathcal{P}(\lambda)^M$ collection of club sets, $|\mathcal{F}| < \mathfrak{c}$, there is $\mathcal{G} \in M$ collection of club sets such that $\mathcal{G}$ is coinitial in $\mathcal{F}$.[8]

Finally, for all $r \in \mathbb{R}$ there is in M a unique sequence of club base numbers starting at a weakly inaccessible[9] which (in the obvious way) code r, and any weakly inaccessible codes some r.

Notice that $\Psi(M)$ is the conjunction of the statement that $|M| = \mathfrak{c}$ and a $\Pi_1(H_{\mathfrak{c}^+}, \in, H_{\omega_1})$ statement about M. Notice as well that M does not have any cardinals above $\mathfrak{c}$. For example, fixing a surjection $\pi : \mathbb{R} \to M$, condition 2 is seen to be Π_1 since the existential quantifier is actually a quantifier over reals, i.e., there is a real r such that, in M, $\pi(r) = D$ is a club set.

For x, y reals, let $\psi(x, y)$ hold iff

There is M such that $\Psi(M)$ holds and in M the sequence coding x appears before the sequence coding y.

Since M is only a model of ZFC^-, $\mathcal{P}(\lambda)^M$ as in 3 above, is to be interpreted as a definable class. This does not affect the desired complexity of ψ.

---

[8]The requirement on the size of $\mathcal{F}$ is not essential. We just include it to ensure the universal quantifier in the definition of the well-ordering we obtain actually ranges over bounded subsets of $\mathfrak{c}$.

[9]Since the ground model satisfies GCH, the weakly inaccessible and the strongly inaccessible cardinals coincide here, and we took care of coding reals at each inaccessible cardinal. It is by no means essential that we decide to code using the inaccessible cardinals, and the coding could have occurred at many other places (say, starting at limit cardinals), with only a straightforward variation in the construction above being required.

The relation ψ just defined can be rendered Σ^2_2 in a straightforward fashion. We are done once we verify that $x <_A y$ holds for reals x, y if and only if $\psi(x, y)$ does. That $x <_A y$ implies $\psi(x, y)$ is easy, $M = V[G_{\mathbb{Q}}][g]_\kappa$ is a witness. To see the converse just observe that any M witnessing $\psi(x, y)$ is correct about cofinalities below κ, and computes correctly club base numbers of cardinals below κ. The uniqueness of the coding of reals by club base numbers ensures that no fake codings (witnessing false relations $x <_A y$) may arise, since every inaccessible below κ is assigned some real this way, and this assignment is a bijection; in particular, $\mathrm{ORD}^M = \mathfrak{c}$ must hold (since no real is coded by $\mathfrak{c}$). Since g was defined precisely to code A using the club base numbers, $\psi(x, y)$ implies $x <_A y$. This completes the proof of Theorem 3.1. □

Remark 3.8. Notice that M as above is not a model of "$\mathbb{R}$ exists". $\mathbb{R}$ and the relation ψ are only *definable* over M. Obviously, any transitive model N of enough set theory that contains all the reals and where there is a well-ordering of $\mathbb{R}$ must satisfy $\mathrm{ORD}^N > \mathfrak{c}$.

Let $(\Sigma^2_2)^+$ denote the class of statements about the reals expressible as a Boolean combination of Σ^2_2 statements. As a consequence of the argument above and Solovay's theorem on preservation of real-valued measurability (Fact 1.27) we obtain that generic invariance of $(\Sigma^2_2)^+$ with respect to real-valued measurability of the continuum[10] is not a theorem of ZFC, even in the presence of projective absoluteness. In effect, the fact that ψ defines a well-ordering of $\mathbb{R}$, for ψ as above, can be expressed as a $(\Sigma^2_2)^+$ statement[11], it can be made true over V as long as there are measurable cardinals in V, and can be made false afterwards simply by adding ω_1 many random reals, see Lemma 2.1.

---

[10] Generic invariance of a class Γ of sentences with respect to a statement ϕ means that whenever $\mathbb{P} * \dot{\mathbb{Q}}$ is a two-step iteration of set forcings such that $V^{\mathbb{P}} \models \phi + 1_{\dot{\mathbb{Q}}} \Vdash \phi$, then for all $\psi \in \Gamma$, $V^{\mathbb{P}} \models \psi$ iff $V^{\mathbb{P}*\dot{\mathbb{Q}}} \models \psi$. For example, it is a theorem of Woodin that if there is a proper class of cardinals which are either measurable Woodin or strongly compact, then generic invariance of Σ^2_1 holds with respect to CH (see for example [33, Theorem 3.2.1]).

[11] It is not accurate to express it in a Σ^2_2 way, even though the relation ψ is Δ^2_2 in $V[G_{\mathbb{Q}}][g]$: Let ψ_1 and ψ_2 be Σ^2_2-formulas such that for all reals r, s, $\psi(r, s) \Leftrightarrow \psi_1(r, s) \Leftrightarrow \neg\psi_2(r, s)$.

The fact that ψ defines a well-ordering of $\mathbb{R}$ can be formalized as follows:

$$\begin{aligned}\forall x, y, z \in \mathbb{R}\, \big[&(\psi(x, y) \vee x = y \vee \psi(y, x)) \wedge (\neg\psi(x, x)) \wedge (\psi(x, y) \to \neg\psi(y, x)) \\ &\wedge (\psi(x, y) \wedge \psi(y, z) \to \psi(x, z)) \wedge \exists n\, \neg\psi(x_{n+1}, x_n)\big],\end{aligned}$$

where $x \mapsto \langle x_n : n < \omega\rangle$ is some recursive bijection between $\mathbb{R}$ and $\mathbb{R}^\omega$. This statement can certainly be expressed in a Σ^2_2 way if ψ_1 and ψ_2 are judiciously used in place of ψ in the displayed formula above. However, we *must* add to it the clause that $\psi_1(x, y) \leftrightarrow \neg\psi_2(x, y)$ (which is not a Σ^2_2 statement), since this equivalence is certainly vital for the validity of the assertion that ψ defines a well-ordering, but it is not a theorem and must therefore be explicitly claimed.

4. A result of Kunen

In this section we sketch a proof (which is probably folklore) of the result of Kunen mentioned in Remark 3.6. We use the technique illustrated in the previous section of finding, in the ground model, filters that are generic over inner models for forcing notions that are sufficiently closed.

Theorem 4.1 (Kunen). *Assuming the consistency of measurable cardinals, it is consistent that there are models $M \subset W \subset V$ and a cardinal κ such that κ is measurable in M and V but not in W.*

Kunen's argument is different from the one to follow. He starts with a ground model N where there is a measurable cardinal κ, and adds a generic S to N for the *Silver preparation forcing* so, in $M = N[S]$, κ is measurable, and it remains measurable after adding to M a Cohen subset of κ. In $W = M[G]$ there is a κ-Suslin tree (so κ is not measurable). G is generic for a forcing like the Prikry-Silver forcing ([9, §7]), but care is taken to ensure that the tree that is added is homogeneous. Finally, in $V = W[H]$ the tree is killed in the usual way. The iterated forcing that first adds G and then adds H is equivalent to adding over M a Cohen subset of κ, so κ is measurable in V. The details of this argument can be found in [30].

In the argument below, we avoid the need for the preparation forcing by working over a nice inner model. Recall:

Definition 4.2. By $L[\mu]$ we mean the smallest proper class inner model of the theory

$$\mathsf{ZFC} + \text{"There exists a measurable,"}$$

in this context, by μ we always mean a witness to measurability, i.e.,

$$L[\mu] \models \mu \text{ is a normal } \kappa\text{-complete measure on some cardinal } \kappa,$$

and by smallest we mean that κ is as small as possible ($L[\mu]$ is sometimes called the *core* ρ*-model*, see [15, Definition 13.8]).

We abuse notation in the usual way, and occasionally talk about the theory $V = L[\mu]$. The minimality assumption is just adopted for definiteness, and not required in the arguments; of course if, for some U and λ,

$$L[U] \models U \text{ is a normal } \lambda\text{-complete measure on } \lambda,$$

then $L[U] \models V = L[\mu]$.

Proof. The idea of the proof is to start with a measurable cardinal κ and a stationary set $S \subset \kappa^+$. We carefully associate to S a sequence

$$\langle S_\delta : \delta < \kappa \text{ inaccessible} \rangle ,$$

$S_\delta \subset \delta^+$ stationary, in such a way that the stationarity of the S_δ can be destroyed by forcing while the stationarity of S is preserved. By a reflection argument, this will contradict the measurability of κ in the extension, but a further extension (destroying the stationarity of S) resurrects its measurability, providing the example we desire.

Recall ([1, Definition 1.1]) that a stationary subset S of a regular cardinal λ is called *fat* iff for every club $C \subseteq \kappa$, $S \cap C$ contains closed sets of ordinals of arbitrarily large order types below κ. It is shown in [1, Theorem 1] that if $\lambda = \rho^+$ where $\rho^{<\rho} = \rho$, $2^\rho = \lambda$, and $S \subseteq \lambda$ is fat, then there is a λ-distributive forcing $\mathbb{P} = \mathbb{P}_S$ such that $|\mathbb{P}| = \lambda$ and $\mathbb{P}$ adds a club $C \subseteq S$. $\mathbb{P}$ is just the set of bounded, closed subsets of S, ordered by end extension. It follows from [1, Lemma 1.2] that if $\lambda = \rho^+$ where ρ is regular, $A \subset \{\, \alpha < \lambda : \operatorname{cf}(\alpha) = \rho \,\}$, and $\{\, \alpha \in \lambda \setminus A : \operatorname{cf}(\alpha) = \rho \,\}$ is stationary, then $\lambda \setminus A$ is fat. In this case, if GCH holds, then $\mathbb{P}_{\lambda\setminus A}$ is η-closed for all $\eta < \rho$.

Work in $L[\mu]$. Let $j_\mu : L[\mu] \to M_1$ be the embedding by the normal measure μ, and let $\kappa = \operatorname{cp}(j_\mu)$, so $\mathcal{P}(\kappa)^{L[\mu]} = \mathcal{P}(\kappa)^{M_1}$ and, in particular, $(\kappa^+)^{L[\mu]} = (\kappa^+)^{M_1}$.

Notice that there is a set $S \in M_1$ such that S is a stationary subset of κ^+ in $L[\mu]$, $\forall \alpha \in S\, (\operatorname{cf}(\alpha) = \kappa)$, and $\{\, \alpha \in \kappa^+ \setminus S : \operatorname{cf}(\alpha) = \kappa \,\}$ is also stationary in $L[\mu]$ (for example, because a $\kappa \times \kappa^+$ Ulam matrix defined in M_1 would still be an Ulam matrix in $L[\mu]$, see [31, Theorem II.6.11]).

Fix a function f such that $j_\mu(f)(\kappa) = S$, so for μ-almost every inaccessible $\delta < \kappa$, $f(\delta)$ is a stationary subset of δ^+ concentrating on the ordinals of cofinality δ such that $\{\, \alpha \in \delta^+ \setminus f(\delta) : \operatorname{cf}(\alpha) = \delta \,\}$ is also stationary. By redefining f on a μ-measure zero set, if necessary, we may assume that this holds for all inaccessible cardinals $\delta < \kappa$.

Consider the Backward Easton support iteration $\mathbb{P}_\kappa$ of forcings $\mathbb{F}_\delta$, $\delta < \kappa$ inaccessible, such that $\mathbb{F}_\delta = \mathbb{P}_{\delta^+ \setminus f(\delta)}$ adds a club subset of $\delta^+ \setminus f(\delta)$. Notice that $\mathbb{P}_\kappa$ is κ-cc ([9, Corollary 2.4]) and that, for every inaccessible δ, $\mathbb{P}_\kappa$ factors as $\mathbb{Q}_\delta * \mathbb{F}_\delta * \mathbb{Q}^\delta$ where $\mathbb{Q}^\delta$ is, say, δ^{++}-closed, and $\mathbb{Q}_\delta$ preserves the stationarity of $f(\delta)$.

By elementarity, in M_1, $j(\mathbb{P}_\kappa) = \mathbb{P}_\kappa * \mathbb{F}_\kappa * \mathbb{Q}^\kappa$, where $\mathbb{F}_\kappa$ adds a club subset of $\kappa^+ \setminus S$ and $\mathbb{Q}^\kappa$ is κ^{++}-closed.

Let G_κ be $\mathbb{P}_\kappa$-generic over $L[\mu]$, and let g be $\mathbb{F}_\kappa$-generic over $L[\mu][G_\kappa]$.

Claim 4.3. *κ is measurable in $L[\mu][G_\kappa][g]$.*

Proof. We find in $L[\mu][G_\kappa][g]$ a lifting

$$j : L[\mu][G_\kappa] \xrightarrow{\prec} M_1[j(G_\kappa)]$$

of j_μ. This suffices because $\mathcal{P}(\kappa)^{L[\mu][G_\kappa]} = \mathcal{P}(\kappa)^{L[\mu][G_\kappa][g]}$, so the $L[\mu][G_\kappa]$-ultrafilter derived from j is an ultrafilter on κ, and it is straightforward to verify that it is non-principal and κ-complete.

The lifting is found arguing as in Claim 3.2: $\mathbb{Q}^\kappa$ is at least κ^+-closed (i.e., closed under extensions of decreasing sequences of length $< \kappa^+$) in $M_1[G_\kappa][g]$ and, in fact, it is κ^+-closed in $L[\mu][G_\kappa][g]$. This follows from standard arguments about Backward Easton iterations, and is almost verbatim as [14, Lemma 11.3 and Lemma 11.6], to which we refer for further details. Since, in $L[\mu][G_\kappa][g]$, $|j(\kappa)| =$

κ^+, there is in $L[\mu][G_\kappa][g]$ an $M_1[G_\kappa][g]$-generic filter H. Defining $j(G_\kappa) = G_\kappa * g * H$, j_μ lifts in the usual way to an embedding j as required.

This concludes the proof of Claim 4.3. □

Claim 4.4. *κ is not measurable in $L[\mu][G_\kappa]$.*

Proof. Suppose otherwise, and let $k : L[\mu][G_\kappa] \to N$ be the corresponding embedding coming from a normal measure on κ. Then $k\restriction(\mathcal{P}(\kappa) \cap L[\mu])$ is the restriction to $\mathcal{P}(\kappa) \cap L[\mu]$ of an iteration of j_μ (see for example [26, Exercise 20.13]), and therefore $k(f)(\kappa) = j_\mu(f)(\kappa) = S$.

But then, by elementarity, $k(G_\kappa)$ adds a club set killing the stationarity of S. This contradicts that $\mathbb{P}_\kappa$ is κ-cc. □

Theorem 4.1 follows at once, taking $M = L[\mu]$, $W = L[\mu][G_\kappa]$, and $V = L[\mu][G_\kappa][g]$. □

Remark 4.5. The use of $L[\mu]$ in the previous argument is by no means essential: An additional preparation forcing (ensuring that in any future extension, for any embedding k with critical point κ, $k(f)(\kappa) = S$) would allow us to start with an arbitrary ground model (of GCH) instead of $L[\mu]$. See [22, Theorem 2] for an elaboration on the argument above that produces an example where κ is atomlessly measurable in the final model V, but the reals of V are not obtained by adding random reals to any inner model where κ is measurable.

5. Anticoding results

The results in this section are folklore, although our presentation may be novel. Of course, the complexity Δ^2_2 of the well-ordering we obtained in section 3 is an overkill; notice the third-order universal quantifier only ranges over *bounded* subsets of κ. It is natural to wonder whether we can improve the complexity of the well-ordering to be Σ^2_1. The problem with following a strategy similar to the one described in 3 is that we need to ensure correctness of the model M with respect to the combinatorial structure of the universe that carries out the coding (the club base numbers, for example). This level of correctness needs to be attained via projective and (at most) third-order existential statements. This seems to suggest that we need to be able to code (suitable) bounded subsets of κ by reals. In general (as in the arguments of [3] and [4]), this is done by arranging that the universe satisfies something like a sufficiently strong fragment of MA to be able to use the coding provided by almost-disjoint forcing.

Unfortunately (as it is well known, see [37]) MA itself fails after adding even one random real, so it is incompatible with real-valued measurability of the continuum. For example:

Theorem 5.1. *If $\mathsf{RVM}(\kappa)$ holds and $\kappa \le \mathfrak{c}$, then there is a ccc partial order $\mathbb{P}$ such that $\mathbb{P} \times \mathbb{P}$ is not ccc.*

Corollary 5.2. *If $\mathsf{RVM}(\kappa)$ holds and $\kappa \le \mathfrak{c}$, then MA_{ω_1} fails.* □

The hypothesis we display is not ideal, but there is some subtlety here, since Prikry showed that MA is compatible with quasi-measurability of the continuum, see [19, Proposition 9G].

Remark 5.3. Corollary 5.2 has been shown in many ways independently of Theorem 5.1. Arguments more in the spirit of forcing axioms are possible: For example, if κ is atomlessly measurable, then

- $\mathrm{non}(\mathbb{R}, \mathcal{N}) = \mathrm{cov}(\mathbb{R}, \mathcal{M}) = \mathrm{add}(\mathcal{M}) = \mathrm{add}(\mathcal{N}) = \mathfrak{p} = \omega_1$. Here, $\mathcal{N}$ is the ideal of Lebesgue null sets and $\mathcal{M}$ is the ideal of meager sets.
- $\mathfrak{b} < \kappa$.

See [19] and references within. The particular case $\mathsf{RVM}(\mathfrak{c}) \Rightarrow \mathfrak{b} < \mathfrak{c}$ is due to Banach-Kuratowski [7].

It is a well-known result (due to Bell) that $\mathfrak{p}$ is the smallest cardinal λ such that $\mathsf{MA}_\lambda(\sigma\text{-centered})$ fails, see [17, §14]. Recall that almost disjoint forcing is σ-centered.

Proof of Theorem 5.1. This is a corollary of the following result of Roitman[12] [37]:

Lemma 5.4. *In $V^{\mathtt{Random}_\omega}$ there is a ccc partial order whose square is not ccc.* □

Corollary 5.5. *Roitman's result* 5.4 *holds in $V^{\mathtt{Random}_\lambda}$ and not just $V^{\mathtt{Random}_\omega}$.*

Proof. Since $\mathtt{Random}_\lambda/\mathtt{Random}_\omega \cong \mathtt{Random}_\lambda$, Corollary 5.5 follows from Corollary 1.32. □

Assume $\mathsf{RVM}(\kappa)$ where $\kappa \leq \mathfrak{c}$, and let λ be such that in $V^{\mathtt{Random}_\lambda}$ there is an embedding $j : V \to N$ with $\mathrm{cp}(j) = \kappa$. By Corollary 5.5 there is in $V^{\mathtt{Random}_\lambda}$ a ccc partial order $\mathbb{P}$ whose square is not ccc. By taking Skolem hulls, we may assume $|\mathbb{P}| \leq \aleph_1$. By Remark 1.21, we may as well assume N is closed under ω_1-sequences, so we can take $\mathbb{P} \in N$ and

$$N \models \mathbb{P} \text{ is ccc but } \mathbb{P} \times \mathbb{P} \text{ is not.}$$

But then, by elementarity, there is such a partial order in V. □

Question 5.6 (Fremlin). *Suppose κ is atomlessly measurable. Are there two ccc posets $\mathbb{P}$ and $\mathbb{Q}$ such that $\mathbb{P} \times \mathbb{Q}$ has an antichain of size κ?*

Another forcing axiom that is used to code information about subsets of reals is the Open Coloring Axiom OCA, see [20].

Theorem 5.7. *If* $\mathsf{RVM}(\kappa)$ *holds and $\kappa \leq \mathfrak{c}$, then* OCA *fails.*

This is essentially due to Todorčević.

Proof. The key to this result is the notion of an entangled linear order, see [41]. The following is [41, Theorem 2]:

Lemma 5.8. *If E is a set of random reals, then E is ω_1-entangled.* □

[12]In [8, Theorem 3.2.30], this is erroneously attributed to Galvin. Galvin devised a general method to construct such posets. Roitman showed that the construction works in $V^{\mathbb{F}}$, where $\mathbb{F} = \mathtt{Add}(\omega, 1)$ or $\mathbb{F} = \mathtt{Random}_\omega$.

I think the following is due to Todorčević and Baumgartner, see [20] for a proof.

Fact 5.9. *If there is an uncountable ω_1-entangled subset of $\mathbb{R}$, then* OCA *fails*[13]. □

Theorem 5.7 now follows as before: For some λ, in $V^{\mathtt{Random}_\lambda}$ there is an embedding $j : V \to N$ with $\mathrm{cp}(j) = \kappa$ and ${}^{\omega_1}N \subseteq N$, so in N there is an uncountable ω_1-entangled subset of $\mathbb{R}$ and, by elementarity, there is such a set also in V. By Fact 5.9, OCA fails in V.[14] □

These arguments should make it clear that any statement sufficiently fragile in the sense that random forcing destroys it and sufficiently absolute in the sense that it transfers to the generic ultrapower of the ground model, is bound to fail if there are atomlessly measurable cardinals. Thus, any naive attempt to improve the complexity of the well-ordering obtained in Section 3 by coding bounded subsets of κ by reals (where κ was measurable in the ground model and turns atomlessly measurable in the extension), say by including into the product we were calling $\mathbb{P}$ small factors that will do the coding of bounded subsets, runs into the immediate difficulty that we are adding random reals by homogeneous forcing (by the poset we were calling $\mathbb{Q}$, which is just $\mathtt{Random}_\kappa$), which most likely will undo our coding. We would then have to do the coding in such a way that no initial segment of the iteration would suffice to code a bounded set of κ in the final model, but this seems difficult as well, because bounded sets of κ would most likely appear in initial segments of the iteration.

This section has highlighted inherent difficulties that a proof of the consistency of $\mathsf{RVM}(\mathfrak{c})$ together with a Σ^2_1-well-ordering of the reals must face.

Woodin's result in the following section solves them in an indirect manner, by restricting in a very serious way the universe over which the argument takes place. The question of whether measurability of κ and GCH (or for that matter, any set of hypotheses which do not carry anti-large cardinal restrictions, or smallness requirements on the universe) suffice to force a model of $\mathsf{RVM}(\mathfrak{c})$ with a Σ^2_1-well-ordering of $\mathbb{R}$ is still open. In section 7 we discuss an alternative approach.

6. Σ^2_1-well-orderings

The result of this section is due to Woodin.

Assume that $V \models \kappa$ is measurable and $2^\kappa = \kappa^+$, and let $j : V \to N$ be a normal ultrapower embedding with $\mathrm{cp}(j) = \kappa$.

Let $\mathbb{Q} = \mathtt{Random}_\kappa$ and $\mathbb{P}$ be the Easton product over the inaccessible cardinals $\lambda < \kappa$ of $\mathtt{Add}(\lambda^+, 1) \times \mathtt{Add}(\lambda^{++}, 1)$.

Force over V with $\mathbb{P} \times \mathbb{Q}$, and let $G_{\mathbb{P}} \times G_{\mathbb{Q}}$ be generic.

---

[13]The existence of uncountable ω_1-entangled subsets of $\mathbb{R}$ also contradicts MA_{ω_1} (see [41]), thus giving yet another proof of Corollary 5.2.

[14]This argument actually shows that if κ is atomlessly measurable, then for every $\lambda < \kappa$ there is an ω_1-entangled subset of $\mathbb{R}$ of size λ.

As before:

- If $j(\mathbb{P}) = \mathbb{P} \times \mathbb{P}_{\text{tail}}$, then there is $G^* \in V$, $\mathbb{P}_{\text{tail}}$-generic over N, such that j lifts to $j_1 : V[G_{\mathbb{P}}] \to N[G_{\mathbb{P}}][G^*]$.
- $j(\mathbb{Q})/\mathbb{Q}$ is isomorphic to an appropriate random forcing in any intermediate model between $V[G_{\mathbb{Q}}]$ and $V_1 := V[G_{\mathbb{Q}}][G_{\mathbb{P}}]$, inclusive, and $\mathfrak{c} = \kappa$ is real-valued measurable in V_1. In fact if H is $j(\mathbb{Q})/\mathbb{Q}$-generic over V_1 then, in $V_1[H]$, j lifts to $j_2 : V_1 \to N[G_{\mathbb{P}}][G^*][G_{\mathbb{Q}}][H]$, thus showing $\mathsf{RVM}(\mathfrak{c})$ in V_1, by Solovay's Theorem 1.6.
- Similarly, in $V[G_{\mathbb{Q}}][H]$, j lifts to $j_3 : V[G_{\mathbb{Q}}] \to N[G_{\mathbb{Q}}][H]$.
- $\mathbb{R}^{V[G_{\mathbb{Q}}]} = \mathbb{R}^{V[G_{\mathbb{Q}}][G_{\mathbb{P}}]}$.

In $V[G_{\mathbb{Q}}]$, let $A \subset \kappa$ code a well-ordering of $\mathbb{R}$ in order type κ.

Let $\langle \delta_\alpha : \alpha < \kappa \rangle$ be the increasing enumeration of the inaccessible cardinals in V below κ. For $\alpha < \kappa$, let G_α be the αth component of $G_{\mathbb{P}}$, so G_α is the product of an $\mathtt{Add}(\delta_\alpha^+, 1)$-generic and an $\mathtt{Add}(\delta_\alpha^{++}, 1)$-generic over V. Let G_α^* be the $\mathtt{Add}(\delta_\alpha^+, 1)$-generic, if $\alpha \in A$, and the $\mathtt{Add}(\delta_\alpha^{++}, 1)$-generic, if $\alpha \notin A$. Finally, let

$$g = \prod_{\alpha<\kappa} G_\alpha^*.$$

Notice A is definable from g.

The same argument as in Claim 3.5 shows G^* and $j_3(A)$ suffice to define $j_2(g)$ (and recall $G^* \in V$ and $j_3(A) \in V[G_{\mathbb{Q}}][H]$). It follows as in that claim that $\mathfrak{c} = \kappa$ is real-valued measurable in $V[G_{\mathbb{Q}}][g]$, and that a lifting of j to $j^* : V[G_{\mathbb{Q}}][g] \to N[G_{\mathbb{Q}}][H][j_2(g)]$ definable in $V[G_{\mathbb{Q}}][g][H]$ serves as a witness.

Theorem 6.1 (Woodin). *If $V = L[\mu]$, then in $V[G_{\mathbb{Q}}][g]$, $\mathsf{RVM}(\mathfrak{c})$ and there is a Δ^2_1-well-ordering of $\mathbb{R}$.*

Proof. Let $V_1 = L[\mu][G_{\mathbb{Q}}][g]$, so $\mathbb{R}^{V_1} = \mathbb{R}^{L[\mu][G_{\mathbb{Q}}]}$ and $\mathsf{RVM}(\mathfrak{c})$ holds in V_1. We claim that the well-ordering coded by A is Σ^2_1 in V_1. This we verify by "guessing" the ground model. What the following claim formalizes is our intuition that any structure which resembles $L[\mu]$ sufficiently close must coincide with $L[\mu]$. This resemblance we indicate in terms of a covering property.

Definition 6.2. Let N be a transitive structure that models enough set theory. We say that N *satisfies countable covering* iff

$$\forall \sigma \in \mathcal{P}_{\omega_1}(N)\, \exists \tau \in N\, (\sigma \subseteq \tau \text{ and } N \models |\tau| \leq \aleph_0).$$

Once again, we use $\mathsf{ZFC}^{-\varepsilon}$ to denote a sufficiently strong fragment of ZFC, say (as before), $\mathsf{ZFC}\restriction\Sigma_{200}$, ZFC with the replacement schema restricted to Σ_{200} statements. Obviously, much less suffices.

Claim 6.3. *In V_1, suppose M is transitive, $|M| = \mathfrak{c}$, $M \models \mathsf{ZFC}^{-\varepsilon} + V = L[\mu]$. Let κ_M be the measurable cardinal in the sense of M, and $\kappa = \mathfrak{c}$. Suppose $\kappa_M \geq \mathfrak{c}$, M is iterable and satisfies countable covering. Then $M_\kappa = L[\mu]_\kappa$.*[15]

The hypothesis of Claim 6.3 requires some expansion. The point of the claim is that we have identified the ground model (or, better, the part of the ground model relevant to our argument) in a projective fashion.

See [15] for a careful exposition of iterability at this level and for the necessary background on the argument to follow. What we refer to as K^{DJ} is just called K in [15], and $L[\mu]$ is called there $L[U]$. K^{DJ} is the Dodd-Jensen core model.

Proof. First notice that an initial segment of $L[\mu]$ itself satisfies the requirements: Iterability is clear, and countable covering holds because $\mathbb{Q}$ is ccc and $\mathbb{P}$ is ω_1-closed.

Assume M satisfies the requirements of the claim. Notice that (provably in ZFC + "$L[\mu]$ exists"), $K^{DJ}_\kappa = L[\mu]_\kappa$. It follows that $M_\kappa \models \mathsf{ZFC} + V = K^{DJ}$ (ZFC holds in M_κ because GCH holds in M, so $M \models \kappa$ is strongly inaccessible). So $M_\kappa = (K^{DJ})^{M_\kappa} = K^{DJ} \cap M_\kappa \subseteq K^{DJ} \cap V_\kappa = L[\mu]_\kappa$, where we use [15, Lemma 14.18] to justify the equality $(K^{DJ})^{M_\kappa} = K^{DJ} \cap M_\kappa$ (namely, K^{DJ} is the union of all the mice in the sense of [15] if $0^\sharp$ exists, but being a mouse relativizes downwards).

If $M_\kappa \subsetneq K^{DJ}_\kappa$, then there is a least *sharplike* mouse $\bar{M} \notin M_\kappa$ such that $M_\kappa \subsetneq L[\bar{M}]_\kappa$ (see [15, Chapter 15]). Notice that $K^{DJ}_\kappa \models L[\mu]$ does not exist, because κ is a cardinal in V_1 (otherwise, in V_1, $L\big[L[\mu]^{K^{DJ}_\kappa}\big]$ would really be a model $L[U]$ with U a normal measure in $L[U]$ on some cardinal $\lambda < \kappa$, contradicting the minimality of κ in V_1. It follows from [15, Chapter 16] that there is a nontrivial $j : M_\kappa \xrightarrow{\prec} M_\kappa$ in $L[\bar{M}]_\kappa$, and this certainly contradicts the countable covering property of M considering, for example, the first ω terms of the critical sequence derived from j.

This completes the proof of Claim 6.3. □

We are basically done now: To require iterability of a model M as in Claim 6.3 is a projective requirement; for example, if $M \models V = K^{DJ}$, iterability of M states that every countable premouse (that we can code with a real) that embeds into M in a Σ_1-elementary way is iterable (if M is coded by a set of reals, the existence of this embedding is an assertion about an ω-sequence of reals, coding the range of the embedding, and about the satisfaction relation between a universal Σ_1 formula and the elements of M; all of this can be expressed in a projective fashion; see [15, Lemma 8.7] for a proof of the claimed characterization of iterability). The iterability of a countable premouse is in turn a Π^1_2 statement (uniformly in a real

[15] The notation we use here is ambiguous. For N a model and α an ordinal,

$$N_\alpha = \{\, x \in N : \mathrm{rk}\,(x) < \alpha \,\},$$

where $\mathrm{rk}\,(x)$ is the set-theoretic rank of x. In particular, $L[\mu]_\kappa$ is *not* the κth-stage in the (classical) constructible hierarchy of $L[\mu]$.

coding the premouse as a parameter), see [15, Lemma 13.21]. Hence, to define A in a Σ^2_1-way following the approach explained at the end of section 2 it suffices to notice the following claim, whose proof concludes the proof of Theorem 6.1.

Claim 6.4. *In V_1 suppose $\hat{\delta} < \kappa$ and $a \subseteq \hat{\delta}^+$ is such that $a \notin L[\mu]$ is $\mathtt{Add}(\hat{\delta}^+, 1)$-generic over $L[\mu]$. Then $\hat{\delta}$ is an inaccessible cardinal δ_β or its successor, and $\beta \in A$ iff $\hat{\delta} = \delta_\beta$.*

It follows that A can be defined by refering to those cardinals $\hat{\delta}$ for which there is a set a as above.

Proof. This follows quite easily by what is essentially the decoding argument given during the proof of Theorem 3.1. □

This completes the proof of Theorem 6.1. □

Notice that essentially the same argument provides models of a Σ^2_1-well-ordering together with $\mathsf{RVM}(\mathfrak{c})$, as long as the ground model is fine structural, and the iterability condition for countable mice is projective[16].

Following this approach, granting large cardinals, and starting with a definable fine structural model, the construction produces a model of $\mathsf{RVM}(\mathfrak{c})$ together with a $\Sigma^2_1(\mathrm{Hom}_\infty)$-well-ordering of $\mathbb{R}$. Here, $\Sigma^2_1(\mathrm{Hom}_\infty)$ is the pointclass of sets of reals A such that for some projective formula ψ and some real parameter r, A can be defined by: For all $s \in \mathbb{R}$,

$$s \in A \iff \exists B\, (\psi(s, r, B) \text{ and } B \in \mathrm{Hom}_\infty).$$

The pointclass Hom_∞ consists of all ∞-Homogeneous sets of reals. Under the background assumption that there are unboundedly many Woodin cardinals (in V, not necessarily in the fine structural model), it coincides with the pointclass of all Universally Baire sets of reals. See [39] for definitions, details and references.

7. Real-valued measurability and the Ω-conjecture

This section announces an improvement due to Woodin of the result in section 6. We include enough definitions to make the statement meaningful.

Recall we have shown inherent difficulties to a straightforward attempt to obtain (without anti-large cardinal assumptions) extensions of the universe where $\mathfrak{c}$ is real-valued measurable and there are Δ^2_1-well-orderings of $\mathbb{R}$. The specific technical difficulty that must be resolved is whether it is possible to devise a coding of bounded subsets of $\mathfrak{c}$ by reals. The usual way of obtaining such coding

[16] If M_κ is the model the corresponding version of Claim 6.3 tries to identify, a fake candidate would give rise to a club of inaccessible cardinals below the distinguished measurable κ, again violating covering. Recall that iterability of a fine structural premouse M is in essence a condition about its countable elementary substructures, and that, in the presence of only finitely many Woodin cardinals, this condition is a projective requirement. See for example the introduction to [36].

is by ensuring that some kind of forcing axiom holds. However, we have shown that real-valued measurability contradicts even very general schema toward such forcing axioms. The way this difficulty was dealt with in the previous section was by circumventing it, by working within a "thin" ground model which could therefore be identified in a projective fashion in the relevant forcing extension.

Woodin's idea is to exploit this "thinness" within a broader context. Specifically, instead of trying to establish directly that a Δ^2_1-well-ordering of $\mathbb{R}$ and $\mathsf{RVM}(\mathfrak{c})$ can be added by forcing, he settles for showing the Ω-consistency of this assumption. We proceed now to present a brief summary of Ω-logic, of Ω-consistency, of its connection with the problem of showing consistency via forcing, and close with the statement of Woodin's result and the question of possible generalizations. The reader may also want to look at [5], in this same volume, where Ω-logic is studied in some detail and its basic theorems are established.

In [44], Woodin introduces Ω-logic as a strong logic extending first-order logic (in fact, extending β-logic), and uses it to argue for a negative solution to Cantor's continuum problem. His argument would justify the adoption of $\neg\mathsf{CH}$ if a particular conjecture holds. This conjecture would show that Ω-logic is in a sense as strong as possible for a wide class of statements (including CH). We advise the interested reader to consult [44] for more details. All the results and definitions presented here, unless otherwise explicitly stated, are due to Woodin. However, it must be pointed out that since the appearance of [44] and even [45], the basic definitions have changed somewhat, see [46]. In particular, the definition of Ω-logic we state below is purely semantic, and corresponds to what [45] calls Ω^*-logic. This move requires a slight change in the definition of *proofs* in Ω-logic, as we will explain.

The concept of strong logic is defined in [45]. We do not need it here, but it is useful to mention that we are only interested in it with respect to theories (in a first-order language) extending ZFC. Ω-logic and first-order logic are both examples of strong logics, at opposite ends of the spectrum, first-order logic being the most generous strong logic there is, in the sense that it allows as many structures as possible, and we regard this generosity as a weakness. On the other hand, Ω-logic is the strongest possible logic, allowing only those structures that pass for acceptable models of set theory, under reasonable requirements of acceptability. For example, while first-order logic allows any structure of the form (M, E) as a possible model, ω-logic only allows those structures that "compute V_ω correctly" and β-logic only allows those structures that are correct about well-foundedness. Ω-logic goes as far in this direction as possible, subject to natural requirements that we list below.

Recall that if M is a transitive structure, $M_\alpha = \{ x \in M : \mathrm{rk}\,(x) < \alpha \}$, see also [5, §1.1]. The following is also [5, Definition 1.7].

Definition 7.1 (Ω-logic). Let $T \supseteq \mathsf{ZFC}$ and let ϕ be a sentence. Then

$$T \models_\Omega \phi$$

iff for all $\mathbb{P}$ and all λ, if $V^{\mathbb{P}}_\lambda \models T$, then $V^{\mathbb{P}}_\lambda \models \phi$.

Remark 7.2. According to this definition, an Ω-*satisfiable* sentence ϕ, i.e., a sentence ϕ such that $\neg\phi$ is not Ω-valid, is one such that for some $\mathbb{P}$ and α, $V_\alpha^{\mathbb{P}} \models \mathsf{ZFC} + \phi$, see [5, Definition 3.1]. It is easy to see that if ϕ is Σ_2 and Ω-satisfiable, then in fact ϕ is forceable *over* V, i.e., for some $\mathbb{P}$, $V^{\mathbb{P}} \models \phi$. In effect, let $\phi \equiv \exists x\, \psi(x)$ be a Σ_2 sentence, where $\psi(x)$ is Π_1. Suppose ϕ is Ω-satisfiable, and let $\alpha, \mathbb{P}$ be such that $V_\alpha^{\mathbb{P}} \models \mathsf{ZFC}+\phi$. Let u be such that $V_\alpha^{\mathbb{P}} \models \psi(u)$ and let $\omega < \kappa < \alpha$ be a cardinal in $V_\alpha^{\mathbb{P}}$ (and therefore in $V^{\mathbb{P}}$), sufficiently large so $u \in H_\kappa^{\mathbb{P}}$. A well-known result of Levy (see [35]) asserts that whenever $\lambda > \omega$ is a cardinal, $H_\lambda \prec_1 V$. Relativizing Levy's result to $V_\alpha^{\mathbb{P}}$, it follows that $H_\kappa^{\mathbb{P}} \models \psi(u)$. Applying Levy's result in $V^{\mathbb{P}}$, we see that $V^{\mathbb{P}} \models \phi$, as wanted.

A logic (in the sense of a satisfaction relation between first-order structures and first-order statements) satisfying the definition of Ω-logic (and, perhaps, being more restrictive) is said to be *generically sound.*

An important difference between first-order logic and Ω-logic is that the latter requires a healthy large cardinal structure on the background universe for certain absoluteness requirements to hold; this absoluteness is essential for a reasonable study of Ω-logic. For this section, let us define:

Definition 7.3. By *our Base Theory* we mean

$$\mathsf{ZFC} + \text{“There is a proper class of Woodin cardinals.”}$$

The following is proven in [5, Theorem 1.8].

Theorem 7.4 (Generic Invariance). *Assume our Base Theory. Let* $T \supseteq \mathsf{ZFC}$ *and let* ϕ *be a sentence. Then* $T \models_\Omega \phi$ *iff for all* $\mathbb{P}$, $V^{\mathbb{P}} \models T \models_\Omega \phi$. □

Corresponding to the semantic notion of satisfiability we want to develop a syntactic counterpart, $\vdash_\Omega$. Recall that proofs in first-order logic can be construed as certain trees. Similarly, for Ω-logic, we develop a notion of *certificate* that plays this role.

The certificates in this case are more specialized, and it is better to present first the sets in terms of which we are to define them, the Universally Baire sets, which we introduce directly in the way we need them, by what is usually stated as a corollary of their standard definition. See also [5, §2.1].

Definition 7.5 (Feng, Magidor, Woodin [16]). Let λ be an infinite cardinal. A set $A \subseteq \omega^\omega$ is λ-*Universally Baire* iff there are λ-*absolutely complementing* trees for A, i.e., a pair T, T^* of trees on $\omega \times X$ for some X, such that

1. $A = p[T]$ and $\omega^\omega \setminus A = p[T^*]$.
2. $1 \Vdash_{\mathbb{P}} p[T] \cup p[T^*] = \omega^\omega$ for any forcing $\mathbb{P}$ of size at most λ.

A is ∞-Universally Baire or, simply, Universally Baire, iff it is λ-Universally Baire for all λ.

Notice that if A is λ-Universally Baire, and $T, T^*, \mathbb{P}$ are as above, then $1 \Vdash_{\mathbb{P}} p[T] \cap p[T^*] = \emptyset$.

The Universally Baire sets generalize the Borel sets and have all the usual regularity properties.

Under reasonable large cardinal assumptions, the pointclass of Universally Baire sets is quite closed. For example (see [16] for the case $A = \mathbb{R}$ or [34] for the general case):

Fact 7.6. *Assume our Base Theory. Suppose A is Universally Baire. Then every set of reals in $L(A, \mathbb{R})$ is Universally Baire.* □

There are somewhat cleaner ways of stating this fact. For example, since our Base Theory grants that every set has a sharp, Fact 7.6 is equivalent to (see [34]):

Fact 7.7. *Assume our Base Theory. Suppose A is Universally Baire. Then $A^\sharp$ is Universally Baire.* □

Given such a set A, it makes sense to talk about its *interpretation* in extensions of the universe, in what generalizes the idea of Borel codes for Borel sets.

Definition 7.8. Let A be Universally Baire. Let $\mathbb{P}$ be a forcing notion, and let G be $\mathbb{P}$-generic over V. Then the interpretation A_G of A in $V[G]$ is

$$A_G = \bigcup \{ p[T] : T \in V \text{ and } V \models A = p[T] \}.$$

This is the natural notion we would expect: If T, T^* are λ-complementing trees such that $p[T] = A$, if $|\mathbb{P}| \leq \lambda$ and G is $\mathbb{P}$-generic over V, then $V[G] \models A_G = p[T]$.[17]

The certificates for Ω-logic are issued in terms of Universally Baire sets, and thus we arrive at the concept of A-closed structures. See also [5, §2.2].

Definition 7.9. Let $A \subseteq \omega^\omega$ be Universally Baire. A transitive set M is *A-closed* iff for all $\mathbb{P} \in M$ and all $\mathbb{P}$-terms $\tau \in M$,

$$\{ p \in \mathbb{P} : V \models p \Vdash \tau \in A_G \} \in M.$$

Remark 7.10. In practice, countable transitive A-closed models M are those admitting a pair of "absolutely complementing with respect to M" trees $T, T^* \in M$ such that the interpretation of A (which needs not be in M) would be in forcing extensions of M by forcing notions in M given by the projection of T, and such that in V, $p[T] \subseteq A$ and $p[T^*] \subseteq \mathbb{R} \setminus A$. Notice that M-generics for forcing notions in M exist in V, since M is countable.

Even though the official definition restricts the A-closed structures from the beginning to transitive sets, it may be helpful to point out that β-logic can be characterized in terms of A-closure: An ω-model $(M, E) \models \mathsf{ZFC}$ is well founded iff,

[17] A word of warning is in order: Suppose $A = \{ r \in \mathbb{R} : \varphi(r) \}$ is Universally Baire, where φ is, say, Σ^1_3. It does *not* follow that

$$A_G = \{ r \in \mathbb{R}^{V[G]} : \varphi(r) \}. \qquad (*)$$

Universally Baire sets figure prominently in generic absoluteness arguments but, in addition, equalities like $(*)$ need to be ensured. See for example [39].

under the proper interpretation, it is A-closed for each Π^1_1-set A, see [5, Theorem 2.23] for a proof.

The following is [44, Lemma 10.143], see [5, Proposition 2.9] for a proof.

Theorem 7.11 (Woodin). *Let $M \models \mathsf{ZFC}$ be transitive, and let A be Universally Baire. Then the following are equivalent:*

1. *M is A-closed.*
2. *Suppose $\mathbb{P} \in M$ and G is $\mathbb{P}$-generic over V. Then*

$$V[G] \models A_G \cap M[G] \in M[G]. \qquad \square$$

With the concept of A-closed structures at hand, we are ready to define provability in Ω-logic. That our discussion is not vacuous is the content of the following fact; in practice more delicate results are required. See [34, §4] for techniques that can easily be adapted to prove strengthened versions of Fact 7.12.

Fact 7.12. *Assume our Base Theory. Let A be a Universally Baire set. Then there are A-closed countable transitive models of* ZFC. $\square$

See [5, §2.4] for basic results about the following notion.

Definition 7.13 ($\vdash_\Omega$). Let $T \supseteq \mathsf{ZFC}$ be a theory, and let ϕ be a sentence. Then

$$T \vdash_\Omega \phi$$

iff there exists a Universally Baire set A such that

1. $L(A, \mathbb{R}) \models \mathsf{AD}^+$.
2. $A^\sharp$ exists and is Universally Baire.
3. Whenever M is a countable, transitive, A-closed model of ZFC and $\alpha \in \mathrm{ORD}^M$ is such that $M_\alpha \models T$, then $M_\alpha \models \phi$.

See [44, Chapter 10] or [47] for an introduction to AD^+.

In [44], the notion now called $\models_\Omega$ was denoted $\vdash_{\Omega^*}$ and called Ω^*-logic. Ω-logic was defined by a slight variation of Definition 7.13, namely instead of requiring that if $M_\alpha \models T$ then $M_\alpha \models \phi$ for initial segments M_α of M, this was required of M itself. The change allows for a cleaner version of the Ω-conjecture, see Conjecture 7.17. Originally, the Ω-conjecture needed to be stated in terms of Π_2 statements. The other difference between the definition given here and the one in [44] is due to the fact that Definition 7.13 is stated in ZFC and not in our Base Theory. Under our Base Theory, assumptions 1 and 2 hold automatically. These assumptions are what is required to prove the existence of appropriate A-closed structures, see [34].

One of the nicest features of $\vdash_\Omega$ is that it does not depend on the particular universe where it is considered, at least if we restrict our attention to possible generic extensions. This is the content of [44, Theorem 10.146], see [5, Theorem 2.35] for a proof.

Theorem 7.14 (Generic Invariance). *Assume our Base Theory. Let $T \supseteq$ ZFC and let ϕ be a sentence. Then $T \vdash_\Omega \phi$ iff for all $\mathbb{P}$, $V^{\mathbb{P}} \models T \vdash_\Omega \phi$.* $\square$

See [5, Theorem 3.3] for a proof of the following under the additional assumption of the existence of a proper class of strongly inaccessible cardinals.

Theorem 7.15 (Generic Soundness). *Let $T \supseteq \mathsf{ZFC}$ and let ϕ be a sentence. Suppose $T \vdash_\Omega \phi$. Then $T \models_\Omega \phi$.* □

Remark 7.16. The previous definition of $\vdash_\Omega$ required the background assumption of our Base Theory in order for Theorem 7.15 to hold. Notice that with the new definition it is stated as a ZFC result.

The Ω-conjecture is the statement that $\vdash_\Omega$ is the notion of provability associated to $\models_\Omega$ in the sense that the completeness theorem for Ω-logic holds. See [5, §3] for an interesting discussion of this conjecture.

Conjecture 7.17 (Ω-Conjecture). *Assume our Base Theory and let ϕ be a sentence. Then $\mathsf{ZFC} \models_\Omega \phi$ iff $\mathsf{ZFC} \vdash_\Omega \phi$.*

Woodin has shown that the Ω-conjecture is true unless (in a precise sense) there are large cardinal hypothesis implying a strong failure of iterability, see [44] and [45].

Definition 7.18 (Ω-consistency). Assume our Base Theory. Let $T \supseteq \mathsf{ZFC}$ and let ϕ be a sentence. Then ϕ is *Ω-consistent relative to T* (and if $T = \mathsf{ZFC}$, we just say ϕ is Ω-consistent) iff for any Universally Baire set A there is an A-closed countable transitive $M \models T + \phi$.

Hence, at least as far as we can see nowadays, in order to prove that a proper class model of a Σ_2-sentence ϕ can be achieved (from large cardinals) by forcing, it suffices to show that for any Universally Baire set A, ϕ holds in an appropriate A-closed model M of ZFC. The intention of this comment is that it is not the same to prove that a sentence ϕ is forceable from an inner model than from the ground model itself. After all, ϕ may hold in forcing extensions of an inner model because that model is not sufficiently correct. For a trivial example, L admits a projective well-ordering of the reals, but such well-orderings are impossible in the presence of mild large cardinals. However, if the Ω-conjecture holds, and ϕ is Ω-consistent, then in fact ϕ can be forced over V.

Notice that any statement of the form $\exists\alpha\,(V_\alpha \models \phi)$, where ϕ is a sentence, is Σ_2, and any statement of the form $\forall\alpha\,(V_\alpha \models \phi)$, for ϕ a sentence, is Π_2. The following follows immediately:

Fact 7.19. *The statement* "$\mathsf{RVM}(\mathfrak{c})+$ There is a Δ^2_1-well-ordering of $\mathbb{R}$" *can be rendered in a Δ_2-way.* □

The reader should appreciate by now how powerful the Ω-conjecture is, since the witnesses to Ω-consistency of a sentence ϕ can be "fine structural-like" models, their fine structural features may be used in essential ways to establish the validity of ϕ, and nonetheless we can conclude that ϕ can be forced over the universe, without the need of any fine structural of anti-large cardinal requirements.

Since we do not know how to force a Δ^2_1-well-ordering of the reals together with $\mathsf{RVM}(\mathfrak{c})$, unless we have some nice control over the ground model itself, it was natural to attempt a proof of the Ω-consistency of this assumption. Woodin has succeeded in this attempt, and we close this section with his result and a few comments.

Theorem 7.20 (Woodin). *Assume our Base Theory. Then it is Ω-consistent that $\mathfrak{c}$ is real-valued measurable and there is a Σ^2_1-well-ordering of $\mathbb{R}$.* □

This result is proved in [47]. The idea is to use the large cardinal assumption to produce, given a Universally Baire set A, A-closed and sufficiently "fine structure-like" inner models of strong versions[18] of AD^+ over which forcing with $\mathbb{Q}_{\max}$ produces ZFC-models with a distinguished measurable cardinal. The measurable is used to produce a further extension, by forcing as in section 6. This provides us, combined with the fine structural features of the ground model, with an appropriate covering argument that can be used in place of Claim 6.3 to correctly identify enough of HOD of the ground model to obtain the desired Σ^2_1-definition. The ground model can in fact be chosen so the forcing extension itself is A-closed, and this gives the result. The covering argument rests on factoring properties of the generic embeddings derived from forcing with the nonstationary ideal, using the features that $\mathbb{Q}_{\max}$ provides.

It follows immediately that granting large cardinals, if the Ω-conjecture holds then the conclusion of Theorem 7.20 can actually be forced. The following, however, remains open (from any large cardinal assumptions).

Question 7.21. *Assume κ is measurable and* GCH *holds. Is there a forcing extension where $\kappa = \mathfrak{c}$ is real-valued measurable, and there is a Δ^2_1-well-ordering of $\mathbb{R}$?*

8. Real-valued huge cardinals

The result of this section serves a two-fold goal. It shows that $\mathsf{RVM}(\mathfrak{c})$ and a Σ^2_1, or even Σ^2_n-well-ordering of $\mathbb{R}$ for some $n < \omega$, cannot be obtained for free. It also shows that there are limits to how far the techniques of this paper can generalize.

Definition 8.1. A cardinal κ is *real-valued huge* iff there is $\lambda \geq \omega_1$ such that in $V^{\mathtt{Random}_\lambda}$ there exists an elementary embedding $j : V \xrightarrow{\prec} N$ with $\mathrm{cp}(j) = \kappa$ and such that ${}^{j(\kappa)}N \subseteq N$.

The following is clear:

Lemma 8.2. *If κ is huge, then $V^{\mathtt{Random}_\kappa} \models \kappa = \mathfrak{c}$ is real-valued huge.*

[18] These models have the form $N = L_\Gamma(\mathbb{R}, \mu)$, where μ is the restriction to N of some normal measure ν on some cardinal κ, $\mu = \nu \cap N \in N$, and Γ is a particular closure operator which also plays the role of the tree for Σ^2_1 inside the model.

Proof. Let $j : V \to M$ in V witness hugeness of κ, so ${}^{j(\kappa)}M \subseteq M$ and $\mathrm{cp}(j) = \kappa$. Set $\mathbb{Q} = \mathtt{Random}_\kappa$. Let G be $\mathbb{Q}$-generic over V, and let H be $j(\mathbb{Q})/\mathbb{Q}$-generic over $V[G]$. By Theorem 1.34, we just need to verify that in $V[G][H]$, j lifts to

$$j^* : V[G] \to M[G][H]$$

and that $V[G][H] \models {}^{j(\kappa)}M[G][H] \subseteq M[G][H]$. As usual, the lifting j^* is given by $j^*(\tau_G) = j(\tau)_{G^\frown H}$. This is well defined and elementary.

Given a sequence of names $\vec{\tau} = \langle \tau_\alpha : \alpha < j(\kappa) \rangle$ with each τ a $j(\mathbb{Q})$-name in M, the whole sequence $\vec{\tau}$ belongs to M and, therefore, $\langle (\tau_\alpha)_{G^\frown H} : \alpha < j(\kappa) \rangle \in M[G][H]$. From this the result follows. □

Having shown the consistency of real-valued hugeness of the continuum, we now point out the following observation due to Woodin:

Fact 8.3 (Woodin). *Suppose $\mathfrak{c}$ is real-valued huge. Then there are no third-order definable well-orderings of the reals.*

Proof. The same argument as for $L(\mathbb{R})$ in Theorem 2.5 works:

Towards a contradiction, let $\varphi(x, y, z)$ be a third-order formula in the language of arithmetic, and let $t \in \mathbb{R}$ be such that for some well-ordering $<$ of $\mathbb{R}$, $\varphi(r, s, t)$ holds of reals r, s iff $r < s$.

Let λ and G a $\mathtt{Random}_\lambda$-generic over V be such that in $V[G]$ there is an embedding $j : V \to N$ with $\mathrm{cp}(j) = \mathfrak{c}^V$ and ${}^{j(\mathfrak{c}^V)}N \subseteq N$. Then $\mathcal{P}(\mathbb{R})^{V[G]} \subseteq N$, since $|\mathbb{R}| = j(\mathfrak{c}^V)$ holds in N (and $\mathbb{R}^N = \mathbb{R}^{V[G]}$, since ${}^\omega N \subseteq N$.) But this means that third-order statements in the language of arithmetic, with parameters from N, are absolute between N and $V[G]$.

We are done, because by elementarity $\varphi(\cdot, \cdot, t)$ would be a third-order definition of a well-ordering of the reals in $V[G]$, but this is impossible by Corollary 2.2. □

Remark 8.4. Notice that what the proof actually shows is that if $\mathfrak{c}$ is real-valued huge and λ is as in Definition 8.1, then $V \equiv_{\underset{\sim}{\Sigma}^2_\omega} V^{\mathtt{Random}_\lambda}$, where boldface indicates that real parameters from V are allowed.

The argument of Theorem 3.1 breaks down very early when trying to adapt it to the case where κ is huge. For example, the existence of the N-generic object we called G^* cannot be ensured due to the strong closure of N.

Remark 8.4 suggests the natural question of whether generic invariance of Σ^2_ω with respect to "$\mathfrak{c}$ is real-valued huge" holds. This seems somewhat delicate, since there does not seem to be a natural counterpart to Solovay's Fact 1.27 for preservation of real-valued hugeness. The hypothesis is by no means intended to be optimal. For example, it is not clear whether the natural real-valued version of $\mathcal{P}^2(\kappa)$-measurability of κ for $\kappa = \mathfrak{c}$ suffices to rule out the existence of third-order definable well-orderings of $\mathbb{R}$.

As expected, real-valued hugeness is a serious large cardinal assumption, strictly stronger than real-valued measurability. Here we content ourselves with some easy observations and a remark:

Fact 8.5. *If κ is real-valued huge, then there are weakly inaccessible cardinals larger than κ.*

Proof. Let λ be as in Definition 8.1, and in $V^{\mathtt{Random}_\lambda}$, let $j : V \to N$ be the witnessing embedding. Then $N \models j(\kappa)$ is real-valued measurable, so in particular $N \models j(\kappa)$ is weakly inaccessible.

But $V[G] \models {}^{j(\kappa)}N \subseteq N$, so $j(\kappa)$ *is* weakly inaccessible in $V[G]$, and therefore in V.

As usual, the proof actually shows that there are fixed points of the weakly Mahlo hierarchy (see [26, after Proposition 1.1]), etc., above κ. □

Theorem 8.6. *If $\kappa \leq \mathfrak{c}$ is real-valued huge, then the real-valued measurable cardinals are unbounded below κ. In fact, for a witnessing probability ν,*

$$\nu(\{\, \alpha < \kappa : \mathsf{RVM}(\alpha)\,\}) = 1.$$

Proof. As before, let λ be as in Definition 8.1. Let $\varphi : \mathtt{Random}_\lambda \to [0,1]$ be the 'probability measure' associated to $\mathtt{Random}_\lambda$, fix a $\mathtt{Random}_\lambda$-generic G over V and, in $V[G]$, let $j : V \to N$ witness real-valued hugeness of κ.

By Fact 1.27, $\mathsf{RVM}(\kappa)$ holds in $V[G]$. Let $\hat{\nu} : \mathcal{P}(\kappa) \to [0,1]$ be a witness. Notice that $[0,1] \in N$. Since in $V[G]$, ${}^{j(\kappa)}N \subseteq N$ and $|\mathcal{P}(\kappa)| = 2^\kappa \leq j(\kappa)$, then in particular $\hat{\nu} \in N$. Thus, $N \models \mathsf{RVM}(\kappa)$.

Since G was arbitrary, $\varphi[\![\kappa \in j(\{\, \alpha : \mathsf{RVM}(\alpha)\,\})]\!] = \mathbb{1}$, where j denotes a term for an embedding witnessing real-valued hugeness of κ.

In V, let $\nu : \mathcal{P}(\kappa) \to [0,1]$ be defined as usual by $\nu(A) = \varphi[\![\kappa \in j(A)]\!]$. Then ν is as required.

As usual, this proof actually gives that κ is limit of real-valued measurable cardinals that also concentrate on real-valued measurable cardinals that concentrate on real-valued measurable cardinals, etc. □

Real-valued huge cardinals imply the existence of inner models for Woodin cardinals. In the presence of measurable cardinals this is an immediate consequence of the following result of Steel. It appears as [40, Theorem 7.1] under the stronger assumption that Ω is measurable.

Theorem 8.7 (Steel). *Suppose $V_\Omega^\sharp$ exists, and let G be $\mathbb{P}$-generic over V for some $\mathbb{P} \in V_\Omega$. Suppose that in $V[G]$ there is a transitive class M and an elementary embedding*

$$j : V \to M \subseteq V[G]$$

with $\mathrm{cp}(j) = \kappa$ and such that $V[G] \models {}^{<j(\kappa)}M \subseteq M$. Then the K^c-construction reaches a non-1-small level[19]*.* □

[19] I.e., $M_1^\sharp$, the sharp for a proper class fine structural inner model with a Woodin cardinal, exists.

In fact, much more follows from this hypothesis. For example, it is straightforward to improve the argument leading to Theorem 8.6 to a proof of the fact that there is a 'probability measure' $\nu : \mathcal{P}(\mathfrak{c}) \to [0,1]$ such that $\nu(\{\, \alpha : \alpha \text{ is real-valued almost huge} \,\}) = 1$. Here, a cardinal κ is called real-valued almost huge iff there is a $\lambda \geq \omega_1$ such that in $V^{\mathtt{Random}_\lambda}$ there is an embedding $j : V \to N$ with $\mathrm{cp}(j) = \kappa$ and such that $V^{\mathtt{Random}_\lambda} \models {}^{<j(\kappa)}N \subseteq N$.

Using his technique of the core model induction, Woodin has shown:

Theorem 8.8 (Woodin). *If there is a real-valued almost huge cardinal, then* $\mathsf{AD}^{L(\mathbb{R}\cup\{\mathbb{R}^\sharp\})}$ *holds.* □

For more on real-valued huge cardinals and a strengthening of the above result, see [12].

Remark 8.9. Anti-definability results can also be achieved by fine structural arguments starting with $V = L[\mu]$.

References

[1] U. Abraham, S. Shelah. *Forcing closed unbounded sets*, The Journal of Symbolic Logic **48 (3)** (1983), 643–657 ([AbSh 146]).

[2] U. Abraham, S. Shelah. *A Δ^2_2 well-order of the reals and incompactness of $L(Q^{MM})$*, Annals of Pure and Applied Logic **59 (1)** (1993), 1–32 ([AbSh 403]).

[3] U. Abraham, S. Shelah. *Martin's Axiom and Δ^2_1 well-ordering of the reals*, Archive for Mathematical Logic **35 (5–6)** (1996), 287–298 ([AbSh 458]).

[4] U. Abraham, S. Shelah. *Coding with ladders a well ordering of the reals*, The Journal of Symbolic Logic **67 (2)** (2002), 579–597 ([AbSh 485]).

[5] J. Bagaria, N. Castells, and P. Larson. *An Ω-logic primer*, this volume.

[6] B. Balcar, T. Jech, and J. Zapletal. *Semi-Cohen Boolean algebras*, Annals of Pure and Applied Logic **87** (1997), 187–208.

[7] S. Banach, K. Kuratowski. *Sur une généralisation du problème de la mesure*, Fundamenta Mathematicae **14** (1929), 127–131.

[8] T. Bartoszyński, H. Judah. **Set Theory. On the structure of the real line**, A.K. Peters (1995).

[9] J. Baumgartner. *Iterated Forcing*, in **Surveys in Set Theory**, A.D.R. Mathias, ed., Cambridge University Press (1983), 1–59.

[10] J. Baumgartner. *Applications of the Proper Forcing Axiom*, in **Handbook of Set-Theoretic Topology**, K. Kunen and J.E. Vaughan, eds., Elsevier Science Publishers (1984), 913–959.

[11] A. Caicedo. *Simply definable well-orderings of the reals*, Ph.D. Dissertation, Department of Mathematics, University of California, Berkeley (2003).

[12] A. Caicedo. *The strength of real-valued huge cardinals*, in preparation.

[13] J. Cummings. *A model where* GCH *holds at successors but fails at limits*, Transactions of the American Mathematical Society **329 (1)** (1992), 1–39.

[14] J. Cummings. *Iterated Forcing and Elementary Embeddings*, to appear in **Handbook of Set Theory**, M. Foreman, A. Kanamori, and M. Magidor, eds., to appear.

[15] A. Dodd. **the Core Model**, Cambridge University Press (1982).

[16] Q. Feng, M. Magidor, and H. Woodin. *Universally Baire sets of reals*, in **Set Theory of the continuum**, H. Judah, W. Just, and H. Woodin, eds., Springer-Verlag (1992), 203–242.

[17] D. fremlin. **Consequences of Martin's Axiom**, Cambridge University Press (1984).

[18] D. Fremlin. *Measure Algebras*, in **Handbook of Boolean Algebras**, volume 3, J. Monk and R. Bonnet, eds., North-Holland (1989), 877–980.

[19] D. Fremlin. *Real-valued measurable cardinals* in **Set Theory of the Reals**, H. Judah, ed., Israel Mathematical Conference Proceedings **6**, Bar-Ilan University (1993), 151–304.

[20] S. Fuchino. *Open Coloring Axiom and Forcing Axioms*, preprint.

[21] M. Gitik, S. Shelah. *Forcing with ideals and simple forcing notions*, Israel Journal of Mathematics **68** (1989), 129–160 ([GiSh 357]).

[22] M. Gitik, S. Shelah. *More on Real-valued measurable cardinals and forcing with ideals*, Israel Journal of Mathematics **124** (2001), 221–242 ([GiSh 582]).

[23] J. Hamkins. *Lifting and extending measures by forcing: fragile measurability*, Ph.D. Dissertation, Department of Mathematics, University of California, Berkeley (1994).

[24] T. Jech. **Set Theory**, Academic Press (1978).

[25] R. Jensen. *Measurable cardinals and the* GCH, in **Axiomatic set theory (Proc. Sympos. Pure Math., Vol. XIII, Part II, Univ. California, Los Angeles, Calif., 1967)**, American Mathematical Society (1974), 175–178.

[26] A. Kanamori. **The Higher Infinite**, Springer-Verlag (1994).

[27] S. Koppelberg. **General Theory of Boolean Algebras**, volume 1 of **Handbook of Boolean Algebras**, J. Monk and R. Bonnet, eds., North-Holland (1989).

[28] S. Koppelberg. *Characterizations of Cohen algebras*, in **Papers on general topology and applications: Seventh Conference at the University of Wisconsin**, S. Andima, R. Kopperman, P. Misra, M.E. Rudin, and A. Todd, eds., Annals of the New York Academy of Sciences **704** (1993), 222–237.

[29] S. Koppelberg, S. Shelah. *Subalgebras of Cohen algebras need not be Cohen*, in **Logic: From Foundations to Applications. European Logic Colloquium**, W. Hodges, M. Hyland, C. Steinhorn, and J. Truss, eds., Oxford University Press (1996) ([KpSh 504]).

[30] K. Kunen. *Saturated ideals*, The Journal of Symbolic Logic **43 (1)** (1978), 65–76.

[31] K. Kunen. **Set Theory. An introduction to independence proofs**, Elsevier Science Publishers (1980).

[32] K. Kunen. *Random and Cohen reals*, in **Handbook of Set-Theoretic Topology**, K. Kunen and J.E. Vaughan, eds., Elsevier Science Publishers (1984), 887–911.

[33] P. Larson. **The Stationary Tower: Notes on a Course by W. Hugh Woodin**, American Mathematical Society (2004).

[34] P. Larson. *Forcing over models of determinacy*, preprint. To appear in **Handbook of Set Theory**, M. Foreman, A. Kanamori, and M. Magidor, eds.

[35] A. Levy. *A hierarchy of formulas in set theory*, Memoirs of the American Mathematical Society **57** (1965).

[36] D. Martin, J. Steel. *Iteration trees*, Journal of the American Mathematical Society **7 (1)** (1994), 1–73.

[37] J. Roitman. *Adding a random or a Cohen real: topological consequences and the effect on Martin's axiom*, Fundamenta Mathematicae **103 (1)** (1979), 47–60.

[38] R. Solovay. *Real-valued measurable cardinals*, in **Axiomatic set theory, Part I**, American Mathematical Society (1971), 397–428.

[39] J. Steel. *The derived model theorem*, preprint.

[40] J. Steel. **The core model iterability problem**, Springer-Verlag (1996).

[41] S. Todorčević. *Remarks on chain conditions in products*, Compositio Mathematica **55 (3)** (1985), 295–302.

[42] S. Todorčević. *Trees and linearly ordered sets* in **Handbook of Set-Theoretic Topology**, K. Kunen and J.E. Vaughan, eds., Elsevier Science Publishers (1984), 235–294.

[43] S. Ulam. *Zur Masstheorie in der allgemeinen Mengenlehre*, Fundamenta Mathematicae **16** (1930), 140–150.

[44] H. Woodin. **The Axiom of Determinacy, Forcing Axioms, and the Nonstationary Ideal**, Walter de Gruyter (1999).

[45] H. Woodin. *The Continuum Hypothesis and the Ω-conjecture*, slides of a talk given at the Fields Institute during the **Thematic Program on Set Theory and Analysis, September–December 2002**, A. Dow, A. Kechris, M. Laczkovich, C. Laflamme, J. Steprans, and S. Todorčević, organizers.

[46] H. Woodin. *Set Theory after Russell; the journey back to Eden*, in **One hundred years of Russell's paradox**, G. Link, ed., Walter de Gruyter (2004).

[47] H. Woodin, A. Caicedo. *Real-valued measurable cardinals and Σ^2_1-well-orderings of the reals*, preprint.

Andrés Eduardo Caicedo
Department of Mathematics
Mail code 253-37
California Institute of Technology
Pasadena, CA 91125, USA
e-mail: caicedo@caltech.edu

Set Theory
Trends in Mathematics, 121–147

Complexity of Sets and Binary Relations in Continuum Theory: A Survey

Alberto Marcone

Contents

In the last few years there have been many applications of descriptive set theory to the study of continua. In this paper we focus on classification results for the complexity of natural sets of continua and binary relations between continua. Recent papers in this area include [Dar00, Kru02, Kru03, Kru04, CDM05, DM04]. However the subject is much older, as witnessed by papers such as [Kur31, Maz31]. On one hand continuum theory provides natural examples for many phenomena of descriptive set theory (e.g., natural sets requiring the difference hierarchy for their classification occur quite frequently in this area – see Theorems 2.35, 2.36 and 2.41 below). On the other hand descriptive set theory sheds some light on continuum theory, e.g., by explaining why some classes of continua do not have simple topological characterizations. In this paper we will survey both classical and recent results, and state some open problems. The present paper can be viewed as an update of a portion (pp. 9–11) of the survey by Becker ([Bec92]), which was much broader in scope.

The relationship between descriptive set theory and continuum theory involves many other topics not covered in this survey: these include universal sets and pairs of inseparable sets. Other striking applications of descriptive set theory to the study of continua include the results by Solecki ([Sol02b], see [Sol02a, §4] for a survey) about the space of composants of an indecomposable continuum, and by Becker and others ([Bec98, BP01]) about path components.

Nadler's monograph [Nad92] and Kechris's textbook [Kec95] are our main references for continuum theory and descriptive set theory, respectively.

Let us recall the basic notions of continuum theory (as in [Nad92], we will be concerned exclusively with continua that are metrizable, i.e., with *metric continua*):

- A *continuum* is a compact and connected metric space.
- A *subcontinuum* of a continuum C is a subset of C which is also a continuum.
- A continuum is *nondegenerate* if it contains more than one point (and hence it has the cardinality of the continuum).
- A continuum is *planar* if it is homeomorphic to a subset of $\mathbb{R}^2$.

Section 1 contains the necessary background in descriptive set theory: the reader already familiar with the basics of the subject can safely skip it and refer back to it when needed. In Section 2 we consider the complexity of natural classes of continua and sketch a few proofs illustrating some basic techniques. Subsection 2.4 includes full details of some unpublished proofs of H. Becker. Section 3 deals with binary relations among continua, and in particular with quasi-orders and equivalence relations; this section includes also results about the complexity of equivalence classes of an equivalence relation and of initial segments of a quasi-order.

1. Descriptive set theory

1.1. Spaces of continua

We start by describing how descriptive set theory deals with continua. If X is a compact metric space with (necessarily complete) metric d, we denote by $\mathsf{K}(X)$ the hyperspace of nonempty compact subsets of X, equipped with the Vietoris topology which is generated by the Hausdorff metric, denoted by d_H. Recall that if $K, L \in \mathsf{K}(X)$ we have

$$d_H(K, L) = \max\left\{\max_{x \in K} d(x, L), \max_{x \in L} d(x, K)\right\}.$$

$\mathsf{K}(X)$ is a compact metric space ([Kec95, §4.F] or [Nad92, chapter IV]). We denote by $\mathsf{C}(X)$ the subset of $\mathsf{K}(X)$ which consists of all connected elements, i.e., of all continua included in X. $\mathsf{C}(X)$ is closed in $\mathsf{K}(X)$ and, therefore, it is a compact metric space. In particular $\mathsf{C}(X)$ is separable and completely metrizable, i.e., a *Polish space*, and thus the typical ambient space for descriptive set theory.

Let I be the closed interval $[0, 1]$. Every compact metric space, and in particular every continuum, is homeomorphic to a closed subset of the Hilbert cube I^ω. Hence $\mathsf{C}(I^\omega)$ is a compact metric space containing a homeomorphic copy of every continuum. We say that $\mathsf{C}(I^\omega)$ is the *Polish space of continua.* Similarly, $\mathsf{C}(I^2)$ is a compact metric space containing a homeomorphic copy of every planar continuum, and we say that $\mathsf{C}(I^2)$ is the *Polish space of planar continua.* Now we can study subsets of $\mathsf{C}(I^\omega)$, $\mathsf{C}(I^2)$, $\mathsf{C}(I^3)$, $\mathsf{C}(I^\omega) \times \mathsf{C}(I^\omega)$, etc., with the tools and techniques of descriptive set theory.

Suppose we are given a class of continua which is topological, i.e., invariant under homeomorphisms (a typical example is the class of continua which are locally connected). In light of the previous discussion it makes sense to identify the class with the set $\mathcal{P} \subseteq \mathsf{C}(I^\omega)$ of all subcontinua of I^ω belonging to the class, so that $\mathcal{P}$ can be studied with the tools and techniques of descriptive set theory. Similarly, by considering $\mathcal{P} \cap \mathsf{C}(I^2)$ as a subset of $\mathsf{C}(I^2)$ we can study the set of planar continua belonging to the class.

In a similar fashion we can translate a relationship between continua (typical examples are homeomorphism and continuous embeddability) into a binary relation on $\mathsf{C}(I^\omega)$ or $\mathsf{C}(I^2)$.

In some situations we are interested in studying metric, rather than topological, properties of continua (see, e.g., §3.6 below). Since the space $\mathsf{C}(I^\omega)$ obviously does not contain an isometric copy of every continuum, it is no longer the appropriate setting for this study.

We denote by M the Urysohn space: it is the unique, up to isometry, Polish space which contains an isometric copy of every Polish space. $\mathsf{C}(M)$ is Polish and contains an isometric copy of every continuum: we say that it is the *Polish space of metric continua.* We now view classes of continua which are isometric, i.e., invariant under isometries, as subsets of $\mathsf{C}(M)$.

We are interested in classes and relations which have been studied for their own sake in continuum theory (rather than being built *ad hoc* so that the corresponding set exhibits certain descriptive set theoretic features) and in this case we often say that the class or the relation is natural (this is a sociological, rather than mathematical, notion).

1.2. Descriptive set theoretic hierarchies

The main goal of the research surveyed in this paper is to establish the position in the descriptive set theoretic hierarchies of sets of continua arising from natural classes. We recall the basic definitions of the hierarchies of descriptive set theory (for more details see, e.g., [Kec95]).

If X is a separable metric space we denote by $\mathbf{\Sigma}^0_1(X)$ the family of open subsets of X. Then for an ordinal $\alpha > 0$, $\mathbf{\Pi}^0_\alpha(X)$ is the family of all complements of sets in $\mathbf{\Sigma}^0_\alpha(X)$, while, for $\alpha > 1$, $\mathbf{\Sigma}^0_\alpha(X)$ is the class of countable unions of elements of $\bigcup_{\beta<\alpha} \mathbf{\Pi}^0_\beta(X)$. At the lowest stages we have that $\mathbf{\Pi}^0_1(X)$ is the family of closed subsets of X, while sets in $\mathbf{\Sigma}^0_2(X)$ and $\mathbf{\Pi}^0_2(X)$ are respectively the F_σ and G_δ subsets of X. Moving a bit further in the hierarchy the $\mathbf{\Pi}^0_4$ sets are the $G_{\delta\sigma\delta}$ sets (i.e., countable intersections of countable unions of G_δ sets). It is straightforward to check that $\mathbf{\Sigma}^0_\alpha \cup \mathbf{\Pi}^0_\alpha \subseteq \mathbf{\Sigma}^0_\beta \cap \mathbf{\Pi}^0_\beta$ whenever $\alpha < \beta$. If X is an uncountable Polish space and $0 < \alpha < \beta < \omega_1$ we have $\mathbf{\Sigma}^0_\alpha \cup \mathbf{\Pi}^0_\alpha \neq \mathbf{\Sigma}^0_\beta \cap \mathbf{\Pi}^0_\beta$.

The fact that $\bigcup_{\alpha<\omega_1} \mathbf{\Sigma}^0_\alpha(X) = \bigcup_{\alpha<\omega_1} \mathbf{\Pi}^0_\alpha(X)$ is exactly the collection of all Borel subsets of X leads to the name *Borel hierarchy*.

We then denote by $\mathbf{\Sigma}^1_1(X)$ the family of subsets of X which are continuous images of a Polish space. For $n > 0$, $\mathbf{\Pi}^1_n(X)$ is the class of all complements of sets in $\mathbf{\Sigma}^1_n(X)$, and $\mathbf{\Sigma}^1_{n+1}(X)$ is the family of continuous images of a set in $\mathbf{\Pi}^1_n(Y)$ for some Polish space Y. Again $\mathbf{\Sigma}^1_n \cup \mathbf{\Pi}^1_n \subseteq \mathbf{\Sigma}^1_m \cap \mathbf{\Pi}^1_m$ whenever $n < m$, and for uncountable Polish spaces the inclusion is strict. This hierarchy is the *projective hierarchy*. $\mathbf{\Sigma}^1_1$ and $\mathbf{\Pi}^1_1$ sets are called resp. *analytic* and *coanalytic* sets.

We will also use the *difference hierarchy*: we restrict our discussion to the very first step of this hierarchy (see [Kec95, §22.E] for a complete definition of the hierarchy). A set is in $\mathrm{D}_2(\mathbf{\Sigma}^i_\alpha)$ if it is the intersection of a $\mathbf{\Sigma}^i_\alpha$ set and a $\mathbf{\Pi}^i_\alpha$ set. A set is in $\check{\mathrm{D}}_2(\mathbf{\Sigma}^i_\alpha)$ if it is the complement of a set in $\mathrm{D}_2(\mathbf{\Sigma}^i_\alpha)$ or, equivalently, if it is the union of a $\mathbf{\Sigma}^i_\alpha$ set and a $\mathbf{\Pi}^i_\alpha$ set.

By establishing the position of a natural class of continua in the Borel and projective hierarchies (i.e., the smallest family to which the set belongs), we obtain some information about the complexity of the class. This gives lower limits for the possibility of characterizing the class: e.g., if the class is $\mathbf{\Pi}^0_4$ and not $\mathbf{\Sigma}^0_4$, then any proposed characterization of the class as a countable union of countable intersections of F_σ sets is bound to be incorrect. In some cases the classification of a class of continua has immediate continuum theoretic consequences. To state an instance of this phenomena recall that a *model* for a class of continua is a continuum C_0 such that the continua in the class are exactly the continuous images of C_0 (the typical example here is the class of locally connected continua, which –

by the Hahn-Mazurkiewicz theorem – has I as a model). Then a class of continua which is not $\boldsymbol{\Sigma}^1_1$ cannot have a model.

The main tool for establishing lower bounds on the complexity of a set is Wadge reducibility.

Definition 1.1. If X and Y are metric spaces, $A \subseteq X$, and $B \subseteq Y$, we say that A is *Wadge reducible* to B (and write $A \leq_{\mathrm{W}} B$) if there exists a continuous function $f : X \to Y$ such that for every $x \in X$, $x \in A$ if and only if $f(x) \in B$.

Notice that if, e.g., B is $\boldsymbol{\Sigma}^0_\alpha$ and $A \leq_{\mathrm{W}} B$ then A is also $\boldsymbol{\Sigma}^0_\alpha$. Thus, proving that $A \leq_{\mathrm{W}} B$ for some A of known complexity yields a lower bound on the complexity of B.

Definition 1.2. If $\boldsymbol{\Gamma}$ is a class of sets in Polish spaces (like the classes $\boldsymbol{\Sigma}^0_\alpha$, $\boldsymbol{\Pi}^0_\alpha$, $\boldsymbol{\Sigma}^1_n$ and $\boldsymbol{\Pi}^1_n$ introduced above), Y is a Polish space and $A \subseteq Y$, we say that A is $\boldsymbol{\Gamma}$*-hard* if $B \leq_{\mathrm{W}} A$ for every $B \subseteq X$ which belongs to $\boldsymbol{\Gamma}$, where X is a zero-dimensional Polish space. We say that A is $\boldsymbol{\Gamma}$*-complete* if, in addition, $A \in \boldsymbol{\Gamma}$.

If A is $\boldsymbol{\Gamma}$-hard and $A \leq_{\mathrm{W}} B$, then B is $\boldsymbol{\Gamma}$-hard: this is the typical way to prove $\boldsymbol{\Gamma}$-hardness.

It turns out that a set is $\boldsymbol{\Sigma}^0_\alpha$-complete if and only if it is $\boldsymbol{\Sigma}^0_\alpha$ but not $\boldsymbol{\Pi}^0_\alpha$, and similarly interchanging $\boldsymbol{\Sigma}^0_\alpha$ and $\boldsymbol{\Pi}^0_\alpha$. If a set is $\boldsymbol{\Pi}^1_n$-complete then it is not $\boldsymbol{\Sigma}^1_n$, and similarly interchanging $\boldsymbol{\Sigma}^1_n$ and $\boldsymbol{\Pi}^1_n$. If a set is $\mathrm{D}_2(\boldsymbol{\Sigma}^i_\alpha)$-complete then it is not $\check{\mathrm{D}}_2(\boldsymbol{\Sigma}^i_\alpha)$.

Most results we will survey in §2 state that a natural set of continua is $\boldsymbol{\Gamma}$-complete for some $\boldsymbol{\Gamma}$, and thus pinpoint the complexity of that particular set by showing that it belongs to $\boldsymbol{\Gamma}$ and not to any simpler class, in particular those of the complements of elements of $\boldsymbol{\Gamma}$.

We are not including in this survey sharpenings of the classification results, as those showing that a set of continua which is $\boldsymbol{\Gamma}$-complete for some $\boldsymbol{\Gamma}$, is actually homeomorphic to a well-known $\boldsymbol{\Gamma}$-complete set. Results of this kind are included, e.g., in [DR94, CDGvM95, Sam03, KS].

Krupski (e.g., in [Kru02, Kru04]) and others have also studied the complexity of classes of continua within $\mathsf{C}(X)$, for X a compact metric space. A typical result of this kind states that, for a set of continua $\mathcal{P}$, a certain topological condition on X is sufficient for $\mathcal{P} \cap \mathsf{C}(X)$ to have in $\mathsf{C}(X)$ the same complexity that $\mathcal{P}$ has in $\mathsf{C}(I^\omega)$. These results are also not included in this survey, where we confine ourselves to subsets of $\mathsf{C}(I^n)$ for $2 \leq n \leq \omega$.

1.3. Descriptive set theory and binary relations

A binary relation on a Polish space X can be viewed as a subset of $X \times X$, and can be studied as such, e.g., by establishing its position in the descriptive set theoretic hierarchies described above. However this approach is much too crude, in that it neglects to take into account the particular features of a binary relation (actually what follows applies as well to n-ary relations for any $n > 1$). The following definition has been introduced and studied in depth in the context of equivalence relations, giving rise to the rich subject of "Borel reducibility for equivalence relations" ([FS89] is a pioneering paper, [BK96], [Hjo00] and [Kec02] are more recent accounts). More recently Louveau and Rosendal ([LR05]) extended its use to arbitrary binary relations, focusing in particular on quasi-orders.

Recall that a *quasi-order* is a binary relation which is reflexive and transitive. Therefore an equivalence relation is a quasi-order which is also symmetric. Moreover a quasi-order R on a set X induces naturally an equivalence relation $\sim_R$ on X defined by $x \sim_E y$ if and only if $x \mathrel{R} y$ and $y \mathrel{R} x$.

Definition 1.3. If R and S are binary relations on sets X and Y respectively, *a reduction of R to S* is a function $f : X \to Y$ such that

$$\forall x_0, x_1 \in X (x_0 \mathrel{R} x_1 \iff f(x_0) \mathrel{S} f(x_1)).$$

If X and Y are Polish spaces and f is Borel we say that *R is Borel reducible to S*, and we write $R \leq_{\mathrm{B}} S$. If $R \leq_{\mathrm{B}} S \leq_{\mathrm{B}} R$ then we say that R and S are *Borel bireducible.*

In practice we do not need the ambient spaces X and Y of Definition 1.3 to be Polish: it suffices that they are standard Borel spaces, i.e., that their Borel sets coincide with the Borel sets of a Polish topology on the same space. The basic fact we will implicitly use is that any Borel subset of a Polish space is standard Borel. This allows to study the behavior of a binary relation restricted to a Borel subset of $\mathsf{C}(I^\omega)$.

It is easy to see that if S is a quasi-order (resp. an equivalence relation) and $R \leq_{\mathrm{B}} S$, then R is a quasi-order (resp. an equivalence relation) as well. Moreover a reduction of the quasi-order R to the quasi-order S is also a reduction of the induced equivalence relation $\sim_R$ to the induced equivalence relation $\sim_S$.

If E and F are equivalence relations such that $E \leq_{\mathrm{B}} F$ we also say that the effective cardinality of the quotient space X/E is less than or equal to the effective cardinality of the quotient space Y/F: this is because there is a one-to-one function from X/E to Y/F that can be lifted to a Borel map from X to Y, i.e., the reduction of E to F. Another way of describing the fact that $E \leq_{\mathrm{B}} F$ is the following: we can assign in a Borel way F-equivalence classes as complete invariants for the equivalence relation E. Therefore the classification problem for F is at least as complicated as the classification problem for E.

As we already said, the research on Borel reducibility for equivalence relations has focused mainly on equivalence relations induced by continuous Polish group actions (or Borel bireducible to such an equivalence relation), under the headline of "descriptive dynamics". It is immediate to check that an equivalence relation of this kind on the Polish space X is $\mathbf{\Sigma}^1_1$ (as a subset of $X \times X$), and Miller proved (see [Kec95, Theorem 15.14]) that each equivalence class is Borel. Even in this restricted setting the structure of $\leq_{\mathrm{B}}$ is rich and complicated (e.g., see [AK00]).

Here we list only the definitions we will need to state the results surveyed in this paper.

Definition 1.4. An equivalence relation on a standard Borel space is *smooth* (or *concretely classifiable*, or *tame*) if it is Borel reducible to equality on some Polish space.

A smooth equivalence relation is considered to be very simple, since it admits "concrete" objects (i.e., elements of a Polish space) as complete invariants.

Definition 1.5. For $\mathcal{L}$ a countable relational language, let $X_{\mathcal{L}}$ be the Polish space of (codes for) $\mathcal{L}$-structures with universe $\mathbb{N}$ (see [Kec95, §16.C]). Let $\cong_{\mathcal{L}}$ denote isomorphism on $X_{\mathcal{L}}$. If E is an equivalence relation on a standard Borel space, E is *classifiable by countable structures* if $E \leq_{\mathrm{B}} \cong_{\mathcal{L}}$ for some $\mathcal{L}$; E is S_∞*-universal* if, in addition, $\cong_{\mathcal{L}} \leq_{\mathrm{B}} E$ for every $\mathcal{L}$.

The reason for the terminology "S_∞-universal" is that such an equivalence relation is as complicated as any equivalence relation induced by a continuous action of the infinite symmetric group S_∞ can be. An example of an equivalence relation which is S_∞-universal is homeomorphism on compact subsets of the Cantor space ([CG01]).

In [LR05] Louveau and Rosendal started the study of Borel reducibility for equivalence relations induced by natural $\mathbf{\Sigma}^1_1$ quasi-orders and showed that several of these are $\mathbf{\Sigma}^1_1$-complete or K_σ-complete in the following sense. (Recall that a subset of a Polish space is K_σ if it is the countable union of compact sets.)

Definition 1.6. A quasi-order R on a Polish space is $\mathbf{\Sigma}^1_1$*-complete* (resp. K_σ*-complete*) if it is $\mathbf{\Sigma}^1_1$ (resp. K_σ) and $S \leq_{\mathrm{B}} R$ for any $\mathbf{\Sigma}^1_1$ (resp. K_σ) quasi-order S.

An equivalence relation E on a Polish space is $\mathbf{\Sigma}^1_1$*-complete* (resp. K_σ*-complete*) if it is $\mathbf{\Sigma}^1_1$ (resp. K_σ) and $F \leq_{\mathrm{B}} E$ for any $\mathbf{\Sigma}^1_1$ (resp. K_σ) equivalence relation F.

Since every equivalence relation is a quasi-order, if the quasi-order R is $\mathbf{\Sigma}^1_1$-complete then the induced equivalence relation $\sim_R$ is $\mathbf{\Sigma}^1_1$-complete among equivalence relations, and similarly for K_σ-complete. A $\mathbf{\Sigma}^1_1$-complete equivalence relation is immensely more complicated than any equivalence relation induced by any Polish group action (e.g., uncountably many of its equivalence classes are not Borel). An example of a $\mathbf{\Sigma}^1_1$-complete quasi-order is isometric embeddability between Polish spaces ([LR05]). The same quasi-order, restricted to Heine-Borel Polish spaces (a metric space is Heine-Borel if its closed bounded subsets are compact), is K_σ-complete ([LR05]).

2. Sets of continua

2.1. Decomposability and Baire category arguments

The following notions provide a basic distinction between continua.

Definition 2.1. A continuum C is *decomposable* if $C = C_1 \cup C_2$ where C_1 and C_2 are proper subcontinua of C. If a continuum is not decomposable then it is *indecomposable*.

Indecomposable continua form a dense $\mathbf{\Pi}_2^0$ in $\mathsf{C}(I^n)$ for $2 \leq n \leq \omega$ ([Nad92, Exercise 1.17]). Since the set of decomposable continua is also dense, a simple Baire category argument shows:

Fact 2.2. *The set of indecomposable continua is $\mathbf{\Pi}_2^0$-complete in $\mathsf{C}(I^n)$ for $2 \leq n \leq \omega$. The set of decomposable continua is $\mathbf{\Sigma}_2^0$-complete in $\mathsf{C}(I^n)$ for $2 \leq n \leq \omega$.*

Definition 2.3. A continuum is *hereditarily decomposable*, if all its nondegenerate subcontinua are decomposable. A continuum is *hereditarily indecomposable*, if all its subcontinua are indecomposable.

There exists a (necessarily unique) continuum (the *pseudoarc*) such that its homeomorphism class is dense $\mathbf{\Pi}_2^0$ in $\mathsf{C}(I^n)$ for $2 \leq n \leq \omega$ ([Bin51], see [Nad92, Exercise 12.70]). The pseudoarc has a fascinating history, briefly sketched in [Nad92, p. 228–229]. The pseudoarc is hereditarily indecomposable, and hereditarily indecomposable continua form also a dense $\mathbf{\Pi}_2^0$ set in $\mathsf{C}(I^n)$ for $2 \leq n \leq \omega$ ([Maz30], see [Nad92, Exercise 1.23 d]). Again using Baire category we immediately obtain:

Fact 2.4. *The set of hereditarily indecomposable continua is $\mathbf{\Pi}_2^0$-complete in $\mathsf{C}(I^n)$ for $2 \leq n \leq \omega$.*

Analogous Baire category arguments also establish the following easy results which can be considered more or less folklore (e.g., [CDM05] suggests the proofs, and [Kru03] contains a detailed proof of (b)):

Fact 2.5. *The following sets of continua are $\mathbf{\Pi}_2^0$-complete in $\mathsf{C}(I^n)$ for $2 \leq n \leq \omega$:*

(a) *the set of unicoherent continua (a continuum C is unicoherent if $C_1 \cap C_2$ is a continuum whenever C_1 and C_2 are subcontinua of C such that $C = C_1 \cup C_2$);*
(b) *the set of hereditarily unicoherent continua (a continuum C is hereditarily unicoherent if every subcontinuum of C is unicoherent or, equivalently, if the intersection of any two subcontinua of C is connected);*
(c) *for $n \geq 2$ the set of irreducible continua between n points (a continuum is irreducible between n points if it contains a set of n points which is not contained in any proper subcontinua);*
(d) *the set of hereditarily irreducible continua (a continuum is hereditarily irreducible if all its nondegenerate subcontinua are irreducible between 2 points).*

2.2. Hereditarily decomposable continua and generalizations of Darji's argument

While the classification of the sets of decomposable, indecomposable and hereditarily indecomposable continua is quite old, the precise classification of the set of hereditarily decomposable continua was obtained more recently by Darji ([Dar00]). There exists "nice" characterizations for indecomposable continua ([IC68]) and for hereditarily indecomposable continua ([Pro72]), but Theorem 2.6 below implies that nothing similar is possible for hereditarily decomposable continua.

Theorem 2.6. *The set $\mathcal{HD}$ of hereditarily decomposable continua is $\mathbf{\Pi}_1^1$-complete in $\mathsf{C}(I^n)$ for $2 \leq n \leq \omega$.*

Sketch of proof. We sketch Darji's proof that $\mathcal{HD}$ is $\mathbf{\Pi}^1_1$-hard (it follows immediately from Fact 2.2 that it is $\mathbf{\Pi}^1_1$). Let $D = \{ \alpha \in 2^\omega \mid \exists m \, \forall n > m \; \alpha(n) = 0 \}$. D is a countable dense subset of the Cantor space 2^ω and it is a classical result due to Hurewicz (see, e.g., [Kec95, Theorem 27.5]) that $A = \{ C \in \mathsf{K}(2^\omega) \mid C \subseteq D \}$ is $\mathbf{\Pi}^1_1$-complete. Darji's proof consists in showing that $A \leq_{\mathrm{W}} \mathcal{HD}$.

To this end for every $\alpha \in 2^\omega$ we construct, using the basic continuum theoretic technique of nested intersections, a continuum $X_\alpha \subseteq I^2$ with the property that if $\alpha \in D$ then X_α is an arc (i.e., homeomorphic to I), while if $\alpha \notin D$ then X_α is nondegenerate indecomposable. Moreover there exists a point p_0 such that $\alpha \neq \alpha'$ implies $X_\alpha \cap X_{\alpha'} = \{p_0\}$, and if $\alpha \in D$ then p_0 is one of the endpoints of the arc X_α. The whole construction is done in such a way that the map $\alpha \mapsto X_\alpha$ is continuous, and this implies that if $C \in \mathsf{K}(2^\omega)$ then $X_C = \bigcup_{\alpha \in C} X_\alpha$ is a continuum and that the map $\mathsf{K}(2^\omega) \to \mathsf{C}(I^2)$, $C \mapsto X_C$ is continuous.

To show that the latter map is the desired reduction, we need to show that $C \in A$ if and only if $X_C \in \mathcal{HD}$. The backward direction is obvious (if $C \notin A$ fix $\alpha \in C \setminus D$: then X_α is a nondegenerate indecomposable subcontinuum of X_C). For the forward direction it suffices to notice that if $C \subseteq D$ then X_C is a countable union of arcs pairwise intersecting in a common endpoint: such a continuum is hereditarily decomposable. □

The construction sketched above has the property that when $C \in A$ then X_C is a quite simple continuum which enjoys more properties than just being hereditarily decomposable, and when $C \notin A$ then X_C is quite complicated. This immediately shows that many sets of continua are $\mathbf{\Pi}^1_1$-hard and hence, when they are $\mathbf{\Pi}^1_1$, are indeed $\mathbf{\Pi}^1_1$-complete. Some of these fairly easy consequences of Darji's proof where noticed in [CDM05] and in [Kru03]:

Corollary 2.7. *The following sets of continua are $\mathbf{\Pi}^1_1$-complete in $\mathsf{C}(I^n)$ for $2 \leq n \leq \omega$:*

(a) *the set of continua which have no nondegenerate hereditarily indecomposable subcontinua;*
(b) *the set of uniquely arcwise connected continua (a continuum C is uniquely arcwise connected if for all distinct $x, y \in C$ there exists a unique arc contained in C with end points x and y);*
(c) *the set of dendroids (a continuum is a dendroid if it is arcwise connected and hereditarily unicoherent);*
(d) *the set of λ-dendroids (a continuum is a λ-dendroid if it is hereditarily decomposable and hereditarily unicoherent).*

The technique of Darji's proof of Theorem 2.6 has also been generalized to obtain other results. One result of this kind is due to Darji and Marcone ([DM04]):

Theorem 2.8. *The set $\mathcal{HLC}$ of hereditarily locally connected continua is $\mathbf{\Pi}^1_1$-complete in $\mathsf{C}(I^n)$ for $2 \leq n \leq \omega$ (a continuum is hereditarily locally connected if all its subcontinua are locally connected).*

Sketch of proof. We use the notation of the sketch of proof of Theorem 2.6 and sketch the proof that $\mathcal{HLC}$ is $\mathbf{\Pi}^1_1$-hard (it follows from Theorem 2.10 below that it is $\mathbf{\Pi}^1_1$). We show that $A \leq_W \mathcal{HLC}$ by constructing, using nested intersections, for every $\alpha \in 2^\omega$ a continuum X_α and considering the map $C \mapsto X_C$ as before. Again X_α is an arc when $\alpha \in D$, and a nondegenerate indecomposable continuum otherwise. Since nondegenerate indecomposable continua are not locally connected, if $C \notin A$ then $X_C \notin \mathcal{HLC}$.

However now we must make sure that $X_C \in \mathcal{HLC}$ when $C \in A$. The construction of the proof of Theorem 2.6 does not work for this, since in that case we have $X_C \notin \mathcal{HLC}$ whenever C is infinite. To solve this problem we must make sure that $X_\alpha \cap X_{\alpha'}$ is quite large, yet X_α and $X_{\alpha'}$ are still distinct. This goal is achieved by putting quite detailed requirements on the nested intersections used to define the X_α's. □

Another generalization of Darji's construction has been introduced by Krupski in [Kru03, §3]. Krupski proves a general lemma stating that any set of continua enjoying some properties with respect to a construction made with inverse limits of polyhedra is $\mathbf{\Pi}^1_1$-hard, and then considers some specific examples of his construction. Here is one of his results:

Theorem 2.9. *The set of strongly countable-dimensional continua is $\mathbf{\Pi}^1_1$-complete in $\mathsf{C}(I^\omega)$ (a continuum is strongly countable-dimensional if it is the countable union of compact finite-dimensional spaces).*

2.3. Continua with strong forms of connectedness

Sets of continua enjoying some strong form of connectedness (as the one classified by Corollary 2.7.(b) and Theorem 2.8) are obviously quite important.

The classification of locally connected continua (also called Peano continua, because of the well-known Hahn-Mazurkiewicz theorem stating that a continuum C is locally connected if and only if there exists $f : I \to C$ continuous and onto) is a classical result due independently to Kuratowski ([Kur31]) and Mazurkiewicz ([Maz31]). A modern proof is included in [CDM05].

Theorem 2.10. *The set of locally connected continua is $\mathbf{\Pi}^0_3$-complete in $\mathsf{C}(I^n)$ for $2 \leq n \leq \omega$.*

The following theorem is due independently to Ajtai and Becker ([Bec92], see [Kec95, Theorem 37.11] for a proof: there the results are stated for compacta, but the proofs actually deal with continua).

Theorem 2.11. *The set of arcwise connected continua is $\mathbf{\Pi}^1_2$-complete in $\mathsf{C}(I^n)$ for $3 \leq n \leq \omega$.*

Theorem 2.11 provides a quite rare example of a natural set whose classification involves the second level of the projective hierarchy, but fails to classify the set of planar arcwise connected continua. The following problem is of obvious importance:

Problem 2.12. Classify the complexity of the set of arcwise connected continua in $\mathsf{C}(I^2)$.

This set is clearly $\mathbf{\Pi}^1_2$ and Becker's proof of Theorem 2.11 shows that it is $\mathbf{\Sigma}^1_1$-hard and hence not $\mathbf{\Pi}^1_1$; Darji's proof of Theorem 2.6 shows that the set is $\mathbf{\Pi}^1_1$-hard and hence not $\mathbf{\Sigma}^1_1$ (the latter result was obtained independently by Just, unpublished). Putting together these proofs (e.g., with the technique used in the proof of Theorem 2.14 below) it can be shown that the set of arcwise connected continua is $\mathrm{D}_2(\mathbf{\Sigma}^1_1)$-hard.

2.4. Simply connected continua

In the 1980's Becker also studied the set $\mathcal{SC}$ of simply connected continua (a continuum C is simply connected if it is arcwise connected and has no holes, i.e., every continuous function from the unit circle into C can be extended to a continuous function from the closed unit disk into C). It is immediate that $\mathcal{SC}$ is $\mathbf{\Pi}^1_2$ in $\mathsf{C}(I^n)$ for every n. It turns out that the precise classification of $\mathcal{SC}$ depends on the dimension of the ambient space.

Theorem 2.13. *The set of simply connected continua $\mathcal{SC}$ is $\mathbf{\Pi}^1_1$-complete in $\mathsf{C}(I^2)$.*

A proof that $\mathcal{SC}$ is $\mathbf{\Pi}^1_1$-hard can be found in [Kec95, Theorem 33.17]. On the other hand Becker's original proof that the set of planar simply connected continua is $\mathbf{\Pi}^1_1$ involves a quite delicate argument based on the techniques of so-called *effective* descriptive set theory ([Bec86], see [Bec92] for a sketch of the proof). It has however been suggested (e.g., by the anonymous referee of this paper) that the ideas and results of [BP01] could be used to obtain a completely *classical*, and probably simpler, proof that $\mathcal{SC}$ is $\mathbf{\Pi}^1_1$ in $\mathsf{C}(I^2)$. The details of this new proof are still to be worked out.

The exact classification of $\mathcal{SC}$ in $\mathsf{C}(I^3)$ is unknown. The best known upper bound is the obvious one, namely $\mathbf{\Pi}^1_2$. The following theorem of Becker ([Bec87]) establishes a lower bound and implies that $\mathcal{SC}$ is more complex in $\mathsf{C}(I^3)$ than in $\mathsf{C}(I^2)$. We include here Becker's proof, which has never been published in print.

Theorem 2.14. *The set $\mathcal{SC}$ of simply connected continua is $\mathrm{D}_2(\mathbf{\Sigma}^1_1)$-hard in $\mathsf{C}(I^3)$.*

We first establish the following weaker result:

Lemma 2.15. *$\mathcal{SC}$ is $\mathbf{\Sigma}^1_1$-hard in $\mathsf{C}(I^3)$.*

Proof. The proof is based on the construction (also due to Becker, see [Bec84]) which shows that the set of arcwise connected continua is $\mathbf{\Sigma}^1_1$-hard in $\mathsf{C}(I^2)$. This construction is essentially contained in the proof that the set of arcwise connected continua is $\mathbf{\Pi}^1_2$-hard in $\mathsf{C}(I^3)$ published in [Kec95, Theorem 37.11]. Actually we do not need all details of the construction, and we summarize in the next paragraph the ones we will use.

We have a continuous function $T \mapsto L_T$ from the set of descriptive set-theoretic trees to $\mathsf{C}(I^2)$. We can assume that $L_T \subset I \times [0, 1/2]$, that the line segments $I \times \{0\}$ and $\{1\} \times [0, 1/2]$ (corresponding to l and $l_\emptyset$ in the notation of [Kec95]) are included in L_T, and that the point $(0, 1/2)$ (corresponding to r in Kechris's notation) belongs to L_T for any T. If T is well founded L_T has exactly two arc components: the points $(0, 1/2)$ and $(0, 0)$ belong to these different arc components. If T is not well founded then L_T is arcwise connected.

Let $M_T = L_T \cup (\{0\} \times [1/2, 1]) \subset I^2$: as far as arcwise connectedness is concerned, M_T has the same properties of L_T, so that in particular if T is well founded $(0, 1)$ and $(0, 0)$ belong to different arc components of M_T, and there are no other arc components of M_T.

Now we work in three-dimensional space and let $Z_0 = I^2 \times \{0\}$, $Z_1 = I^2 \times \{1\}$, $Y_0 = I \times \{0\} \times I$, $Y_1 = I \times \{1\} \times I$, $X_0 = \{0\} \times I^2$, and $X_1 = \{1\} \times I^2$. The set $A = Z_0 \cup Z_1 \cup Y_0 \cup Y_1$ is a cube with the interior of the opposite faces X_0 and X_1 removed. Let $B_T = M_T \times I$ and notice that half of each X_0 and X_1 is included in B_T for each T. Let $C_T = B_T \cup A$: the function $T \mapsto C_T$ from the set of descriptive set-theoretic trees to $\mathsf{C}(I^3)$ is clearly continuous and we claim that T is not well founded if and only if $C_T \in \mathcal{SC}$. Since the set of well-founded descriptive set-theoretic trees is $\mathbf{\Pi}^1_1$-complete the claim completes the proof.

First of all notice that C_T is arcwise connected for every T, and hence we need to show that T is not well founded if and only if C_T has no holes. It is clear that A has holes because, e.g., a homeomorphism of the unit circle onto $A \cap X_0$ cannot be extended to a continuous function from the unit disk to A. Notice also that any continuous function from the unit circle to B_T can be extended to a continuous function from the unit disk to C_T with range intersecting Z_1 (the idea is that we can "lift" the circle in B_T to Z_1, where we can contract it to a single point without problems). Therefore the only possible reason for C_T being not simply connected is the hole in A, i.e., if a continuous function from the unit circle to C_T has no extension to a continuous function from the unit disk to C_T, it "goes around the hole in A".

If T is not well founded then M_T is arcwise connected and let $X \subseteq M_T$ be an arc with endpoints $(0, 1)$ and $(0, 0)$. Now $X \times I \subseteq B_T$ "fills the hole" in A: e.g., the homeomorphism of the unit circle onto $A \cap X_0$ can be extended to the unit disk by a continuous function with range included in $X \times I$. Any continuous function from the unit circle to C_T which "goes around the hole in A" can be extended to a continuous function from the unit disk to C_T using $X \times I$, and therefore C_T has no holes.

If T is well founded there is no arc in M_T with endpoints $(0, 1)$ and $(0, 0)$ and hence there is no arc in B_T with endpoints $(0, 1, 0)$ and $(0, 0, 0)$. This implies that each arc in C_T with endpoints $(0, 1, 0)$ and $(0, 0, 0)$ intersects either Z_0 or Z_1. It follows that the homeomorphism of the unit circle onto $A \cap X_0$ cannot be extended to a continuous function from the unit disk to C_T. Therefore C_T has holes. □

Proof of Theorem 2.14. By Lemma 2.15 and the part of Theorem 2.13 proved in [Kec95, Theorem 33.17] $\mathcal{SC}$ is both $\mathbf{\Sigma}^1_1$-hard and $\mathbf{\Pi}^1_1$-hard in $\mathsf{C}(I^3)$. This means that there exist continuous functions $T \mapsto C_T$ and $T \mapsto C'_T$ from the set of descriptive set-theoretic trees to $\mathsf{C}(I^3)$ such that T is well founded if and only if $C_T \notin \mathcal{SC}$, if and only if $C'_T \in \mathcal{SC}$. Moreover we can assume $C_T \subseteq [0,1/3] \times I^2$, $C'_T \subseteq [2/3,1] \times I^2$, $(1/3,0,0) \in C_T$, and $(2/3,0,0) \in C'_T$ for every T.

We define a continuous function $(T,S) \mapsto D_{T,S}$ from the product of the set of descriptive set-theoretic trees with itself to $\mathsf{C}(I^3)$ by setting $D_{T,S} = C_T \cup ([1/3,2/3] \times \{0\} \times \{0\}) \cup C'_S$ (i.e., we are joining with a segment C_T and C'_S). $D_{T,S}$ is a continuum and it is simply connected if and only if T is not well founded and S is well founded. Since the set $\{ (T,S) \mid T$ is not well founded and S is well founded$\}$ is easily seen to be $\mathrm{D}_2(\boldsymbol{\Sigma}^1_1)$-complete, this completes the proof. □

The following problem is therefore still open.

Problem 2.16. Classify the complexity of $\mathcal{SC}$ in $\mathsf{C}(I^3)$.

Becker ([Bec87]) showed that the obvious upper bound for $\mathcal{SC}$ is sharp in $\mathsf{C}(I^n)$ for $n > 3$. Again we include the proof, which has not appeared in print elsewhere.

Theorem 2.17. *The set $\mathcal{SC}$ of simply connected continua is $\boldsymbol{\Pi}^1_2$-complete in $\mathsf{C}(I^n)$ for $4 \leq n \leq \omega$.*

Proof. We already noticed that $\mathcal{SC}$ is $\boldsymbol{\Pi}^1_2$ in $\mathsf{C}(I^n)$ for any n. The idea of the proof that $\mathcal{SC}$ is $\boldsymbol{\Pi}^1_2$-hard in $\mathsf{C}(I^4)$ is to lift up one dimension the proof of Lemma 2.15. (This is analogous to the technique used to prove that the set of arcwise connected continua is $\boldsymbol{\Pi}^1_2$-hard in $\mathsf{C}(I^3)$ starting from the proof that the set of arcwise connected continua is $\boldsymbol{\Sigma}^1_1$-hard in $\mathsf{C}(I^2)$.)

It suffices to show that $A \leq_{\mathrm{W}} \mathcal{SC} \cap \mathsf{C}(I^4)$ for any $\boldsymbol{\Pi}^1_2$ subset A of the Baire space $\mathbb{N}^\omega$. Any such A is of the form

$$\{ x \in \mathbb{N}^\omega \mid \forall y \in 2^\omega \; \exists z \in \mathbb{N}^\omega \; \forall n \; (x[n], y[n], z[n]) \in S \}$$

for some descriptive set-theoretic tree S on $\mathbb{N} \times 2 \times \mathbb{N}$ (here $x[n]$ is the initial segment of x with length n). If $x \in \mathbb{N}^\omega$ and $y \in 2^\omega$ let

$$S(x,y) = \{ s \in \mathbb{N}^{<\omega} \mid (x[\mathrm{lh}\, s], y[\mathrm{lh}\, s], s) \in S \}$$

(here $\mathrm{lh}\, s$ is the length of the finite sequence s). The function $(x,y) \mapsto S(x,y)$ from $\mathbb{N}^\omega \times 2^\omega$ to the set of descriptive set-theoretic trees is continuous, and we have $x \in A$ if and only if for every $y \in 2^\omega$ the tree $S(x,y)$ is not well founded.

By Lemma 2.15 there exists a continuous function $T \mapsto C_T$ from the set of descriptive set-theoretic trees to $\mathsf{C}(I^3)$ such that T is not well founded if and only if $C_T \in \mathcal{SC}$. We may assume $(0,0,0) \in C_T$ for every T. Moreover we can identify the Cantor space 2^ω with Cantor's middle third set, which is a subset of I. For each $x \in \mathbb{N}^\omega$ and $y \in 2^\omega$ let $U_{x,y} = C_{S(x,y)} \times \{y\}$. The function $\mathbb{N}^\omega \times 2^\omega \to \mathsf{C}(I^4)$, $(x,y) \mapsto U_{x,y}$ is clearly continuous, and $U_{x,y} \in \mathcal{SC}$ if and only if $S(x,y)$ is not well founded.

For every $x \in \mathbb{N}^\omega$ let $E_x = \bigcup_{y \in 2^\omega} U_{x,y} \cup (\{0\}^3 \times I)$. To check that E_x is a continuum, notice that compactness follows from the continuity of $y \mapsto U_{x,y}$, and E_x is connected because the various $U_{x,y}$'s are joined by a segment. The function

$x \mapsto E_x$ is easily seen to be continuous from $\mathbb{N}^\omega$ to $\mathsf{C}(I^4)$. Since $E_x \in \mathcal{SC}$ if and only if each $U_{x,y} \in \mathcal{SC}$, it is straightforward to check that $x \in A$ if and only if $E_x \in \mathcal{SC}$. □

Theorem 2.17 provides another example of a natural set which is $\mathbf{\Pi}^1_2$-complete.

2.5. Continua which do not contain subcontinua of a certain kind

The technique used to prove Theorem 2.9 was used by Krupski in [Kru03, §3] to prove results dealing with sets of continua which do not contain a copy of a fixed continuum:

Theorem 2.18. *The set of continua which do not contain the pseudo-arc is* $\mathbf{\Pi}^1_1$-*complete in* $\mathsf{C}(I^n)$ *for* $2 \leq n \leq \omega$.

The pseudo-arc can be characterized as the unique hereditarily indecomposable continuum which is the inverse limit of a sequence of arcs. A hereditarily indecomposable continuum which is the inverse limit of a sequence of circles but not the inverse limit of a sequence of arcs is called a *pseudo-solenoid*. There exists a unique (up to homeomorphism) planar pseudo-solenoid, which is called the *pseudo-circle*. Krupski used the technique discussed above to prove:

Theorem 2.19. *The set of continua which do not contain any pseudo-solenoid is* $\mathbf{\Pi}^1_1$-*complete in* $\mathsf{C}(I^n)$ *for* $2 \leq n \leq \omega$. *The set of continua which do not contain the pseudo-circle is* $\mathbf{\Pi}^1_1$-*complete in* $\mathsf{C}(I^n)$ *for* $2 \leq n \leq \omega$.

A similar result is the following, which is stated explicitly in [Kru03], but whose proof is based on Becker's construction for the hardness part of the proof of Theorem 2.13 (see [Kec95, p. 256–257]).

Theorem 2.20. *The set of continua which do not contain any circle is* $\mathbf{\Pi}^1_1$-*complete in* $\mathsf{C}(I^n)$ *for* $2 \leq n \leq \omega$.

In [Kru03, §4] Krupski proves, using techniques not directly related to Darji's construction, another result of this kind:

Theorem 2.21. *The set of continua which do not contain any arc is* $\mathbf{\Pi}^1_1$-*complete in* $\mathsf{C}(I^n)$ *for* $2 \leq n \leq \omega$.

This theorem is sharpened by the following two results, again proved in [Kru03, §4]:

Theorem 2.22. *The set of* λ-*dendroids which do not contain any arc is* $\mathbf{\Pi}^1_1$-*complete in* $\mathsf{C}(I^n)$ *for* $2 \leq n \leq \omega$.

Theorem 2.23. *The set of hereditarily decomposable continua which do not contain any arc is* $\mathbf{\Pi}^1_1$-*complete in* $\mathsf{C}(I^n)$ *for* $2 \leq n \leq \omega$.

In the same vein the following problem is interesting:

Problem 2.24. Classify the complexity of the set of continua which do not contain any hereditarily decomposable subcontinuum in $\mathsf{C}(I^n)$ for $2 \leq n \leq \omega$.

A straightforward computation shows that this set is $\mathbf{\Pi}^1_2$, and, as noticed by Krupski in [Kru03], the proof of Theorem 2.21 shows that it is $\mathbf{\Pi}^1_1$-hard.

A classification result concerning continua which do not contain point exhibiting a certain behavior is the following theorem of Krupski ([Kru02]):

Theorem 2.25. *The set of locally connected continua which do not contain any local cut point is $\mathbf{\Pi}^0_3$-complete in $\mathsf{C}(I^n)$ for $2 \leq n \leq \omega$ ($x \in C$ is a local cut point of the continuum C if there exists an open neighborhood U of x such that $U \setminus \{x\}$ is not connected).*

2.6. More results by Krupski

In [Kru04] Krupski studies other natural classes of continua. Here are some of his results:

Theorem 2.26. *The set of continua with the property of Kelley is $\mathbf{\Pi}^0_3$-complete in $\mathsf{C}(I^n)$ for $2 \leq n \leq \omega$ (a continuum C with compatible metric d has the property of Kelley if for every $\epsilon > 0$ there exists $\delta > 0$ such that whenever $x, y \in C$ are such that $d(x, y) < \delta$ and D is a subcontinuum of C with $x \in D$ there exists a subcontinuum E of C with $y \in E$ and $d_H(D, E) < \epsilon$).*

Theorem 2.27. *The set of arc continua is $\mathbf{\Pi}^1_1$-complete in $\mathsf{C}(I^n)$ for $3 \leq n \leq \omega$ (an arc continuum is a continuum such that all its proper nondegenerate subcontinua are arcs).*

Since planar arc continua do exist, the following problem is natural:

Problem 2.28. Classify the complexity of the set of arc continua in $\mathsf{C}(I^2)$.

Krupski studied also the following problem, which is still open:

Problem 2.29. Classify the complexity of the set of solenoids in $\mathsf{C}(I^n)$ for $3 \leq n \leq \omega$ (a solenoid is a continuum which is homeomorphic to the inverse limit of unit circles with maps $z \mapsto z^n$ for $n > 1$).

Solenoids are non-planar continua, so the problem above is not interesting in $\mathsf{C}(I^2)$. In [Kru04] it is shown that the set of solenoids is Borel and $\mathbf{\Pi}^0_3$-hard in $\mathsf{C}(I^n)$ for $3 \leq n \leq \omega$.

2.7. Curves

An important part of continuum theory is the theory of curves: a curve is a 1-dimensional continuum. Since a theorem of Mazurkiewicz (see, e.g., [Nad92, Theorem 13.57]) asserts that every compact metric space of dimension at least 2 contains a nondegenerate indecomposable continuum, every hereditarily decomposable continuum is a curve. We already dealt with some sets of continua which are actually sets of curves: beside hereditarily decomposable continua, these include hereditarily locally connected continua and solenoids. We will now list some classification results (and an open problem) for sets of curves which are included in the set of hereditarily decomposable continua. Each statement (from Theorem 2.30 to Theorem 2.35 included) deals with a set of curves properly included in the ones considered in the statements preceding it.

Theorem 2.30. *The set of Suslinian continua is* $\mathbf{\Pi}^1_1$*-complete in* $\mathsf{C}(I^n)$ *for* $2 \leq n \leq \omega$ *(a continuum is Suslinian if each collection of its pairwise disjoint nondegenerate subcontinua is countable).*

This result is proved in [DM04], and it follows fairly easily from the proof of Theorem 2.6 and from the fact that a non Suslinian continuum contains a Cantor set of pairwise disjoint nondegenerate subcontinua ([CL78]).

Problem 2.31. Classify the complexity of the set of rational continua (a continuum C is *rational* if every point of C has a neighborhood basis consisting of sets with countable boundary).

The set of rational continua is easily seen to be $\mathbf{\Sigma}^1_2$ and $\mathbf{\Pi}^1_1$-hard. If it turns out to be $\mathbf{\Sigma}^1_2$-complete would be the first example of a natural set of continua with this classification. Notice that every hereditarily locally connected continuum is rational, but not viceversa.

Theorem 2.32. *The set of finitely Suslinian continua is* $\mathbf{\Pi}^1_1$*-complete in* $\mathsf{C}(I^n)$ *for* $2 \leq n \leq \omega$ *(a continuum* C *is finitely Suslinian if for every* $\varepsilon > 0$ *each collection of pairwise disjoint subcontinua of* C *with diameter* $\geq \varepsilon$ *is finite).*

This result is proved by Darji and Marcone in [DM04]: the hardness part follows from Theorem 2.8 since a planar continuum is hereditarily locally connected if and only if it is finitely Suslinian ([Lel71]). In general each finitely Suslinian continuum is hereditarily locally connected, but not viceversa.

The following theorem was also proved in [DM04]. We sketch the proof of the lower bound because it shares some common features with many other proofs in the area. To prove that a set of continua $\mathcal{P}$ is $\mathbf{\Gamma}$-hard we start from a continuum $C \notin \mathbf{\Gamma}$. Usually C was originally defined as a counterexample showing that $\mathcal{P}$ does not enjoy some property and/or does not coincide with some other set. We look for modifications to the construction of C which lead to a continuum belonging to $\mathcal{P}$. The modifications should depend continuously on some parameter ranging in a Polish space, and yield a member of $\mathcal{P}$ if and only if the parameter belongs to a $\mathbf{\Gamma}$-complete set. When this strategy succeeds, we have shown that $\mathcal{P}$ is $\mathbf{\Gamma}$-hard.

Theorem 2.33. *The set* $\mathcal{R}$ *of regular continua is* $\mathbf{\Pi}^0_4$*-complete in* $\mathsf{C}(I^n)$ *for* $2 \leq n \leq \omega$ *(a continuum* C *is regular if every point of* C *has a neighborhood basis consisting of sets with finite boundary).*

Sketch of proof. Recall the following characterization of regular continua due to Lelek ([Lel71]): a continuum C is regular if and only if for every $\varepsilon > 0$ there exists n such that every collection of n subcontinua of C of diameter $\geq \varepsilon$ is not pairwise disjoint. Using this characterization it can be shown fairly easily that $\mathcal{R}$ is $\mathbf{\Pi}^0_4$.

To prove that $\mathcal{R}$ is $\mathbf{\Pi}^0_4$-hard we apply the strategy discussed before the statement of the theorem. Let C be the continuum described by Nadler in [Nad92, Example 10.38] and partially drawn in Figure 1 (the construction should be continued by adding infinitely many horizontal segments above the lowest segment,

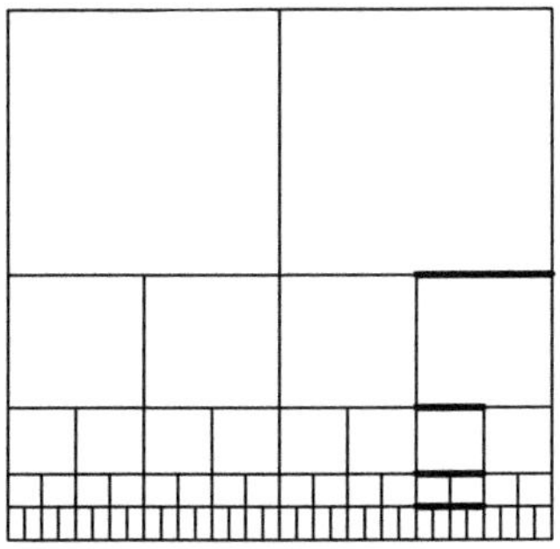

FIGURE 1. The continuum C of the proof of Theorem 2.33

and 2^{n-1} new vertical segments from the nth horizontal segment to the lowest segment). It is easy to check using Lelek's characterization that C is not regular (the horizontal segments are an infinite pairwise disjoint collection of subcontinua of the same diameter). C was originally introduced to show that a continuum which is the union of two regular continua (or even of two dendrites) need not to be regular, or even hereditarily locally connected.

If we modify C by deleting all but finitely many of the horizontal segments (and leaving all vertical segments and the bottom horizontal segment untouched), the resulting continuum will be regular. If we view the nth horizontal segment (counting from the top down) as consisting of 2^n subsegments of equal length, to achieve regularity it suffices that for n sufficiently large we delete a portion of each of these subsegments (to make sure that the resulting set is a continuum we should delete a nonempty open connected subset of each subsegment).

We can further assign each of the subsegments to a column: a subsegment is in the first column if at least its leftmost part is in the left half of the figure, in the second column if at least its leftmost part is in the third quarter (counting from the left) of the figure, and in general belongs to the kth column if at least its leftmost part is in the $2^k - 1$th 2^{-k} piece (counting from the left) of the figure. In Figure 1 the subsegments belonging to the third column are thicker (notice that for $n \geq 3$, 2^{n-3} subsegments of the nth horizontal segment belong to the third column). With this notation and using the characterization mentioned above, it can be shown that to obtain regularity it suffices that in each column only finitely many subsegments are not affected by our deletion process.

This leads to a way of reducing the $\mathbf{\Pi}^0_4$-complete set

$$A = \{\, \alpha \in I^{\mathbb{N}\times\mathbb{N}} \mid \forall k\ \exists m\ \forall n > m\ \alpha(k,n) < 1 \,\}$$

to $\mathcal{R}$. For any $\alpha \in I^{\mathbb{N}\times\mathbb{N}}$ we define a continuum C_α obtained by deleting a (possibly empty) open connected subset of every horizontal subsegment belonging to column k and horizontal segment n. The size of the open subset we delete is dictated by

$\alpha(k,n)$, so that in particular if $\alpha(k,n) = 1$ the subsegment is left untouched. This process can be set up so that the map $I^{\mathbb{N}\times\mathbb{N}} \to \mathsf{C}(I^2)$, $\alpha \mapsto C_\alpha$ is continuous.

If $\alpha \in A$ then in defining C_α we are deleting a nonempty subset of all but finitely many subsegments of each column, so that $C_\alpha \in \mathcal{R}$. If $\alpha \notin A$ there is at least a column where infinitely many subsegments are not affected by our deletion process. It follows that a subcontinuum of C_α is homeomorphic to C, and therefore $C_\alpha \notin \mathcal{R}$. Therefore $A \leq_{\mathrm{W}} \mathcal{R}$ and the proof is complete. □

This result is particularly interesting from a descriptive set theoretic viewpoint because natural sets appearing at the fourth level of the Borel hierarchy are quite rare (and the only claim to an example appearing at later levels is in [Sof02]).

Theorem 2.34. *The set of dendrites is $\mathbf{\Pi}^0_3$-complete in $\mathsf{C}(I^n)$ for $2 \leq n \leq \omega$ (a dendrite is a locally connected continuum which does not contain any circle).*

This result is proved by Camerlo, Darji and Marcone in [CDM05], essentially in the same way used in that paper for proving Theorem 2.10.

Dendrites are quite important planar continua (e.g., Nadler devotes a whole chapter of [Nad92] to their study, and the cover of the same book depicts a universal dendrite) and will play a role also in Section 3, where often the complexity of a binary relation will be studied on the set of dendrites.

Theorem 2.35. *The set of trees is $\mathrm{D}_2(\mathbf{\Sigma}^0_3)$-complete in $\mathsf{C}(I^n)$ for $2 \leq n \leq \omega$ (a tree is a dendrite which can be written as the union of finitely many arcs which pairwise intersect only in their end points).*

This theorem is also due to Camerlo, Darji and Marcone ([CDM05]) and is also particularly interesting from a descriptive set theoretic viewpoint: natural sets whose classification involves the difference hierarchy are uncommon. Recalling the definition of the difference hierarchy, Theorem 2.35 asserts that the set of trees is the intersection of a $\mathbf{\Sigma}^0_3$ and a $\mathbf{\Pi}^0_3$ set, but cannot be written as the union of a $\mathbf{\Sigma}^0_3$ and a $\mathbf{\Pi}^0_3$ set.

By removing from the definition of tree the requirement that they are dendrites, we obtain the important notion of a graph (a whole chapter of [Nad92] deals with graphs).

Theorem 2.36. *The set of graphs is $\mathrm{D}_2(\mathbf{\Sigma}^0_3)$-complete in $\mathsf{C}(I^n)$ for $2 \leq n \leq \omega$.*

This theorem is again proved in [CDM05] in the same way of Theorem 2.35, and again provides a rare example of a natural set whose classification needs the difference hierarchy.

The proof that the sets of trees and graphs are $\mathrm{D}_2(\mathbf{\Sigma}^0_3)$ is sketched after Theorem 3.11 below.

2.8. Retracts

Recall the following definition:

Definition 2.37. A separable metric space C is an *absolute retract* if whenever C is embedded as a closed subset in a separable metric space Y, there exists a continuous function $f : Y \to C$ which is the identity on C.

A compact absolute retract is a continuum, so any classification result about the set of absolute retracts in $\mathsf{K}(X)$ holds also in $\mathsf{C}(X)$.

Cauty, Dobrowolski, Gladdines, and van Mill ([CDGvM95]) studied the set of compacta which are absolute retracts in $\mathsf{K}(I^2)$.

Theorem 2.38. *The set of absolute retracts is* $\mathbf{\Pi}^0_3$*-complete in* $\mathsf{C}(I^2)$.

Dobrowolski and Rubin ([DR94]) studied the same set in $\mathsf{K}(I^n)$ for $3 \leq n \leq \omega$, and found the situation to be quite different.

Theorem 2.39. *The set of absolute retracts is* $\mathbf{\Pi}^0_4$*-complete in* $\mathsf{C}(I^n)$ *for* $3 \leq n \leq \omega$.

Let us also recall the definition of absolute neighborhood retract:

Definition 2.40. A separable metric space C is an *absolute neighborhood retract* if whenever C is embedded as a closed subset in a separable metric space Y, there exist a neighborhood U of C in Y and a continuous function $f : U \to C$ which is the identity on C.

Cauty, Dobrowolski, Gladdines, and van Mill in [CDGvM95] studied also the set of compacta which are absolute neighborhood retracts in $\mathsf{K}(I^2)$. Since there exist compact absolute neighborhood retracts which are not connected, one cannot immediately translate to our case results obtained in $\mathsf{K}(X)$. However in Remarque 3.10 of [CDGvM95] it is suggested how to adapt the proof of the result on compacta to obtain:

Theorem 2.41. *The set of continua which are absolute neighborhood retracts is* $\mathrm{D}_2(\mathbf{\Sigma}^0_3)$*-complete in* $\mathsf{C}(I^2)$.

Hence we have another example of a natural set of continua whose classification needs the difference hierarchy.

Dobrowolski and Rubin in [DR94] explicitly studied the set of absolute neighborhood retracts which are continua in dimension > 2 and found again the situation to be different than in the planar case:

Theorem 2.42. *The set of continua which are absolute neighborhood retracts is* $\mathbf{\Pi}^0_4$*-complete in* $\mathsf{C}(I^n)$ *for* $3 \leq n \leq \omega$.

2.9. σ-ideals of continua

In [Cam05] Camerlo studied σ-ideals of continua. Here is the definition:

Definition 2.43. Let X be a Polish space and $\emptyset \neq \mathcal{I} \subseteq \mathsf{C}(X)$. Then $\mathcal{I}$ is a *σ-ideal of continua* if:

(a) any subcontinuum of an element of $\mathcal{I}$ belongs to $\mathcal{I}$;
(b) if $C_n \in \mathcal{I}$ for all $n \in \mathbb{N}$ and $\bigcup_{n\in\mathbb{N}} C_n$ is a continuum, then $\bigcup_{n\in\mathbb{N}} C_n \in \mathcal{I}$.

Some of the sets of continua we mentioned before (e.g., the set of Suslinian continua and the set of strongly countable-dimensional continua) are actually σ-ideals.

The basic classification result obtained by Camerlo mirrors the dichotomy for $\mathbf{\Pi}^1_1$ σ-ideals of compact sets proved by Kechris, Louveau and Woodin ([KLW87]), stating that a $\mathbf{\Pi}^1_1$ σ-ideal of compact sets in a Polish space is either $\mathbf{\Pi}^0_2$ or $\mathbf{\Pi}^1_1$-complete. Here is Camerlo's theorem:

Theorem 2.44. *Let $\mathcal{I} \subseteq \mathsf{C}(I^n)$ for $2 \leq n \leq \omega$ be a $\mathbf{\Pi}^1_1$ σ-ideal of continua which is closed under homeomorphisms. Then $\mathcal{I}$ is either $\mathbf{\Pi}^0_2$ or $\mathbf{\Pi}^1_1$-complete.*

Using Theorem 2.44 the proofs of Theorems 2.9 and 2.30 can be simplified: to prove that the σ-ideal under consideration is $\mathbf{\Pi}^1_1$-complete it suffices to show that it is $\mathbf{\Pi}^1_1$ and $\mathbf{\Sigma}^0_2$-hard.

Kechris, Louveau and Woodin in [KLW87] also proved that a $\mathbf{\Sigma}^1_1$ σ-ideal of compact sets is $\mathbf{\Pi}^0_2$, and Camerlo mirrored also this result:

Theorem 2.45. *Let $\mathcal{I} \subseteq \mathsf{C}(I^n)$ for $2 \leq n \leq \omega$ be a $\mathbf{\Sigma}^1_1$ σ-ideal of continua which is closed under homeomorphisms. Then $\mathcal{I}$ is $\mathbf{\Pi}^0_2$.*

Camerlo remarks that the hypothesis in Theorems 2.44 and 2.45 that the σ-ideal is closed under homeomorphisms can be relaxed but cannot be totally deleted. [Cam05] contains several other results on σ-ideals of continua.

3. Binary relations between continua

3.1. Homeomorphism

The most natural equivalence relation between continua is that of homeomorphism. Kechris and Solecki proved that homeomorphism between compact metric spaces is induced by a Polish group action ([Hjo00, §4.4]). On the other hand, Hjorth ([Hjo00, §4.3]) proved the following theorem:

Theorem 3.1. *The equivalence relation of homeomorphism between locally connected continua, and a fortiori between arbitrary continua, in $\mathsf{C}(I^n)$ for $3 \leq n \leq \omega$ is strictly more complicated (in the sense of $\leq_{\mathrm{B}}$) than any equivalence relation classifiable by countable structures.*

(Hjorth actually talks about compacta, but his construction produces locally connected continua.) The most difficult part of the proof of Theorem 3.1 is showing that homeomorphism is not classifiable by countable structures: this is an application of Hjorth's theory of turbulence.

In contrast with Theorem 3.1, Camerlo, Darji and Marcone ([CDM05]) proved the following result:

Theorem 3.2. *The equivalence relation of homeomorphism between dendrites in $\mathsf{C}(I^n)$ for $2 \leq n \leq \omega$ is classifiable by countable structures.*

Theorem 3.2 shows that the homeomorphism type of a dendrite is easier to recognize than the homeomorphism type of an arbitrary locally connected continuum. Camerlo, Darji and Marcone showed also that the classification given by Theorem 3.2 is optimal, by showing that homeomorphism between dendrites is S_∞-universal and hence can still be considered very complicated. Their result is actually sharper and to state it we need the following definition.

Definition 3.3. If C is a continuum and $x \in C$ the *order of x in C*, denoted by $\mathrm{ord}(x, C)$, is the smallest cardinal number κ such that there exists a neighborhood-base for x in C consisting of open sets each with boundary of cardinality less than or equal to κ. A point $x \in C$ is a *branching point of C* if $\mathrm{ord}(x, C) > 2$.

Theorem 3.4. *The equivalence relation of homeomorphism between dendrites with all branching points belonging to an arc is S_∞-universal in $\mathsf{C}(I^n)$ for $2 \leq n \leq \omega$.*

The dendrites mentioned in this theorem are quite simple (see the remark after Fact 3.12 below). Theorem 3.4 (combined with Theorem 3.2) states that already at this level homeomorphism is as complicated as it can get when dendrites are concerned.

3.2. Continuous embeddability

A natural $\mathbf{\Sigma}^1_1$ quasi-order between continua is that of continuous embeddability. The sharpest result about it is due to Camerlo ([Cam05b]), and improves previous results by Louveau and Rosendal ([LR05]) and Marcone and Rosendal ([MR04]):

Theorem 3.5. *The quasi-order of continuous embeddability restricted to dendrites whose points have order at most 3 is $\mathbf{\Sigma}^1_1$-complete in $\mathsf{C}(I^n)$ for $2 \leq n \leq \omega$.*

As explained in §1.3, this implies that the equivalence relation of mutual continuous embeddability between dendrites (and, a fortiori, between arbitrary continua) is $\mathbf{\Sigma}^1_1$-complete (as an equivalence relation) and hence immensely more complicated than the equivalence relations considered in §3.1.

Theorem 3.5 implies also that for continuous embeddability the situation is different from the one illustrated in §3.1: restricting the quasi-order of continuous embeddability between continua to dendrites does not lead to a simpler quasi-order.

3.3. Continuous surjections

In §3.2 we considered embeddings, i.e., injective maps between continua. However Darji pointed out that in continuum theory continuous surjections (epimorphisms) are at least as important as continuous injections (see, e.g., [Nad92, Theorem 3.21]), and raised the question of the complexity of the corresponding quasi-order. Let us define $C \preceq_e D$ if and only if there exists $f : D \to C$ continuous and onto. Camerlo ([Cam05a]) answered Darji's question by proving:

Theorem 3.6. *The quasi-order of epimorphism between continua is $\mathbf{\Sigma}^1_1$-complete in $\mathsf{C}(I^n)$ for $2 \leq n \leq \omega$.*

Actually Camerlo's proof shows that the result holds also if we restrict the kind of surjective maps we are considering (the definition of $\leq_B$ is such that neither of these corollaries of the proof of Theorem 3.6 implies its statement or follows from it):

Corollary 3.7. *The quasi-order of monotone epimorphism between continua is $\mathbf{\Sigma}^1_1$-complete in $\mathsf{C}(I^n)$ for $2 \leq n \leq \omega$ (a continuous function between continua is monotone if the preimage of every point is a continuum).*

Corollary 3.8. *The quasi-order of weakly confluent epimorphism between continua is $\mathbf{\Sigma}^1_1$-complete in $\mathsf{C}(I^n)$ for $2 \leq n \leq \omega$ (a continuous function between continua is weakly confluent if every subcontinua of the image is image of a subcontinuum of the domain).*

As far as $\preceq_e$ is concerned it does not make sense to study dendrites (as we did in Theorems 3.2, 3.4, and 3.5). In fact the Hahn-Mazurkiewicz theorem and the fact that every continuum is homeomorphic to a subset of I^ω imply that all nondegenerate locally connected continua (and in particular all dendrites) are equivalent with respect to the equivalence relation induced by $\preceq_e$. However if we require that the surjective function is not only continuous, but also open, dendrites become interesting once more, and Camerlo ([Cam05a]) proved:

Theorem 3.9. *The quasi-order of open epimorphism between dendrites (and, a fortiori, between continua) is $\mathbf{\Sigma}^1_1$-complete in $\mathsf{C}(I^n)$ for $2 \leq n \leq \omega$.*

3.4. Likeness and quasi-homeomorphism

An important quasi-order on continua is the relation of likeness (e.g., a whole chapter of [Nad92] deals with arc-like continua).

Definition 3.10. If $\mathcal{C}$ is a class of continua and D is a continuum we say that D is *$\mathcal{C}$-like* if for every $\varepsilon > 0$ there exist $C \in \mathcal{C}$ and a continuous map $f : D \to C$ such that f is onto and $\{ x \in D \mid f(x) = y \}$ has diameter less than ε for each $y \in C$.

When $\mathcal{C} = \{C\}$ we say that D is *C-like* and write $D \preceq C$. The equivalence relation induced by likeness is called *quasi-homeomorphism*.

Camerlo, Darji and Marcone ([CDM05]) studied extensively likeness, mainly when $\mathcal{C}$ is a set of dendrites or graphs. They obtained complexity results for many initial segments of $\preceq$, including the following:

Theorem 3.11.

(a) *If C is a nondegenerate dendrite or a graph, then the set of C-like continua is $\mathbf{\Pi}^0_2$-complete in $\mathsf{C}(I^n)$ for $2 \leq n \leq \omega$;*
(b) *the set of graph-like continua is $\mathbf{\Pi}^0_2$-complete in $\mathsf{C}(I^n)$ for $2 \leq n \leq \omega$;*
(c) *the set of tree-like continua is $\mathbf{\Pi}^0_2$-complete in $\mathsf{C}(I^n)$ for $2 \leq n \leq \omega$.*

In particular Theorem 3.11.(a) implies that the sets of arc-like continua and of circle-like continua are both $\mathbf{\Pi}^0_2$-complete in $\mathsf{C}(I^n)$ for $2 \leq n \leq \omega$.

Theorem 3.11 is used to obtain the upper bounds for Theorems 2.35 and 2.36. In fact, by a result of Kato and Ye ([KY00]), if C is a locally connected continuum, D a graph (resp. tree) and $C \preceq D$ then C is also a graph (resp. tree). Hence if $\{ D_n \mid n \in \mathbb{N} \}$ is a sequence of graphs (resp. trees) which contains at least a member from each quasi-homeomorphism class of graphs (resp. trees), then a continuum C is a graph (resp. tree) if and only if C is locally connected and $C \preceq D_n$ fo some n. By Theorems 2.10 and 3.11.a this shows that the set of graphs (resp. trees) is $\mathrm{D}_2(\mathbf{\Sigma}^0_3)$.

3.5. Some homeomorphism and quasi-homeomorphism classes

The fact that each homeomorphism class of a continuum is a Borel subset of $\mathsf{C}(I^\omega)$ is a classical result due to Ryll-Nardzewski ([RN65]). The following fact is a consequence of Theorem 3.4:

Fact 3.12. *For every $\alpha < \omega_1$ there exists a dendrite C with all branching points belonging to an arc such that the homeomorphism class of C in $\mathsf{C}(I^n)$ for $2 \leq n \leq \omega$, is not $\mathbf{\Pi}^0_\alpha$.*

This means that the set of dendrites with all branching points belonging to an arc (and a fortiori the sets of dendrites and of continua) is partitioned into homeomorphism classes of unbounded Borel complexity. (The intuitive fact that the dendrites with all branching points belonging to an arc – already mentioned in Theorem 3.4 – are "simple" is made precise by the following result of [CDM05]: there exist exactly two $\preceq$-minimal quasi-homeomorphism classes of dendrites with infinitely many branching points, and one of them consists precisely of all dendrites with infinitely many branching points which all belong to an arc.)

If we fix a specific continuum we can study the complexity of the homeomorphism and quasi-homeomorphism class of that continuum. To this end the following theorem, which combines results of Segal ([Seg68], rediscovered by Kato and Ye in [KY00]) and Camerlo, Darji and Marcone ([CDM05]) is useful:

Theorem 3.13. *Let C be either a graph or a dendrite with finitely many branching points. A continuum is homeomorphic to C if and only if it is quasi-homeomorphic to C.*

In general quasi-homeomorphism is coarser than homeomorphism, but Theorem 3.13 states that for fairly simple continua the two equivalence relation coincide.

Camerlo, Darji and Marcone ([CDM05]) proved:

Theorem 3.14. *Let C be either a graph or a nondegenerate dendrite. The quasi-homeomorphism class of C is $\mathbf{\Pi}^0_3$-complete in $\mathsf{C}(I^n)$ for $3 \leq n \leq \omega$. If C is planar (e.g., if it is a dendrite) then this holds also in $\mathsf{C}(I^2)$.*

The proof of Theorem 3.14 uses the results of Theorem 3.11.

Combining the last two Theorems we obtain the following result ([CDM05]):

Theorem 3.15. *Let C be either a graph or a dendrite with finitely many branching points. The homeomorphism class of C is $\mathbf{\Pi}^0_3$-complete in $\mathsf{C}(I^n)$ for $3 \leq n \leq \omega$. If C is planar then this holds also in $\mathsf{C}(I^2)$.*

Some instances of Theorem 3.15 are much older: e.g., $\mathbf{\Pi}^0_3$-completeness of the set of all arcs follows from the results in [Kur31] and [Maz31].

The following fact follows by a Baire category argument from the remarks before Fact 2.4:

Fact 3.16. *The homeomorphism class of the pseudo-arc is $\mathbf{\Pi}^0_2$-complete in $\mathsf{C}(I^n)$ for $2 \leq n \leq \omega$.*

In [Kru02] Krupski studied the homeomorphism class of two important continua: Sierpiński universal curve M^2_1 (see [Nad92, Example 1.11]) and Menger universal curve M^3_1 (see, e.g., [Eng78, p.122]). The former contains a homeomorphic copy of every 1-dimensional planar compacta, while the latter contains a homeomorphic copy of every 1-dimensional compacta. Krupski proved:

Theorem 3.17. *The homeomorphism class of the Sierpiński universal curve is $\mathbf{\Pi}^0_3$-complete in $\mathsf{C}(I^n)$ for $2 \leq n \leq \omega$.*

Theorem 3.18. *The homeomorphism class of the Menger universal curve is $\mathbf{\Pi}^0_3$-complete in $\mathsf{C}(I^n)$ for $3 \leq n \leq \omega$.*

3.6. Isometry and Lipschitz isomorphism

The equivalence relation of isometry between Polish metric spaces is very complicated and has been studied in depth by Gao and Kechris ([GK03]). However if we restrict ourselves to compact Polish metric spaces, Gromov ([Gro99], see [Hjo00, §4.4] for a sketch) has shown that isometry is smooth. This immediately yields (recall from page 123 that M denotes the Urysohn space):

Theorem 3.19. *The equivalence relation of isometry between continua in $\mathsf{C}(M)$ is smooth.*

Rosendal ([Ros05]) studied the metric equivalence relation of Lipschitz isomorphism: two metric spaces C and C' with metrics d and d' are *Lipschitz isomorphic* if there exist a bijection $f : C \to C'$ and $c \geq 1$ such that $\frac{1}{c}d(x,y) \leq d'(f(x), f(y)) \leq c\,d(x,y)$ for every $x, y \in C$ (obviously such an f is a homeomorphism). Rosendal proved:

Theorem 3.20. *The equivalence relation of Lipschitz isomorphism between continua in $\mathsf{C}(M)$ is K_σ-complete.*

Rosendal's proof uses continua which are not locally connected, but Rosendal himself pointed out that a straightforward modification yields also:

Theorem 3.21. *The equivalence relation of Lipschitz isomorphism between dendrites in $\mathsf{C}(M)$ is K_σ-complete.*

Acknowledgment

I thank Riccardo Camerlo and Udayan B. Darji for many fruitful conversations on the topic of this survey. Howard Becker kindly allowed the inclusion of some of his unpublished proofs. I am in debt with P. Krupski and with the anonymous referee for several bibliographic suggestions.

References

[AK00] Scot Adams and Alexander S. Kechris, *Linear algebraic groups and countable Borel equivalence relations*, J. Amer. Math. Soc. **13** (2000), no. 4, 909–943.

[Bec84] Howard Becker, *Some theorems and questions about some natural examples in descriptive set theory*, manuscript, 1984.

[Bec86] ———, *Simply connected sets and hyperarithmetic paths*, manuscript, 1986.

[Bec87] ———, *On classifying simple connectedness in the projective hierarchy*, manuscript, 1987.

[Bec92] ———, *Descriptive set theoretic phenomena in analysis and topology*, Set Theory of the Continuum (H. Judah, W. Just, and H. Woodin, eds.), Springer-Verlag, 1992, pp. 1–25.

[Bec98] ———, *The number of path-components of a compact subset of* $\mathbf{R}^n$, Logic Colloquium '95 (Haifa), Lecture Notes in Logic, vol. 11, Springer-Verlag, Berlin, 1998, pp. 1–16.

[Bin51] R.H. Bing, *Concerning hereditarily indecomposable continua*, Pacific J. Math. **1** (1951), 43–51.

[BK96] Howard Becker and Alexander S. Kechris, *The descriptive set theory of Polish group actions*, Cambridge University Press, Cambridge, 1996.

[BP01] Howard Becker and Roman Pol, *Note on path-components in complete spaces*, Topology Appl. **114** (2001), no. 1, 107–114.

[Cam05a] Riccardo Camerlo, *Universal analytic preorders arising from surjective functions*, Fund. Math., **187** (2005), 193–212.

[Cam05b] ———, *Universality of embeddability relations for coloured total orders*, Order **22** (2005), 289–300.

[Cam05] ———, *Continua and their* σ*-ideals*, Topology Appl. **150** (2005), 1–18.

[CDGvM95] Robert Cauty, Tadeusz Dobrowolski, Helma Gladdines, and Jan van Mill, *Les hyperespaces des rétractes absolus et des rétractes absolus de voisinage du plan*, Fund. Math. **148** (1995), no. 3, 257–282 (French).

[CDM05] Riccardo Camerlo, Udayan B. Darji, and Alberto Marcone, *Classification problems in continuum theory*, Trans. Amer. Math. Soc. **357** (2005), 4301–4328.

[CG01] Riccardo Camerlo and Su Gao, *The completeness of the isomorphism relation for countable Boolean algebras*, Trans. Amer. Math. Soc. **353** (2001), no. 2, 491–518.

[CL78] H. Cook and A. Lelek, *Weakly confluent mappings and atriodic Suslinian curves*, Canad. J. Math. **30** (1978), no. 1, 32–44.

[Dar00] Udayan B. Darji, *Complexity of hereditarily decomposable continua*, Topology Appl. **103** (2000), no. 3, 243–248.

[DM04] Udayan B. Darji and Alberto Marcone, *Complexity of curves*, Fund. Math. **182** (2004), no. 1, 79–93.

[DR94] Tadeusz Dobrowolski and Leonard R. Rubin, *The space of ANR's in* $\mathbb{R}^n$, Fund. Math. **146** (1994), no. 1, 31–58.

[Eng78] Ryszard Engelking, *Dimension theory*, North-Holland Publishing Co., Amsterdam, 1978, Translated from the Polish and revised by the author, North-Holland Mathematical Library, 19.

[FS89] Harvey Friedman and Lee Stanley, *A Borel reducibility theory for classes of countable structures*, J. Symbolic Logic **54** (1989), no. 3, 894–914.

[GK03] Su Gao and Alexander S. Kechris, *On the classification of Polish metric spaces up to isometry.*, Mem. Am. Math. Soc. **766** (2003), 78 p.

[Gro99] Misha Gromov, *Metric structures for Riemannian and non-Riemannian spaces*, Progress in Mathematics, v. 152, Birkhäuser, Boston, 1999.

[Hjo00] Greg Hjorth, *Classification and orbit equivalence relations*, American Mathematical Society, Providence, RI, 2000.

[IC68] W.T. Ingram and Howard Cook, *A characterization of indecomposable compact continua*, Topology Conference (Arizona State Univ., Tempe, Ariz., 1967), Arizona State Univ., Tempe, Ariz., 1968, pp. 168–169.

[Kec95] Alexander S. Kechris, *Classical descriptive set theory*, Graduate Texts in Mathematics, no. 156, Springer-Verlag, New York, 1995.

[Kec02] ———, *Actions of Polish groups and classification problems*, Analysis and logic (Mons, 1997), London Math. Soc. Lecture Note Ser., vol. 262, Cambridge Univ. Press, Cambridge, 2002, pp. 115–187.

[KLW87] Alexander S. Kechris, Alain Louveau, and W. Hugh Woodin, *The structure of σ-ideals of compact sets*, Trans. Amer. Math. Soc. **301** (1987), no. 1, 263–288.

[Kru02] Paweł Krupski, *Hyperspaces of universal curves and 2-cells are true $F_{\sigma\delta}$-sets*, Colloq. Math. **91** (2002), no. 1, 91–98.

[Kru03] ———, *More non-analytic classes of continua*, Topology Appl. **127** (2003), no. 3, 299–312.

[Kru04] ———, *Families of continua with the property of Kelley, arc continua and curves of pseudo-arcs*, Houston J. Math. **30** (2004), no. 2, 459–482.

[KS] Paweł Krupski and Alicja Samulewicz, *Strongly countable dimensional compacta form the Hurewicz set*, Topology Appl., to appear.

[Kur31] Kazimierz Kuratowski, *Évaluation de la classe borélienne ou projective d'un ensemble de points à l'aide des symboles logiques*, Fund. Math. **17** (1931), 249–272.

[KY00] Hisao Kato and Xiangdong Ye, *On Burgess's theorem and related problems*, Proc. Amer. Math. Soc. **128** (2000), no. 8, 2501–2506.

[Lel71] A. Lelek, *On the topology of curves. II*, Fund. Math. **70** (1971), no. 2, 131–138.

[LR05] Alain Louveau and Christian Rosendal, *Complete analytic equivalence relations*, Trans. Amer. Math. Soc. **357** (2005), 4839–4866.

[Maz30] Stefan Mazurkiewicz, *Sur le continus absolutement indécomposables*, Fund. Math. **16** (1930), 151–159.

[Maz31] ———, *Sur l'ensemble des continus péaniens*, Fund. Math. **17** (1931), 273–274.

[MR04] Alberto Marcone and Christian Rosendal, *The complexity of continuous embeddability between dendrites*, J. Symbolic Logic **69** (2004), no. 3, 663–673.

[Nad92] Sam B. Nadler, Jr., *Continuum theory*, Marcel Dekker Inc., New York, 1992.

[Pro72] C. Wayne Proctor, *A characterization of hereditarily indecomposable continua*, Proc. Amer. Math. Soc. **34** (1972), 287–289.

[RN65] C. Ryll-Nardzewski, *On a Freedman's problem*, Fund. Math. **57** (1965), 273–274.

[Ros05] Christian Rosendal, *Cofinal families of Borel equivalence relations and quasiorders*, J. Symbolic Logic **70** (2005), 1325–1340.

[Sam03] Alicja Samulewicz, *The hyperspace of hereditarily decomposable subcontinua of a cube is the Hurewicz set*, preprint, 2003.

[Seg68] Jack Segal, *Quasi dimension type. II: Types in 1-dimensional spaces*, Pac. J. Math. **25** (1968), no. 2, 353–370.

[Sof02] Nikolaos Efstathiou Sofronidis, *Natural examples of $\mathbf{\Pi}^0_5$-complete sets in analysis*, Proc. Amer. Math. Soc. **130** (2002), 1177–1182.

[Sol02a] Sławomir Solecki, *Descriptive set theory in topology*, Recent progress in general topology, II, North-Holland, Amsterdam, 2002, pp. 485–514.

[Sol02b] ———, *The space of composants of an indecomposable continuum*, Adv. Math. **166** (2002), no. 2, 149–192.

Alberto Marcone
Dipartimento di matematica e informatica
Università di Udine
Via delle Scienze 208
I-33100 Udine, Italy
e-mail: `marcone@dimi.uniud.it`

Set Theory
Trends in Mathematics, 149–224

Weak Systems of Gandy, Jensen and Devlin

A.R.D. Mathias

Abstract. In Part I, we formulate and examine some systems that have arisen in the study of the constructible hierarchy; we find numerous transitive models for them, among which are supertransitive models containing all ordinals that show that Devlin's system BS lies strictly between Gandy's systems PZ and BST'; and we use our models to show that BS fails to handle even the simplest rudimentary functions, and is thus inadequate for the use intended for it in Devlin's treatise. In Part II we propose and study an enhancement of the underlying logic of these systems, build further models to show where the previous hierarchy of systems is preserved by our enhancement; and consider three systems that might serve Devlin's purposes: one the enhancement of a version of BS, one a formulation of Gandy-Jensen set theory, and the third a subsystem common to those two. In Part III we give new proofs of results of Boffa by constructing three models in which, respectively, TCo, AxPair and AxSing fail; we give some sufficient conditions for a set not to belong to the rudimentary closure of another set, and thus answer a question of McAloon; and we comment on Gandy's numerals and correct and sharpen other of his observations.

Contents

Part III

0. Introduction

During the 1960's, as knowledge of the constructible hierarchy advanced, pre-eminently through the work of Jensen [J1] [J2], there was a drive to study various weak systems of set theory, all weaker than that of Kripke–Platek. Those systems included Δ_0 separation but weakened Δ_0 collection in various ways, and their purpose was to give a finer account of the growth of the constructible hierarchy. As is well known, this move has been extraordinarily fruitful.

Gandy [G] proposed four systems which he called PZ (for "predicative Zermelo"), BST', BRT and PZF. and which he proved to be strictly ascending in strength. Devlin in his treatise [Dev] proposed a further system, which he called BS. We shall, starting in Section 1, introduce new names for those five systems and others which have suggested themselves, but shall use the old in this introduction.

So, roughly, PZ is a weak base theory plus Δ_0 separation. BS adds Cartesian product to that. BST' is the result of adding an axiom of infinity to Gandy's theory BST, of which the transitive models are precisely the rudimentarily closed sets. BRT has what Gandy calls the bounded replacement axiom; and PZF has Δ_0 replacement, making it weaker than but close to and equiconsistent with the system of Kripke–Platek with an axiom of infinity. We shall also look briefly at what Gandy would have called the bounded collection axiom, and at our preferred formulation of the system of Kripke and Platek.

When, in the next section and later, we give precise formulations of systems, we shall put names of systems and axioms in nine-point sans-serif type to indicate that it is our particular formulations that are being discussed, as defined either in this paper or in [M2]. In our formulations we shall change some of Gandy's terminology and notation, since Gandy uses the term "basic" for the functions that Jensen called "rudimentary"; and further Gandy studies two versions of the axiom of replacement, calling the one "basic" and the other "bounded", an unfortunate combination of adjectives as both begin with 'b'. Therefore we shall follow Jensen's usage, often shortening "rudimentary" to "rud", and shall use "RR" to name what Gandy called the basic replacement axiom. We shall reserve the word "basic" for a proper subclass $\mathcal{B}$ of the class $\mathcal{R}$ of rudimentary functions, namely those generated by composition from Gödel's functions $\mathcal{F}_1, \dots, \mathcal{F}_8$, and we shall use "flat" where Gandy used "bounded" in naming axioms.

In discussing these systems it will, as in *The Strength of Mac Lane Set Theory* [M2], at times be necessary to maintain a careful distinction between three levels of language, which we call the *metalanguage*, which is English, the *language of discourse*, which is a language of set theory formulated with atomic predicates $\in$ and $=$, and various *object languages*, again set-theoretical in nature, with atomic predicates symbolised by ϵ and $\mathtt{=}$. We use Fraktur lower case letters $\mathfrak{k}, \mathfrak{l}, \mathfrak{m}, \ldots$ for *concrete* integers, which are quantified only in the metalanguage, and the corresponding terms for them in the language of discourse. This visual aid may be used to mark the distinction between a system T being able, for each $\mathfrak{k}$, to prove some statement $\Phi(\mathfrak{k})$ and being able to prove $\forall k \Phi(k)$.

Three areas of uncertainty in the choice of axioms

The above authors differ in their treatment of the scheme of foundation: Gandy makes no mention of foundation in his formulations, whereas Devlin calls for the full scheme of foundation in his. Without foundation, his system is intermediate between PZ and BST'. The question of the amount of foundation possessed by a system is not idle: in our paper [M2] we showed that in terms of consistency strength Π_2 foundation is in some cases strictly stronger than Π_1 foundation – see Metacorollary 9.21 and Metatheorem 9.34 of [M2] – and there is evidence that Π_1 foundation is the "right" amount to have in formulating the system of Kripke–Platek; see Corollary 1.22(ii) and Proposition 3.14 (ii') of [M2]. The investigations of the present paper suggest that Π_1 foundation is also the "right" amount to have in these weaker systems.

A second area of uncertainty is the axiom of transitive containment, TCo, which asserts that every set is a member of a transitive set. It was shown by Boffa [B1], [B2] that TCo is not provable in Zermelo set theory: we give a new proof of that result in Section 12. TCo is, however, provable in our formulation of Kripke-Platek.

Finally it is of interest to see to what extent the axiom of infinity can be avoided.

So our policy will be, at least initially, to exclude the "special" axioms of infinity and transitive containment from the general axioms of our systems, and explicitly to note each use of those special axioms as it occurs. As for foundation, we shall include the scheme of Π_1 foundation in our systems, and draw attention to areas where foundation can be avoided, and where the full scheme of foundation is required.

In many sections of the paper, our focus will be chiefly not on the consistency strength of the various theories but on constructing transitive models for them; and in such models, the full scheme of foundation will be inherited from our ambient set theory. Further TCo and AxInf will be true in most of our models. We remark that we are not in this paper concerned to find the minimal ambient set theory in which our examples can be built. ZF is certainly too strong; $\mathsf{Z} + \mathsf{KP}$ is usually enough, apart from the occasional appeal to the existence of $V_{\omega+\omega}$ and similar sets. The axiom of choice is used only in a very few peripheral remarks.

Some differences

In the calibration of these systems, certain sets function as litmus paper:

0.0. Definition. We write $\mathcal{S}(x)$ for the set of finite subsets of x; for each $\mathfrak{k} > 0$, $[\omega]^{\mathfrak{k}}$ for the class of subsets of ω of size $\mathfrak{k}$; **HF** for the class of hereditarily finite sets, which in appropriate set theories will coincide with the classes notated V_ω, L_ω and J_1; EVEN for the class of even numbers, ACK for the Ackermann relation on ω, defined as $\{(m, n)_2 \mid 2^m$ is one of the summands in the expression of n as a sum of powers of $2\}$; and $\mathcal{G}_+$ for the graph of integer addition, defined as the class $\{(p, m, n)_3 \mid m + n = p\}$.

We shall see that PZ cannot prove the existence even of $[\omega]^1$; BS can prove the existence of $[\omega]^1$ and $[\omega]^2$ but not of $[\omega]^3$; BST' can prove the existence of each $[\omega]^{\mathfrak{k}}$; BST' with Π_1 foundation can prove that $\forall k\, [\omega]^k \in V$ but cannot prove the existence of $\mathcal{S}(\omega)$; BRT can prove the existence of $\mathcal{S}(\omega)$ but not of **HF**; and PZF proves the existence of **HF**. Further, we shall see that BRT proves that $\mathcal{G}_+$ is a set but that BST' fails to do so.

The contents of the paper

In the first of the three parts of the paper, we shall formulate, in Section 1, eight systems, with variants, and note in Section 2 various results provable in them. In Section 3, we review some simple techniques for building transitive models of weak systems. In the next four sections, we work through the systems in order of increasing strength, summarising Gandy's model-theoretic constructions and giving new ones of our own; our models will demonstrate the unprovability of various results.

The second part begins with the heavily syntactic Section 8, in which we examine the result of strengthening our previous systems by uniformly adding an axiom of infinity and the principle that the class of all finite subsets of any given set is a set; and study the effect of enhancing those strengthened systems by adding limited quantifiers of the form "for some finite subset of a" and "for all finite subsets of a". In Section 9, we give further models illustrating the limitations of our strengthened systems. Then in Section 10 we turn to an examination of Devlin's book *Constructibility*, of which certain passages have been known since its publication to be problematical; we use our models to shed light on those passages, and draw attention to three of our systems that might serve Devlin's purposes better than his system BS.

We begin the final part of the paper by showing in Section 11 that Gandy's remarks concerning certain variants of his systems are not correct. In Section 12, we return to model-building and give a new proof of the result of Boffa that TCo is not provable in Zermelo set theory; Section 13 looks briefly at the axiom of pairing; in Section 14 we find an answer to a question raised by McAloon in the 1970's by giving criteria for one set not to lie in the rudimentary closure of another; finally in Section 15 we apply the technique of Section 14 to show that the set of Gandy numerals is not in the rudimentary closure of ω.

Part I

1. Formulations of the various systems

We outline the syntactical development of our systems: various aspects will be discussed in the projected sequel [M4] in greater detail.

We start with enough syntax to introduce the axioms of our first, very weak system, and to define for each $\mathfrak{n}$ the ordered $\mathfrak{n}$-tuple; then we shall enlarge the syntax to include some convenient extensions of the class-forming operator, and shall then be able to enunciate in the language of discourse the axioms of the systems we intend to study.

We begin therefore with two undefined binary relations $\in$, $=$; propositional connectives $\neg$, $\vee$, & , $\Longrightarrow$, $\Longleftrightarrow$; unrestricted quantifiers $\forall x$, $\exists x$; and restricted quantifiers $\forall x_{\in y}$, $\exists x_{\in y}$, where x and y are not permitted to be the same letter, and the quantifier binds x but not y, in harmony with the axioms that express their intended meaning:

$$\exists x_{\in y}\,\mathfrak{A} \Longleftrightarrow \exists x[x \in y \ \&\ \mathfrak{A}]; \qquad \forall x_{\in y}\,\mathfrak{A} \Longleftrightarrow \forall x[x \in y \Longrightarrow \mathfrak{A}].$$

The rules of formation are the usual ones of classical logic.

We then define a Δ_0 formula or a Δ_0 class to be one containing no unrestricted quantifiers; a Π_1 formula is one of the form $\forall x\mathfrak{A}$ where $\mathfrak{A}$ is Δ_0; a Σ_2 formula is one of the form $\exists y\mathfrak{B}$ where $\mathfrak{B}$ is Π_1; a Σ_1 formula is one of the form $\exists x\mathfrak{A}$ where $\mathfrak{A}$ is Δ_0, and so on.

We have the usual axioms of classical propositional and predicate logic; we should (but do not) define the result of substituting one variable for another, indicated informally by such usages as $\mathfrak{A}(x)$ and $\mathfrak{A}(y)$.

It is convenient to permit the use of the class-forming operator $\{\cdot \mid \ldots\}$, but we emphasize that our language is unramified and that our logic includes the Church conversion schema

$$x \in \{y \mid \mathfrak{A}(y)\} \Longleftrightarrow \mathfrak{A}(x)$$

so that all occurrences of the class-forming operator are in principle eliminable. We adopt appropriate axioms interpreting the result of substituting a class for a variable in a formula, which we summarise in these three equivalences, in which A

is a class and t a class or a variable:

$$\exists x_{\in A}\mathfrak{A} \iff \exists x[x \in A \ \&\ \mathfrak{A}]; \qquad t = A \iff \forall y_{\in t}\, y \in A \ \&\ \forall y_{\in A}\, y \in t;$$
$$A \in t \iff \exists z_{\in t}\, z = A.$$

With this syntax, we may give axioms for our first, very weak, system:

S_0 The axiom of extensionality, $[\forall x_{\in a}\, x \in b \ \&\ \forall x_{\in b}\, x \in a] \Longrightarrow a = b$, and axioms of empty set, pair set, difference and sumset (or union):

$$\varnothing \in V, \quad \{x,\, y\} \in V, \quad x \setminus y \in V, \quad \bigcup x \in V.$$

In this system we introduce, successively, ordered $\mathfrak{k}$-tuples, in the Wiener-Kuratowski manner:

$$\begin{aligned}
(y_1)_1 &=_{\mathrm{df}} y_1\\
(y_1, y_2)_2 &=_{\mathrm{df}} \{\{y_1\},\, \{y_1,\, y_2\}\};\\
(y_1, y_2, y_3)_3 &=_{\mathrm{df}} (y_1, (y_2, y_3)_2)_2\\
(y_1, y_2, y_3, y_4)_4 &=_{\mathrm{df}} (y_1, (y_2, y_3, y_4)_3)_2\\
(y_1, y_2, y_3, y_4, y_5)_5 &=_{\mathrm{df}} (y_1, (y_2, y_3, y_4, y_5)_4)_2\\
&\ \ldots
\end{aligned}$$

1.0. Remark. Thus all WK-tuples are generated from the single binary function $\{x,\, y\}$.

We may now develop the usual theory of relations, $\mathfrak{k}$-ary functions and so on: we treat functions as a subclass of their image $\times$ their domain. We shall see that these weak systems are sensitive to the choice of implementation of function, and so it is necessary to distinguish notationally between concepts that "the working mathematician" would often conflate. Thus we adopt a policy of writing ${}^3X$ for the set of 3-sequences of members of X, reserving X^3 for the set of WK 3-tuples of members of X; thus $\omega^3 = \omega \times (\omega \times \omega)$.

1.1. It is convenient further to enlarge the syntax to permit certain classes with quantified terms, namely those where the terms are WK-tuples: where there might otherwise be ambiguity, we indicate the variables to be quantified in a list placed subscript to the vertical bar, for example:

$$\{(x,y)_2 \mid_{x,y} \mathfrak{A}(x,y)\}; \quad \{(x,a)_2 \mid_x \mathfrak{A}(x,a)\}.$$

The first of those will equal $\{z \mid \exists x \exists y[z = (x,y)_2 \ \&\ \mathfrak{A}(x,y)]\}$; the second, $\{z \mid \exists x[z = (x,a)_2 \ \&\ \mathfrak{A}(x,a)]\}$ for the given a: such equalities are accomplished by adding the following scheme to our system:

$$x \in \{(y_1, y_2, \ldots, y_{\mathfrak{k}})_{\mathfrak{k}} \mid_{y_1, y_2, \ldots, y_{\mathfrak{k}}} \mathfrak{A}\} \iff \exists y_1 \exists y_2 \cdots \exists y_{\mathfrak{k}}[x = (y_1, y_2, \ldots, y_{\mathfrak{k}})_{\mathfrak{k}} \ \&\ \mathfrak{A}].$$

1.2. We informally permit classes with other quantified terms, for example $\{\bigcup x \mid_x x \in a\}$.

1.3. Definition. Foundation, the axiom of (set) foundation, is

$$x \neq \varnothing \Longrightarrow \exists y_{\in x}\, x \cap y = \varnothing.$$

S_0' S_0 + Foundation

A calculus of Δ_0 terms

1.4. Definition. We call a term A, possibly with free variables, T-*semi-suitable*, where T is some system of set theory, if whenever Φ is Δ_0, and the variable w is not free in A, then $\forall w_{\in A}\,\Phi$ is Δ_0^{T}, meaning "equivalent, provably in the system T, to a Δ_0 formula". If in addition, T proves that A is a set, we call A T-*suitable*.

1.5. Remark. S_0 is adequate for the development of a surprisingly large number of suitable terms. In particular, $\bigcup x$ is S_0-suitable, as is each $\bigcup^{\mathfrak{k}} x$, where we define inductively $\bigcup^{\mathfrak{k}+1} x =_{\mathrm{df}} \bigcup(\bigcup^{\mathfrak{k}} x)$. S_0 easily proves that if $x = (y, z)_2$, then $y \in \bigcup^2 x$ and $z \in \bigcup^2 x$; hence if $\mathfrak{A}$ is Δ_0 then the $\mathfrak{k}$-class $\{(y_1, y_2, \ldots, y_{\mathfrak{k}})_{\mathfrak{k}} \mid_{y_1, y_2, \ldots, y_{\mathfrak{k}}} \mathfrak{A}\}$ is equal, provably in S_0, to a Δ_0 1-class.

With Foundation added, the formulation of "ordinal" becomes Δ_0 and much of the elementary theory of ordinals can then be developed.

1.6. Remark. Gandy in [G] proves that the term ω is S_0'-semi-suitable in that if Φ is Δ_0 then the formula $\exists y_{\in \omega}\,\Phi$ is equivalent in S_0' to a Δ_0 formula. His proof will work for appropriate terms for each ordinal strictly less than ω^ω, an interesting ordinal shown by Delhommé [Del] to be the first non-automatic ordinal, but, by [DoMT, page 44, Theorem 38], no further.

1.7. Remark. Gandy [G] and Dodd [Do] have a concept of "substitutable" which is similar to our "suitable" but formulated semantically rather than syntactically. Jensen [J2] and Devlin [Dev] have the same concept but call it "simple". In the present author's opinion, that concept has the danger of blurring the levels of language. If one considers a rudimentary function to be defined by a class of the language of discourse, then implicitly there is a quantification taking place in the meta-language whenever one uses such phrases as "rud closed" or "the class of rud functions". That is scarcely satisfactory, though the situation is saved by defining a rud closed set to be one closed under, say, the explicit list of nine functions given in **2.61**. What would be better would be to resort to some mild recursion theory, and to list terms of an object language defining certain (set-theoretical) computations, and then when one speaks of closure the quantification will indeed be going on in the language of discourse.

Thus it would seem that the axiom TCo, not adopted by Mac Lane, expresses a characteristic of set theory, namely that it is often concerned with computations going on in small portions of the universe, perhaps the transitive sets, or else the transitive sets closed under pairing functions. Not adopting TCo is a sop to the structuralists; but adopting it is what set theorists should do if they are to be true to their underlying intuitions. The point is linked to the meaning of Δ_0 and will recur in Remark **10.1**.

Names of systems

Our policy will be this: if we have a system X, X_0 will mean the variant of that system with no axiom of foundation, no TCo, and no axiom of infinity. Without that subscript, Π_1 foundation will be customary. We use "restricted" to mean Δ_0. We use "flat", where Gandy used "bounded", to mean that a certain quantifier limits its variable to subsets of a named set.

Four of our names will reflect the fact that a significant part of the system is the scheme of restricted separation, flat restricted replacement, flat restricted collection or restricted replacement: ReS, fReR, fReC, ReR.

We shall add the letter I to indicate the adjunction of an axiom of infinity, usually in the form $\omega \in V$. In Section 8 we shall add the letter S, either in upper or lower case, to existing names to indicate the adjunction of both the axiom of infinity and the axiom $\mathcal{S}(x) \in V$. TCo will be listed by name when needed.

Gandy's first system

Gandy called his weakest system PZ, for "predicative Zermelo", and his strongest PZF, for "predicative Zermelo–Fraenkel". They are both something of a misnomer as he overlooked the power-set axiom; and without that axiom, as shown by Zarach [Z], the difference between replacement and separation-with-collection becomes significant. We use ReS, for "restricted separation".

ReS_0 S_0 plus the Δ_0 separation axiom: $x \cap A \in V$ for A a Δ_0 class.

ReS ReS_0 plus the scheme of Π_1 foundation: $A \neq \varnothing \implies \exists x_{\in A}\, x \cap A = \varnothing$ for A a Π_1 class.

We shall call functions of the form $x \mapsto x \cap A$ *separators.*

Devlin's system and variant

The next system, which we call DB for "Devlin Basic", adds the existence of Cartesian product to ReS_0, but as it thereby becomes finitely axiomatisable, by a result of which many variants are found in the literature, and presumably going back to Bernays, we give it officially as that finite axiomatisation.

DB_0 The system of which the set-theoretic axioms are Extensionality and the following nine set-existence axioms:

$$\begin{array}{rrr} \varnothing \in V & \bigcup x \in V & a \cap \{(x,y)_2 \mid x \in y\} \in V \\ \{x, y\} \in V & \mathrm{Dom}\,(x) \in V & \{(y,x,z)_3 \mid (x,y,z)_3 \in b\} \in V \\ x \smallsetminus y \in V & x \times y \in V & \{(y,z,x)_3 \mid (x,y,z)_3 \in c\} \in V \end{array}$$

DB DB_0 plus Π_1 foundation.

1.8. Remark. All those nine are theorems of ReS_0 + Cartesian product, without foundation.

1.9. Definition. Although in one model that we consider, we must use a different formulation, we shall usually take the axiom of infinity in the form $\omega \in V$, ω being defined as the class of all von Neumann ordinals such that they and all their predecessors are either 0 or successor ordinals.

1.10. Remark. If we add the axiom of infinity plus the scheme of foundation for all classes to DB we obtain the system BS as formulated on page 36 of Devlin's book *Constructibility*:

BS $\qquad$ ReS_0 + Cartesian product + full foundation + $\omega \in V$.

The Gandy–Jensen system

The next system, called BST by Gandy, represents a considerable step forward, in that it involves the class of rudimentary functions. Foundation apart, it is finitely axiomatisable, and indeed needs only one axiom beyond those of DB_0. We give first the scheme of Gandy, and in Remark **1.12** shall indicate why all instances of it are derivable in the finitely-axiomatisable version.

GJ_0 $\qquad$ S_0 + the rudimentary replacement axiom:

(*RR*) $$\forall x \exists w \forall \vec{v}_{\in x} \exists t_{\in w} \forall u (u \in t \iff . u \in x \;\&\; \phi[u, \vec{v}]).$$

for ϕ any Δ_0 formula.

1.11. Remark. At first glance, it might seem more appropriate to call that a collection axiom, since it says that a certain family of sets is included in a set, rather than being a set. But if φ is Δ_0, x a set and $\vec{v}$ parameters, not necessarily in x, then a term x_1 and a Δ_0 formula φ_1 are readily found so that S_0 proves that x_1 is a set containing each parameter in the list $\vec{v}$, that $x \cap \{u \mid \varphi\} = x_1 \cap \{u \mid \varphi_1\}$ and that the latter is a set. So GJ_0 indeed proves Δ_0 separation.

1.12. Remark. GJ_0 is the result of adding a single axiom, which I call R_8, to DB_0:

(R_8) $$\{x``\{w\} \mid w \in y\} \in V$$

To see that, use the fact remarked above and reformulated again as Proposition **2.65** that DB_0 generates all Δ_0 separators; each instance of (RR) then follows by taking the F of the Gandy-Jensen Lemma **2.72** to be an appropriate such separator.

GJ $\qquad$ GJ_0 + the scheme of Π_1 foundation.

Flat restricted replacement

The next system has what Gandy called the bounded replacement axiom, but we shall prefer to use the adjective "flat".

$\mathsf{fReR_0}$ $\mathsf{S_0}$ plus the flat Δ_0 replacement axiom: namely, for any ϕ in Δ_0,

(*Flat* Δ_0 *Replacement*)

$$\forall x_{\in u}\, \exists! y(\phi(x,y)\ \&\ y \subseteq z) \Longrightarrow \exists v \forall y[y \in v \Longleftrightarrow \exists x_{\in u}\, (\phi(x,y)\ \&\ y \subseteq z)].$$

In words, the image of a set by a function whose values are all included in a set is itself a set.

fReR $\mathsf{fReR_0}$ + the scheme of Π_1 foundation.

Flat restricted collection

$\mathsf{fReC_0}$ $\mathsf{S_0}$ plus Δ_0 separation plus the following scheme, for ϕ any Δ_0 formula:

(*Flat* Δ_0 *Collection*)

$$\forall x_{\in u}\, \exists y(\phi(x,y)\ \&\ y \subseteq z) \Longrightarrow \exists v \forall x_{\in u}\, \exists y_{\in v}\, (\phi(x,y)\ \&\ y \subseteq z)].$$

fReC $\mathsf{fReC_0}$ + the scheme of Π_1 foundation.

1.13. Remark. Π_1 Foundation aside, the axioms of the above systems are all provable in the system $\mathsf{M_0}$ studied in [M2], which is the system $\mathsf{ReS_0}$ + the power set axiom, $\mathcal{P}(x) \in V$ and is a subsystem of Mac Lane's system ZBQC, which in turn, shorn of the axiom of choice, is a subsystem of Zermelo's system Z.

Restricted replacement

We depart now from a linearly ordered set of systems: we shall see that ReR is not a subsystem of fReC, and I suspect that methods of Zarach will show that fReC is not a subsystem of ReR.

ReR_0 S_0 + the following scheme, for ϕ any Δ_0 formula:

(Δ_0 *Replacement*)

$$\forall x_{\in u} \exists! y\, \phi(x, y) \Longrightarrow \exists v \forall y[y \in v \Longleftrightarrow \exists x_{\in u}\ \phi(x, y)].$$

ReR ReR_0 + the scheme of Π_1 foundation.

Kripke–Platek

Finally we arrive at Kripke–Platek set theory, KP which we formulate with Π_1 foundation.

KP Δ_0 separation, Π_1 foundation, and Δ_0 collection, in the formulation of which u and v are to be variables having no occurrence in the Δ_0 formula ϕ:

(Δ_0 *Collection*)

$$\forall x \exists y \phi \Longrightarrow \forall u \exists v \forall x_{\in u} \exists y_{\in v}\, \phi(x, y).$$

We shall indicate the addition of the axiom of infinity to one of the above systems by adding the letter I: thus $\mathsf{DB}_0\mathsf{I}$, KPI.

1.14. Remark. By a result of Boffa, TCo, the statement that every set is a member of a transitive set, is not provable in Z, and therefore not in its subsystems. It is, however, provable in KP when that system is formulated, as here, to include Π_1 foundation, and in ReRI: see Proposition **2.108** and Problem **2.107**.

2. Theorems of the various systems

On ReS and finite sets

We shall work with two definitions of *finite*: we get an easy Σ_1 definition of **HF** by taking "finite" to mean "in bijection with a member of ω"; we shall get an easy proof that the union of two finite sets is finite by taking "finite" to mean "possesses a double well-ordering"; and we need Π_1 foundation to prove the equivalence of the two definitions (or to develop the arithmetic necessary were we to work only with the "member of ω" definition).

2.0. Definition. x is *finite* if x carries a double well-ordering, that is, a linear ordering such that every non-empty subset has both a least and a greatest element.

The natural ordering of any member of ω is a double well-ordering.

2.1. Proposition. (ReS) *If a set is finite then it is in bijection with some member of ω.*

Proof. Let X be a set with a double well-ordering $\leqslant_X$. We say that f is *an attempt at x in X* if $\mathrm{Dom}(f) = \{y \mid y \leqslant_X x\}$ and for all y in $\mathrm{Dom}(f)$, $f(y) = \{f(z) \mid z <_X y\}$. The class

$$\{x \mid x \in X \;\&\; \neg\exists f \; f \text{ is an attempt at } x\}$$

is Π_1 and if non-empty, has a $\leqslant_X$-least element $\bar{x}$. $\bar{x}$ is not the first member of X, as an attempt at that point is easily built; nor can $\bar{x}$ be a successor, as an attempt at its predecessor is easily extended. So that class is in fact empty. Let ϖ be an attempt at the largest element of X: then a further induction shows that ϖ maps $(X, <_X)$ bijectively to a member of ω. (2.1) □

As we are working without assuming that Cartesian products exist in general, the converse, which is true, requires some preparation.

2.2. Lemma. (ReS) *For all m and k in ω, $\{m\} \times k$ is a set.*

Proof. Fix m. Use the fact that

$$\{(m,n)_2 \restriction_n n < k+1\} = \{(m,n)_2 \restriction_n n < k\} \cup \{(m,k)_2\}. \qquad (2.2)\ \square$$

2.3. Lemma. (ReS) *For all m and k in ω, $k \times \{m\}$ is a set.*

Proof. Fix m. Use the fact that

$$\{(n,m)_2 \restriction_n n < k+1\} = \{(n,m)_2 \restriction_n n < k\} \cup \{(k,m)_2\}. \qquad (2.3)\ \square$$

2.4. Proposition. (ReS) *For all m and n in ω, $m \times n$ is a set.*

Proof. Use the fact that

$$(m+1) \times (m+1) = (m \times m) \cup (\{m\} \times m) \cup (m \times \{m\}) \cup (\{m\} \times \{m\}).$$

(2.4) □

2.5. Remark. Note that that cannot lead to a proof that $\omega \times \omega$ is a set. We cannot form the collection of attempts.

2.6. Corollary. (ReS) *The Cartesian product of two sets, each in bijection with a member of ω, is a set.*

Proof. First, reason thus: if $g : m \longleftrightarrow a$ and $h : n \longleftrightarrow b$, define the function f with domain $m \times n$ by

$$f((i,j)_2) = (g(i), h(j))_2.$$

Then the image of that function is $a \times b$.

But that reasoning, though sound in GJ, is not available in BS or ReS_0. Hence we must do an induction structured as above: first for $m = 1$ prove, by induction on n, that for any n, and g and h as above, the Cartesian product exists. Then do an induction on m. (2.6) □

2.7. Remark. In systems without the Axiom of Cartesian Products, it cannot be assumed that the inverse of an injective function will always exist: see the variant of Model 4 mentioned in **4.8**.

2.8. Proposition. (ReS) *If X is in bijection with some member of ω, then it is finite.*

Proof. From the above we know that $X \times X$ and each $n \times X$ exist. Now given $f : n \longleftrightarrow X$, we may form its inverse g thus:

$$g := (n \times X) \cap \{(a,b)_2 \mid_{a,b} (b,a)_2 \in f\}$$

and we may then form the set $X \times X \cap \{(x,y)_2 \mid_{x,y} g(x) \leqslant g(y)\}$, which will be a double well-ordering. (2.8) □

2.9. Proposition. *Every subset of a finite set is finite.*

Proof. A restriction of a double well-ordering is ditto. (2.9) □

2.10. Proposition. *If x and y are finite, so is $x \cup y$.*

Proof. A double well-ordering of $x \cup y$ can easily be constructed given ones of x and of $y \setminus x$. (2.10) □

2.11. Lemma. *Let z be a finite set, and $a \notin \bigcup z$. Then $\{y \cup \{a\} \mid_y y \in z\}$ is a set and is finite.*

Proof. Let $f : n \longleftrightarrow z$. Define

$$g(0) = \{f(0) \cup \{a\}\}$$
$$g(k+1) = g(k) \cup \{f(k) \cup \{a\}\}$$

Then $g(n)$ will be defined – appeal to Π_1 foundation if not! – and will be the desired set, which is evidently in bijection with z and therefore finite. (2.11) □

2.12. Lemma. (S_0) *Let z be a set, and $a \notin z$. Then*

$$\mathcal{P}(z \cup \{a\}) = \mathcal{P}(z) \cup \{y \cup \{a\} \mid_y y \in \mathcal{P}(z)\}.$$

2.13. Proposition. *Let w be finite. Then $\mathcal{P}(w)$ is a set and is finite.*

Proof. Write $F(a,z)$ for $\{y \cup \{a\} \mid_y y \in z\}$. Let $f : n \longleftrightarrow w$. Define

$$g(0) = \{\varnothing\}$$
$$g(k+1) = g(k) \cup F(f(k), g(k))$$

As before, we consider the least m for which there is no attempt at m for this recursion; and obtain a contradiction. So $g(n)$ will be the desired set $\mathcal{P}(w)$.

To see that $\mathcal{P}(w)$ is finite, argue, again by induction on $k \leq n$, and using **2.10**, **2.11** and **2.12**, that each $g(k)$ is finite, (the class of failures being again Π_1, the argument succeeds); so $g(n)$ is finite. (2.13) □

2.14. Proposition. *The Cartesian product of two finite sets is finite.*

Proof. By a similar argument, starting from the observation that

$$x \times (z \cup \{a\}) = (x \times z) \cup (x \times \{a\}).$$ (2.14) □

2.15. Proposition. *A surjective image of a finite set is finite.*

[trivial if the surjection is a set; if it is defined by some formula, we may need full foundation.]

2.16. Definition. $\mathcal{S}(x) =_{\mathrm{df}} \{y \mid y \subseteq x \ \&\ y \text{ is finite}\}$.

[It is not assumed that $\mathcal{S}(x)$ is a set.]

2.17. Definition. Let $\Psi_{\mathcal{S}}(q, y)$ be the Δ_0 formula $\varnothing \in q \ \&\ \forall w_{\in q} \forall x_{\in y}\, w \cup \{x\} \in q$.

2.18. Lemma. (ReS) $\Psi_{\mathcal{S}}(q, y) \Longrightarrow q \supseteq \mathcal{S}(y)$.

2.19. Lemma. (ReS) $x \in \mathcal{S}(y) \Longleftrightarrow \exists f (x \subseteq y \ \&\ \exists n_{\in \omega} \mathrm{Fn}(f) \ \&\ f : n \longleftrightarrow x)$.

Hence, using the semi-suitablility of the constant ω recorded in Remark **1.6**:

2.20. Corollary. *"$x \in \mathcal{S}(y)$" is Σ_1^{ReS}.*

2.21. Lemma. (ReS) $\mathcal{S}(y) \in V \Longrightarrow \forall x[x \in \mathcal{S}(y) \Longleftrightarrow \forall q(\Psi_{\mathcal{S}}(q, y) \Longrightarrow x \in q)]$.

2.22. Lemma. (ReS) $\mathcal{S}(y) \in V \Longrightarrow [z \subseteq \mathcal{S}(y) \Longleftrightarrow \forall q(\Psi_{\mathcal{S}}(q, y) \Longrightarrow q \supseteq z)]$.

2.23. Lemma. (ReS) $\mathcal{S}(y) \in V \Longrightarrow [z = \mathcal{S}(y) \Longleftrightarrow z \subseteq \mathcal{S}(y) \ \&\ \Psi_{\mathcal{S}}(z, y)]$.

Next, a principle of collection for finite sets.

2.24. Metatheorem. *Let $\mathfrak{A}$ be a $\Pi_{\mathfrak{k}}$ wff; then it is provable in ReS_0 with $\Pi_{\mathfrak{k}+1}$ foundation that for v finite, $\forall x_{\in v} \exists y \mathfrak{A} \Longrightarrow \exists w \forall x_{\in v} \exists y_{\in w} \mathfrak{A}$.*

Proof. Let $f : n \longleftrightarrow v$. Let $P(k)$ say that there is a function g with domain k such that $\forall i < k\ \mathfrak{A}(f(i), g(i))$. Find the least $k \leqslant n$ such that $P(k)$ fails. By taking cases on k, we see that it cannot exist. So $P(n)$ holds. Take the image of a corresponding g for w. (2.24) □

2.25. Remark. The above result is self-strengthening to the case that $\mathfrak{A}$ is $\Sigma_{\mathfrak{k}+1}$.

Proof that HF models ZF minus infinity

2.26. Definition. We define **TF** to be the class of all finite transitive sets, and **HF** to be its union.

2.27. Remark. In a set theory without an axiom of foundation, **HF** might be strictly greater than V_ω; for example, any *Quine atom*, that is, a set x which equals its own singleton $\{x\}$, would be in **HF** as we have defined it. To exclude such ill-founded sets we should define **HF** as the union of transitive finite sets u which are well founded in the sense that $\forall x_{\subseteq u}(x \neq \varnothing \Longrightarrow \exists y_{\in x}\, y \cap x = \varnothing)$; and would then have to add occasional remarks to the discussion below. But as our chief focus is on contexts where the axiom of foundation is true, we may leave our definition of **HF** as it is.

2.28. Metatheorem. *Let $\mathfrak{A}$ be any axiom of* ZF *other than that of infinity. Then $(\mathfrak{A})^{\mathbf{HF}}$ is a theorem of ReS_0 + full foundation.*

We begin a sequence of verifications. We frequently use the fact that for Δ_0 concepts it suffices to prove that the object in question is in **HF** as its definition will relativise without difficulty.

2.29. Lemma. **HF** *is transitive.*

2.30. Lemma. $(\mathsf{Extensionality})^{\mathbf{HF}}$.

Proof. Assured by the transitivity of **HF**. (2.30) □

2.31. Lemma. $\mathbf{TF} \subseteq \mathbf{HF}$.

Proof. Since u transitive and finite implies $u \cup \{u\}$ is too; and hence u is in **HF**. (2.31) □

2.32. Corollary. $(\mathsf{TCo})^{\mathbf{HF}}$.

2.33. Lemma. $(\mathsf{Emptyset})^{\mathbf{HF}}$

Proof. $\{\varnothing\}$ is transitive and finite. (2.33) □

2.34. Lemma. $(\mathsf{Pairing})^{\mathbf{HF}}$

Proof. By Proposition **2.10** and the fact that the union of two transitive sets is transitive. (2.34) □

2.35. Lemma. $(\mathsf{Sumset})^{\mathbf{HF}}$

Proof. If $x \in u \in \mathbf{TF}$, then $\bigcup x \subseteq u$ and $\bigcup x \in u \cup \{\bigcup x\} \in \mathbf{TF}$. (2.35) □

2.36. Lemma. $(\Delta_0\ \mathsf{Separation})^{\mathbf{HF}}$

Proof. Δ_0 separation will relativise to any transitive set. (2.36) □

2.37. Remark. Indeed an "external" version of Δ_0 separation holds, in that $x \cap A \in \mathbf{HF}$ whenever $x \in \mathbf{HF}$ and A is a Δ_0 class, possibly with parameters that are not in **HF**.

2.38. Lemma. $(\mathsf{Powerset})^{\mathbf{HF}}$

Proof. By Proposition **2.13** and the fact that if u is transitive and $\forall x_{\in a}\, x \subseteq u$ then $u \cup a$ is transitive. (2.38) □

2.39. Lemma (set foundation). $(\mathsf{Foundation})^{\mathbf{HF}}$

2.40. Remark. Foundation is definitely needed here: the result would be false if **HF** contained Quine atoms. (2.39) □

At this point we have proved that all of M_1 is true in **HF**.

2.41. Definition. $u^\star =_{\mathrm{df}} u \cup [u]^1 \cup [u]^2 \cup (u \times u)$.

2.42. Lemma. *If u is finite and transitive then so is $u^\star$.*

Proof. $[u]^1 \cup [u]^2$ is a Δ_0 subclass of $\mathcal{P}(u)$, $u \times u$ is finite by what we have seen, and the transitivity is easily verified. (2.42) □

2.43. Proposition. *"all sets are finite" is true in* **HF**.

Proof. If $x \in u \in \mathbf{TF}$ and $f : n \longleftrightarrow u$, then $f \subseteq u \times n$; $(u \cup n)^\star$ is in **TF**, and so is $(u \cup n)^\star \cup u \times n \cup \{u \times n\}$. (2.43) □

2.44. Lemma. *"$x \in \mathbf{HF}$" is Σ_1^{ReS}.*

Proof. $x \in \mathbf{HF} \iff \exists u \exists f \exists n[n \in \omega \;\&\; \bigcup u \subseteq u \;\&\; f : n \longleftrightarrow u]$. (2.44) □

2.45. Remark. Here we benefit from the "simplified" definition of **HF**: if we had to say that u is well founded, that would introduce a Π_1 clause.

2.46. Lemma. (ReS) $((\Pi_1 \text{ foundation}))^{\mathbf{HF}}$.

Proof. Let Φ be Δ_0 and $B = (\{x \mid \forall b\, \Phi\})^{\mathbf{HF}}$.

Let $C = \{x \mid \forall b[b \in \mathbf{HF} \implies \Phi]\}$. Then C is Π_1 and $B \subseteq C$; indeed $B = C \cap \mathbf{HF}$. Suppose that B is non-empty and that x is a member. Then there is $u \in \mathbf{TF}$ with $x \in u$. Then $C \cap u$ is Π_1 and non-empty; let $\bar{x}$ be a minimal element. Then $\bar{x}$ is a minimal element of B. (2.46) □

2.47. Corollary. (ReS) $(\Delta_0 \text{ collection})^{\mathbf{HF}}$.

Thus ReS proves the relative consistency of the system MOST (as defined in [M2]) less infinity.

2.48. Remark. The above sheds some light on relative consistency strengths: reasoning in ReS we have shown the relative consistency of adding the power set axiom.

With Full Foundation

By results of [M3] we could now conclude that all axioms of ZF save that of infinity are true in **HF** provided we established the truth of the principle called Repcoll in [M3] and shown there to imply all the axioms of ZF in the system M_1, which is M_0 + TCo+ set Foundation. M_0 is ReS_0 plus $\mathcal{P}(x) \in V$.

2.49. Lemma. (ReS + full Foundation) $(\text{Repcoll})^{\mathbf{HF}}$

We shall not give the proof, because we shall derive the truth of ZF$-\infty$ in **HF** by another route.

2.50. Lemma. *Let A be any class: then* ReS + full Foundation *proves* $A \cap \mathbf{HF} \neq \varnothing \implies \exists x_{\in A \cap \mathbf{HF}}\, x \cap A = \varnothing$.

2.51. Remark. Here we definitely need the "simplied" version of **HF** that does not mention well-foundedness.

If we use full foundation we can establish an "external" form of full separation, as in the following scheme:

2.52. Lemma. (ReS + full Foundation) $x \in \mathbf{HF} \implies x \cap A \in \mathbf{HF}$ *for A any class.*

Proof. Let $f : n \longleftrightarrow x$. Consider the class

$$B := \{k \leqslant n \mid_k \neg\exists y[y \subseteq x \ \& \ \forall m :\leqslant k(f(m) \in y \Longleftrightarrow f(m) \in A\}.$$

By full Foundation, that, if non-empty has a minimal element, $\bar{k}$, say. The case $\bar{k} = 0$ is easily dismissed; if $\bar{k} = k + 1$, we know that $z =_{\rm df} \{f(i) \mid i \leqslant k\} \cap A$ is a set, and $\{f(i) \mid i \leqslant \bar{k}\} \cap A$ will be either z or $z \cup \{f(\bar{k})\}$; as both are sets, we have a contradiction; so the class B is empty and the theorem is proved. □

2.53. Theorem. (ReS + full Foundation) (full Collection)$^{\mathbf{HF}}$.

Proof. From the above, since we know from Lemma **2.50** that **HF** models full foundation and from Proposition **2.43** that **HF** thinks that all sets are finite. (2.53) □

With HF $\in V$

2.54. Lemma. (ReS_0 + **HF** $\in V$) (full Separation)$^{\mathbf{HF}}$

Proof. By re-writing the formula relativising all quantifiers to the set **HF**, and then applying Δ_0 Separation. □

2.55. Lemma. (ReS_0 + **HF** $\in V$ + set Foundation) (full Foundation)$^{\mathbf{HF}}$

Proof. By Lemma **2.54** and Corollary **2.32**. (2.55) □

Another example of the amount of foundation needed for a proof being reduced by the assumption that **HF** $\in V$ is furnished by the next sub-section.

Do graphs of recursive functions exist?

2.56. Consider the following argument, intended to prove that addition on ω is total:

Let $\phi(m, n)$ say that there is no function with domain $(m + 1) \times (n + 1)$ which satisfies the definition of addition for $m' + n'$ for $m' \leqslant m$ and $n' \leqslant n$. [We call such functions *attempts at integer addition.* "f is an attempt at integer addition" is Δ_0, and therefore rudimentary.]

Consider the class of $m \in \omega$ such that there is some $n \in \omega$ for which $\phi(m, n)$ is true. If non-empty, use Π_1 foundation to find its least member, $\bar{m}$, which cannot be 0, as the function $f(0, n) \equiv n$ would work: a subset of $(n+1) \times (\{0\} \times (n+1))$, and so is some $m + 1$. Now minimise n. Again it cannot be 0. So it is some $n + 1$. But we have a function h defined up to $m+1, n$, and can extend it to g by setting $g(m + 1, n + 1) = h(m + 1, n) \dot{+} 1$, a contradiction. We have proved the following:

2.57. Proposition. (ReS) *Every pair* (m, n) *of integers is in the domain of some attempt at integer addition.*

2.58. Now comes the great task of putting all the attempts together: what does it take to prove that the graph of integer addition is a set? The axiom of infinity is certainly necessary, but not sufficient: we shall see in Proposition **2.95** that fReRI would do this very well, and in Remarks **5.21** and **6.0**, that neither BS nor GJI can do it, though see also Remark **5.24** for a fine point. Happily, our system DS does prove it. **HF** $\in V$ would also do it.

2.59. Remark. Proposition **2.1**, taken with Propositions **2.10** and **2.14**, suggests the possibility of using ideas from cardinal rather than ordinal arithmetic to define addition and multiplication within ReS.

On $\mathsf{DB_0I}$:

2.60. Proposition. ($\mathsf{DB_0I}$) *$[\omega]^1$ and $[\omega]^2$ exist.*

Proof. $\omega \in V$ is an axiom of $\mathsf{DB_0I}$. By the definition of ordered pair, $[\omega]^1 \cup [\omega]^2 \subseteq \bigcup(\omega \times \omega)$, and the result follows by Δ_0 separation. (2.60) □

On GJ and the class of rudimentary functions

The companion papers *Rudimentary recursion* and *Rudimentary forcing* will contain more detailed material on rudimentary functions and related topics. Here we merely give a summary, drawing on but in places differing from the material in Jensen [J2], Gandy [G], Devlin [Dev] and Dodd [Do].

2.61. Corresponding to the systems of $\mathsf{DB_0}$ and $\mathsf{GJ_0}$, we introduce the rudimentary functions $R_0, \dots, R_8$ and certain auxiliary functions $A_0 \dots A_{15}$ generated by them: this is not the shortest possible list, but one that conveniently extends the list that generates the Δ_0 separators. Of the auxiliaries, we list only the most important, A_{14}.

$$
\begin{aligned}
R_0(x,y) &= \{x,y\}\\
R_1(x,y) &= x \setminus y\\
R_2(x) &= \bigcup x\\
R_3(x) &= \mathrm{Dom}\,(x)\\
R_4(x,y) &= x \times y\\
R_5(x) &= x \cap \{(a,b)_2 \mid a \in b\}\\
R_6(x) &= \{(b,a,c)_3 \mid (a,b,c)_3 \in x\}\\
R_7(x) &= \{(b,c,a)_3 \mid (a,b,c)_3 \in x\}\\
A_{14}(x,y) &= x``\{y\}\ [= \mathrm{Dom}\,((x \cap ([\textstyle\bigcup\bigcup x] \times \{y\}))^{-1})]\\
R_8(x,y) &= \{x``\{w\} \mid w \in y\}
\end{aligned}
$$

2.62. Proposition. *Each of $R_0 \dots R_7$ and $A_0, \dots, A_{14}$ is $\mathsf{DB_0}$-suitable; R_8 is $\mathsf{GJ_0}$-suitable.*

2.63. Definition. Let $\mathcal{B}$ be the closure of $R_0 \dots R_7$ under composition.

2.64. Proposition. *Each function in $\mathcal{B}$ is $\mathsf{DB_0}$-suitable.*

2.65. Proposition. *For each Δ_0 class A the map $x \mapsto x \cap A$ is in $\mathcal{B}$.*

2.66. Remark. That corresponds to the derivability of Δ_0 separation in $\mathsf{DB_0}$.

2.67. Definition. Let $\mathcal{R}$ be the closure of $R_0 \dots R_8$ under composition.

2.68. Proposition. *Each function in $\mathcal{R}$ is GJ_0-suitable.*

The collection of functions in $\mathcal{R}$ is also closed under formation of images: by which is meant that if F is in $\mathcal{R}$ so is $x \mapsto F``x$. To prove this we introduce the notion of a *companion.* We will actually have two such notions.

Let T be some system of set theory extending DB, and let G and F be Δ_0 classes such that T proves that both G and F are total functions.

2.69. Definition. G is a 1-*companion* of F in T if G is T-suitable and

$$\vdash_{\mathsf{T}} \vec{x} \in \vec{u} \Longrightarrow F(\vec{x}) \downarrow\in G(\vec{u})$$

2.70. Definition. H is a 2-*companion* of F in T if H is T-suitable and

$$\vdash_{\mathsf{T}} \vec{x} \in \vec{u} \Longrightarrow F(\vec{x}) \downarrow\subseteq H(\vec{u})$$

where $\vec{x} \in \vec{u}$ abbreviates $x_1 \in u_1$ & $\ldots x_n \in u_n$ for an appropriate n.

The collection of functions with a 1-companion is easily seen to be closed under composition; but usually it is much easier to spot a 2-companion of a function. The following is easily verified by inspection.

2.71. Proposition. *Each of the functions $R_0, \ldots, R_7$ and A_{14} has a 2-companion in DB_0.*

Generation of 1-companions from 2-companions and separators

The Gandy-Jensen Lemma is the core of the proof that $\mathcal{R}$ is closed under formation of images. Versions of it are to be found in the papers of Gandy [G] and Jensen [J2]. We discuss it only for 1-ary functions.

2.72. Lemma (Gandy-Jensen Lemma). *Suppose that H is a 2-companion of F, and that '$a \in F(b)$' is Δ_0. Then F is generated by composition from H and members of $\mathcal{B}$; further $F``x \in V$ and $F``$ (as a function) is generated by H and members of $\mathcal{R}$ and (as a term) is $\mathcal{S}$-suitable and is a 1-companion of F in $\mathcal{S}$.*

Proof. We have

$$x \in u \Longrightarrow F(x) \subseteq H(u).$$

Form

$$h(u) =_{\text{df}} \big(H(u) \times u\big) \cap \{(a,b)_2 | b \in u \,\&\, a \in F(b)\}.$$

Actually, we could just take

$$h(u) =_{\text{df}} \big(H(u) \times u\big) \cap \{(a,b)_2 | a \in F(b)\}.$$

Since $a \in F(b)$ is Δ_0 and for each Δ_0 A, the separator $x \mapsto x \cap A$ is in $\mathcal{F}$ and is DB-suitable, we have that h is generated by H and functions in $\mathcal{F}$.

Now note that for $b \in u$, $F(b) = h(u)``\{b\} = A_{13}(h(u), b)$, so $F$ is built from $H$ and functions in $\mathcal{F}$; if $R_8$ is available, we may argue further that $F``u = R_8(h(u), u)$ so $F``$ is built from $H$ and rudimentary functions; hence $F``u \in V$, and this function $F``$ now forms a 1-companion of F. (2.72) □

Proofs that $\mathcal{R}$ is closed under the rudimentary schemata may be found in the cited works on fine structure.

A single generating function for rud(u)

Following Jensen, we define rud(u) to be the rud closure of $u \cup \{u\}$. Various functions with properties similar to those of the following may be found in the literature.

2.73. Definition.

$$\begin{aligned}\mathbb{T}(u) = {} & u \cup \{u\} \\ & \cup [u]^1 \cup [u]^2 \\ & \cup \{x \setminus y \mid_{x,y} x, y \in u\} \\ & \cup \{\bigcup x \mid_x x \in u\} \\ & \cup \{\mathrm{Dom}\,(x) \mid_x x \in u\} \\ & \cup \{u \cap (x \times y) \mid_{x,y} x, y \in u\} \\ & \cup \{x \cap \{(a,\, b)_2 \mid_{a,b} a \in b\} \mid_x x \in u\} \\ & \cup \{u \cap \{(b, a, c)_3 \mid_{a,b,c} (a, b, c)_3 \in x\} \mid_x x \subset u\} \\ & \cup \{u \cap \{(b, c, a)_3 \mid_{a,b,c} (a, b, c)_3 \in x\} \mid_x x \in u\} \\ & \cup \{x``\{w\} \mid_{x,w} x \in u, w \in u\} \\ & \cup \Big\{u \cap \{x``\{w\} \mid_w w \in y\} \Big|_{x,y} x, y \in u\Big\}.\end{aligned}$$

2.74. Remark. The successive lines of the definition of $\mathbb{T}$, after the first, may be written more prosaically as $R_0``(u \times u)$, $R_1``(u \times u)$, $R_2``u$, $R_3``u$, $\{u \cap R_4(x, y) \mid_{x,y} x, y \in u\}$, $R_5``u$, $\{u \cap R_6(x) \mid_x x \in u\}$, $\{u \cap R_7(x) \mid_x x \in u\}$, $A_{14}``(u \times u)$ and $\{u \cap R_8(x, y) \mid_{x,y} x, y \in u\}$. It will be notationally convenient to treat all these functions as having three variables, so let us define $S_i(u; x, y) := R_i(x, y)$ for $i = 0, 1$; $S_i(u; x, y) := R_i(x)$ for $i = 2, 3, 5$; $S_i(u; x, y) := u \cap R_i(x, y)$ for $i = 4, 8$; $S_i(u; x, y) := u \cap R_i(x)$ for $i = 6, 7$; and $S_9(u; x, y) := A_{14}(x, y)$.

Then each of those lines is of the form $S_i``(\{u\} \times (u \times u))$ for some i. If we further define $S_{10}(u; x, y) := u$ and $S_{11}(u; x, y) := x$, then we are still within the class of rudimentary functions, as $\varnothing = R_1(x, x)$, $S_{10}(u; x, y) = R_1(u, \varnothing)$ and $S_{11}(u; x, y) = R_1(x, \varnothing)$, and, easily, $S_{11}``(\{u\} \times (u \times u)) = u$ and for non-empty $u$, $S_{10}``(\{u\} \times (u \times u)) = \{u\}$, so that $\mathbb{T}(\varnothing) = \{\varnothing\}$ and for u non-empty, $\mathbb{T}(u) = \bigcup_{i<12} S_i``(\{u\} \times (u \times u))$.

We have proved the first clause of the following, and the others are easy.

2.75. Proposition. *$\mathbb{T}$ is rudimentary, $u \subseteq \mathbb{T}(u)$ and $u \in \mathbb{T}(u)$. Further, if u is transitive, then $\mathbb{T}(u)$ is a set of subsets of u, and hence $\mathbb{T}(u)$ is transitive.*

2.76. Remark. It will not in general be true that $u \subseteq v \implies \mathbb{T}(u) \subseteq \mathbb{T}(v)$, the problem being that $u \in \mathbb{T}(u)$, but if v is countably infinite, so is $\mathbb{T}(v)$ which therefore cannot contain all the subsets of v. Fortunately, $u \subseteq \mathbb{T}(u) \subseteq \mathbb{T}^2(u) \ldots$

2.77. Lemma. *For x, y in u, $R_4(x, y) = x \times y \subseteq u \times u \subseteq \mathbb{T}^2(u)$.*

2.78. Corollary. *For x, y in u, $R_4(x, y) \in \mathbb{T}^3(u)$.*

2.79. Lemma. *For a, b c in u, $(a, c)_2 \in \mathbb{T}^2(u)$ and $(b, a, c)_3 \in \mathbb{T}^4(u)$.*

2.80. Corollary. *For $x \in u$, $R_6(x)$ and $R_7(x)$ are in $\mathbb{T}^5(u)$.*

2.81. Lemma. *For x, $y \in u$, $R_8(x, y) \in \mathbb{T}^2(u)$.*

Proof. For x, w in u, x"$w \in \mathbb{T}(u)$, so $R_8(x, y) = \mathbb{T}(u) \cap \{x\text{``}w \mid_w w \in y\}$; $x, y \in \mathbb{T}(u)$, so $R_8(x, y) \in \mathbb{T}^2(u)$. (2.81) □

2.82. Proposition. *For any transitive u, $\bigcup_{n \in \omega} \mathbb{T}^n(u)$ is the rudimentary closure of $u \cup \{u\}$, and in it,* TCo *holds.*

2.83. Problem. I do not see how to form a single rud function which will in similar fashion give the rud closure of u. Perhaps this has something to do with the question of MacAloon and Stanley discussed in Section 14.

Other remarks on GJ

2.84. Remark. RR produces a collection of subsets of x.

2.85. Proposition (Gandy; Jensen). *A transitive set is rud closed (= basically closed) iff it models* GJ_0.

2.86. Remark. GJ_0 proves that the Cartesian product of two sets is a set.

2.87. Remark. Δ_0 separation is a theorem scheme of GJ_0.

2.88. Proposition. *RR is self-strengthening to*

$$(RR^+) \qquad \forall x_1 \forall x_2 \exists w \forall \vec{v}_{\in x_1} \exists t_{\in w} \forall u (u \in t \iff .u \in x_2 \;\&\; \phi[u, \vec{v}]).$$

for ϕ any Δ_0 formula.

2.89. Problem. Does GJ prove the existence of a bijection between ω and $\omega \times \omega$? I suspect that BS does, as everything necessary is in **HF**.

The next result is a scheme of theorems:

2.90. Proposition. (GJ_0) *Each $[\omega]^{\mathfrak{k}}$ exists; indeed, each $[a]^{\mathfrak{k}}$ exists for any set a.*

Proof. $[a]^0 = \{\varnothing\} \in V$. $[a]^1 = A_0\text{``}a \in V$. $[a]^{\mathfrak{k}+1} = \{s \cup \{x\} \mid (s, x)_2 \in ([a]^{\mathfrak{k}} \times a) \cap \{(s, x)_2 \mid x \notin s\}\}$, which is in V, being of the form h"b for some set b and rudimentary function h. (2.90) □

2.91. Theorem. (GJ) $\forall a \forall k_{\in \omega}\ [a]^k \in V$.

We omit the proof, it being similar to that of Theorem **2.93**.

2.92. Problem. Is the quantified form provable without Π_1 foundation?

2.93. Theorem. (GJ) $\forall a \forall m_{\in \omega}\ {}^m a \in V$.

Proof. Fix a, and consider the Π_1 class

$$\omega \cap \{m \mid \neg\exists x[\forall y_{\in x}\,(y : m \longrightarrow a \;\&\; \forall k_{\in m} \forall t_{\in a} \exists z_{\in x}\,(z \restriction k = y \restriction k \;\&\; z(k) = t))]\}.$$

The theorem states that that class is empty: if it is not, let m be its minimal element. But then m is either 0 or a successor; if 0, nothing to prove; if $m = k+1$, then ${}^k a$ exists and we can then form ${}^m a$ as the image of a rudimentary function applied to ${}^k a \times a$, since

$${}^{k+1}a = \{f \cup \{(t,k)_2\} \mid_{f,t} f \in {}^k a \;\&\; t \in a\}. \qquad (2.93) \;\square$$

2.94. Problem. Is ${}^m a$ suitable in any sense? What seems to be true is that each ${}^{\mathfrak{k}}a$ is rud, and each $[a]^{\mathfrak{k}}$ but that $[b]^n$ is not a rud function of two variables, as, if it were, $\mathcal{S}(b,x) \;=_{\mathrm{df}}\; \bigcup_{n \in x} [b]^n$ would be a rud function; but by Gandy the rud closure of $\omega + 1$ omits $\mathcal{S}(\omega) = \mathcal{S}(\omega, \omega)$.

On fReR

That GJ is a subsystem of fReR would follow from the theory of companions.

2.95. Proposition. (fReRI) *The graph of addition, and indeed of every primitive recursive function is a set.*

Proof. We prove first that $\forall n \exists f\; f \subseteq \omega \times (\omega \times \omega)$ with $\mathrm{Dom}\,(f) = n \times n$ and

$$\forall m :< n \forall k :< n[f(m,0) = m \;\&\; f(m, k \dotplus 1) = f(m,k) \dotplus 1].$$

The collection of all such f's is a set, of which the union will be the graph of addition. (2.95) $\square$

2.96. Corollary. *The Ackermann relation may be proved to exist in* fReR.

2.97. Corollary. (fReRI) $\mathcal{S}(\omega) \in V$.

For another proof, one may reflect that every finite set of natural numbers is of the form

$$\{i \mid \mathsf{p}_i \text{ divides } n\}$$

for some n, where p_i is the ith rational prime.

2.98. Corollary. (fReRI) Even *is a set.*

2.99. Proposition. (fReRI) *If x is countable then $\mathcal{S}(x)$ exists.*

2.100. Problem. Does fReR prove that each $\mathcal{S}(x)$ is a set? or at least that each $\mathcal{S}(\zeta)$ exists?

It may be that in a model with amorphous sets in the sense of Truss, there will be difficulties.

2.101. Proposition. fReR *is self-strengthening to allowing ϕ in (BdR) to have further free variables.*

Proof. Note that if $\mathrm{Rel}(s)$ and $\mathrm{Dom}\, s \neq \varnothing$ and $s \subseteq z \times \{w\}$, then $s = y \times \{w\}$ for some $y \subseteq z$; further, $\mathrm{Dom}\, s = \{w\}$, $\bigcup \mathrm{Dom}\, s = w$ and $\mathrm{Im}\, s = y$.

Let $\psi(x, s) \iff_{\mathrm{df}} \mathrm{Rel}(s)\ \&\ \mathrm{Dom}\, s \neq \varnothing\ \&\ \phi(x, \mathrm{Im}\, s, \bigcup \mathrm{Dom}\, s)$. Then ψ is Δ_0. Let $z_1 = z \times \{w\}$, and suppose that $\forall x_{\in u} \exists! y[\phi(x, y, w)\ \&\ y \subseteq z.]$ That tells us that

$$\forall x_{\in u} \exists! s[\psi(x, s)\ \&\ s \subseteq z_1],$$

so applying (BdR), we deduce that the class $\{y \times \{w\} \mid \exists x_{\in u} \phi(x, y, w)\ \&\ y \subseteq z\}$ is a set, v, say. Then applying an appropriate rudimentary function, we see that the class $\{\mathrm{Im}\, t \mid t \in v\}$ is a set; but that class is $\{y \mid \exists x_{\in u} \phi(x, y, w)\ \&\ y \subseteq z\}$, as desired. (2.101) □

On ReRI

2.102. Proposition. (ReRI) $\omega + \omega \in V$.

Proof. $\forall n_{\in \omega} \exists f[\mathrm{Fn}(f)\ \&\ \mathrm{Dom}\,(f) = n + 1\ \&\ (f(0) = \omega)\ \&\ \forall m :< n f(m \dot{+} 1) = f(m) \dot{+} 1]$, by an easy application of Π_1 foundation, and for each n there cannot be two distinct such f's. Hence by Δ_0 replacement, the set F of those f's exists, and $\omega + \omega$ will be $\mathrm{Im}\,(\bigcup F)$. (2.102) □

2.103. Proposition. (ReRI) $\mathcal{S}(x) \in V$

Proof. Fix x. Let G be the rudimentary function given by $G(y, z) = \{a \cup b \mid a \in y\ \&\ b \in [z]^1\}$. We seek to define a function $f : \omega \longrightarrow V$ by the following recursion:

$$f(0) = [x]^1; f(n + 1) = G(f(n), x).$$

We call f a *G-attempt at n* if

$$Fn(f)\ \&\ \mathrm{Dom}\,(f) = n + 1\ \&\ f(0) = [x]^1\ \&\ \forall k_{\in n} f(k + 1) = G(f(k), x).$$

Using set foundation it is easily seen that any two G-attempts agree on their common domain, so that there is at most one attempt at n; and, using Π_1 foundation to obtain a minimal element of the class of those n at which there is no attempt, we see that that class in fact must be empty, and hence that there is a unique attempt at each n.

Since being an attempt is Δ_0 in our present system, ReRI proves that there is a set containing (exactly) the attempts for each n. The union of that set is therefore a set and a function, and the union of its image is $\mathcal{S}(x)$. (2.103) □

2.104. Remark. A similar argument will show in ReRI that the transitive closure of any set exists.

2.105. Proposition. (ReRI) $\mathbf{HF} \in V$

We omit the proof as the Proposition is a special case of Proposition **8.28**.

2.106. Remark. I would guess that ReRI suffices to define the relation $u \models \varphi$, and the constructible hierarchy; and that the L of a model of ReRI is a model of KPI, so that indeed the two theories are equiconsistent.

2.107. Problem. Does ReR prove TCo?

On KP

2.108. Proposition. (KP)TCo

Proof. Let $A = \{x \mid \forall u \bigcup u \subseteq u \Longrightarrow x \notin u\}$. By Π_1 foundation, A, if non-empty, has an $\in$-minimal element $\bar{x}$. So $\forall x_{\in \bar{x}} \exists u \bigcup u \subseteq u \;\&\; x \in u$. By Δ_0 Collection there is a v such that $\forall x_{\in \bar{x}} \exists u_{\in v} \bigcup u \subseteq u \;\&\; x \in u$. Let $w = v \cap \{u \mid \bigcup u \subseteq u\}$. w is a set by Δ_0 separation; let $\bar{u} = \bigcup w$. The $\bar{u}$ is transitive and $\bar{x} \subseteq \bar{u}$. Hence $\bar{x}$ is a member of the transitive set $\bar{u} \cup \{\bar{x}\}$, and is therefore in A, a contradiction. (2.108) □

3. Remarks on transitive models

Many of our models are of the following simple kind. We define a class $\mathbf{A}$ of transitive sets, and take $\mathbf{M} = \bigcup \mathbf{A}$.

3.0. Proposition.

(i) *Such an* $\mathbf{M}$ *will always be transitive, and will model the Axiom of Extensionality and the full scheme of Foundation for all classes, and be absolute for all* Δ_0 *formulæ.*

(ii) *If* $\mathbf{A}$ *is non-empty, the axiom* $\varnothing \in V$ *will be true in* $\mathbf{M}$*; if* $\omega + 1 \in \mathbf{A}$ *then* $\mathbf{M}$ *will model* $\omega \in V$.

(iii) *If* $u \in \mathbf{A}$ *and* $y \subseteq u$ *implies* $u \cup \{y\} \in \mathbf{A}$*, then* $\mathbf{M}$ *will model the sumset axiom; further* $\mathbf{M}$ *will be supertransitive and will therefore model the full separation scheme; and* $\mathbf{A}$ *will be a subclass of* $\mathbf{M}$*, which will therefore model* TCo*, and indeed the transitive closure of any member of* $\mathbf{M}$ *will also be a member of* $\mathbf{M}$.

(iv) *If the hypothesis of* (iii) *holds and, additionally,* $u \in \mathbf{A}$ *and* $v \in \mathbf{A}$ *implies* $u \cup v \in \mathbf{A}$*, then* $\mathbf{M}$ *will model* AxPair.

The proof is straightforward. Models of that kind, therefore, are always models of Gandy's system ReS_0 with TCo, and with full foundation and full separation.

3.1. Remark. Just to clarify that last remark: to prove full foundation in the model, we require (if the model be a proper class) full foundation in the background theory; and similarly for full separation.

Slim models of weak systems

Many such classes $\mathbf{A}$ can be found by modifying a definition to be found in *Slim Models of Zermelo Set Theory* [M1]:

3.2. Definition. $\mathcal{T}$ is *weakly fruitful* if

(i) every x in $\mathcal{T}$ is transitive;

(iii) $x \in \mathcal{T} \;\&\; y \in \mathcal{T} \Longrightarrow x \cup y \in \mathcal{T}$;

(iv′) $x \in \mathcal{T} \;\&\; a \subseteq x \Longrightarrow x \cup \{a\} \in \mathcal{T}$.

The missing condition (ii) lists three possible conditions on the ordinals in the class $\mathcal{T}$:

(ii) $1 \in \mathcal{T}$; $\omega + 1 \in \mathcal{T}$; $ON \subseteq \mathcal{T}$,

respectively;

So our theorem above gives the following:

3.3. Proposition. *If $\mathcal{T}$ is weakly fruitful, then $\bigcup\mathcal{T}$ will be a supertransitive model of* ReS_0 *with* TCo, *full separation and full foundation, and if $1 \in \mathcal{T}$, of Empty Set; if $\omega+1 \in \mathcal{T}$, the axiom of infinity will hold in $\bigcup\mathcal{T}$ in the form $\omega \in V$, and in the third case, the model $\bigcup\mathcal{T}$ will contain all ordinals.*

There is a simple further requirement on $\mathbf{A}$ that ensures that $\bigcup\mathbf{A}$ is closed under Cartesian products. Recall our definition from Section 2:

Definition. $u^\star =_{\mathrm{df}} u \cup [u]^1 \cup [u]^2 \cup (u \times u)$.

3.4. Lemma. *$u^\star$ is* BS *suitable; if u is transitive, so is $u^\star$, and $u \times u \subseteq u^\star$.*

3.5. Proposition. *If $\mathbf{A}$ is a collection of transitive sets closed under *, union of two elements, adding a subset to an element, and containing the set $\omega+1$, then $\bigcup\mathbf{A}$ will model* BS *with* TCo *and* full Separation.

As in *Slim Models*, we may obtain some interesting examples of such models by estimating the rate of growth of various transitive sets. Given a function $Q : \omega \longrightarrow V$, set $f_x^Q(n) = \overline{\overline{x \cap Q(n)}}$. For $\mathcal{G}$ a class of functions, form $\mathcal{T}^{Q,\mathcal{G}} =_{\mathrm{df}} \{x \mid \bigcup x \subseteq x \;\&\; f_x^Q \in \mathcal{G}\}$.

3.6. Proposition. *If $\mathcal{G}$ has these properties then $\mathcal{T}^{Q,\mathcal{G}}$ will be weakly fruitful:*

$$f \leqslant g \in \mathcal{G} \Longrightarrow f \in \mathcal{G};$$
$$f, g \in \mathcal{G} \Longrightarrow f+g \in \mathcal{G};$$
$$\bigcup x \subseteq x \;\&\; f_x^Q \in \mathcal{G} \Longrightarrow f_x^Q + 1 \in \mathcal{G}.$$

The three conditions on ordinals considered correspond to the three requirements

$$f_1^Q \in \mathcal{G};\; f_{\omega+1}^Q \in \mathcal{G};\; \forall \zeta f_\zeta^Q \in \mathcal{G}.$$

3.7. Proposition. *A sufficient further condition on $\mathcal{G}$ for Cartesian products to exist in $\bigcup\mathcal{T}^{Q,\mathcal{G}}$, when $Q(n) = V_n$, is this:*

$$(f \in \mathcal{G} \;\&\; g \in \mathcal{G} \;\&\; C \in \omega) \Longrightarrow C.f.g \in \mathcal{G}$$

Proof. We must show that in these circumstances, $u \in \mathcal{T} \Longrightarrow u^\star \in \mathcal{T}$. Note that for $n \geqslant 2$,

$$\overline{\overline{[u]^1 \cap V_n}} = \overline{\overline{u \cap V_{n-1}}};\quad \overline{\overline{[u]^2 \cap V_n}} \leqslant (\overline{\overline{u \cap V_{n-1}}})^2;\quad \overline{\overline{(u \times u) \cap V_n}} = (\overline{\overline{u \cap V_{n-2}}})^2.$$

Hence

$$f_{u^\star}^Q(n) = \overline{\overline{u^\star \cap V_n}} \leqslant f_u^Q(n) + f_u^Q(n-1) + (f_u^Q(n-1))^2 + (f_u^Q(n-2))^2.$$

Since $V_n \subseteq V_{n+1}$ each f_u^Q is monotonic; the proposition now follows by elementary analysis. (3.7) □

Of our collection, Models 3, 5 and 8 are obtained by the above rate-of-growth method, of which the last two model the Axiom of Cartesian Products. Models 1, 2, 4, 6, 7, 9, and 10 are obtained by a different method, which we now describe.

3.8. Proposition. *Let X be a class. Put $\mathbf{A}^X$ = the class of those transitive u whose intersection with X is finite. Then $\mathbf{M}^X =_{\text{df}} \bigcup \mathbf{A}^X$ will be supertransitive and will model extensionality; foundation; full separation, difference and $\bigcup$; pairing; and* TCo, *since $\mathbf{A}^X \subseteq \mathbf{M}^X$; as long as X contains only finitely many ordinals, $\mathbf{M}^X$ will model infinity; if u in $\mathbf{A}^X$ implies $u^\star$ is in $\mathbf{A}^X$ then $\mathbf{M}^X$ will be closed under Cartesian products.*

Models 11–15 are obtained by yet other methods. TCo holds in all these models; all are supertransitive save for Model 14 and some variants of Model 11.

4. Models of ReS

Gandy: A set which models PZ but not BST.

We take $\mathbf{G}_1$ to be the class of all x such that everything in $tcl(\{x\})$ is either finite or differs from ω by a finite set. Gandy remarks that (a) $\mathbf{G}_1$ is transitive; (b) if x is in $\mathbf{G}_1$ $\bigcup x$ is a subset of $\mathbf{G}_1$; (c) $\omega \in \mathbf{G}_1$; (d) $\mathbf{G}_1$ contains every finite subset of itself, and every x in $\mathbf{G}_1$ is a substitutable constant in his sense. (e) $\mathbf{G}_1$ satisfies Δ_0 separation, the proof of which uses the fact that every Δ_0 subset of ω is finite or cofinite, by his quantifier elimination lemma. (f) $\omega \times \omega$ is not in A.

It follows from those remarks that $\mathbf{G}_1$ is not supertransitive and that $\mathbf{G}_1 \cap ON = \omega + \omega$. We verify the following in detail:

4.0. Proposition. *If $x \in \mathbf{G}_1$ then so are $\bigcup x$ and* $\text{tcl}(x)$.

Now $\text{tcl}(\{\bigcup x\}) = \{\bigcup x\} \cup \text{tcl}(\bigcup x)$ and $\text{tcl}(\bigcup x) \subseteq \text{tcl}(x) \subseteq \text{tcl}(\{x\})$, so it is enough to prove that if x is in $\mathbf{G}_1$, $\bigcup x$ is either finite or almost ω.

First note that if x is finite and in $\mathbf{G}_1$, then $x = y \cup z$, where y is the set of finite members of x and z is the set of members of x which are infinite and therefore almost equal to ω. If z is empty, then $\bigcup x = \bigcup y$, and is thus finite. If z is non-empty, then $\bigcup x = \bigcup y \cup \bigcup z$; y and z are both finite, and so $\bigcup y$ will be finite, and $\bigcup z$ will be almost equal to ω. Hence $\bigcup x$ is almost equal to ω.

Thus we have verified that if x is a finite member of $\mathbf{G}_1$ then $\bigcup x \in \mathbf{G}_1$.

If on the other hand, x almost equals ω, then we can write $x = y \cup z$ where z is a cofinite subset of ω, and y is a finite set disjoint from $\omega = \varnothing$. As $\mathbf{G}_1$ is transitive, y is a a finite subset of it, and therefore a member of it, and therefore $\bigcup y \in \mathbf{G}_1$, by the previous paragraph. So $\bigcup x = \bigcup y \cup \omega$; $\bigcup y$ is either finite or almost ω; either way, $\bigcup x$ is almost ω.

To show that $x \in \mathbf{G}_1 \Longrightarrow \text{tcl}(x) \in \mathbf{G}_1$, suppose that x is a counterexample of minimal rank. It is enough to show that $\text{tcl}(x)$ is either finite or almost ω.

$$\text{tcl}(x) = x \cup \bigcup_{t \in x} \text{tcl}(t),$$

where by the minimality of x each $\text{tcl}(t)$ is in $\mathbf{G}_1$.

4.1. Remark. The displayed formula implies easily that $\text{tcl}(a \cup b) = \text{tcl}(a) \cup \text{tcl}(b)$.

So if x is finite, $\mathrm{tcl}(x)$ is the union of a finite set and finitely many sets each either finite or almost ω, so that $\mathrm{tcl}(x)$ itself must be either finite or almost ω, and therefore in $\mathbf{G}_1$. Thus the minimal counterexample must be almost ω.

But now we may write x as the union of a finite set y disjoint from ω and a cofinite subset z of ω. We know that $\mathrm{tcl}(y) \in \mathbf{G}_1$ by the argument of the previous paragraph, the rank of y not exceeding that of x, and that $\mathrm{tcl}(z) = \omega$, so that again $\mathrm{tcl}(x)$, being the union of a pair of elements of $\mathbf{G}_1$ is itself in $\mathbf{G}_1$.

Model 1: A model of ReS with full separation in which Cartesian products are absent

Consider, working in some suitable theory such as ZF, the class $\mathbf{A}_1$ of all transitive sets which contain but finitely many ordered pairs.

Then $\mathbf{M}_1 = \bigcup \mathbf{A}_1$, which is the same as the class of all sets x such that $\mathrm{tcl}(x)$ contains but finitely many ordered pairs,is supertransitive and contains all ordinals, and models Extensionality, AxPair, Sum Set, Infinity and full Separation, full foundation and TCo. $\omega \in \mathbf{M}_1$ but $\omega \times \omega$ is not. Indeed the Cartesian product of an infinite set and a non-empty set is never there; but the Cartesian product of two finite sets is there, so in this model a set a is finite if and only if $a \times a \in V$.

4.2. Remark. Note also that the graph of addition is not present in this model, since its domain would be $\omega \times \omega$, and the domain can be recovered using the axioms of union and Δ_0 separation.

4.3. Remark. $\mathcal{S}(\omega) \in \mathbf{M}_1$; indeed for each ordinal ζ, $\mathcal{S}(\zeta) \in \mathbf{M}_1$.

4.4. Remark. $\mathbf{M}_1$ contains no bijection between ω and $\mathcal{S}(\omega)$. For a bijection would be an infinite set of ordered pairs. Indeed, $\mathbf{M}_1$ contains no functions with infinite domain!

Model 1a

Write $\mathcal{S}(x)$ for the set of finite subsets of x. Then in $\mathbf{M}_1$, $\mathcal{S}(\omega)$ exists, but $\mathcal{S}(\mathcal{S}(\omega))$ does not. Indeed if a is infinite, $\mathcal{S}(\mathcal{S}(a))$ never exists. So let $\mathbf{M}_{1a}$ be the set of members x of $\mathbf{M}_1$ such that $\mathcal{S}(y)$ exists in $\mathbf{M}_1$ for each member y of $\mathrm{tcl}(\{x\})$. Then the model $\mathbf{M}_{1a}$ contains all ordinals but not $\mathcal{S}(\omega)$, and in it, a is finite iff $\mathcal{S}(a)$ exists iff $\mathcal{P}(a)$ exists. What else is true there?

Model 2: A model of ReS with full separation in which $[\omega]^1$ and $[\omega]^2$ do not exist

Take $\mathbf{A}_2$ to be the class of those transitive u such that $\{x \in u \mid \overline{\overline{x}} \leqslant 2\}$ is finite, and $\mathbf{M}_2$ to be $\bigcup \mathbf{A}_2$.

4.5. Remark. If we look at $\mathbf{C}$, the class of those x such that $\mathrm{tcl}(x)$ contains only finitely many sets of cardinality 2, we get a model that is nearly the same as the model $\mathbf{M}_1$; the chief difference seems to be that $[\omega]^\omega$ is not a member of $\mathbf{C}$, but is a member of $\mathbf{M}_1$.

4.6. Remark. We shall return to this mode of construction for Model 6.

Model 3: ringing the changes

Consider for any given $\mathfrak{k}$ the set $\mathbf{A}_{3,\mathfrak{k}}$ of those u with f_u $O(n^{\mathfrak{k}})$. This gives a model $\mathbf{M}_{3,\mathfrak{k}}$ of full separation in which Cartesian product will fail. $[\omega]^{\mathfrak{k}}$ will be in the model but not $[\omega]^{\mathfrak{k}+1}$.

The arguments are modifications of those of [M1]: a similar argument is worked in detail below.

Model 4: asymmetry of Cartesian product

Let $\mathbf{A}_4 = \{u \mid u \text{ is transitive and } (V \times \{\omega\}) \cap u \text{ is finite }\}$.

Put $\mathbf{M}_4 = \bigcup \mathbf{A}_4$. Then $\omega \times \{\omega\} \notin \mathbf{M}_4$, but both $\{\omega\} \times \omega$ and $\omega \times \{\omega + 1\}$ are in $\mathbf{M}_4$.

4.7. In one of our later systems we would be able to define the *right Wiener–Kuratowski rank* of a set by this rudimentary recursion:

$$\varrho_{rWK}(x) = \begin{cases} 0 & \text{if } x \text{ is not an ordered pair} \\ 1 + \varrho_{rWK}(\text{right}(x)) & \text{otherwise} \end{cases}$$

and prove that for any x, $\varrho_{rWK}(x) < \omega$.

For the moment we content ourselves with a weak form, for which S_0 is adequate, and which will be useful for some of our model-building:

4.8. Definition. The *weak right Wiener–Kuratowski rank* is defined by cases:

$$\varrho_{wrWK}(x) = \begin{cases} 0 & \text{if } x \text{ is not an ordered pair} \\ 1 & \text{if } x \text{ is an ordered pair but right}(x) \text{ is not} \\ 2 & \text{if both } x \text{ and right}(x) \text{ are ordered pairs} \end{cases}$$

Now, for a variant of Model 4, take X to be the class of those sets of weak right WK rank 2. Then $\omega \times (\omega \times \omega)$ will not be in M_X, whereas $(\omega \times \omega) \times \omega$ will be.

Hence we have the curiosity that in this model, there will be a bijection one way but not the other.

5. Models of DB

Model 5: A slim model for Devlin

5.0. Proposition. *There is a supertransitive model of* DB *containing all ordinals but omitting the set of finite sets of natural numbers.*

Write f_u for the map $n \mapsto \overline{\overline{u \cap V_n}}$. Write g_k for the map $n \mapsto n^k$.

5.1. Definition. Let $\mathbf{A}_5$ be the class of transitive sets u such that the map f_u is dominated (i.e. eventually majorised) by some g_k. Let $\mathbf{M}_5 = \bigcup \mathbf{A}_5$.

5.2. Lemma. $\mathbf{A}_5 \subseteq \mathbf{M}_5$.

Proof. If $u \in \mathbf{A}_5$, then $u \in u \cup \{u\} \in \mathbf{A}_5$. (5.2) □

5.3. Lemma. $\mathbf{M}_5$ *is transitive, being the union of transitive sets.*

5.4. Lemma. $\mathbf{M}_5$ *is supertransitive.*

Proof. If $x \subseteq y \in u \in \mathbf{A}_5$ then $x \subseteq u$; put $v = u \cup \{x\}$. v is transitive and for each n $\overline{\overline{v \cap V_n}} \leqslant \overline{\overline{u \cap V_n}} + 1$, so $v \in \mathbf{A}_5$. (5.4) □

5.5. Corollary. (Z) $\mathbf{M}_5$ *models extensionality, difference, full foundation and full separation.*

5.6. Lemma. $\omega \in \mathbf{M}_5$*: indeed,* $\mathbf{A}_5$ *contains all ordinals.*

5.7. Lemma. *For each* k*,* $[\omega]^k$ *is in* $\mathbf{M}_5$*.*

Proof. $u_k =_{\text{df}} \omega \cup [\omega]^k \cup \{[\omega]^k\}$ is transitive. $\overline{\overline{(u_k \cap V_n)}} = \binom{n}{k} < n^k$. (5.7) □

5.8. Remark. Indeed for each $x \in \mathbf{M}_5$ and each $k \in \omega$, $[x]^k \in \mathbf{M}_5$.

Proof. Fix $x \in u \in \mathbf{A}_5$ and $k \in \omega$. It is enough to show that $[u]^k$ is in $\mathbf{M}_5$. Let $v = u \cup [u]^k \cup \{[u]^k\}$: then v is transitive. We shall show that v is in $\mathbf{A}_5$. Note that $a \in [u]^k \cap V_{n+1} \iff a \subseteq V_n \;\&\; a \in [u]^k$, so that $[u]^k \cap V_{n+1} = [V_n \cap u]^k$. So if $V_n \cap u$ is of size $O(n^\ell)$, then $V_{n+1} \cap [u]^k$ is of size $O(n^{k\ell})$. (5.8) □

5.9. Lemma. $[\omega]^{<\omega}$ *is not in* $\mathbf{M}_5$*.*

Proof. Suppose $[\omega]^\omega \in u$, a transitive set. Then $\overline{\overline{u \cap V_n}} \geqslant 2^n$, and the map $n \mapsto 2^n$ eventually strictly dominates all the $n \mapsto n^k$'s. (5.9) □

5.10. Corollary. $\mathcal{P}(\omega) \notin \mathbf{M}_5$.

5.11. Lemma. $\varnothing \in \mathbf{M}_5$.

5.12. Lemma. *If* a *and* b *are in* $\mathbf{M}_5$ *so is* $\{a, b\}$*.*

Proof. Let $a \in u \in \mathbf{A}_5$ and $b \in v \in \mathbf{A}_5$. Put $w = u \cup v$. Then f_w is dominated by $f_u + f_v$, so if f_u is dominated by g_k and f_v by g_ℓ, then f_w is dominated by $g_{\max(k,\ell)+1}$. (5.12) □

5.13. Lemma. *If* a *is in* $\mathbf{M}_5$*, so is* $\bigcup a$*.*

Proof. Let $a \in u \in \mathbf{A}_5$. Then $a \subseteq u$, so $\bigcup a \subseteq \bigcup u \subseteq u$; as before $\{\bigcup a\} \cup u$ will be in $\mathbf{A}_5$. (5.13) □

5.14. Lemma. TCo *holds in* $\mathbf{M}_5$*; indeed* $x \in \mathbf{M}_5 \Longrightarrow tcl(x) \in \mathbf{M}_5$*.*

Proof. Let $v = tcl(x)$ where $x \in u \in \mathbf{A}_5$. Then $v \subseteq u$ and is therefore in $\mathbf{M}_5$ by supertransitivity. (5.14) □

5.15. Lemma. *If* a *and* b *are in* $\mathbf{M}_5$ *so is* $a \times b$*.*

Proof. It is enough to show that if u is in $\mathbf{A}_5$, then $u^\star \in \mathbf{A}_5$. By the reasoning in the proof of Proposition **3.7**, if f_u is dominated by g_k then $f_{u^\star}(n)$ for sufficiently large n is at most $n^k + (n-1)^k + (n-1)^{2k} + (n-2)^{2k}$ which in turn is at most $4g_{2k}(n)$; thus $f_{u^\star}$ is dominated by g_{2k+1} and $u^\star$ is accordingly in $\mathbf{A}_5$. (5.15) □

5.16. Lemma. *If* $x \in u \in \mathbf{A}_5$, *then* $\operatorname{Dom} x \subseteq u$ *and is thus in* $\mathbf{M}_5$.

The following verifications are related to the finite axiomatisation of DB. We check that for a in $\mathbf{M}_5$,

$$a \cap \{(p,q)_2 \mid p \in q\} \in V$$
$$\{\langle q,p,r\rangle \mid \langle p,q,r\rangle \in a\} \in \mathbf{M}_5$$
$$\{\langle q,r,p\rangle \mid \langle p,q,r\rangle \in a\} \in \mathbf{M}_5$$

The first is immediate by supertransitivity, and for the other two, if $a \in u \in \mathbf{A}_5$, both the given classes are contained in $u \times (u \times u)$, and are thus in $\mathbf{M}_5$ by supertransitivity.

5.17. Remark. The model being supertransitive, the set of even numbers is in it. That is of interest, because that was Gandy's test set, studied in Section 2. His arguments use quantifier elimination; our examples do not.

We show that $\mathbf{M}_5$ is not a model of GJ. Recall the definition of the Ackermann relation $\text{ACK} \subseteq \omega \times \omega$: $m \,\text{ACK}\, n$ if and only if 2^m is one of the summands in the binary expression of n as a sum of powers of 2.

5.18. Lemma. $\text{ACK} \in \mathbf{M}_5$.

Proof. $\omega \times \omega \in \mathbf{M}_5$ and $\mathbf{M}_5$ is supertransitive. (5.18) □

5.19. Proposition. $\mathbf{M}_5$ *is not a model of* GJ.

Proof. $\{\text{ACK}``\{n\} \mid n \in \omega\} = [\omega]^{<\omega}$. By Lemmata **5.9** and **5.18**, Axiom R_8 fails in $\mathbf{M}_5$. (5.19) □

5.20. Remark. The graph of addition is present in this model, as it will be in any supertransitive model of DB_0 containing ω; one may also argue directly that if u is the transitive closure of the singleton of that graph, f_u is dominated by g_3.

5.21. Remark. Gandy's model $\mathbf{G}_2$, given below, is a model of GJI without the graph of addition; the submodel $\bigcup(\mathbf{G}_2 \cap \mathbf{A}_5)$ will be supertransitive relative to $\mathbf{G}_2$, and will be a transitive model of DB, indeed of BS, in which GJ fails and in which the graph of addition is absent.

Model 6

We consider a variant of the construction $\mathbf{M}_2$ of section 2.

Here we wish to study the extent to which DB proves the existence of the sets $[\omega]^{\mathfrak{k}}$

5.22. Proposition. *For any* $\mathfrak{k} \geqslant 3$, DB, *if consistent, fails to prove that* $[\omega]^{\mathfrak{k}}$ *exists.*

Fix $\mathfrak{k} \geqslant 3$. We shall exhibit a supertransitive model $\mathbf{M}_{6,\mathfrak{k}}$ of DB in which $[\omega]^{\ell}$ exists iff $\ell \neq \mathfrak{k}$.

5.23. Remark. Indeed the existence of $[\omega]^{\ell}$ for different ℓ is independent. So we can code an arbitrary subset of ω into the theory of such a model.

Guided by Proposition **3.8**, we let $X_{6,\mathfrak{k}}$ be the class of all sets of cardinality $\mathfrak{k}$, we take $\mathbf{A}_{6,\mathfrak{k}}$ to be the class of all transitive u such that $u \cap X_{6,\mathfrak{k}}$ is finite, and $\mathbf{M}_{6,\mathfrak{k}}$ to be $\bigcup \mathbf{A}_{6,\mathfrak{k}}$. Then that will model S_0 with full separation and full foundation; for $\mathfrak{k} \geqslant 3$, it will model Cartesian Product, since then for u transitive, $X_{6,\mathfrak{k}} \cap \big([u]^1 \cup [u]^2 \cup (u \times u)\big) = \varnothing$, and so $u \in \mathbf{A}_{6,\mathfrak{k}} \Longrightarrow u^\star \in \mathbf{A}_{6,\mathfrak{k}}$.

If $\mathfrak{l} \neq \mathfrak{k}$, then for each x in $\mathbf{M}_{6,\mathfrak{k}}$, $[x]^\mathfrak{l}$ will be in $\mathbf{M}_{6,\mathfrak{k}}$: if $x \in u \in \mathbf{A}_{6,\mathfrak{k}}$, $[x]^\mathfrak{l} \subseteq [u]^\mathfrak{l}$; $u \cup [u]^\mathfrak{l}$ is transitive, and its intersection with $X_{6,\mathfrak{k}}$ equals $u \cap X_{6,\mathfrak{k}}$, and is therefore finite. By the supertransitivity of $\mathbf{M}_{6,\mathfrak{k}}$, $[x]^\mathfrak{l} \in \mathbf{M}_{6,\mathfrak{k}}$.

On the other hand for no infinite member x of $\mathbf{M}_{6,\mathfrak{k}}$ will $[x]^\mathfrak{k}$ be in $\mathbf{M}_{6,\mathfrak{k}}$, as no member of $\mathbf{M}_{6,\mathfrak{k}}$ can have infinitely many members of cardinality $\mathfrak{k}$.

So it will also be true that ${}^\mathfrak{k}\omega$ is not in the model, although $\omega \times (\omega \times (\ldots))$ ($\mathfrak{k}$ times) will be.

5.24. Remark. Consider the case $\mathfrak{k} = 3$: the graph of addition, implemented (as we do) as a subset of $\omega \times (\omega \times \omega)$, is a member of $\mathbf{M}_{6,3}$, but implemented as a set of 3-tuples is not, since in that model, no infinite subset of ${}^3\omega$ exists. Thus these weak theories are extremely sensitive to the implementation of functions, a point that is touched on by Stanley in his review [St] of Devlin's book [De].

5.25. Remark. If we ask that for each k u contains only finitely many sets of size k, the resulting model, though containing all the ordinals, will contain none of the sets $[\omega]^k$; if we ask for u to contain only finitely many finite sets, the resulting model will be **HF**, given that we are using the Axiom of Foundation. In a universe with Quine atoms, of course, the situation would be different.

A variant of Model 6

Let $\mathbf{A} = \{u \mid \bigcup u \subseteq u \;\&\; u \cap {}^3[\omega, \omega+\omega) \text{ is finite}\}$, and let $\mathbf{M} = \bigcup \mathbf{A}$. Then $\mathbf{HF} \in \mathbf{M}$ but ${}^3[\omega, \omega+\omega)$ is not. $\mathbf{M}$ contains all ordinals and is a supertransitive model of BS.

Model 7: a failure of $\bigcup$ "

Here we shall exhibit a transitive model of BS in which the following failure of GJ occurs: there is a set B such that $\{\bigcup x \mid x \in B\}$ is not a set.

Following Proposition **3.8**, take X to be the class of transitive sets of limit rank, $\mathbf{A}_7$ to be A^X, the class of all transitive sets u such that only finitely many transitive sets of limit rank are members of u, and $\mathbf{M}_7$ to be $\bigcup \mathbf{A}_7$.

Then $\mathbf{M}_7$ is a supertransitive model of ReS_0 + full Foundation + TCo; "$x \times y \in V$" will be true in it since for u transitive, $u^\star \cap X = u \cap X$, as all members of $[u]^1 \cup [u]^2 \cup (u \times u)$ are non-empty finite sets and therefore of successor rank; and it contains all the ordinals below ω^2, and thus models the axiom of infinity. To prove the failure of GJ, we turn to the idea of a *Zermelo tower* from [M1], which is defined thus:

5.26. Definition. For a any set, put $Z_0(a) = \varnothing$; $Z_1(a) = \{a\}$; $Z_{n+1}(a) = \{a\} \cup (\mathcal{P}(Z_n(a)) \smallsetminus \{\varnothing\})$; $Z(a) = \bigcup_{n \in \omega} Z_n(a)$.

If one thinks of **HF** as a collection of words in $\varnothing$, $\{$ and $\}$ then $Z(a)$ is the collection of the corresponding words with a substituted for $\varnothing$ throughout. Thus every member either is a finite non-empty set or equals a.

Now let $\mathcal{X}$ be the set of those subsets a of $\omega+1$ of which ω is a member.

For each such a let $x(a) =_{\mathrm{df}} \{Z_n(a) \mid n \in \omega\}$. The rank of $x(a)$ is $\omega+\omega$.

Let $x^*(a) = x_a \cup \{\omega+1\}$. All the members of $x^*(a)$ are of successor rank, and so $x^*(a)$ is not transitive, but $\bigcup x^*(a) = Z(a) \cup (\omega+1)$ which is transitive, and of rank $\omega+\omega$; its only transitive member of limit rank is ω; thus each $x^*(a)$ is in $\mathbf{M}_7$.

Take B to be $\{x^*(a) \mid a \in \mathcal{X}\}$. Note that

$$\mathrm{tcl}(\{B\}) = \{B\} \cup B \cup \{Z_n(a) \mid n \in \omega \;\&\; a \in \mathcal{X}\} \cup \{\omega+1\} \cup \omega+1,$$

a transitive set of which the sole transitive member of limit rank is ω. Hence $B \in \mathbf{M}_7$; but $\{\bigcup x \mid x \in B\}$ will not be, since it is an infinite set of transitive sets of limit rank.

Model 8: in which $\mathcal{S}(\omega)$ exists but not $\mathcal{S}(\omega\times\omega)$

Note that the cardinality of $\mathcal{S}(\omega\times\omega)\cap V_n$ is about $2^{(n-2)^2}$, an order of magnitude higher than that of $\mathcal{S}(\omega)\cap V_n$; we have to take the transitive closure of course, but that will only make it higher.

So take $\mathbf{A}_8$ to be the class of all transitive u such that the map f_u defined by $f_u(n) = \overline{\overline{u \cap V_n}}$ is eventually dominated, for some k, by $n \mapsto 2^{kn}$, and $\mathbf{M}_8$ to be $\bigcup \mathbf{A}_8$.

By Proposition **3.3** and Proposition **3.7**, $\mathbf{M}_8$ models BS.

5.27. Remark. By estimating the number of ordered triples in V_n, and considering those transitive u with f_u dominated by $n \mapsto 2^{kn^2}$ for some k, we would obtain a model containing $\mathcal{S}(\omega\times\omega)$ but omitting $\mathcal{S}(\omega\times(\omega\times\omega))$.

Model 9: a failure of Seq

The importance of this example will be explained in our discussion in **10.6**: it provides a model of BS containg **HF** that refutes Devlin's claim that BS proves $\forall a \forall n_{\in\omega}\, \exists u\, \mathrm{Seq}(u,a,n)$.

Let $\mathbf{A}_9$ be $\{u \mid \bigcup u \subseteq u \;\&\; u \cap {}^3A \text{ is finite}\}$, where we have yet to choose A.

5.28. Lemma. $\mathbf{HF} \cap {}^3A = {}^3(\mathbf{HF} \cap A)$.

So take A to be $\{\omega\}\times\omega$. The resulting model $\mathbf{M}_9 = \bigcup \mathbf{A}_9$ will have **HF** as a member; ${}^3(\{\omega\}\times\omega)$ will not be there, but ${}^3(\omega\times\{\omega\})$ will be. The model will contain a bijection between the two sets $\omega\times\{\omega\}$ and $\{\omega\}\times\omega$, and therefore will fail to model GJ.

We should check that $\star$-closure holds in Model 9. Recall that $u^\star = u \cup [u]^1 \cup [u]^2 \cup u \times u$.

The members of ${}^3A$ are 3-sequences, which are neither singletons nor doubletons nor ordered pairs. So in this case

$$u^\star \cap {}^3A = u \cap {}^3A,$$

and all is well.

5.29. Remark. In the next section we give Gandy's model of GJI, which thus contains for each a and n a u such that $\mathrm{Seq}(u, a, n)$ but which does not, for $a = \omega$, contain the set of all finite sequences of members of a.

Model 10: from sheer perversity

Let P be an almost disjoint family of infinite subsets of ω; for X in P, consider the class A_X of all transitive sets having finite intersection with ${}^3X$. Take for Q any subset of P, $\mathbf{A}_Q$ to be the intersection of all the A_X for $X \in Q$. Then, for X in P, $\bigcup \mathbf{A}_Q$ will contain ${}^3X$ iff X is not in Q, and will model BS.

6. Models of GJ

Gandy: A set that models GJ but not fReR

Take $\mathbf{G}_2$ to be the rudimentary closure of $\{\omega\}$.

The set of even numbers is not in $\mathbf{G}_2$, not being Δ_0. Π_1, indeed full, foundation is true in $\mathbf{G}_2$; TCo will be true there as ω is transitive, by Proposition **2.82**. But as we saw in Section 2, fReR proves the existence of EVEN.

The next two remarks are semantical versions of [G, Theorems 2.2.2(ii) and 3.1.1].

6.0. Remark. It follows that the graph G of addition is not a member of this model, for

$$\text{EVEN} = \omega \cap \{n \mid n = 0 \vee \exists m_{\in n}\, (n, m, m) \in G\}.$$

6.1. Remark. The graph of concatenation is not in this model.

The unprovability of $\mathcal{S}(\omega) \in V$ in GJI

6.2. Remark. If Δ_0 separation is true and $\mathcal{S}(\omega) \in V$, then the set of even numbers can be built as

$$\bigcup\bigl(\mathcal{S}(\omega) \cap \{x \mid x \subseteq \omega \;\&\; 0 \in x \;\&\; \forall n :< \textstyle\bigcup x\, (n \in x \iff n+1 \notin x)\}\bigr)$$

6.3. Corollary. *"$\mathcal{S}(\omega) \in V$" is false in the rud closure of $\{\omega\}$.*

Proof. By Gandy, who showed that EVEN is not there. □

6.4. Corollary. *"$\mathcal{S}(\omega) \in V$" is not provable in* GJI.

6.5. Corollary. *Since the existence of $\mathcal{S}(\omega)$ is derivable in* GJ *from the existence of* ACK, *the existence of* ACK *is not provable in* GJI.

7. Models of fReR and beyond

Gandy: A set that models fReR but not ReR

Take $\mathbf{G}_3$ to be $V_{\omega+\omega}$.

Model 11:

Write **HC** for the union of all countable transitive sets. Then, assuming choice for countable families, $\mathbf{M}_{11} =_{\mathrm{df}} V_{\omega+\omega} \cap \mathbf{HC}$, that is, the union of all countable transitive sets of rank less than $\omega + \omega$, will be a model of fReRI but not, by Proposition **2.102**, ReR.

Variants of Model 11:

As often in this paper, we can obtain further models by carrying out one construction within another. Let **N** be an admissible set of height $\kappa > \omega$. For $0 < \eta = \bigcup \eta < \kappa$, let $\mathbf{N}_{11,\eta}$ be the union of transitive sets in **N** of rank less than η. Then that will be a model of fReC, and of AxInt if $\eta > \omega$. For a second example, assume that AC holds in **N** and consider the union **P** of all transitive sets which are members of **N** and countable there. Then **P** will be a model of fReC. Further **P** will be a model of $\mathcal{S}(x) \in V$.

Model 12: of fReR omitting HF

Since fReR_0 is a subtheory of Z, it is enough to find a transitive model of Z in which **HF** is not a set. The construction of one such model is sketched in Remark **14.24**; for others, see [M1] and the further references there.

7.0. Problem. For which λ and α are L_λ and J_α a model of fReR or fReC? Material in Section 9 suggests that a necessary condition will be that $\alpha = \omega\alpha$. Is that also sufficient?

Zarach: a set that models ReR but not KPI

See [Z], Theorem 6.4.

Model 13: a model of Z + TCo in which rank is not everywhere defined

Let λ be a limit ordinal. Define

$$\mathbf{A}_{13,\lambda} =_{\mathrm{df}} \{u \mid \bigcup u \subseteq u \;\&\; u \cap \lambda < \lambda\}; \qquad \mathbf{M}_{13,\lambda} = \bigcup \mathbf{A}_{13,\lambda};$$

Note that if u and v are members of $\mathbf{A}_{13,\lambda}$ then $u \cup v \in \mathbf{A}_{13,\lambda}$, and $u \cup \mathcal{P}(u) \cup \{\mathcal{P}(u)\} \in \mathbf{A}_{13,\lambda}$; so $\mathbf{M}_{13,\lambda}$ will be a supertransitive model of all of Z except (in the case $\lambda = \omega$) the axiom of infinity. As $\mathbf{A}_{13,\lambda} \subseteq \mathbf{M}_{13,\lambda}$, $\mathbf{M}_{13,\lambda}$ will also model TCo. V_λ will be a subclass but not a member of $\mathbf{M}_{13,\lambda}$; $ON \cap \mathbf{M}_{13,\lambda} = \lambda$. V_λ will be definable over $\mathbf{M}_{13,\lambda}$ as the class of those sets which lie in the domain of an attempt at the rank function. The union of those attempts will be a class but not a set of $\mathbf{M}_{13,\lambda}$.

We show that $\mathbf{M}_{13,\lambda}$ will contain sets of all ranks. Let u be any member of $\mathbf{A}_{13,\lambda}$ which is not an ordinal. Define the sequence

$$u_0 = u; \quad u_{\nu+1} = u_\nu \cup \{u_\nu\}; \quad u_\eta = \bigcup_{\nu<\eta} u_\nu \quad \text{for } 0 < \eta = \bigcup \eta.$$

Then it is easily shown by induction on ν that no u_ν is an ordinal; that each u_ν is transitive; that each u_ν is a member of each $u_{\nu'}$ with $\nu < \nu'$; that $\varrho(u_\nu) = \varrho(u_0) + \nu$; that $u_\nu \cap ON = u_0 \cap ON$; and hence that each u_ν is in $\mathbf{A}_{13,\lambda}$ and therefore in $\mathbf{M}_{13,\lambda}$.

The case $\lambda = \omega$ gives us a model of Z which has infinite members but for which the axiom of infinity in the form $\omega \in V$ is false.

Variants of Model 13 will be studied in *Rudimentary Recursion* [M4].

Part II

8. Adding $\mathcal{S}(x) \in V$ to these systems

Devlin in his book [Dev] had the aim of finding a theory that would hold in all structures I_λ for λ a limit ordinal, and in all structures J_α for α an arbitrary non-zero ordinal, be strong enough for a unified development of both hierarchies, and yet not require the introduction of rudimentary functions at too early a stage; and proposed BS as such a theory. Alas, it proves to be too weak, as we shall see in Section 10 through the use of the models that we have built in earlier sections. Devlin's treatment is further flawed by other mistakes such as those mentioned by Stanley in his review (*Journal of Symbolic Logic* **53** pp 864–8) of Devlin's book *Constructibility*, where Solovay (unpublished) is quoted as declaring [Dev, I.9.5] to be false "as can be seen by a forcing argument," and [Dev I.9.3] to be refutable "by the use of Ehrenfeucht games."

Stanley concludes his review of [Dev] by asking whether such a theory might be found. We have three candidates: our first proposal, which we call DS, for "Devlin strengthened", is to add to the axioms of DB the axioms

$$\omega \in V \text{ and } \mathcal{S}(x) \in V,$$

where $\mathcal{S}(x)$ is to mean the set of finite subsets of x. Call ReSs, GJs, fReRs the result of adding, to ReS, GJ and fReR respectively, the same two principles. Note that whereas BS had full foundation, we allow DS and our other systems to have only Π_1 foundation.

8.0. Proposition. *The existence of Cartesian products is provable in* ReSs*: so* DS *is the same as* ReSs.

Proof. Given a, $\mathcal{S}(a)$ will contain all 1- and 2-element subsets of a; hence $a \times a$ is a Δ_0 subclass of the set $\mathcal{S}(\mathcal{S}(a))$; to form $b \times c$, take $a = b \cup c$ and apply Δ_0 Separation. (8.0) □

At the stronger end of our lattice of theories, the enhancement amounts to no more than adding the axiom of infinity, since by Proposition **2.103**, ReRI proves that $\forall x\, \mathcal{S}(x) \in V$.

8.1. Problem. Is TCo derivable from the other axioms of ReR?

8.2. Remark. It is tempting to add a further axiom,

$$\mathbf{HF} \in V,$$

which in many ways makes life easier, because **HF** is a model of ZF – Infinity, and therefore a large number of functions become automatically available. But a feeling, that doing so does not address the chief problem with BS, is reinforced by the variant of Model 6 mentioned after Remark **5.25**, in which **HF** exists but some ${}^3x$ not.

Our aim, in this section and the next, is to study these systems, and we shall begin by enlarging our syntax to treat a class of formulæ that is slightly more general than Δ_0 but still limited in a specific sense.

A syntactical enhancement

We examine the consequences of allowing limited quantifiers $\forall y_{\in \mathcal{S}(x)}$, $\exists y_{\in \mathcal{S}(x)}$. The paradigm for our discussion is section 6 of "The Strength of Mac Lane Set Theory" where the quantifiers $\forall y_{\in \mathcal{P}(x)}$, there written as $\forall y{:}\subseteq x$ and in the present paper as $\forall y_{\subseteq x}$, were discussed.

We call a formula $\Delta_{0,\mathcal{S}}$ if all its quantifiers are of the form $Qx_{\in \mathcal{S}(y)}$ or $Qx_{\in y}$ where Q is $\forall$ or $\exists$, and x and y are distinct variables. We preserve "restricted" as a description of the quantifiers $Qx_{\in y}$, and speak of the occurrences of y in $Qx_{\in \mathcal{S}(y)}$ or $Qx_{\in y}$ as *limiting* the range of the bound variable x. It is tempting, indeed, to adopt a different presentation of the language by declaring the class of atomic formulæ to consist of every formula of one of the three forms

$$x \in y \qquad x = y \qquad x \in \mathcal{S}(y)$$

and to have three kinds of quantifiers, $\forall x$, $\forall x_{\in y}$ and $\forall x_{\in \mathcal{S}(y)}$ in the language; but we shall not formally adopt this approach here. Gandy in his paper [G] suggests considering the ancestral $\in^*$ of $\in$, where $x \in^* y$ iff $x \in \mathrm{tcl}(y)$, which will become easily available in our system.

8.3. Proposition. (DS) *"$x \in \mathcal{S}(y)$", "$x = \mathcal{S}(y)$" and "$\mathcal{S}(y) \in x$" are all $\Delta_{0,\mathcal{S}}$.*

Normal forms for $\Delta_{0,\mathcal{S}}$ formulæ

8.4. We sketch a method of rewriting a $\Delta_{0,\mathcal{S}}$ formula so that all variables are limited by terms constructed from the free variables of the original formula using only $\bigcup$; thus ultimately the terms limiting variables contain no variables that are themselves bound by other quantifiers.

Unlike $\in$, $\subseteq$ is transitive. Hence the following reduction is available:

$$\exists x_{\in \mathcal{S}(t)} \forall y_{\in \mathcal{S}(x)} \mathfrak{A} \Longleftrightarrow \exists x_{\in \mathcal{S}(t)} \forall y_{\in \mathcal{S}(t)} [y \subseteq x \Longrightarrow \mathfrak{A}].$$

Note here that on the left-hand side the x limiting y in the quantifier $\forall y_{\in \mathcal{S}(x)}$ is itself bound by the preceding quantifier $\exists x_{\in \mathcal{S}(t)}$, whereas on the right-hand side the t that limits both quantifiers is itself free. We may speak of t in the above displayed formula or $\bigcup t$ in the next as a *free term*.

We thus obtain these reductions:

$$\forall x_{\in a}\, \exists y_{\in x}\, \mathfrak{A} \iff \forall x_{\in a}\, \exists y_{\in \bigcup a}\, [y \in x \ \&\ \mathfrak{A}];$$

$$\forall x_{\in \mathcal{S}(a)}\, \exists y_{\in x}\, \mathfrak{A} \iff \forall x_{\in \mathcal{S}(a)}\, \exists y_{\in a}\, [y \in x \ \&\ \mathfrak{A}];$$

$$\begin{aligned} \forall x_{\in a}\, \exists y_{\in \mathcal{S}(x)}\, \mathfrak{A} &\iff \forall x_{\in a}\, \exists y_{\in \mathcal{S}(\bigcup a)}\, [y \subseteq x \ \&\ \mathfrak{A}] \\ &\iff \forall x_{\in a}\, \exists y_{\in \mathcal{S}(\bigcup a)}\, [\forall s_{1 \in \bigcup a}\, (s_1 \in y \Longrightarrow y_1 \in x) \ \&\ \mathfrak{A}]; \end{aligned}$$

$$\begin{aligned} \forall x_{\in \mathcal{S}(a)}\, \exists y_{\in \mathcal{S}(x)}\, \mathfrak{A} &\iff \forall x_{\in \mathcal{S}(a)}\, \exists y_{\in \mathcal{S}(a)}\, [y \subseteq x \ \&\ \mathfrak{A}] \\ &\iff \forall x_{\in \mathcal{S}(a)}\, \exists y_{\in \mathcal{S}(a)}\, [\forall s_{2 \in a}\, (s_2 \in y \Longrightarrow s_2 \in x) \ \&\ \mathfrak{A}]. \end{aligned}$$

Those equivalences, which are all valid in S_0, and, where applicable, preserve the stratifiability of the formula under consideration, show that one may progressively rewrite the formula to one in which all limitations are of the form ${}_{\in \mathcal{S}(\bigcup^{\mathfrak{k}} a)}$ or ${}_{\in \bigcup^{\mathfrak{k}} a}$ with a a free variable. We call such a formula one in *free form.* Our expansion of $y \subseteq x$ in the fourth and sixth lines, which would be unnecessary if we treated $y \subseteq x$ as atomic, helps to secure free form. We call the bound variables s, t introduced in those expansions *subsidiary variables*: we shall suppress mention of them in our discussion below, so that when we speak of "every quantifier", we mean "every quantifier binding other than a subsidiary variable".

Given a formula in free form, we replace each limiting free term by a new variable and add a clause expressing the equality of the term and the variable.

We have reached the

8.5. First Limited Normal Form. *Let Φ be a $\Delta_{0,\mathcal{S}}$ formula with free variables $a_0, \dots, a_{\mathfrak{n}}$. Let $\mathfrak{m}+1$ be the number of quantifiers occurring in Φ. Then for $0 \leqslant j \leqslant \mathfrak{m}$, there are numbers $0 \leqslant \mathfrak{k}(j) \leqslant \mathfrak{n}$, $0 \leqslant \mathfrak{l}(j)$, determined by the quantifier structure of Φ, new variables $y_0, \dots, y_{\mathfrak{m}}$, and a $\Delta_{0,\mathcal{S}}$ formula Ψ_1 with free variables $a_0, \dots, a_{\mathfrak{n}}, y_0, \dots, y_{\mathfrak{m}}$, in which every quantifier is limited by one of the parameters y_i, such that, abbreviating $\forall y_0, \dots, \forall y_{\mathfrak{m}}$ by $\forall \vec{y}$, we have*

$$\vdash_{\mathsf{DB}_0} \forall \vec{a} \forall \vec{y} \Big[\bigwedge_{0 \leqslant j \leqslant \mathfrak{m}} y_j = \bigcup{}^{\mathfrak{l}(j)} a_{\mathfrak{k}(j)} \Longrightarrow \big[\Phi(\vec{a}) \iff \Psi_1(\vec{a}, \vec{y})\big]\Big]$$

To take things to a second stage, if we know that we intend using the formula $\Phi(a)$ in a context where a_i will be constrained to be a member of b_i, we may replace the restriction ${}_{\in \bigcup^{\mathfrak{l}} a_i}$ by the restriction ${}_{\in \bigcup^{\mathfrak{l}+1} b_i}$; and each limitation ${}_{\in \mathcal{S}(\bigcup^{\mathfrak{l}} a_i)}$ by the limitation ${}_{\in \mathcal{S}(\bigcup^{\mathfrak{l}+1} b_i)}$, since if $a \in b$, $\bigcup^{\mathfrak{l}} a \subseteq \bigcup^{\mathfrak{l}+1} b$, and make a corresponding adjustment to the matrix.

We could also consider intended limitations $a_i \subseteq b_i$ instead of restrictions $a_i \in b_i$: the replacements to be made then would be ${}_{\in \bigcup^{\mathfrak{l}} a_i}$ by ${}_{\in \bigcup^{\mathfrak{l}} b_i}$ and ${}_{\in \mathcal{S}(\bigcup^{\mathfrak{l}} a_i)}$ by ${}_{\in \mathcal{S}(\bigcup^{\mathfrak{l}} b_i)}$, since if $a \subseteq b$ then $\bigcup^{\mathfrak{l}} a \subseteq \bigcup^{\mathfrak{l}} b$.

Further, we could mix our intentions, and also leave some a_i untouched, which is tantamount to saying $a_i = b_i$.

We thus have the

8.6. Second Limited Normal Form. *Continuing the notation of the First Limited Normal Form, let R, S and U be disjoint sets partitioning $[0, \mathfrak{n}]$, and let $b_0, \dots, b_\mathfrak{n}$ be variables not occurring in Φ. Then for the same numbers $\mathfrak{k}(j)$, $\mathfrak{l}(j)$, there is a $\Delta_{0,\mathcal{S}}$ formula Ψ_2 with free variables $a_0, \dots, a_\mathfrak{n}, y_0, \dots, y_\mathfrak{m}$, in which every quantifier is limited to one of the parameters y_i, such that*

$$\vdash_{\mathsf{DB}_0} \forall\vec{b}\forall\vec{a}\forall\vec{y}\Bigg[\Bigg[\bigwedge_{i \text{ in } R} a_i \in b_i \;\&\; \bigwedge_{i \text{ in } S} a_i \subseteq b_i \;\&\; \bigwedge_{i \text{ in } U} a_i = b_i \;\&\; \bigwedge_{\mathfrak{k}(j) \text{ in } R} y_j = \bigcup{}^{\mathfrak{l}(j)+1} b_{\mathfrak{k}(j)} \;\&\; \bigwedge_{\substack{\mathfrak{k}(j) \text{ in} \\ S \text{ or } U}} y_j = \bigcup{}^{\mathfrak{l}(j)} b_{\mathfrak{k}(j)}\Bigg] \implies \Big[\Phi(\vec{a}) \iff \Psi_2(\vec{a}, \vec{y})\Big]\Bigg]$$

8.7. Example. Let $\mathfrak{A}$ be quantifier-free, with six variables a, b, x, y, z, w. Suppose we want to re-write the formula

$$\Phi(a, b) \iff_{\mathrm{df}} \exists x_{\in a} \forall y_{\in \mathcal{S}(x)} \exists z_{\in x} \forall w_{\in \mathcal{S}(z)} \mathfrak{A}(a, b, x, y, z, w).$$

Let

$$\mathfrak{B}(a, b, x, y, z, w) \iff_{\mathrm{df}} \big(y \subseteq x \implies [z \in x \;\&\; (w \subseteq z \implies \mathfrak{A}(a, b, x, y, z, w))]\big).$$

Notice that $\mathfrak{B}$ is Δ_0, or indeed quantifier-free if we count $s \subseteq t$ as atomic. Then

$$\exists x_{\in a} \forall y_{\in \mathcal{S}(x)} \exists z_{\in x} \forall w_{\in \mathcal{S}(z)} \mathfrak{A}(a, b) \iff$$
$$\iff \exists x_{\in a} \forall y_{\in \mathcal{S}(\cup a)} \exists z_{\in \cup a} \forall w_{\in \mathcal{S}(\cup\cup a)} \big[\mathfrak{B}(a, b, x, y, z, w)\big]$$

In order not to use $\mathcal{S}$ applied to a term that is not a variable, we introduce further variables z_j.

8.8. First Restricted Normal Form. *Continuing the notation of the First Limited Normal Form, for the same numbers $\mathfrak{k}(j)$, $\mathfrak{l}(j)$, there is a partition of $\{j \mid 0 \leqslant j \leqslant \mathfrak{m}\}$ into disjoint sets L_Φ, R_Φ; there are new variables y_j, z_j for $0 \leqslant j \leqslant \mathfrak{m}$; and there is a Δ_0 formula Ψ_3, with free variables the a's and the z's; such that every quantifier in Ψ_3 is restricted to one of the parameters z_i, and*

$$\vdash_{\mathsf{DB}_0} \forall\vec{a}\forall\vec{y}\forall\vec{z}\Bigg[\Bigg[\bigwedge_{j \text{ in } R_\Phi}\big(z_j = y_j \;\&\; y_j = \bigcup{}^{\mathfrak{l}(j)} a_{\mathfrak{k}(j)}\big) \;\&\; \bigwedge_{j \text{ in } L_\Phi}\big(z_j = \mathcal{S}(y_j) \;\&\; y_j = \bigcup{}^{\mathfrak{l}(j)} a_{\mathfrak{k}(j)}\big)\Bigg] \Rightarrow \Big[\Phi(\vec{a}) \iff \Psi_3(\vec{a}, \vec{z})\Big]\Bigg]$$

Taking that to the corresponding second stage, and noting that if $a \subseteq b$ then $\mathcal{S}(\bigcup{}^{\mathfrak{l}} a) \subseteq \mathcal{S}(\bigcup{}^{\mathfrak{l}} b)$, whereas if $a \in b$, $\mathcal{S}(\bigcup{}^{\mathfrak{l}} a) \subseteq \mathcal{S}(\bigcup{}^{\mathfrak{l}+1} b)$, we reach the

8.9. Second Restricted Normal Form. *Let Φ be a $\Delta_{0,\mathcal{S}}$ formula with free variables $a_0, \dots, a_\mathfrak{n}$. Let R, S and U be disjoint sets partitioning $[0, \mathfrak{n}]$, and let $b_0, \dots, b_\mathfrak{n}$ be variables not occurring in Φ. Let $\mathfrak{m} + 1$ be the number of quantifiers occurring in*

Φ. Then there is a partition of $\{j \mid 0 \leqslant j \leqslant \mathfrak{m}\}$ into disjoint sets L_Φ, R_Φ; for $0 \leqslant j \leqslant \mathfrak{m}$, there are numbers $0 \leqslant \mathfrak{k}(j) \leqslant \mathfrak{n}$, $0 \leqslant \mathfrak{l}(j)$, determined by the quantifier structure of Φ, there are new variables y_j, z_j for $0 \leqslant j \leqslant \mathfrak{m}$; and there is a Δ_0 formula Ψ_4 with free variables the a's and the z's, in which every quantifier is restricted to one of the parameters z_i; such that,

$$\vdash_{\mathsf{DB}_0} \forall\vec{b}\forall\vec{a}\forall\vec{y}\forall\vec{z}\Bigg[\Bigg[\bigwedge_{i \text{ in } R} a_i \in b_i \;\&\; \bigwedge_{i \text{ in } S} a_i \subseteq b_i \;\&\; \bigwedge_{i \text{ in } U} a_i = b_i$$

$$\& \bigwedge_{\substack{j \text{ in } R_\Phi,\\ \mathfrak{k}(j) \text{ in } R}} \Big(z_j = y_j \;\&\; y_j = \bigcup{}^{\mathfrak{l}(j)+1} b_{\mathfrak{k}(j)}\Big)$$

$$\& \bigwedge_{\substack{j \text{ in } L_\Phi,\\ \mathfrak{k}(j) \text{ in } R}} \Big(z_j = \mathcal{S}(y_j) \;\&\; y_j = \bigcup{}^{\mathfrak{l}(j)+1} b_{\mathfrak{k}(j)}\Big)$$

$$\& \bigwedge_{\substack{j \text{ in } R_\Phi,\\ \mathfrak{k}(j) \text{ in } S \text{ or } U}} \Big(z_j = y_j \;\&\; y_j = \bigcup{}^{\mathfrak{l}(j)} b_{\mathfrak{k}(j)}\Big)$$

$$\& \bigwedge_{\substack{j \text{ in } L_\Phi,\\ \mathfrak{k}(j) \text{ in } S \text{ or } U}} \Big(z_j = \mathcal{S}(y_j) \;\&\; y_j = \bigcup{}^{\mathfrak{l}(j)} b_{\mathfrak{k}(j)}\Big)\Bigg]$$

$$\Longrightarrow \Big[\Phi(\vec{a}) \iff \Psi_4(\vec{a}, \vec{z})\Big]\Bigg]$$

Self-strengthening of DS

We may now deduce the

8.10. Metatheorem. DS *proves all instances of the scheme of $\Delta_{0,\mathcal{S}}$ separation.*

Proof. Suppose that there are $\mathfrak{m}+1$ quantifiers in the $\Delta_{0,\mathcal{S}}$ formula $\Phi(x,a)$. By the Second Restricted Normal Form, we know that there are new variables $y_0, \ldots, y_\mathfrak{m}$, $z_0, \ldots, z_\mathfrak{m}$ and a Δ_0 formula $\Psi_4(x, \vec{a}, \vec{z})$ with the free variables shown, such that

$$\mathsf{DB}_0 \vdash x \in d \;\&\; \Big\langle \text{conditions on } \vec{z},\, \vec{y},\, d \text{ and } \vec{a} \Big\rangle \Longrightarrow \big[\Phi(x,\vec{a}) \iff \Psi_4(x,\vec{a},\vec{z})\big],$$

where there are $\mathfrak{m}+1$ conditions, each of one of the four following types, according to the quantifier structure of Φ:

$$[z = y \;\&\; y = \bigcup{}^{\mathfrak{l}+1} d];$$
$$[z = \mathcal{S}(y) \;\&\; y = \bigcup{}^{\mathfrak{l}+1} d];$$
$$[z = y \;\&\; y = \bigcup{}^{\mathfrak{l}} a];\; [z = \mathcal{S}(y) \;\&\; y = \bigcup{}^{\mathfrak{l}} a].$$

In DS we may prove that given d and $\vec{a}$ there are y's and z's satisfying the conditions, and for those z, we have $\forall x_{\in d}\big(\Phi(x,\vec{a}) \iff \Psi_4(x,\vec{a},\vec{z})\big)$, whence

$$d \cap \{x \mid \Phi(x,\vec{a})\} = d \cap \{x \mid \Psi_4(x,\vec{a},\vec{z})\} \in V. \qquad (8.10)\ \square$$

8.11. Proposition. DS *proves that the graph $\mathcal{G}_+$ of integer addition, or indeed of any partial recursive function, is a set.*

Proof. To get the graph of addition, we would apply separation to $\omega \times (\omega \times \omega)$ to form the set of all triples such that there exists an attempt: *prima facie* Σ_1 or perhaps just Δ_1 separation, given that attempts are unique (a fact that we have not proved). But the attempts are all in $\mathcal{S}(\omega \times (\omega \times \omega))$, and so with that set as a parameter, only Δ_0 separation is needed. (8.11) □

The results following Definitions **2.16** and **2.17** can be improved:

8.12. Lemma. *"$x \in \mathcal{S}(y)$" is Δ_1^{DS}.*

Proof. By Corollary **2.20** and Lemma **2.21**. □

8.13. Lemma. (DS)

$$z \subseteq \mathcal{S}(y) \iff \exists c[\forall w_{\in z}\, w \subseteq y \;\&\; \forall w_{\in z} \exists f_{\in c} \exists n_{\in \omega}\, f : n \longleftrightarrow w].$$

Proof. Take $c = \mathcal{S}(y \times \omega)$. (8.13) □

8.14. Corollary. *"$z \subseteq \mathcal{S}(y)$" and "$z = \mathcal{S}(y)$" are Δ_1^{DS}.*

Proof. The first part by Lemmata **2.22** and **8.13**; the second then follows by Lemma **2.23**. (8.14) □

8.15. Remark. The above discussion shows that the function $x \mapsto \mathcal{S}(x)$ is Σ_1 in ReR with $\omega \in V$ and Π_1 foundation.

8.16. Metatheorem. *Every $\Pi_{1,\mathcal{S}}$ predicate is Π_1^{DS}.*

Proof. Consider a predicate of the form $\forall c \Phi(c, \vec{a})$ where Φ is $\Delta_{0,\mathcal{S}}$. We again use the Second Restricted Normal Form, which tells us that there is a Δ_0 predicate $\Psi_4(c, a, \vec{z})$ and further variables $\vec{b}$ and $\vec{y}$, such that $\Phi(c, a)$ is equivalent to $\Psi_4(c, a, \vec{z})$ provided finitely many conditions hold, of the form $z = \mathcal{S}(y) \;\&\; y = \bigcup^{\ell} b$ or $z = y \;\&\; y = \bigcup^{\ell} b$, and each a and c is either a member of or a subset of or equal to the corresponding b.

Thus, writing out a sample condition,

$$\forall c \Phi(c, a) \iff \forall c \forall \vec{b} \forall \vec{z} \forall \vec{y} \Big[[[\underbrace{z = \mathcal{S}(y)}_{\Sigma_1} \;\&\; \underbrace{y = \textstyle\bigcup^{\ell} b \;\&\; a \subseteq b]}_{\Delta_0} \;\&\; \ldots$$

$$\ldots \;\&\; \underbrace{[\ldots]}_{\Sigma_1}] \implies \underbrace{\Psi_4(c, a, \vec{z})}_{\Delta_0} \Big],$$

which is Π_1, as required. (8.16) □

DS with TCo

8.17. Proposition. (DS + TCo) $\mathrm{tcl}(x) \in V$.

Proof. Fix x, and using TCo, let u be a transitive set of which x is a member. Using $\mathcal{S}(x) \in V$, let a be the set $\mathcal{S}(u \times \omega)$.

Say that f *descends from x to y* if

$$\mathrm{Fn}(f) \ \&\ \mathrm{Dom}\, f \in \omega \ \&\ 2 \leqslant \mathrm{Dom}\, f \ \&\ f(0) = x$$
$$\&\ \forall k :< \mathrm{Dom}\,(f) - 1 f(k+1) \in f(k) \ \&\ f(\mathrm{Dom}\,(f) - 1) = y.$$

That is a Δ_0 predicate of f, and each such f is in a, so the class

$$u \cap \{y \mid \exists f_{\in a} [f \text{ descends from } x \text{ to } y]\}$$

is a set and is the desired transitive closure of x. (8.17) □

Self-strengthening of GJs

8.18. Lemma. (GJs) $\{\mathcal{S}(x) \mid x \in a\} \in V$.

Proof. Fix the set a. If $x \in a$ then $x \subseteq \bigcup a$, so $\mathcal{S}(x) \subseteq \mathcal{S}(\bigcup a)$. The desired set is the class

$$\{\mathcal{S}(\textstyle\bigcup a) \cap \{y \mid y \subseteq x\} \mid_x x \in a\},$$

which is a set by an application of RR^+. (8.18) □

8.19. Corollary. (GJs) $\{\langle \mathcal{S}(\bigcup w), \mathcal{S}(w)\rangle \mid_w w \in b\} \in V$.

Proof. Consider

$$\{\mathcal{S}(v) \mid_v v \in a\} \times \{\mathcal{S}(w) \mid_w w \in b\} \cap \{(c,d)_2 \mid_{c,d} \textstyle\bigcup c = \bigcup\bigcup d\},$$

taking $a = \{\bigcup w \mid_w w \in b\}$. (8.19) □

8.20. Proposition. GJs *proves* $\Delta_{0,\mathcal{S}}$ *rud replacement.*

Proof. Aiming, in fact, for the extended form corresponding to RR^+, defined in **2.88**, we must show that

$$\forall x_2 \forall x_1 \exists w \forall \vec{v}_{\in x_1} \exists t_{\in w} \forall u (u \in t \Longleftrightarrow u \in x_2 \ \&\ \Phi(u, \vec{v}),$$

where Φ is a $\Delta_{0,\mathcal{S}}$ formula with the free variables shown.

Suppose that there are $\mathfrak{m}+1$ quantifiers in Φ. By the Second Restricted Normal Form, we know that there are new variables $y_0, \dots, y_{\mathfrak{m}}$, $z_0, \dots, z_{\mathfrak{m}}$ and a Δ_0 formula $\Psi_4(u, \vec{v}, \vec{z})$ with the free variables shown, such that

$$\mathsf{DB}_0 \vdash u \in x_2 \ \&\ v \mathrel{\vec{\in}} x_1 \ \&\ \big\langle \text{conditions on } \vec{z},\ \vec{y},\ x_1, \text{ and } x_2 \big\rangle$$
$$\Longrightarrow \big[\Phi(u, \vec{v}) \Longleftrightarrow \Psi_4(u, \vec{v}, \vec{z})\big],$$

where there are $\mathfrak{m}+1$ conditions, each of one of the four following types, according to the quantifier structure of Φ:

$$[z = y \ \&\ y = \textstyle\bigcup^{\mathfrak{l}+1} x_2]; \quad [z = \mathcal{S}(y) \ \&\ y = \textstyle\bigcup^{\mathfrak{l}+1} x_2];$$
$$[z = y \ \&\ y = \textstyle\bigcup^{\mathfrak{l}+1} x_1]; \quad [z = \mathcal{S}(y) \ \&\ y = \textstyle\bigcup^{\mathfrak{l}+1} x_1].$$

A slight extension of RR^+ would tell us that

$$\forall x_2 \forall x_1 \exists w \forall \vec{z}_{\in A} \forall \vec{v}_{\in x_1} \exists t_{\in w} \forall u \big(u \in t \iff u \in x_2 \ \&\ \Psi_4(u, \vec{v}, \vec{z})\big),$$

where A is a certain class, provably a set containing at most $\mathfrak{m}+1$ elements, namely the values of the form $\bigcup^{\mathfrak{l}+1} x_2$ or $\mathcal{S}(\bigcup^{\mathfrak{l}+1} x_2)$ given to the z's by the conditions.

To show that, fix x_2. If we write x_3 for $x_1 \cup A$, then by RR^+, we may deduce that

$$\exists w \forall \vec{v}_{\in x_3} \forall z_{\in x_3} \exists t_{\in w} \forall u \big(u \in t \iff u \in x_2 \ \&\ \Psi_4(u, \vec{v}, \vec{z})\big),$$

whence

$$\exists w \forall \vec{v}_{\in x_3} \forall z_{\in x_3} \exists t_{\in w} \forall u \big(u \in t \iff u \in x_2 \ \&\ \Phi(u, \vec{v})\big).$$

We may now cut this w down to exactly the one we want by applying $\Delta_{0,\mathcal{S}}$ separation. (8.20) □

Self-strengthening of fReRs

8.21. Proposition. fReRs *proves flat* $\Delta_{0,\mathcal{S}}$ *replacement.*

Proof. We must show that

$$\forall x_{\in u} \exists! d[\Phi(x, d) \ \&\ d \subseteq e] \implies \exists v \forall d\big[d \in v \iff \exists x_{\in u} [\Phi(x, d) \ \&\ d \subseteq e]\big],$$

where Φ is a $\Delta_{0,\mathcal{S}}$ formula with the two free variables shown.

Suppose that there are $\mathfrak{m}+1$ quantifiers in Φ. By the Second Restricted Normal Form, we know that there are new variables $y_0, \dots, y_{\mathfrak{m}}$, $z_0, \dots, z_{\mathfrak{m}}$ and a Δ_0 formula $\Psi_4(x, d, \vec{z})$ with $\mathfrak{m}+3$ free variables, such that

$$\begin{aligned} \mathsf{DB}_0 \vdash x \in u \ \&\ d \subseteq e \ \&\ \big\langle \text{conditions on } \vec{z}, \vec{y}, u, \text{ and } e\big\rangle \\ \implies \big[\Phi(x, d) \iff \Psi_4(x, d, \vec{z})\big], \end{aligned}$$

where there are $\mathfrak{m}+1$ conditions, each of one of the four following types, according to the quantifier structure of Φ:

$$[z = y \ \&\ y = \textstyle\bigcup^{\mathfrak{l}+1} u]; \quad [z = \mathcal{S}(y) \ \&\ y = \textstyle\bigcup^{\mathfrak{l}+1} u];$$
$$[z = y \ \&\ y = \textstyle\bigcup^{\mathfrak{l}} e]; \quad [z = \mathcal{S}(y) \ \&\ y = \textstyle\bigcup^{\mathfrak{l}} e].$$

Fix u and e; then, using $\forall x\, \mathcal{S}(x) \in V$, the conditions will give fixed values to the y's and z's; for those values we shall have that for $x \in u$ and $d \subseteq e$, $\Phi(x, d) \iff \Psi_4(x, d, \vec{z})$.

Suppose now that $\forall x_{\in u} \exists! d[\Phi(x, d) \ \&\ d \subseteq e]$; then $\forall x_{\in u} \exists! d[\Psi_4(x, d, \vec{z}) \ \&\ d \subseteq e]$. We appeal to the extended form of (BdR) proved as Proposition **2.101**, to deduce that

$$\exists v \forall d\big[d \in v \iff \exists x_{\in u} [\Psi_4(x, d, \vec{z}) \ \&\ d \subseteq e]\big],$$

whence

$$\exists v \forall d\big[d \in v \iff \exists x_{\in u} [\Phi(x, d) \ \&\ d \subseteq e]\big]. \qquad (8.21)\ \square$$

8.22. Remark. In [M5] it will be seen that the system fReRs proves appropriate for the development of the definition of forcing, and that fReCs might be the weakest system persistent under set-generic extensions.

8.23. Remark. In [M4] we shall study rudimentary recursions on the ancestral and related relations.

Self-strengthening of ReR

8.24. Lemma. (ReR) *All instances of Δ_0 replacement where, as in Proposition* **2.101**, *φ is allowed to have further free variables.*

Proof. Suppose that $\mathfrak{A}$ is Δ_0 and that $\forall x_{\in u} \exists! y \mathfrak{A}(x, y, w)$. Let $u_1 = u \times \{w\}$. Then

$$\forall x_{\in u_1} \exists! y \underbrace{\mathfrak{A}(\text{left}(x), y, \text{right}(x))}_{\Delta_0^{\mathsf{S}_0}}.$$

So applying Δ_0 replacement, we get $\exists v \forall y (y \in v \text{ iff } \exists x_{\in u_1} \mathfrak{A}(\text{left}(x), y, \text{right}(x))$, which in turn is equivalent to $\exists x_{\in u} \mathfrak{A}(x, y, w)$, as required. (8.24) □

8.25. Proposition. ReRI *proves each instance of $\Delta_{0,\mathcal{S}}$ replacement.*

Proof. The argument given for **8.21** adapts easily, using the Lemma. (8.25) □

8.26. Problem. Does ReR prove $\mathcal{S}(x) \in V$? The idea being that if there is an infinite set, then one ought to be able to prove that ω exists, and thence that $\mathcal{S}(x) \in V$; and if all sets are finite a proof of $\mathcal{S}(x) \in V$ will be provided by Proposition **2.13**.

We pause to establish two results concerning the sets $Z(a)$ defined in [M1], whose definition was recalled in our discussion of Model 7.

8.27. Definition. We write "f attempts $Z(a)$ at n" for the $\Delta_{0,\mathcal{S}}$ formula

$$\text{Fn}(f) \;\&\; \text{Dom}\,(f) = n + 1 \;\&\; f(0) = \varnothing \;\&\; \forall k_{\in n} \,(f(k+1) = \mathcal{S}(f(k)) \cup \{a\} \setminus \{\varnothing\}).$$

8.28. Proposition. (ReRI) *$\forall a : \omega \longrightarrow 2$, $Z(a)$ exists.*

Proof. Fix a. Note that if $\text{Fn}(f)$ then

$$x = \mathcal{S}(f(k)) \iff \underbrace{\exists y_{\in \bigcup\bigcup(f)} \, (y, k)_2 \in f \;\&\; x = \mathcal{S}(y)}_{\Delta_{0,\mathcal{S}}}.$$

Hence we may assert that

$$\forall n_{\in \omega} \exists f (f \text{ attempts } Z(a) \text{ at } n);$$

for the class of n for which the assertion fails is $\Pi_{1,\mathcal{S}}$ and therefore by Metatheorem **8.16** has, if non-empty, a minimal element, necessarily a successor; which can rapidly be refuted.

For each n, there can be at most one such f, so by $\Delta_{0,\mathcal{S}}$ replacement, the set of such f exists; its union will be a function, of which the class $Z(a)$ is the image and therefore a set. (8.28) □

8.29. Definition. Let $\Psi(x, a)$ be the $\Delta_{0,\mathcal{S}}$ formula

$$a \in x \;\&\; \forall b_{\in x} \,[\{b\} \in x \;\&\; (b \in \mathcal{S}(x) \vee b = a) \;\&\; (b = \varnothing \Longrightarrow b = a)]$$
$$\&\; \forall s_{\in \mathcal{S}(x)} \,[s \neq \varnothing \Longrightarrow s \in x].$$

8.30. Lemma. (ReRI) $Z(a) \in V \Longrightarrow [x = Z(a) \Longleftrightarrow \Psi(x, a)]$.

Proof. It is readily checked that $x = Z(a) \Longrightarrow \Psi(x, a)$.

Suppose that $Z(a) \in V$ and that $\Psi(x, a)$. Let $c = \mathcal{S}(Z(a) \times \omega)$. Then

$$\{n \mid Z_n(a) \not\subseteq x\} = \{n \mid \underbrace{\exists f_{\in b} f \text{ attempts } Z(a) \text{ at } n \,\&\, f(n) \not\subseteq x}_{\Delta_{0,\mathcal{S}}}\};$$

Π_1 foundation would yield a minimal element of that class, if non-empty; but $Z_0(a) = \varnothing \subseteq x$, and it is easily checked that $\Psi(x, a) \;\&\; Z_n(a) \subseteq x \Longrightarrow Z_{n+1} \subseteq x$. Thus $Z(a) \subseteq x$.

If $x \not\subseteq Z(a)$, let y be an $\in$-minimal element of $x \setminus Z(a)$. Then $y \neq \varnothing$, $y \in \mathcal{S}(x)$ and $y \subseteq Z(a)$. Hence $\forall z_{\in y} \exists! n_{\in \omega} \underbrace{(z \in Z_{n+1}(a) \;\&\; z \notin Z_n(a))}_{\Delta_{0,\mathcal{S}}}$; the class of such n's is therefore a set, which is finite and therefore bounded in ω; so $\exists m_{\in \omega}\, y \subseteq Z_m(a)$, whence $y \in Z_{m+1}(a)$, contradicting $y \notin Z(a)$. (8.30) □

8.31. Corollary. (ReRI) *"$x = Z(a)$" is $\Delta_{0,\mathcal{S}}$.*

8.32. Proposition. (ReRI) $\forall b_{\subseteq {}^{\omega}2} \{Z(a) \mid a \in b\} \in V$.

Proof. Fix b. Then

$$\forall a_{\in b} \exists! x \; \underbrace{x = Z(a)}_{\Delta_{0,\mathcal{S}}}.$$

Apply $\Delta_{0,\mathcal{S}}$ replacement to complete the proof. (8.32) □

Self-strengthening of KPI

8.33. Proposition. KPI *proves every instance of $\Delta_{0,\mathcal{S}}$ collection.*

Proof. We may either use Remark **8.30** or else Metatheorem **8.31**, which implies that in the context of KPI, every $\Delta_{0,\mathcal{S}}$ formula is equivalent to a Σ_1 one; but it is well known that KP is self-strengthening to Σ_1 collection. (8.33) □

8.34. Problem. In Proposition 8.33, can KPI be reduced to KP? In KP rank is definable and the rank of an infinite set must be at least ω; but with infinity $\mathcal{S}(x) \in V$ becomes provable.

9. The Gandy sequence

In this section we wish to assess the relative strength of the enhanced theories DS, etc.

9.0. Proposition. *There is a model of* DS *plus* $\mathbf{HF} \in V$ *in which* GJ *is false.*

Proof. The model $\mathbf{M}_7$ will do. We have to prove that "$\mathcal{S}(x) \in V$" is true in $\mathbf{M}_7$. Note that any non-empty finite set must have successor rank. So if u is transitive and contains only finitely many transitive sets of limit rank, then $u \cup \mathcal{S}(u) \cup \{\mathcal{S}(u)\}$ will have the same property. That suffices. (9.0) □

GJs in L and J

Now we wish to verify that GJs is true in every L_λ ($\lambda = \bigcup\lambda > \omega$)and J_α ($\alpha > 1$).

9.1. Proposition. *"$\mathcal{S}(x) \in V$" is true in every L_λ.*

Proof. Evidently so for $\lambda = \omega$; thereafter we have languages. Given $x \in L_\zeta$, all its finite subsets will be in $L_{\zeta+1}$, and the set of them will be in $L_{\zeta+2}$. (9.2) □

9.2. Proposition. *"$\mathcal{S}(x) \in V$" is true in every J_α.*

9.3. Lemma. *The sequence $\langle [\zeta]^{<\omega} \mid \zeta < \omega\alpha \rangle$ is uniformly Σ_1 over every J_α.*

Proof. By a rudimentary recursion, as discussed in [M4]. (9.3) □

The $S_{\omega\beta+k}$ used in the next proof may be defined as in Dodd's book, or one might use the sets corresponding to the T_n defined in the proof of Proposition **9.7**.

9.4. Lemma. *In each J_α, to every set x there is an ordinal λ and a surjection $f : \lambda \xrightarrow{\text{onto}} x$.*

Proof. In J_α each set is a member of some $S_{\omega\beta+k}$, with $\beta < \alpha$, so we may derive the lemma from [Do], chapter 1, section 2, Lemma 2.42 on page 20, which Dodd proves within his theory R^+_ω that he introduces on page 12. In our terms that is the theory GJ plus TCo (in view of his Lemma 2.6) plus a version of "V = L" plus certain instances of the scheme of full foundation. He shows though that each J_α models this theory: see his Lemma 2.21 on page 14. (9.4) □

Proof of the proposition. Let $f \in J_\alpha$ be a surjection from ζ to x. Then

$$\mathcal{S}(x) = \{f``a \mid a \in \mathcal{S}(\zeta)\}. \qquad (9.2)\ \square$$

9.5. Proposition. *Let λ be a limit ordinal. Then L_λ models (RR).*

Proof. For if x is in L_ζ each of the $x \cap \{u|\phi(u, \vec{v})\}$ is in $L_{\zeta+1}$ and the set of them is in $L_{\zeta+2}$. (9.5) □

9.6. Proposition. **HF** $= L_\omega = J_1$, *and hence is a member of $L_{\omega+\nu}$ and of $J_{1+\nu}$ for each $\nu > 0$.*

Model 14: of GJs without fReR

9.7. Theorem. *There is a model of* GJs *plus* **HF** $\in V$ *in which* fReR *is false.*

Proof. Such a model is J_2. Here we shall use the existence of our single rudimentary function $\mathbb{T}$ of Definition **2.73** that for any transitive set u generates the rudimentary closure of $u \cup \{u\}$. It has these properties: the elements of $\mathbb{T}(u)$ are subsets of u and, for non-empty u, are precisely the sets of the form $S(u; x, y)$, where S is one of our list $S_0, \ldots, S_{11}$ of twelve rudimentary functions, and $x, y \in u$. Similarly the elements of $\mathbb{T}(\mathbb{T}(u))$ are the sets $S(\mathbb{T}(u); x, y)$, where x and y are members of $\mathbb{T}(u)$, and are subsets of $\mathbb{T}(u)$.

Our function $\mathbb{T}$ differs slightly from those used by Jensen, Devlin and Dodd, and so we make a corresponding change of notation. We write T_0 for J_1, and successively T_{n+1} for $\mathbb{T}(T_n)$. Then $T_0 \subseteq T_1 \subseteq \ldots$ and $J_2 = \bigcup_{n\in\omega} T_n$.

Our intention is to build a calculus of terms, using names $\dot{S}_i$ for S_i in that finite list, and allowing as arguments names for the various T_n and their members. We define the class of terms recursively. $\mathcal{W}_0$ is to comprise symbols for the members of J_1. Having formed $\mathcal{W}_n$, we take a new symbol τ_n for T_n, and let $\mathcal{W}_{n+1}$ be the set of words of the form $\dot{S}_i\dot{(}\tau_n; v, w\dot{)}$ where v and w are words in $\mathcal{W}_n$, $0 \leqslant i \leqslant 9$, and $\dot{(}$ and $\dot{)}$ are the parentheses of the formal language we are developing.

Thus $\mathcal{W}_1$ comprises words of the form $\dot{S}\dot{(}\tau_1; x, y\dot{)}$ where x and y are in $\mathcal{W}_0$.

We suppose that our symbols are coded so that $\mathcal{W}_n \subseteq \omega \subseteq J_1 = \mathbf{HF}$, and that the $\mathcal{W}_n$ are pairwise disjoint, and that the coding has been done in some reasonable recursive way, so that in particular the map $k \mapsto \ulcorner k \urcorner$ is recursive with recursive inverse, and that there is a recursive map $(n, k)_2 \mapsto w^n_k$ such that for each n, $(w^n_k)_k$ is a recursive enumeration of the words in $\mathcal{W}_n$.

Let $\mathcal{E}_n$ be the evaluation function of these words: so that the set of evaluations, $\mathcal{E}_n[\mathcal{W}_n]$, is precisely the set T_n just defined.

Let $\mathcal{M}_n$ be the relation on ω defined by

$$\mathcal{M}_n(w, v) \iff w \in \mathcal{W}_n \;\&\; v \in \mathcal{W}_n \;\&\; \mathcal{E}_n(w) \in \mathcal{E}_n(v).$$

Let $\mathcal{Q}_n$ be the relation on $\mathcal{W}_n$ defined by

$$\mathcal{Q}_n(w, v) \iff w \in \mathcal{W}_n \;\&\; v \in \mathcal{W}_n \;\&\; \mathcal{E}_n(w) = \mathcal{E}_n(v).$$

9.8. Remark. In our context, of full extensionality, $\mathcal{Q}_n$ will of course be rudimentary in $\mathcal{M}_n$, and might therefore be dropped from this discussion; but with possible applications of the present argument in a non-extensional context in mind, we keep both predicates in play.

9.9. Proposition. *There are rudimentary functions G and H such that*

$$\mathcal{M}_{n+1} = G(\mathcal{M}_n, \mathcal{Q}_n) \;\&\; \mathcal{Q}_{n+1} = H(\mathcal{M}_n, \mathcal{Q}_n)$$

Proof. We examine the passage from one stage to the next in greater detail. We have a non-empty set W of words and an evaluation $\mathcal{E}$ for those words, such that $\mathcal{E}[W] = U$, a non-empty transitive set. We add a term τ to the language to denote U. We define a new set of words thus:

$$W^+ = \{\dot{S}_i\dot{(}\tau; v, w\dot{)} \mid 0 \leqslant i \leqslant 11, v \in W, w \in W\}.$$

We define an evaluation $\mathcal{E}^+$ of the words in W^+ thus:

$$\mathcal{E}^+(\dot{S}_i(\tau; v, w)) = S_i(U; \mathcal{E}(v), \mathcal{E}(w)).$$

The evaluation of course takes place in the set theoretical universe. We wish to show that it can be carried out at a more formal level.

We define relations $\mathcal{M}$, $\mathcal{Q}$ on W, and $\mathcal{M}^+$, $\mathcal{Q}^+$ on W^+, and we shall show that the second pair are uniformly rudimentary in the first pair.

9.10. Definition.

$$\mathcal{M}(v,w) \iff_{\text{df}} \mathcal{E}(v) \in \mathcal{E}(w)$$
$$\mathcal{Q}(v,w) \iff_{\text{df}} \mathcal{E}(v) = \mathcal{E}(w)$$

and similarly

$$\mathcal{M}^+(v^+,w^+) \iff_{\text{df}} \mathcal{E}^+(v^+) \in \mathcal{E}^+(w^+)$$
$$\mathcal{Q}^+(v^+,w^+) \iff_{\text{df}} \mathcal{E}^+(v^+) = \mathcal{E}^+(w^+)$$

9.11. Remark. Let $U^+ = \mathcal{E}^+(W^+)$: then $U^+ = \mathbb{T}(U)$. Thus each evaluation $\mathcal{E}^+(v^+)$ of a word in W^+ will be a subset of U, and therefore quantification over U suffices for comparing one evaluation with another; the finitely many functions involved being rudimentary, describing the evaluations will always be Δ_0.

9.12. Lemma. *For $z \in W$ and w^+ a word in W^+, the relation $\mathcal{E}(z) \in \mathcal{E}^+(w^+)$ is (uniformly) rudimentary in W, $\mathcal{M}$ and $\mathcal{Q}$.*

Proof. Essentially because the class of rudimentary relations is closed under definition by rudimentarily distinguishable cases. Let w^+ be $\dot{S}_p(\tau; w_1, w_2)$. If, say, $p = 2$, we shall have

$$\mathcal{E}(z) \in \mathcal{E}^+(w^+) \iff \exists w_{3 \in W}\, (\mathcal{M}(z, w_3)\ \&\ \mathcal{M}(w_3, w_1)).$$

For the general case, the function S_i being rudimentary, the predicate $z \in S_i(u; x, y)$ will be a Δ_0 predicate of z, u, x and y; rewrite that predicate by requiring all bound variables to be restricted to members of W, and as for atomic formulæ, replace $a = b$ by $\mathcal{Q}(a,b)$ and $a \in b$ by $\mathcal{M}(a,b)$. Note that for $i = 0, 1, 2, 3, 5, 9$ and 11, u does not occur; otherwise u only occurs in contexts such as $u \cap R_i(x)$ (for $i = 6$ or 7), $u \cap R_j(x,y)$ (for $i = 4$ or 8), and $z \in u$ (for $i = 10$); and so in all cases when the formula is written out, u will occur only in atomic formulæ of the form $a \in u$; of which the formal counterparts will always be evaluated as **true**, as τ denotes U, the set of evaluations of the variables. (9.12) □

Given that lemma, the relation $\mathcal{Q}^+(v^+, w^+)$ being equivalent to $\forall z_{\in W}\, (\mathcal{E}(z) \in \mathcal{E}^+(v^+) \iff \mathcal{E}(z) \in \mathcal{E}^+(w^+)$, will be rudimentary in W, $\mathcal{M}$ and $\mathcal{Q}$.

Now for $\mathcal{M}^+$.

9.13. Lemma. *For $z \in W$ and w^+ a word in W^+, the relation $\mathcal{E}(z) = \mathcal{E}^+(w^+)$ is (uniformly) rudimentary in W, $\mathcal{M}$ and $\mathcal{Q}$.*

Proof. With Remark **9.11** in mind, we see that $\mathcal{E}(z) = \mathcal{E}^+(w^+) \iff \forall y_{\in W} \left[\mathcal{E}(y) \in \mathcal{E}^+(w^+) \iff \mathcal{M}(y,z)\right]$, since $\mathcal{M}(y,z) \iff \mathcal{E}(y) \in \mathcal{E}(z)$. (9.13) □

Now $\mathcal{M}^+(v^+,w^+) \iff \exists z_{\in W}\, \mathcal{E}^+(v^+) = \mathcal{E}(z)\ \&\ \mathcal{E}(z) \in \mathcal{E}^+(w^+)$, and so $\mathcal{M}^+$ is rudimentary in W, $\mathcal{M}$ and $\mathcal{Q}$ by the last two lemmata.

Our Proposition is now established by the uniformity of the above discussion.

(9.9) □

Hence we may write a Δ_0 formula $\Phi(n, Z)$ which says that Z, a subset of ω, codes the sequences $\langle \mathcal{M}_m \mid 0 \leqslant m \leqslant n \rangle$ and $\langle \mathcal{Q}_m \mid 0 \leqslant m \leqslant n \rangle$; once we have fixed our coding, there will be a unique Z, call it Z_n, that does that.

$\mathcal{M}_0$ and $\mathcal{Q}_0$ will be in J_2, since $J_1 \in J_2$ and J_1 is an admissible set, and hence terms for the members of J_1, and the corresponding evaluation function, can be set up very easily in a way that is definable over J_1. Thus $\mathcal{M}_0$ and $\mathcal{Q}_0$ can be obtained by applying Δ_0 separation (with J_1 as a parameter) inside J_2.

Then repeated application of the Proposition, together with the fact that J_2 is rud closed, will show that each $\mathcal{M}_n$ and $\mathcal{Q}_n$ is in J_2; and by the uniformity of the progression, J_2 will model the statement that $\forall n \exists ! Z \Phi(n, Z)$.

Suppose that fReR were true in J_2. Then there would be a set containing all the Z_n's, and therefore a set containing all the $\mathcal{M}_n$'s. But uniformly from $\mathcal{M}_n$ we can form the set X_n defined by

$$X_n =_{\text{df}} \{k \in \omega \mid \neg \mathcal{M}_n(\ulcorner k \urcorner, w^n_k)\},$$

where $\ulcorner k \urcorner$ is our canonical symbol for k (so that $\mathcal{E}_n(\ulcorner k \urcorner) = k$ for every n) and $(w^n_k)_k$ is our recursive enumeration of $\mathcal{W}_n$. Hence there will be some ℓ such that T_ℓ contains all the X_n's. We now get a contradiction, for X_ℓ itself cannot be a member of T_ℓ. If it were, it would for some k be the evaluation $\mathcal{E}_\ell(w^\ell_k)$ of some word w^ℓ_k. But then for that k,

$$k \in X_\ell \Longleftrightarrow \mathcal{M}_\ell(\ulcorner k \urcorner, w^\ell_k) \Longleftrightarrow k \notin X_\ell. \qquad (9.7) \square$$

9.14. Proposition. *There is a model of* fReCs *in which* ReR *is false.*

Proof. $V_{\omega+\omega}$; alternatively, $V_{\omega+\omega} \cap \mathbf{HC}$. (9.14) □

Model 15: of Z without restricted rank-bounded replacement

We apply the pivotal idea of Zarach [Z] to the model-building of [M1, section 4]. We have above recalled the definition of $Z(a)$; we shall use these further definitions from [M1]:

9.15. Definition. $b_0(n) = n; b_{k+1}(n) = 2^{b_k(n)}$; $\mathcal{F}$ is the family of functions from ω to ω that are dominated by some b_k; for u transitive, $f^a_u(n) = \overline{\overline{u \cap Z_n(a)}}$; $\mathcal{T}^a = \{u \mid \bigcup u \subseteq u \;\&\; f^a_u \in \mathcal{F}\}$. $T(a) = \mathrm{tcl}(a) \cup Z(a) \cup \{Z(a)\}$.

9.16. Lemma.

(i) *If* $Z(b)$ *is in* u, *transitive, then* f^b_u *is not in* $\mathcal{F}$, *so* u *is not in* $\mathcal{T}^b$.

(ii) *For* $a \neq b$, $Z(b) \in T(b) \in \mathcal{T}^a$.

Proof. As in the proof of [M1, Theorem **4.8**], but note that (ii) of the present lemma corrects a slip in the last sentence of the first paragraph of that proof. (9.16) □

Now let A be an infinite subset of ${}^\omega 2$. Let I be a proper ideal on A extending the Fréchet ideal of all finite subsets of A. For $s \in I$, let $A^s = \bigcap\{\mathcal{T}^a \mid a \in A \setminus s\}$, and let $M^s = \bigcup A^s$. Finally, set $\mathbf{M}_{15} = \bigcup_s M^s$.

9.17. Now $M^s \cup M^t \subseteq M^{s\cup t}$, since $s_1 \subseteq s_2 \Longrightarrow A^{s_1} \subseteq A^{s_2}$, so AxPair will hold in $\mathbf{M}$. Further, $b \in s \Longrightarrow T(b) \in A^s$, so $Z(b) \in M^s$, and so each $Z(b)$ is in $\mathcal{M} = \bigcup_s M^s$.

Indeed, $\mathbf{M}_{15}$ is a supertransitive model of Z containing all ordinals, in which full flat collection holds, and TCo; and in which every set has a rank.

But $\{Z(b) \mid b \in A\}$ is not in $\mathbf{M}_{15}$; if it were a member of u, transitive and in A^s, take $a \in A \setminus s$; then f^a_u is not in $\mathcal{F}$ so $u \notin \mathcal{T}^a$ and therefore not in A^s. Hence by Proposition **8.32**, $\mathbf{M}_{15}$ is not a model of ReRl; and indeed the failure is one of rank-bounded replacement in that all the $Z(a)$ are of rank $\omega + \omega$. (9.17) □

9.18. As ReRl proves $\mathcal{S}(x) \in V$ and $\mathbf{HF} \in V$, Zarach's model suffices to show that that theory does not prove restricted collection.

10. Mending the flaws in Devlin's book

We turn now to a discussion of the flaws in Devlin's book *Constructibility* to which attention was drawn in Stanley's review mentioned in Section 8.

We begin with some notes on Devlin's notation, which is not always identical with ours; in this section unexplained notation will be as defined in [Dev]. We then mention a general problem, not, alas, confined to Devlin's book; then we work through Section 9 of Chapter I, where the system BS is introduced as the intended vehicle for the stream of thought in that section: we point out places where BS is inadequate, and places where, with some correction, it suffices; as we go, we suggest various revisions of Devlin's definitions; we mention passages in Chapters II and VI that are affected by those errors in Chapter I; then we introduce a system, which we call MW, that forms a mild strengthening of DBI and furnishes a framework within which the desired Σ_1 definition of the satisfaction relation $\models_u \varphi$ can be given; finally we suggest that the systems DS and GJI, each of them a strengthening of MW, offer possibly smoother treatments than that available in MW itself.

Some notes on Devlin's notation

On page 9: an n-tuple is introduced as a Wiener-Kuratowski one. In a familiar tradition, a function is treated as a subset of its image × its domain. On page 11: a *sequence* is defined as a function whose domain is an ordinal; so a finite sequence is one whose domain is a finite ordinal; a natural number is a finite ordinal.

Thus an n-sequence is an object of cardinality n consisting of ordered pairs of which the second elements form a finite initial segment of the ordinals. The 4-sequence $\langle 5,1,4,2\rangle$ is written thus to distinguish it from the (WK) 4-tuple $(5,1,4,2)_4$.

We maintain our policy of writing ${}^3X$ for the set of 3-sequences of members of X; X^3 for the set of WK 3-tuples of members of X; thus $\omega^3 = \omega \times (\omega \times \omega)$.

10.0. Remark. Devlin makes no distinction between $(X \times X) \times X$ and $X \times (X \times X)$, writing both as X^3. With weak systems that is scarcely satisfactory, since the variant given of Model 4, using weak right WK-rank, is a model of ReS_0 which contains $(\omega \times \omega) \times \omega$ but not $\omega \times (\omega \times \omega)$; and, following the lead of Model 9, we can get models of BS containing either, but not both, of ${}^3(\omega \times (\omega \times \omega))$ and ${}^3((\omega \times \omega) \times \omega)$.

As for abbreviations of lists of variables, Devlin follows the useful convention that $\vec{x} \in A$ abbreviates $x_1 \in A$ & ... & $x_n \in A$, whereas $(\vec{x}) \in A$ indicates that the corresponding WK n-tuple is in A.

We shall make a slight change to his notation: we shall use the letters φ, ϑ and χ for formal formulæ, ψ and θ for building sequences, or similar sequences of formulæ, and α, β and γ for (finite) attempts at addition. The reader will be able to distinguish a reference to his Lemma 9.4 from one to our Lemma **9.4** by the boldface font.

The problem of levels of language

There is an ambiguity over the meaning of Δ_0 (which Devlin calls Σ_0). Devlin on page 230 writes:

> "In class terms a function is Σ_0 if of the form $\{(y, \vec{x}) \mid \Phi(y, \vec{x})\}$ where Φ is a Σ_0 formula of LST. In set-theoretic terms a function f is said to be Σ_0 if there is a Σ_0 formula φ of $\mathcal{L}$ such that for any $\vec{x}$, y, if M is a transitive set such that $\vec{x}$, $y \in M$, then
>
> $$f(\vec{x}) = y \Longleftrightarrow \models_M \varphi(\mathring{y}, \mathring{\vec{x}}).\text{"}$$

10.1. Remark. The second definition has the advantage that one can then legitimately quantify over all φ; but the disadvantage that the definition collapses if TCo is false; whereas the first definition is still operational. Thus Devlin's remark that the two definitions are "equivalent" is dangerous.

Errors in Chapter I

Definition of Finseq

10.2. Remark. The definition, on page 33 of [Dev], of Finseq might not be as intended; what is written is that members of Finseq are functions with domain a non-empty bounded subset of ω (possibly not a proper initial segment of ω).

We shall suppose that the definition has been corrected to mean that members of Finseq are functions with domain a non-empty bounded initial segment of ω; that is still Δ_0, so no harm has been done.

Lemmata 9.1 and 9.2 are correct.

The trouble starts on page 34, with the formula $F_\wedge(\theta, \phi, \psi)$: in its definition the clause "$\mathrm{Dom}\,(\theta) = \mathrm{Dom}\,(\phi) + \mathrm{Dom}\,(\psi) + 3$" occurs, and thus addition of natural numbers is being used to define concatenation.

Lemma 9.3 "$F_\wedge$ is Δ_0"

Though the other parts of Lemma 9.3 are correct as stated, that statement is false – Solovay has remarked that that can be seen by Ehrenfeucht-Fraissé games.

Its falsehood may indeed be established by arguments from Gandy's paper, where he proves (by a quantifier elimination argument, which is what, presumably, Solovay had in mind) that every Δ_0 subset of ω is finite or cofinite; from that he shows that the graph of addition is not Δ_0, and further deduces that the graph of concatenation is not Δ_0.

Suppose we consider a language which accepts as atomic formulæ all finite constant sequences of $*$'s. Note that each such sequence is expressible as $\{*\} \times n$ for some n.

Let $\tau_{n,k}$ be the term

$$(\{*\} \times (k+3) \setminus \{(*,0)_2, (*, n+1)_2, (*, k+2)_2\}) \cup \{(\dot{(}, 0)_2, (\wedge, n+1)_2, (\dot{)}, k+2)_2\},$$

where $\dot{(}$, $\wedge$ and $\dot{)}$ code the left parenthesis, conjunctive connective and right parenthesis of the formal language. Then $k = n + m \iff F_\wedge(\tau_{n,k}, \{*\} \times n, \{*\} \times m)$, and thus $F_\wedge$ cannot be Δ_0 as the graph of addition is not.

Complexity of $F_\wedge$

10.3. However, the Lemma is nearly correct in that one might say that $F_\wedge$ is Δ_0 in any sufficiently long attempt at integer addition. We therefore propose to revise the definition of $F_\wedge$, by making explicit the attempt at integer addition that is being used, as follows:

$$\begin{aligned}
\mathrm{At}_+(\vartheta;\alpha) \iff_{\mathrm{df}} & \big[\mathrm{Fn}(\alpha) \mathbin{\&} \mathrm{Dom}\,(\alpha) \supseteq \mathrm{Dom}\,(\vartheta) \times \mathrm{Dom}\,(\vartheta) \\
& \quad \mathbin{\&} \alpha \text{ is an attempt at integer addition } \big]; \\
F^0_\wedge(\vartheta,\varphi,\chi;\alpha) \iff_{\mathrm{df}} & \big[[\mathrm{Dom}\,(\varphi) < \mathrm{Dom}\,(\vartheta)] \mathbin{\&} [\mathrm{Dom}\,(\chi) < \mathrm{Dom}\,(\vartheta)] \\
& \quad \mathbin{\&} [\mathrm{Dom}\,(\vartheta) = \alpha(\mathrm{Dom}\,(\varphi)+1, \mathrm{Dom}\,(\chi)+1)+1] \\
& \quad \mathbin{\&} [\vartheta(0) = 0] \mathbin{\&} [\vartheta(1) = 6] \mathbin{\&} [\vartheta(\|\vartheta\|) = 1] \\
& \quad \mathbin{\&} \forall i_{\in \mathrm{Dom}\,(\varphi)} [\vartheta((i+1)+1) = \varphi(i)] \\
& \quad \mathbin{\&} \forall i_{\in \mathrm{Dom}\,(\chi)} [\vartheta(\alpha(\mathrm{Dom}\,(\varphi)+1, i+1)) = \chi(i)]\big]; \\
F_\wedge(\vartheta,\varphi,\chi) \iff_{\mathrm{df}} & \mathrm{Finseq}(\vartheta) \mathbin{\&} \mathrm{Finseq}(\varphi) \mathbin{\&} \mathrm{Finseq}(\chi) \mathbin{\&} \exists\alpha\big[\mathrm{At}_+(\vartheta;\alpha) \\
& \quad \mathbin{\&} F^0_\wedge(\vartheta,\varphi,\chi;\alpha)\big].
\end{aligned}$$

Proposition. At_+ and $F^0_\wedge$ are Δ_0^{ReS}; $F_\wedge$ is Δ_1^{ReS}.

Proof. At_+ and $F^0_\wedge$ are composed entirely of S_0-suitable terms; therefore $F_\wedge$ is Σ_1^{ReS}; with Propositions **2.14** and **2.57** in mind, and because there is no disagreement between two attempts at addition where both are defined, we see that $F_\wedge$ is equivalent in ReS to the formula

$$\mathrm{Finseq}(\vartheta) \mathbin{\&} \mathrm{Finseq}(\varphi) \mathbin{\&} \mathrm{Finseq}(\chi) \mathbin{\&} \forall\alpha\big[\mathrm{At}_+(\vartheta;\alpha) \implies F^0_\wedge(\vartheta,\varphi,\chi;\alpha)\big]$$

which is Π_1^{ReS}. (10.3) □

The definition of Build

The trouble caused by $F_\wedge$ continues with the next Lemma:

Lemma 9.4 "Build(φ, ψ) is Δ_0"

The proof is certainly invalid since it uses 9.3. The statement is suspect: suppose we add to the definition of Build extra clauses admitting the "formulæ" $\{*\} \times n$, as atomic: that would not change the Δ_0 character of Build, as those clauses would be Δ_0, even (by Gandy's proof that ω is S_0-semi-suitable) when quantified over $n \in \omega$. Then for $\tau_{n,k}$ the term defined above,

$$k = n + m \iff \mathrm{Build}(\tau_{n,k}, \langle \{*\} \times n, \{*\} \times m, \tau_{n,k} \rangle),$$

and therefore Build (in the form modified to allow atomic formulæ of the form $\{*\} \times n$) cannot be Δ_0 as the graph of addition is not.

Complexity of Build

As one might again say that Build is Δ_0 in any sufficiently long attempt at addition, we shall make a similar revision of its definition by introducing a name, β, for the attempt at addition on which the formula implicitly relies; but first there is a further danger to be noted. Suppose that Build(φ, ψ). Now let θ result from ψ by adding various formulæ to the sequence, keeping φ always the last and observing the other rules of Build ; for example one might add many atomic formulæ and build up long conjunctions of atomic formulæ or one might interpolate the terms of some ψ' that builds some other formula, subject only to the condition on variables, which is that the only variables with bound occurrences are those with such occurrences in φ. Then θ also builds φ according to Devlin's definition of Build, but might easily list formulæ that contain free variables not occurring in φ or that are actually longer than φ and therefore beyond the domain of competence of the attempt at addition being used. Ideally one would like to require every formula listed to be actually a subformula of the formula being built, but we have not yet defined the notion of formula, let alone subformula. We shall therefore, in our reformulation of the definition of Build, impose the milder requirement that no finite sequence listed by ψ is strictly longer than φ.

10.4. Here is our revised definition:

$$\begin{aligned}
\mathrm{Build}^0(\varphi, \psi) &\iff_{\mathrm{df}} \mathrm{Finseq}(\varphi) \;\&\; \mathrm{Finseq}(\psi) \;\&\; [\psi_{\|\psi\|} = \varphi] \\
&\qquad \& \; \forall i_{\in \mathrm{Dom}\,(\psi)} \, [\mathrm{Finseq}(\psi_i) \;\&\; \mathrm{Dom}\,(\psi_i) \leqslant \mathrm{Dom}\,(\varphi)] \\
\mathrm{Build}^1(\varphi, \psi; \beta) &\iff_{\mathrm{df}} \forall i_{\in \mathrm{Dom}\,(\psi)} \big[\mathrm{PFml}(\psi_i) \lor \exists j, k_{\in i} \, F^0_\wedge(\psi_i, \psi_j, \psi_k; \beta) \\
&\qquad \lor \exists j_{\in i} \, F_\neg(\psi_i, \psi_j) \\
&\qquad \lor \exists j_{\in i} \, \exists u_{\in \mathrm{ran}(\varphi)} \, (\mathrm{Vbl}(u) \;\&\; F_\exists(\psi_i, u, \psi_j))\big]; \\
\mathrm{Build}(\varphi, \psi) &\iff_{\mathrm{df}} \mathrm{Build}^0(\varphi, \psi) \;\&\; \exists \beta \, [\mathrm{At}_+(\varphi; \beta) \;\&\; \mathrm{Build}^1(\varphi, \psi; \beta)]
\end{aligned}$$

Proposition. $\mathrm{Build}^0(\varphi, \psi)$ and $\mathrm{Build}^1(\varphi, \psi; \beta)$ are Δ_0^{ReS}; $\mathrm{Build}(\varphi, \psi)$ is Δ_1^{ReS}.

Proof. The first part by inspection; for the second, note that Proposition **2.57** implies that

$$\vdash_{\mathsf{ReS}} \mathrm{Build}(\varphi, \psi) \iff \left[\mathrm{Build}^0(\varphi, \psi) \;\&\; \forall\beta\,[\mathrm{At}_+(\varphi;\beta) \implies \mathrm{Build}^1(\varphi,\psi;\beta)]\right]$$

(10.4) □

10.5. Problem. Does the absence of uniqueness matter? One might try for a minimality condition of the form "ψ builds φ and no proper subsequence of ψ does". But that hardly seems worth the effort, as the redundancy in such formulæ as $\varphi \wedge (\vartheta \wedge \varphi)$ is liable to reappear in the corresponding building functions.

The formula Seq

At the bottom of page 36 of [Dev] a formula $\mathrm{Seq}(u, a, n)$ is defined which expresses the statement that u is the set of all finite sequences, of length less than n, of elements of a, and is correctly stated to be Σ_1. But this formula gives trouble in the proof of the next Lemma.

Lemma 9.5 "Seq is Δ_1^{BS}"

According to Solovay, the statement is false, "as may be seen using a forcing argument". I have been unable to demonstrate the falsity of the assertion using my present methods, but the model-building of Section 5 will pin-point flaws in the argument as printed.

In Model 6, there is no u such that $\mathrm{Seq}(u, \omega, 4)$; so in that model the proposed Π_1 form of the definition is true of everything, and the proposed Σ_1 form is false of everything. So the equivalence is not a theorem of BS, and the proposed proof of I.9.5 cannot succeed.

In greater detail:

10.6. The first displayed formula in the proof of 9.5 asserts that

"it is clear from the definition of BS that:

$$\mathrm{BS} \vdash (\forall a)(\forall n \in \omega)(\exists u)\mathrm{Seq}(u, a, n)."$$

But that statement, on lines 5 and 6 of page 37, is not a theorem of BS, as is shown by Model 9, in which there is no u with $\mathrm{Seq}(u, \{\omega\} \times \omega, 4)$, or, indeed, by Model 6, in which for no infinite a is there a u with $\mathrm{Seq}(u, a, 4)$.

10.7. Devlin wishes to bound the quantifier f by the set of n-sequences of finite sequences from a.

First problem: is it a set? No, even if a has only two members: if A is the class of n-sequences of finite sequences of members of a, the class B of finite sequences of members of a is a subclass of $\bigcup\bigcup\bigcup A$; and Model 5 is a supertransitive model of BS not containing the set BIN of finite binary sequences, the reason being that $\overline{\overline{BIN \cap V_n}} = 2^{n-3}$ for all $n \geqslant 3$; and hence in Model 5, the class A is not a set.

Second problem: would B be a bounding class for the quantifier $\exists f$? No; it is the wrong type. The values of f are not finite sequences but sets of finite sequences.

However, the faulty proof of [Dev] Lemma I.9.5 becomes true if we confine a to being finite. First, a general lemma:

10.8. Proposition. *Let G be a Δ_0 class. Then*

$$\vdash_{\mathsf{ReS}} \mathrm{Fn}(G) \,\&\, \mathrm{Dom}\,(G) = V \Longrightarrow \forall a\big(a \text{ finite } \Longrightarrow G\text{“}a \in V\big).$$

Proof. Let $f : n \longleftrightarrow a$. Consider the class $n \cap \{k \mid G\text{“}\{f(i) \mid_i i < k\} \notin V\}$. That is Π_1, and so if not empty, a minimal element exists, which, trivially, is > 0, and hence equals $k+1$ for some k. Thus $G\text{“}\{f(i) \mid_i i < k\} \in V$; to that we must add $\{G(f(k))\}$. (10.8) □

10.9. Remark. Under the hypotheses of the Proposition, $G\text{“}a$ will be finite.

10.10. Lemma. (ReS) *If a is finite, then for each n there is a u such that Seq(u, a, n). Hence for a finite, Seq$(u, a, n) \Longleftrightarrow \forall u' \neq u\, \neg$Seq$(u', a, n)$.*

Proof. An induction on n. The induction step will require us to form $\{x \cup y \mid_{x,y} x \in A \,\&\, y \in B\}$, where A and B are finite; but that is of the form $g\text{“}(A \times B)$ where g is rudimentary and provably total in ReS, and thus satisfies the hypotheses of Proposition **10.7**. $A \times B$ will be finite by Proposition **2.14**. (10.10) □

Lemma 9.6 “Fml(x) is Δ_1^{BS}”

This result is actually true, indeed it can be sharpened to “Fml(x) is Δ_1^{ReS}”, but the proof given is seriously flawed.

There is a slight error in the definition of $A(x)$; replace the third occurrence of ‘n’ by ‘m’.

At the bottom of page 37, in the proof of Lemma 9.6, the claim, said to be “easily checked”, that

$$\text{“BS} \vdash \forall x \exists y[y = A(x)].\text{”}$$

is untrue, as is shown by Model 9 for appropriate infinite x.

However, this claim is needed only in the case that x is a finite sequence, when the result is indeed provable in the following form:

10.11. Lemma. (ReS) *If x is a finite sequence, then $A(x)$ is a finite set.*

Proof. Let x be a finite set, and k a finite ordinal. Then the set $B(k, x)$ of functions from k to x is a Δ_0 subclass of $\mathcal{P}(x \times k)$, which as we have seen is, provably in ReS, a finite set.

This principle, applied twice, will yield the Lemma. (10.11) □

To complete the proof of 9.6, we appeal twice to our Proposition **10.4**, that Build is Δ_1^{ReS}: first, it implies that Fml(x), being of the form $\exists f \mathrm{Build}(x, f)$, is Σ_1; and secondly, in view of our Metatheorem **2.24**, it implies that, v being the finite set $A(x)$, the subformula $(\exists f_{\in v})\mathrm{Build}(x, f)$ is (taking the Π_1 form of Build), Π_1^{ReS}, and thus that the given alternative form of Fml is indeed Π_1.

Lemma 9.7

The above arguments, appropriately modified, will prove Lemma 9.7, with Δ_1^{BS} sharpened to Δ_1^{ReS}. The restriction in Fml(x, u) of the formal constants to those for members of u is Δ_0 and causes no difficulty.

The definition of Fr

Devlin now writes

"Our next task is to write down an LST formula Fr(φ, x) such that

$$\mathrm{Fr}(\varphi, x) \leftrightarrow \mathrm{Fml}(\varphi) \wedge [x \text{ is the set of variables occurring free in } \varphi]."$$

But the formula that he proposes does not work: given the fact that a ψ with Build(φ, ψ) may contain many formulæ with free variables not among those of φ, the truth of his formula Fr(φ, x) only guarantees that x *contains* all the variables with at least one free occurrence in φ. That invalidates the proof of his Lemma 9.8.

But really one wishes to know whether a particular occurrence is free or not. So it would be better to aim at achieving that. We shall be able to do so by using the relation Sub that Devlin is, without using Fr, about to define; so let us go on to that and postpone the present definition.

The definition of Sub

First, two minor points: in the fifth line from the bottom of page 39 of [Dev], for $F_\in$ one should read $F_\exists$; and in the build-up to Lemma 9.9, the phrase "the scope of this quantifier" is used but not defined.

Lemma 9.9 "Sub is Δ_1^{BS}"

The Lemma is essentially correct, and indeed admits a sharpening of Δ_1^{BS} to Δ_1^{ReS}, but there is a problem with Devlin's suggestion for Sub: as $F_\wedge$ is used, it is not immediately clear that Sub will be Σ_1. We could appeal to Metatheorem **2.24** since the domain of ψ is a finite set, but it will be better to follow the style of our earlier revisions and first formulate a Δ_0 version of Sub with explicit names for the various supporting characters. Here it is, where $S(\cdot,\cdot,\cdot,\cdot)$ is the Δ_0 formula given by Devlin on his page 39.

$$\begin{aligned}
&\mathrm{Sub}^0(\varphi', \varphi, v, t;\ \psi, \theta;\ \beta) \iff_{\mathrm{df}} \\
&\qquad \underbrace{\mathrm{Vbl}(v)\ \&\ \mathrm{Const}(t)}_{\mathfrak{A}(v,t)}\ \& \\
&\qquad \underbrace{\mathrm{At}_+(\varphi;\beta)\ \&\ \mathrm{Build}^0(\varphi,\psi)\ \&\ \mathrm{Build}^1(\varphi,\psi;\beta)}_{\mathfrak{B}(\varphi,\psi;\beta)}\ \& \\
&\qquad \underbrace{\mathrm{At}_+(\varphi';\beta)\ \&\ \mathrm{Build}^0(\varphi',\theta)\ \&\ \mathrm{Dom}\,(\theta) = \mathrm{Dom}\,(\psi)}_{\mathfrak{C}(\varphi';\psi,\theta;\beta)}\ \&\ \underbrace{\theta_{\|\theta\|} = \varphi'}_{\mathfrak{D}(\varphi';\theta)}\ \& \\
&\qquad \underbrace{\begin{aligned}&\forall i_{\in \mathrm{Dom}\,(\psi)} \big[\exists j, k_{\in i}\,(F^0_\wedge(\psi_i,\psi_j,\psi_k;\beta)\ \&\ F^0_\wedge(\theta_i,\theta_j,\theta_k;\beta)) \\
&\qquad\qquad \vee\ \exists j_{\in i}\,(F_\neg(\psi_i,\psi_j)\ \&\ F_\neg(\theta_i,\theta_j)) \\
&\qquad \vee\ \exists j_{\in i}\,\exists u_{\in \mathrm{ran}(\varphi)}\,(\mathrm{Vbl}(u)\ \&\ u \neq v\ \&\ F_\exists(\psi_i,u,\psi_j)\ \&\ F_\exists(\theta_i,u,\theta_j)) \\
&\qquad \vee\ \exists j_{\in i}\,(F_\exists(\psi_i,v,\psi_j)\ \&\ (\theta_i = \psi_i)) \\
&\qquad \vee\ S(\theta_i,\psi_i,v,t)\big]\end{aligned}}_{\mathfrak{E}(\varphi,v,t;\psi,\theta;\beta)}
\end{aligned}$$

Then we define, omitting the listing of free variables given in the underbraces to the above display,

$$\mathrm{Sub}(\varphi,\varphi',v,t) \iff_{\mathrm{df}} \mathfrak{A}\ \&\ \exists\psi\exists\theta\exists\beta\left[\mathfrak{B}\ \&\ \mathfrak{C}\ \&\ \mathfrak{E}\ \&\ (\theta_{\|\theta\|} = \varphi')\right]$$

and prove in ReS that a φ' with $\mathrm{Sub}(\varphi,\varphi',v,t)$ always exists (by a recursion of finite length); whence

$$\vdash_{\mathsf{ReS}} \mathrm{Sub}(\varphi,\varphi',v,t) \iff \mathfrak{A}\ \&\ \forall\psi\forall\beta\forall\theta\Big[[\mathfrak{B}\ \&\ \mathfrak{C}\ \&\ \mathfrak{E}] \implies (\theta_{\|\theta\|} = \varphi')\Big]$$

10.12. Proposition. *$\mathfrak{A}, \mathfrak{B}, \mathfrak{C}, \mathfrak{D}$ and $\mathfrak{E}$ are all Δ_0;* Sub *is* Δ_1^{ReS}.

10.13. Remark. We should (but won't) prove that ϕ' is a formula, by modifying θ to give a building sequence for it, and that the outcome of these tests is independent of the building sequence used.

We may now characterise *bound occurrences* of a given variable in a formula as those for which no change results in the formula when the above procedure is followed for substituting some constant for that variable, and then we may define *sentences* to be those formulæ whose every occurrence of a variable is bound.

With trifling loss of generality we take that constant to be $\mathring{\varnothing}$, the constant denoting the empty set, which will usually be a member of the sets in which we shall wish to interpret formulae, and may now give our definition of Sen^0 and Sen.

10.14. Definition.

i) $\mathrm{Sen}^0(\varphi; v; \psi, \theta, \gamma) \iff_{\mathrm{df}} \mathrm{Sub}^0(\varphi, \varphi; v, \mathring{\varnothing}; \psi, \theta, \gamma)$.

ii) $\mathrm{Sen}(\varphi) \iff_{\mathrm{df}} \mathrm{Fml}(\varphi)\ \&\ \forall v_{\in \mathrm{ran}(\varphi)}\big[\mathrm{Vbl(v)} \implies \exists\psi\exists\theta\exists\gamma\mathrm{Sen}^0(\varphi; v; \psi, \theta, \gamma)\big]$.

iii) Let v be a formal variable. If $\varphi(i) = v$, that occurrence of v at i in φ is *bound* $\Longleftrightarrow_{\rm df}$ whenever $\mathrm{Sub}(\varphi, \varphi', v, \mathring{\varnothing})$, $\varphi'(i) = v$.

10.15. Remark. It is necessary to include $\mathrm{Fml}(\varphi)$ in the definition of $\mathrm{Sen}(\varphi)$, lest φ have no variables at all in its range.

10.16. Remark. In a manner to which we have become accustomed, the above concepts will be Δ_0 in any appropriate parameter, and Δ_1^{ReS} if no parameters are mentioned.

10.17. Definition. $\mathrm{Sen}(\varphi, u) \Longleftrightarrow_{\rm df} \mathrm{Fml}(\varphi, u)$ & all occurrences of its variables are bound.

10.18. Lemma. $\mathrm{Sen}(\varphi, u)$ *is* Δ_1^{ReS}.

The definition of Fr reconsidered

10.19. Definition. $\mathrm{Fr}(\varphi, x) \Longleftrightarrow_{\rm df} x = \mathrm{Vbl} \cap \{\varphi(i) \mid$ that occurrence is bound$\}$

10.20. Remark. Such an x will be a Δ_1^{ReS} subclass of a bounded subset of ω, and therefore can be proved in ReS to be a set, by an argument reminiscent of the proof of Lemma **2.52**.

The above wffs are Δ_0 in any w containing sufficiently many building sequences (and their attendant attempts), so we could give an alternative prove of the existence of an x with $\mathrm{Fr}(\varphi, x)$ by using Δ_0 separation with w as a parameter.

Lemma 9.8: "Fr is Δ_1^{BS}"

The Lemma is true, and can be sharpened to "Fr is Δ_1^{ReS}".

The definition of the precursor $S(u, \varphi)$ to Sat

At the bottom of page 40, Devlin introduces a formula $S(u, \varphi)$ and alleges that it defines the satisfaction relation. There is a minor slip in the last line of page 40: for $F_\in$ read $F_\exists$; but there is a more substantial error in the formula. Devlin's strategy is to build two finite sequences f and g of sets of formulæ; roughly at stage i, $f(i)$ is to comprise all formulæ of $\mathcal{L}_u$ built up within i steps from atomic formulae; and $g(i)$ is to comprise the sentences of $\mathcal{L}_u$ which are both members of $f(i)$ and true in u. But let ϑ be a member of $f(i)$ which has a free occurrence of a variable, and therefore is not a sentence; then $\vartheta \notin g(i)$; let χ be $\neg\vartheta$; then according to his definition χ should be placed in $g(i+1)$; but it is not a sentence. Thus his definition should be amended by adding the requirement that the members of each $g(i)$ are sentences.

We shall also require a bound for the length of formulæ to be considered when evaluating the truth of φ. Atomic formulæ are of length 5; by inspection, the length of formulæ in $f(i+1)$ will be at most three times the length of the longest formula in $f(i)$; if φ is of length ℓ it will be in $f(\ell)$; thus a bound for the length of any other formula in $f(\ell)$ is 5.3^ℓ, and we should therefore establish in ReS that every integer is in the domain of an attempt at the function $n \mapsto 3^n$. Arguments similar to those we have given for attempts at addition will suffice for that, and will show in addition that the property of being such an attempt is Δ_0.

Let us now revise the definition of $S(u, \varphi)$ in the light of these remarks and our previous revisions. The predicate E used in the definition of S^3 is that defined in [Dev] in the lower half of page 40.

$$S^0(u,\varphi) \iff_{\mathrm{df}} u \neq \varnothing \;\&\; \mathrm{Sen}(\varphi, u);$$

$$\begin{aligned} S^1(\varphi; f, g) \iff_{\mathrm{df}} & \ \mathrm{Finseq}(f) \;\&\; \mathrm{Finseq}(g) \;\&\; \mathrm{Dom}\,(f) = \mathrm{Dom}\,(g) \\ & \;\&\; \forall i_{\in \mathrm{Dom}\,(f)}\, \forall x_{\in f(i) \cup g(i)}\, [\mathrm{Finseq}(x) \\ & \;\&\; \mathrm{Dom}\,(x) \leqslant \mathrm{Dom}\,(\varphi)] \end{aligned}$$

$$\begin{aligned} S^2(u;\chi;f;\alpha) \iff_{\mathrm{df}} & \ \mathrm{At}_+(\chi;\alpha) \;\&\; \Big[[\chi \in f(0) \iff \mathrm{PFml}(\chi, u)] \;\&\; \\ & \forall j_{\in \mathrm{Dom}\,(f)}\, \forall i_{\in j}\, [\chi \in f(i+1) \iff \\ & \quad (\chi \in f(i) \vee \exists \vartheta, \vartheta'_{\in f(i)}\, F^0_\wedge(\chi, \vartheta, \vartheta'; \alpha) \\ & \quad\quad \vee \exists \vartheta_{\in f(i)}\, F_\neg(\chi, \vartheta) \\ & \quad\quad \vee \exists \vartheta_{\in f(i)}\, \exists v_{\in \mathrm{ran}(\chi)}\, [\mathrm{Vbl}(v) \;\&\; F_\exists(\chi, v, \vartheta)])]\Big]; \end{aligned}$$

$$\begin{aligned} S^3(u;\chi;f,g;\alpha;\psi,\theta) \iff_{\mathrm{df}} & \ \mathrm{At}_+(\chi;\alpha) \;\&\; \Big[[\chi \in g(0) \iff E(\chi, u)] \;\&\; \\ & \forall j_{\in \mathrm{Dom}\,(f)}\, \forall i_{\in j}\, [\chi \in g(i+1) \iff \\ & \quad (\chi \in g(i) \vee \exists \vartheta, \vartheta'_{\in g(i)}\, F^0_\wedge(\chi, \vartheta, \vartheta'; \alpha) \\ & \quad \vee \exists \vartheta_{\in f(i)}\, \mathrm{Sen}^0(\vartheta; v; \psi, \theta; \alpha) \;\&\; (\vartheta \notin g(i) \;\&\; F_\neg(\chi, \theta)) \\ & \quad \vee \exists \vartheta_{\in f(i)}\, \exists v_{\in \mathrm{ran}(\chi)}\, \exists x_{\in u}\, \exists \vartheta'_{\in g(i)} \\ & \quad\quad [\mathrm{Vbl}(v) \;\&\; F_\exists(\chi, v, \vartheta) \\ & \quad\quad \;\&\; \mathrm{Sub}^0(\vartheta', \vartheta; v, \mathring{x}; \psi, \theta; \alpha)])]\Big] \end{aligned}$$

$$S^4(\varphi; g) \iff_{\mathrm{df}} \varphi \in g(\|g\|)$$

10.21. Proposition. *Each $S^{\mathfrak{k}}$ is Δ_0.*

We are getting warm: we may now show that

$$\begin{aligned} \models_u \varphi \iff & S^0(u,\varphi) \;\&\; \\ & \exists f, \exists g \Big[S^1(\varphi; f, g) \\ & \quad \;\&\; [\text{for all appropriate } \chi \text{ and} \\ & \quad\quad \text{for all sufficiently long } \alpha\ S^2(u;\chi;f;\alpha)] \\ & \quad \;\&\; [\text{for all appropriate } \chi \text{ and} \\ & \quad\quad \text{for all sufficiently long } \alpha, \psi, \theta,\ S^3(u;\chi;f,g;\alpha;\psi,\theta)] \\ & \quad \;\&\; S^4(\varphi; g) \Big] \end{aligned}$$

Here "appropriate" is to mean the Δ_0 requirement that χ is a finite sequence all of whose terms are either symbols of our formal language or constants for members of u, and whose length is at most $p =_{\mathrm{df}} 5.3^{\mathrm{Dom}(\varphi)}$; and "sufficiently long" is, in the case of α, an attempt at integer addition, to mean that its domain includes $p \times p$.

We might remark here that a further restraint on the possible values of χ is possible whilst preserving the above equivalence, namely by requiring the formal variables occurring in χ to be among those occurring in φ.

The definition of Sat

At this point, Devlin's strategy (in our revised context) is to convert the above universal quantifications, which we have qualified with phrases such as "appropriate" and "sufficiently long", to restricted ones by finding a set w which will contain sufficiently many possible values of the variables χ, α, ψ, θ to preserve the intended meaning of $S(u,\varphi)$ and, as his candidate for w, defines, on his page 41, a class $w(u,\varphi)$. But there is a final problem: as is shown by Model 9, the class $w(u,\phi)$ is not provable in BS to be a set. Even if we adopt the further restraint on variables mentioned above, and give a correspondingly restrained definition of a class we might call $w^*(u,\varphi)$, its set-hood, for arbitrary u, would not be provable in BS.

Lemma 9.10 "the LST formula Sat(u,ϕ) is Δ_1^{BS}"

The statement is false, so this time there is no hope of saving the proof. In Model 6, for no infinite set x does there exist a y with $\mathrm{Seq}(y,x,4)$; for u infinite, the set a of names of members of u will be infinite, and so the given Σ_1 formula for $\mathrm{Sat}(u,\varphi)$ will always be false; but then so is the Σ_1 version of $\mathrm{Sat}(u,\neg\varphi)$; but one of them ought to be true !

10.22. Remark. We have just used the axiom of infinity to build our counterexample, and necessarily so, for we could indeed, without invoking the axiom of infinity, give a Σ_1^{ReS} definition of $\models_u \varphi$ for *finite* u by adopting the above restraint, so that the set-hood (and finiteness) of the correspondingly restrained class, $w^*(u,\varphi)$, would be provable in ReS. Thus Lemma 9.10 holds in sharpened form for u finite.

But as we wish to use and to define truth in infinite sets, we must seek a set theory, including the axiom of infinity, sufficiently strong to prove that Devlin's classes $w(u,\varphi)$ are indeed sets, even when u is infinite; for if they are, the rest of his argument is correct and we shall finally have reached a Δ_1 definition of Sat.

Before discussing possible candidates for such a theory, we comment briefly on some other passages in Chapters I, II and VI of Devlin's book.

Lemma 9.12

The amended proofs of Lemmata 9.6 and 9.7 will now yield Lemma 9.12, with Δ_1^{BS} sharpened to Δ_1^{ReS}. **Lemma 9.14** is in error named Lemma 9.4.

Errors in Chapter II

Amenability

On page 45, in section 10, a set M is defined to be *amenable* if it is transitive and satisfies five conditions: closed under pairing, sumsets, and Cartesian products; contains ω; and closed under $\dot{\Delta}_0(M)$ separators, though Devlin writes "Σ_0."

Given the ambiguity in the meaning of Δ_0 discussed in Remark **10.1**, I would suggest defining an amenable set as a transitive set containing ω and closed under the functions in the finite set $R_0, \ldots, R_7$, listed in paragraph **2.61**, of generators of the class $\mathcal{B}$.

On page 65, in section 2 of Chapter 2, Devlin writes

> "by repeating the proof of I.9.10 for $\mathcal{L}$ in place of LST, we obtain a proof of the fact that the class Sat $(= \{(u, \varphi) \mid \text{Sat}(u, \varphi)\})$ is uniformly Δ_1^M for amenable sets M. That is, there is a Σ_1 formula $\psi(x, y)$ of $\mathcal{L}$ and a Π_1 formula $\theta(x, y)$ of $\mathcal{L}$ such that for any amenable set M, if $u, \varphi \in M$ then
>
> $$\text{Sat}(u, \varphi) \Longleftrightarrow \models_M \psi(\mathring{u}, \mathring{\varphi}) \Longleftrightarrow \models_M \theta(\mathring{u}, \mathring{\varphi}).$$
>
> (The formulas ψ and θ are just the $\mathcal{L}$ analogues of the LST formulas described in I.9.10.)"

With Model $\mathbf{M}_{6,5}$ in mind, we give a counterexample to the alleged uniformity for the specific formulation of Sat given by Devlin.

Let u be an infinite transitive set containing only finitely many sets of cardinality 5. Let M be the rud closure of $u \cup \{u\}$. Let N be the union of the class of all transitive members of M which have only finitely many sets of cardinality 5. So $u \in N$ and N is amenable. Suppose we wish to evaluate the truth in u of the sentence $\bigwedge x \bigvee y\ x \epsilon y$: readers will recognise that that is true in many u and also false in many others. M can correctly make that evaluation; so the Π_1 form holds in M; therefore in N; therefore, if Devlin's assertion were correct, the Σ_1 form would hold in N. But it is false in N, because all atomic formulae such as $(x \epsilon y)$ are sequences of length 5, and therefore, u being infinite, the set of atomic sentences of $\mathcal{L}_u$ is infinite and therefore not a member of N; and therefore not available to be the $f(0)$ of Devlin's formulation.

10.23. Remark. This argument suggests that no other pair of Π_1 and Σ_1 formulæ will work for amenable sets such as N, as information concerning the infinitely many atomic formulæ must be coded in some way into any truth-evaluation, which cannot therefore lie in N if the said information can be recovered by some rudimentary function.

If one calls a set M *$\mathcal{S}$-amenable* if it is amenable and for each $x \in M$ $\mathcal{S}(x) \in M$, then Sat will indeed be uniformly Δ_1^M for $\mathcal{S}$-amenable sets M. By the remark following Proposition **10.26**, the same will be true for amenable sets M that are *weakly $\mathcal{S}$-amenable* in the sense that for each x in M and each k in ω, $[x]^k$ is in M.

Errors in Chapter VI

Lemma VI.1.13 "Sat^A is Δ_1^{BS}"

The statement is false, being a generalisation of the false Lemma I.9.10.

Lemma VI.1.14 "truth for Δ_0 wffs is uniformly Σ_1 for transitive rud-closed structures $\langle M, A\rangle$."

This ought to be correct, and it is of the greatest importance. We make some minor comments, but *Rudimentary Recursion* a full discussion of the proof.

On page 242, in the proof of Lemma VI.1.14, the displayed formula in the middle of the page is incomplete as 't' does not occur on the right-hand side. I suggest that the clause $f(\mathrm{Dom}\,(f) - 1) = t$ should be added.

There is a delicate visual confusion of the meaning of brackets in the following subformula of that same displayed formula:

$$\big(f(i) = \mathring{F}_0\dot{(}f(j), f(k)\dot{)} \Longrightarrow g(i) = F_0(g(j), g(k))\big)$$

where the two parentheses that I have dotted are part of the syntax of the object language, not the language of discourse; but in Devlin's text no visual difference is made between them. Normally of course such confusion would cause no trouble, but in this particular context, greater exactitude might be desirable.

Lower on page 242, in line -7, there is a typo: t^φ should be t_φ.

Finally on page 243, some correction will be needed as the troublemaker $F_\wedge$ recurs here and appeal is made to the false Lemma I.9.3.

The definition of $G_\exists$ oscillates between two and three variables.

On page 243, line -5, reference to 1.7 should perhaps be to 1.8.

Taking stock

Much of the problem with Chapter I Section 9 has now been repaired, but the proposed definition of Sat is not possible in BS, and no other seems likely to succeed.

In the Introduction we spoke of three systems that might work in place of BS. One is our suggestion DS; the second is GJI; and we now introduce the third system, which we call MW, for "Middle Way":

MW $\quad$ DBI + $\forall a \forall k_{\in\omega}\, [a]^k \in V$

Of those, GJI is the longest established candidate, being, apart from the restriction to Π_1 foundation, the system RUD discussed in Stanley's review; DS emerged as the present author's first response to Stanley's call for a replacement for BS that does not use the theory of rudimentary functions; and then at a late stage in the writing of the present paper, the system MW, which is a proper subsystem both of GJI and of DS, revealed itself, and might now be thought to be the "right" answer to Stanley's call.

The system MW proves Theorem **2.93**, and is a proper extension of DBI. Model 5 provides an example of a structure where MW is true but both DS and GJI fail; in Model 7, DS is true but not GJI, and in Gandy's model $\mathbf{G}_2$, GJI is true but not DS.

We shall discuss the definition of Sat first in the system MW and then in GJI and in DS. The reader might wonder what is to be gained by considering the problem of defining Sat within the latter two systems, once one knows that a definition of Sat in MW is possible and that MW is a subsystem of both.

Our answer would be that defining Sat in MW is laborious, whereas it seems possible that each of the other two systems can supply a more elegant treatment, the one drawing on the theory of rudimentary functions, and the other on the enhanced logic of limited quantitifers discussed in Section 8.

The cure in MW

A first step is to collect the $[a]^k$ for given a and a bounded set of k's. The proof seems not to be trivial: Model 6 provides examples where the existence of $[a]^{\mathfrak{k}}$ for one value of $\mathfrak{k}$ does not imply its existence for another.

10.24. Definition. $P(g,k,a) \iff_{\mathrm{df}} g$ is a function with domain $k+1$ and $g(0)=\varnothing$ and for all $i<k$, $g(i+1)=\{x\cup\{p\} \mid x\in g(i) \,\&\, p\in a\smallsetminus x\}$. P is Δ_0.

10.25. Lemma. (MW)
(i) *If* $P(g,k,a)$, *then for each* $i\leqslant k$, $g(i)=[a]^i$.
(ii) $\forall a \forall k_{\in\omega} \exists g\, P(g,k,a))$.

Proof of (ii). Fix a; use Π_1 foundation to find the least k such that there is no g with $P(g,k,a)$; show that k is not 0; $[a]^k$ exists, so if $\ell+1=k$ and $P(h,\ell,a)$ we can create g with $g\restriction k=h$ and $g(k)=[a]^k$, after all. So no failure k exists. (10.25) □

10.26. Proposition. (MW) $\forall a \forall n_{\in\omega} \exists t(t=[a]^{\leqslant n})$.

Proof. Lemma **10.24** shows that $x=[a]^i$ is a Σ_1^{MW} predicate. Since $\forall i_{\in n}\exists x\, x=[a]^i$, Metatheorem **2.24** coupled with Remark **2.25** implies that there is a w such that $\forall i_{\in n}\exists x_{\in w}\, x=[a]^i$. Then the desired t is a Δ_0 subclass of $\bigcup w$ and therefore a set. (10.26) □

That argument readily extends to give the set-hood of the classes $w(u,\varphi)$. We may now implement Devlin's definition of Sat and show that it is Δ_1^{MW}; by working with the restrained versions $w^*(u,\varphi)$, we could avoid appeal to the axiom of infinity in defining Sat; though of course if we want our languages to be sets we must use it.

The cure in GJ

10.27. Lemma. (GJ) $\forall n_{\in\omega} \forall a \exists u\, \mathrm{Seq}(u,a,n)$.

Proof. Fix a; least failed n is given by Π_1 foundation. then piece things together using appropriate rudimentary functions. (10.27) □

10.28. Proposition. *The LST formula Seq(u,a,n) is* Δ_1^{GJ}.

10.29. Lemma. (GJ) $w^*(u,\varphi)\in V$; *if* $\omega\in V$, *then* $w(u,\varphi)\in V$.

Proof. Use the result and reasoning behind Theorem **2.93**. (10.29) □

10.30. Remark. The natural proof of Devlin I.9.6 would use Π_2 foundation to reduce the problem to showing that $\{\bigcup x \mid x\in a\}$ is a set, which is possible in GJ, but, by Model $\mathbf{M}_7$, not in DB.

With the existence of $w(u,\varphi)$ and $w^*(u,\varphi)$ now established, we could follow the structure of Devlin's argument; but the present author's inclination would now be to adopt a slightly different approach to the definition of Sat. Fix u and φ. For any sentence ϑ of $\mathcal{L}_u$, let $B(\vartheta)$ be the set of "simpler sentences" to which the computation of $\models_u \vartheta$ is naturally referred; thus if ϑ is atomic, $B(\vartheta)$ will be empty; if ϑ is $\vartheta_1\wedge\vartheta_2$ then $B(\vartheta)=\{\vartheta_1,\vartheta_2\}$; if ϑ is $\neg\vartheta_1$, then $B(\vartheta)=\{\vartheta_1\}$; and if ϑ is $\forall x\vartheta_1(x)$, then $B(\vartheta)$ will be the set of all substitution instances $\vartheta_1(\mathring{a})$ for $a\in u$.

The first step would then be to define the function that unfolds the formula φ as a tree $\mathcal{T}_\phi$ with φ as its top point; immediately below φ one would place all the members of $B(\varphi)$; immediately below each such formula ϑ one would place all the members of $B(\vartheta)$, and so on, so that the bottom points of the tree are atomic sentences of $\mathcal{L}_u$. GJI is strong enough to do that, for the length of φ gives an upper bound to the (finite) number of steps required.

Then by recursion on the tree $\mathcal{T}_\varphi$ one can compute the truth of $\models_u \vartheta$ for each node ϑ of the tree, culminating with the computation of the truth of $\models_u \varphi$. Thus we would arrive at a proof of

10.31. Proposition. *The LST formula Sat(u,ϕ) is* Δ_1^{GJ}.

We should remark that Gandy in developing (his variant of) the system GJI was specifically aiming at an elegant framework for treating the syntax of formalised languages.

The cure in DS

We recall that DS is the theory $\mathsf{S}_0+\Delta_0$ separation $+\ \Pi_1$ foundation $+\ \omega\in V\ +$ $\mathcal{S}(x)\in V$. The following remarks are intended to suggest that in DS, given a greater knowledge of the behaviour of limited quantifiers with respect to rudimentary substitution, we might arrive at a third proof.

$F_\wedge$ is Δ_0 in the parameter $\mathcal{S}(\omega\times\omega)$, by the result, given as Proposition **8.11**, that in DS the graph of each partial recursive function is a set. Further,

corrected Lemma I.9.3:

10.32. Lemma. $F_\wedge$ *is* $\Delta_{0,\mathcal{S}}^{\mathsf{DS}}$

corrected Lemma I.9.4:

10.33. Lemma. *Build is* $\Delta_{0,\mathcal{S}}^{\mathsf{DS}}$.

10.34. Lemma. (DS)
(i) $\forall a \; {}^{<\omega}a \in V$.
(ii) $\forall n_{\in\omega} \forall a \exists u \mathrm{Seq}(u, a, n)$.

Proof. By two applications of Δ_0 separation, as

$$ {}^{<\omega}a = \mathcal{S}(a \times \omega) \cap \{x \mid \underbrace{\mathrm{Fn}(x) \;\&\; \mathrm{Dom}\,(x) \in \omega}_{\Delta_0}\} $$

and the desired u with $\mathrm{Seq}(a, u, n)$ is

$$ {}^{<\omega}a \cap \{x \mid \underbrace{\mathrm{Fn}(x) \;\&\; \mathrm{Dom}\,(x) \in n}_{\Delta_0}\}. \qquad (10.34)\ \Box $$

corrected Lemma I.9.5:

10.35. Proposition. *The LST formula Seq*(u, a, n) *is* Δ_1^{DS}.

Corrected I.9.10:

10.36. Lemma. (DS) $w(u, \varphi) \in V$.

Proof. By arguments similar to those of Lemma **10.33**. (10.36) □

10.37. Proposition. *The LST formula Sat*(u, ϕ) *is* Δ_1^{DS}

Proof. Apply Lemma **10.35** and Proposition **10.34**. (10.37) □

Conclusion

As each of the three systems holds in all J_ν and L_λ with λ a limit ordinal $> \omega$ each might be claimed to be a good replacement for BS.

Each of the three is open to criticism: whilst DS is perhaps closest to Devlin's original conception, and the enhancement of its logic studied in Section 8 gives it a certain smoothness, it might be felt that the axiom $\mathcal{S}(x) \in V$ is too strong for its intended use; MW avoids that problem, but at the cost of a certain austerity; whether it will lend itself to an enhancement of its logic of the kind studied in Section 8 and enjoyed by DS must remain a question for another time. GJI is open to the pedagogical criticism that it relies on too early an introduction of the notion of rudimentary function.

The proof of VI.1.14 rests on a different idea, unrelated to the problems of defining Sat. The proof given by Devlin is tainted by its appeal to the false Lemma I.9.3, and therefore I intend in [M4] to rework the proof.

I cannot claim to have checked through the whole book, but my remarks reassure me, if no-one else, that the errors are not catastrophic. A modest strengthening of the meaning of BS and all seems to be well.

Part III

11. Gandy's inexact remarks

Gandy in [G] says of his four weak set theories PZ, BST', BRT and PZF, that were one to drop the requirement of Δ_0 the four would stretch from Zermelo to Zermelo–Fraenkel, and continues "presumably these are also all distinct". His first remark is *prima facie* false as he makes no mention of the power set axiom (nor of the axiom of foundation) and the power set axiom is certainly independent of the others as (working say in ZFC) **HC** satisfies all other axioms of ZF.

We insert BS in the sequence and comment on the effect on the five of dropping the restriction to Δ_0, of adding the power set axiom, and of doing both.

The full systems without power set

The first system will have axioms of extensionality, pairset, sumset and infinity, and the full separation scheme. The second system will add Cartesian product to that.

The model $\mathbf{M}_2$ satisfies full separation but not Cartesian product.

Corresponding to GJ, we have the "full rudimentary" replacement scheme:

(*full RR*) $\forall x \exists w \forall v_{\in x} \exists t_{\in w} \forall u (u \in w \iff . u \in x \;\&\; \phi[u, v])$.
for ϕ any formula.

The model $\mathbf{M}_7$ satisfies full separation and Cartesian product, but witnesses a failure of (restricted) rudimentary replacement.

Corresponding to fReR we have the full flat replacement axiom: namely, for any ϕ,

(*full flat repl.*)

$$\forall x_{\in u} \exists ! y (\phi(x, y) \;\&\; y \subseteq z) \Longrightarrow \exists u \forall y [y \in v \iff \exists x_{\in u} (\phi(x, y) \;\&\; y \subseteq z)]$$

But full flat replacement is derivable from "full rudimentary" replacement, using the self-strengthening of full RR corresponding to that noted in Proposition **2.88** for RR, by remarking that the set promised by an instance of full flat replacement is of the form

$$\{Z \cap \{y \mid \exists Y \Phi(X, Y) \;\&\; y \in Y\} \mid X \in U\}.$$

So in fact the distinction between the two systems will collapse already at Σ_1.

As for full flat collection, full replacement and full collection, Gandy's choice $\mathbf{G}_3 = V_{\omega+\omega}$ gives a model of full flat collection in which replacement fails – but since gfReR is a subsystem of Z, we may also find a model for it in which **HF** does not exist – and Zarach's model, [Z] Theorem 6.4, gives a model of full replacement in which collection, possibly even flat collection, fails.

Gandy's systems with added power set

PZ + P is the system M_0, in which Cartesian product is provable, as are Rudimentary Replacement, and flat Δ_0 Replacement and Collection. PZF + P is strictly stronger, as it builds $\omega + \omega$.

11.0. Problem. Is KPI + P the same as ReR + P?

The full systems with foundation and power set added

We have just Z in the first case; and the first four cases now coincide, for full flat replacement is provable in Z, just as fReR is provable in M_0 using power set plus Δ_0 separation. The fifth is ZF.

12. A model of Z plus full Foundation in which TCo fails

Boffa [B1] [B2] has constructed two other models of $\mathsf{Z} + \neg\mathsf{TCo}$; ours appears to be a third.

12.0. Definition. $\iota^0(x) =_{\mathrm{df}} x$; $\iota^{n+1}(x) =_{\mathrm{df}} \{\iota^n(x)\}$.

12.1. Definition. ϱ is the set-theoretical rank of x.

12.2. Definition. $V_n =_{\mathrm{df}} \{x \mid \varrho(x) < n\}$; $b_n =_{\mathrm{df}} \iota^n(V_n)$.

12.3. Definition. For each $n \in \omega$, set $c_n =_{\mathrm{df}} \{\bigcup^n b_m \mid n \leqslant m < \omega\}$.

12.4. Example. $c_0 = \{V_0, \{V_1\}, \{\{V_2\}\}, \ldots\}$; $c_1 = \{V_1, \{V_2\}, \{\{V_3\}\}, \ldots\}$; $c_2 = \{V_2, \{V_3\}, \{\{V_4\}\}, \ldots\}$.

12.5. Proposition. $\bigcup c_n = V_n \cup c_{n+1}$.

12.6. Definition. $K_0 =_{\mathrm{df}} \omega \cup \{c_0\}$; $K_{n+1} =_{\mathrm{df}} \mathcal{P}(K_n) \cup K_n \cup c_n$; $K =_{\mathrm{df}} \bigcup_{n\in\omega} K_n$.

12.7. Theorem. *K is a supertransitive model of Zermelo set theory Z in which some set is a member of no transitive set.*

12.8. Lemma. *$K_n \subseteq K_{n+1}$, and $K_n \in K_{n+1} \subseteq K$, so that each $K_n \in K$.*

12.9. Corollary. *K models Pairing.*

12.10. Lemma. *$V_n \subseteq K_n$.*

Proof. Induction on n. $V_0 = \varnothing$; if $V_n \subseteq K_n$, $V_{n+1} = \mathcal{P}(V_n) \subseteq \mathcal{P}(K_n) \subseteq K_{n+1}$. (12.10) □

12.11. Corollary. *K includes all of $V_\omega =$* **HF**; *in particular K contains all finite ordinals. Moreover $\omega \in K_1 \subseteq K$.*

12.12. Lemma. *K is transitive:*

Proof. Let $x \in y \in K_0$. Then either $y \in \omega$ when $x \in K$ or $y = c_0$ when $x \in \mathbf{HF} \subseteq K$.

Let $x \in y \in K_{n+1}$. Then either $y \subseteq K_n$, when $x \in K_n$, or $y \in K_n$, when inductively we have already shown that $x \in K$; or $y \in c_n \subseteq \mathbf{HF}$, when $x \in \mathbf{HF} \subseteq K$. (12.12) □

12.13. Corollary. *K models Extensionality, Null Set, Infinity and (full) Foundation.*

12.14. Lemma. *K is supertransitive.*

Proof. $x \subseteq y \in K_n \Longrightarrow x \in K_{n+1} \in K$. (12.14) □

12.15. Corollary. *K is a model of full Separation.*

12.16. Lemma. *Each $\bigcup K_n$ is a subset of K_{n+1} and thus is in K by supertransitivity.*

Proof. $\bigcup K_0 = \omega \cup c_0 \subseteq K_1 \in K$. If $\bigcup K_n \subseteq K_{n+1}$, $\bigcup K_{n+1} = K_n \cup \bigcup K_n \cup V_n \cup c_{n+1} \subseteq K_{n+2}$. (12.16) □

12.17. Corollary. *K models Union.*

Proof. If $y \in K_n$, then $y \subseteq \bigcup K_n \subseteq K_{n+1}$, so $\bigcup y \subseteq \bigcup K_{n+1} \in K$, so $\bigcup y$ is in K. (12.17) □

12.18. Lemma. *K models Power set.*

Proof. If $x \in K_n$, $x \subseteq K_{n+1}$ so $\mathcal{P}(x) \subseteq \mathcal{P}(K_{n+1}) \subseteq K_{n+2}$. (12.18) □

Thus we have shown that K models Z.

12.19. Proposition. $\forall n \forall m[m \geqslant n+3 \Longrightarrow V_m \notin K_n]$.

Proof. $V_0 = 0$; $V_1 = 1$, $V_2 = 2$ but for $m \geqslant 3$, V_m is not an ordinal and is therefore not in ω, nor is it, a finite set, equal to c_0, an infinite set. Hence $V_3 \notin K_0$.

Suppose that $V_m \notin K_n$, for any $m \geqslant n+3$. If $V_{m+1} \in K_{n+1}$, then either $V_{m+1} \subseteq K_n$, so that $V_m \in K_n$, contradicting the inductive hypothesis; or $V_{m+1} \in K_n$, again contrary to the inductive hypothesis; or $V_{m+1} \in c_n = \{V_n, \{V_{n+1}\}, \{\{V_{n+2}\}\} \ldots\}$, again impossible by inspection. (12.19) □

12.20. Proposition. TCo *fails in K.*

Proof. $c_0 \in K$. Suppose that $c_0 \in u \in K$ with u transitive. Then $\mathbf{HF} \subseteq u$, so $\mathbf{HF} \in K$, and hence $HF \in K_n$ say, so that $\mathbf{HF} \subseteq K_{n+1}$. But K_{n+1} contains at most $n+4$ of the sets V_m. (12.20) □

Other constructions of models of Zermelo are given in *Slim Models*. The constructions there furnish an entertaining independence argument for the axiom of pairing, which we shall give in the next section.

13. AxPair and AxSing

Let Z be Zermelo set theory, including the axioms of infinity and foundation. Let TCo be the assertion that every set is a member of a transitive set. Let TIn be the assertion that every set is a subset of a transitive set. Let AxSing be the assertion that for each set x, $\{x\}$ is a set. Let AxPair be the assertion that for all sets x and y, $\{x, y\}$ is a set.

13.1. Remark. TCo trivially (in the strict sense) implies TIn; TIn + AxSing implies TCo. AxSing is usually derived from AxPair, either by taking $x = y$ or if AxPair is confined to the strict case, by using separation. Indeed AxSing is provable using separation and power set, since each set x is a member of its power set, should the latter exist.

We shall exhibit a model of almost all of Zermelo, in which AxSing is true but AxPair is false, and a model of a substantial amount of set theory in which TIn holds but AxSing and TCo fail.

It is amusing to note that in the system of Bourbaki, the pairing axiom has been proved to be redundant. see Sonner [S]. That it is not redundant in Z was first shown by Boffa [B3].

Failure of AxPair

Let T be the theory Z + TCo + WO, – WO being the statement "every set has a well-ordering" – and let T^- be the theory T with the axiom of pairing replaced by its negation: $\exists x \exists y \{x, y\} \notin V$, and with the addition of AxSing.

13.2. Remark. The scheme of foundation for all classes is provable in T^-.

We find a model for T^-: indeed we show that if Consis(Z) then Consis(T^-).

It follows from the last part of Theorem 5 of *The Strength of Mac Lane Set Theory* [M2], proved in Section 5 of that paper, that if Z is consistent, so is Z + KP + WO.

A set or class **M** is said to be *supertransitive* if it is transitive and, further, $x \subseteq y \in \mathbf{M} \Longrightarrow x \in \mathbf{M}$.

As in the proof of Theorem **4.8** of *Slim Models of Zermelo Set Theory* [M1] one can, working in the theory Z + KP + WO, build two supertransitive models **M** and **N** of Z + TCo + WO, with neither a subset of the other: e.g. take **M** to contain $Z(0)$ but not $Z(\omega)$ and **N** to contain $Z(\omega)$ but not $Z(0)$, in the notation of that paper.

Theorem. *Let* **M** *and* **N** *be supertransitive models of* T, *neither included in the other; then* $\mathbf{M} \cup \mathbf{N}$ *is a model of* T^-, *and* $\mathbf{M} \cap \mathbf{N}$ *is a model of* T.

Proof. Note first that $\mathbf{M} \cup \mathbf{N}$ is supertransitive, and hence absolute for most of the set-theoretical concepts used in the axioms; therefore it will be a model of Extensionality, Sum Set, Power Set, full Separation, Foundation, TCo (whence also Foundation for all classes), and WO.

[For power set, use supertransitivity; otherwise there would be a risk of **N** containing subsets of some element of **M** which were not in **M**. Supertransitivity also gives the truth of full separation in **P**. For the other axioms the transitivity of **P** is enough.]

Pairing fails, for if $a \in \mathbf{M} \setminus \mathbf{N}$ and $b \in \mathbf{N} \setminus \mathbf{M}$, then $\{a, b\} \notin \mathbf{M} \cup \mathbf{N}$. But AxSing holds.

The verification of the second assertion is straightforward. (13.2) □

Metacorollary. *If* Z *is consistent so is* T′.

13.3. Remark. The **M** and **N** just used can be chosen to contain all ordinals, all sequences of ordinals and all sets of sequences of ordinals, and to be such that for all limit ordinals $\lambda > \omega$, neither of $\mathbf{P}_\lambda =_{\text{df}} \mathbf{M} \cap V_\lambda$ nor $\mathbf{Q}_\lambda =_{\text{df}} \mathbf{N} \cap V_\lambda$ is contained in the other. In such a case, each of $\mathbf{P}_\lambda$, $\mathbf{Q}_\lambda$ and $\mathbf{R}_\lambda =_{\text{df}} \mathbf{P}_\lambda \cap \mathbf{Q}_\lambda$ will be a supertransitive model of Z+ WO, each being the intersection of two such. If $p \in \mathbf{P}_\lambda \setminus \mathbf{R}_\lambda$, then for $x \in \mathbf{R}_\lambda$ $\{p, x\}$ will be in $\mathbf{P}_\lambda \setminus \mathbf{R}_\lambda$; so the three sets $\mathbf{P}_\lambda \setminus \mathbf{R}_\lambda$, $\mathbf{Q}_\lambda \setminus \mathbf{R}_\lambda$ and $\mathbf{R}_\lambda$ will all be of cardinal $\beth_\lambda$.

13.4. Remark. Boffa in [B3] shows of every member a of **HF** that it is provable in Z that for any x, the pair $\{a, x\}$ exists: for example both the empty set and x are in $\mathcal{P}(x)$, and therefore the pair $\{\varnothing, x\}$ can be recovered using Separation. Thus a set which might not form a pair with something must be of rank at least ω, and Boffa shows that the set $\{\varnothing, \{\varnothing\}, \{\{\varnothing\}\} \ldots\}$ of Zermelo integers indeed has that property.

Failure of AxSing

Consider, working in some suitable theory such as ZF, the class **C** of all sets x such that $\mathrm{tcl}(x)$ contains at most one strict pair, that is, a set of the form $\{b, c\}$ with $b \neq c$. **C** is supertransitive, and models "much" of Z: namely Extensionality, full separation, sum set, and infinity; and it contains all the ordinals, of which $2 = \{0, 1\}$ is the only strict pair. AxSing fails since $\{5, 6\}$ is a member of **C** but $\{\{5, 6\}\}$ is not.

Moreover TIn holds in **C** , since the transitive closure of an element of **C** is itself an element of **C** ; but TCo is false, since for example $\{5, 6\}$ cannot be a member of any transitive element of **C**.

Inadequate axioms in a French textbook

The well-established textbook *Tome* 1, *Algèbre,* of the *Cours de mathématiques* by Jacqueline Lelong-Ferrand and Jean-Marie Arnaudiès, [L-F,A], in its opening chapter gives some axioms for what is in effect a subsystem of Z. They follow Bourbaki in giving axioms for ordered pairs, but not for unordered pairs. But a model for the axioms that they state is furnished by any $\mathbf{P}_\lambda \cup \mathbf{Q}_\lambda$ as discussed in Remark **13.2**; for let $f : \mathbf{R}_\lambda \longleftrightarrow \mathbf{R}_\lambda \times \{0\}$, $g : \mathbf{P}_\lambda \setminus \mathbf{R}_\lambda \longleftrightarrow \mathbf{R}_\lambda \times \{1\}$ and $h : \mathbf{Q}_\lambda \setminus \mathbf{R}_\lambda \longleftrightarrow \mathbf{R}_\lambda \times \{2\}$, let f_x be f if $x \in \mathbf{R}_\lambda$, g if $x \in \mathbf{P}_\lambda \setminus \mathbf{R}_\lambda$ and h if $x \in \mathbf{Q}_\lambda \setminus \mathbf{R}_\lambda$, and interpret the formal ordered pair of x and y as $(f_x(x), f_y(y))_2$;

and in that model whenever $x \in \mathbf{P}_\lambda \setminus \mathbf{R}_\lambda$ and $y \in \mathbf{Q}_\lambda \setminus \mathbf{R}_\lambda$, their union $x \cup y$ will not be a set.

The reader will find in [M6] a more detailed scrutiny of the account of logic and set theory in [L-F,A].

14. A remark on rud closure answering a question of MacAloon

Let $\mathbb{T}$ be the rudimentary function of Definition **2.73**.

14.0. Lemma. *Let $\langle u_n \mid n \in \omega \rangle$ be any sequence of transitive sets. Define*

$$K_0 = u_0; \quad K_{n+1} = \mathbb{T}^5(K_n) \cup u_n \quad K_\omega = \bigcup_{n<\omega} K_n.$$

Then K_ω is rud closed.

Proof. We show that K_ω is closed under each of the functions R_0 to R_8. By the properties of $\mathbb{T}$ established in Section 2, x, y in u implies $R_i(x) \in \mathbb{T}(u)$ for i = 2,3,5; and x, y, in u implies $R_i(x,y) \in \mathbb{T}(u)$ for i = 0,1; x in u implies $R_i(x) \in \mathbb{T}^5(u)$ for i = 6, 7; x, y in u implies $R_4(x,y) \in \mathbb{T}^3(u)$; and x, y in u implies $R_8(x,y) \in \mathbb{T}^2(u)$. As $u \subset \mathbb{T}(u) \subset \mathbb{T}^2(u) \ldots$, it follows that for each n, $K_n \subseteq \mathbb{T}^5(K_n) \subseteq K_{n+1}$. (14.0) □

14.1. Definition. $\iota(x) =_{\mathrm{df}} \{x\}$

14.2. Lemma. *If $x \notin u$ then $\iota(x) \notin \mathbb{T}(u)$; and hence $\iota^4(x) \notin \mathbb{T}^4(u)$.*

Proof. Every member of $\mathbb{T}(u)$ is a subset of u. (14.2) □

14.3. Proposition. *Suppose that u is a transitive set closed under pairing. Then whenever w is a transitive set of which u is not a subset, u is not a member of the rud closure of $u \cup w$.*

Proof. u must be of limit rank λ say. Suppose first that u is countable, so that λ is of cofinality ω. Let $\lambda_n \nearrow_n \lambda$. We fix an enumeration of x and use it to make the following choices.

- Pick $x_0 \in u \setminus w$. Let $u_0 = u \cap V_{\max\{\lambda_0, \varrho(x_0)+1\}}$.
- Pick $x_1 \in u \setminus u_0$, with $\iota^4 x_0 \in x_1$. Let $u_1 = u \cap V_{\max\{\lambda_1, \varrho(x_1)+1\}}$.
- Pick $x_{n+1} \in u \setminus u_n$, with $\iota^4 x_n \in x_{n+1}$. Let $u_{n+1} = u \cap V_{\max\{\lambda_{n+1}, \varrho(x_{n+1})+1\}}$.
- Finally let $K_0 = w$; $K_{n+1} = \mathbb{T}^5(K_n) \cup u_n$; $K_\omega = \bigcup_n K_n$.

Then every K_n is transitive and by the Lemma, K_ω is rud closed, and includes $w \cup \{w\} \cup u$. If $x_{n+1} \in K_{n+1}$, it cannot, by construction, be a member of u_n and so must be a subset of $\mathbb{T}^4(K_n)$, so $\iota^4(x_n) \in \mathbb{T}^4(K_n)$, which by Lemma **14.2** implies $x_n \in K_n$. But $x_0 \notin K_0$; so by induction no $x_n \in K_n$. Hence no superset of $\{x_n \mid n \in \omega\}$ can be a member of K_ω. In particular, u cannot be.

The Proposition is now proved for the case that u is countable. In the general case, go to a generic extension of the universe in which u is countable; the hypotheses will still hold; hence in the generic extension, u is not in the rud closure of $u \cup w \cup \{w\}$; but that latter statement is absolute and therefore true in the ground model. (14.3) □

14.4. Corollary. *Let u be transitive and closed under pairing; then u is not in the rud closure of $ON \cup u$.*

A particular case answers a question posed by McAloon in the 1970's:

14.5. Corollary. *For any $\alpha > 0$, $J_\alpha \notin$ rud cl$(J_\alpha \cup \{\omega\alpha\})$*

I thank Lee Stanley for telling me of McAloon's question.

14.6. Remark. So far as the definition of K_ω goes, other functions T could be used instead of $\mathbb{T}$, provided they had the property that the members of $T(u)$ are subsets of u: for example, if we instead use $u \mapsto \mathcal{P}(u)$, K_ω will be a model of Zermelo set theory, probably including the axiom of infinity, though possibly not in the form $\omega \in V$; we adopt this strategy in the following variant.

14.7. Proposition. *Suppose that $(x_n)_n$ and $(u_n)_n$ are two sequences of sets such that for each $n < \omega$:*

(14.7.0) $x_n \in u_n$;

(14.7.1) $u_n \subseteq u_{n+1}$;

(14.7.2) u_n *is transitive;*

(14.7.3) $x_n \in \mathrm{tcl}(x_{n+1})$;

(14.7.4) $x_{n+1} \notin u_n$.

Then $\bar{u} =_{\mathrm{df}} \bigcup_n u_n$ is transitive and if w is a transitive set with $x_0 \notin w$, the set $\bar{x} =_{\mathrm{df}} \{x_n \mid n \in \omega\}$ is not a member of the rud closure of $\bar{u} \cup w \cup \{w\}$. If in addition $\omega \subseteq w$, then there is a supertransitive model of Zermelo set theory of which $\bar{u} \cup w \cup \{w\}$ is a subset but $\bar{x}$ and $\bar{u}$ are not members.

Proof. Let K be the model formed as follows:

$$K_0 = w; \qquad K_{n+1} = \mathcal{P}(K_n) \cup u_n; \qquad K = \bigcup_n K_n.$$

Then each K_n is transitive. (14.7) □

14.8. Lemma. *Each K_n is a member of K_{n+1}.*

14.9. Lemma. *$K_0 \subseteq K_1$; if $K_n \subseteq K_{n+1}$ then $K_{n+1} \subseteq K_{n+2}$.*

Proof. As K_0 is transitive, its members are also subsets of it and therefore members of K_1. Under the hypotheses of the second statement, $\mathcal{P}(K_n) \subseteq \mathcal{P}(K_{n+1}) \subseteq K_{n+2}$ and $u_n \subseteq u_{n+1} \subseteq K_{n+2}$. (14.9) □

14.10. Lemma. *$\bigcup K_0 \subseteq K_0$; $\bigcup K_{n+1} = K_n \cup \bigcup u_n$.*

14.11. Lemma. *If $x \in K$, then for some ℓ, $x \subseteq K_\ell$.*

14.12. Lemma. *K is transitive.*

Proof. If $y \in x \in K$ then for some ℓ, $y \in x \subseteq K_\ell$, so $y \in K_\ell \subseteq K$. (14.12) □

14.13. Lemma. *K is supertransitive.*

Proof. If $y \subseteq x \in K$ then for some ℓ, $y \subseteq x \subseteq K_\ell$, so $y \in \mathcal{P}(K_\ell) \subseteq K_{\ell+1} \subseteq K$. (14.13) □

14.14. Corollary. *K models the full separation scheme.*

14.15. Lemma. $x \in K \Longrightarrow \mathcal{P}(x) \in K$.

Proof. By Lemma **14.11**, x is a subset of some K_ℓ; by the proof of Lemma **14.13**, any subset of x is in $K_{\ell+1}$, and so $\mathcal{P}(x)$ is a subset of $K_{\ell+1}$ and therefore a member of $K_{\ell+2}$. (14.15) □

14.16. Lemma. *Each $\bigcup K_n$ is in K.*

Proof. By supertransitivity, as each $K_n \in K$. (14.16) □

14.17. Lemma. $x \in K \Longrightarrow \bigcup x \in K$.

Proof. If $x \subseteq K_\ell$, then $\bigcup x \subseteq \bigcup K_\ell$, which is in K; as K is supertransitive, $\bigcup x \in K$. (14.17) □

14.18. Lemma. *For no n is x_n a member of K_n; hence $\bar{x}$ is a subset of no K_n; hence neither it not $\bar{u}$ can be a member of K.*

Proof. $x_0 \notin K_0$ by hypothesis. Suppose that $x_{n+1} \in K_{n+1}$, then either $x_{n+1} \subseteq K_n$, giving $x_n \in K_n$, (since K_n is transitive) or else $x_{n+1} \in u_n$, contrary to hypothesis.

So $x_n \notin K_n \Longrightarrow x_{n+1} \notin K_{n+1}$; by induction, for no n is x_n a member of K_n; as $x_n \in \bar{x}$, $\bar{x} \not\subseteq K_n$. Lemma **14.11** now implies that $\bar{x}$ is not a member of K; as it is a subset of $\bar{u}$ and K is supertransitive, $\bar{u}$ cannot be a member of K. (14.18) □

14.19. Lemma. $\bar{u} \cup w \cup \{w\} \subseteq K$.

14.20. Lemma. *If $x \in K_m$ and $y \in K_n$ then for $\ell = \max(m, n)$, $\{x, y\} \subseteq K_\ell$ and so is in K.*

14.21. Proposition. $\omega \in K \Longleftrightarrow \omega \subseteq w$.

14.22. Proposition. *K is a model of all axioms of Zermelo set theory except possibly the axiom of infinity.*

14.23. Corollary. *K is rud closed.*

14.24. Remark. If we take $u = \mathbf{HF}$ and $w = \omega$, K_ω will be a set model of Zermelo of which **HF** is not a member. Thus our argument generalises constructions to be found in the texts of Moschovakis and Enderton.

A third possibility is in the proof of the next remark.

14.25. Proposition. *Let u be transitive and be the strictly increasing union of a sequence u_n of transitive sets with u_0 not an ordinal and $u_n \in u_{n+1}$. Let $\zeta = ON \cap u$. Then the rud closure of $u \cup \{\zeta\}$ is a proper subset of the rud closure of $u \cup \{u\}$.*

Proof. Define $K_0 = \zeta$; $K_{n+1} = \mathrm{Def}(K_n) \cup u_n$; $K = \bigcup_n K_n$.

K is rud closed and includes $u \cup \{\zeta\}$; but one may show that each $u_n \notin K_n$; hence $u \notin K$. (14.25) □

15. An application to Gandy numerals

The method of Section 14 casts some light on the proposal made by Gandy in [G] for discarding the von Neumann ordinals as numerals for the purpose of developing formal syntax. Their problem is that the rank of n is n. His method makes use of ideas of Smullyan [Sm].

First step: $\hat{\omega}$

15.0. Definition. We assign to each $n \in \omega$ a hereditarily finite set $\hat{n}$ and a level $\lambda(n) \in \omega$. $\hat{0} = 0; \hat{1} = \{0\}; \lambda(0) = \lambda(1) = 0$. For $n > 0$ let $n - 1 = \Sigma_{\ell<k} a_\ell 2^\ell$, where $a_\ell \in \{1, 2\}$. Then put

$$\hat{n} = \{\{\hat{\ell} \mid \ell < k \ \&\ a_\ell = 2\}, \{\hat{\ell} \mid \ell < k\}\}; \lambda(n) = k.$$

15.1. Example. $\hat{2} = \{0, \{0\}\}$; $\hat{3} = \{\{0\}\}$; $\lambda(2) = \lambda(3) = 1$

$\hat{4} = \{0, \{0, \{0\}\}\}$; $\hat{5} = \{\{0\}, \{0, \{0\}\}\}$; $\hat{6} = \{\{\{0\}\}, \{0, \{0\}\}\}$; $\hat{7} = \{\{0, \{0\}\}\}$; $\lambda(4) = \lambda(5) = \lambda(6) = \lambda(7) = 2$.

Set $\hat{\omega} = \{\hat{n} \mid n \in \omega\}$.

To get λ we need the graph of exponentiation.

Second step: $\bar{\omega}$

Then set $\bar{n} =_{\mathrm{df}} \{\hat{m} \mid m < n\}$ and $\bar{\omega} =_{\mathrm{df}} \{\bar{n} \mid n \in \omega\}$.

It is the members of $\bar{\omega}$ that Gandy proposes, and which we shall call Gandy numerals. He proves that

> the predicate $x \in \bar{\omega}$ is Δ_0; addition and multiplication of Gandy numerals are rudimentary; concatenation of sequences of Gandy numerals is rudimentary; but exponentiation of Gandy numerals is not rudimentary.

His reason for not remaining with $\hat{\omega}$ is that he was unable to prove that $x \in \hat{\omega}$ is Δ_0, and he speculated that $x \in \hat{\omega}$ is in fact not.

15.2. Proposition. *Neither $\hat{\omega}$ nor $\bar{\omega}$ is in* rud cl$(\{\omega\})$.

Proof. We apply Proposition **14.7**. $\hat{0} = 0$, $\hat{1} = 1$, $\hat{2} = 2$ but $\hat{3} = \{1\}$ which is not an ordinal. Therefore let $x_0 = \hat{3}$, and $u_0 = \mathrm{tcl}(\{x_0\})$. Let x_{n+1} be $\hat{k}$ for k the least such that $\hat{k} \notin u_n$ and $x_n \in \mathrm{tcl}\hat{k}$; take $u_{n+1} = u_n \cup \mathrm{tcl}(\{x_{n+1}\})$. The resulting supertransitive model K is rud closed and does not contain $\bar{x}$; therefore it does not contain $\hat{\omega}$, of which $\bar{x}$ is a subset. But it does include the rudimentary closure of $\{\omega\}$.

Since $\bigcup \bar{\omega} = \hat{\omega}$, $\bar{\omega}$, too, cannot be in K. (15.2) □

15.3. Remark. We can define a version, $\widehat{\text{ACK}}$, of the Ackermann relation by $\hat{m}\widehat{\text{ACK}}\hat{n} =_{\text{df}} \hat{m} \in \bigcap \hat{n}$.

By the Proposition, $\hat{\omega}$ is not provably a set in GJ. But in GJ, we can show that if $\hat{\omega}$ is a set, then so is the relation $\widehat{\text{ACK}}$, and therefore the set of all finite subsets of $\hat{\omega}$ will be obtainable as $\{\widehat{\text{ACK}}``\{x\} \mid x \in \hat{\omega}\}$.

Acknowledgments

This article owes its existence to my participation in the Set Theory Year, 2003–4, at the Centre de Recerca Matemàtica at Bellaterra outside Barcelona: I express my gratitude to Joan Bagaria and the organisers of the Set Theory Year for their invitation; to my fellow-members of the research group ERMIT at the Université de la Réunion, who enabled me to maximise my stay at the CRM; to the Director and staff of the CRM for the excellent working conditions that they continue to provide; to the members of the set theory seminars in Réunion and at the CRM who patiently heard successive versions of this material; and to Thomas Forster, Kai Hauser, Ronald Jensen, Robert Lubarsky, Colin McLarty, Thoralf Räsch and Lee Stanley for their stimulating encouragement and helpful observations.

References

[B1] Maurice Boffa, Axiome et schéma de fondement dans le système de Zermelo, *Bull. Acad. Polon. Sci. Sér. Math. Astron. Phys.* **17** 1969 113–5. MR 40 # 38.

[B2] Maurice Boffa, Axiom and scheme of foundation, *Bull. Soc. Math. Belg.* **22** (1970) 242–247; MR 45 #6618.

[B3] Maurice Boffa, L'axiome de la paire dans le système de Zermelo, *Arch. Math. Logik Grundlagenforschung* **15** (1972) 97–98. MR 47 #6486

[Del] Ch. Delhommé, Automaticité des ordinaux et des graphes homogènes (Automaticity of ordinals and of homogeneous graphs) *C.R. Acad. Sci. Paris*, Ser. I **339** pp. 5–10 (2004). (*French with abridged English version*)

[Dev] K. Devlin, *Constructibility*, Perspectives in Mathematical Logic, *Springer-Verlag, Berlin*, 1984.

[Do] A.J. Dodd, *The Core Model*, London Mathematical Society Lecture Note Series, 61, Cambridge University Press, 1982. MR **84a:**03062.

[DoMT] J.E. Doner, A. Mostowski and A. Tarski, The elementary theory of well-ordering: a metamathematical study, in *Logic Colloquium '77*, eds. A. Macintyre, L. Pacholski, J. Paris, North-Holland Publishing Company, 1978, 1–54.

[G] R.O. Gandy, Set-theoretic functions for elementary syntax, in *Proceedings of Symposia in Pure Mathematics*, **13**, Part II, ed. T. Jech, American Mathematical Society, 1974, 103–126.

[J1] R.B. Jensen, Stufen der konstruktiblen Hierarchie. Habilitationsschrift, Bonn, 1968.

[J2] R.B. Jensen, The fine structure of the constructible hierarchy, with a section by Jack Silver, *Annals of Mathematical Logic,* **4** (1972) 229–308; erratum ibid **4** (1972) 443.

[L-F,A] Jacqueline Lelong-Ferrand and Jean-Marie Arnaudiès, *Cours de mathématiques, Tome* 1, *Algèbre*, Editions Dunod, Paris; first edition 1971; often reprinted.

[M1] A.R.D. Mathias, Slim models of Zermelo Set Theory, *Journal of Symbolic Logic* **66** (2001) 487–496; MR 2003a:03076.

[M2] A.R.D. Mathias, The Strength of Mac Lane Set Theory, *Annals of Pure and Applied Logic,* **110** (2001) 107–234; MR 2002g:03105.

[M3] A.R.D. Mathias, A note on the schemes of replacement and collection, *to appear in the Archive for Mathematical Logic.*

[M4] A.R.D. Mathias, Rudimentary recursion, *in preparation.*

[M5] A.R.D. Mathias, Rudimentary forcing, *in preparation.*

[M6] A.R.D. Mathias, The banning of formal logic from a French national examination, *in preparation.*

[Sm] R.M. Smullyan, *Theory of formal systems*, Ann. of Math. Studies, no, 3, Princeton University Press, Princeton N.J., 1940 . MR **2**, 66.

[So] Johann Sonner, On sets with two elements, Arch. Math (Basel) **20** (1969) 225–7. MR **40** #36.

[St] Lee Stanley, review of [Dev], *Journal of Symbolic Logic* **53** (1987) 864–8.

[Z] Andrzej M. Zarach, Replacement $\nrightarrow$ Collection, in Gödel '96, ed. Petr Hájek, (Springer Lecture Notes in Logic, Volume 6).

Centre de Recerca Matemàtica
Bellaterra, Catalonia

and

ERMIT
Université de la Réunion

Set Theory
Trends in Mathematics, 225–255

Some New Directions in Infinite-combinatorial Topology

Boaz Tsaban

Abstract. We give a light introduction to selection principles in topology, a young subfield of infinite-combinatorial topology. Emphasis is put on the modern approach to the problems it deals with. Recent results are described, and open problems are stated. Some results which do not appear elsewhere are also included, with proofs.

Contents

0. Introduction

The modern era of what we call *infinite combinatorial topology*, or *selection principles in mathematics* began with Scheepers' paper [40] and the subsequent work [23]. In these works, a unified framework was given that extends many particular investigations carried on in the classical era. The current paper aims to give the reader a taste of the field by telling six stories, each shedding light on one specific theme. The stories are short, but in a sense never ending, since each of them poses several open problems, and more are expected to arise when these are solved.

This is not intended to be a systematic exposition to the field, not even when we limit our scope to selection principles in *topology*. For that see Scheepers' survey [44] as well as Kočinac's [24]. Rather, we describe themes and results with which we are familiar. This implies the disadvantage that we are often quoting our own results, which only form a tiny portion of the field.[1]

Some open problems are presented here. For many more see [56]. After reading this introduction, the reader can proceed directly to some of the works of the ex-

[1]To partially compensate for this, the name of the present author is never explicitly mentioned in the paper.

perts in the field (or in closely related fields), such as:[2] Liljana Babinkostova, Taras Banakh, Tomek Bartoszyński, Lev Bukovský, Krzysztof Ciesielski, David Fremlin, Fred Galvin, Salvador García-Ferreira, Janos Gerlits, Cosimo Guido, Istvan Juhasz, Ljubisa Kočinac, Adam Krawczyk, Henryk Michalewski, Arnold Miller, Zsigmond Nagy, Andrzej Nowik, Janusz Pawlikowski, Roman Pol, Ireneusz Recław, Miroslav Repický, Masami Sakai, Marion Scheepers, Lajos Soukup, Paul Szeptycki, Stevo Todorcevic, Tomasz Weiss, Lyubomyr Zdomskyy, and many others.

Notation. In most cases, the notation and terminology we use is Scheepers' modern one, and we do not pay special attention to the historical predecessors of the notations we use. The online version of this paper [59] has an index that can be used to locate the definitions the reader is missing.

By *set of reals* we usually mean a zero-dimensional, separable metrizable space, though some of the results hold in more general situations.

Apology. Some of the results might be miscredited or misquoted (or both). Please let us know of any mistake you find and we will correct it in the online version of this paper [59].

1. The Menger-Hurewicz conjectures

1.1. The Menger and Hurewicz properties

In 1924, Menger [31] introduced the following basis covering property for a metric space X:

> For each basis $\mathcal{B}$ of X, there exists a sequence $\{B_n\}_{n\in\mathbb{N}}$ in $\mathcal{B}$ such that $\lim_{n\to\infty} \operatorname{diam}(B_n) = 0$ and $X = \bigcup_n B_n$.

It is an amusing exercise to show that every compact, and even σ-compact space has this property. Menger conjectured that this property characterizes σ-compactness. In 1925, Hurewicz [21] introduced two properties of the following prototype. For collections $\mathscr{A}, \mathscr{B}$ of covers of a space X, define

$\mathsf{U}_{fin}(\mathscr{A}, \mathscr{B})$: For each sequence $\{\mathcal{U}_n\}_{n\in\mathbb{N}}$ of members of $\mathscr{A}$ *which do not contain a finite subcover*, there exist finite (possibly empty) subsets $\mathcal{F}_n \subseteq \mathcal{U}_n$, $n \in \mathbb{N}$, such that $\{\cup\mathcal{F}_n\}_{n\in\mathbb{N}} \in \mathscr{B}$.

Hurewicz proved that for $\mathcal{O}$ the collection of *all* open covers of X, $\mathsf{U}_{fin}(\mathcal{O}, \mathcal{O})$ is equivalent to Menger's basis property. Hurewicz did not settle Menger's conjecture, but he suggested a more modest one: Call an open cover $\mathcal{U}$ of X a γ-*cover* if

[2] This *very* incomplete list is ordered alphabetically. We did not give references to works not explicitly mentioned in this paper, but the reader can find some of these in the bibliographies of the given references, most notably, in [44, 24]. We should also comment that not all works of the authors are formulated using the systematic notation of Scheepers which we use here, and probably some of the mentioned experts do not consider themselves as working in the discussed field (but undoubtly, each of them made significant contributions to the field).

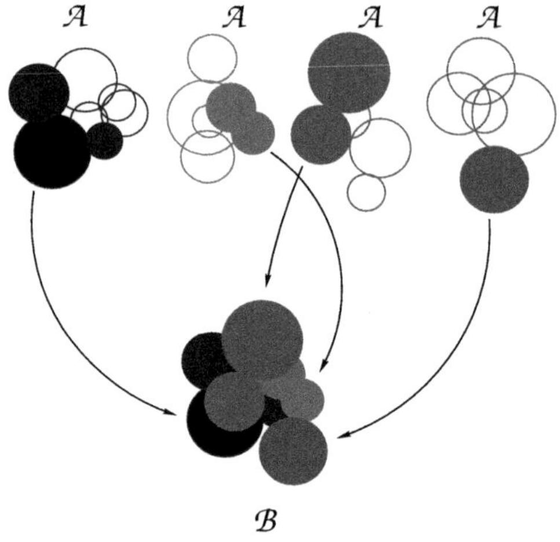

FIGURE 1. $\mathsf{U}_{fin}(\mathscr{A},\mathscr{B})$

$\mathcal{U}$ is infinite, and each $x \in X$ belongs to all but finitely many members of $\mathcal{U}$. Let Γ denote the collection of all open γ-covers of X. Clearly,

$$\sigma\text{-compact} \Rightarrow \mathsf{U}_{fin}(\mathcal{O},\Gamma) \Rightarrow \mathsf{U}_{fin}(\mathcal{O},\mathcal{O}),$$

and Hurewicz conjectured that for metric spaces, $\mathsf{U}_{fin}(\mathcal{O},\Gamma)$ (now known as the *Hurewicz property*) characterizes σ-compactness.

1.2. Consistent counter examples

It did not take long to find out that these conjectures are false assuming the Continuum Hypothesis: Observe that every uncountable F_σ set of reals contains an uncountable perfect set, which in turn contains an uncountable set which is both meager (i.e., of Baire first category) and null (i.e., of Lebesgue measure zero). A set of reals L is a *Luzin set* if it is uncountable, but for each meager set M, $L \cap M$ is countable. It was proved by Mahlo (1913) and Luzin (1914) that the Continuum Hypothesis implies the existence of a Luzin set. Sierpiński pointed out that each Luzin set has Menger's property $\mathsf{U}_{fin}(\mathcal{O},\mathcal{O})$ (hint: If we cover a countable dense subset of the Luzin set by open sets, then the uncovered part is meager and therefore countable), and is therefore a counter example to Menger's conjecture.[3]

Similarly, a set of reals S is a *Sierpiński set* if it is uncountable, but for each null set N, $S \cap N$ is countable. Sierpiński showed that the Continuum Hypothesis implies the existence of such sets, and it can be shown that Sierpiński sets have the Hurewicz property, and is therefore a counter-examples to Hurewicz' (and therefore Menger's) conjecture.

[3]This observation was "added in proof" just after footnote 1 on page 196 of Hurewicz' 1927 paper [22].

Why must a Sierpiński set satisfy Hurewicz' property $\mathsf{U}_{fin}(\mathcal{O},\Gamma)$? A classical proof can be carried out using Egoroff's Theorem, but let us see how a modern, combinatorial proof goes [45]. The *Baire space* $\mathbb{N}^\mathbb{N}$ (a Tychonoff power of the discrete space $\mathbb{N}$) carries an interesting combinatorial structure: For $f, g \in \mathbb{N}^\mathbb{N}$, write

$$f \leq^* g \quad \text{if} \quad f(n) \leq g(n) \text{ for all but finitely many } n.$$

$B \subseteq \mathbb{N}^\mathbb{N}$ is *bounded* if there exists $g \in \mathbb{N}^\mathbb{N}$ such that $f \leq^* g$ for all $f \in B$. $D \subseteq \mathbb{N}^\mathbb{N}$ is *dominating* if for each $g \in \mathbb{N}^\mathbb{N}$ there exists $f \in D$ such that $g \leq^* f$. It is easy to see that a countable union of bounded sets in $\mathbb{N}^\mathbb{N}$ is bounded, and that compact (and therefore σ-compact) subsets of $\mathbb{N}^\mathbb{N}$ are bounded. The following theorem is essentially due to Hurewicz, who proved a variant of it in [22]. In the form below, the theorem was stated and proved in Recław [35] in the zero-dimensional case, and then extended by Zdomskyy [62] to arbitrary subsets of $\mathbb{R}$.[4]

Theorem 1.1 (Hurewicz). *For a set of reals X:*

1. *X satisfies $\mathsf{U}_{fin}(\mathcal{O},\Gamma)$ if, and only if, all continuous images of X in $\mathbb{N}^\mathbb{N}$ are bounded.*
2. *X satisfies $\mathsf{U}_{fin}(\mathcal{O},\mathcal{O})$ if, and only if, all continuous images of X in $\mathbb{N}^\mathbb{N}$ are not dominating.*

Assume that $S \subseteq [0,1]$ is a Sierpiński set and $\Psi : S \to \mathbb{N}^\mathbb{N}$ is continuous. Then Ψ can be extended to all of $[0,1]$ as a Borel function. By a theorem of Luzin, there exists for each n a closed subset C_n of $[0,1]$ such that $\mu(C_n) \geq 1 - 1/n$, and such that Ψ is continuous on C_n. Since C_n is compact, $\Phi[C_n]$ is bounded in $\mathbb{N}^\mathbb{N}$. The set $N = [0,1] \setminus \bigcup_n C_n$ is null, and so its intersection with S is countable. Consequently, $\Psi[S]$ is contained in a union of countably many bounded sets in $\mathbb{N}^\mathbb{N}$, and is therefore bounded.

1.3. Counter examples in ZFC

But are the conjectures *consistent*? It turns out that the answer is negative. Surprisingly, it was only recently that this question was clarified, again using a combinatorial approach. Let $\mathfrak{b}$ denote the minimal size of an unbounded subset of $\mathbb{N}^\mathbb{N}$, and $\mathfrak{d}$ denote the minimal size of a dominating subset of $\mathbb{N}^\mathbb{N}$. The *critical cardinality* of a (nontrivial) collection $\mathcal{J}$ of sets of reals is

$$\mathsf{non}(\mathcal{J}) = \min\{|X| : X \subseteq \mathbb{R},\ X \notin \mathcal{J}\}.$$

By Hurewicz' Theorem 1.1, $\mathsf{non}(\mathsf{U}_{fin}(\mathcal{O},\Gamma)) = \mathfrak{b}$, and $\mathsf{non}(\mathsf{U}_{fin}(\mathcal{O},\mathcal{O})) = \mathfrak{d}$.

In 1988, Fremlin and Miller [17] used their celebrated dichotomic argument to refute Menger's conjecture (in ZFC): By the last observation, if $\aleph_1 < \mathfrak{d}$ then any set of reals of size $\aleph_1$ will do; and if $\aleph_1 = \mathfrak{d}$, then one can use a sophisticated combinatorial construction. Chaber and Pol [14], exploiting a celebrated topological technique due to Michael, extended the dichotomic argument to show that there always exists a counter example of size $\mathfrak{b}$ to Menger's conjecture.

---

[4]In the current form, Theorem 1.1 does not hold for subsets of $\mathbb{R}^2$ [62].

Hurewicz' conjecture was refuted in 1996 [23], this time using the dichotomy $\aleph_1 < \mathfrak{b}$ or $\aleph_1 = \mathfrak{b}$ and even more sophisticated combinatorial arguments in the second case. This was improved by Scheepers [43], who showed (again on a dichotomic basis) that there always exists a counter example of size $\mathfrak{t}$ ($\mathfrak{t}$, to be defined in Section 5.2, is an uncountable cardinal which is consistently greater than $\aleph_1$).

A simple construction was very recently found [7, 9] to refute both conjectures, and not on a dichotomic basis: There exists a non σ-compact set of reals H of size $\mathfrak{b}$ which has the Hurewicz property. The construction does not use any special hypothesis: Let $\mathbb{N} \cup \{\infty\}$ be the one point compactification of $\mathbb{N}$, and $\mathcal{Z}$ be the nondecreasing functions $f \in (\mathbb{N} \cup \{\infty\})^{\mathbb{N}}$. For a finite nondecreasing sequence s of natural numbers, let q_s be the element of $\mathcal{Z}$ extending s and being equal to ∞ on all new n's. Then the collection Q of all these elements q_s is dense in $\mathcal{Z}$. Define a $\mathfrak{b}$-*scale* to be an unbounded set $\{f_\alpha : \alpha < \mathfrak{b}\} \subseteq \mathbb{N}^{\mathbb{N}}$ of increasing functions, such that $f_\alpha \leq^* f_\beta$ whenever $\alpha < \beta$. It is an easy exercise to construct a $\mathfrak{b}$-scale in ZFC.

Theorem 1.2 (Bartoszyński, et al. [9]). *Let H be a union of a $\mathfrak{b}$-scale and Q. Then all finite powers of H satisfy $\mathsf{U}_{fin}(\mathcal{O}, \Gamma)$ (but H is not σ-compact).*

Consequently, there exists a counter example G_H to the Hurewicz conjecture, such that $|G_H| = \mathfrak{b}$ and G_H is a *subgroup* of $\mathbb{R}$ [55].

Similarly, it was shown in [9] that there exists a counter example of size $\mathfrak{d}$ to the Menger conjecture. However it is open whether the group theoretic version also holds.

Problem 1.3 ([55]). *Does there exist (in ZFC) a subgroup G_M of $\mathbb{R}$ such that $|G_M| = \mathfrak{d}$ and G_M has Menger's property $\mathsf{U}_{fin}(\mathcal{O}, \mathcal{O})$?*

2. The Borel conjecture

2.1. Strong measure zero

Recall that a set of reals X is *null* if for each positive ϵ there exists a cover $\{I_n\}_{n\in\mathbb{N}}$ of X such that $\sum_n \operatorname{diam}(I_n) < \epsilon$. In his 1919 paper [12], Borel introduced the following stronger property: A set of reals X is *strongly null* (or: has *strong measure zero*) if, for each sequence $\{\epsilon_n\}_{n\in\mathbb{N}}$ of positive reals, there exists a cover $\{I_n\}_{n\in\mathbb{N}}$ of X such that $\operatorname{diam}(I_n) < \epsilon_n$ for all n. But Borel was unable to construct a nontrivial (that is, an uncountable) example of a strongly null set. He therefore conjectured that there exist no such examples. Sierpiński (1928) observed that every Luzin set is strongly null (see the hint on page 228 for the reason), thus the Continuum Hypothesis implies that *Borel's Conjecture* is false. Sierpiński asked whether the property of being strongly null is preserved under taking homeomorphic (or even continuous) images. The answer, given by Rothberger (1941) in [36], is negative under the Continuum Hypothesis.

If we carefully check Rothberger's argument, we can obtain a slightly stronger result without making the proof more complicated, and with the benefit of understanding the underlying combinatorics better. Theorem 2.1 and Proposition 2.2 below are probably folklore, but we do not know of a satisfactory reference for them so we give complete proofs. (The proof of Theorem 2.1 is a modification of the proof of [5, Theorem 2.9].) Let SMZ denote the collection of strongly null sets of reals. A subset A of $\mathbb{N}^\mathbb{N}$ is *strongly unbounded* if for each $f \in \mathbb{N}^\mathbb{N}$, $|\{g \in A : g \leq^* f\}| < |A|$. Observe that there exist strongly unbounded sets of sizes $\mathfrak{b}$ and $\mathfrak{d}$, thus any of the hypotheses $\mathsf{non}(\mathrm{SMZ}) = \mathfrak{b}$ or $\mathsf{non}(\mathrm{SMZ}) = \mathfrak{d}$ (in particular, the Continuum Hypothesis) implies the assumption in the following theorem.

Theorem 2.1. *Assume that there exists a strongly unbounded set of size* $\mathsf{non}(\mathrm{SMZ})$. *Then there exist a strongly null set of reals X and a continuous image Y of X such that Y is not strongly null.*

Proof. Let $\kappa = \mathsf{non}(\mathrm{SMZ})$. Then there exist: A strongly unbounded set $A = \{f_\alpha : \alpha < \kappa\}$, and a set of reals $Y = \{y_\alpha : \alpha < \kappa\}$ that is not strongly null. By standard translation arguments (see, e.g., [58]) we may assume that $Y \subseteq [0,1]$ and therefore think of Y as a subset of $\{0,1\}^\mathbb{N}$ ($\{0,1\}^\mathbb{N}$ is *Cantor's space*, which is endowed with the product topology). Consequently, the set $A' = \{f'_\alpha : \alpha < \kappa\}$, where for each α, $f'_\alpha(n) = 2f_\alpha(n) + y_\alpha(n)$ for all n, is also strongly unbounded, and the mapping $A' \to Y$ defined by $f(n) \mapsto f(n) \bmod 2$ is continuous and surjective.

It remains to show that A' is a continuous image of a strongly null set of reals X. Let $\Psi : \mathbb{N}^\mathbb{N} \to \mathbb{R} \setminus \mathbb{Q}$ be a homeomorphism (e.g., taking continued fractions), and let $X = \Psi[A']$. We claim that X is strongly null. Indeed, assume that $\{\epsilon_n\}_{n\in\mathbb{N}}$ is a sequence of positive reals. Enumerate $\mathbb{Q} = \{q_n : n \in \mathbb{N}\}$, and choose for each n an open interval I_{2n} of length less than ϵ_{2n} such that $q_n \in I_{2n}$. Let $U = \bigcup_n I_{2n}$. Then $\mathbb{R} \setminus U$ is σ-compact, thus $B = \Psi^{-1}[\mathbb{R} \setminus U] = \mathbb{N}^\mathbb{N} \setminus \Psi^{-1}[U]$ is a σ-compact and therefore bounded subset of $\mathbb{N}^\mathbb{N}$. As A' is strongly unbounded, $|A' \setminus \Psi^{-1}[U]| = |A' \cap B| < \kappa = \mathsf{non}(\mathrm{SMZ})$, thus $\Psi[A' \cap B] = \Psi[A'] \setminus U$ is strongly null, so we can find intervals I_{2n+1} of diameter at most ϵ_{2n+1} covering this set, and notice that

$$X = \Psi[A'] \subseteq (\Psi[A'] \setminus U) \cup U \subseteq \bigcup_{n\in\mathbb{N}} I_n. \qquad \square$$

From Theorem 2.1 it is possible to deduce that SMZ is not provably closed under taking homeomorphic images. The argument in the following proof, that is probably similar to Rothberger's, was pointed out to us by T. Weiss. Observe that SMZ is hereditary (that is, if X is strongly null and Y is a subset of X, then Y is strongly null too), and that it is preserved under taking uniformly continuous images.

Proposition 2.2. *If a hereditary property P is not preserved under taking continuous images, but is preserved under taking* uniformly *continuous images, then it is not preserved under taking homeomorphic images.*

Proof. Assume that X satisfies P, Y does not satisfy P, and $f : X \to Y$ is a continuous surjection. Let $\tilde{X} \subseteq X$ (so that $\tilde{X}$ satisfies P) be such that $f : \tilde{X} \to Y$ is a (continuous) bijection. Then the set $f \subseteq X \times Y$ (we identify f with its graph) is homeomorphic to $\tilde{X}$ which satisfies P, but the projection of f on the second coordinate, which is a uniformly continuous image of f, is equal to Y. Thus f does not satisfy P. □

2.2. Rothberger's property

This lead Rothberger to introduce the following topological version of strong measure zero (which is preserved under taking continuous images). Again, let $\mathscr{A}$ and $\mathscr{B}$ be collections of open covers of a topological space X. Consider the following prototype of a selection hypothesis.

$\mathsf{S}_1(\mathscr{A}, \mathscr{B})$: For each sequence $\{\mathcal{U}_n\}_{n\in\mathbb{N}}$ of members of $\mathscr{A}$, there exist members $U_n \in \mathcal{U}_n$, $n \in \mathbb{N}$, such that $\{U_n\}_{n\in\mathbb{N}} \in \mathscr{B}$.

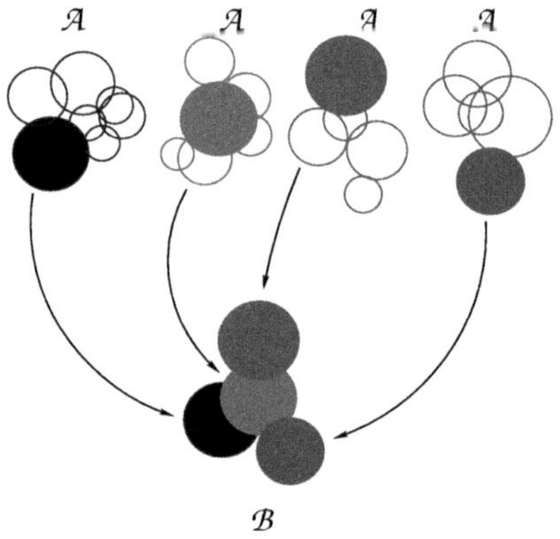

FIGURE 2. $\mathsf{S}_1(\mathscr{A}, \mathscr{B})$

Then Rothberger introduced the case $\mathscr{A} = \mathscr{B} = \mathcal{O}$ (the collection of all open covers).[5] Clearly, *Rothberger's property* $\mathsf{S}_1(\mathcal{O}, \mathcal{O})$ implies being strongly null, and the usual argument shows that every Luzin set L satisfies $\mathsf{S}_1(\mathcal{O}, \mathcal{O})$. Moreover, Fremlin and Miller [17] proved that for a metric space $\langle X, d\rangle$, $\mathsf{S}_1(\mathcal{O}, \mathcal{O})$ is the same as having strong measure zero with respect to all metrics which generate the same topology as the one defined by d.

The question of the consistency of Borel's Conjecture was settled in 1976, when Laver in his deep work [28] showed that Borel's Conjecture is consistent. We will return to Borel's Conjecture in Section 3.6.

[5]Originally, Rothberger denoted this property C'', the reason being as follows. In his 1919 paper [12], Borel considered several properties, which he enumerated as A, B, C, and so on. The property that was numbered C was that of strong measure zero. . Thus, Rothberger used C' to denote continuous images of elements of C, and C'' to be what we now call $\mathsf{S}_1(\mathcal{O}, \mathcal{O})$, since it implies C'.

3. Classification

3.1. More properties

Having the terminology introduced thus far, we can also consider the properties $\mathsf{S}_1(\Gamma,\mathcal{O})$, $\mathsf{S}_1(\Gamma,\Gamma)$, and $\mathsf{S}_1(\mathcal{O},\Gamma)$. The last property turns out trivial (consider an open cover with no γ-subcover), but the first two make sense even in the restricted setting of sets of reals. These properties turn out much more restrictive than Menger's property $\mathsf{U}_{fin}(\mathcal{O},\mathcal{O})$, but they do not admit an analogue of the Borel conjecture. In fact, the set H from Theorem 1.2 satisfies $\mathsf{S}_1(\Gamma,\mathcal{O})$ (by an argument similar to that in Theorem 2.1) [9], and in fact there always exist uncountable elements satisfying $\mathsf{S}_1(\Gamma,\Gamma)$ [23, 43].

Problem 3.1.

1. (*Bartoszyński, et al.* [9]) *Does the set H from Theorem* 1.2 *satisfy* $\mathsf{S}_1(\Gamma,\Gamma)$*?*
2. *Does there always exist a set of size* $\mathfrak{b}$ *satisfying* $\mathsf{S}_1(\Gamma,\Gamma)$*?*

3.2. ω-covers

We need not stop here, and may wish to consider other important types of covers which appeared in the literature. An open cover $\mathcal{U}$ of X is an *ω-cover* of X if no single member of $\mathcal{U}$ covers X, but for each finite $F \subseteq X$ there exists a single member of $\mathcal{U}$ covering F. Let Ω denote the collection of open ω-covers of X. Then $\mathsf{S}_1(\Omega,\Gamma)$ is equivalent to the *γ-property* introduced by Gerlits and Nagy (1982) in [19], and $\mathsf{S}_1(\Omega,\Omega)$ was studied by Sakai (1988) in [38], both properties naturally arising in the study of the space of continuous real valued functions on X (we will return to this in Section 6.5).

3.3. Arkhangel'skiǐ's property

Another prototype for a selection hypothesis generalizes a property studied by Arkhangel'skiǐ, also in the context of function spaces, in 1986 [1]. The prototype is similar to $\mathsf{U}_{fin}(\mathscr{A},\mathscr{B})$, but we do not "glue" the finite subcollections.

$\mathsf{S}_{fin}(\mathscr{A},\mathscr{B})$: For each sequence $\{\mathcal{U}_n\}_{n\in\mathbb{N}}$ of members of $\mathscr{A}$, there exist finite (possibly empty) subsets $\mathcal{F}_n \subseteq \mathcal{U}_n$, $n \in \mathbb{N}$, such that $\bigcup_{n\in\mathbb{N}}\mathcal{F}_n \in \mathscr{B}$.

Then the property studied by Arkhangel'skiǐ is equivalent to $\mathsf{S}_{fin}(\Omega,\Omega)$.

3.4. The Scheepers diagram

Thus far we have a selection hypothesis corresponding to each member of the 27 element set $\{\mathsf{S}_1,\mathsf{S}_{fin},\mathsf{U}_{fin}\}\times\{\Gamma,\Omega,\mathcal{O}\}^2$. Fortunately, it suffices to consider only some of them. First, observe that in the cases we consider,

$$\mathsf{S}_1(\mathscr{A},\mathscr{B}) \Rightarrow \mathsf{S}_{fin}(\mathscr{A},\mathscr{B}) \Rightarrow \mathsf{U}_{fin}(\mathscr{A},\mathscr{B}),$$

and we have the following monotonicity property: For $\Pi \in \{\mathsf{S}_1,\mathsf{S}_{fin},\mathsf{U}_{fin}\}$, if $\mathscr{A} \subseteq \mathscr{C}$ and $\mathscr{B} \subseteq \mathscr{D}$, then:

$$\begin{array}{ccc} \Pi(\mathscr{A},\mathscr{B}) & \rightarrow & \Pi(\mathscr{A},\mathscr{D}) \\ \uparrow & & \uparrow \\ \Pi(\mathscr{C},\mathscr{B}) & \rightarrow & \Pi(\mathscr{C},\mathscr{D}) \end{array}$$

After removing trivial properties and proving equivalences among the remaining ones (see [23] for a summary of these), we get the *Scheepers Diagram* (Figure 3). In this diagram, as in the ones to follow, an arrow denotes implication, and below each property we wrote its critical cardinality. ($\mathsf{cov}(\mathcal{M})$ denotes the minimal cardinality of a cover of $\mathbb{R}$ by meager sets, and $\mathfrak{p}$ is the *pseudo-intersection number* to be defined in Section 5.2. See [11] for information on these cardinals as well as other cardinals which we mention later.)

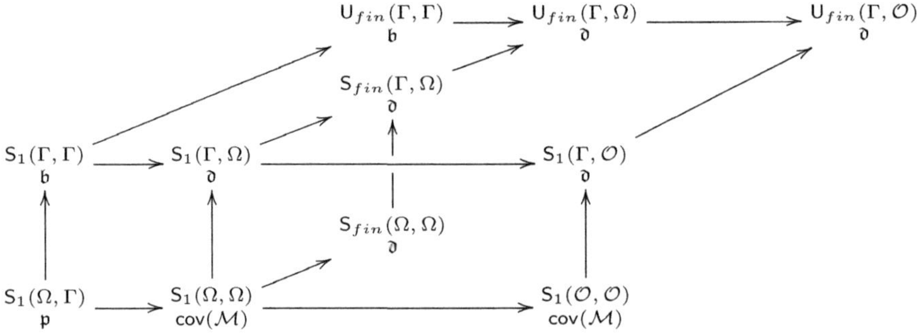

FIGURE 3. The Scheepers Diagram

There remain only two problems concerning this diagram.

Problem 3.2 (Just, Miller, Scheepers, Szeptycki [23]).

1. *Does* $\mathsf{U}_{fin}(\Gamma,\Omega)$ *imply* $\mathsf{S}_{fin}(\Gamma,\Omega)$*?*
2. *If not, does* $\mathsf{U}_{fin}(\Gamma,\Gamma)$ *imply* $\mathsf{S}_{fin}(\Gamma,\Omega)$*?*

All other implications are settled in [40, 23], using two methods. One approach uses consistency results concerning the values of the critical cardinalities. For example, it is consistent that $\mathfrak{b} < \mathfrak{d}$, thus none of the properties with critical cardinality $\mathfrak{d}$ can imply any of those with critical cardinality $\mathfrak{b}$. Another approach is by transfinite constructions under the Continuum Hypothesis (such as special kinds of Luzin and Sierpiński sets). Recently, an approach combining these two approaches was investigated – see, e.g., [13, 4, 16].

3.5. Borel covers

There are other natural types of covers which appear in the literature, but probably the first natural question is: What happens if we replace "open" by "countable Borel" in the types of covers which we consider? Let $\mathcal{B}, \mathcal{B}_\Omega, \mathcal{B}_\Gamma$ denote the collections of *countable Borel* covers, ω-covers, and γ-covers of the given space, respectively. It turns out that the same analysis is applicable when we plug in these families instead of $\mathcal{O}, \Omega, \Gamma$, and in fact one gets more equivalences. Moreover, some of the resulting properties turn out equivalent to properties which appeared in the literature in other guises. This is shown in [45], where it is also shown that no arrows can be added (except perhaps those corresponding to Problem 3.2) to the extended diagram (Figure 4).

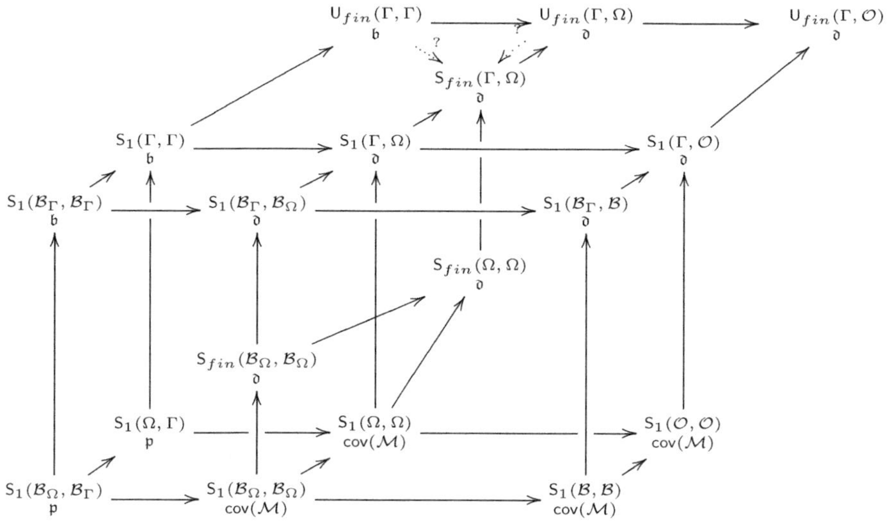

FIGURE 4. The extended Scheepers Diagram

3.6. Borel's conjecture revisited

It is easy to verify that every countable set of reals X satisfies the strongest property in the extended Scheepers Diagram 4, namely, $\mathsf{S}_1(\mathcal{B}_\Omega,\mathcal{B}_\Gamma)$. By Laver's result mentioned in Section 2.2, we know that it is consistent that all properties between $\mathsf{S}_1(\mathcal{B}_\Omega,\mathcal{B}_\Gamma)$ and $\mathsf{S}_1(\mathcal{O},\mathcal{O})$ (inclusive) are consistent to hold only for countable sets of reals.

All other classes in the original Scheepers Diagram 3 contain uncountable elements: Recall that every σ-compact set satisfies the Hurewicz property $\mathsf{U}_{fin}(\mathcal{O},\Gamma)$. Now, $\mathsf{S}_{fin}(\Omega,\Omega)$ is equivalent to satisfying Menger's property $\mathsf{U}_{fin}(\mathcal{O},\mathcal{O})$ in all finite powers [23]. Thus, every σ-compact set satisfies $\mathsf{S}_{fin}(\Omega,\Omega)$. As for the remaining properties, recall from Section 1.3 that $\mathsf{S}_1(\Gamma,\Gamma)$ always contains an uncountable element. In other words, none of the properties except those mentioned in the previous paragraph can satisfy an analogue of Borel's Conjecture. However, by a result of Miller, Borel's Conjecture for $\mathsf{S}_1(\mathcal{B}_\Gamma,\mathcal{B}_\Gamma)$ is consistent [9].

Problem 3.3.

1. *(folklore) Is it consistent that every set of reals which satisfies* $\mathsf{S}_1(\mathcal{B}_\Gamma,\mathcal{B})$ *is countable?*
2. *What about* $\mathsf{S}_{fin}(\mathcal{B}_\Omega,\mathcal{B}_\Omega)$ *and* $\mathsf{S}_1(\mathcal{B}_\Gamma,\mathcal{B}_\Omega)$*?*

A combinatorial formulation of the first question in Problem 3.3 is obtained by replacing $\mathsf{S}_1(\mathcal{B}_\Gamma,\mathcal{B})$ with the equivalent property "every Borel image in $\mathbb{N}^\mathbb{N}$ is not dominating".

Other interesting investigations in this direction are of the form: Is it consistent that a certain property in the diagram satisfies Borel's Conjecture, whereas another one does not? Some results in this direction are the following.

Theorem 3.4 (Miller [34]).

1. *Borel's Conjecture for* $\mathsf{S}_1(\mathcal{O},\mathcal{O})$ *implies Borel's Conjecture (for strong measure zero);*
2. *Borel's Conjecture for* $\mathsf{S}_1(\Omega,\Gamma)$ *does not imply Borel's Conjecture.*

The proofs use, of course, combinatorial arguments (and forcing in the second case: The model is obtained by adding $\aleph_2$ dominating reals with a finite support iteration to a model of the Continuum Hypothesis). Using yet more combinatorial arguments, it is possible to extend Theorem 3.4.

Theorem 3.5 (Weiss, et al. [58]). *Borel's Conjecture for* $\mathsf{S}_1(\Omega,\Omega)$ *implies Borel's Conjecture. Consequently, Borel's Conjecture for* $\mathsf{S}_1(\Omega,\Gamma)$ *does not imply Borel's Conjecture for* $\mathsf{S}_1(\Omega,\Omega)$.

This settles completely this investigation when we restrict attention to the original Scheepers Diagram 3.

4. Preservation of properties

4.1. Continuous images

It is easy to see that all properties in the Scheepers Diagram 3 (as well as all other selection properties in this paper) are preserved under taking continuous images [23]. Similarly, the selection properties involving Borel covers are preserved under taking Borel images [45]. The situation is not as good concerning other types of preservation...

4.2. Additivity

In [23] (1996), Just, Miller, Scheepers, and Szeptycki raised the following *additivity problem*: It is easy to see that some of the properties in the Scheepers diagram are (provably) preserved under taking finite and even countable unions (i.e., they are *σ-additive*). What about the remaining ones? Figure 4.2(a) summarizes the knowledge that was available at the point the question was posed, where the positions are according to Figure 3.

In 1999, Scheepers [43] proved that the answer is positive for $\mathsf{S}_1(\Gamma,\Gamma)$.

The key observation towards solving the problem for the remaining properties was the following analogue of Hurewicz' Theorem 1.1. According to Blass, a family $Y \subseteq \mathbb{N}^\mathbb{N}$ is *finitely dominating* if for each $g \in \mathbb{N}^\mathbb{N}$ there exist k and $f_1,\dots,f_k \in Y$ such that $g(n) \le \max\{f_1(n),\dots,f_k(n)\}$ for all but finitely many n.

Theorem 4.1 ([51]). *A set of reals* X *satisfies* $\mathsf{U}_{fin}(\Gamma,\Omega)$ *if, and only if, each continuous image of* X *in* $\mathbb{N}^\mathbb{N}$ *is not finitely dominating.*

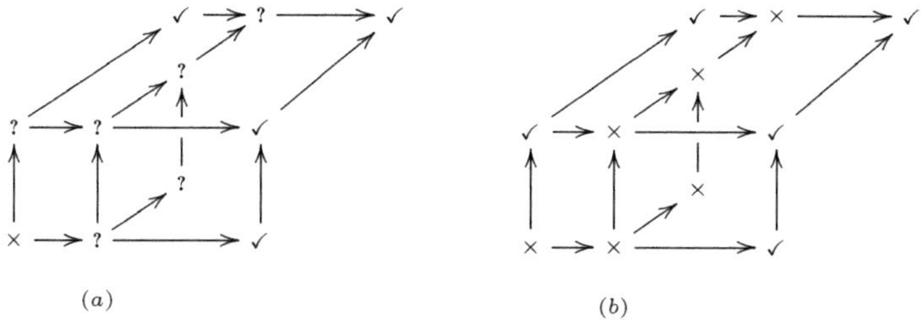

FIGURE 5. The additivity problem (a) and its solution (b)

Motivated by this observation, the following theorem was proved. (As usual, $\mathfrak{c} = 2^{\aleph_0}$ denotes the size of the continuum.)

Theorem 4.2 (Bartoszyński, Shelah, et al. [8]). *Assume the Continuum Hypothesis (or just* $\mathsf{cov}(\mathcal{M}) = \mathfrak{c}$*). Then there exist sets of reals* L_0*, and* L_1 *satisfying* $\mathsf{S}_1(\mathcal{B}_\Omega, \mathcal{B}_\Omega)$ *such that* $L_0 \cup L_1$ *is finitely dominating.*

The proof used a tricky "power sharing" between L_0 and L_1 during their transfinite-inductive construction.

Consequently, *none of the remaining properties* is provably preserved under taking finite unions (Figure 4.2(b)). This also settled all corresponding problems in the Borel case (see the extended diagram 4). Interestingly, the simple observation in Theorem 4.1 was also the key behind proving that consistently (namely, assuming NCF), $\mathsf{U}_{fin}(\mathcal{O}, \Omega)$ is σ-additive.

Bad transmission of knowledge. The transmission of knowledge concerning the additivity problem was very poor (take a deep breathe): It posteriorly turns out that if we restrict attention to the open case only, then the additivity problem was already implicitly solved earlier. In 1999, Scheepers [42] constructed sets of reals L_0, and L_1 satisfying $\mathsf{S}_1(\Omega, \Omega)$ such that $L_0 + L_1$ is finitely dominating. It is easy to see that this implies that $L_0 \cup L_1$ is finitely dominating, which settles the problem if we add the missing ingredient Theorem 4.1, which, funnily, seems to be the simplest part of the solution. Scheepers was unaware of this observation and consequently of that solving the additivity problem completely, but he did point out that his construction implied that the properties between $\mathsf{S}_1(\Omega, \Omega)$ and $\mathsf{S}_{fin}(\Omega, \Omega)$ (inclusive) are not provably additive. In turn, we were not aware of this when we proved Theorem 4.2. On top of that, Theorem 4.1, the open part of Theorem 4.2, and the result concerning NCF were independently proved by Eisworth and Just in [16], and similar results were also independently obtained by Banakh, Nickolas, and Sanchis in [3]. As if this is not enough, the main ingredient of the result concerning NCF was also independently obtained by Blass [10].

This is a good point to recommend the reader announce his new results in the $\mathcal{SPM}$ **Bulletin** (see [57]), so as to avoid similar situations.

4.3. Hereditarity

A property (or a class of topological spaces) is *hereditary* if it is preserved under taking subsets. Despite the fact that the properties in the Scheepers diagram are (intuitively) notions of smallness, none of them is (provably) hereditary. Define a topology on the space $P(\mathbb{N})$ of all sets of natural numbers by identifying it with $\{0,1\}^{\mathbb{N}}$. Note that $[\mathbb{N}]^{<\aleph_0}$, the collection of finite subsets of $\mathbb{N}$, is dense in $P(\mathbb{N})$. By taking increasing enumerations, we have that the *Rothberger space* $[\mathbb{N}]^{\aleph_0} = P(\mathbb{N}) \setminus [\mathbb{N}]^{<\aleph_0}$ is identified with the space of increasing sequences of natural numbers, which in turn is homeomorphic to the Baire space $\mathbb{N}^{\mathbb{N}}$.

Theorem 4.3 (Babinkostova, Kočinac, and Scheepers [2]; Bartoszyński, et al. [9]). *Assuming (a portion of) the Continuum Hypothesis, there exists a set of reals X satisfying $\mathsf{S}_1(\Omega, \Gamma)$ and a countable subset Q of X, such that $X \setminus Q$ does not satisfy $\mathsf{U}_{fin}(\mathcal{O}, \mathcal{O})$.*

The proof for this assertion is a modification of the construction of Galvin and Miller [18], with $X \subseteq P(\mathbb{N})$, $Q = [\mathbb{N}]^{<\aleph_0}$, so that $X \setminus [\mathbb{N}]^{<\aleph_0} \subseteq [\mathbb{N}]^{\aleph_0}$ is dominating (when viewed as a subset of $\mathbb{N}^{\mathbb{N}}$) which is what we look for in light of Hurewicz' Theorem 1.1.

However, in the Borel case, $\mathsf{S}_1(\mathcal{B}, \mathcal{B})$ and all properties of the form $\Pi(\mathcal{B}_\Gamma, \mathscr{B})$ are hereditary. This is immediate from the combinatorial characterizations [45], but is also easy to prove directly [9]. However, not all properties in the Borel case are hereditary, e.g., Miller [33] proved that assuming the Continuum Hypothesis, $\mathsf{S}_1(\mathcal{B}_\Omega, \mathcal{B}_\Gamma)$ is not hereditary.

Problem 4.4 (Bartoszyński, et al. [9]). *Is any of $\mathsf{S}_1(\mathcal{B}_\Omega, \mathcal{B}_\Omega)$ or $\mathsf{S}_{fin}(\mathcal{B}_\Omega, \mathcal{B}_\Omega)$ hereditary?*

5. The Minimal Tower Problem and τ-covers

5.1. "Rich" covers of spaces

[6] To avoid trivialities, let us decide that when we say that $\mathcal{U}$ is a *cover* of X, we mean that $X = \cup\mathcal{U}$ (the usual requirement), and in addition no single member of $\mathcal{U}$ covers X. We also always assume that the space X is infinite.

$\mathcal{U}$ is a *large cover* of X if for each $x \in X$, there exist infinitely many $U \in \mathcal{U}$ such that $x \in U$. Let Λ denote the collection of large open covers of X. Then

$$\Gamma \subseteq \Omega \subseteq \Lambda,$$

the last implication being a cute exercise.

$\mathcal{U}$ is a *τ-cover* of X if it is a large cover of X, and for each $x, y \in X$, at least one of the sets $\{U \in \mathcal{U} : x \in U \text{ and } y \notin U\}$ or $\{U \in \mathcal{U} : y \in U \text{ and } x \notin U\}$ is finite. This is a reminiscent of the negation of the T_1 property, where here the

---

[6] Richness can be viewed as some sort of redundancy, and some people are richer than others. This is precisely the case with *rich covers*.

notion is applied "modulo finite". Let T (uppercase τ) denote the collection of open τ-covers of X. Then

$$\Gamma \subseteq \mathrm{T} \subseteq \Omega.$$

The motivation behind the definition of τ-covers is the Minimal Tower Problem.

5.2. The Minimal Tower Problem

Recall that $[\mathbb{N}]^{\aleph_0}$ is the collection of infinite subsets of $\mathbb{N}$. $A \subseteq^* B$ means that $A \setminus B$ is finite. $\mathcal{F} \subseteq [\mathbb{N}]^{\aleph_0}$ is *centered* if for each finite $\mathcal{A} \subseteq \mathcal{F}$, $\cap\mathcal{A}$ is infinite. $A \in [\mathbb{N}]^{\aleph_0}$ is a *pseudo-intersection* of $\mathcal{F}$ if $A \subseteq^* B$ for all $B \in \mathcal{F}$. Let $\mathfrak{p}$ denote the minimal size of a centered family $\mathcal{F} \subseteq [\mathbb{N}]^{\aleph_0}$ with no pseudo-intersection, and $\mathfrak{t}$ denote the minimal size of a $\subseteq^*$-linearly ordered family $\mathcal{F} \subseteq [\mathbb{N}]^{\aleph_0}$ which has no pseudo-intersection. Then $\mathfrak{p} \leq \mathfrak{t}$, and the *Minimal Tower Problem* is:

Problem 5.1. *Is it provable that* $\mathfrak{p} = \mathfrak{t}$*?*

This is one of the major and oldest problems of infinitary combinatorics. Allusions to this problem can be found in Rothberger's 1940's works (see, e.g., [37]). The problem is explicitly mentioned in van Douwen's survey [15], and quoted in Vaughan's survey [61, Problem 333], where it is considered the most interesting open problem in the field. Extensive work of Shelah and others in the field solved all problems of this sort which were mentioned in [15], except for the Minimal Tower Problem.

5.3. A dictionary

How is this problem related to τ-covers? Each type of rich cover corresponds to some combinatorial notion (of richness), as follows. If we confine attention to sets of reals, then we may assume that all open covers we consider are countable. Thus, assume that $\mathcal{U} = \{U_n\}_{n\in\mathbb{N}}$ is a countable (not necessarily open) cover of X, and define the *Marczewski characteristic function* of $\mathcal{U}$ [30] by

$$h(x) = \{n : x \in U_n\},$$

so that $h : X \to P(\mathbb{N})$. h is continuous if the sets U_n are clopen, and Borel if the sets U_n are Borel.

We have the following *dictionary* translating properties of $\mathcal{U}$ to properties of the image $h[X]$.

$\{U_n\}_{n\in\mathbb{N}}$	$h[X]$
large cover	subset of $[\mathbb{N}]^{\aleph_0}$
ω-cover	centered
τ-cover	linearly ordered by $\subseteq^*$
γ-cover	cofinite sets
contains a γ-cover	has a pseudo-intersection

(Note that, since we assume that $X \not\subseteq U_n$ for all n, we have that $h[X]$ is centered if, and only if, it is a base for a nonprincipal filter on $\mathbb{N}$.)

5.4. Topological approximations of the Minimal Tower Problem

For families $\mathscr{B} \subseteq \mathscr{A}$ of covers of a space X, define the property $\mathscr{A}$ *choose* $\mathscr{B}$ as follows.

$\binom{\mathscr{A}}{\mathscr{B}}$: For each $\mathcal{U} \in \mathscr{A}$ we can choose a subset $\mathcal{V} \subseteq \mathcal{U}$ such that $\mathcal{V} \in \mathscr{B}$.

This is a prototype for many classical topological notions, most notably compactness and being Lindelöf.

In 1982, Gerlits and Nagy [19] introduced the γ-*property* $\binom{\Omega}{\Gamma}$, and proved that it is equivalent to $\mathsf{S}_1(\Omega, \Gamma)$.

By the dictionary (Section 5.3), we have that

1. $\mathsf{non}\binom{\Omega}{\Gamma} = \mathsf{non}\binom{\mathcal{B}_\Omega}{\mathcal{B}_\Gamma} = \mathfrak{p}$; and
2. $\mathsf{non}\binom{\mathrm{T}}{\Gamma} = \mathsf{non}\binom{\mathcal{B}_\mathrm{T}}{\mathcal{B}_\Gamma} = \mathfrak{t}$.

Clearly, $\binom{\Omega}{\Gamma} \subseteq \binom{\mathrm{T}}{\Gamma}$, and $\binom{\mathcal{B}_\Omega}{\mathcal{B}_\Gamma} \subseteq \binom{\mathcal{B}_\mathrm{T}}{\mathcal{B}_\Gamma}$. We therefore have the following topological problems related to the Minimal Tower Problem.

1. Is $\binom{\Omega}{\Gamma} = \binom{\mathrm{T}}{\Gamma}$?
2. Is $\binom{\mathcal{B}_\Omega}{\mathcal{B}_\Gamma} = \binom{\mathcal{B}_\mathrm{T}}{\mathcal{B}_\Gamma}$?

Observe that the answer is "No" for both problems if $\mathfrak{p} < \mathfrak{t}$ is consistent.

But it turns out that both problems can be solved outright in ZFC: The first problem was solved by Shelah [50], who proved that $\{0,1\}^\mathbb{N}$ satisfies $\binom{\mathrm{T}}{\Gamma}$. A set satisfying $\binom{\Omega}{\Gamma}$ must satisfy $\mathsf{S}_1(\mathcal{O}, \mathcal{O})$ and therefore have strong measure zero. Obviously, $\{0,1\}^\mathbb{N}$ does not have strong measure zero (it has $[0,1]$ as a uniformly continuous image). A modification of Shelah's construction yields a negative solution to the second question as well: Indeed, $\mathbb{N}^\mathbb{N}$ satisfies $\binom{\mathrm{T}}{\Gamma}$ [54], and Borel images of $\mathbb{N}^\mathbb{N}$ are analytic, and can therefore be represented as continuous images of $\mathbb{N}^\mathbb{N}$, so Borel images of $\mathbb{N}^\mathbb{N}$ satisfy $\binom{\mathrm{T}}{\Gamma}$, and therefore $\mathbb{N}^\mathbb{N}$ satisfies $\binom{\mathcal{B}_\mathrm{T}}{\mathcal{B}_\Gamma}$ [54].

5.5. Tougher topological approximations

It is easy to see that $\mathsf{non}(\mathsf{S}_1(\Omega, \Gamma)) = \mathsf{non}(\mathsf{S}_1(\mathcal{B}_\Omega, \mathcal{B}_\Gamma)) = \mathfrak{p}$, and $\mathsf{non}(\mathsf{S}_1(\mathrm{T}, \Gamma)) = \mathsf{non}(\mathsf{S}_1(\mathcal{B}_\mathrm{T}, \mathcal{B}_\Gamma)) = \mathfrak{t}$ [50], and we have the following implications:

$$\begin{array}{ccc} \mathsf{S}_1(\Omega, \Gamma) & \rightarrow & \mathsf{S}_1(\mathrm{T}, \Gamma) \\ \uparrow & & \uparrow \\ \mathsf{S}_1(\mathcal{B}_\Omega, \mathcal{B}_\Gamma) & \rightarrow & \mathsf{S}_1(\mathcal{B}_\mathrm{T}, \mathcal{B}_\Gamma) \end{array}$$

We therefore arrive at the following "tighter" approximations.

1. Is $\mathsf{S}_1(\Omega, \Gamma) = \mathsf{S}_1(\mathrm{T}, \Gamma)$?
2. Is $\mathsf{S}_1(\mathcal{B}_\Omega, \mathcal{B}_\Gamma) = \mathsf{S}_1(\mathcal{B}_\mathrm{T}, \mathcal{B}_\Gamma)$?

Again, $\mathfrak{p} < \mathfrak{t}$ implies a negative answer for both problems. It turns out that the answer is indeed negative, at least assuming the Continuum Hypothesis [54]. The main ingredients in the proof of this are the following: $\mathsf{S}_1(\mathcal{B}_\mathrm{T}, \mathcal{B}_\Gamma) = \mathsf{S}_1(\mathcal{B}_\Gamma, \mathcal{B}_\Gamma) \cap \binom{\mathcal{B}_\mathrm{T}}{\mathcal{B}_\Gamma}$, and is therefore countably additive. Now, the Continuum Hypothesis implies that there exists a hereditary $\mathsf{S}_1(\mathcal{B}_\Omega, \mathcal{B}_\Gamma)$-set $X \subseteq [0,1]$ (e.g., Brendle [13] or Miller

[33]). By a result of Galvin and Miller (1984) [18], if $Y \subseteq X$ is not F_σ or not G_δ, then $(X \setminus Y) \cup (Y+1) \notin \mathsf{S}_1(\Omega, \Gamma)$. This solves both problems at once.

The striking observation we get from this journey of topological approximations to the Minimal Tower Problem is that, despite the fact that the purely combinatorial problem is very difficult, if we add to it topological structure we get a negative answer quite easily.

It is surprising, though, that the following related problem remains open.

Problem 5.2 ([54]). *Is* $\binom{\Omega}{\Gamma} = \binom{\Omega}{\mathrm{T}}$*?*

The critical cardinality of $\binom{\Omega}{\mathrm{T}}$ is $\mathfrak{p}$ [46], so both properties have the same critical cardinality.

Remark 5.3. The reader interested in another topological study which was inspired by (and is related to) the Minimal Tower Problem is referred to Machura's [29].

5.6. Known implications and critical cardinalities

Having this new notion of rich covers, namely τ-covers, it is interesting to try and add it to Scheepers' framework of selection principles. This yields many more properties, but again some of them can be proved to be equivalent [54].

As already mentioned, a basic tool to prove *nonimplications* is the computation of critical cardinalities: If P and Q are properties with $\mathsf{non}(P) < \mathsf{non}(Q)$ consistent, then Q does not imply P. The critical cardinalities of most of the new properties were found in [54] and in Shelah, et al. [46]. Still, 6 critical cardinalities remained unsettled. These remaining cardinalities were addressed by Mildenberger, Shelah, et al., [32]. We give here one example of that treatment, and quote the main results. (Everything in the remainder of this subsection is quoted, without further notice, from [32]).

Let C_Γ, C_T, and C_Ω denote the collections of *clopen* γ-covers, τ-covers, and ω-covers of X, respectively. Recall that, since we are dealing with sets of reals, we may assume that all open covers are countable. Restricting attention to countable covers, we have the following, where an arrow denotes inclusion:

$$\begin{array}{ccccccc}
\mathcal{B}_\Gamma & \rightarrow & \mathcal{B}_\mathrm{T} & \rightarrow & \mathcal{B}_\Omega & \rightarrow & \mathcal{B} \\
\uparrow & & \uparrow & & \uparrow & & \uparrow \\
\Gamma & \rightarrow & \mathrm{T} & \rightarrow & \Omega & \rightarrow & \mathcal{O} \\
\uparrow & & \uparrow & & \uparrow & & \uparrow \\
C_\Gamma & \rightarrow & C_\mathrm{T} & \rightarrow & C_\Omega & \rightarrow & C
\end{array}$$

As each of the properties $\Pi(\cdot\ ,\cdot)$, $\Pi \in \{\mathsf{S}_1, \mathsf{S}_{fin}, \mathsf{U}_{fin}\}$, is monotonic in its first variable, we have that for each $x, y \in \{\Gamma, \mathrm{T}, \Omega, \mathcal{O}\}$,

$$\Pi(\mathcal{B}_x, \mathcal{B}_y) \rightarrow \Pi(x, y) \rightarrow \Pi(C_x, C_y)$$

(here $C_\mathcal{O} := C$ and $\mathcal{B}_\mathcal{O} := \mathcal{B}$). Consequently,

$$\mathsf{non}(\Pi(\mathcal{B}_x, \mathcal{B}_y)) \leq \mathsf{non}(\Pi(x, y)) \leq \mathsf{non}(\Pi(C_x, C_y)).$$

Definition 5.4. We use the short notation $\forall^\infty$ for "for all but finitely many" and $\exists^\infty$ for "there exist infinitely many".

1. $A \in \{0,1\}^{\mathbb{N}\times\mathbb{N}}$ is a *γ-array* if $(\forall n)(\forall^\infty m)\ A(n,m)=1$.
2. $\mathcal{A} \subseteq \{0,1\}^{\mathbb{N}\times\mathbb{N}}$ is a *γ-family* if each $A \in \mathcal{A}$ is a γ-array.
3. A family $\mathcal{A} \subseteq \{0,1\}^{\mathbb{N}\times\mathbb{N}}$ is *finitely τ-diagonalizable* if there exist finite (possibly empty) subsets $F_n \subseteq \mathbb{N}$, $n \in \mathbb{N}$, such that:
   (a) For each $A \in \mathcal{A}$: $(\exists^\infty n)(\exists m \in F_n)\ A(n,m)=1$;
   (b) For each $A, B \in \mathcal{A}$:
   Either $(\forall^\infty n)(\forall m \in F_n)\ A(n,m) \leq B(n,m)$,
   or $(\forall^\infty n)(\forall m \in F_n)\ B(n,m) \leq A(n,m)$.

Using the dictionary of Section 5.3, one can prove the following. (Notice that $\{0,1\}^{\mathbb{N}\times\mathbb{N}}$ is topologically the same as the Cantor space $\{0,1\}^{\mathbb{N}}$.)

Theorem 5.5. *For a set of reals X, the following are equivalent:*

1. *X satisfies $\mathsf{S}_{fin}(\mathcal{B}_\Gamma, \mathcal{B}_\mathrm{T})$; and*
2. *For each Borel function $\Psi : X \to \{0,1\}^{\mathbb{N}\times\mathbb{N}}$, if $\Psi[X]$ is a γ-family, then it is finitely τ-diagonalizable.*

The corresponding assertion for $\mathsf{S}_{fin}(C_\Gamma, C_\mathrm{T})$ holds when "Borel" is replaced by "continuous".

Lemma 5.6. *The minimal cardinality of a γ-family which is not finitely τ-diagonalizable is $\mathfrak{b}$.*

(To appreciate the result in Lemma 5.6, the reader may wish to try proving it for a while.) Having Theorem 5.5 and Lemma 5.6, we get that

$$\mathfrak{b} = \mathsf{non}(\mathsf{S}_{fin}(\mathcal{B}_\Gamma, \mathcal{B}_\mathrm{T})) \leq \mathsf{non}(\mathsf{S}_{fin}(\Gamma, \mathrm{T})) \leq \mathsf{non}(\mathsf{S}_{fin}(C_\Gamma, C_\mathrm{T})) = \mathfrak{b},$$

and therefore $\mathsf{non}(\mathsf{S}_{fin}(\Gamma, \mathrm{T})) = \mathfrak{b}$.

Notice that Theorem 5.5 reduced the original topological question into a purely combinatorial question. Its solution in Lemma 5.6 may be of independent interest to those working on this sort of pure combinatorics.

It follows that $\mathsf{non}(\mathsf{S}_1(\Gamma, \mathrm{T})) = \mathfrak{b}$, and using a similar approach, it can be proved that $\mathsf{non}(\mathsf{S}_1(\mathrm{T}, \mathrm{T})) = \mathfrak{t}$, and $\mathsf{non}(\mathsf{S}_{fin}(\mathrm{T}, \mathrm{T})) = \min\{\mathfrak{b}, \mathfrak{s}\}$.

What about the remaining two cardinals? It is not difficult to see that $\mathsf{non}(\mathsf{S}_{fin}(\mathrm{T}, \Omega)) = \mathsf{non}(\mathsf{S}_{fin}(\mathrm{T}, \mathcal{O}))$, call this joint cardinal $\mathfrak{od}$, the *o-diagonalization number*; the reason for this to be explained soon. The surviving properties (in the open case) appear in Figure 6, with their critical cardinalities, and serial numbers (for later reference). The new cardinalities computed in [32] are framed.

By Figure 6,

$$\mathsf{cov}(\mathcal{M}) = \mathsf{non}(\mathsf{S}_1(\mathcal{O}, \mathcal{O})) \leq \mathsf{non}(\mathsf{S}_1(\mathrm{T}, \mathcal{O})) \leq \mathsf{non}(\mathsf{S}_1(\Gamma, \mathcal{O})) = \mathfrak{d},$$

thus $\mathsf{cov}(\mathcal{M}) \leq \mathfrak{od} \leq \mathfrak{d}$.

Definition 5.7. A family $\mathcal{A} \subseteq \{0,1\}^{\mathbb{N}\times\mathbb{N}}$ is a *τ-family* if:

1. For each $A \in \mathcal{A}$: $(\forall n)(\exists^\infty m)\ A(n,m)=1$;

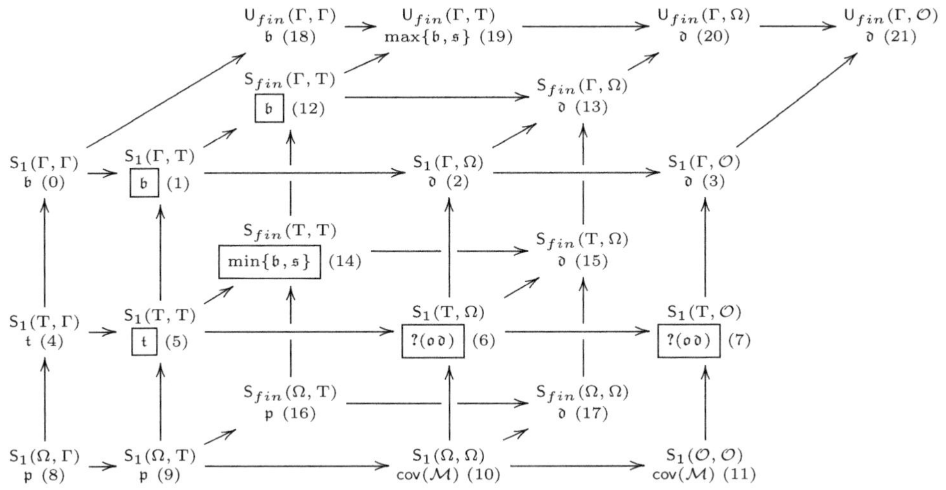

FIGURE 6. The Scheepers diagram, enhanced with τ-covers

2. For each $A, B \in \mathcal{A}$ and each n:
   Either $(\forall^\infty m)\ A(n,m) \leq B(n,m)$,
   or $(\forall^\infty m)\ B(n,m) \leq A(n,m)$.

A τ-family $\mathcal{A}$ is *o-diagonalizable* if there exists a function $g : \mathbb{N} \to \mathbb{N}$, such that:

$$(\forall A \in \mathcal{A})(\exists n)\ A(n, g(n)) = 1.$$

(Equivalently, $(\forall A \in \mathcal{A})(\exists^\infty n)\ A(n, g(n)) = 1$.)

As in Theorem 5.5, $\mathsf{S}_1(\mathcal{B}_\mathrm{T}, \mathcal{B})$ and $\mathsf{S}_1(C_\mathrm{T}, C)$ have a natural combinatorial characterization. This characterization implies that $\mathfrak{od}$ is equal to the minimal cardinality of a τ-family that is not o-diagonalizable. A detailed study of $\mathfrak{od}$ is initiated in [32], where it is shown that consistently $\mathfrak{od} < \min\{\mathfrak{h}, \mathfrak{s}, \mathfrak{b}\}$.

Problem 5.8 (Mildenberger, Shelah, et al. [32]). *Is it consistent (relative to ZFC) that* $\mathsf{cov}(\mathcal{M}) < \mathfrak{od}$*?*

This problem, which originated from the topological studies of the minimal tower problem, is of similar flavor: It is well known that if $\mathfrak{p} = \aleph_1$, then $\mathfrak{t} = \aleph_1$ too. We have a similar assertion for $\mathsf{cov}(\mathcal{M})$ and $\mathfrak{od}$: If $\mathsf{cov}(\mathcal{M}) = \aleph_1$, then $\mathfrak{od} = \aleph_1$, either.

5.7. A table of open problems

It is possible that the diagram in Figure 6 is incomplete: There are many unsettled possible implications in it. After [54, 46], there remained 76 (!) potential implications which were not proved or ruled out. The study made recently in [32] and described in the previous section ruled out 21 of these implications, so that

55 implications remain unsettled. The situation is summarized in Table 1, which updates the corresponding table given in [56].

Each entry (i, j) (ith row, jth column) contains a symbol. ✓means that property (i) in Figure 6 implies property (j) in Figure 6. × means that property (i) does not (provably) imply property (j), and **?** means that the corresponding implication is still unsettled.

	0	1	2	3	4	5	6	7	8	9	10	11	12	13	14	15	16	17	18	19	20	21
0	✓	✓	✓	✓	×	×	×	×	×	×	×	×	✓	✓	×	?	×	×	✓	✓	✓	✓
1	?	✓	✓	✓	×	×	×	×	×	×	×	×	✓	✓	×	?	×	×	?	✓	✓	✓
2	×	×	✓	✓	×	×	×	×	×	×	×	×	×	✓	×	?	×	×	×	×	✓	✓
3	×	×	×	✓	×	×	×	×	×	×	×	×	×	×	×	×	×	×	×	×	×	✓
4	✓	✓	✓	✓	✓	✓	✓	✓	×	×	?	?	✓	✓	✓	✓	×	?	✓	✓	✓	✓
5	?	✓	✓	✓	?	✓	✓	✓	×	×	?	?	✓	✓	✓	✓	×	?	?	✓	✓	✓
6	×	×	✓	✓	×	×	✓	✓	×	×	?	?	×	✓	×	✓	×	?	×	×	✓	✓
7	×	×	×	✓	×	×	×	✓	×	×	×	?	×	×	×	×	×	×	×	×	×	✓
8	✓	✓	✓	✓	✓	✓	✓	✓	✓	✓	✓	✓	✓	✓	✓	✓	✓	✓	✓	✓	✓	✓
9	?	✓	✓	✓	?	✓	✓	✓	?	✓	✓	✓	✓	✓	✓	✓	✓	✓	?	✓	✓	✓
10	×	×	✓	✓	×	×	✓	✓	×	×	✓	✓	×	✓	×	✓	×	✓	×	×	✓	✓
11	×	×	×	✓	×	×	×	✓	×	×	×	✓	×	×	×	×	×	×	×	×	×	✓
12	?	?	?	?	×	×	×	×	×	×	×	×	✓	✓	×	?	×	×	?	✓	✓	✓
13	×	×	×	×	×	×	×	×	×	×	×	×	×	✓	×	?	×	×	×	×	✓	✓
14	?	?	?	?	×	×	×	×	×	×	×	×	✓	✓	✓	✓	×	?	?	✓	✓	✓
15	×	×	×	×	×	×	×	×	×	×	×	×	×	✓	×	✓	×	?	×	×	✓	✓
16	?	?	?	?	?	?	?	?	?	?	?	?	✓	✓	✓	✓	✓	✓	?	✓	✓	✓
17	×	×	×	×	×	×	×	×	×	×	×	×	×	✓	×	✓	×	✓	×	×	✓	✓
18	×	×	×	×	×	×	×	×	×	×	×	×	×	?	×	?	×	×	✓	✓	✓	✓
19	×	×	×	×	×	×	×	×	×	×	×	×	×	?	×	?	×	×	×	✓	✓	✓
20	×	×	×	×	×	×	×	×	×	×	×	×	×	?	×	?	×	×	×	×	✓	✓
21	×	×	×	×	×	×	×	×	×	×	×	×	×	×	×	×	×	×	×	×	×	✓

TABLE 1. Known implications and nonimplications

Problem 5.9 (Mildenberger, Shelah, et al. [54, 32]). *Settle any of the remaining* 55 *implication in Table* 1.

5.8. The Minimal Tower Problem revisited

The study of τ-covers was motivated by a combinatorial problem. Interestingly, it rewarded us with a closely related *purely combinatorial* problem. One of the difficulties with treating τ-covers is the following: Unlike the case of ω-covers or γ-covers, it could be that $\mathcal{U}$ is a τ-cover of X which refines another cover $\mathcal{V}$ of X, but $\mathcal{V}$ is not a τ-cover of X. This led to the introduction of the following close relative [54].

A family $Y \subseteq [\mathbb{N}]^{\aleph_0}$ is *linearly refinable* if for each $y \in Y$ there exists an infinite subset $\hat{y} \subseteq y$ such that the family $\hat{Y} = \{\hat{y} : y \in Y\}$ is linearly (quasi)ordered by $\subseteq^*$. A cover $\mathcal{U}$ of X is a τ^**-cover* of X if it is large, and $h[X]$ (where h is the Marczewski characteristic function of $\mathcal{U}$, see Section 5.3) is linearly refinable. Let T^* denote the collection of all *countable* open τ^*-covers of X.

If we restrict attention to countable covers (this does not make a difference for sets of reals), then $\mathrm{T} \subseteq \mathrm{T}^* \subseteq \Omega$. Let $\mathfrak{p}^* = \mathsf{non}(\binom{\Omega}{\mathrm{T}^*})$. Then by the dictionary (Section 5.3),

$$\mathfrak{p}^* = \min\{|Y| : Y \subseteq [\mathbb{N}]^{\aleph_0} \text{ is centered but not linearly refineable}\},$$

and we have the following interesting theorem (exercise):

Theorem 5.10. $\mathfrak{p} = \min\{\mathfrak{p}^*, \mathfrak{t}\}$.

Thus $\mathfrak{p} < \mathfrak{t}$ implies the somewhat more pathological situation $\mathfrak{p}^* < \mathfrak{t}$. If $\mathfrak{p}^*$ is not provably equal to $\mathfrak{p}$ then Theorem 5.10 may be regarded as a partial (but easy) solution to the Minimal Tower Problem.

Problem 5.11 (Shelah, et al. [54, 46]). *Is $\mathfrak{p} = \mathfrak{p}^*$?*

As we have mentioned before, the closely related problem whether $\mathfrak{p} = \mathsf{non}(\binom{\Omega}{\mathrm{T}})$ has a positive answer in [46].

6. Some connections with other fields

6.1. Ramsey theory

Recall that *Ramsey's Theorem*, often written as

$$(\forall n, k)\ \aleph_0 \to (\aleph_0)^n_k\,,$$

asserts that for each n, k, and a countably infinite set I: For each coloring $f : [I]^n \to \{1, \dots, k\}$, there exists a color j and an infinite $J \subseteq I$ such that $f \restriction [J]^n \equiv j$.

This motivates the following prototype for Ramsey theoretic hypotheses [40].

$\mathscr{A} \to (\mathscr{B})^n_k$: For each $\mathcal{U} \in \mathscr{A}$ and $f : [\mathcal{U}]^n \to \{1, \dots, k\}$, there exists j and $\mathcal{V} \subseteq \mathcal{U}$ such that $\mathcal{V} \in \mathscr{B}$ and $f \restriction [\mathcal{V}]^n \equiv j$.

Using this notation, Ramsey's Theorem is $(\forall n, k)\ [\mathbb{N}]^{\aleph_0} \to ([\mathbb{N}]^{\aleph_0})^n_k$. This prototype can be applied to the case where $\mathscr{A}$ and $\mathscr{B}$ are collections of rich covers, in accordance to the major theme of Ramsey theory that often, when rich enough structures are split into finitely many pieces, one of the pieces contains a rich substructure (Furstenberg).

The simplest case of the mentioned Ramseyan property is where $n = 1$ (so that we color the elements of the given member of $\mathscr{A}$ rather than finite subsets of it). This case is well understood: Fix a space X. Since a finite partition of an infinite set must contain an infinite element, and since every infinite subset of a γ-cover of X is again a γ-cover of X, we have that

$$(\forall k)\ \Gamma \to (\Gamma)^1_k.$$

It is less trivial but still not difficult to show that

$$(\forall k)\ \Omega \to (\Omega)^1_k$$

also holds, and since each τ-cover is an ω-cover (which in turn is a large cover), and each large subcover of a τ-cover is again a τ-cover, we have that

$$(\forall k)\ \mathrm{T} \to (\mathrm{T})^1_k.$$

The corresponding assertions for Borel or clopen (instead of open) covers also hold for the same reasons, and it can be shown that nothing more can be said (except for what trivially follows from the above assertions), concerning the partition property $\mathscr{A} \to (\mathscr{B})^1_k$ where $\mathscr{A}, \mathscr{B} \in \{\mathcal{O}, \Omega, \mathrm{T}, \Gamma\}$. We now turn to the case that $n > 1$.

Ramsey's Theorem is exactly the same as

$$(\forall n, k)\ \Gamma \to (\Gamma)^n_k.$$

However, if we substitute other classes of covers for $\mathscr{A}$ and $\mathscr{B}$ the property may not hold for all X. An elegant example is the following.

Theorem 6.1 (Just-Miller-Scheepers-Szeptycki [40, 23]). *The following properties are equivalent:*

1. $\mathsf{S}_1(\Omega, \Omega)$,
2. $\Omega \to (\Omega)^2_2$; *and*
3. $(\forall n, k)\ \Omega \to (\Omega)^n_k$.

In other words, X satisfies $\mathsf{S}_1(\Omega, \Omega)$ if, and only if, whenever we color with 2 colors the edges of a complete graph whose vertices are elements of an ω-cover of X, we can find a complete monochromatic subgraph whose vertices form an ω-cover of X.

Corollary 6.2.

1. $\mathsf{S}_1(\Omega, \Gamma)$ *is equivalent to* $\Omega \to (\Gamma)^2_2$.
2. $\mathsf{S}_1(\Omega, \mathrm{T})$ *implies* $\Omega \to (\mathrm{T})^2_2$.

And in both cases we can use any n and k as a superscript and a subscript.

Proof. ($\Rightarrow$) for (1) and (2): Assume that $\Omega \supseteq \mathscr{B}$ and X satisfies $\mathsf{S}_1(\Omega, \mathscr{B})$. Then X satisfies $\mathsf{S}_1(\Omega, \Omega)$ as well as $\binom{\Omega}{\mathscr{B}}$. By Theorem 6.1, X satisfies $\Omega \to (\Omega)^2_2$, but since X satisfies $\binom{\Omega}{\mathscr{B}}$ and a (complete) subgraph of a monochromatic graph is monochromatic, X satisfies $\Omega \to (\mathscr{B})^2_2$.

($\Leftarrow$) for (1): Clearly, $\Omega \to (\Gamma)^2_2$ implies $\binom{\Omega}{\Gamma}$, which we know is the same as $\mathsf{S}_1(\Omega, \Gamma)$. □

Problem 6.3. *Is* $\mathsf{S}_1(\Omega, \mathrm{T})$ *equivalent to* $\Omega \to (\mathrm{T})^2_2$*?*

Even for $n = k = 2$, Ramsey's Theorem cannot be extended to $\aleph_1$. To see this, consider a set of reals X of size $\aleph_1$, and a wellordering $\prec$ of X. Give a pair $\{x, y\} \subseteq X$ the color 1 if $<$ and $\prec$ agree on (x, y), and 0 otherwise. Then a monochromatic subgraph would give rise to an $<$-chain of reals of size $\aleph_1$, which is impossible, since between each two elements of the chain there is a rational number. (This nice argument is due to Sierpiński.)

There are some other ways to extend Ramsey's Theorem to larger cardinals, sometimes taking a quick route to large cardinals, see [20] for a survey. But Theorem 6.1 and its relatives suggest other generalizations. If we forget about the topology and consider arbitrary (but countable) covers of the given space (that is, set), then Theorem 6.1 implies the following. For an infinite cardinal κ, let Ω_κ denote the collection of all *countable* ω-covers of κ.

Theorem 6.4 (Scheepers [41]). *If* $\kappa < \mathsf{cov}(\mathcal{M})$*, then* $(\forall n, k)\ \Omega_\kappa \to (\Omega_\kappa)^n_k$*, where* Ω_κ *is the collection of countable* ω*-covers of* κ*. (Moreover, for each n and k the bound* $\mathsf{cov}(\mathcal{M})$ *is tight.)*

This is so because $\mathsf{non}(\mathsf{S}_1(\Omega,\Omega)) = \mathsf{cov}(\mathcal{M})$. This is a nice and typical case where the results in the topological studies, which were motivated by results from pure infinite combinatorics, often project to new results in pure infinite combinatorics.

Remark 6.5. For a much more comprehensive survey of this subject, see [25].

6.2. Countably distinct representatives and splittability

Among the main tools for proving Ramsey theoretic results of the flavor of the previous section one can find the properties of the following form:

$\mathsf{CDR}(\mathscr{A},\mathscr{B})$: For each sequence $\{\mathcal{U}_n\}_{n\in\mathbb{N}}$ of elements of $\mathscr{A}$, there exist countable, pairwise disjoint elements $\mathcal{V}_n \subseteq \mathcal{U}_n$, $n \in \mathbb{N}$, such that $\mathcal{V}_n \in \mathscr{B}$ for all n.

$\mathsf{Split}(\mathscr{A},\mathscr{B})$: For each $\mathcal{U} \in \mathscr{A}$, there exist disjoint $\mathcal{V}_1, \mathcal{V}_2 \subseteq \mathcal{U}$ such that $\mathcal{V}_1, \mathcal{V}_2 \in \mathscr{B}$.

Clearly, $\mathsf{CDR}(\mathscr{A},\mathscr{B})$ implies $\mathsf{Split}(\mathscr{A},\mathscr{B})$, of which an almost complete classification was carried in [53], when $\mathscr{A},\mathscr{B} \in \{\Omega, \mathrm{T}, \Gamma, \Lambda\}$. As before, some of the properties are trivial, and several equivalences are provable among the remaining properties.

The following dictionary is the key behind the classification, where the negation of each property corresponds to some combinatorial property of $h[X]$ where h is the Marczewski characteristic function (Section 5.3) of $\mathcal{U}$ for a cover $\mathcal{U}$ witnessing the failure of that property.

Property	$h[X]$
$\neg\mathsf{Split}(\Lambda,\Lambda)$	reaping family
$\neg\mathsf{Split}(\Omega,\Lambda)$	ultrafilter base
$\neg\mathsf{Split}(\Omega,\Omega)$	ultrafilter subbase
$\neg\mathsf{Split}(\mathrm{T},\mathrm{T})$	simple P-point base

The results of the classification are summarized in the following theorem.

Theorem 6.6 ([53]). *No additional implication (except perhaps the dotted ones) is provable.*

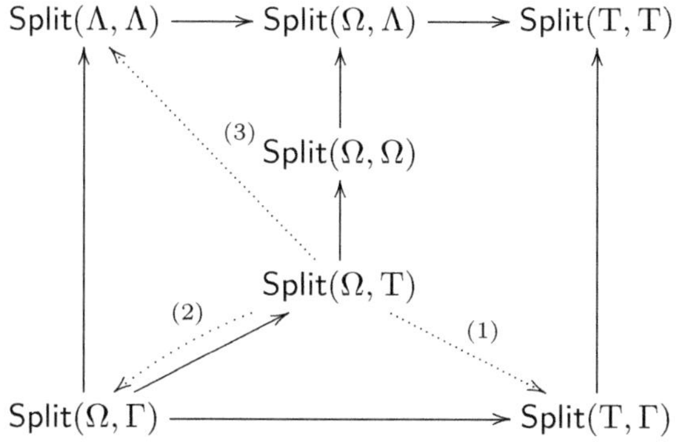

Problem 6.7 ([53]). *Is the dotted implication* (1) (*and therefore* (2) *and* (3)) *in the diagram true? If not, then is the dotted implication* (3) *true?*

6.3. An additivity problem

With regards to the additivity (preservation under taking finite unions) and σ-additivity (countable unions), the following is known (✓means that, in the figure of Theorem 6.6, the property in this position is σ-additive, and $\times$ means that it is not additive).

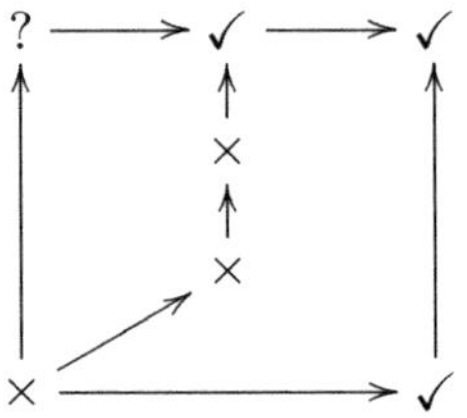

Thus, the only unsettled problem is the following:

Problem 6.8 ([53]). *Is* $\mathsf{Split}(\Lambda, \Lambda)$ *additive?*

In Proposition 1.1 of [53] it is shown that for a set of reals X (in fact, for any hereditarily Lindelöf space X), each large open cover of X contains a *countable* large open cover of X. Consequently, using standard arguments [53], the problem is closely related to the following one (where $[\mathbb{N}]^{\aleph_0}$ is the space of all infinite sets of natural numbers, with the topology inherited from $P(\mathbb{N})$, the latter identified with $\{0,1\}^{\mathbb{N}}$).

Problem 6.9. *If* R *denotes the sets of reals* X *such that each continuous image of* X *in* $[\mathbb{N}]^{\aleph_0}$ *is not reaping, then is* R *additive?*

Zdomskyy has proved the following surprising result concerning the Hurewicz and Menger properties.

Theorem 6.10 (Zdomskyy [63]**).** *Assume that* $\mathfrak{u}<\mathfrak{g}$. *Then* $\mathsf{U}_{fin}(\mathcal{O},\Gamma)=\mathsf{Split}(\Lambda,\Lambda)$, *and* $\mathsf{U}_{fin}(\mathcal{O},\Omega) = \mathsf{U}_{fin}(\mathcal{O},\mathcal{O})$.

Since $\mathsf{U}_{fin}(\mathcal{O},\Gamma)$ is easily seen to be countably additive, it follows that the answer to Problem 6.8 is consistently positive.

6.4. Topological games

Historically, the bridge from general to infinite-combinatorial topology was through topological games related to covering properties, introduced by Telgársky in [47, 48], and extensively studied by him and his colleagues (see Telgársky's survey [49]). This sort of game theory is still an important tool in proving Ramsey-theoretic results as those in Section 6.1. The games appear under various guises, but we will focus on their form which is motivated by the selection hypotheses.

$\mathsf{G}_1(\mathscr{A},\mathscr{B})$ is the game-theoretic version of $\mathsf{S}_1(\mathscr{A},\mathscr{B})$. In this game, ONE chooses in the nth inning an element $\mathcal{U}_n \in \mathscr{A}$ and then TWO responds by choosing $U_n \in \mathcal{U}_n$. They play an inning per natural number.

This is illustrated in the following figure.

$$\begin{array}{lcccccc} \text{ONE:} & \mathcal{U}_1 \in \mathscr{A} & & & \mathcal{U}_2 \in \mathscr{A} & & \dots \\ & & \searrow & \nearrow & & \searrow & \\ \text{TWO:} & & & U_1 \in \mathcal{U}_1 & & & U_2 \in \mathcal{U}_2 \quad \dots \end{array}$$

TWO wins if $\{U_1, U_2, \dots\} \in \mathscr{B}$, otherwise ONE wins.

The game $\mathsf{G}_{fin}(\mathscr{A}, \mathscr{B})$ is played similarly, where TWO responds with finite subsets $\mathcal{F}_n \subseteq \mathcal{U}_n$ and wins if $\bigcup_n \mathcal{F}_n \in \mathscr{B}$.

Observe that if ONE does not have a winning strategy in $\mathsf{G}_1(\mathscr{A}, \mathscr{B})$ (respectively, $\mathsf{G}_{fin}(\mathscr{A}, \mathscr{B})$), then $\mathsf{S}_1(\mathscr{A}, \mathscr{B})$ (respectively, $\mathsf{S}_{fin}(\mathscr{A}, \mathscr{B})$) holds. The converse is not always true; when it is true, the game is a powerful tool for studying the combinatorial properties of $\mathscr{A}$ and $\mathscr{B}$. Fortunately, this is often the case. In fact, this is always the case for the properties in the Scheepers Diagram (see, e.g., [27, 2, 60], and references therein).

There is, though, a well-known property of similar flavor for which the question of equivalence is still open. Consider the following generalized selection hypothesis.

$\mathsf{S}_1(\{\mathscr{A}_n\}_{n\in\mathbb{N}}, \mathscr{B})$: For each sequence of elements $\mathcal{U}_n \in \mathscr{A}_n$, $n \in \mathbb{N}$, there are elements $U_n \in \mathcal{U}_n$, $n \in \mathbb{N}$, such that $\{U_n : n \in \mathbb{N}\} \in \mathscr{B}$.

Define its game theoretic version $\mathsf{G}_1(\{\mathscr{A}_n\}_{n\in\mathbb{N}}, \mathscr{B})$ in the natural way. For most instances of $\{\mathscr{A}_n\}_{n\in\mathbb{N}}$ and $\mathscr{B}$, we do not get anything new by considering these properties [60]. However, this is not always the case.

A cover $\mathcal{U}$ of X is an *n-cover* of X if each subset F of X of cardinality at most n is contained in some member of $\mathcal{U}$. For each n denote by $\mathcal{O}_n$ the collection of all open n-covers of X. Then the property $\mathsf{S}_1(\{\mathcal{O}_n\}_{n\in\mathbb{N}}, \Gamma)$, introduced by Galvin and Miller in [18] (where it was called *strong γ-property*),[7] is strictly stronger than $\mathsf{S}_1(\Omega, \Gamma)$ (which in turn is the strongest property in the Scheepers Diagram) [6]. The strong γ-property $\mathsf{S}_1(\{\mathcal{O}_n\}_{n\in\mathbb{N}}, \Gamma)$ never had a game theoretic characterization. A natural candidate is the following. Let $\mathsf{G}_1(\{\mathscr{A}_n\}_{n\in\mathbb{N}}, \mathscr{B})$ be the game theoretic version of $\mathsf{S}_1(\{\mathscr{A}_n\}_{n\in\mathbb{N}}, \mathscr{B})$ (so it is like $\mathsf{G}_1(\mathscr{A}, \mathscr{B})$, but in the nth inning ONE chooses $\mathcal{U}_n \in \mathscr{A}_n$ instead of $\mathcal{U}_n \in \mathscr{A}$).

Problem 6.11 ([60]). *Is the strong γ-property $\mathsf{S}_1(\{\mathcal{O}_n\}_{n\in\mathbb{N}}, \Gamma)$ equivalent to ONE not having a winning strategy in $\mathsf{G}_1(\{\mathcal{O}_n\}_{n\in\mathbb{N}}, \Gamma)$?*

What about the case that TWO has a winning strategy in the game $\mathsf{G}_1(\mathscr{A}, \mathscr{B})$ or $\mathsf{G}_{fin}(\mathscr{A}, \mathscr{B})$? It turns out that in the cases corresponding to the properties in the Scheepers Diagram, these questions have interesting and elegant solutions. In this interpretation, the conjectures made by Menger, Hurewicz, and Borel are all correct!

Theorem 6.12 (Telgársky). *For a metrizable space X: TWO has a winning strategy in the game $\mathsf{G}_{fin}(\mathcal{O}, \mathcal{O})$ if, and only if, the space X is σ-compact.*

[7] Actually, Galvin and Miller defined the property as $(\exists k_n \nearrow \infty)\ X \in \mathsf{S}_1(\{\mathcal{O}_{k_n}\}_{n\in\mathbb{N}}, \Gamma)$, but it was observed in [60] that the quantifier can be eliminated from their definition.

Notice that $\mathsf{S}_{fin}(\mathcal{O},\mathcal{O})$ is the same as the Menger property $\mathsf{U}_{fin}(\mathcal{O},\mathcal{O})$, so $\mathsf{G}_{fin}(\mathcal{O},\mathcal{O})$ is the game theoretic version of the Menger property.

Theorem 6.13 (Galvin; Telgársky). *For a metrizable space X: TWO has a winning strategy in the game $\mathsf{G}_1(\mathcal{O},\mathcal{O})$ if, and only if, the space X is countable.*

$\mathsf{G}_1(\mathcal{O},\mathcal{O})$ is the game theoretic version of Rothberger' property $\mathsf{S}_1(\mathcal{O},\mathcal{O})$. Recall from Section 3.6, that the Borel Conjecture is equivalent to the assertion that all spaces satisfying $\mathsf{S}_1(\mathcal{O},\mathcal{O})$ are countable.

Another way to interpret these results is as follows:

If we assume that $\mathsf{G}_1(\mathcal{O},\mathcal{O})$ is determined, then Borel's Conjecture is true.

Corollary 6.14. *$\mathsf{G}_1(\mathcal{O},\mathcal{O})$ is determined if, and only if, Borel's Conjecture is true.*

Note that, by Section 1.3, we have to give up the Axiom of Choice in order to make the corresponding assertion for $\mathsf{G}_{fin}(\mathcal{O},\mathcal{O})$ and the Menger Conjecture meaningful. Scheepers told us that the Axiom of Determinacy (which rules out the Axiom of Choice) implies that the above-mentioned games are determined.

6.5. Arkhangel'skiĭ duality theory

Consider $C(X)$, the space of continuous real-valued functions $f : X \to \mathbb{R}$, as a subspace of $\mathbb{R}^X$ (the Tychonoff product of X many copies of $\mathbb{R}$). This is often called the *topology of pointwise convergence* and sometimes denoted $C_p(X)$ rather than $C(X)$.

The object $C(X)$ is very complicated from a topological point of view, in fact, even the behavior of the closure operator in this space is complicated. To this end, a comprehensive duality theory was developed by Arkhangel'skiĭ and his followers, which translates properties of $C(X)$ into properties of X, which are easier to work with.

For example, recall that a space Y is *Fréchet-Urysohn* if for each $A \subseteq Y$ and $x \in \overline{A}$, there is a sequence $\{a_n\}_{n\in\mathbb{N}}$ in A such that $\lim_{n\to\infty} a_n = x$. In their celebrated 1982 paper [19], Gerlits and Nagy proved that $C(X)$ is Fréchet-Urysohn if, and only if, X satisfies $\binom{\Omega}{\Gamma}$ (or, equivalently, $\mathsf{S}_1(\Omega,\Gamma)$).

An interesting example of the applicability of infinite-combinatorial methods for these questions is the following. In 1992, at a seminar in Moscow, Reznicenko introduced the following property: A space Y is *weakly Fréchet-Urysohn* if, for each $A \subseteq Y$ and $x \in \overline{A} \setminus A$, there exist finite disjoint sets $F_n \subseteq A$, $n \in \mathbb{N}$, such that for each neighborhood U of x, $F_n \cap U \neq \emptyset$ for all but finitely many n.

In [26], Kočinac and Scheepers made the following conjecture, which is now a theorem.

Theorem 6.15 ([52]). *The minimal cardinality of a set of reals X such that $C(X)$ does not have the weak Fréchet-Urysohn property is $\mathfrak{b}$.*

The surprising thing is that its proof only required a translation into the language of combinatorics and an application of an existing result: It was known that this minimal cardinality is at least $\mathfrak{b}$. A result of Sakai [39] can be used to prove that if $C(X)$ is weakly Fréchet-Urysohn, then a continuous image of X cannot be a subbase for a non-feeble filter on $\mathbb{N}$ (see [11] for the definition of non-feeble filter). Now it remained to apply a result of Petr Simon which tells that there exists a non-feeble filter with base of size $\mathfrak{b}$.

7. Conclusions

The theory emerging from the systematic study of diagonalizations of covers in a unified framework is not only aesthetically pleasing, but is also useful in turning otherwise ingenious ad hoc arguments into natural explanations. For this a good terminology is required, and the major part of this was already suggested by Scheepers' selection prototypes.

This study has connections and applications in several related fields, like Ramsey theory, function spaces, and topological groups. The usage of infinite-combinatorial methods in this theory has proved successful, and is often the "correct" tool to investigate these problems. This approach sometimes rewords by implying interesting results in the field of pure infinite-combinatorics.

While with regards to some of the investigations concerning the classical types of covers the picture is rather complete now, there remains much to be explored with regards to the new types of covers and their connection to the related fields.

We have only given a tiny sample of each theme. The reader is referred to [44, 24, 56] to have a more complete picture of the framework and the open problems it poses.

7.1. Acknowledgments

We thank Tomasz Weiss for the proof of Proposition 2.2. We also thank Rastislav Telgársky, Lyubomyr Zdomskyy, and the referee, for making several interesting comments.

References

[1] A.V. Arkhangel'skiĭ, *Hurewicz spaces, analytic sets, and fan tightness of function spaces*, Soviet Mathematical Doklady **33** (1986), 396–399.

[2] L. Babinkostova, Lj.D.R. Kočinac, and M. Scheepers, *Combinatorics of open covers (VIII)*, Topology and its Applications **140** (2004), 15–32.

[3] T. Banakh, P. Nickolas, and M. Sanchis, *Filter games and pathological subgroups of a countable product of lines*, Journal of the Australian Mathematical Society, to appear.

[4] T. Bartoszyński, *Invariants of measure and category*, in: **Handbook of Set Theory** (M. Foreman, A. Kanamori, and M. Magidor, eds.), Kluwer Academic Publishers, Dordrecht, to appear.

[5] T. Bartoszyński and H. Judah, *Strong measure zero sets*, Israel Mathematical Conference Proceedings **6** (1993), 13–62.

[6] T. Bartoszyński and I. Recław, *Not every γ-set is strongly meager*, Contemporary Mathematics **192** (1996), 25–29.

[7] T. Bartoszyński and S. Shelah, *Continuous images of sets of reals*, Topology and its Applications **116** (2001), 243–253.

[8] T. Bartoszyński, S. Shelah, and B. Tsaban, *Additivity properties of topological diagonalizations*, The Journal of Symbolic Logic **68** (2003), 1254–1260. (Fuller treatment: `http://arxiv.org/abs/math.GN/0604451`)

[9] T. Bartoszyński and B. Tsaban, *Hereditary topological diagonalizations and the Menger-Hurewicz Conjectures*, Proceedings of the American Mathematical Society **134** (2006), 605–615.

[10] A.R. Blass, *Nearly adequate sets*, Logic and Algebra (Yi Zhang, ed.), Contemporary Mathematics **302** (2002), 33–48.

[11] A.R. Blass, *Combinatorial cardinal characteristics of the continuum*, in: **Handbook of Set Theory** (M. Foreman, A. Kanamori, and M. Magidor, eds.), Kluwer Academic Publishers, Dordrecht, to appear.

[12] E. Borel, *Sur la classification des ensembles de mesure nulle*, Bulletin de la Société Mathématique de France **47** (1919), 97–125.

[13] J. Brendle, *Generic constructions of small sets of reals*, Topology and it Applications **71** (1996), 125–147.

[14] J. Chaber and R. Pol, *A remark on Fremlin-Miller theorem concerning the Menger property and Michael concentrated sets*, unpublished notes.

[15] E.K. van Douwen, *The integers and topology*, in: **Handbook of Set Theoretic Topology** (eds. K. Kunen and J. Vaughan), North-Holland, Amsterdam: 1984, 111–167.

[16] T. Eisworth and W. Just, *NCF and the combinatorics of open covers*, unpublished notes.

[17] D.H. Fremlin and A.W. Miller, *On some properties of Hurewicz, Menger and Rothberger*, Fundamenta Mathematica **129** (1988), 17–33.

[18] F. Galvin and A.W. Miller, *γ-sets and other singular sets of real numbers*, Topology and it Applications **17** (1984), 145–155.

[19] J. Gerlits and Zs. Nagy, *Some properties of $C(X)$, I*, Topology and its Applications **14** (1982), 151–161.

[20] A. Hajnal and J. Larson, *A survey of partition relations*, in: **Handbook of Set Theory** (M. Foreman, A. Kanamori, and M. Magidor, eds.), Kluwer Academic Publishers, Dordrecht, to appear.

[21] W. Hurewicz, *Über eine Verallgemeinerung des Borelschen Theorems*, Mathematische Zeitschrift **24** (1925), 401–421.

[22] W. Hurewicz, *Über Folgen stetiger Funktionen*, Fundamenta Mathematicae **9** (1927), 193–204.

[23] W. Just, A.W. Miller, M. Scheepers, and P.J. Szeptycki, *The combinatorics of open covers II*, Topology and its Applications **73** (1996), 241–266.

[24] Lj.D.R. Kočinac, *Selected results on selection principles*, in: **Proceedings of the 3rd Seminar on Geometry and Topology** (Sh. Rezapour, ed.), July 15-17, Tabriz, Iran, 2004, 71–104.

[25] Lj.D.R. Kočinac, *Generalized Ramsey Theory and topological properties: A survey*, **Proceedings of the International Symposium on Graphs, Designs and Applications, Messina** (September 30–October 4, 2003), Rendiconti del Seminario Matematico di Messina, Serie II, **9** (2003), 119–132.

[26] Lj.D.R. Kočinac and M. Scheepers, *Function spaces and a property of Reznichenko*, Topology and it Applications **123** (2002), 135–143.

[27] Lj.D.R. Kočinac and M. Scheepers, *Combinatorics of open covers (VII): Groupability*, Fundamenta Mathematicae **179** (2003), 131–155.

[28] R. Laver, *On the consistency of Borel's conjecture*, Acta Mathematica **137** (1976), 151–169.

[29] M. Machura, *Cardinal invariants* $\mathfrak{p}$, $\mathfrak{t}$ *and* $\mathfrak{h}$ *and real functions*, Tetra Mountains Journal **28** (2004), 97–108.

[30] E. Marczewski (Szpilrajn), *The characteristic function of a sequence of sets and some of its applications*, Fundamenta Mathematicae **31** (1938), 207–233.

[31] K. Menger, *Einige Überdeckungssätze der Punktmengenlehre*, Sitzungsberichte der Wiener Akademie **133** (1924), 421–444.

[32] H. Mildenberger, S. Shelah, and B. Tsaban, *The combinatorics of* τ*-covers*, Topology and its Applications, to appear. `http://arxiv.org/abs/math.GN/0409068`

[33] A.W. Miller, *A Nonhereditary Borel-cover* γ*-set*, Real Analysis Exchange **29** (2003/4), 601–606.

[34] A.W. Miller, *The* γ*-Borel conjecture*, Archive for Mathematical Logic **44** (2005), 425–434.

[35] I. Recław, *Every Luzin set is undetermined in point-open game*, Fundamenta Mathematicae **144** (1994), 43–54.

[36] F. Rothberger, *Sur des familles indénombrables de suites de nombres naturels, et les problèmes concernant la propriété C*, Proceedings of the Cambridge Philosophical Society **37** (1941), 109–126.

[37] F. Rothberger, *On some problems of Hausdorff and of Sierpiński*, Fundamenta Mathematicae **35** (1948), 29–46.

[38] M. Sakai, *Property* C'' *and function spaces*, Proceedings of the American Mathematical Society **104** (1988), 917–919.

[39] M. Sakai, *The Pytkeev property and the Reznichenko property in function spaces*, Note di Matematica **22** (2003), 43–52.

[40] M. Scheepers, *Combinatorics of open covers I: Ramsey theory*, Topology and its Applications **69** (1996), 31–62.

[41] M. Scheepers, *The least cardinal for which the Baire Category Theorem fails*, Proceedings of the American Mathematical Society **125** (1997), 579–585.

[42] M. Scheepers, *The length of some diagonalization games*, Arch. Math. Logic **38** (1999), 103–122.

[43] M. Scheepers, *Sequential convergence in* $\mathsf{C}_p(X)$ *and a covering property*, East-West Journal of Mathematics **1** (1999), 207–214.

[44] M. Scheepers, *Selection principles and covering properties in topology*, Note di Matematica **22** (2003), 3–41.

[45] M. Scheepers and B. Tsaban, *The combinatorics of Borel covers*, Topology and its Applications **121** (2002), 357–382.

[46] S. Shelah and B. Tsaban, *Critical cardinalities and additivity properties of combinatorial notions of smallness*, Journal of Applied Analysis **9** (2003), 149–162.

[47] R. Telgársky, *Spaces defined by topological games*, Fundamenta Mathematicae **88** (1975), 193–223.

[48] R. Telgársky, *Spaces defined by topological games II*, Fundamenta Mathematicae **116** (1983), 189–207.

[49] R. Telgársky, *Topological Games: On the 50th Anniversary of the Banach-Mazur Game*, Rocky Mountain Journal of Mathematics **17** (1987), 227–276.

[50] B. Tsaban, *A topological interpretation of* $\mathfrak{t}$, Real Analysis Exchange **25** (1999/2000), 391–404.

[51] B. Tsaban, *A diagonalization property between Hurewicz and Menger*, Real Analysis Exchange **27** (2001/2002), 757–763.

[52] B. Tsaban, *The minimal cardinality where the Reznichenko property fails*, Israel Journal of Mathematics **140** (2004), 367–374.

[53] B. Tsaban, *The combinatorics of splittability*, Annals of Pure and Applied Logic **129** (2004), 107–130.

[54] B. Tsaban, *Selection principles and the minimal tower problem*, Note di Matematica **22** (2003), 53–81.

[55] B. Tsaban, *o-bounded groups and other topological groups with strong combinatorial properties*, Proceedings of the American Mathematical Society **134** (2006), 881–891.

[56] B. Tsaban, *Selection principles in Mathematics: A milestone of open problems*, Note di Matematica **22** (2003), 179–208.

[57] B. Tsaban (ed.),
SPM Bulletin **1** (2003), `http://arxiv.org/abs/math.GN/0301011`
SPM Bulletin **2** (2003), `http://arxiv.org/abs/math.GN/0302062`
SPM Bulletin **3** (2003), `http://arxiv.org/abs/math.GN/0303057`
SPM Bulletin **4** (2003), `http://arxiv.org/abs/math.GN/0304087`
SPM Bulletin **5** (2003), `http://arxiv.org/abs/math.GN/0305367`
SPM Bulletin **6** (2003), `http://arxiv.org/abs/math.GN/0312140`
SPM Bulletin **7** (2004), `http://arxiv.org/abs/math.GN/0401155`
SPM Bulletin **8** (2004), `http://arxiv.org/abs/math.GN/0403369`
SPM Bulletin **9** (2004), `http://arxiv.org/abs/math.GN/0406411`
SPM Bulletin **10** (2004), `http://arxiv.org/abs/math.GN/0409072`
SPM Bulletin **11** (2005), `http://arxiv.org/abs/math.GN/0412305`
SPM Bulletin **12** (2005), `http://arxiv.org/abs/math.GN/0503631`
SPM Bulletin **13** (2005), `http://arxiv.org/abs/math.GN/0508563`
SPM Bulletin **14** (2005), `http://arxiv.org/abs/math.GN/0509432`
SPM Bulletin **15** (2005), `http://arxiv.org/abs/math.GN/0512275`
SPM Bulletin **16** (2006), `http://arxiv.org/abs/math.GN/0603290`

[58] B. Tsaban and T. Weiss, *Products of special sets of real numbers*, Real Analysis Exchange **30** (2004/5), 819–836.

[59] B. Tsaban, *Some new directions in infinite-combinatorial topology*, online version, `http://arxiv.org/abs/math.GN/0409069`

[60] B. Tsaban, *Strong γ-sets and other singular spaces*, Topology and its Applications **153** (2005), 620–639.

[61] J. Vaughan, *Small uncountable cardinals and Topology*, in: **Problems in Topology** (eds. Jan van Mill and G.M. Reed), North-Holland Pub. Co., Amsterdam: 1990, 195–218.

[62] L. Zdomskyy, *A characterization of the Menger and Hurewicz properties of sets of reals*, Mathematychni Studii **24** (2005), 115–119.

[63] L. Zdomskyy, *A semifilter approach to selection principles*, Commentarii Mathematicae Universitatis Carolinae **46** (2005), 525–540.

Boaz Tsaban
Department of Mathematics
The Weizmann Institute of Science
Rehovot 76100, Israel
e-mail: `boaz.tsaban@weizmann.ac.il`
URL: `http://www.cs.biu.ac.il/~tsaban`

Set Theory
Trends in Mathematics, 257–273

The Number of Near-Coherence Classes of Ultrafilters is Either Finite or $2^{\mathfrak{c}}$

Taras Banakh and Andreas Blass

Abstract. We prove that the number of near-coherence classes of non-principal ultrafilters on the natural numbers is either finite or $2^{\mathfrak{c}}$. Moreover, in the latter case the Stone-Čech compactification $\beta\omega$ of ω contains a closed subset C consisting of $2^{\mathfrak{c}}$ pairwise non-nearly-coherent ultrafilters. We obtain some additional information about such closed sets under certain assumptions involving the cardinal characteristics $\mathfrak{u}$ and $\mathfrak{d}$.

Applying our main result to the Stone-Čech remainder $\beta\mathbb{R}_+ - \mathbb{R}_+$ of the half-line $\mathbb{R}_+ = [0, \infty)$ we obtain that the number of composants of $\beta\mathbb{R}_+ - \mathbb{R}_+$ is either finite or $2^{\mathfrak{c}}$.

1. Introduction

In this paper, all filters are on the set ω of natural numbers, and they contain all cofinite subsets of ω. In particular, all ultrafilters are non-principal ultrafilters on ω. If $\mathcal{F}$ is a filter and $f : \omega \to \omega$ is a finite-to-one function, then $f(\mathcal{F})$ is defined to be $\{X \subseteq \omega : f^{-1}(X) \in \mathcal{F}\}$. This is a filter, and it is an ultrafilter if $\mathcal{F}$ is.

As in [5], we call two filters $\mathcal{F}_1$ and $\mathcal{F}_2$ *coherent* if $\mathcal{F}_1 \cup \mathcal{F}_2$ has the finite intersection property, i.e., if there is a filter that includes them both. We call $\mathcal{F}_1$ and $\mathcal{F}_2$ *nearly coherent* if there is a finite-to-one $f : \omega \to \omega$ such that $f(\mathcal{F}_1)$ and $f(\mathcal{F}_2)$ are coherent. Notice that a filter and an ultrafilter are coherent if and only if the former is included in the latter; in particular, two ultrafilters are coherent only if they are equal.

Near-coherence is an equivalence relation, introduced and extensively studied in [5]. It is natural to ask how many equivalence classes it has. Since the number of ultrafilters is $2^{2^{\aleph_0}}$ by a theorem of Pospíšil [17], the number of near-coherence classes of ultrafilters is obviously between 1 and $2^{2^{\aleph_0}}$, inclusive. Its exact value, however, is independent of the usual (ZFC) axioms of set theory. The known consistency results are, in chronological order, using the standard notation $\mathfrak{c}$ for the cardinality $2^{\aleph_0}$ of the continuum:

1. It is consistent relative to ZFC, and in fact it is a consequence of the continuum hypothesis (CH) or of Martin's axiom (MA), that the number of near-coherence classes is $2^{\mathfrak{c}}$.
2. It is consistent relative to ZFC that there is only one near-coherence class of ultrafilters.

The first of these consistency results follows from the fact that among selective ultrafilters (those such that every function on ω is either constant or one-to-one on a set in the ultrafilter) near coherence is the same as isomorphism (via permutations of ω). In particular, any selective ultrafilter is nearly coherent with only $\mathfrak{c}$ others. On the other hand, CH (or just MA) implies that there are $2^{\mathfrak{c}}$ selective ultrafilters, and therefore that there are $2^{\mathfrak{c}}$ near-coherence classes. The history of this result is a bit obscure. Booth [10] writes that Galvin was the first to prove the existence of selective ultrafilters under CH, but doesn't say that Galvin made the slight extension to get $2^{\mathfrak{c}}$ selective ultrafilters. Booth himself weakens the hypothesis to MA and shows that the selective ultrafilters are dense in the Stone-Čech remainder $\omega^* = \beta\omega - \omega$. Rudin [20] proves that CH yields $2^{\mathfrak{c}}$ selective ultrafilters, but she describes the result as well known. In the second author's thesis [4], the assumption here is reduced to MA (and in fact to the hypothesis there called FRH(ω) but nowadays expressed as $\mathfrak{p} = \mathfrak{c}$ or as MA(σ-centered)), but the proof is essentially the same as under CH. In this paper we shall show that the number of near-coherence classes of ultrafilters is $2^{\mathfrak{c}}$ under the weaker hypothesis $\mathfrak{u} \geq \mathfrak{d}$ (see Section 2 for the definitions of $\mathfrak{u}$ and $\mathfrak{d}$).

The statement that there is only one near-coherence class of ultrafilters is known as the principle of near coherence of filters (NCF) and is proved consistent in [8]. For more information about it, see [5, 6].

It is shown in [7] that the statement "there are exactly two near-coherence classes of ultrafilters" follows from the statement "there are simple P_κ-points for two different cardinals κ." The consistency of the latter statement is the content of Section 6 of [8], but Dow has found an error in that section. So it is, for the time being, an open problem whether it is consistent to have exactly two near-coherence classes of ultrafilters. Shelah has proposed a new construction of a model with simple $P_{\aleph_1}$-points and $P_{\aleph_2}$-points, but the details of the construction remain to be written down carefully and checked. If correct, it will restore the previously believed result that there can be exactly two near-coherence classes of ultrafilters.

For all cardinals κ strictly between 2 and $2^{\mathfrak{c}}$, it has always been an open question whether there could be exactly κ near-coherence classes of ultrafilters. In this paper, we present the first negative result about this question, eliminating all infinite cardinals except $2^{\mathfrak{c}}$. In particular, the number of near-coherence classes of ultrafilters cannot be $\aleph_0$ or $\mathfrak{c}$.

Our principal result is:

Theorem 1. *If there are infinitely many near coherence classes of ultrafilters, then $\beta\omega$ contains a closed subset C consisting of $2^{\mathfrak{c}}$ pairwise non-nearly-coherent ultra-*

filters. Consequently, the number of near-coherence classes of ultrafilters is either finite or equal to $2^{\mathfrak{c}}$. *This number is equal to* $2^{\mathfrak{c}}$ *if* $\mathfrak{u} \geq \mathfrak{d}$.

Remark 2. J.Mioduszewski [15, 16] has established that the number of near-coherence classes of ultrafilters is the same as the number of composants of the indecomposable continuum $\mathbb{R}_+^* = \beta\mathbb{R}_+ - \mathbb{R}_+$, the Stone-Čech remainder of a closed half-line $\mathbb{R}_+ = [0, \infty)$. (See also [6] for a discussion and a proof with less machinery.) Thus, our result implies that the number of composants of $\mathbb{R}_+^*$ is either finite or $2^{\mathfrak{c}}$. In particular, $\mathbb{R}_+^*$ cannot have exactly $\aleph_0$ or $\mathfrak{c}$ composants. Moreover, if the number of composants is infinite then there is a closed subset $\mathsf{C} \subset \mathbb{R}_+^*$ of size $2^{\mathfrak{c}}$ having at most one-point intersection with each composant. This result can be viewed as an analog of Mazurkiewicz's classical theorem [14] that a non-degenerate metrizable indecomposable continuum has an uncountable closed subset containing at most one point from each composant.

The proof of Theorem 1 is rather lengthy and requires some preparatory work. We shall give different proofs for the cases $\mathfrak{u} \geq \mathfrak{d}$ (Section 3) and $\mathfrak{u} < \mathfrak{d}$ (Section 4). Section 2 reviews some known results that we shall need and obtains some immediate consequences of them. Section 5 gives some additional information, under the assumption $\mathfrak{u} > \mathfrak{d}$, about the closed set whose existence Theorem 1 asserts. Finally, Section 6 presents some problems that remain open.

2. Preliminaries

This section is a review of known information that will be needed in our proofs.

2.1. Cardinals

We shall need three of the standard cardinal characteristics of the continuum, in addition to the cardinality $\mathfrak{c}$ of the continuum already used in the introduction.

The *dominating number*, $\mathfrak{d}$, is defined as the smallest cardinality of any family $\mathcal{D}$ of functions $\omega \to \omega$ such that every function $\omega \to \omega$ is eventually majorized by some member of $\mathcal{D}$. The *ultrafilter number*, $\mathfrak{u}$, is defined as the smallest size of any base for an ultrafilter. The *unsplitting number* $\mathfrak{r}$, sometimes called the *refining* or *reaping number*, is the smallest cardinality of a family $\mathcal{R}$ of infinite subsets of ω that is unsplittable in the sense that, for any $S \subseteq \omega$, there is some $R \in \mathcal{R}$ with either $R - S$ or $R \cap S$ finite, i.e., R is almost included in S or in $\omega - S$.

It is easy to see that all of these cardinals are between $\aleph_1$ and $\mathfrak{c}$, inclusive, and that $\mathfrak{r} \leq \mathfrak{u}$ (because any ultrafilter base is unsplittable). We shall need the theorem of Aubrey [1] that $\mathfrak{r} \geq \min\{\mathfrak{u}, \mathfrak{d}\}$. In other words, although each of the inequalities $\mathfrak{r} < \mathfrak{u}$ and $\mathfrak{r} < \mathfrak{d}$ is consistent (the former by [11] and the latter by [8] or [9]), their conjunction is inconsistent. We shall use Aubrey's result in the following form.

Lemma 3. *If* $\mathfrak{u} \geq \mathfrak{d}$ *then also* $\mathfrak{r} \geq \mathfrak{d}$.

If $\mathcal{F}$ is a filter, then we denote by $\chi(\mathcal{F})$ the smallest cardinality of any base for $\mathcal{F}$. Notice that this is also the smallest cardinality of any family of sets that generates $\mathcal{F}$, because closing such a family under finite intersections to produce a base will not increase the cardinality. Notice also that $\mathfrak{u} = \min\{\chi(\mathcal{U}) : \mathcal{U}$ an ultrafilter$\}$.

2.2. Topology in $\beta\omega$

The ultrafilters (non-principal and on ω, as always) can be identified with the points of the Stone-Čech remainder $\omega^* = \beta\omega - \omega$ of the discrete space ω. The topology of ω^* has as a basis of open sets (and also a basis of closed sets) the sets of the form $[A] = \{\mathcal{U} : A \in \mathcal{U}\}$ for infinite $A \subseteq \omega$. These sets $[A]$ are exactly the nonempty clopen subsets of ω^*. The nonempty closed subsets of ω^* are exactly those of the form $[\mathcal{F}] = \{\mathcal{U} : \mathcal{F} \subseteq \mathcal{U}\}$ for filters $\mathcal{F}$. The smallest number $\chi(\mathcal{F})$ of generators for $\mathcal{F}$ is also the smallest number of open sets whose intersection is $[\mathcal{F}]$.

By a *discrete sequence* (in ω^*), we mean a sequence $\langle \mathcal{U}_n : n \in \omega\rangle$ of distinct ultrafilters whose range is a discrete set. Equivalently, for each $n \in \omega$, there is a set that belongs to $\mathcal{U}_n$ but not to $\mathcal{U}_m$ for any $m \neq n$.

Because ω^* is a compact Hausdorff space, any sequence $\langle \mathcal{U}_n : n \in \omega\rangle$ has a limit along any ultrafilter $\mathcal{V}$. We denote the limit by $\mathcal{V}\text{-}\lim_n \mathcal{U}_n$; this ultrafilter consists of those $A \subseteq \omega$ for which $\{n \in \omega : A \in \mathcal{U}_n\} \in \mathcal{V}$.

The following result of Rudin [19, Lemma 2] will be important in our proofs.

Lemma 4. *If $\langle \mathcal{U}_n : n \in \omega\rangle$ is a discrete sequence of ultrafilters, then distinct ultrafilters $\mathcal{V}$ yield distinct limits $\mathcal{V}\text{-}\lim_n \mathcal{U}_n$.*

Since any infinite set in ω^* (or indeed in any regular space) has an infinite discrete subset, it follows that every infinite closed subset of ω^* has the same cardinality as the whole space ω^*. By a theorem of Pospíšil [17], that cardinality is $2^{\mathfrak{c}}$. We summarize for future reference:

Lemma 5. *If a filter $\mathcal{F}$ is included in infinitely many ultrafilters, then it is included in $2^{\mathfrak{c}}$ ultrafilters, and these constitute a closed subset of ω^*.*

In connection with this result, it will be useful to also have the following information about the contrary case.

Lemma 6. *If a filter $\mathcal{F}$ is included in only finitely many ultrafilters, then each of those ultrafilters $\mathcal{U}$ has $\chi(\mathcal{U}) \leq \chi(\mathcal{F})$.*

Proof. As $[\mathcal{F}]$ is finite, each of its points $\mathcal{U}$ is isolated in $[\mathcal{F}]$. This means that there is a set $A \subseteq \omega$ such that $\mathcal{U}$ is the only ultrafilter extending $\mathcal{F}$ and containing A. Then $\mathcal{U}$ is generated by A together with any system of generators for $\mathcal{F}$. □

If f is any finite-to-one map $\omega \to \omega$, then the induced map $\mathcal{U} \mapsto f(\mathcal{U})$ of ultrafilters is a continuous function from ω^* to itself. In particular,

$$f(\mathcal{V}\text{-}\lim_n \mathcal{U}_n) = \mathcal{V}\text{-}\lim_n f(\mathcal{U}_n).$$

2.3. P-points

An ultrafilter $\mathcal{U}$ is called a *P-point* if, whenever $\{A_n : n \in \omega\}$ is a countable family of elements of $\mathcal{U}$, then there is some $B \in \mathcal{U}$ that is almost included in each of them, i.e., $B - A_n$ is finite for every n. Such a B is called a *pseudointersection* of the A_n's.

A P-point $\mathcal{U}$ cannot be in the closure of a countable set of other ultrafilters in ω^*. Indeed, we could find, for each of those other ultrafilters, a set in $\mathcal{U}$ but not in the other ultrafilter. A pseudointersection of those sets in $\mathcal{U}$ gives a neighborhood of $\mathcal{U}$ containing none of the countably many other ultrafilters.

If $\mathcal{U}$ is a P-point, then so is $f(\mathcal{U})$ for any finite-to-one f.

Shelah has shown [23] that the existence of P-points is not provable in ZFC, though it is a consequence of CH ([21]). We shall need the following lemma and corollary, due to Ketonen [12] though a version of the lemma was already in [18].

Lemma 7. *Suppose $\mathcal{F}$ is a filter with $\chi(\mathcal{F}) < \mathfrak{d}$, and suppose we are given a decreasing ω-sequence of sets $A_0 \supseteq A_1 \supseteq \dots$ such that each A_n intersects every set in $\mathcal{F}$. Then the A_n's have a pseudointersection B that also intersects every set in $\mathcal{F}$.*

In the special case where $\mathcal{F}$ is an ultrafilter, so that "intersects every set in $\mathcal{F}$" is synonymous with "is in $\mathcal{F}$," the lemma reduces to the following.

Corollary 8. *Every ultrafilter $\mathcal{U}$ with $\chi(\mathcal{U}) < \mathfrak{d}$ is a P-point.*

It is well known that, if $\mathcal{U}$ is a P-point and f is a one-to-one function from ω into ω^* (or into any compact Hausdorff space), then there is a set $A \in \mathcal{U}$ whose image $f(A)$ is discrete. We shall need the following slightly more general result, whose proof, though essentially the same as for the result just quoted, we give for the sake of completeness.

Lemma 9. *Suppose X is a compact Hausdorff space, $f : \omega \to X$ is a one-to-one map, and $\mathcal{U}$ is an ultrafilter. Then there is a decreasing ω-sequence $A_0 \supseteq A_1 \supseteq \dots$ of sets in $\mathcal{U}$ such that, if $B \subseteq A_0$ is a pseudointersection of the A_n's, then $f(B)$ is discrete and has at most one limit point in the image of f.*

Proof. Since X is a compact Hausdorff space, the sequence f has a limit x with respect to $\mathcal{U}$. For each $k \in \omega$ choose a closed neighborhood N_k of x that does not contain $f(k)$ unless $f(k) = x$. (Note that the exceptional situation $f(k) = x$ arises for at most one k.) Define

$$A_n = \bigcap_{k \leq n} f^{-1}(N_k - \{x\}).$$

Clearly, these A_n's form a decreasing sequence of sets in $\mathcal{U}$. Let B be any pseudointersection of them. To finish the proof, we show that no $f(k)$ is a limit point of $f(B)$ unless $f(k) = x$. Indeed, if $f(k) \neq x$ then $f(k) \notin N_k$. As N_k is closed, $X - N_k$ is a neighborhood of $f(k)$. As B is a pseudointersection of the A_n's, there are only finitely many elements $m \in B$ for which $f(m) \in X - N_k$, so $f(k)$ cannot be a limit point of $f(B)$. □

2.4. Discrete and ω^*-discrete ultrafilters

Baumgartner generalized the notion of P-point to the notion of discrete ultrafilter. We shall need the further generalization to the notion of Y-discrete ultrafilter, defined as follows.

Definition 10. An ultrafilter $\mathcal{U}$ is a *discrete ultrafilter* if for any one-to-one map $f : \omega \to \mathbb{R}$ there is a set $A \in \mathcal{U}$ such that $f(A)$ is discrete. More generally, for any topological space Y, an ultrafilter $\mathcal{U}$ is *Y-discrete* if, whenever $f : \omega \to Y$ is one-to-one, the image of some $U \in \mathcal{U}$ is discrete.

Thus, "discrete" is the same as $\mathbb{R}$-discrete. We shall be interested primarily in ω^*-discreteness.

It follows immediately from Lemma 9 that all P-points are discrete ultrafilters. We shall need the fact that the converse fails, unless both classes are empty; see [3, Corollary 2.9].

Lemma 11. *If there exists a P-point then there also exists a discrete ultrafilter $\mathcal{U}$ that is not a P-point.*

The following lemma will allow us to transfer information about discreteness to the more general notion of Y-discreteness, in particular for $Y = \omega^*$.

Proposition 12. *Every discrete ultrafilter is ω^*-discrete and in fact Y-discrete for every functionally Hausdorff space Y.*

A space is called functionally Hausdorff if its points can be separated by real-valued continuous functions.

Proof. Since we shall need this proposition only for the ω^* case, we prove this case in detail and then indicate briefly the proof for arbitrary functionally Hausdorff spaces.

Let $f : \omega \to \omega^*$ be a one-to-one map. There is a continuous function g from ω^* to the product $\{0,1\}^\omega$ of discrete two-point spaces, such that the restriction of g to $f(\omega)$ is one-to-one. To construct such a g, it suffices to choose countably many sets $A_k \subseteq \omega$ such that, for each pair $m \neq n$ in ω, there is at least one A_k that belongs to $f(m)$ but not to $f(n)$. Then define the kth component of $g(\mathcal{V})$, for any ultrafilter $\mathcal{V}$, to be 1 or 0 according to whether A_k is or is not in $\mathcal{V}$. Since $\{0,1\}^\omega$ is homeomorphic to the Cantor set $\subseteq \mathbb{R}$, any discrete ultrafilter contains an X such that $g(f(X))$ is discrete. But any space with a one-to-one continuous map to a discrete space is itself discrete, so $f(X)$ is discrete.

The argument with Y in place of ω^* is similar. The function g will now map Y into $\mathbb{R}^\omega$, and its components are provided by the assumption that Y is functionally Hausdorff. Unlike the Cantor set above, this $\mathbb{R}^\omega$ cannot be regarded as a subspace of $\mathbb{R}$, but all we really need to embed into $\mathbb{R}$ is the range of $g \circ f$. As a countable, metrizable space, this can be embedded in $\mathbb{R}$, and then the proof can be completed as above. $\square$

2.5. Testing near coherence

The definition of near-coherence says, for ultrafilters $\mathcal{U}$ and $\mathcal{V}$, that $f(\mathcal{U}) = f(\mathcal{V})$ for some finite-to-one $f : \omega \to \omega$. We shall need to know that, under certain circumstances, the search for such an f need not extend to all finite-to-one functions.

Definition 13. Let $\mathcal{F}$ be a filter. A family $\mathcal{T}$ of finite-to-one functions $\omega \to \omega$ is called a *test family over* $\mathcal{F}$ if, whenever two ultrafilters $\mathcal{U}$ and $\mathcal{V}$ in $[\mathcal{F}]$ are nearly coherent, then there is an $f \in \mathcal{T}$ with $f(\mathcal{U}) = f(\mathcal{V})$. When $\mathcal{F}$ is the filter of cofinite sets, then we abbreviate "test family over the cofinite filter" to "test family."

We shall need the following result, which is Lemma 10 of [5].

Lemma 14. *There is a test family of cardinality* $\mathfrak{d}$.

In Section 4, we shall obtain even smaller test families over certain filters under the hypothesis $\mathfrak{u} < \mathfrak{d}$. The use of test families in our proofs will be by way of the following main lemma.

Lemma 15. *Assume*

- $\mathcal{F}$ *and* $\mathcal{G}$ *are filters,*
- $\mathcal{U}_n$ *are ultrafilters extending* $\mathcal{G}$, *for* $n \in \omega$,
- $\mathcal{T}$ *is a test family over* $\mathcal{G}$, *and*
- *for each* $f \in \mathcal{T}$, *there is* $A \in \mathcal{F}$ *for which the sequence* $\langle f(\mathcal{U}_n) : n \in A \rangle$ *is discrete.*

Then the ultrafilters $\mathcal{V}\text{-}\lim_n \mathcal{U}_n$ *for* $\mathcal{V} \in [\mathcal{F}]$ *are pairwise distinct and not nearly coherent. These ultrafilters* $\mathcal{V}\text{-}\lim_n \mathcal{U}_n$ *form a closed set in* ω^*, *whose cardinality is that of* $[\mathcal{F}]$.

Proof. For any $f \in \mathcal{T}$, let A be as in the fourth assumption. Then the ultrafilters

$$f(\mathcal{V}\text{-}\lim_n \mathcal{U}_n) = \mathcal{V}\text{-}\lim_n f(\mathcal{U}_n)$$

for varying $\mathcal{V} \in [\mathcal{F}]$ are distinct by Lemma 4, because $\langle f(\mathcal{U}_n) : n \in A \rangle$ is discrete and $A \in \mathcal{F} \subseteq \mathcal{V}$. Since $\mathcal{T}$ is a test family over $\mathcal{G}$, and all the ultrafilters $\mathcal{V}\text{-}\lim_n \mathcal{U}_n$ (for $\mathcal{V} \in [\mathcal{F}]$) extend $\mathcal{G}$, we have that these ultrafilters are pairwise distinct and not nearly coherent. Furthermore, these ultrafilters are exactly the images of the ultrafilters $\mathcal{V} \in [\mathcal{F}]$ under the continuous extension $\beta\omega \to \beta\omega$ of the map $\omega \to \beta\omega$ sending each n to $\mathcal{U}_n$. So they constitute the image of the compact set $[\mathcal{F}]$ under a continuous map, and such an image is, of course, closed. Finally, the lemma's assertion about cardinality follows from the previous assertion of pairwise distinctness. $\square$

3. Proof when $\mathfrak{u} \geq \mathfrak{d}$

This section is devoted to the proof of Theorem 1 under the assumption that $\mathfrak{u} \geq \mathfrak{d}$. Notice that in this situation the theorem's last sentence means that the hypothesis in the first sentence, that there are infinitely many near-coherence classes of ultrafilters, must be proved rather than being assumed. The following proposition summarizes the part of the theorem to be proved in this section.

Proposition 16. *Assume $\mathfrak{u} \geq \mathfrak{d}$. There is an infinite closed set $\mathsf{C} \subseteq \omega^*$ no two of whose elements are nearly coherent. Thus, the number of near-coherence classes of ultrafilters is $2^{\mathfrak{c}}$.*

Proof. By Lemma 14, fix a test family $\mathcal{T}$ of cardinality $\mathfrak{d}$. We intend to construct ultrafilters $\mathcal{U}_n$ for $n \in \omega$ so that the assumptions of Lemma 15 are satisfied by the $\mathcal{U}_n$'s, $\mathcal{T}$, and the filter of cofinite sets (in the roles of both the $\mathcal{F}$ and the $\mathcal{G}$ of the lemma). In fact, we shall arrange that the whole sequence $\langle f(\mathcal{U}_n) : n \in \omega\rangle$ is discrete for every $f \in \mathcal{T}$. Then Lemma 15 will complete the proof of the proposition.

The ultrafilters $\mathcal{U}_n$ will be constructed simultaneously in a transfinite induction of length $\mathfrak{d}$, in which the stages are indexed by pairs $(f, k) \in \mathcal{T} \times \omega$. Since there are only $\mathfrak{d}$ of these pairs, such an indexing is possible. At each stage, we shall have constructed filters $\mathcal{G}_n$, each of which will be included in the corresponding $\mathcal{U}_n$. Although $\mathcal{G}_n$ will change – in fact grow – from one stage to the next, it will not be necessary to clutter the notation with an explicit mention of the stage. From each stage to the next, at most one new generator will be added to each $\mathcal{G}_n$, so that, since the induction has length $\mathfrak{d}$, we shall have $\chi(\mathcal{G}_n) < \mathfrak{d}$ at each stage. Indeed, since only countably many generators are added at any one stage (at most one new generator for each $\mathcal{G}_n$), there will be, at each stage, a set of size $< \mathfrak{d}$ that contains generators for all the $\mathcal{G}_n$. (In fact, this also follows directly from $\chi(\mathcal{G}_n) < \mathfrak{d}$, since $\mathfrak{d}$ cannot have cofinality ω.)

At the initial stage of the induction, we let each $\mathcal{G}_n$ be the filter of cofinite sets. At limit stages, we take, for each n, the union of the filters $\mathcal{G}_n$ from all the previous stages. It remains to describe the successor stages of the induction, so consider the step from the stage indexed by (f, k) to the next stage, and consider the filters $\mathcal{G}_n$ produced by all the previous stages. As noted above, there is a set $\mathcal{X}$ of size $< \mathfrak{d}$ containing generators for all the $\mathcal{G}_n$. Since we are assuming $\mathfrak{u} \geq \mathfrak{d}$, Lemma 3 implies that there is a set $S \subseteq \omega$ such that both $f(X) \cap S$ and $f(X) - S$ are infinite for all $X \in \mathcal{X}$. This means that we can adjoin $f^{-1}(S)$ as a new generator to $\mathcal{G}_k$ and adjoin $\omega - f^{-1}(S)$ as a new generator to all $\mathcal{G}_n$ for $n \neq k$ and still have filters, i.e., the new generators do not destroy the finite intersection property. Adjoining these generators gives the filters $\mathcal{G}_n$ for the next stage. Now, with these updated filters, we have that $f(\mathcal{G}_k)$ contains a set, namely S, whose complement is in $f(\mathcal{G}_n)$ for all $n \neq k$. Thus, this stage of the construction ensures that, for the ultimately constructed ultrafilters $\mathcal{U}_n \supseteq \mathcal{G}_n$, the point $f(\mathcal{U}_k)$ is isolated in $\{f(\mathcal{U}_n) : n \in \omega\}$.

After all $\mathfrak{d}$ of the stages are completed, take, for each n, the union of the filters $\mathcal{G}_n$ constructed during the induction, and extend this filter (arbitrarily) to an ultrafilter $\mathcal{U}_n$. Since there was a stage for every pair (f, k), we have ensured that, for each $f \in \mathcal{T}$, every element of $\{f(\mathcal{U}_n) : n \in \omega\}$ is isolated, i.e., this set is discrete. Thus, Lemma 15 applies, and the proof of the proposition is complete. □

The following is a corollary not of the proposition but of its proof; we shall obtain an improvement later, in Proposition 24.

Corollary 17. *Assume* $\mathfrak{u} \geq \mathfrak{d}$. *Let* $\mathcal{F}$ *be a filter with* $\chi(\mathcal{F}) < \mathfrak{r}$. *Then there is an infinite closed set* C *of pairwise not nearly coherent ultrafilters with* $\mathsf{C} \subseteq [\mathcal{F}]$. *Thus, if a closed set in* ω^* *is an intersection of fewer than* $\mathfrak{r}$ *open sets, then it includes a set* C *as in the proposition.*

Proof. Repeat the proof of the proposition, using $\mathcal{F}$ in place of the cofinite filter as the initial $\mathcal{G}_n$ for all n. The induction still works, because we still have, at each stage, a family $\mathcal{X}$ of size $< \mathfrak{r}$ containing generators for all the $\mathcal{G}_n$. At the end, all the $\mathcal{U}_n$ will, by construction, be in $[\mathcal{F}]$. As $[\mathcal{F}]$ is closed, it will also contain all the limits of the $\mathcal{U}_n$, which, according to the proof of Lemma 15, are the elements of the C that we finally obtain. □

4. Proof when $\mathfrak{u} < \mathfrak{d}$

To prove our main theorem in the case $\mathfrak{u} < \mathfrak{d}$, we shall need three preliminary propositions, which may be of some interest in their own right. Two of them involve the dominating number of a filter, defined as the cofinality of the reduced power of ω. More explicitly:

Definition 18. For any filter $\mathcal{F}$, its *dominating number* $\mathfrak{d}(\mathcal{F})$ is defined as the smallest cardinality of any family $\mathcal{D}$ of functions $\omega \to \omega$ such that every function $\omega \to \omega$ is majorized by some function from $\mathcal{D}$ on some set in $\mathcal{F}$.

Notice that, when $\mathcal{F}$ is the filter of cofinite sets, then $\mathfrak{d}(\mathcal{F})$ is simply $\mathfrak{d}$, and that, for all filters, $\mathfrak{d}(\mathcal{F}) \leq \mathfrak{d}$.

The first of our three preliminary propositions can be deduced from Theorem 5.5.3 and Corollary 10.3.2 of [2], but for convenience we give a self-contained proof here.

Proposition 19. *If a filter* $\mathcal{F}$ *and an ultrafilter* $\mathcal{U}$ *are not nearly coherent, then* $\mathfrak{d}(\mathcal{F}) \leq \chi(\mathcal{U})$.

Proof. Let $\mathcal{F}$ and $\mathcal{U}$ be as in the hypothesis of the proposition, and fix a base $\mathcal{B}$ for $\mathcal{U}$ with cardinality $|\mathcal{B}| = \chi(\mathcal{U})$. For each $X \in \mathcal{B}$ and each $n \in \omega$, define $\text{next}(X, n)$ to be the first element of X that is $> n$.

Suppose, toward a contradiction, that $\chi(\mathcal{U}) < \mathfrak{d}(\mathcal{F})$. In particular, the functions $\text{next}(X, -)$ for $X \in \mathcal{B}$ do not form a family $\mathcal{D}$ as in the definition of $\mathfrak{d}(\mathcal{F})$. So fix $h : \omega \to \omega$ that is not majorized on any set in $\mathcal{F}$ by any of the functions $\text{next}(X, -)$ for $X \in \mathcal{B}$. Partition ω into intervals $I_0 = [i_0, i_1)$, $I_1 = [i_1, i_2)$, etc., where $0 = i_0 < i_1 < i_2 < \ldots$ and where each i_{k+1} is chosen to be larger than $h(n)$ for all $n \leq i_k$. Let $g : \omega \to \omega$ be the function that sends all elements of any interval I_k to k. Then $g(\mathcal{F})$, not being coherent with $g(\mathcal{U})$, cannot be the cofinite filter. Thus, we can find a set $A \in \mathcal{F}$ that is the union of some of the intervals I_k but that omits infinitely many of the I_k.

From the sequence of intervals $\langle I_k\rangle$, extract the subsequence $\langle J_k = [l_k, r_k) : k \in \omega\rangle$ consisting of those intervals that are outside A. Define $f : \omega \to \omega$ to map all elements of $[r_{k-1}, r_k)$ to k (where r_{-1} means 0). Less formally, f takes the value k on J_k, the kth of the intervals omitted by A, and also on the block of I-intervals in A that immediately precede this J_k.

By assumption, $f(\mathcal{F})$ and $f(\mathcal{U})$ are not coherent. So we can find $F \in \mathcal{F}$ and $X \in \mathcal{U}$ with $f(F) \cap f(X) = \varnothing$. Shrinking these sets if necessary, we can assume that $F \subseteq A$ and that $X \in \mathcal{B}$. By our choice of h, there is some $n \in F$ with $\mathrm{next}(X, n) < h(n)$. Letting k be the (unique) index with $n \in I_k$, we have that $I_k \subseteq A$ because $n \in F \subseteq A$. Thus, f takes the same value on I_k and I_{k+1}. Also, since $n \le i_{k+1}$, we have $h(n) < i_{k+2}$, and so

$$i_k \le n < \mathrm{next}(X, n) < h(n) < i_{k+2}.$$

Thus, both the element n of F and the element $\mathrm{next}(X, n)$ of X lie in the set $I_k \cup I_{k+1}$ on which f is constant. This contradicts the fact that $f(F)$ and $f(X)$ are disjoint, and so the proof is complete. □

Proposition 20. *Suppose $\mathcal{V}$ is an ultrafilter with $\chi(\mathcal{V}) < \mathfrak{d}$, and suppose $\mathcal{U}_n$, for $n \in \omega$, are ultrafilters that are not nearly coherent with $\mathcal{V}$. Then the filter $\bigcap_{n\in\omega} \mathcal{U}_n$ is also not nearly coherent with $\mathcal{V}$.*

Proof. Suppose the contrary, and let f be a finite-to-one function such that $f(\bigcap_{n\in\omega} \mathcal{U}_n)$ and $f(\mathcal{V})$ are coherent. Since the latter is an ultrafilter, we have $f(\bigcap_{n\in\omega} \mathcal{U}_n) \subseteq f(\mathcal{V})$. That is, $f(\mathcal{V})$ belongs to $[f(\bigcap_{n\in\omega} \mathcal{U}_n)]$, which is the closure of the set $\{f(\mathcal{U}_n) : n \in \omega\}$ in ω^*. But by Corollary 8, we know that $f(\mathcal{V})$ is a P-point, so it cannot be in the closure of a countable set of other ultrafilters. Therefore $f(\mathcal{V}) = f(\mathcal{U}_n)$ for some n, contrary to the assumption that no $\mathcal{U}_n$ is nearly coherent with $\mathcal{V}$. □

Proposition 21. *For any filter $\mathcal{F}$, there is a test family of cardinality $\mathfrak{d}(\mathcal{F})$ over $\mathcal{F}$.*

Proof. We first handle the case that, for some finite-to-one $f : \omega \to \omega$, $f(\mathcal{F})$ is the cofinite filter. In this case, we have $\mathfrak{d}(\mathcal{F}) = \mathfrak{d}$. Indeed, if $\mathcal{D}$ is as in the definition of $\mathfrak{d}(\mathcal{F})$, then it is straightforward to check that the family of functions

$$x \mapsto \max\{g(y) : f(y) = x\}$$

for $g \in \mathcal{D}$ is a dominating family. Since there always exists a test family of size $\mathfrak{d}$, and since a test family serves as a test family over any filter, the proposition is proved in this case.

Assume from now on that no finite-to-one function sends $\mathcal{F}$ to the cofinite filter. Fix a family $\mathcal{H}$ of functions $\omega \to \omega$ such that $|\mathcal{H}| = \mathfrak{d}(\mathcal{F})$ and such that every function $\omega \to \omega$ is majorized, on some set in $\mathcal{F}$, by some $h \in \mathcal{H}$.

For each $h \in \mathcal{H}$, proceed as in the proof of Proposition 19 to construct a partition of ω into intervals I_k, a function g, a set $A \in \mathcal{F}$, a subsequence of intervals J_k, bigger intervals $[r_{k-1}, r_k)$, and the function f that is constant on just these bigger intervals. The key property of f that we shall need is that, whenever

$a \le b \le h(a)$ and $a \in A$, then $f(a) = f(b)$. To verify this property, first use the definition of the intervals I_k to see that a and b are either in the same one of these intervals or in consecutive ones. Then use the assumption that $a \in A$ to see that, even if they are in consecutive intervals, the first of these intervals (the one containing a) is not among the J_k's, and so a and b lie in the same $[r_{k-1}, r_k)$. Thus $f(a) = f(b)$.

We have associated to each $h \in \mathcal{H}$ a function f, and we shall show that these f's constitute a test family over $\mathcal{F}$. Since the number of these f's is at most $|\mathcal{H}| = \mathfrak{d}(\mathcal{F})$, this will complete the proof of the proposition.

Suppose, therefore, that $\mathcal{U}$ and $\mathcal{V}$ are nearly coherent ultrafilters in $[\mathcal{F}]$. Fix a finite-to-one function $j : \omega \to \omega$ such that $j(\mathcal{U}) = j(\mathcal{V})$. By the defining property of $\mathcal{H}$, fix some $F \in \mathcal{F}$ and some $h \in \mathcal{H}$ such that, for all $n \in F$, $h(n) > \max(j^{-1}(j([0, n])))$. If we view j as partitioning ω into finite pieces $j^{-1}(n)$, then the requirement on $h(n)$, for $n \in F$, is that it be larger than the right endpoints of all the (finitely many) pieces whose left endpoints are $\le n$. Let f be the function constructed from h above. We shall complete the proof by showing that $f(\mathcal{U}) = f(\mathcal{V})$.

We may assume, by shrinking F if necessary, that $F \subseteq A$. To prove that $f(\mathcal{U}) = f(\mathcal{V})$ it suffices, since these are ultrafilters, to show that $f(U) \cap f(V) \ne \varnothing$ for all $U \in \mathcal{U}$ and $V \in \mathcal{V}$. So consider any such U and V. Shrinking them if necessary, we may assume, since $F \in \mathcal{F} \subseteq \mathcal{U}, \mathcal{V}$, that $U, V \subseteq F \subseteq A$. As $j(\mathcal{U}) = j(\mathcal{V})$, the sets $j(U)$ and $j(V)$ must intersect. Fix some $a \in U$ and $b \in V$ such that $j(a) = j(b)$. Without loss of generality, suppose $a \le b$. It follows, by our choice of h, that $b \le h(a)$. Thanks to the key property of f verified above, we conclude that $f(a) = f(b)$, and the proof is complete. □

After these preparatory results, we are ready to prove the case $\mathfrak{u} < \mathfrak{d}$ of the main theorem; we state the result, with some additional information, as a separate proposition.

Proposition 22. *Assume that $\mathfrak{u} < \mathfrak{d}$ and that $\{\mathcal{U}_n : n \in \omega\}$ is a set of distinct ultrafilters that are pairwise not nearly coherent. Then the closure of $\{\mathcal{U}_n : n \in \omega\}$ in ω^* has an infinite closed subset (hence of size $2^{\mathfrak{c}}$) whose members are pairwise not nearly coherent.*

In contrast to Proposition 16, here we must assume, rather than prove, that infinitely many non-nearly-coherent ultrafilters are given. As is shown in [8], it is consistent to have only a single near-coherence class of ultrafilters (with $\mathfrak{u} < \mathfrak{d}$).

Proof of the Proposition. Assume the hypotheses. We intend to find filters $\mathcal{F}$ and $\mathcal{G}$ and a family $\mathcal{T}$ of functions satisfying the hypotheses of Lemma 15. Furthermore, we shall ensure that $[\mathcal{F}]$ is infinite. Since the closed set of non-nearly-coherent ultrafilters produced by the lemma consists of ultrafilters of the form $\mathcal{V}\text{-}\lim_n \mathcal{U}_n$, which are in the closure of $\{\mathcal{U}_n : n \in \omega\}$, and since this closed set has the same cardinality as $[\mathcal{F}]$, this will suffice to complete the proof of the proposition.

Let $\mathcal{W}$ be an ultrafilter with $\chi(\mathcal{W}) = \mathfrak{u} < \mathfrak{d}$. Recall that, by Corollary 8, $\mathcal{W}$ is a P-point.

Since the $\mathcal{U}_n$ are pairwise not nearly coherent, at most one of them is nearly coherent with $\mathcal{W}$. Discarding one of the $\mathcal{U}_n$'s, we assume from now on that none of them are nearly coherent with $\mathcal{W}$. Define the filter $\mathcal{G}$ to be the intersection of the $\mathcal{U}_n$'s, and note that, by Proposition 20, $\mathcal{G}$ is not nearly coherent with $\mathcal{W}$. By Proposition 19, it follows that $\mathfrak{d}(\mathcal{G}) \leq \chi(\mathcal{W}) = \mathfrak{u}$. By Proposition 21, fix a test family $\mathcal{T}$ over $\mathcal{G}$ with $|\mathcal{T}| = \mathfrak{u}$. We have defined $\mathcal{G}$ and $\mathcal{T}$ in such a way that the hypotheses of Lemma 15 are satisfied insofar as they do not mention $\mathcal{F}$. It remains to define a filter $\mathcal{F}$ such that, for each $f \in \mathcal{T}$, there is $A \in \mathcal{F}$ for which the sequence $\langle f(\mathcal{U}_n) : n \in A\rangle$ is discrete.

Since $\mathcal{W}$ is a P-point, Lemma 11 gives us a discrete ultrafilter $\mathcal{V}$ that is not a P-point. By Proposition 12, $\mathcal{V}$ is ω^*-discrete, and by Corollary 8 it has character $\chi(\mathcal{V}) \geq \mathfrak{d} > \mathfrak{u}$. We shall use $\mathcal{V}$ to help us build the required $\mathcal{F}$, in a transfinite induction of length $\mathfrak{u}$, constructing an increasing sequence of filters. The stages of the induction will be indexed by the functions $f \in \mathcal{T}$; this can be arranged since $|\mathcal{T}| = \mathfrak{u}$. We begin with the filter of cofinite sets, and at limit stages we form the union of the previously built filters. It remains to describe the successor steps. At each of these steps, we shall add at most one new generator to our filter, and this generator will be a set in $\mathcal{V}$. This ensures that, at each stage during the construction, the current filter is generated by fewer than $\mathfrak{u}$ sets and that the filter obtained at the end of the construction is generated by at most $\mathfrak{u}$ sets and is included in $\mathcal{V}$. Furthermore, by always using sets from the ultrafilter $\mathcal{V}$, we avoid any concern about whether the added sets preserve the finite intersection property and so generate a proper filter; they will automatically do so since $\mathcal{V}$ has the finite intersection property.

Consider the step from the stage indexed by f to the next stage, and let $\mathcal{F}'$ be the filter constructed so far. Consider the ultrafilters $f(\mathcal{U}_n)$. They are distinct, because the $\mathcal{U}_n$ are pairwise not nearly coherent. Since $\mathcal{V}$ is ω^*-discrete, there is a set $A \in \mathcal{V}$ such that $\{f(\mathcal{U}_n) : n \in A\}$ is discrete. Choose one such A and adjoin it to $\mathcal{F}'$ as a new generator. This ensures that, at the end of our induction, $\mathcal{F}$, which is the union of all the filters $\mathcal{F}'$ built during the inductive stages, will satisfy the hypothesis of Lemma 15.

According to that lemma, we obtain, in the closure of $\{\mathcal{U}_n : n \in \omega\}$, a closed set of pairwise not nearly coherent ultrafilters, and this closed set has the same cardinality as $[\mathcal{F}]$. It remains to verify that this cardinality is infinite.

Suppose, toward a contradiction, that $[\mathcal{F}]$ were finite. By Lemma 6, we infer that $\mathcal{V}$, being an ultrafilter extending $\mathcal{F}$, has $\chi(\mathcal{V}) \leq \chi(\mathcal{F}) \leq \mathfrak{u}$, whereas we saw, immediately after choosing $\mathcal{V}$, that its character is $> \mathfrak{u}$. □

Propositions 16 and 22 together complete the proof of Theorem 1.

5. Closed sets when $\mathfrak{u} > \mathfrak{d}$

Comparing Propositions 16 and 22, we see that the latter has not only an additional hypothesis, namely that infinitely many non-nearly-coherent ultrafilters must be given, but also an additional conclusion, namely that $2^{\mathfrak{c}}$ non-nearly-coherent ultrafilters (in fact an infinite closed set of them) can be found in the closure of the given ultrafilters. It is natural to ask whether this additional conclusion can also be obtained, of course under the same additional hypothesis, when $\mathfrak{u} \geq \mathfrak{d}$. We do not know the answer to this question, but we have an affirmative answer under the stronger assumption that $\mathfrak{u} > \mathfrak{d}$. The result is exactly like Proposition 22 except that the inequality between $\mathfrak{u}$ and $\mathfrak{d}$ is reversed.

Proposition 23. *Assume that $\mathfrak{u} > \mathfrak{d}$ and that $\{\mathcal{U}_n : n \in \omega\}$ is a set of distinct ultrafilters that are pairwise not nearly coherent. Then the closure of $\{\mathcal{U}_n : n \in \omega\}$ in ω^* has an infinite closed subset (hence of size $2^{\mathfrak{c}}$) whose members are pairwise not nearly coherent.*

Proof. Assume the hypotheses of the proposition. As in previous proofs, we shall produce $\mathcal{F}$, $\mathcal{G}$, and $\mathcal{T}$ to satisfy the hypotheses of Lemma 15. Furthermore, we shall ensure that $[\mathcal{F}]$ is infinite. Then Lemma 15 will complete the proof of the proposition. We take $\mathcal{G}$ to be the filter of cofinite sets, and, invoking Lemma 14, we take $\mathcal{T}$ to be a test family of cardinality $\mathfrak{d}$. It remains to construct a filter $\mathcal{F}$ such that $[\mathcal{F}]$ is infinite and, for each $f \in \mathcal{T}$, there is $A \in \mathcal{F}$ for which the sequence $\langle f(\mathcal{U}_n) : n \in A\rangle$ is discrete. As in previous proofs, the construction is a transfinite induction, starting with the cofinite filter, taking unions at limit steps, and adding at most one new generator to our filter at each successor step. There will be $\mathfrak{d}$ steps in the induction, indexed by the elements of $\mathcal{T}$. So at every stage, the part of $\mathcal{F}$ already constructed will be a filter $\mathcal{F}'$ with $\chi(\mathcal{F}') < \mathfrak{d}$. The final result will therefore be a filter $\mathcal{F}$ with $\chi(\mathcal{F}) \leq \mathfrak{d} < \mathfrak{u}$. In view of Lemma 6, it follows that there are infinitely many ultrafilters in $[\mathcal{F}]$.

It remains to carry out the successor step, from the stage indexed by $f \in \mathcal{T}$ to the next stage, adding, to the filter $\mathcal{F}'$ produced by the previous stages, some set A such that $\langle f(\mathcal{U}_n) : n \in A\rangle$ is discrete. To obtain an appropriate A, first apply Lemma 9 to the compact Hausdorff space ω^*, the function $\omega \to \omega^*$ that sends n to $f(\mathcal{U}_n)$, and an arbitrary ultrafilter $\mathcal{V} \supseteq \mathcal{F}'$. The result is a decreasing sequence of sets $A_0 \supseteq A_1 \supseteq \dots$ in $\mathcal{V}$ such that, if $B \subseteq A_0$ is a pseudointersection of the A_n's, then $\langle f(\mathcal{U}_n) : n \in B\rangle$ is discrete. So we need only obtain such a pseudointersection B that can be added to $\mathcal{F}'$, i.e., that meets every set in $\mathcal{F}'$. Noting that each A_n, being in $\mathcal{V}$, meets every set in $\mathcal{F}'$, and remembering that $\chi(\mathcal{F}') < \mathfrak{d}$, we obtain the required B from Lemma 7. □

Using the preceding result, we can improve Corollary 17 as follows, replacing the bound $\mathfrak{r}$ for $\chi(\mathcal{F})$ by the possibly larger cardinal $\mathfrak{u}$.

Proposition 24. *Assume $\mathfrak{u} \geq \mathfrak{d}$. Let $\mathcal{F}$ be a filter with $\chi(\mathcal{F}) < \mathfrak{u}$. Then there is an infinite closed set C of pairwise not nearly coherent ultrafilters with $\mathsf{C} \subseteq [\mathcal{F}]$. Thus,* *such a closed set, of size $2^{\mathfrak{c}}$, can be found inside any non-empty subset of ω^* that is the intersection of fewer than $\mathfrak{u}$ open sets.*

Proof. Assume that $\mathfrak{u} \geq \mathfrak{d}$ and that $\mathcal{F}$ is as in the proposition. To obtain the required C, we may assume $\mathfrak{r} < \mathfrak{u}$, for otherwise we get C immediately from Corollary 17. By Lemma 3 and our assumption that $\mathfrak{u} \geq \mathfrak{d}$, we have $\mathfrak{d} \leq \mathfrak{r} < \mathfrak{u}$. Thus, Proposition 23 applies, and we need only show, given any filter $\mathcal{F}$ with $\chi(\mathcal{F}) < \mathfrak{u}$, that there are infinitely many, pairwise not nearly coherent ultrafilters extending $\mathcal{F}$. It therefore suffices, given any finite number n (possibly zero) of ultrafilters $\mathcal{U}_1, \dots, \mathcal{U}_n$, to find another ultrafilter $\mathcal{U} \supseteq \mathcal{F}$ that is not nearly coherent with any of them.

Let such $\mathcal{U}_i$ be given, and, by Lemma 14, let $\mathcal{T}$ be a test family of size $\mathfrak{d}$. We shall construct the required $\mathcal{U}$ in an induction of length $\mathfrak{d}$, indexed by the functions $f \in \mathcal{T}$. We begin the induction with the given filter $\mathcal{F}$, and at limit stages we take unions. At any successor step, we shall add at most one new generator to the filter under construction. Since $\chi(\mathcal{F}) < \mathfrak{u}$ and since there are only $\mathfrak{d} < \mathfrak{u}$ steps in the induction, we shall have at each stage of the induction a filter generated by fewer than $\mathfrak{u}$ sets.

At the step from the stage indexed by some $f \in \mathcal{T}$ to the next stage, the generator to be added to the current filter, say $\mathcal{G}$, is obtained as follows. Since $\mathcal{G}$ is generated by fewer than $\mathfrak{u}$ sets, so is $f(\mathcal{G})$. By Lemma 6, there are infinitely many ultrafilters extending $f(\mathcal{G})$. Let $\mathcal{V}$ be an ultrafilter extending $f(\mathcal{G})$ and distinct from the finitely many ultrafilters $f(\mathcal{U}_i)$. Let B be a set that is in $\mathcal{V}$ but in none of the $f(\mathcal{U}_i)$. Then adjoin $f^{-1}(B)$ to $\mathcal{G}$ as a new generator.

At the end of the induction, we have produced a filter $\mathcal{G} \supseteq \mathcal{F}$ such that, for each $f \in \mathcal{T}$, none of the $f(\mathcal{U}_i)$ extend $f(\mathcal{G})$, because the latter filter contains the set B obtained during the induction step for f, while all of the former contain $\omega - B$. Thus, we can get the required $\mathcal{U}$ by extending $\mathcal{G}$ arbitrarily to an ultrafilter.

Finally, to prove the last assertion in the proposition, we first observe that it follows from what is already proved if the given set is closed. To establish the general case, let X be a non-empty intersection of fewer than $\mathfrak{u}$ open subsets G_i of ω^*, and let $\mathcal{U}$ be an arbitrary element of X. For each i, choose a basic open set $[A_i]$ such that $\mathcal{U} \subseteq [A_i] \subseteq \mathsf{G}_i$. Since all the A_i are in $\mathcal{U}$, they have the finite intersection property and so generate a filter $\mathcal{F}$. By the part of the proposition already proved, there is an infinite closed set C of pairwise not nearly coherent ultrafilters with

$$\mathsf{C} \subseteq [\mathcal{F}] = \bigcap_i [A_i] \subseteq \bigcap_i \mathsf{G}_i = \mathsf{X}. \qquad \square$$

6. Some open problems

As indicated before Proposition 23, there is a gap between it and the very similar Proposition 22, namely the case $\mathfrak{u} = \mathfrak{d}$.

Question 25. Given an infinite set of ultrafilters that are pairwise not nearly coherent, must its closure include an infinite closed subset no two of whose members are nearly coherent?

By Propositions 22 and 23, the answer is affirmative as long as $\mathfrak{u} \neq \mathfrak{d}$. Inspecting the proof of Proposition 22, we can see that the answer to Question 25 is also affirmative under the assumption that there is an ω^*-discrete ultrafilter $\mathcal{V}$ with character $\chi(\mathcal{V}) > \min\{\mathfrak{u}, \mathfrak{d}\}$.

In the light of this observation it is interesting to mention that discrete ultrafilters need not exist in ZFC: according to [22] there is a model of ZFC without nowhere dense ultrafilters, i.e., a model where every ultrafilter has an image, under a map $\omega \to \mathbb{R}$, containing no nowhere dense subset of $\mathbb{R}$, and therefore certainly no discrete set.

Question 26. Is the existence of a $\beta\omega$-discrete ultrafilter provable in ZFC?

We may pose this question is a more general form.

Question 27. Describe Tychonov spaces Y for which Y-discrete ultrafilters exist in ZFC. Given such a space Y describe the structure of Y-discrete ultrafilters.

It can be shown that the class of spaces Y described in Question 27 includes scattered spaces (for a scattered space Y each OK-point, as defined in [13], will be Y-discrete) and excludes all spaces containing a copy of the space $\mathbb{Q}$ of rationals (because $\mathbb{Q}$-discrete ultrafilters are discrete, and thus need not exist in ZFC).

There are natural questions concerning the relation of ω^*-discrete ultrafilters with discrete or nowhere dense ultrafilters; the following seems particularly basic.

Question 28. Is each ω^*-discrete ultrafilter discrete? Is it nowhere dense?

Recall that a *weak P-point* is an ultrafilter that is not in the closure of any countable set of other ultrafilters. Kunen [13] showed (in ZFC) that weak P-points exist. One can show that an ω^*-discrete ultrafilter has the additional "at most one limit point not in the range" property (as in Lemma 9) if and only if it is a weak P-point. Baumgartner's proof of Lemma 11 produces, from any such ultrafilter, another ω^*-discrete ultrafilter that is not a weak P-point. Given this, and given the connection between discrete ultrafilters and P-points, it is natural to ask about further connections.

Question 29. Is there a weak P-point that is not ω^*-discrete? Can an ω^*-discrete ultrafilter be a weak P-point without being a P-point?

The next question is motivated by the circumstance that we proved that the number of near-coherence classes is infinite if $\mathfrak{u} \geq \mathfrak{d}$ (Proposition 16) but had to assume it when $\mathfrak{u} < \mathfrak{d}$ (Proposition 22).

Question 30. Is $\mathfrak{u} < \mathfrak{d}$ provably equivalent to the assertion that there are only finitely many near-coherence classes of ultrafilters?

Of course, Proposition 16 gives the implication in one direction, so the question reduces to asking whether a model of $\mathfrak{u} < \mathfrak{d}$ can have infinitely many near-coherence classes of ultrafilters. The difficulty of answering this question arises from the paucity of known models of $\mathfrak{u} < \mathfrak{d}$. Apart from models of NCF (i.e., models with only one near-coherence class) and the model recently proposed by Shelah to replace the erroneous one in [8, Section 6] (with two near-coherence classes), there are the models constructed in [9]. Michael Canjar (unpublished) has shown that the latter models do not satisfy NCF, but we do not know whether they have infinitely many near-coherence classes.

Finally, we repeat the central open question motivating this work.

Question 31. What cardinals can consistently be the number of near-coherence classes of ultrafilters?

The known results – those quoted in the introduction and those established in this paper – say that the cardinals in question include 1 and $2^{\mathfrak{c}}$ but no other infinite cardinals. Once Shelah's replacement for the model of [8, Section 6] is checked, the cardinal 2 can be added to the list. So the remaining question will concern finite cardinals ≥ 3.

References

[1] Jason Aubrey, "Combinatorics for the dominating and unsplitting numbers," *J. Symbolic Logic* 69 (2004) 482–498.

[2] Taras Banakh and Lyubomyr Zdomskyy, *Coherence of Semifilters*, http://www.franko.lviv.ua/faculty/mechmat/Departments/Topology/booksite.html

[3] James Baumgartner, "Ultrafilters on ω," *J. Symbolic Logic* 60 (1995) 624–639.

[4] Andreas Blass, *Orderings of Ultrafilters*, Ph.D. thesis, Harvard University (1970).

[5] Andreas Blass, "Near coherence of filters, I: Cofinal equivalence of models of arithmetic," *Notre Dame J. Formal Logic* 27 (1986) 579–591.

[6] Andreas Blass, "Near coherence of filters, II: Applications to operator ideals, the Stone-Čech remainder of a half-line, order ideals of sequences, and slenderness of groups," *Trans. Amer. Math. Soc.* 300 (1987) 557–581.

[7] Andreas Blass and Heike Mildenberger, "On the cofinality of ultrapowers," *J. Symbolic Logic* 64 (1999) 727–736.

[8] Andreas Blass and Saharon Shelah, "There may be simple $P_{\aleph_1}$ and $P_{\aleph_2}$ points and the Rudin-Keisler order may be downward directed," *Ann. Pure Appl. Logic* 33 (1987) 213–243.

[9] Andreas Blass and Saharon Shelah, "Ultrafilters with small generating sets," *Israel J. Math.* 65 (1989) 259–271.

[10] David Booth, "Ultrafilters on a countable set," *Ann. Math. Logic* 2 (1970) 1–24.

[11] Martin Goldstern and Saharon Shelah, "Ramsey ultrafilters and the reaping number – Con($\mathfrak{r} < \mathfrak{u}$)," *Ann. Pure Appl. Logic* 49 (1990) 121–142.

[12] Jussi Ketonen, "On the existence of P-points in the Stone-Čech compactification of integers," *Fund. Math.* 92 (1976) 91–94.

[13] Kenneth Kunen, "Weak P-points in $\mathbb{N}^*$," *Topology*, Colloq. Math. Soc. János Bolyai 23, ed. Á. Császár, North-Holland (1980) 741–749.

[14] Stefan Mazurkiewicz, "Sur les continus indécomposables," *Fund. Math.* 10 (1927) 305–310.

[15] Jerzy Mioduszewski, "On composants of $\beta R - R$," *Proc. Conf. Topology and Measure (Zinnowitz, 1974)*, ed. J. Flachsmeyer, Z. Frolík, and F. Terpe, Ernst-Moritz-Arndt-Universität zu Greifswald (1978) 257–283.

[16] Jerzy Mioduszewski, "An approach to $\beta\mathbf{R} - \mathbf{R}$," *Topology*, Colloq. Math. Soc. János Bolyai 23, ed. Á. Császár, North-Holland (1980) 853–854.

[17] Bedřich Pospíšil, "Remark on bicompact spaces," *Ann. Math. (2)* 38 (1937) 845–846.

[18] Fritz Rothberger, "On some problems of Hausdorff and Sierpiński," *Fund. Math.* 35 (1948) 29–46.

[19] Mary Ellen Rudin, "Types of ultrafilters," *Topology Seminar Wisconsin, 1965*, eds. R.H. Bing and R.J. Bean, Annals of Mathematics Studies 60 (1966) 147–151.

[20] Mary Ellen Rudin, "Composants and βN," *Proc. Washington State Univ. Conf. on General Topology* (1970) 117–119.

[21] Walter Rudin, "Homogeneity problems in the theory of Čech compactifications," *Duke Math. J.* 23 (1956) 409–419

[22] Saharon Shelah, "There may be no nowhere dense ultrafilter," *Logic Colloquium '95 (Haifa)*, eds. J.A. Makowsky and E.V. Ravve, Lecture Notes in Logic 11, Springer-Verlag (1998) 305–324.

[23] Edward Wimmers, "The Shelah P-point independence theorem," *Israel J. Math.* 43 (1982) 28–48.

Taras Banakh
Lviv University (Lviv, Ukraine)
Nipissing University (North Bay, Canada)
Akademia Swiętokrzyska (Kielce, Poland)
e-mail: `tbanakh@yahoo.com`

Andreas Blass
Mathematics Department
University of Michigan
Ann Arbor, MI 48109–1109, USA
e-mail: `ablass@umich.edu`

Set Theory
Trends in Mathematics, 275–283

Stable Axioms of Set Theory

Sy-David Friedman

Abstract. I discuss criteria for the choice of axioms to be added to ZFC, introducing the criterion of *stability*. Then I examine a number of popular axioms in light of this criterion and propose some new axioms.

1. Criteria for new axioms

The incompleteness phenomenon is particularly evident in the field of set theory: The standard axiom system ZFC for set theory has a vast range of different types of models. Some people have suggested that this is an essential feature of set theory, because ZFC exhausts our set-theoretic intuition. A more optimistic view is that by increasing our knowledge of set theory, we will arrive at new axioms which are so compelling in their naturalness and in their ability to clarify the structure of the set-theoretic universe that we can assert that our intuition is in fact strong enough to justify adding them to ZFC as standard axioms.

By adopting new axioms, we narrow our view of set theory. Therefore it is important to suggest criteria for doing so. Below are some criteria to consider.

Naturalness. Axioms should be directly concerned with the structure of the set-theoretic universe V.

Natural axioms typically make assertions about the height or width of the universe, or assert the existence of certain types of elementary embeddings between inner models.

Power. Axioms should explain a lot.

Powerful axioms give us more detail about the structure of V than ZFC alone can provide.

Consistency. Axioms should be consistent (with ZFC).

The author wishes to thank the Austrian Research Fund (FWF) for its generous support through grants P16334-NO5 and P16790-NO4.

This criterion presents a problem: How are we to know whether or not a proposed axiom is consistent? By Gödel's second incompleteness theorem there is no way to definitively establish the consistency of an axiom. This leads to:

An axiom is deemed to be consistent as long as no proof of its inconsistency is currently known.

Thus what we accept as consistent can change with time. I see no alternative to this.

The above three criteria are in my view essential to the choice of any new axiom. The next criterion however is not.

Stability. (a) (Syntactic stability) Axioms should be unaffected by small changes. In particular, small changes should not lead to inconsistency. (b) (Semantic stability) A small extension of a model of the axioms should also be a model.

As with naturalness or power, I do not attempt to give here a rigorous definition of stability. In particular, I offer no precise definition of the notion of small extension used to define semantic stability. But surely the addition of one Cohen real must be viewed as a small extension. And the notion of small extension cannot be restricted to just set-generic extensions, as class-forcing provides consistent ways to enlarge the set-theoretic universe in the same way that set-forcing does. Although proper classes do not themselves belong to the universe of sets, via forcing they do produce new sets with important properties that cannot fairly be excluded from consideration.

As we shall see, stability is very restrictive. Indeed, it is violated by almost all of the axioms that have been proposed as candidates for addition to ZFC. Stability is however appealing, as it rules out the choice of axioms which are obtained by slightly weakening inconsistent principles. And as we will see below, there are attractive proposals for stable axioms of considerable naturalness and strength.

2. Examples

I first examine some well-known axioms in terms of the criteria of naturalness, power and stability.

a. $V = L$

Of course this axiom is natural and very powerful. But by the work of Cohen we know that $V = L$ is easily contradicted by forcing, and therefore the criterion of stability is violated. The same problem exists with any axiom of the form

$$V = L[G]$$

where G is P-generic over L for an L-definable forcing P, as one can similarly violate this easily by further forcing.

b. Large cardinals

Typically these are of the form

There exists $j : V \to M$, where M is "close" to V.

Certainly such axioms are natural and very powerful. They are however unstable. If we require M to equal V, we have a contradiction, by Kunen's theorem [5]. If we only require M to agree with V up to $j(\kappa)$ where κ is the critical point of j, then by stability, we must also allow agreement up to arbitrary iterates of j applied to κ, another contradiction to Kunen's theorem.

c. Determinacy

I am not referring to the full axiom AD, as this contradicts the axiom of choice, but rather to determinacy for sets of reals that are "definable" in some sense. This axiom has proved to be powerful, and in addition appears to be stable.

Unfortunately the existence of strategies for infinite games is not directly concerned with the structure of V, in violation of naturalness. I will however argue below that some definable determinacy is a consequence of natural and stable axioms, even though determinacy itself does not qualify as one.

d. Generic absoluteness

Absoluteness asserts that the truth of certain formulas is not affected by enlarging the universe in certain ways. The classical example of this is Lévy-Shoenfield absoluteness, which says that $\Sigma_1(H(\omega_1))$ formulas (with parameters) are absolute for arbitrary extensions.

Typically, one considers only *set-generic* absoluteness, in which only set-generic extensions of the universe are allowed, and only formulas which are first-order over $H(\omega_2)$. This is to enable the formulation of consistent principles. However such principles fail as soon as more general extensions, such as class-generic extensions, are allowed. Even when restricted to set-generic extensions, instability is present: $\Sigma_1(H(\kappa))$ absoluteness for ccc forcing extensions is inconsistent when κ is greater than $2^{\aleph_0}$. As one typically has $2^{\aleph_0} = \aleph_2$ in the context of set-generic absoluteness principles, inconsistency already occurs when κ is $\aleph_3$.

e. Forcing axioms

The most common such axioms assert that for certain set-forcings P and certain collections X of dense subsets of P, there is a compatible subset G of P which intersects all elements of X. The classical example is Martin's axiom (at ω_1), which asserts this for ccc P and collections X of cardinality ω_1.

As with the set-generic absoluteness principles, these axioms are unstable. For example, one cannot have this forcing axiom for κ-many dense sets with respect to even ccc forcings when κ is at least $2^{\aleph_0}$.

Other types of forcing axioms have also been considered. Foreman, Magidor and Shelah proposed:

Every set-forcing either adds a real or collapses a cardinal.

Little is known about this interesting axiom. A stable version would however require the consideration of more than just set-generic extensions.

An axiom of Chalons (as modified by Larson and then Hamkins) states:

> "If a statement with real parameters holds in a set-forcing extension and all further set-forcing extensions, then it holds in V; moreover this property is not only true in V, but also in all set-generic extensions of V."

Woodin proved the consistency of this axiom from large cardinals. Unfortunately, even a weak form of this axiom is inconsistent when "set-forcing" is replaced by "class-forcing". A consistent class-forcing version of this axiom is not known.

f. Strong logics

These are logics whose set of validities is large and remains unchanged by set-forcing. One can obtain such a logic as follows: Say that φ is $**$-provable iff for some set-forcing P, if P belongs to V_α and V_α satisfies ZFC, then V_α^{P*Q} satisfies φ for all Q in V_α^P. Woodin proposes the use of such a strong logic, together with the existence of a proper class of Woodin cardinals. This gives a $**$-complete theory of $H(\omega_1)$ and, assuming that $H(\omega_2)$ is obtained by forcing with Woodin's forcing P_{max} over $L(R)$, gives a $**$-complete theory of $H(\omega_2)$. Therefore under Woodin's assumptions, the theory of $H(\omega_2)$ cannot be changed by set-forcing.

There are several difficulties with this approach.

i. The assumption of the existence of a proper class of Woodin cardinals is left unjustified.

However I will propose below some natural and stable axioms which lead to an *inner model* for this assumption.

ii. Although strong logics are immune to set-forcing, they are not immune to class-forcing.

As mentioned earlier, class-forcing methods provide consistent ways to enlarge the set-theoretic universe in the same way that set-forcing methods do. Therefore adopting as new axioms the validities of a logic with only set-generic absoluteness does not achieve stability. One needs at least a plausible notion of "acceptable class forcing" and a corresponding property of absoluteness for such class forcings.

iii. The axiom asserting that $H(\omega_2)$ is obtained by set-forcing over $L(R)$ is easily contradicted by class-forcing, and therefore as in ii. leads to instability.

3. Some stable axioms of strong absoluteness

As mentioned earlier, the typical absoluteness principles which generalise Lévy-Shoenfield absoluteness refer exclusively to set-generic extensions, and are unstable. The Lévy-Shoenfield absoluteness principle itself, however, applies to arbitrary extensions. The *strong absoluteness principles* discussed below are in the tradition of Lévy-Shoenfield and impose no genericity requirement on the extensions considered. This leads to the possibility of obtaining stable axioms.

By *extension of* V I shall mean a ZFC model V^* which contains V and has the same ordinals as V. This is best formalised by regarding V as a countable transitive model of ZFC and allowing V^* to range over countable transitive ZFC models which contain V and have the same ordinal height as V.

Any consistent generalisation of Lévy-Shoenfield absoluteness must deal with the following two obstacles:

Counterexample 1. There is a Σ_1 formula with parameters from $H(\omega_2)$ which holds in some set-generic extension V^* of V but not in V.

Counterexample 2. There is a Σ_1 formula with parameters from $H((2^{\aleph_0})^+)$ which holds in some ccc set-generic extension V^* of V but not in V.

Counterexample 1 is witnessed by the formula "ω_1^V is countable". Counterexample 2 is witnessed by the formula "There is a real not in $\mathcal{P}(\omega)^V$".

Let us say that a Σ_1 *absoluteness principle* is a principle asserting the absoluteness of certain Σ_1 formulas with certain parameters with respect to certain extensions of V. Our counterexamples imply that a consistent Σ_1 absoluteness principle must impose some restriction either on the choice of formulas, on the choice of parameters, on the choice of extensions, or a combination of the three.

I offer three proposals. The first allows arbitrary parameters, at the cost of restricting the choice of extensions. The second allows arbitrary extensions, at the cost of restricting the allowable parameters. And the third weakens the parameter restrictions of the second proposal, at the cost of restricting the choice of formulas.

a. Σ_1 absoluteness with arbitrary parameters

A first attempt to avoid Counterexample 1 is to require that V and V^* have the same ω_1. But Σ_1 absoluteness with parameters from $H(\omega_2)$ even for ω_1-preserving extensions is also inconsistent: Let A be a stationary subset of ω_1. Then the formula which asserts that A contains a CUB subset is Σ_1 and true in a cardinal-preserving (set-generic) extension; therefore Σ_1 absoluteness with parameters from $H(\omega_2)$ for ω_1-preserving extensions implies that A contains a CUB subset. But there are disjoint stationary subsets of ω_1, giving disjoint CUB subsets of ω_1, a contradiction.

Even requiring stationary-preservation at ω_1 (i.e, that stationary subsets of ω_1 in V remain stationary in V^*) results in inconsistency:

Theorem 1. *There exists an extension V^* of V which is stationary-preserving at ω_1 such that some Σ_1 sentence with parameters from $H(\omega_2)^V$ true in V^* is false in V.*

Proof. By a theorem of Beller-David (see [1]) there is an extension V^* with the same ω_1 as V containing a real R such that $L_\alpha[R]$ fails to satisfy ZFC for each ordinal α. Moreover, V^* is stationary-preserving at ω_1 (see [2]). Now suppose that the Theorem fails. Then there is such a real R in V, as this property of R can be expressed by a Σ_1 sentence with parameters R and ω_1. In particular, ω_1 is not inaccessible to reals. It is easy to see that the failure of the Theorem implies that Σ_3^1-absoluteness holds between V and its stationary-preserving at ω_1 extensions.

It then follows from Lemma 7 of [2] that ω_1 is inaccessible to reals after all, contradiction. □

One could continue to make further restrictions on the extension V^*, such as stationary-preservation at ω_1 together with full cardinal-preservation, in the hope of achieving the consistency of $\Sigma_1(H(\omega_2))$ absoluteness (without imposing the requirement that V^* be a set-generic extension of V). But we must also reckon with Counterexample 2.

A possible solution is described by the following. I say that an extension V^* of V *strongly preserves* $H(\kappa)$ iff the $H(\kappa)$ of V^* equals the $H(\kappa)$ of V and all cardinals of V less than or equal to card $(H(\kappa)) = 2^{<\kappa}$ remain cardinals in V^*.

Σ_1 absoluteness with arbitrary parameters

Suppose that κ is an infinite cardinal and a Σ_1 formula φ with parameters from $H(\kappa^+)$ holds in an extension V^* of V which strongly preserves $H(\kappa)$. Then φ holds in V.

When κ is ω, this is Lévy-Shoenfield absoluteness. When κ is ω_1, this asserts $\Sigma_1(H(\omega_2))$ absoluteness for extensions which do not add reals and which preserve cardinals up to $2^{\aleph_0}$. Note that in the presence of $\sim$ CH, this axiom does rule out the two standard set-forcings for destroying the stationarity of a subset of ω_1.

b. Σ_1 absoluteness for arbitrary extensions

Counterexample 2 implies that to obtain a consistent version of absoluteness for arbitrary Σ_1 formulas with respect to arbitrary extensions, we must impose some restriction on our choice of parameters. A suitable restriction is perhaps provided by the following definition.

Definition. An *absolute cardinal-formula* is a parameter-free formula of the form

$$\varphi(\kappa) \text{ iff } L(H(\kappa)) \vDash \psi(\kappa),$$

where κ ranges over cardinals. We say that the cardinal κ is *absolute* between V and an extension V^* iff there is an absolute cardinal-formula which has κ as its unique solution in both V and V^*.

Σ_1 absoluteness for arbitrary extensions

Suppose that the cardinals $\kappa_1 < \cdots < \kappa_n$ are absolute between V and V^*, where V and V^* have the same cardinals $\leq \kappa_n$. Then any Σ_1 formula with parameters $\kappa_1, \ldots, \kappa_n$ which holds in V^* also holds in V.

Remark. David Asperó and I showed that if one drops the requirement that cardinals up to κ_n are preserved, then the above principle is inconsistent.

c. Cardinality and cofinality absoluteness principles

Other forms of strong absoluteness result by considering special types of Σ_1 formulas. First I introduce a variant of the notion of absolute parameter.

Definition. Suppose that α is an ordinal, P is a subset of V and V^* is an extension of V. Then α is *weakly absolute relative to parameters in P between V and V^** iff there is a formula with parameters from P which defines α not only in V, but also in V^*.

For cardinality and cofinality we have the following absoluteness principles.

Cardinal absoluteness. Suppose that α is an ordinal, V^* is an extension of V and α is weakly absolute relative to bounded subsets of α between V and V^*. Then if α is collapsed (i.e., not a cardinal) in V^*, it is also collapsed in V.

Cofinality absoluteness. Suppose that α is an ordinal, V^* is an extension of V and α is weakly absolute relative to bounded subsets of α between V and V^*. Then if α is singular in V^*, it is also singular in V.

The consistency strength of strong absoluteness principles

I do not know if any of the above principles of strong absoluteness are provably consistent relative to large cardinals. In this subsection I provide some lower bounds on their consistency strength.

Theorem 2. *Σ_1 absoluteness with arbitrary parameters implies that the GCH fails at every infinite cardinal, and for regular uncountable κ, there is no κ-Suslin tree.*

Proof. Suppose that the GCH held at the infinite cardinal κ. Choose $S \subseteq \kappa^+$ to be a fat-stationary subset of κ^+ which does not contain a CUB subset. (S is *fat-stationary* iff $S \cap C$ contains closed subsets of any ordertype less than κ^+, for each CUB $C \subseteq \kappa^+$.) The existence of such a set is guaranteed by a result of Krueger [4]. Then the forcing P that adds a CUB subset to S using closed subsets of S ordered by end-extension has cardinality κ^+ and, using the fatness of S, is κ^+-distributive. It follows that $H(\kappa^+)$ is strongly preserved by P. But a CUB subset of S witnesses a Σ_1 formula with parameter S not true in the ground model, in contradiction to our hypothesis.

Suppose that there were a κ-Suslin tree T for an uncountable regular cardinal κ. Then forcing with this tree strongly preserves $H(\kappa)$ and adds a witness to a Σ_1 formula with parameter T not witnessed in the ground model, in contradiction to our hypothesis. □

By work of Mitchell [6]:

Corollary 3. *Σ_1 absoluteness with arbitrary parameters implies the consistency of a measurable cardinal κ of Mitchell order κ^{++}.*

A lower bound for the strength of Σ_1 absoluteness for arbitrary extensions follows from the arguments of [3]:

Theorem 4. *Suppose that Σ_1 absoluteness for arbitrary extensions holds. Then there is an inner model with a strong cardinal.*

For cardinal absoluteness we have:

Theorem 5. *Cardinal absoluteness implies that for each infinite cardinal κ, κ^+ is greater than (κ^+ of HOD).*

Proof. If G is generic for the Lévy collapse of κ^+ to ω, then HOD is the same in V and in $V[G]$, by the homogeneity of the forcing. This contradicts our absoluteness hypothesis. □

By [7] and [8]:

Corollary 6. *Cardinal absoluteness implies that there is an inner model with a strong cardinal, and, if there is a proper class of subtle cardinals, there is an inner model with a Woodin cardinal.*

It is possible to extend Corollary 6 to obtain inner models with a proper class of Woodin cardinals containing any given set, under the assumption of cardinal absoluteness and a proper class of subtle cardinals. This is more than enough to imply Projective Determinacy.

Corollary 6 also holds for cofinality absoluteness, as the latter implies cardinality absoluteness.

4. Some final thoughts

The most important axioms of set theory that have been explored until now have arisen unavoidably out of the need to solve central problems in the field. This is especially true of the large cardinal axioms, which have even provided a measure for the consistency strength of virtually all set-theoretic statements. In my view we should not however impatiently assert that any axiom is "correct" until we can derive it from axioms which meet criteria like the ones discussed above.

I believe that the axioms of finite set theory "capture" the first-order theory of $H(\omega)$, as $H(\omega)$ is the unique well-founded model of finite set theory and no clear examples of ill-founded models are known. PD (projective determinacy) provides attractive axioms for the first-order theory of $H(\omega_1)$. The strong absoluteness axioms of the previous section are stable and natural axioms which lead to inner models with Woodin cardinals, and therefore to PD. Therefore PD follows from natural and stable axioms, and in my view can be judged to be "correct". Although I have not seen a convincing argument that PD "captures" the first-order theory of $H(\omega_1)$, I do believe this to be the case.

Though the axioms of strong absoluteness lead to the existence of *inner models* with Woodin cardinals, they do not produce large cardinals in V. Fortunately, large cardinals in V do not appear to be necessary to obtain "correct" axioms which capture the first-order theory of $H(\omega_2)$, a goal which in my view is still well beyond our reach.

References

[1] Beller, A., Jensen, R. and Welch, P. *Coding the Universe*, London Math Society Lecture Note Series, **47** (1982) Cambridge University Press.

[2] Friedman, S. Generic Σ^1_3 absoluteness, Journal of Symbolic Logic, Vol. 69, No. 1, pp. 73–80, 2004.

[3] Friedman, S. Internal consistency and the inner model hypothesis, to appear, Bulletin of Symbolic Logic, 2006.

[4] Krueger, J. Fat sets and saturated ideals, Journal of Symbolic Logic, vol. 68, pp. 837–845, 2003.

[5] Kunen, K. Elementary embeddings and infinitary combinatorics. J. Symbolic Logic 36, pp. 407–413, 1971.

[6] Mitchell, W. The core model for sequences of measures I, Math. Proc. Cambridge Phil. Soc. 95, pp. 229–260, 1984.

[7] Mitchell, W., Schimmerling, E. and Steel, J. The Weak Covering Lemma up to a Woodin Cardinal, Annals of Pure and Applied Logic , vol. 84, pp. 219–255, 1997.

[8] Steel, J. *The core model iterability problem*, Lecture Notes in Logic 8, Springer-Verlag, Berlin, 1996.

Sy-David Friedman
Kurt Gödel Research Center for Mathematical Logic
Universität Wien

Set Theory
Trends in Mathematics, 285–295

Forcing with Finite Conditions

Sy-David Friedman

Abstract. We present a generalisation to ω_2 of Baumgartner's forcing for adding a CUB subset of ω_1 with finite conditions.

The following well-known result appears in Baumgartner, Harrington, Kleinberg [2]. For the reader's convenience we provide a proof here.

Theorem 1. *Suppose that $X \subseteq \omega_1$. Then the following are equivalent:*

a. *X contains a CUB subset in an outer model which preserves ω_1.*
b. *X is stationary.*

Proof. (a) implies (b) because any two CUB sets must intersect. Conversely, consider the forcing P whose conditions are closed, countable subsets of X, ordered by end-extension. Clearly P adds a CUB subset to X; we must show that ω_1 is preserved.

First a general comment about ω_1-preservation. We say that D is *predense below* p iff every condition below p is compatible with an element of D. Then ω_1-preservation is a consequence of the following:

($*$) For any p and D_i, $i < \omega$ with each D_i predense below p, there are $q \leq p$ and *countable* d_i, $i < \omega$ with $d_i \subseteq D_i$ and d_i predense below q for each $i < \omega$.

For if ($*$) holds and p forces σ to be a function from ω to ω_1, then we can consider $D_i = \{q \mid \text{for some } \alpha < \omega_1,\ q \text{ forces } \sigma(i) = \alpha\}$; by ($*$), there is $q \leq p$ and a countable β such that q forces $\sigma(i) < \beta$ for each $i < \omega$, and therefore q forces that σ is bounded.

Now to see that P preserves ω_1, suppose that $\langle D_i \mid i < \omega\rangle$ is a sequence of sets which are predense below p and choose a continuous elementary chain $\langle M_j \mid j < \omega_1\rangle$ of countable elementary submodels of H_θ, θ large, so that $X, p, \langle D_i \mid i < \omega\rangle$ belong to M_0 and $M_j \in M_{j+1}$ for each j. As $C = \{M_j \cap \omega_1 \mid j < \omega_1\}$ is CUB, we

This article was prepared during research visits to the Centre de Recerca Matematica, Bellaterra during the months of September 2003 and February–March 2005. Research support was provided by Forschungsprojekt Nr. P16334-NO5 des österreichischen Fonds zur Förderung der wissenschaftlichen Forschung. The author also wishes to thank Bill Mitchell, who independently obtained a similar result, for sharing his insights into this problem.

can choose j so that $\alpha = M_j \cap \omega_1 \in X$. Then as each $D_i \cap M_j$ is predense below p on $P \cap M_j$, we can choose $p = p_0 \geq p_1 \geq \dots$ so that p_{i+1} belongs to M_j and extends an element r_i of D_i for each $i < \omega$, and in addition $\bigcup_i p_i$ has supremum α. Then $q = \bigcup_i p_i \cup \{\alpha\}$ is a condition extending p, and for each i, $\{r_i\} \subseteq D_i$ is predense below q, proving $(*)$. □

Next we ask the following.

Question. Suppose that X is a stationary subset of ω_1. Then is there a *cardinal-preserving* forcing P which adds a CUB subset to X?

The difficulty with the forcing used to prove Theorem 1 is that it will collapse $2^{\aleph_0}$ to ω_1, and therefore not preserve cardinals if CH fails. However, Baumgartner found a way of adding CUB sets with "finite conditions" which yields a positive answer to the above question (see [1]).

Theorem 2. *Let X be a stationary subset of ω_1. Then there is a forcing P which adds a CUB subset to X which preserves cofinalities.*

Proof. We use Uri Avraham's variant of Baumgartner's original technique (see [1]). A condition is a finite set p of disjoint closed intervals in ω_1 whose left endpoints belong to X. (We allow the one-point intervals $[\alpha, \alpha]$, $\alpha \in X$.) A condition q extends p iff q contains p.

It is clear that for generic G, C_G = the set of all left endpoints of intervals in $\cup G$ is an unbounded subset of X. Each countable ordinal either belongs to some interval in G or fails to be a limit point of X; it follows that C_G is closed. As there are only ω_1 conditions, it only remains to show that ω_1 is preserved.

Suppose that p is a condition and D_i, $i < \omega$ are predense below p. Choose a continuous elementary chain $\langle M_j \mid j < \omega_1 \rangle$ of countable elementary submodels of H_θ, θ large, so that $X, p, \langle D_i \mid i < \omega \rangle$ belong to M_0 and $M_j \in M_{j+1}$ for each j. As $C = \{M_j \cap \omega_1 \mid j < \omega_1\}$ is CUB, we can choose j so that $\alpha = M_j \cap \omega_1 \in X$. Let q be the condition $p \cup \{[\alpha, \alpha]\}$. If r extends q and $r_0 = r \restriction \alpha$ then every extension s_0 of r_0 in $P \cap M_j$ is compatible with r. This is because $[\alpha, \alpha]$ belongs to q. It follows that $d_i = D_i \cap M_j$ is predense below q for each i, as if r extends q then we can choose $s_0 \leq r_0$ which extends a condition in d_i, and therefore since s_0 is compatible with r, r is compatible with an element of d_i. Hence ω_1 is preserved. □

Now we look at the situation for ω_2. Unfortunately there is no analogue for Theorem 1.

Theorem 3. *(See [3].) Suppose that $0^\#$ exists. Then*

- *$\{X \subseteq \omega_2^L \mid X \in L$ and X has a CUB subset in an inner model where $\omega_2 = \omega_2^L\}$*
- *is not constructible, and indeed has L-degree $0^\#$. In particular, there are X which belong to the above set but have no CUB subset in any set-generic extension of L in which $\omega_2 = \omega_2^L$.*

However (under CH) there is a nice sufficient condition for a subset of ω_2 to contain a CUB in an extension preserving ω_1 and ω_2: $X \subseteq \omega_2$ is *fat stationary* iff $X \cap \mathrm{cof}\,\omega_1$ is stationary and for all α in $X \cap \mathrm{cof}\,\omega_1$, $X \cap \alpha$ contains a CUB subset of α.

Theorem 4. *Assume CH. If $X \subseteq \omega_2$ is fat stationary then there is a set-forcing extension preserving both ω_1 and ω_2 in which X contains a CUB subset.*

Proof. In analogy with the proof of Theorem 1, force with closed subsets of X of ordertype less than ω_2, ordered by end-extension. Countably closed models of size ω_1 and the fat stationarity of X are used as in the proof of Theorem 1 to show that if p is a condition and D_i, $i < \omega_1$, are predense below p then there is $q \leq p$ extending an element of D_i for each i. It follows that no new ω_1-sequences are added by the forcing and therefore both ω_1 and ω_2 are preserved. □

The forcing of Theorem 4 will collapse cardinals if GCH fails at ω_1. Avraham discovered a way of avoiding this problem, but still assuming CH. Is there a version for ω_2 of Baumgartner's forcing (as modified by Avraham) to add a CUB subset of a fat stationary set using finite conditions, without collapsing cardinals and without assuming CH? The following result provides a positive answer under the assumption of the existence of a thin stationary subset of $P_{\omega_1}(\omega_2)$ (an assumption weaker than CH).

Definition. $P_{\omega_1}(\omega_2)$ denotes the collection of countable subsets of ω_2. A subset S of $P_{\omega_1}(\omega_2)$ is *thin* iff for each $\alpha < \omega_2$, the set $\{x \cap \alpha \mid x \in S\}$ has cardinality at most ω_1.

Theorem 5. *Assume that there exists a thin stationary subset of $P_{\omega_1}(\omega_2)$ and that $D \subseteq \omega_2$ is fat stationary. Then there is a forcing P that preserves cofinalities and adds a CUB subset of D.*

Remark. The following results appear in [4]: Thin *cofinal* subsets of $P_{\omega_1}(\omega_2)$ exist provably in ZFC. The existence of a thin stationary subset of $P_{\omega_1}(\omega_2)$ is strictly weaker than that of a special Aronszajn tree on ω_2. Thin stationary subsets of $P_{\omega_1}(\omega_2)$ do not exist if Martin's Maximum (MM) holds.

Proof of Theorem 5. Let D_1 denote $D \cap \mathrm{cof}\,\omega_1$. We can assume that D consists exclusively of limit ordinals and that $\alpha + \omega$ belongs to D whenever α belongs to D. Let T be a thin stationary subset of $P_{\omega_1}(\omega_2)$ and assume that T is closed under initial segments. Choose $B \subseteq \omega_2$ such that $T \subseteq L[B]$ and ω_2 equals (ω_2 of $L[B]$). An ordinal α is *good* iff it is a limit ordinal between ω_1 and ω_2 and for every $\beta < \alpha$, cof β equals (cof β in $L_\alpha[B]$). The set of good ordinals forms a CUB subset of ω_2.

For an ordinal α and a set x with $\alpha < \sup(x \cap \mathrm{Ord})$, let α_x denote the least ordinal $\geq \alpha$ in x. Note that if $\alpha < \alpha_x$ and x is Σ_1 elementary in some $L_\beta[B]$, β good, then α_x must have uncountable cofinality.

A condition is a pair $p = (A, S)$, where:

1. A is a finite set of disjoint closed intervals whose left endpoints belong to D. (We allow the one-point intervals $[\alpha, \alpha]$, $\alpha \in D$.) Let L_A denote the set of left endpoints of intervals in A.
2. S is a finite set of countable Σ_1 elementary submodels x of some $L_\beta[B]$, β good, such that $x \cap \mathrm{Ord}$ belongs to T and $\sup(x \cap \alpha)$ belongs to D whenever α belongs to $(x \cap D_1) \cup \{\omega_2\}$.
3. For each interval $I = [\alpha, \beta]$ in A and each $x \in S$:
   3a. If I intersects x then I belongs to x.
   3b. If $I = [\alpha, \beta]$ does not intersect x and $\alpha < \sup(x \cap \mathrm{Ord})$ then α_x belongs to L_A.
4. Let F_A be the set of all elements of L_A of cofinality ω_1, together with ω_2. For $x \in S$, the *F_A-height of x* is the least element of F_A greater than $\sup(x \cap \mathrm{Ord})$.
   4a. If x belongs to S and α belongs to F_A then $x \cap L_\alpha[B]$ belongs to S.
   4b. Suppose that $x, y \in S$ have the same F_A-height. Then $x \in y$, $y \in x$ or $x = y$.

Write $p = (A_p, S_p)$ and let L_p, F_p denote the L_A, F_A of 1, 4 above. q extends p iff A_q contains A_p and S_q contains S_p. For any condition q^* and $\alpha < \omega_2$ we define $q^* \restriction \alpha$ to be the pair $q = (A_q, S_q)$ where:

$$A_q \text{ is } A_{q^*} \cap L_\alpha[B], \qquad S_q \text{ is } S_{q^*} \cap L_\alpha[B].$$

Claim 1. Suppose that p belongs to P.

(a) If C is a CUB subset of ω_2 then there exists $\alpha \in D_1 \cap C$ such that p belongs to $L_\alpha[B]$ and every subset of α in T belongs to $L_\alpha[B]$. For such α, obtain p^* from p by adding the interval $[\alpha, \alpha]$ to A_p (and otherwise leaving p unchanged). Then p^* is a condition extending p.

(b) Let α and p^* be as in part (a) and suppose that q^* extends p^*. Then $q^* \restriction \alpha = q$ is a condition in $L_\alpha[B]$ extending p such that every extension of q in $L_\alpha[B]$ is compatible with q^*.

Proof of Claim 1: (a) Such α exist since D_1 is stationary and $T \subseteq L_{\omega_2}[B]$ is thin. Property 1 is satisfied by p^* as α is greater than the right endpoint of any interval in A_p. Property 2 is the same for p^* as for p. Property 3a is the same for p^* as for p, as α does not belong to any element of S_p. Property 3b is the same for p^* as for p, as α is greater than $\sup(x \cap \mathrm{Ord})$ for any $x \in S_p$. And property 4 holds for p^* as $F_{p^*} = F_p \cup \{\alpha\}$, $x \cap L_\alpha[B] = x$ for all $x \in S_p$ and x, $y \in S_p$ have the same F_{p^*}-height iff they have the same F_p-height.

(b) Suppose that q^* extends p^* and set $q = q^* \restriction \alpha$.

Subclaim 1. q is a condition in $L_\alpha[B]$ which extends p.
Clearly q, if a condition, extends p since q^* does and p belongs to $L_\alpha[B]$. To verify that q is a condition, we need only verify properties 3b and 4.

3b. Assume that $I \cap x = \emptyset$ and the left endpoint β of $I = [\beta, \gamma]$ is less than $\sup(x \cap \mathrm{Ord})$, where I belongs to $A_{q^*} \cap L_\alpha[B]$ and x belongs to $S_{q^*} \cap L_\alpha[B]$. Then

since q^* is a condition, β_x is the left endpoint of some interval J in S_{q^*}. But since $[\alpha, \alpha]$ belongs to A_{q^*}, the right endpoint of J is less than α and therefore J belongs to $A_{q^*} \cap L_\alpha[B] = A_q$.
For property 4, first note that $F_q = F_{q^*} \cap \alpha$, together with ω_2.

4a. If x is in S_q and $\beta \in F_q$ then $x \cap L_\beta[B]$ is in S_{q^*} and therefore also in $S_q = S_{q^*} \cap L_\alpha[B]$, since, using our hypothesis on α, $x \cap L_\beta[B]$ is an element of $L_\alpha[B]$.

4b. If $x, y \in S_q$ have the same F_q-height then since they both belong to $L_\alpha[B]$, they have the same F_{q^*}-height. Thus the desired conclusion follows as $x, y \in S_{q^*}$ and q^* is a condition.

Now suppose that r is an extension of q, and r belongs to $L_\alpha[B]$. We must find a common extension of r and q^*. We define t by

$$A_t = A_r \cup A_{q^*}, \qquad S_t = S_r \cup S_{q^*}.$$

Subclaim 2. t is a condition extending both r and q^*.
Clearly t, if a condition, extends both r and q^*. We now verify that t is a condition, by verifying properties 1–4.

1. The intervals in A_t are disjoint, as r is a condition extending q, all intervals in A_r have right endpoint less than α and all intervals in A_{q^*} not in A_q have left endpoint at least α.

2. Clear.

3. Suppose that I is an interval in $A_t - A_r$ and x belongs to S_r. Then $\sup(x \cap \mathrm{Ord})$ is less than α and the left endpoint of I is at least α. So property 3 is vacuous in this case. Suppose that I belongs to A_r and x belongs to $S_t - S_r$. Then $x \cap L_\alpha[B]$ belongs to $S_q \subseteq S_r$ and therefore property 3 holds for I and $x \cap L_\alpha[B]$. It follows that 3a holds for I and x, since if I intersects x it must also intersect $x \cap L_\alpha[B]$. And 3b holds for I and x: If I is disjoint from x and the left endpoint β of I is less than $\sup(x \cap \mathrm{Ord})$ then I is also disjoint from $x \cap L_\alpha[B]$ and either (i) β is less than $\sup(x \cap \alpha)$, in which case $\beta_x = \beta_{x \cap \alpha}$ and therefore the result follows since r is a condition, (ii) $\beta_x = \alpha$, in which case the result follows since $[\alpha, \alpha]$ belongs to A_{q^*}, or (iii) $\beta_x = \alpha_x$, in which case the result follows since q^* is a condition. The remaining cases, where I belongs to A_r and x belongs to S_r, or where I belongs to $A_t - A_r$ and x belongs to $S_t - S_r$, are immediate since r and q^* are conditions.

4a. We must show that if x belongs to S_t and $\beta \in F_t$ then $x \cap L_\beta[B]$ belongs to S_t. If x belongs to S_r then either β is in F_r, in which case $x \cap L_\beta[B]$ belongs to $S_r \subseteq S_t$ since r is a condition, or β is at least α, in which case $x \cap L_\beta[B] = x \in S_r \subseteq S_t$. If x belongs to S_{q^*} then either β is in F_{q^*}, in which case the result follows since q^* is a condition, or β is in F_r, in which case $x \cap L_\beta[B] = (x \cap L_\alpha[B]) \cap L_\beta[B] \in S_r \subseteq S_t$, since $x \cap L_\alpha[B] \in S_q \subseteq S_r$ and r is a condition.

4b. We must show that if $x, y \in S_t$ have the same F_t-height, then $x \in y, y \in x$ or $x = y$. If x belongs to S_r then the F_t height of x is at most α and therefore y also belongs to S_r; thus x, y have the same F_r-height and the result follows since r is a condition. If x belongs to $S_{q^*} - S_r$ then the F_t-height of x is greater than α, and therefore y also belongs to S_{q^*}; thus x, y have the same F_{q^*}-height and the desired conclusion follows since q^* is a condition.

This completes the proof of Claim 1.

Claim 2. Suppose that p belongs to P.

(a) For any CUB $C \subseteq P_{\omega_1}(\omega_2)$ there exists a countable elementary submodel x of $L_{\omega_2}[B]$ such that $x \cap \mathrm{Ord}$ belongs to $C \cap T$, p belongs to x and whenever α belongs to $(x \cap D_1) \cup \{\omega_2\}$, $\sup(x \cap \alpha)$ belongs to D. For such x obtain p^* from p by adding $x \cap L_\alpha[B]$ to S_p for all $\alpha \in F_p$ (and otherwise leaving p unchanged). Then p^* is a condition extending p.

(b) Let x and p^* be as in part (a). Then if q^* extends p^* there is q in x extending p such that every extension of q in x is compatible with q^*.

Proof of Claim 2: (a) To see that such x exist, argue as follows. Choose β in D_1 such that $C \cap P_{\omega_1}(\beta)$ is CUB in $P_{\omega_1}(\beta)$. Also choose $y \in T$ such that $y \cap \beta$ belongs to $C \cap P_{\omega_1}(\beta)$ and $\sup(y \cap \alpha)$ belongs to D whenever α belongs to $(y \cap \beta \cap D_1) \cup \{\beta\}$. As T is closed under initial segments, $x = y \cap \beta$ belongs to T and has the desired properties.

Clearly p^*, if a condition, extends p. To verify that p^* is a condition we need only check properties 3 and 4.

3a. As p belongs to x, each interval in A_p belongs to x and therefore the conclusion of 3a holds for x. It follows easily that 3a also holds for $x \cap L_\alpha[B]$ whenever α belongs to F_p. 3a holds for other elements of S_{p^*} since p is a condition.

3b. This is vacuous for $x \cap L_\alpha[B]$, $\alpha \in F_p$, and holds for other elements of S_{p^*} since p is a condition.

4a. This is true for $x \cap L_\alpha[B]$, $\alpha \in F_p$, by definition of p^*, and for other elements of S_{p^*} since p is a condition.

4b. Suppose that $y, z \in S_{p^*}$ have the same F_{p^*}-height ($= F_p$-height). If both y, z belong to S_p then the desired conclusion follows since p is a condition. Assume that $y = x \cap L_\alpha[B]$ where α belongs to F_p. If z belongs to S_p then z belongs to x and since it has the same F_p-height as y, must also belong to $L_\alpha[B]$; hence z belongs to y. If z is of the form $x \cap L_\beta[B]$, $\beta \in F_p$, and has the same F_p-height as y then $z = y$, since the F_p-height of $x \cap L_\beta[B]$ equals β for any $\beta \in F_p$.

(b) Let q^* extend p^* and define q as follows:

$$A_q \text{ is } A_{q^*} \cap x, \qquad S_q \text{ is } S_{q^*} \cap x.$$

Subclaim 1. q is a condition in x extending p.

Clearly q, if a condition, extends p since q^* extends $p^* \leq p$ and p belongs to x. To check that q is a condition we need only verify properties 3b and 4.

3b. Suppose that I belongs to A_q, I is disjoint from y and the left endpoint α of I is less than $\sup(y \cap \mathrm{Ord})$, where y belongs to S_q. Then α_y is the left endpoint of some $J \in A_{q^*}$ since q^* is a condition. Since J intersects y, J must belong to y and therefore also to x, since y belongs to x. Thus J belongs to A_q.

4a. If y belongs to S_q and α belongs to $F_q \cap \omega_2$ then $y \cap L_\alpha[B]$ belongs to S_{q^*} since q^* is a condition. Since both y and α belong to x, we get $y \cap L_\alpha[B] \in S_q$. If y belongs to S_q then $y \cap L_{\omega_2}[B] = y$ and therefore belongs to S_q.

4b. Suppose that $y \in S_q$ has F_q-height α and F_{q^*}-height β. Suppose that β is less than $\sup(x \cap \mathrm{Ord})$. Then either β equals α or is the left endpoint of some interval in A_{q^*} disjoint from x. In the latter case, β_x is the left endpoint of some interval in $A_{q^*} \cap x = A_q$ and therefore β_x belongs to F_q, since it must have uncountable cofinality. Thus $\beta_x = \alpha$. So we conclude that the F_{q^*}-height of y is the least $\beta \in F_{q^*}$ such that either β is less than $\sup(x \cap \mathrm{Ord})$ and $\beta_x = \alpha$, or β is greater than $\sup(x \cap \mathrm{Ord})$. Therefore the F_{q^*}-height of $y \in S_q$ is uniquely determined by the F_q-height of y. If $y, z \in S_q$ have the same F_q-height then they therefore also have the same F_{q^*}-height, and since q^* is a condition, either $y \in z$, $z \in y$ or $y = z$, as desired.

Now suppose that r in x extends q. We must find a common extension t of r and q^*. We define t by:

$$A_t = A_r \cup A_{q^*} \qquad S_t = S_r \cup \{y \cap L_\alpha[B] \mid y \in S_{q^*}, \quad \alpha \in F_r\}.$$

Subclaim 2. t is a condition extending both r and q^*.

Clearly t, if a condition, extends both r and q^*. We show now that t is a condition.

1. Suppose that I is an interval in A_{q^*} but not in A_r. Then I is disjoint from x. If the left endpoint α of I is at least $\sup(x \cap \mathrm{Ord})$, then I is disjoint from all intervals in A_r, since the latter belong to x. Otherwise α_x is the left endpoint of some interval J in A_q. It follows that the intervals in A_r are disjoint from I, as they belong to x and are either equal to or disjoint from J. Thus A_t consists of pairwise disjoint intervals.

2. We must show that if y belongs to S_t and $\alpha \in (y \cap D_1) \cup \{\omega_2\}$ then $\sup(y \cap \alpha)$ belongs to D. This is clear if y belongs to S_r since r is a condition. It also holds if y belongs to S_{q^*} since q^* is a condition. This implies the result for arbitrary $y \in S_t$ when α is not ω_2. It remains to show: If $y \in S_{q^*}$, $\alpha \in F_r$ then $\sup(y \cap \alpha)$ belongs to D. Let β be least in $F_{q^*} - \alpha$. If β is not the F_{q^*}-height of $y \cap L_\beta[B]$ then $y \cap \alpha = y \cap \beta$ and therefore $\sup(y \cap \alpha)$ belongs to D since q^* is a condition. Otherwise, $y \cap L_\beta[B]$ and $x \cap L_\beta[B]$ have the same F_{q^*}-height, since α belongs to x. Since q^* is a condition, either $y \cap L_\beta[B] \in x \cap L_\beta[B]$, $x \cap L_\beta[B] \in y \cap L_\beta[B]$ or $y \cap L_\beta[B] = x \cap L_\beta[B]$. In the first case, $y \cap L_\beta[B]$ belongs to S_r so $y \cap \alpha = (y \cap \beta) \cap \alpha$ belongs to D since r is a condition. In the latter two cases, α belongs to $y \cap D_1$, and therefore the result follows since q^* is a condition.

3a. Suppose that I is an interval in A_t, y belongs to S_t and I intersects y. We must show that I belongs to y. First we consider the case where I belongs to A_r and y belongs to $S_t - S_r$. Write $I = [\alpha, \beta]$ and $y = z \cap L_\gamma[B]$, where z belongs to $S_{q^*} - S_r$ and γ belongs to F_r. Let β^* be the least element of F_{q^*} greater than α. Since we have shown that A_t consists of pairwise disjoint intervals, it follows that β^* is greater than β. Therefore the F_{q^*}-heights of $x \cap L_{\beta^*}[B]$ and $z \cap L_{\beta^*}[B]$ are both β^*, the former since β belongs to x and the latter since z intersects $I = [\alpha, \beta]$. Thus either $z \cap L_{\beta^*}[B] \in x \cap L_{\beta^*}[B]$, $x \cap L_{\beta^*}[B] \in z \cap L_{\beta^*}[B]$ or $x \cap L_{\beta^*}[B] = z \cap L_{\beta^*}[B]$. The first possibility implies that I intersects $y \cap L_{\beta^*}[B] = (z \cap L_{\beta^*}[B]) \cap L_\gamma[B] \in S_r$, and therefore since r is a condition, I belongs to $y \cap L_{\beta^*}[B] \subseteq y$, as desired. The second and third possibilities imply that y contains $x \cap L_{\beta^*}[B]$ as a subset and therefore I as an element. Now consider the case where I belongs to $A_t - A_r$ and y belongs to S_r. Then I belongs to A_{q^*} and intersects x, which belongs to S_{q^*}. Thus I belongs to x, contradicting the fact that I does not belong to $A_q \subseteq A_r$. The case where I belongs to A_r and y belongs to S_r follows since r is a condition. Finally, if I belongs to $A_t - A_r$ and y belongs to $S_t - S_r$, write $y = z \cap L_\alpha[B]$ where $z \in S_{q^*}$ and $\alpha \in F_r$. Since q^* is a condition, I belongs to z. If I does not belong to y, then I intersects x and therefore belongs to x, again since q^* is a condition. But this contradicts the hypothesis that I does not belong to A_r.

3b. Suppose that $I = [\alpha, \beta]$ belongs to A_t, y belongs to S_t, I is disjoint from y and α is less than $\sup(y \cap \mathrm{Ord})$. We must show that α_y is the left endpoint of some interval in A_t. First we consider the case where I belongs to A_r and y belongs to $S_t - S_r$. Write $y = z \cap L_\gamma[B]$ where z belongs to S_{q^*} and γ belongs to F_r. Let β^* be the least element of F_{q^*} greater than β. If $\alpha_y = \beta^*$ then α_y is the left endpoint of some interval in A_{q^*} and we are done. If $\alpha_y > \beta^*$, then let J be the interval of A_{q^*} with left endpoint β^*. Since q^* is a condition and $\alpha_y = \alpha_z$, J is not an element of z and therefore is disjoint from z. Since q^* is a condition, $\alpha_y = (\beta^*)_z$ is the left endpoint of some interval of A_{q^*}, as desired. Finally, if $\beta < \alpha_y < \beta^*$, it follows that $x \cap L_{\beta^*}[B]$ and $z \cap L_{\beta^*}[B]$ both have F_{q^*}-height β^*, and therefore $x \cap L_{\beta^*}[B] \in z \cap L_{\beta^*}[B]$, $z \cap L_{\beta^*}[B] \in x \cap L_{\beta^*}[B]$ or $x \cap L_{\beta^*}[B] = z \cap L_{\beta^*}[B]$. The first and third of these possibilities contradict our hypothesis that $I \in x$ is disjoint from y. In the second possibility, $z \cap L_{\beta^*}[B]$ belongs to S_q and since $\alpha_y = \alpha_z$ is less than β^*, we have that α_y is the left endpoint of some interval in A_r since r is a condition. Next we consider the case where I belongs to $A_t - A_r$ and y belongs to S_r. Thus I belongs to A_{q^*} and must be disjoint from x, else it would belong to x and therefore to A_r. As α is less than $\sup(x \cap \mathrm{Ord})$, it follows that α_x is the left endpoint of some interval in A_{q^*}, which in fact belongs to A_r. If $\alpha_y = \alpha_x$ then we are done. Otherwise α_y equals $(\alpha_x)_y$, which must be the left endpoint of an interval in A_r, since r is a condition. The remaining two cases, where either I belongs to A_r and y belongs to S_r, or where I belongs to $A_t - A_r$ and y belongs to $S_t - S_r$, follow since both r and q^* are conditions.

4a. We must show that if y belongs to S_t and α belongs to F_t then $y \cap L_\alpha[B]$ belongs to S_t. This is clear if y belongs to S_r and α belongs to F_r, or if y belongs to F_{q^*} and α belongs to F_{q^*}, since r and q^* are conditions. This is also true if y belongs to S_{q^*} and α belongs to F_r, by definition of S_t. And we may assume that y belongs to $S_r \cup S_{q^*}$. So we need only check the case where y belongs to S_r, α belongs to F_{q^*} and α is less than $\sup(y \cap \mathrm{Ord})$. If α belongs to x then it also belongs to F_r so we are done since r is a condition. Otherwise α_x is defined and belongs to F_r. So since r is a condition, $y \cap L_\alpha[B] = y \cap L_{\alpha_x}[B]$ belongs to S_r.

4b. We must show that if $y, z \in S_t$ have the same F_t-height then either $y \in z$, $z \in y$ or $y = z$. Note that y, z also have the same F_r-height and the same F_{q^*}-height. If y, z both belong to S_r then the desired conclusion follows since r is a condition. Suppose that y, z are of the form $y^* \cap L_\alpha[B]$, $z^* \cap L_\beta[B]$, respectively, where y^*, z^* belong to S_{q^*} and $\alpha, \beta \in F_r$. We may assume that α, β are the F_r-heights of y, z, respectively, and therefore $\alpha = \beta$. Let α^* be the common F_{q^*}-height of y, z. Then $y^* \cap L_{\alpha^*}[B]$, $z^* \cap L_{\alpha^*}[B]$ have F_{q^*}-height α^* and therefore since q^* is a condition, we have $y^* \cap L_{\alpha^*}[B] \in z^* \cap L_{\alpha^*}[B]$, $y^* \cap L_{\alpha^*}[B] = z^* \cap L_{\alpha^*}[B]$ or $z^* \cap L_{\alpha^*}[B] \in y^* \cap L_{\alpha^*}[B]$. The second possibility yields $y = z$. The first possibility implies that y belongs to $z^* \cap L_{\alpha^*}[B]$ since it is an initial segment of $y^* \cap L_{\alpha^*}[B]$; as $y \in L_\alpha[B]$ we get $y \in z^* \cap L_\alpha[B] = z$. The third possibility is handled identically to the first, with y and z switched. Finally assume that y belongs to S_r and $z = z^* \cap L_\alpha[B]$ where $z^* \in S_{q^*}$ and $\alpha \in F_r$. We may assume that α is the F_r-height of z, which is also the F_r-height of y. Let α^* be the common F_{q^*}-height of y and z. Then α^* is also the F_{q^*}-height of $x \cap L_{\alpha^*}[B]$, since x contains y, and is the F_{q^*}-height of $z^* \cap L_{\alpha^*}[B]$. Since q^* is a condition, we have $z^* \cap L_{\alpha^*}[B] \in x \cap L_{\alpha^*}[B]$, $x \cap L_{\alpha^*}[B] \in z^* \cap L_{\alpha^*}[B]$ or $z^* \cap L_{\alpha^*}[B] = x \cap L_{\alpha^*}[B]$. Under the first possibility, $(z^* \cap L_{\alpha^*}[B]) \cap L_\alpha[B] = z$ belongs to S_r, so we are done since r is a condition. The second and third possibilities imply that $y \in z^* \cap L_\alpha[B] = z$.

This completes the proof of Claim 2.

Claim 1 implies that ω_2 is preserved. Claim 2 implies that ω_1 is preserved. As P has cardinality ω_2, all cofinalities are preserved.

Claim 3. Let G be P-generic and $C_G = \{\gamma \mid \gamma$ is a left endpoint of some interval in $\cup\{A_p \mid p \in G\}\}$. Then C_G is a CUB subset of D.

Proof of Claim 3: It follows from *Claim 1 (a)* that C_G is unbounded. We show that C_G is closed.

Suppose that p is a condition and for the sake of contradiction, $p \Vdash (\alpha \in \text{Lim } C_G$ and $\alpha \notin C_G)$. We may assume that for each $y \in S_p$, if α_y is defined and forced by some extension of p to belong to C_G, then α_y is the left endpoint of some interval in A_p; otherwise we can enlarge A_p without changing S_p to guarantee this property. Note that for $q \leq p$, α does not belong to any interval in A_q, else q forces either that α belongs to C_G or is not the limit of elements of C_G.

Suppose that y belongs to S_p, α is not in Lim $(y \cap \text{Ord})$ but α is less than $\sup(y \cap \text{Ord})$. Then observe that α_y must be a left endpoint of some interval in A_p, else by requirement 3b on conditions, no extension of p can introduce a new interval with left endpoint between $\sup(y \cap \alpha)$ and α, and hence p cannot force that α is a limit point of C_G. Let β be the least element of F_p greater than α and consider $S = \{y \in S_p \mid \alpha \leq \sup(y \cap \text{Ord}) < \beta\}$. Then by requirement 4b on conditions, the elements of S form an $\in$-chain.

Assume first that $y \cap \alpha$ is cofinal in α for some $y \in S$, and let y_0 be the $\in$-least such. Note that if α_{y_0} is defined and is the left endpoint of some interval in A_p then α must belong to D, by requirement 2 on conditions. We show that we can extend p to force either that α belongs to C_G or that α is not a limit point of C_G, achieving the desired contradiction. Note that $D \cap y_0 \cap \alpha$ must be cofinal in α, as there are cofinally many $\gamma < \alpha$ which are forced by extensions of p into C_G and for any such $\gamma \notin y_0$, γ_{y_0} belongs to D by requirement 3b on conditions. Since $\gamma + \omega$ belongs to D whenever γ does, it follows that $D \cap y_0 \cap \alpha \cap \operatorname{cof} \omega$ is also cofinal in α.

If α_{y_0} is defined and not the left endpoint of some interval in A_p, then let γ be an element of $D \cap y_0 \cap \alpha \cap \operatorname{cof} \omega$ greater than the right endpoint of any interval of A_p with left endpoint less than α, and larger than $\sup(y \cap \alpha)$ for all $y \in S_p$ with $\sup(y \cap \alpha) < \alpha$. We claim a condition results when the interval $I = [\gamma, \alpha_{y_0}]$ is added to p: I is disjoint from intervals of A_p with left endpoint less than α by choice of γ. And it is disjoint from intervals of A_p with left endpoint greater than α since by assumption, α_{y_0} is not the left endpoint of an interval of A_p, and therefore by 3a, 3b neither is any ordinal between α and α_{y_0}. I does not intersect any $y \in S_p$ with $\sup(y \cap \beta) < \alpha$ by choice of γ. I does not intersect any $y \in S_p$ with $\sup(y \cap \alpha) < \alpha < \sup(y \cap \beta)$: For such y we have $y \cap \beta \in y_0$ and therefore $y \cap \beta \subseteq y_0$; also $\alpha_y > \alpha_{y_0}$ since, as observed earlier, α_y must be a left endpoint of some interval of A_p and by assumption α_{y_0} is not. Any other $y \in S_p$ contains y_0 as an element and therefore as a subset, and therefore also the interval I as an element. For those $y \in S_p$ disjoint from I with $\gamma < \sup(y \cap \mathrm{Ord})$, we have $\gamma_y = \alpha_y$, and as observed earlier, α_y is the left endpoint of an interval of A_p. This completes the verification that adding I to p results in a condition.

If α_{y_0} is defined and the left endpoint of some interval in A_p, then let $I = [\alpha, \alpha]$. We claim that a condition results when we add I to p: Of course I is disjoint from all intervals of A_p since α does not belong to any such interval. Trivially, if I intersects $y \in S_p$ then it belongs to y. If I is disjoint from $y \in S_p$ and $\alpha < \sup(y \cap \mathrm{Ord})$, then $\alpha_y \geq \alpha_{y_0}$, as otherwise $y \cap L_\beta[B] \in S$, $y_0 \in y \cap L_\beta[B]$ and therefore $\alpha = \sup(y_0 \cap \alpha_{y_0}) \in y$, against our hypothesis. So α_y must be a left endpoint of some interval of A_p, else by requirements 3a,3b on conditions, α_{y_0} could not be. This completes the verification that adding I to p results in a condition.

If α_{y_0} is undefined then again set $I = [\alpha, \alpha]$. We claim that a condition results when we add I to p: By the argument of the previous paragraph, it suffices to show that if I is disjoint from $y \in S_p$ and $\alpha < \sup(y \cap \mathrm{Ord})$ then α_y is the left endpoint of some interval of A_p. If α_y is at least β, then this must be the case, as otherwise β could not be the left endpoint of such an interval. If α_y is less than β then $y \cap L_\beta[B]$ belongs to S and therefore either y_0 is an element of y, contradicting $\alpha \notin y$, or $\alpha_y \geq \alpha_{y_0}$. If α_y equals α_{y_0} then α_y is the left endpoint of some interval of A_p by hypothesis, and if α_y is greater than α_{y_0} then it must be the left endpoint of an interval of A_p, else α_{y_0} could not be the left endpoint of such an interval. This completes the verification that adding I to p results in a condition.

Lastly, we treat the case where $y \cap \alpha$ is not cofinal in α for all $y \in S$. In this case we choose $I = [\gamma, \alpha]$, where γ is an element of $D \cap \alpha \cap \operatorname{cof} \omega$ greater than the right endpoint of any interval of A_p with left endpoint less than α, and larger than $\sup(y \cap \alpha)$ for all $y \in S_p$ with $\sup(y \cap \alpha) < \alpha$. We claim that a condition results when we add I to p: I is disjoint from all intervals in A_p by choice of γ. I is disjoint from each $y \in S_p$, as $y \cap \alpha$ is contained in γ by choice of γ and the case hypothesis, and if α belongs to y, we have $\alpha < \sup(y \cap \mathrm{Ord})$, which, as observed earlier, implies that $\alpha_y = \alpha$ the left endpoint of an interval of A_p, an impossibility. If y belongs to S_p and $\gamma < \sup(y \cap \mathrm{Ord})$ then α_y must be the left endpoint of an interval of A_p, as observed earlier. This completes the verification that adding I to p results in a condition.

This completes the proof of Theorem 5. □

Some remarks and questions. The hypothesis that D is fat stationary is not necessary for Theorem 5. The proof only uses that there is a thin stationary subset S of $P_{\omega_1}(\omega_2)$ such that for $x \in S$, $\sup(x \cap \alpha)$ belongs to D whenever α belongs to $x \cap D \cap \mathrm{Lim}\ D$ or $\alpha = \omega_2$. However this hypothesis is not substantially weaker than the one stated in Theorem 5 as, at least under CH, there is a countably distributive, cofinality-preserving forcing that adds a fat stationary subset to such a set D. In [4] it is shown that the assumption of a thin stationary subset of $P_{\omega_1}(\omega_2)$ cannot be eliminated from the statement of Theorem 5. Is Theorem 5 still true if CH is added to both its hypothesis and conclusion? And is there a version of Theorem 5 for ω_3?

References

[1] Abraham, U. and Shelah, S. *Forcing closed unbounded sets*, Journal of Symbolic Logic, Vol. 48, pp. 643–657, 1983.

[2] Baumgartner J., Harrington L. and Kleinberg E. *Adding a closed unbounded set*, Journal of Symbolic Logic, Vol. 41, pp. 481–482, 1976.

[3] Friedman, S. *Cardinal-preserving extensions*, Journal of Symbolic Logic, Vol. 68, No. 4, pp. 1163–1170, 2003.

[4] Friedman, S. and Krueger, J. *Thin stationary sets and disjoint club sequences*, to appear, Transactions of the American Mathematical Society, 2006.

Sy-David Friedman
Kurt Gödel Research Center for Mathematical Logic
University of Vienna

Set Theory
Trends in Mathematics, 297–308

Subgroups of Abelian Polish Groups

Greg Hjorth

1. Introduction

Ilijas Farah and Slawek Solecki showed in [2] that every Polish group contains arbitrarily complicated Borel subgroups and go on to ask whether the same can be shown for *Polishable* subgroups. As a partial answer, this notes shows that an uncountable abelian Polish group contains Polishable subgroups of unbounded Borel complexity.

2. Notation

In what follows, G is an uncountable abelian Polish group and $d(\cdot,\cdot)$ a compatible, complete, two sided-invariant metric. We write the group operations on G with reference to it being abelian – thus $+$ is the group operation and $n \cdot g$ stands for

$$g + g + \cdots (n \text{ times}) \cdots + g.$$

0 is the group identity in G.

We will find it convenient to use the notation of an infinite sum. Thus we say that

$$g = \sum_{n=0}^{\infty} g_n \qquad \text{if} \qquad d(\sum_{n=0}^{N} g_n, g) \to 0.$$

Since d is two sided-invariant metric we obtain that

$$\sum_{n=0}^{\infty} g_n = \sum_{n=0}^{\infty} h_n \qquad \text{if and only if} \qquad \sum_{n=0}^{\infty} (g_n - h_n) = 0.$$

For g a group element we use $o(g)$ to denote the *order* – that is to say, the least ℓ with $\ell \cdot g = 0$ if such an ℓ exists, and infinity otherwise.

3. The subgroups

We choose a sequence of non-zero elements

$$(g_i)_i$$

in G such that:

(I) for any $k \in \mathbb{N}$ and $(n_\ell)_{\ell \le k}$ chosen from $\mathbb{Z}$, with each

$$|n_\ell| \le 4\ell k$$

and

$$h = n_0 \cdot g_0 + n_1 \cdot g_1 \cdots + n_k \cdot g_k \in G$$
$$h \neq 0$$

we have that for any choice of $(\hat{n}_\ell)_{\ell > k}$ with each $|\hat{n}_\ell| < 4\ell k$ that

$$h \neq \sum_{\ell > k} \hat{n}_\ell \cdot g_\ell;$$

(II) for any ℓ and n with $|n| \le \ell^2$ we have

$$d(0, n \cdot g_\ell) < 2^{-\ell};$$

(III) either every g_i has infinite order, or the g_i's all have the same *finite* order, or $o(g_i) \to \infty$ as $i \to \infty$.

We obtain (I) and (II) just by taking any sequence which converges to 0 sufficiently fast. (III) then follows by refining the sequence as needed.

Definition. If $h \in G$ then a *k-representation* of h is an expression of the form

$$\sum_{\ell \in \mathbb{N}} n_\ell \cdot g_\ell \qquad \text{with} \qquad \lim_{M \to \infty} \sum_{\ell=0}^{M} n_\ell \cdot g_\ell = h$$

and each $|n_\ell| \le \ell \cdot k$. A *representation* is a k-representation for some k; a group element is *representable* if it has a representation.

Lemma 3.1. *If*

$$\sum_{\ell \in \mathbb{N}} n_\ell \cdot g_\ell, \qquad \sum_{\ell \in \mathbb{N}} \hat{n}_\ell \cdot g_\ell,$$

are two $2k$*-representations of* h *then for all* $\ell > k$

$$n_\ell \cdot g_\ell = \hat{n}_\ell \cdot g_\ell.$$

Proof. By the definition of $2k$-representation we have $n_\ell, \hat{n}_\ell < 2\ell k$ each ℓ, and then by our assumption (I) on the (g_i) sequence we have that at all $M \ge k$

$$\sum_{\ell \le M} (n_\ell - \hat{n}_\ell) \cdot g_\ell = 0.$$

From this the lemma follows. □

Definition. We let H be the subgroup of all h which have some representation

$$\sum_{\ell \in \mathbb{N}} n_\ell \cdot g_\ell \qquad \text{with} \qquad \frac{|n_\ell|}{\ell} \to 0.$$

Definition. If

$$\sum_{\ell \in \mathbb{N}} n_\ell \cdot g_\ell$$

is a representation of h, then we let

$$\varphi(\sum_{\ell \in \mathbb{N}} n_\ell \cdot g_\ell) = \sup_\ell \frac{|n_\ell|}{\ell}.$$

If $h \in H$ we let $\psi(h)$ be the infinum of

$$\varphi(\sum_{\ell \in \mathbb{N}} n_\ell \cdot g_\ell)$$

as $\sum n_\ell \cdot g_\ell$ ranges over representations of h.

Lemma 3.2. *In the metric*

$$d_H(h, h') = \psi(h - h') + d(h, h')$$

we have that H is a Polish group.

Proof. First we need to check that this is indeed a metric, and that follows from the observation that given any two representations

$$\sum n_\ell \cdot g_\ell, \qquad \sum \hat{n}_\ell \cdot g_\ell,$$

for $h - h'$ and $h' - h''$ respectively, we have

$$\varphi(\sum (n_\ell + \hat{n}_\ell) \cdot g_\ell) \leq \varphi(\sum n_\ell \cdot g_\ell) + \varphi(\sum \hat{n}_\ell \cdot g_\ell).$$

The metric *is* separable since the representations with finite support $(\forall^\infty \ell(n_\ell = 0))$ give rise to a dense subgroup. Continuity of the group operations follows since

$$\psi(h) + \psi(h') \geq \psi(h + h').$$

This only leaves us with a campaign to show that the metric is complete.

For this purpose, consider a Cauchy sequence $(h_i)_i$ in H, with each h_i having

$$\sum_{\ell \in \mathbb{N}} n^i_\ell \cdot g_\ell$$

as a representation. Since

$$|\psi(h_i) - \psi(h_j)| \leq \psi(h_i - h_j)$$

we get that $(\psi(h_i))_{i \in \mathbb{N}}$ is Cauchy and so there is a fixed k such that each is a k-representation. Since the space of k-representations is precompact, we may assume there exist $(n^\infty_\ell)_{\ell \in \mathbb{N}}$ such that

$$\forall^\infty j(n^j_\ell = n^\infty_\ell).$$

Claim (1). As $j \to \infty$ and then $M \to \infty$

$$d_H(h_j, \sum_{\ell=0}^{M} n_\ell^\infty \cdot g_\ell) \to 0.$$

Proof of Claim (1). Fix $\epsilon > 0$. We need to show that for all sufficiently large j we have

$$\forall^\infty M(d_H(h_j, \sum_{\ell=0}^{M} n_\ell^\infty \cdot g_\ell) < \epsilon).$$

Choose j large enough that $d_H(h_j, h_{j'}) < \epsilon$ all $j' \geq j$ and $n_\ell^j = n_\ell^\infty$ all $\ell \leq k$. We may assume without loss of generality that $\epsilon < \frac{1}{k}$. Thus at each $j' \geq j$ we have that if

$$\sum_{\ell \in \mathbb{N}} m_\ell \cdot g_\ell$$

is a $2k$-representation of $h_j - h_j'$ with φ value less than ϵ then

$$m_\ell = 0$$

at all $\ell \leq k$. Since by 3.1 we have

$$\sum (n_\ell^j - n_\ell^{j'}) \cdot g_\ell$$

as the unique $2k$-representation of $h_j - h_{j'}$ with coefficient 0 at the ℓth coordinate all $\ell \leq k$, and thus

$$\sum (n_\ell^j - n_\ell^{j'}) \cdot g_\ell$$

is the unique $2k$-representation of $h_j - h_{j'}$ with φ value less than ϵ.

Therefore if at arbitrarily large M we have

$$d_H(h_j, \sum_{\ell=0}^{M} n_\ell^\infty \cdot g_\ell) > \epsilon$$

then for arbitrarily large M we have

$$d_H(\sum_{\ell=0}^{M} n_\ell^j \cdot g_\ell, \sum_{\ell=0}^{M} n_\ell^\infty \cdot g_\ell) > \epsilon,$$

but then going to some j' with $n_\ell^{j'} = n_\ell^\infty$ for all $\ell \leq M$ we have

$$\varphi(\sum (n_\ell^j - n_\ell^{j'}) \cdot g_\ell) > \epsilon,$$

with a contradiction.

This proves Claim (1).

Claim (2). As $N \to \infty$

$$\varphi(\sum_{\ell=N}^{\infty} n_\ell^\infty \cdot g_\ell) \to 0.$$

Proof of Claim (2). Consider $\epsilon > 0$. Choose M_0 such that at all $j \geq M_0$ and $\ell \leq k$ we have $n_\ell^j = n_\ell^\infty$. Choose $M_1 > M_0$ such that at all $j, j' \geq M_1$ we have $\psi(h_j - h_{j'}) < \epsilon$. Choose N_0 such that at all $\ell \geq N_0$

$$\frac{|n_\ell^{M_1}|}{\ell} \leq \epsilon, \qquad \therefore \varphi(\sum_{\ell=N_0}^{\infty} n_\ell^{M_1} \cdot g_\ell) \leq \epsilon.$$

As in the proof of claim(1) we may assume that $\epsilon < \frac{1}{k}$, and again as in that proof we have that for all sufficiently large j'

$$\sum_0^\infty (n_\ell^{j'} - n_\ell^{M_1}) \cdot g_\ell$$

is the only possible representation of $h_{j'} - h_{M_1}$ with

$$\varphi(\sum_0^\infty (n_\ell^{j'} - n_\ell^{M_1}) \cdot g_\ell) < \epsilon,$$

and hence by 3.1 for j' sufficiently large we have $\sum_0^\infty (n_\ell^{j'} - n_\ell^{M_1}) \cdot g_\ell$ is *the* representation of $h_{j'} - h_j$ with $\varphi(\sum_0^\infty (n_\ell^{j'} - n_\ell^{M_1}) \cdot g_\ell) < \epsilon$. Hence at each $N > N_0$

$$\varphi(\sum_{\ell=N_0}^{N} n_\ell^\infty \cdot g_\ell) \leq \varphi(\sum_{\ell=N_0}^{N} (n_\ell^{j'} - n_\ell^{M_1}) \cdot g_\ell) + \varphi(\sum_{\ell=N_0}^{N} n_\ell^{M_1} \cdot g_\ell)$$

$$\leq \varphi(\sum_0^\infty (n_\ell^{j'} - n_\ell^{M_1}) \cdot g_\ell) + \epsilon = \psi(h_{j'} - h_{M_1}) + \epsilon < 2\epsilon.$$

Letting $\epsilon \to 0$ this proves the claim.

This proves Claim (2).

These two claims between them demonstrate that

$$\sum n_\ell^\infty \cdot g_\ell$$

represents some $h_\infty \in H$ and moreover

$$\psi(h_j - h_\infty) \to 0$$

as $j \to \infty$. □

Definition. We define the notion of α-calibration by induction on the countable ordinal α.

A *0-calibration* is an infinite $A \subset \mathbb{N}$. An $\alpha + 1$-calibration is an infinite $A \subset \mathbb{N}$ equipped with a partition

$$A = \dot{\bigcup} B_n,$$

where each B_n is an α-calibration.

For λ a limit, with a specified[1]

$$\alpha_n \nearrow \lambda,$$

we say that $A \subset \mathbb{N}$ is λ-calibration if it is partitioned

$$A = \dot{\bigcup} B_n,$$

with each B_n an α_n-calibration.

Definition. For A a 0-calibration, a representation

$$\sum n_\ell \cdot g_\ell$$

is A-0-*good* if there is an n_∞ such that

$$\forall^\infty \ell \in A(n_\ell = n_\infty);$$

that is to say, the n_ℓ's, for $\ell \in A$, converge to n_∞. We then let $\pi_{A,0}(\sum n_\ell \cdot g_\ell)$ be this indicated n_∞ and

$$\varphi_{A,0}(\sum n_\ell \cdot g_\ell) = \varphi(\sum n_\ell \cdot g_\ell) + |\pi_{A,0}(\sum n_\ell \cdot g_\ell)| + |\{\ell \in A : n_\ell \neq \pi_{A,0}(\sum n_\ell \cdot g_\ell)\}|;$$

in other words, we add into the norm on representations for H information about the eventual value n_∞ along with a counter for the number of times this value is missed.

For A an $\alpha + 1$-calibration, $A = \dot{\bigcup} B_n$, we say that

$$\sum n_\ell \cdot g_\ell$$

is A-$\alpha + 1$-*good* if it is B_n-α-good at every n and there is an m_∞ such that

$$\forall^\infty n(\pi_{B_n,\alpha}(\sum n_\ell \cdot g_\ell) = m_\infty).$$

We then let $\pi_{A,\alpha+1}(\sum n_\ell \cdot g_\ell)$ be this eventual m_∞ value and let

$$\varphi_{A,\alpha+1}(\sum n_\ell \cdot g_\ell) = \varphi(\sum n_\ell \cdot g_\ell) + \sum_n 2^{-n} \frac{\varphi_{B_n,\alpha}(\sum n_\ell \cdot g_\ell)}{1 + \varphi_{B_n,\alpha}(\sum n_\ell \cdot g_\ell)}$$

$$+ |\pi_{A,\alpha+1}(\sum n_\ell \cdot g_\ell)| + |\{n : \pi_{B_n,\alpha}(\sum n_\ell \cdot g_\ell) \neq \pi_{A,\alpha+1}(\sum n_\ell \cdot g_\ell)\}|;$$

that is to say, we merge into a single norm information about the eventual value m_∞, the set of exceptional n's at which this value is not yet reached, along with a bounded version of norms for the various B_n sets.

For λ a limit ordinal, $\alpha_n \nearrow \lambda$, A a λ-calibration, we can define A-λ-goodness in the exactly parallel manner, replacing each mention of B_n-α-goodness by B_n-α_n-goodness.

[1] For the purpose of this definition, imagine that all our limit ordinals come with a previously decided cofinal sequence.

We want to now define $\pi_{A,\alpha}(h)$ in a way that is independent of the representation. Here there is a split in cases, depending on the situation dictated at (III).

Observe first of all by 3.1 we have that if $\sum n_\ell \cdot g_\ell$ and $\sum n'_\ell \cdot g_\ell$ are two A-α-representations of $h \in H$ then the terms

$$n_\ell \cdot g_\ell, \qquad n'_\ell \cdot g_\ell$$

differ at only finitely many places.

Therefore in the case of every g_ℓ having infinite order we get

$$\pi_{A,\alpha}(\sum n_\ell \cdot g_\ell) = \pi_{A,\alpha}(\sum n'_\ell \cdot g_\ell)$$

for any two distinct representations. Therefore we just do the natural thing and let $\pi_{A,\alpha}(h)$ be the common value realized on all the A-α-good representations.

The next case is when the g_ℓ's *all* have the *same* finite order. Use M to denote this common value. We then let $\pi_{A,\alpha}(h)$ be the unique $m \in \{0, 1, \ldots, M-1\}$ for which there is some A-α-good representation having

$$\pi_{A,\alpha}(\sum n_\ell \cdot g_\ell) = m.$$

In the case that the g_ℓ's have strictly increasing but finite orders, we let $\pi_{A,\alpha}(h)$ be the unique $m \in \mathbb{Z}$ for which there is some A-α-good representation with

$$\pi_{A,\alpha}(\sum n_\ell \cdot g_\ell) = m.$$

Note also that the collection of A-α-good elements is closed under the group operations – a critical fact underpinning the next definition.

Definition. For A an α-calibration we let

$$H_{A,\alpha}$$

be the subgroup of H consisting of all elements which admit an A-α-good representation. For h such an element, we let

$$\psi_{A,\alpha}(h)$$

be the infinum of

$$\varphi_{A,\alpha}(\sum_{\ell \in \mathbb{N}} n_\ell \cdot g_\ell)$$

as $\sum n_\ell \cdot g_\ell$ ranges over all possible representations. We then let $d_{A,\alpha}(\cdot,\cdot)$ be defined by

$$d_{A,\alpha}(h, h') = \psi_{A,\alpha}(h - h') + d_H(h, h').$$

Lemma 3.3. *$d_{A,\alpha}$ defines a complete, separable metric on $H_{A,\alpha}$ for any α-calibration A.*

Proof. That this is a metric follows as in the proof for H. Separability follows by induction on α. For $\alpha = 0$ we observe that the set of elements with a representation

$$\sum n_\ell \cdot g_\ell$$

for which there is some m_∞ with

$$\forall^\infty \ell(n_\ell = m_\infty)$$

forms a countable dense subset. For $A = \dot{\bigcup} B_n$ with each B_n a β_n-calibration, we assume that there are countable dense subsets D_n of each H_{B_n,β_n} and we take as our countable dense subset those representations which on finitely many B_n's are an element of a corresponding D_n, and elsewhere are eventually constant.

Again, these arguments are quickly dismissed, and the main battle is for completeness. So we suppose that we have some $d_{A,\alpha}$-Cauchy sequence $(h_j)_{j\in\mathbb{N}}$, each h_j having

$$\sum_{\ell\in\mathbb{N}} n_\ell^j \cdot g_\ell$$

as a representation. Again by the compactness of the space of k-representations, as in 3.2, we may assume there are $(n_\ell^\infty)_{\ell>k}$ such that

$$\forall^\infty j(n_\ell^j - n_\ell^\infty).$$

Note that for all sufficiently large j, j' we have

$$\psi_{A,\alpha}(h_j - h_{j'}) < 1, \qquad \text{and hence} \qquad \pi_{A,\alpha}(h_j - h_{j'}) = 0$$

by definition of $\psi_{A,\alpha}(\cdot)$.

Thus we may find a single m_∞ such that

$$\forall^\infty j(\pi_{A,\alpha}(h_j) = m_\infty).$$

Let h_∞ be the limit in the $d_H(\cdot,\cdot)$ metric of the $(h_j)_j$ sequence, with representation

$$\sum_\ell n_\ell^\infty \cdot g_\ell.$$

Note that each $|n_\ell^\infty| \leq \ell \cdot k$ since $\forall^\infty j(n_\ell^j = n_\ell^\infty)$ and each $|n_\ell^j| \leq \ell \cdot k$. Thus $\sum n_\ell^\infty \cdot g_\ell$ is a k-representation.

In the case that $\alpha = 0$ we have the following claim:

Claim. There are only finitely many $a \in A$ with

$$n_a^\infty \neq m_\infty.$$

Proof of the Claim. We again use the fact that $2k$-representations are unique on coordinates bigger than k. Thus we obtain that if for some $c \geq 2k$ we have $2c$-many a's with $n_a^\infty \neq m_\infty$, then for all sufficiently large j we have *all* $2k$ representations of h_j have at least c many coordinates which present a value other than

$$m_\infty = \pi_{A,0}(h_j),$$

and hence we are left with $\psi_{A,0}(h_j) > c$.

Thus to recap the argument, we have seen that if at *every* c there exist more than c many $a's$ with $n_a^\infty \neq m_\infty$ then $\psi_{A,0}(h_j) \to \infty$ as $j \to \infty$, contradicting the assumption that $(\psi_j)_{j\in\mathbb{N}}$ is Cauchy.

This proves the Claim.

Thus we have $h_\infty \in H_{A,0}$, $\pi_{A,0}(h_\infty) = m_\infty$; the function

$$j \mapsto \{a \in A : n_a^j \neq m_\infty\}$$

is eventually constant by the definition of $\psi_{A,0}(\cdot)$ and uniqueness of the k-representation on coordinates bigger than k. Therefore with respect to $d_{A,0}(\cdot,\cdot)$ we have

$$h_j \to h_\infty.$$

If $\alpha > 0$, $A = \dot{\bigcup} B_n$, each B_n a β_n-calibration, we may assume inductively that there are (p_n) such that at every n

$$\forall^\infty j(\pi_{B_n,\beta_n}(h_j) = p_n), \quad h_\infty \in H_{B_n,\beta_n}, \quad \text{and} \quad h_j \to h_\infty$$

in $d_{B_n,\beta_n}(\cdot,\cdot)$. By Cauchyness of the (h_j) sequence we must eventually have that for all sufficiently large j, j'

$$\forall n(\pi_{B_n,\beta_n}(h_j) = \pi_{B_n,\beta_n}(h_{j'})),$$

and hence the (p_n)-sequence is eventually constant, with value m_∞. □

Lemma 3.4. *Given any $\underset{\sim}{\Delta}^0_\alpha$ set $D \subset \omega^\omega$ and α-calibration $A = \dot{\bigcup} B_n$, each B_n a β_n-calibration, there is continuous*

$$f_\alpha^{A,D} : \omega^\omega \to G$$

such that

(0) *for every y there is a representation*

$$\sum n_\ell \cdot g_\ell$$

of $f_\alpha^{A,D}(y)$ with each $n_\ell \in \{0,1\}$.

(i) *if $y \in D$ then $\forall^\infty n(\pi_{B_n,\beta_n}(f_\alpha^{A,D}(y)) = 1)$;*

(ii) *if $y \notin D$ then $\forall^\infty n(\pi_{B_n,\beta_n}(f_\alpha^{A,D}(y)) = 0)$;*

(iii) *the representation of each $f_\alpha^{A,D}(y)$ has support in A and each $f_\alpha^{A,D}(y) \in H_{A,\alpha}$.*

Proof. This is proved by induction on α, in a manner entirely parallel to the similar claims from [3]. First of all one sees that for any clopen $D \subset \omega^\omega$ and A a 0-calibration we can certainly find continuous $f^{A,D} : \omega^\omega \to G$ with

$$f^{A,D}(x) = 0$$

if $x \notin D$, but

$$f^{A,D}(x) = \lim_{N\to\infty} \sum_{\ell \in A, \ell \le N} g_\ell,$$

if $x \in D$. This forms the base of our induction.

Now we do the inductive step, and suppose that D is $\underset{\sim}{\Delta}^0_\alpha$, $A = \dot{\bigcup} B_n$ is an α-calibration, each B_n a β_n-calibration. We recall the classical fact (compare [3]) that there must be $\underset{\sim}{\Delta}^0_{\beta_n}$ sets $(E_n)_n$ such that

(a) if $y \in D$ then $\forall^\infty n(y \in E_n)$;

(b) if $y \notin D$ then $\forall^\infty n(y \notin E_n)$.

Appealing to our inductive hypothesis we may find

$$f_n : \omega^\omega \to H_{B_n,\beta_n},$$

continuous with respect to the Polish G-topology, each $f_n(x)$ having support inside B_n, with $\pi_{B_n,\beta_n}(f_n(x)) = 1$ if $x \in E_n$, $\pi_{B_n,\beta_n}(f_n(x)) = 0$ if $x \notin E_n$, and every $f_n(x)$ having a representation with coordinates taken solely from $\{0, 1\}$. We can then simply let

$$f^{A,D}(x) = \lim_{N\to\infty} f_0(x) + f_1(x) + f_2(x) + \cdots + f_N(x).$$

Our inductive assumptions on the f_n's suffice to give that this is in H, and then we obtain it will be in $H_{A,\alpha}$ with either (a) or (b) above by the choice of the E_n sets. It is continuous by condition (II) on the g_ℓ-sequence and the fact that each $f_n(x)$ has support inside B_n. □

Corollary 3.5. *If $A = \dot{\bigcup} B_n$ is an $\alpha + 1$-calibration, then $H_{A,\alpha+1}$ is not $\underset{\sim}{\Pi}^0_\alpha$.*

Proof. Let $C \in \underset{\sim}{\Sigma}^0_\alpha$ be of the form

$$C = \bigcup D_n,$$

where each $D_n \in \underset{\sim}{\Delta}^0_\alpha$. Following the last lemma we may find continuous

$$f_n : \omega^\omega \to H_{B_n,\beta_n}$$

such that

(0) for every y there is a representation

$$\sum_{\ell \in B_n} n_\ell \cdot g_\ell$$

of $f_n(y)$ with each $n_\ell \in \{0, 1\}$.

(i) if $y \in \bigcup_{i\leq n} D_i$ then $\pi_{B_{2n},\beta_{2n}}(f_{2n}(y)) = 1$;

(ii) if $y \notin \bigcup_{i\leq n} D_i$ then $\pi_{B_{2n},\beta_{2n}}(f_{2n}(y)) = 0$;

(iii) the representation of each $f_n(y)$ has its support in B_n;

(iv) for m odd at every y we have $\pi_{B_m,\beta_m}(f_m(y)) = 1$.

After this we can simply take

$$f^C(y) = \lim_{N\to\infty} f_0(y) + f_1(y) + \cdots f_N(y).$$

Then in order for the values of the $\pi_{B_n,\beta_n}(f^C(y))$ to settle down into a limit we must have that $y \in C$, and thus membership C is a necessary condition for $f(y)$ to be in $H_{A,\alpha+1}$. But conversely membership in C is also a sufficient condition for $f(y)$ to be in this subgroup, and thus we have continuously reduced membership in an arbitrary $\underset{\sim}{\Sigma}^0_\alpha$ set to membership in $H_{A,\alpha+1}$. □

4. After thoughts

As observed by Slawek Solecki, this result for uncountable abelian Polish groups implies the same for uncountable locally compact Polish groups.

The main point is that we can use approximation by Lie groups – in particular the theorem that every connected locally compact group has arbitrarily small normal subgroups for which the resulting quotient is Lie. (See [5].)

One case is that the group is totally disconnected, and hence, by local compactness (see [5]) it will have a neighborhood basis of clopen compact subgroups. Appealing to [6] we obtain an uncountable abelian compact subgroup, and finish by the results of the last section.

The other case is that there exists a closed connected subgroup, which in turn will have an uncountable Lie group as a quotient. But any Lie group contains one-parametrized subgroups, arrived at via the exponential map from the tangent bundle at the identity, and hence a continuous image of $\mathbb{R}$ or $\mathbb{T}$, and again we fall back into the cases covered by the last section.

The most optimistic conjecture, then, would be that every uncountable Polish group contains an uncountable abelian subgroup. From this and the results above we could obtain arbitrarily complicated Polishable subgroups of any uncountable Polish group.

However the spirit of fearless scientific research demands that we should never take a position of optimism when pessimism is unrefuted. Thus:

Conjecture 4.1. *There is an uncountable Polish group all of whose abelian subgroups are discrete.*

It may even be that some minor variation of the the surjectively universal two-sided invariant group (see [4], [1]) provides a counterexample. Accordingly it seems unlikely that the construction above will answer the question raised in [2].

Acknowledgements

These results were largely proved during the course of the conference at CRM during the year dedicated to Set Theory, and came about after listening to Farah and Solecki describe their recent work. I am very grateful for a number of conversations with both of them, including one lengthy discussion carried out in the taxi on the way out to the airport for the flight home.

I am also grateful to CRM for making this meeting possible, and I further thank Joan Bagaria and the other members of the organizing committee for inviting me to participate.

I am more than indebted to the anonymous referee for a careful and exacting reading of the earlier version of this paper.

References

[1] H. Becker and A.S. Kechris, *The descriptive set theory of Polish group actions*, London Mathematical Society Lecture Note Series, 232, Cambridge University Press, Cambridge, 1996.

[2] I. Farah and S. Solecki, Borel subgroups of Polish groups, preprint, `http://www.math.yorku.ca/~ifarah/preprints.html`.

[3] G. Hjorth, A.S. Kechris, and A. Louveau, The Borel equivalence relations induced actions of the infinite symmetric group, *Annals of Pure and Applied Logic*, vol. 92(1998), pp. 63–112.

[4] A.S. Kechris, *Definable equivalence relations and Polish group actions*, manuscript, Caltech, 1993.

[5] D. Montgomery and L. Zippin, *Topological transformation groups*, reprint of the 1955 original, Robert E. Krieger Publishing Co., Huntington, N.Y., 1974. xi+289 pp.

[6] E.I. Zelmanov, On periodic compact groups, *Israel Journal of Mathematics*, vol. 77(1992), pp. 83–95.

Greg Hjorth
MSB 6363
UCLA
CA 90095-1555
e-mail: `greg@math.ucla.edu`

Set Theory
Trends in Mathematics, 309–320

On the Strength of Mutual Stationarity

Peter Koepke and Philip Welch

Abstract. For $(\kappa_n)_{n<\omega}$ a strictly increasing sequence of regular cardinals $\geqslant \aleph_2$, Foreman and Magidor showed: if every sequence $(S_n)_{n<\omega}$ of sets S_n, which are stationary in κ_n with $\forall \xi \in S_n$ $\mathrm{cof}(\xi) = \omega_1$, is mutually stationary then $V \neq L$. We show that the existence of a sequence $(\kappa_n)_{n<\omega}$ with this property is equiconsistent with the existence of a measurable cardinal. In case $(\kappa_n)_{n<\omega}=(\aleph_{n+3})_{n<\omega}$ the property implies the existence of inner models with many measurable cardinals .

1. Introduction

The concept of *mutual stationarity* was introduced by M. Foreman and M. Magidor [4] in order to transfer some combinatorial aspects of stationary subsets of regular cardinals to singular cardinals. Together with J. Cummings they further investigated the status of such sequences in [3].

Definition 1. *Let* $(\kappa_n)_{n<\omega}$ *be a strictly increasing sequence of regular cardinals* $\geqslant \aleph_2$ *with* $\kappa_\omega = sup_{n<\omega}\kappa_n$. *A sequence* $(S_n)_{n<\omega}$ *is called* mutually stationary *in* $(\kappa_n)_{n<\omega}$ *if every first-order structure* $\mathfrak{A}$ *of countable type with* $\kappa_\omega \subseteq \mathfrak{A}$ *has an elementary substructure* $\mathfrak{B} \prec \mathfrak{A}$ *such that* $\forall n < \omega \sup |\mathfrak{B}| \cap \kappa_n \in S_n$.

Note that if $(S_n)_{n<\omega}$ is mutually stationary in $(\kappa_n)_{n<\omega}$ then each $S_n \cap \kappa_n$ is stationary in κ_n. In the following we shall denote the class $\{\xi \in \mathrm{Ord} | \mathrm{cf}(\xi) = \lambda\}$ by Cof_λ. For $X \subseteq \mathrm{Ord}$ a set, we write $\mathrm{ot}(X)$ for its order type.

Definition 2. *Let* $(\kappa_n)_{n<\omega}$ *be a strictly increasing sequence of regular cardinals and* $\lambda < \kappa_0$, λ *regular. The mutual stationarity property* $\mathrm{MS}((\kappa_n)_{n<\omega}, \lambda)$ *is the statement: if* $(S_n)_{n<\omega}$ *is a sequence of sets* $S_n \subseteq \mathrm{Cof}_\lambda$ *which are stationary in* κ_n *then* $(S_n)_{n<\omega}$ *is mutually stationary in* $(\kappa_n)_{n<\omega}$.

M. Foreman and M. Magidor [4] proved the following two theorems:

Theorem. *For* $(\kappa_n)_{n<\omega}$ *a strictly increasing sequence of uncountable regular cardinals,* $\mathrm{MS}((\kappa_n)_{n<\omega}, \omega)$ *holds.*

Theorem. $\mathrm{MS}((\kappa_n)_{n<\omega}, \omega_1)$ implies $V \neq L$.

In fact, they proved much more in the latter theorem: assuming $V = L$ they exhibited a double-indexed sequence $S_n^h \subseteq \omega_{n+2}(n < \omega, 1 \leq h < \omega)$, where each $S_n^h = \mathrm{Defcol}(h) \cap \mathrm{Cof}_{\omega_1}$. For α not a cardinal, let $\beta(\alpha) + 1$ be the least level of the L-hierarchy where α is singular, and let $h(\alpha)$ be the least level of definability, so that there is a $\Sigma_h(L_{\beta(\alpha)})$ definable function witnessing the singularity of α. Then $\mathrm{Defcol}(h) = \{\alpha \in \mathrm{Ord} | h(\alpha) = h\}$. Their result then is: For any function $f : \omega \longrightarrow \omega$, $\langle S_n^{f(n)} \rangle$ is mutually stationary if and only if f is eventually constant.

We strengthen this to:

Theorem 1. *The theories* $\mathrm{ZFC} + \exists(\kappa_n)_{n<\omega}\mathrm{MS}((\kappa_n)_{n<\omega}, \omega_1)$ *and* $\mathrm{ZFC} + \exists\kappa$ *(κ measurable) are equiconsistent.*

The implication from right to left was proved by J. Cummings, Foreman, and Magidor [3] via Prikry forcing. Again they proved more than this. They showed that a tail of the Prikry generic sequence satisfies $\mathrm{MS}((\kappa_n)_{n<\omega}, \lambda)$ for any $\lambda < \kappa_0$ (or indeed the mutual stationarity of *any* sequence of stationary sets $S_n \subseteq \kappa_n$ irrespective of the cofinalities of the ordinals in the S_n.) This is essentially obtained by utilising the fact that a tail of the Prikry generic sequence remains *coherently Ramsey* in the generic extension. The converse which we prove here uses the core model K of A.J. Dodd and R.B. Jensen (see [2]). We deduce the existence of $O^\sharp$ from $\mathrm{MS}((\kappa_n)_{n<\omega}, \omega_1)$ in detail. The proof involves the global square principle $\Box$ in L and techniques from the Jensen Covering theorem for L (see [1]). Fine structural details will be presented in the hyperfine structure theory of S.D. Friedman and the first author [5]. Although the hyperfine structure for the Dodd-Jensen Core Model is not yet published we shall nevertheless indicate how to transfer the arguments from L to the Dodd-Jensen K for the proof of the full theorem.

In case $(\kappa_n)_{n<\omega}$ consists of "small" cardinals we can obtain higher consistency strengths:

Theorem 2. *If* $\mathrm{MS}((\aleph_{n+3})_{n<\omega}, \omega_1)$ *holds then there is an inner model with infinitely many measurable cardinals κ of Mitchell order $o(\kappa) = \omega_1$.*

Better results than the above are obtainable, but we leave the precise statement (and a proof of Theorem 2) to a later paper. For these results, the hyperfine structure theory has not been developed, and so there recourse is made to more standard fine structure.

2. Order types of square sequences

Definition 3. *Let* $\mathrm{Sing} = \{\beta \in \mathrm{Ord} \mid \lim(\beta) \wedge \mathrm{cf}(\beta) < \beta\}$ *be the class of singular limit ordinals. Global square ($\Box$) is the assertion: there is a system* $(C_\beta)_{\beta\in\mathrm{Sing}}$ *satisfying:*

(a) C_β *is a closed cofinal subset of* β;
(b) $\mathrm{ot}(C_\beta) < \beta$;
(c) *if* $\overline{\beta}$ *is a limit point of* C_β *then* $\overline{\beta} \in \mathrm{Sing}$ *and* $C_{\overline{\beta}} = C_\beta \cap \overline{\beta}$.

Jensen [6] introduced the principle $\Box$ and proved it in L. The second author [8] proved $\Box$ in the Dodd-Jensen core model K. From the order types of the square sequences C_ξ we shall define stationary sets S_n to which we shall apply the MS-principle.

Theorem 3. *Let* κ *be a regular cardinal* $\geq \aleph_2$ *and* λ *a regular cardinal* $< \kappa$. *Then for every ordinal* θ *such that* $\theta^+ < \kappa$ *the set*

$$\{\beta \in \mathrm{Cof}_\lambda \cap \kappa \mid \mathrm{ot}(C_\beta) \geq \theta\}$$

is stationary in κ.

Proof. Let $C \subseteq \kappa$ be closed unbounded in κ. Let $\mu = \max(\lambda, \theta^+)$ which is an uncountable regular cardinal $< \kappa$. Take a singular limit point γ of C of cofinality μ. Then $C \cap C_\gamma$ is closed unbounded in γ of ordertype $\geq \mu$. Take β to be a singular limit point of $C \cap C_\gamma$ such that $\mathrm{cof}(\beta) = \lambda$ and $\mathrm{ot}(C \cap C_\gamma \cap \beta) \geq \theta$. By the coherency property 3 (c), $C_\beta = C_\gamma \cap \beta$. Thus $\beta \in C \cap \{\beta \in \mathrm{Cof}_\lambda \cap \kappa \mid \mathrm{ot}(C_\beta) \geq \theta\} \neq \emptyset$. $\Box$

Note that $(S_n)_{n<\omega}$ with

$$S_n = \{\beta \in \mathrm{Cof}_{\omega_1} \cap \aleph_{n+3} \mid \mathrm{ot}(C_\beta) \geq \aleph_{n+1}\}$$

is a sequence of stationary sets to which we could apply the MS-principle.

3. Hyperfine singularizations

Let β be a singular ordinal in L. We shall assign to β a level of the fine structural hierarchy and a parameter which canonically witness the singularity of β. We use the hyperfine hierarchy of S.D. Friedman and the first author [5] where the same singularizations were used in the proof of global square.

The hyperfine structural hierarchy refines Gödel's L_α-hierarchy. The levels of the hierarchy are indexed by *locations* $s = (\alpha, \varphi_m, \vec{x})$ where $\alpha \in \mathrm{Ord}$, $\varphi_m(v_0, \ldots, v_{k-1})$ is an $\in$-formula, and $\vec{x} = x_1, \ldots, x_{k-1} \in L_\alpha$. $(\varphi_m)_{m<\omega}$ is an appropriate list of all $\in$-formulas. Then

$$L_s = (L_\alpha, \in, <_L, I, N, S, S^{L_\alpha}_{\varphi_0}, \ldots, S^{L_\alpha}_{\varphi_m} \restriction \vec{x}, \emptyset, \emptyset, \ldots);$$

Here, $<_L$ is the canonical well-ordering of L, I, N, S are an *interpretation function*, a *naming function*, and a *Skolem function* respectively for L; $S^{L_\alpha}_{\varphi_i}$ is a Skolem function for φ_i computed in L_α. Moreover the last function $S^{L_\alpha}_{\varphi_m}$ is restricted to arguments $\vec{y}$ which are lexicographically smaller than $\vec{x}$, where the lexicographical order $<_{\mathrm{lex}}$ is derived from $<_L$.

The locations are well ordered lexicographically by $\tilde{<}$. For each L_s there is a hulling operator $L_s\{.\}$; $L_s\{X\}$ is the smallest substructure of L_s which contains

X. The basic fine structural laws of (L_s) and the associated hulling operations are described in [5].

For a given limit ordinal β which is singular in L we describe its singularization; in view of the intended applications we also assume that $\mathrm{cof}(\beta) \geq \omega_1$. There is a location $s = (\gamma, \varphi, \vec{x})$ and a finite set $p \subseteq L_\gamma$ such that $\gamma \geq \beta$ and

(1) $\{\overline{\beta} < \beta \mid \overline{\beta} = \beta \cap L_s\{\overline{\beta} \cup p\}\}$ *is bounded below* β.

We say that β is *semi-singularized at* (L_s, p). Let $s = s(\beta)$ be the $\tilde{<}$-minimal location such that β is semi-singularized at (L_s, p) for some p. Then let $p = p(\beta)$ be a finite set such that $(L_{s(\beta)}, p)$ semi-singularizes β where p is minimal with respect to the $<_*$-well-ordering of finite subsets of L: $p <_* q \leftrightarrow \exists z \in q \setminus p \forall u (u <_L z \to (u \in p \leftrightarrow u \in q))$.

(2) $(L_{s(\beta)}, p(\beta))$ *exists and semi-singularizes* β*; it is called the* L*-singularization of* β.

We give some more information about the L-singularization. Note that by $\mathrm{cof}(\beta) > \omega_1$ we are in the "generic case" of [5].

(3) $s(\beta) = (\gamma, \varphi, \vec{x}) \neq (\gamma, \varphi_0, \vec{0})$.

(4) *There is* $\alpha_0 = \alpha_0(\beta) < \beta$ *minimal such that* $L_s\{\alpha_0 \cup p\}$ *is unbounded below* s*, i.e., for all* $\vec{y} <_{lex} \vec{x}$*,* $|\vec{y}| = |\vec{x}|$ *there is* $\vec{z} \in L_s\{\alpha_0 \cup p\}, |\vec{z}| = |\vec{x}|$ *such that* $\vec{y} <_{lex} \vec{z}$*; in case that* $\vec{x} = \vec{0}$ *we have to require instead that* $L_s\{\alpha_0 \cup p\}$ *is cofinal in* $(L_s, <_L)$.

In the construction of the canonical $\Box$-sequence C_β some ordinal $\alpha \leq \alpha_0$ will be used as a "steering ordinal". As a brief sketch, we want to define the C_β sequence with reference to a cofinalising sequence in the location s. If α_0 is a limit ordinal, then we shall take α_0 itself as α. Otherwise $\alpha_0 = \alpha_0' + 1$, and we have some $\alpha_1 < \alpha_0'$ so that $L_s\{\alpha_1 \cup \{p, \alpha_0'\}\}$ is unbounded below s; if $\alpha_1 > 0$ but is $\alpha_1' + 1$, we repeat, and see that $L_s\{\alpha_2 \cup \{p, \alpha_0' \, \alpha_1'\}\}$ is unbounded below s for some $\alpha_2 < \alpha_1$. After a finite number k of steps we find that α_k is zero (in which case we deduce that the cofinality of $\beta = \omega$) or a limit. In the latter case, by recursion on $\iota \leq \alpha_k$ we define an increasing sequence of hulls in the location whose suprema below β will be the elements of what will ultimately contain the C_β We thus have bounded the order type of C_β by this "steering ordinal" $\alpha_k(\beta) \leq \alpha_0$. Hence

(5) $\mathrm{otp}(C_\beta) \leq \alpha_0 < \beta$.

This restriction on order types will later conflict with the choice of $(S_n)_{n<\omega}$ described in section 2 and conclude a proof by contradiction.

4. Lifting up singularizations

The following argument is an upward extensions of embeddings construction as known from the proof of Jensen's Covering Theorem:

Theorem 4. *Let* $\pi : (L_\beta, \in) \to (L_{\beta^*}, \in)$ *be an elementary cofinal map between* ZF^-*-models. Let* β *be singular in* L *and* $\mathrm{cof}(\beta) \geq \omega_1$*, let* (L_s, p) *be the* L*-singularization* *of β as described in the previous paragraph. Then there are a uniquely defined structure preserving map $\pi^* : L_s \to L_{s^*}$ and a parameter p^* satisfying:*

a) $\pi^* \upharpoonright L_\beta = \pi$, $\pi^*``p = p^*$;

b) (L_{s^*}, p^*) *is the L-singularization of β^*.*

Proof. The proof of $\square$ in L shows that L_s can be represented as $L_s = \bigcup_{i<\tau} L_{s_i}\{\beta_i \cup p\}$ for strictly increasing sequences $(\beta_i)_{i<\tau}$ and $(s_i)_{i<\tau}$ converging to β and s resp., such that each transitive collapse $\sigma_i : M_i \cong (L_{s_i}\{\beta_i \cup p\}, p)$ is the singularization of β_i.

For $i \leq j < \tau$ let $\sigma_{ij} = \sigma_j^{-1} \circ \sigma_i : M_i \to M_j$. The minimality of s implies that each $M_i \in L_\beta$ and $\sigma_{ij} \in L_\beta$. $(M_i)_{i<\tau}, (\sigma_{ij})_{i\leq j<\tau}$ is a directed system of L-singularizations all of whose components are elements of L_β.

We can now map the directed system pointwise to L_{β^*}: for $i < \tau$ let $M_i^* = \pi(M_i)$ and $\sigma_{ij}^* = \pi(\sigma_{ij})$. $(M_i^*)_{i<\tau}, (\sigma_{ij}^*)_{i\leq j<\tau}$ is a commutative system of L-singularizations for the ordinals $\beta_i^* = \pi(\beta_i)$.

(1) The direct limit of $(M_i^*)_{i<\tau}, (\sigma_{ij}^*)_{i\leq j<\tau}$ is well founded.

Proof. The indexing ordinal τ has cofinality $\geq \omega_1$. So any descending ω-sequence in the direct limit is already represented in some M_j^* with $j < \tau$. But M_j^* is transitive. $\square$(1)

Let $M^*, (\sigma_i^*)_{i<\tau}$ be the direct limit of the system $(M_i^*), (\sigma_{ij}^*)$. An argument similar to the proof of the condensation theorem in [5] shows that M^* is a level of the hyperfine hierarchy, say $M^* = L_{s^*}$. Define the map $\pi^* : L_s \to L_{s^*}$ by $\sigma_i(z) \mapsto \sigma_i^*(\pi(z))$. π^* is a homomorphism by general facts about direct limits. If $z \in L_\beta$, then $\sigma_i(z) = z$ for sufficiently high $i < \tau$, and so

(2) $\pi^* \supseteq \pi$.

Let $p^* = \pi^*``p$.

(3) $\pi^* : L_s \to L_{s^*}$ is cofinal with respect to the well-ordering $\tilde{<}$ of locations.

Proof. The location s^* is determined as the $\tilde{<}$-minimal location such that $\sigma_i^* : M_i^* \to M^*$ is a well-defined homomorphism. This property is equivalent to: for all $i < \tau$ and $M_i^* = L_{s_i^*}$ and for all $t \tilde{<} s^*$ holds $\sigma_i^*(t) \tilde{<} s^*$.

Consider $r^* \tilde{<} s^*$. Then take $i < \tau, M_i^* = L_{s_i^*}$ and some $t_0 \tilde{<} s_i^*$ such that $r^* \tilde{\leq} \sigma_i^*(t_0)$.

Take j, $i < j < \tau$ such that $s_i \in L_{s_j}\{\beta_j \cup p\}$. Let $s_i = \sigma_j(s_j'), M_i = L_{\tilde{s}_i}$. Then

$$\forall t \tilde{<} \tilde{s}_i : \sigma_{ij}(t) \tilde{<} s_i', \quad \forall t \tilde{<} s_i^* : \sigma_{ij}^*(t) \tilde{<} \pi(s_i'); \quad \forall t \tilde{<} s_i^* : \sigma_i^*(t) \tilde{<} \sigma_j^*(\pi(s_i')) = \pi^*(s_i).$$

In particular: $r^* \tilde{\leq} \sigma_i^*(t_0) \tilde{<} \pi^*(s_i)$, as required. $\square$(3)

(4) (L_{s^*}, p^*) is the L-singularization of β^*.

Proof. Take $\delta < \beta$ such that $\{\overline{\beta} < \beta \mid \overline{\beta} = \beta \cap L_s\{\overline{\beta} \cup p\}\} \subseteq \delta$. Set $\delta^* = \pi(\delta)$. We claim that $\{\eta < \beta^* \mid \eta = \beta^* \cap L_{s^*}\{\eta \cup p^*\}\} \subseteq \delta^*$. Let $\eta \geq \delta^*$. Take $\overline{\beta} < \beta$ minimal such that $\pi(\overline{\beta}) \geq \eta$. Then $\overline{\beta} \geq \delta$ and $\overline{\beta} \not\subseteq \beta \cap L_s\{\overline{\beta} \cup p\}$. Take an L_s-term t and $\vec{x} \subseteq \overline{\beta}$ such that $\overline{\beta} \leq t^{L_s}(\vec{x}, p) < \beta$. Since π^* is a homomorphism,

$$\eta \leq \pi(\overline{\beta}) \leq t^{L_{s^*}}(\pi(\vec{x}), p^*) < \beta^*, \text{and } \pi(\vec{x}) \subseteq \pi''\overline{\beta} \subseteq \eta.$$

Hence $\eta \neq \beta^* \cap L_{s^*}\{\eta \cup p^*\}$, and s^* satisfies the semi-singularity property for β^*.

To show that s^* is minimal semi-singularizing β^* consider $r^* \tilde{<} s^*$. By the cofinality property (3) take $r \tilde{<} s$ such that $r^* \tilde{<} \pi^*(r)$. By the minimality of s, $\{\overline{\beta} < \beta \mid \overline{\beta} = \beta \cap L_r\{\overline{\beta} \cup p\}\}$ is unbounded in β. Let $\overline{\beta} < \beta$, $\quad \overline{\beta} = \beta \cap L_r\{\overline{\beta} \cup p\}\}$. Take $i < \tau$ such that $r \tilde{<} s_i$, $\quad \overline{\beta} < \beta$, $\quad r \in L_{s_i}\{\beta_i \cup p\}$. Then:

$$L_{s_i} \models \overline{\beta} = \beta \cap L_r\{\overline{\beta} \cup p\},$$
$$M_i \models \overline{\beta} = \beta_i \cap L_{\sigma_i^{-1}(r)}\{\overline{\beta} \cup \sigma_i^{-1} \text{“} p\},$$
$$M_i^* \models \pi(\overline{\beta}) = \pi(\beta_i) \cap L_{\pi(\sigma_j^{-1}(r))}\{\pi(\overline{\beta}) \cup \pi(\sigma_i^{-1} \text{“} p)\}.$$

Apply σ_i^*:

$$\pi(\overline{\beta}) = \beta^* \cap L_{\pi^*(r)}\{\pi(\overline{\beta}) \cup p^*\}, \qquad \text{and} \qquad \pi(\overline{\beta}) = \beta^* \cap L_{r^*}\{\pi(\overline{\beta}) \cup p^*\}.$$

Since the set of such $\pi(\overline{\beta})$ is cofinal in β^*, r^* does not semi-singularize β^*, as required.

Now we examine the properties of p^*. By construction:

$$(*) L_{s^*} = L_{s^*}\{\beta^* \cup p^*\}.$$

Suppose that some $q^* <_* p^*$ also satisfies $(*)$. Then $p^* = t(\vec{x}, q^*)$ for some term t and $\vec{x} < \beta^*$.

$$L_{s^*} \models \exists \vec{x} < \beta^* \exists q^* <_* p^* \quad p^* = t(\vec{x}, q^*).$$

This existential property can be pulled back to L_s via the directed systems:

$$L_s \models \exists \vec{x} < \beta \exists q <_* p \quad p = t(\vec{x}, q),$$

which contradicts the minimal choice of p. □

The previous proof shows that π^* is cofinal in the locations. This affects the “steering ordinal” α_0 as follows:

Lemma 1. *In the situation of the previous theorem,*

$$\alpha_0(\beta^*) \leq \pi(\alpha_0(\beta)) \quad \textit{and} \quad \operatorname{otp}(C_{\beta^*}) \leq \pi(\alpha_0(\beta)).$$

5. Getting $0^\sharp$

Theorem 5. *If* $\operatorname{MS}((\kappa_n)_{n<\omega}, \omega_1)$ *holds then* $0^\sharp$ *exists.*

Proof. Assume $\neg 0^\sharp$. Without loss of generality we may assume that $\kappa_0 \geq \aleph_3$. Set $\kappa_\omega = \sup_{n<\omega} \kappa_n$. Define a sequence $(S_n)_{n<\omega}$ of stationary sets as in Section 2: $S_0 = \operatorname{Cof}_{\omega_1} \cap \kappa_0$, $S_1 = \operatorname{Cof}_{\omega_1} \cap \kappa_1$, and for $n \geq 2$:

$$S_n = \{\beta \in \operatorname{Cof}_{\omega_1} \cap \kappa_n \mid \operatorname{otp}(C_\beta) \geq \kappa_{n-2}\}.$$

Take a first-order structure $\mathfrak{A} = (L_{\kappa_\omega^+}, \cdots)$ with countable language which has a family of Skolem functions f_i for $L_{\kappa_\omega^+}$, constants $\kappa_0, \kappa_1, \cdots, \kappa_\omega$ and functions g_i, n :

$$g_{i,n}(\vec{x}) = \sup\{f_i(\vec{x}, \vec{y}) \mid \vec{y} < \aleph_2\} \cap \kappa_n.$$

Applying $\mathrm{MS}((\kappa_n)_{n<\omega}, \omega_1)$ to $(S_n)_{n<\omega}$ and the structure $\mathfrak{A}$ yields some $X \prec L_{\kappa_\omega^+}$ such that $\{\kappa_n \mid n \leq \omega\} \subseteq X$, $\forall n < \omega \sup(X \cap \kappa_n) \in S_n$, and $\omega_2 \subseteq X$. Let $\pi : (L_\delta, \in) \cong (X, \in)$, and $\beta_n = \pi^{-1}(\kappa_n)$ for $n \leq \omega$. For each $n < \omega : \beta_n \geq \aleph_2$ and $\mathrm{cf}(\beta_n) = \omega_1$. The Jensen Covering Theorem for L implies that every β_n is a singular ordinal in L. For $n < \omega$ let (L_{s_n}, p_n) be the singularization of β_n.

(1) If $s_n = (\gamma, -, -)$ then $\gamma \geq \beta_\omega$, since inside L_{β_ω}, β_n is a regular cardinal.

(2) $s_{n+1} \tilde{\leq} s_n$.

Proof. We show that s_n singularizes β_{n+1} as well as β_n :

$$L_{s_n} = L_{s_n}\{\beta_n \cup p_n\} \supseteq \beta_\omega \supseteq \beta_{n+1},$$

and so

$$\{\overline{\beta} < \beta_{n+1} \mid \overline{\beta} = \beta_{n+1} \cap L_{s_n}\{\overline{\beta} \cup p_n\}\} \subseteq \beta_n. \qquad \square(2)$$

Since $\tilde{<}$ is a well-order there is $n_0 < \omega$ such that $s_{n_0} = s_{n_0+1} = s_{n_0+2} = \ldots$. Set $s = s_{n_0}$.

(3) For $n_0 \leq n < \omega : p_{n+1} \leq_* p_n$.

Proof. We show that p_n satisfies the property in the definition of p_{n+1}.

$$L_s = L_s\{\beta_n \cup p_n\} \quad \text{and so} \quad L_s = L_s\{\beta_{n+1} \cup p_n\}. \qquad \square(3)$$

Since $<_*$ is a well-order there is some $n_1 < \omega$, $n_1 \geq n_0$ such that $p_{n_1} = p_{n_1+1} = p_{n_1+2} = \ldots$. Set $p = p_{n_1}$. Then (L_s, p) is the L-singularization of $\beta_{n_1}, \beta_{n_1+1}, \ldots$. Let $\alpha = \alpha_0(\beta_{n_1}) < \beta_{n_1}$ as defined in Section 3. As the location s for singularization of the β_m is the same for $m \geq n_1$, the definition of $\alpha_0(\beta_m)$ is independent of $m \geq n_1$. Thus $\alpha = \alpha_0(\beta_m)$ for $n_1 \leq m < \omega$. For $\beta = \beta_{n_1+2}, \beta^* = \sup(X \cap \kappa_{n_1+2})$, we have

$$\pi \restriction L_\beta : L_\beta \to L_{\beta^*}$$

cofinally as required in Theorem 4. Then Lemma 1 yields

$$\mathrm{otp}(C_{\beta^*}) \leq \pi(\alpha_0(\beta)) = \pi(\alpha) < \pi(\beta_{n_1}) = \kappa_{n_1}.$$

But $\beta^* \in S_{n_1+2}$ and $\mathrm{otp}(C_{\beta^*}) \geq \kappa_{n_1}$. Contradiction! $\square$

6. Singularizations in core models

For stronger results, we have to apply *core models* instead of the inner model L. We use models of the form $K = L[E]$ where E is a sequence of measures on ordinals. For Theorem 1 we use the DODD-JENSEN core model below one measurable cardinal [2], (and for Theorem 2 we should have to use core model for sequences of measures [7], where E is a sequence of total and partial measures on K, together with the more usual fine structure – rather than hyperfine structure). Since our proofs are dependent on a range of results and techniques from core model theory the further presentation has to omit many details and tries to convey basic ideas. We are forced to make several simplifying assumptions and have to argue by analogy with the L-case. The general reference to core model theory is the book [9] by Martin Zeman.

We use the fine structure as developed by Jensen. For small core models, where the only measures that appear are of Mitchell order 0, an extender E with critical point κ is a filter indexed by some ν. The latter will be the successor cardinal of κ in the ultrapower of the model by E. For higher core models containing sequences of measures, or extenders proper, then larger indices are used (see [9], Chapter 8 for details).

Subsequently, the letter K stands for the Dodd-Jensen core model. Global Square is proved in K by carefully assigning singularizing sequences to singular ordinals in K. We describe the singularization of an ordinal β in terms of the Jensen fine structure for measure sequences ("mice"). It will consist of a level of the fine structural hierarchy and a parameter(-sequence) which canonically witness the singularity of β.

Definition 4. *Let $M = J_\alpha[E]$ be a mouse and let $p \in M$ be some finite parameter. Then (M, p)* semi-singularizes *β, if $\{\overline{\beta} < \beta \mid \overline{\beta} = \beta \cap M\{\overline{\beta} \cup p\}\}$ is bounded below β. Here $M\{X\}$ denotes the fine structural hull of X in M. For simplicity, we shall say "*singularize*" instead of "semi-singularize". M as above is called a canonical singularization of β if*

a) *$M \models$ "β is regular" or $\beta = \omega\alpha$;*
b) *$M = M\{\beta \cup p_M\}$;*
c) *(M, p_M) singularizes β where p_M is the standard parameter of M.*

Again, we only say "K-singularization*" instead of "canonical singularization".*

From a K-singularization M of β one can readily define a subset C_β of β as in the proof of $\Box$ which is cofinal in β of ordertype $< \beta$. Let us indicate some elements of that definition. In view of the intended applications we also assume that $\mathrm{cof}(\beta) \geq \omega_1$. For simplicity we may assume that the first projectum $\rho^1_M < \beta$ so that we can use the relatively simple Σ_1-finestructure. There is $\alpha_0 = \alpha_0(\beta) < \beta$ minimal such that $M\{\alpha_0 \cup p_M\}$ is unbounded in M. In the construction of the canonical $\Box$-sequence C_β some ordinal $\alpha \leq \alpha_0$ will be used as a "steering ordinal" which will imply that $\mathrm{otp}(C) \leq \alpha_0 < \beta$. This restriction on order types will later conflict with the choice of $(S_n)_{n<\omega}$ described above and conclude a proof by contradiction. The coherency property of $\Box$ is due to the coherency between various K-singularizations.

Lemma 2. *If M and N are K-singularizations of β and $M \restriction \beta = N \restriction \beta$ then $M = N$. Also p_M is the least parameter p such that $\{\overline{\beta} < \beta \mid \overline{\beta} = \beta \cap M\{\overline{\beta} \cup p\}\}$ is bounded below β.*

Proof. Coiterate M and N up to $\tilde{M}$and $\tilde{N}$. As $M \restriction \beta = N \restriction \beta$, and we are here dealing with measure filters, no critical point of any measure used in this coiteration is below β. Thus if $\tilde{M} \in \tilde{N}$ then $\tilde{N}$ contains a code for M and hence $N \models$ "β is singular" which contradicts the definition of a K-singularization. Hence $\tilde{M} = \tilde{N}$. By the preservation of standard parameters, p_M and p_M are both mapped to the standard parameter of $\tilde{M}$. Therefore M and N are both the β-core of $\tilde{M}$ and thus equal. Assume there is some $p < p_M$ such that $\{\overline{\beta} < \beta \mid \overline{\beta} = \beta \cap M\{\overline{\beta} \cup p\}\}$ is bounded below β. Let $(Q, \bar{p}) \cong (M\{\beta \cup p\}, p)$ be the transitivization. Then $(Q, \bar{p})$ singularizes β. The uniqueness argument above shows that $Q = M$ and $M = M\{\beta \cup \bar{p}\}$ with $\bar{p} < p_M$ which contradicts the minimality of p_M. □

7. Lifting up K-singularizations

We transfer the upward extensions of embeddings technique to the core model situation:

Theorem 6. *Let $\pi : (J_\beta[\bar{E}], \in) \to (J_{\beta^*}[E], \in)$ be an elementary cofinal map between* ZF$^-$*-models with* $\mathrm{cof}(\beta) \geq \omega_1$ *. Let $M = J_\alpha[\widetilde{E}]$ be a K-singularization of β which end extends $J_\beta[\bar{E}]$, i.e., $\alpha > \beta$ and $\widetilde{E} \restriction \beta = \bar{E} \restriction \beta$. Then there is a uniquely defined structure preserving map $\pi^* : M \to M^*$, $M^* = J_{\alpha^*}[\widetilde{E}^*]$ satisfying:*

a) $\pi^* \restriction J_\beta[\bar{E}] = \pi$, π^* "$p_M = p_{M^*}$;

b) *M^* is the unique K-singularization of β^* satisfying $\widetilde{E}^* \restriction \beta = E \restriction \beta$.*

Proof. The proof of □ in K shows that M can be represented as

$$M = \bigcup_{i<\tau} J_{\alpha_i}[\widetilde{E}]\{\beta_i \cup p_M\}$$

for strictly increasing sequences $(\beta_i)_{i<\tau}$ and $(\alpha_i)_{i<\tau}$ converging to β and α respectively, such that each transitive collapse $\sigma_i : M_i \cong J_{\alpha_i}[\widetilde{E}]\{\beta_i \cup p_M\}$ is the K-singularization of β_i. For $i \leq j < \tau$ let $\sigma_{ij} = \sigma_j^{-1} \circ \sigma_i : M_i \to M_j$. Since β is a cardinal in M and by acceptability, each $M_i \in J_\beta[\bar{E}]$ and each $\sigma_{ij} \in J_\beta[\bar{E}]$. $(M_i)_{i<\tau}, (\sigma_{ij})_{i \leq j < \tau}$ is a directed system of K-singularizations all of whose components are elements of M.

We can now map the directed system pointwise to $J_{\beta^*}[E]$: for $i < \tau$ let $M_i^* = \pi(M_i)$ and $\sigma_{ij}^* = \pi(\sigma_{ij})$. $(M_i^*)_{i<\tau}, (\sigma_{ij}^*)_{i \leq j < \tau}$ is a commutative system of singularizations for the ordinals $\beta_i^* = \pi(\beta_i)$.

(1) The direct limit of $(M_i^*)_{i<\tau}, (\sigma_{ij}^*)_{i \leq j < \tau}$ is well founded.

Proof. The indexing ordinal τ has cofinality $\geq \omega_1$. So any descending ω-sequence in the direct limit is already represented in some M_j^* with $j < \tau$. But M_j^* is transitive. □(1)

Let $M^*, (\sigma_i^*)_{i<\tau}$ be the direct limit of the system $(M_i^*), (\sigma_{ij}^*)$. M^* is a level of a J-hierarchy, say $M^* = J_{\alpha^*}[\widetilde{E}^*]$.

(2) M^* is a mouse.

Proof. This runs similar to the proof of (1): if M^* were not iterable fine structurally then this would be testified in some M_j^* with $j < \tau$. But M_j^* is iterable since M_j is iterable and π is elementary. $\square$(2)

Define the map $\pi^* : M \to M^*$ by $\sigma_i(z) \mapsto \sigma_i^*(\pi(z))$. π^* is a homomorphism by general facts about direct limits. If $z \in J_\beta$, then $\sigma_i(z) = z$ for sufficiently high $i < \tau$, and so

(3) $\pi^* \supseteq \pi$.

(4) $\pi^* : M \to M^*$ is $\in$-cofinal.

Let $p^* = \pi^* \text{``} p_M$. By the direct limit construction:

(5) $M^* = M^*\{\beta^* \cup p^*\}$.

(6) $p^* = p_{M^*}$.

Proof. $p^* \geq p_{M^*}$ If $p^* > p_{M^*}$ then this would be reflected in some M_j^* with $j < \tau$ but then the elementarity of π would yield the contrary. $\square$(6)

(7) (M^*, p_{M^*}) singularizes β^*.

Proof. Take $\delta < \beta$ such that $\{\overline{\beta} < \beta \mid \overline{\beta} = \beta \cap M\{\overline{\beta} \cup p_M\}\} \subseteq \delta$. Set $\delta^* = \pi(\delta)$. We claim that $\{\eta < \beta^* \mid \eta = \beta^* \cap M^*\{\eta \cup p_{M^*}\}\} \subseteq \delta^*$. Let $\eta \geq \delta^*$. Take $\overline{\beta} < \beta$ minimal such that $\pi(\overline{\beta}) \geq \eta$: then $\overline{\beta} \geq \delta$ and $\overline{\beta} \nsubseteq \beta \cap M\{\overline{\beta} \cup p_M\}$. Take a term t and $\vec{x} \subseteq \overline{\beta}$ such that $\overline{\beta} \leq t^M(\vec{x}, p_M) < \beta$. Since π^* is a homomorphism,

$$\eta \leq \pi(\overline{\beta}) \leq t^{M^*}(\pi(\vec{x}), p_{M^*}) < \beta^*, \text{and } \pi(\vec{x}) \subseteq \pi''\overline{\beta} \subseteq \eta.$$

Hence $\eta \neq \beta^* \cap M^*\{\eta \cup p^*\}$, and M^* satisfies the semi-singularity property for β^*. $\square$(7)

The uniqueness of the K-singularization M^* follows from Lemma 2. $\square$

We saw in the previous proof that $\pi^* : M \to M^*$ is cofinal. This again affects the "steering ordinal" α_0 as follows.

Lemma 3. *In the situation of the previous theorem,* $\alpha_0(\beta^*) \leq \pi(\alpha_0(\beta))$ *and thus* $\text{ot}(C_{\beta^*}) \leq \pi(\alpha_0(\beta))$.

8. Getting an inner model with a measurable cardinal

We modify the proof of Theorem 5 to yield the existence of an inner model with a measurable cardinal. We assume $\text{MS}((\kappa_n)_{n<\omega}, \omega_1)$ and work with the DODD-JENSEN core model K under the assumption that there is no inner model with a measurable cardinal. By the Dodd-Jensen covering theorem for K every ordinal $\beta \geq \omega_2$ with $\text{cof}(\beta) \leq \omega_1$ is singular in K. In particular $\kappa_\omega = \sup_{n<\omega} \kappa_n$ is singular in K. Take the sequence $(S_n)_{n<\omega}$ of stationary sets $S_n \subseteq \kappa_n$ as in the proof of Theorem 5. Define the first-order structure $\mathfrak{A} = (H^K_{\kappa_\omega^+}, \ldots)$ in analogy to that proof. The mutual stationarity property yields some $X \prec H^K_{\kappa_\omega^+}$ such that

$$\{\kappa_n \mid n \leq \omega\} \subseteq X, \ \forall n < \omega \ (\sup X \cap \kappa_n) \in S_n, \text{ and } \omega_2 \subseteq X.$$

Let $\pi : (\bar{K}, \in) \cong (X, \in)$ where $\bar{K}$ is transitive, and $\beta_n = \pi^{-1}(\kappa_n)$ for $n \leq \omega$. $\bar{K}$ is a mouse without a total measure. For $n < \omega$ take $M_n = (J_{s_n}[E], p_n)$ to be the K-singularization of β_n (by which we mean the least location s_n in the K-hierarchy where we take $E = E^K$).

Coiterate the mice $\bar{K}$ and M_n. $\bar{K}$ comes out below M_n because M_n has information for singularizing β_n whereas β_n is regular in $\bar{K}$. So in the coiteration there is no truncation on the $\bar{K}$-side and M_n either is an end-extension of $\bar{K}$, or will coiterate up to one.

(1) If M_n is not an end-extension of $\bar{K}$, let $(\lambda_i \mid i \leq \theta)$ be the sequence of critical points of the M_n-side of the coiteration. Then $\lambda_\theta \geq \beta_\omega$ and $\beta_\omega \notin \{\lambda_i \mid i \leq \theta\}$.

Proof. Suppose M_n is not an end-extension of $\bar{K}$. If $\lambda_\theta < \beta_\omega$ then either the M_n-side had a total measure on λ_θ which $\bar{K}$ does not have, or M_n were a proper initial segment of $\bar{K}$. Both possibilities lead to a contradiction.

If $\beta_\omega = \lambda_i$, then the ith iterate M_n^i of M_n would contain $\mathcal{P}(\beta_\omega) \cap \bar{K}$. Since β_ω is singular in $\bar{K}$, M_n^i would contain a cofinal subset of β_ω of small ordertype. But λ_i is regular in M_n^i. $\square$(1)

So the coiterate M_n^θ is the minimal iterate of M_n whose critical point is $> \beta_\omega$, or is M_n itself. In the former case, by (1) there is some maximal $i < \theta$ such that $\lambda_i < \beta_\omega$. Then the iterate M_n^{i+1} is generated from $\lambda_i + 1$ together with some finite parameter, and the critical point of M_n^{i+1} is $> \beta_\omega$. So in this former case, M_n^{i+1} semi-singularizes all β_m such that $\lambda_{i+1} < \beta_m < \beta_n$. However in the latter case, since $M_n = M_n\{\beta_n \cup p_{M_n}\}$ and $On^{M_n} \geq \beta_\omega$ it is clear that M_n itself semi-singularizes all β_m for $m \geq n$. This implies:

(2) For all $n < \omega$ there exists $n' < \omega, n' \geq n$ such that for all $m, n' \leq m < \omega$: $M_m \leq^* M_n$.

Since $\leq^*$ is a pre-wellorder of mice, one can choose a $\leq^*$-minimal element of $\{M_n \mid n < \omega\}$. Choose $n_0 < \omega$ such that

(3) for all $m, n_0 \leq m < \omega : M_m \leq^* M_{n_0}$ and $M_{n_0} \leq^* M_m$.

By the properties of the $\leq^*$-relation:

(4) M_{m+1} is an iterate of M_m, for $m \geq n_0$.

Then $(M_m)_{m \geq n_0}$ is a subsequence of the M_{n_0}-side of the coiteration of $\bar{K}$ and M_{n_0}.

By (1), β_ω is not a critical point in that iteration of M_{n_0}. So there must be some $n_1 < \omega$, $n_1 \geq n_0$, so that

(5) $M_{m+1} = M_m$, for $m \geq n_1$.

Set $M = M_{n_1}$. As in the L-case:

(6) $p_{m+1} \leq_* p_m$, for $m \geq n_1$.

By the wellfoundedness of $\leq_*$ take $n_2 < \omega$, $n_2 \geq n_1$ such that $p = p_{n_2} = p_{n_2+1} = \ldots$. So (M, p) is a common K-singularization of $\beta_{n_2}, \beta_{n_2+1}, \ldots$. We can then conclude the proof by contradiction as in the proof of Theorem 5.

References

[1] K.J. Devlin and R.B. Jensen. Marginalia to a theorem of Silver. In A. Oberschelp G.H. Müller and K. Potthoff, editors, $\models$ *ISILC Logic Conference, 1974*, number 499 in Lecture Notes in Mathematics, pages 115–142. Springer, 1975.

[2] A.J. Dodd. *The Core Model*, volume 61 of *London Mathematical Society Lecture Notes in Mathematics.* Cambridge University Press, Cambridge, 1982.

[3] J. Cummings, M. Foreman and M. Magidor. Canonical structure for the universe of set theory; part 2. Annals of Pure and Applied Logic, published on-line version: February 2006, pp 21.

[4] M. Foreman and M. Magidor. Mutually stationary sequences of sets and the non-saturation of the non-stationary ideal on $\mathcal{P}_\kappa(\lambda)$. *Acta Math.* 186, no. 2: 271–300, 2001.

[5] S.D. Friedman and P. Koepke. An elementary approach to the fine structure of *L*. *Bull. Symb. Logic*, 3, no. 4:453–468, 1997.

[6] R.B. Jensen. The fine structure of the constructible hierarchy. *Annals of Mathematical Logic*, 4:229–308, 1972.

[7] W.J. Mitchell. The Core Model for Sequences of Measures. I. *Math. Proc. Cambridge Philos. Soc.*, 95, no. 2:229–260, 1984.

[8] P.D. Welch. *Combinatorial Principles in the Core Model.* D.Phil. thesis, Oxford University, 1979.

[9] M. Zeman. *Inner Models and Large Cardinals*, volume 5 of *Series in Logic and its Applications.* de Gruyter, Berlin, New York, 2002.

Peter Koepke
Mathematisches Institut der Universität Bonn
Beringstraße 6
D-53115 Bonn, Germany

Philip Welch
School of Mathematics
University of Bristol
Bristol BS8 1TW, England

Set Theory
Trends in Mathematics, 321–344

Part(κ, λ) and Part$^*(\kappa, \lambda)$

Pierre Matet

Abstract. We show that if κ is $\lambda^{<\kappa}$-ineffable (respectively, almost $\lambda^{<\kappa}$-ineffable) then $(NI_{\kappa,\lambda}^{[\lambda]^{<\kappa}} | A)^+ \to ((NS_{\kappa,\lambda}^{[\lambda]^{<\kappa}})^+)^2$ (respectively, $(NAI_{\kappa,\lambda}^{[\lambda]^{<\kappa}})^+ \to (I_{\kappa,\lambda}^+)^2$) for some A, where $NI_{\kappa,\lambda}^{[\lambda]^{<\kappa}}$ (respectively, $NAI_{\kappa,\lambda}^{[\lambda]^{<\kappa}}$) denotes the projection of the non-ineffable (respectively, non-almost ineffable) ideal on $P_\kappa(\lambda^{<\kappa})$, and $NS_{\kappa,\lambda}^{[\lambda]^{<\kappa}}$ the smallest strongly normal ideal on $P_\kappa(\lambda)$.

Mathematics Subject Classification (2000). 03E02, 03E55, 03E35.

Keywords. $P_\kappa(\lambda)$, Part(κ, λ), Part$^*(\kappa, \lambda)$.

0. Introduction

Let κ be a regular uncountable cardinal, and λ be a cardinal with $\lambda \geq \kappa$. Part$^*(\kappa, \lambda)$ (respectively, Part(κ, λ)) asserts that for every $F : P_\kappa(\lambda) \times P_\kappa(\lambda) \to 2$, there is a stationary (respectively, cofinal) subset A of $P_\kappa(\lambda)$ such that F is constant on the set $\{(a, b) \in A \times A : a \subset b\}$. Kamo [16] established that if κ is λ-ineffable and $cf(\lambda) \geq \kappa$, then Part$^*(\kappa, \lambda)$ holds. We show that if κ is $\lambda^{<\kappa}$-ineffable, then Part$^*(\kappa, \lambda)$ holds and in fact $(NI_{\kappa,\lambda}^{[\lambda]^{<\kappa}} \mid A)^+ \to ((NS_{\kappa,\lambda}^{[\lambda]^{<\kappa}})^+)^2$ for some A, where $NS_{\kappa,\lambda}^{[\lambda]^{<\kappa}}$ denotes the smallest strongly normal ideal on $P_\kappa(\lambda)$ and $NI_{\kappa,\lambda}^{[\lambda]^{<\kappa}}$ the projection of the non-ineffable ideal on $P_\kappa(\lambda^{<\kappa})$. Kamo's result follows from ours since by a result of Johnson [13], if κ is λ-ineffable and $cf(\lambda) \geq \kappa$, then $\lambda^{<\kappa} = \lambda$. Concerning converses, we observe that $\{P_\kappa(\lambda)\} \to ((NS_{\kappa,\lambda}^{[\lambda]^{<\kappa}})^+)^2$ does imply that κ is $\lambda^{<\kappa}$-ineffable, but also that it is consistent that "Part$^*(\kappa, \lambda)$ holds and κ is not $\lambda^{<\kappa}$-ineffable". We use a result of Apter and Shelah [3] to establish the consistency of Part$^*(\kappa, \kappa^+)$ at a nonmeasurable cardinal. As for Part(κ, λ), we show that it holds if κ is almost $\lambda^{<\kappa}$-ineffable. In fact $(NAI_{\kappa,\lambda}^{[\lambda]^{<\kappa}} \mid A)^+ \to (I_{\kappa,\lambda}^+)$ for some A, where $NAI_{\kappa,\lambda}^{[\lambda]^{<\kappa}}$ denotes the projection of the non-almost ineffable ideal on $P_\kappa(\lambda)$.

Our results rely heavily on previous work of Baumgartner, Carr, Johnson, Abe and Kamo.

The paper is organized as follows. Sections 1–5 are concerned with various ideals on $P_\kappa(\lambda)$. In Section 1 we recall the respective definitions of the noncofinal ideal $I_{\kappa,\lambda}$, the nonstationary ideal $NS_{\kappa,\lambda}$, the smallest strongly normal ideal $NS_{\kappa,\lambda}^{[\lambda]^{<\kappa}}$ and the non-Shelah ideal $NSh_{\kappa,\lambda}$. Section 2 contains basic material concerning the non-almost-m-ineffable ideal $NAI_{\kappa,\lambda,m}$ and the non-m-ineffable ideal $NI_{\kappa,\lambda,m}$, where $m = 1, 2, \ldots$ $NAI_{\kappa,\lambda,m}^{[\lambda]^{<\kappa}}$ (respectively, $NI_{\kappa,\lambda,m}^{[\lambda]^{<\kappa}}$) denotes the projection of $NAI_{\kappa,\lambda^{<\kappa},m}$ (respectively, $NI_{\kappa,\lambda^{<\kappa},m}$). That is, for $B \subseteq P_\kappa(\lambda)$, B lies in $NAI_{\kappa,\lambda,m}^{[\lambda]^{<\kappa}}$ (respectively, $NI_{\kappa,\lambda,m}^{[\lambda]^{<\kappa}}$) if and only if $\{x \in P_\kappa(\lambda^{<\kappa}) : x \cap \lambda \in B\}$ lies in $NAI_{\kappa,\lambda^{<\kappa},m}$ (respectively, $NI_{\kappa,\lambda^{<\kappa},m}$). Elementary properties of $NAI_{\kappa,\lambda,m}^{[\lambda]^{<\kappa}}$ and $NI_{\kappa,\lambda,m}^{[\lambda]^{<\kappa}}$ are established in Section 3. In Section 4 we prove that if a subset Z of $P_\kappa(\lambda)$ is large with respect to $NAI_{\kappa,\lambda,m}^{[\lambda]^{<\kappa}}$, then there are many (in the sense of the ideal) $a \in Z$ at which the largeness of Z does not reflect (i.e., $Z \cap P_{|a\cap\kappa|}(a) \in NAI_{|a\cap\kappa|,a,m}^{[a]^{<|a\cap\kappa|}}$). We also establish the corresponding result for $NI_{\kappa,\lambda,m}^{[\lambda]^{<\kappa}}$. In Section 5 it is shown that if κ is almost $\lambda^{<\kappa}$-ineffable, then $|a \cap \kappa|$ is a-Shelah for almost all (in the sense of the ideal $NAI_{\kappa,\lambda}^{[\lambda]^{<\kappa}}$) a. It follows that on the large cardinal scale, "κ is λ-Shelah" is strictly below "κ is almost $\lambda^{<\kappa}$-ineffable". In the same spirit we show that if κ is $\lambda^{<\kappa}$-ineffable and almost $(\lambda^{<\kappa}, m)$-ineffable, then $| a \cap \kappa |$ is almost $(| a |^{<|a\cap\kappa|}, m)$-ineffable for almost all (in the sense of $NI_{\kappa,\lambda}^{[\lambda]^{<\kappa}}$) a. This concludes the preliminaries. In Section 6 we prove that there is a set $Q_{\kappa,\lambda}$ in $(NAI_{\kappa,\lambda,m}^{[\lambda]^{<\kappa}})^+$ with the property that if κ is almost $(\lambda^{<\kappa}, m)$-ineffable, then $(NAI_{\kappa,\lambda,m}^{[\lambda]^{<\kappa}} \mid Q_{\kappa,\lambda})^+ \to (I_{\kappa,\lambda}^+)^{m+1}$. In Section 7 we prove that κ is supercompact if and only if $NS_{\kappa,\tau}^* \to (I_{\kappa,\tau}^+)^2$ for every cardinal $\tau \geq \kappa$. In Section 8 we show that there are cases when $(NAI_{\kappa,\lambda,m}^{[\lambda]^{<\kappa}} \mid Q_{\kappa,\lambda})^+ \to (NSS_{\kappa,\lambda}^+)^{m+1}$, where $NSS_{\kappa,\lambda}$ denotes the smallest seminormal ideal on $P_\kappa(\lambda)$, or even $(NAI_{\kappa,\lambda,m}^{[\lambda]^{<\kappa}} \mid Q_{\kappa,\lambda})^+ \to (NS_{\kappa,\lambda}^+)^{m+1}$. In Section 9 we discuss the possibility of replacing $Q_{\kappa,\lambda}$ by a bigger subset of $P_\kappa(\lambda)$. It is shown that if κ is almost $(\lambda^{<\kappa}, m)$-ineffable and there is no Mahlo cardinal τ with $\kappa < \tau \leq \lambda$, then $(NAI_{\kappa,\lambda,m}^{[\lambda]^{<\kappa}})^+ \to (I_{\kappa,\lambda}^+)^{m+1}$. In Section 10 we prove that if κ is $(\lambda^{<\kappa}, m)$-ineffable, then $(NI_{\kappa,\lambda,m}^{[\lambda]^{<\kappa}} \mid Q_{\kappa,\lambda})^+ \to ((NS_{\kappa,\lambda}^{[\lambda]^{<\kappa}})^+)^{m+1}$. We also show that if κ is $(\lambda^{<\kappa}, m)$-ineffable and λ is a strong limit cardinal of cofinality less than κ, then $(NI_{\kappa,\lambda,m}^{[\lambda]^{<\kappa}})^+ \to (NS_{\kappa,\lambda}^+)^{m+1}$. Finally, in Section 11 we establish the consistency of "$\text{Part}^*(\kappa, \kappa^+)^r$ holds for any r with $1 < r < \omega$" at a nonmeasurable cardinal.

1. Ideals

Definition. Given a set A and a cardinal μ, we set $P_\mu(A) = \{B \subseteq A : \mid B \mid < \mu\}$.

Throughout Sections 1, 2 and 3 $\mathcal{L}$ will denote a set with $\kappa \subseteq \mathcal{L}$.

In this section we review basic material concerning the four ideals $I_{\kappa,\mathcal{L}}$, $NS_{\kappa,\mathcal{L}}$, $NS_{\kappa,\mathcal{L}}^{[\mathcal{L}]^{<\kappa}}$ and $NSh_{\kappa,\mathcal{L}}$.

Definition. $I_{\kappa,\mathcal{L}}$ is the set of all $A \subseteq P_\kappa(\mathcal{L})$ such that $\{a \in A : b \subseteq a\} = \phi$ for some $b \in P_\kappa(\mathcal{L})$.

Definition. An ideal on $P_\kappa(\mathcal{L})$ is a collection J of subsets of $P_\kappa(\mathcal{L})$ such that

(i) $P_\kappa(\mathcal{L}) \notin J$,
(ii) $I_{\kappa,\mathcal{L}} \subseteq J$,
(iii) $P(A) \subseteq J$ for every $A \in J$, and
(iv) $A \cup B \in J$ whenever $A, B \in J$.

J is *κ-complete* if $\cup X \in J$ for every $X \subseteq J$ with $\mid X \mid < \kappa$.

Note that $I_{\kappa,\mathcal{L}}$ is a κ-complete ideal on $P_\kappa(\mathcal{L})$.

Definition. Given an ideal J on $P_\kappa(\mathcal{L})$, we let $J^+ = P(P_\kappa(\mathcal{L})) \setminus J$ and $J^* = \{P_\kappa(\mathcal{L}) \setminus B : B \in J\}$.

For $A \in J^+$, we let $J|A = \{B \subseteq P_\kappa(\mathcal{L}) : B \cap A \in J\}$.

Note that $J|A$ is an ideal on $P_\kappa(\mathcal{L})$.

Definition. For an ideal J on $P_\kappa(\mathcal{L})$, $\mathrm{cof}(J)$ is the least cardinality of any $X \subseteq J$ such that $J = \bigcup_{A \in X} P(A)$.

It is well known (see, e.g., [22]) that if $2^{<\kappa} = \kappa$, then $\mathrm{cof}(I_{\kappa,\mathcal{L}}) = |\mathcal{L}|^{<\kappa}$.

Definition. An *ideal* on $P_\kappa(\mathcal{L})$ is *normal* if $\{a \in P_\kappa(\mathcal{L}) : \forall q \in a\ (a \in A_q)\} \in J^*$ whenever $A_q \in J^*$ for $q \in \mathcal{L}$.

Note that every normal ideal on $P_\kappa(\mathcal{L})$ is κ-complete.

Definition. $NS_{\kappa,\mathcal{L}}$ is the set of all $A \subseteq P_\kappa(\mathcal{L})$ such that $\{a \in A : \forall p, q \in a (f(p,q) \subseteq a)\} \subseteq \{\phi\}$ for some $f : \mathcal{L} \times \mathcal{L} \to P_\kappa(\mathcal{L})$.

Lemma 1.1 (Carr [5], Menas [24]). *$NS_{\kappa,\mathcal{L}}$ is the smallest normal ideal on $P_\kappa(\mathcal{L})$.*

Definition. We let $<$ denote the partial order on $P_\kappa(\mathcal{L})$ defined by $a < b$ if and only if $a \in P_{|b \cap \kappa|}(b)$.

Definition. An ideal J on $P_\kappa(\mathcal{L})$ is *strongly normal* if $\{a \in P_\kappa(\mathcal{L}) : \forall e < a (a \in A_e)\} \in J^*$ whenever $A_e \in J^*$ for $e \in P_\kappa(\mathcal{L})$.

Note that every strongly normal ideal on $P_\kappa(\mathcal{L})$ is normal.

Definition. $NS_{\kappa,\mathcal{L}}^{[\mathcal{L}]^{<\kappa}}$ is the set of all $A \subseteq P_\kappa(\mathcal{L})$ such that $\{a \in A : \forall e \in \{\phi\} \cup P_{|a \cap \kappa|}(a)(g(e) \subseteq a)\} = \phi$ for some $g : P_\kappa(\mathcal{L}) \to P_\kappa(\mathcal{L})$.

Lemma 1.2 (Matet [20]). *Suppose κ is Mahlo. Then $NS_{\kappa,\mathcal{L}}^{[\mathcal{L}]^{<\kappa}}$ is a strongly normal ideal on $P_\kappa(\mathcal{L})$. Moreover, it is the smallest such ideal.*

Definition. $E_{\kappa,\mathcal{L}}$ is the set of all $a \in P_\kappa(\mathcal{L})$ such that $a \cap \kappa$ is an uncountable inaccessible cardinal.

Lemma 1.3 (Carr-Pelletier [10]). $P_\kappa(\mathcal{L}) \setminus E_{\kappa,\mathcal{L}} \in NS_{\kappa,\mathcal{L}}^{[\mathcal{L}]^{<\kappa}}$.

Definition. $NSh_{\kappa,\mathcal{L}}$ is the set of all $A \subseteq P_\kappa(\mathcal{L})$ with the property that one can find $g_a : a \to a$ for $a \in A$ so that for every $f : \mathcal{L} \to \mathcal{L}$, there is $b \in P_\kappa(\mathcal{L})$ with $\{a \in A : b \subseteq a \text{ and } f \restriction b = g_a \restriction b\} = \phi$.

κ is *$\mathcal{L}$-Shelah* if $P_\kappa(\mathcal{L}) \notin NSh_{\kappa,\mathcal{L}}$.

Note that κ is $\mathcal{L}$-Shelah if and only if κ is $|\mathcal{L}|$-Shelah. By a result of Carr [6], if κ is $|\mathcal{L}|$-Shelah, then κ is μ-Shelah for every cardinal μ with $\kappa \leq \mu < |\mathcal{L}|$.

Lemma 1.4.

(i) **(folklore)** $NS_{\kappa,\mathcal{L}}^{[\mathcal{L}]^{<\kappa}} \subseteq NSh_{\kappa,\mathcal{L}}$.

(ii) **(Carr [7])** *If κ is $\mathcal{L}$-Shelah, then $NSh_{\kappa,\mathcal{L}}$ is a normal ideal on $P_\kappa(\mathcal{L})$.*

2. $NAI_{\kappa,\mathcal{L},m}$ and $NI_{\kappa,\mathcal{L},m}$

The non-almost-m-ineffable ideal on κ and the non-m-ineffable ideal on κ were introduced by Baumgartner in [4]. In this section we consider two-cardinal versions of these ideals.

Definition. For $A \subseteq P_\kappa(\mathcal{L})$ and $0 < m < \omega$, we let $[A]^m_< = \{(a_1, \dots, a_m) \in A \times \cdots \times A : a_1 < \cdots < a_m\}$.

Definition. Given $0 < m < \omega$, $NAI_{\kappa,\mathcal{L},m}$ (respectively, $NI_{\kappa,\mathcal{L},m}$) is the set of all $A \subseteq P_\kappa(\mathcal{L})$ with the property that one can find $s_{a_1 \dots a_m} \subseteq a_1$ for $(a_1, \dots, a_m) \in [A]^m_<$ so that there does not exist $S \subseteq \mathcal{L}$ and B in $I^+_{\kappa,\mathcal{L}} \cap P(A)$ (respectively, $NS^+_{\kappa,\mathcal{L}} \cap P(A)$) such that $S \cap a_1 = s_{a_1 \dots a_m}$ for every $(a_1, \dots, a_m) \in [B]^m_<$.

κ is *almost $(\mathcal{L}, m)$-ineffable* (respectively, is *$(\mathcal{L}, m)$-ineffable*) if $P_\kappa(\mathcal{L})$ does not lie in $NAI_{\kappa,\mathcal{L},m}$ (respectively, $NI_{\kappa,\mathcal{L},m}$).

κ is *almost $\mathcal{L}$-ineffable* (respectively, is *$\mathcal{L}$-ineffable*) if it is almost $(\mathcal{L}, 1)$-ineffable (respectively, is $(\mathcal{L}, 1)$-ineffable). We let $NAI_{\kappa,\mathcal{L}} = NAI_{\kappa,\mathcal{L},1}$ and $NI_{\kappa,\mathcal{L}} = NI_{\kappa,\mathcal{L},1}$.

The following is essentially due to Carr [7].

Lemma 2.1.

(i) $NSh_{\kappa,\mathcal{L}} \subseteq NAI_{\kappa,\mathcal{L}} \subseteq NAI_{\kappa,\mathcal{L},2} \subseteq NAI_{\kappa,\mathcal{L},3} \subseteq \cdots$

(ii) $NI_{\kappa,\mathcal{L}} \subseteq NI_{\kappa,\mathcal{L},2} \subseteq NI_{\kappa,\mathcal{L},3} \subseteq \cdots$

(iii) $NAI_{\kappa,\mathcal{L},m} \subseteq NI_{\kappa,\mathcal{L},m}$ *for* $m = 1, 2, 3, \dots$

(iv) *Let $0 < m < \omega$. If κ is almost $(\mathcal{L}, m)$-ineffable (respectively, is $(\mathcal{L}, m)$-ineffable), then $NAI_{\kappa,\mathcal{L},m}$ (respectively, $NI_{\kappa,\mathcal{L},m}$) is a normal ideal on $P_\kappa(\mathcal{L})$.*

(v) *Let $0 < m < \omega$. If κ is almost $(\mathcal{L}, m)$-ineffable (respectively, is $(\mathcal{L}, m)$-ineffable), then κ is almost $(\mathcal{L}', m)$-ineffable (respectively, is $(\mathcal{L}', m)$-ineffable) for every $\mathcal{L}'$ with $\kappa \subseteq \mathcal{L}' \subseteq \mathcal{L}$.*

It is simple to see that if $j : \mathcal{L} \to |\mathcal{L}|$ is a bijection and $A \subseteq P_\kappa(\mathcal{L})$, then A lies in $NAI_{\kappa,\mathcal{L},m}$ (respectively, $NI_{\kappa,\mathcal{L},m}$) if and only if $\{j''a : a \in A\}$ lies in $NAI_{\kappa,|\mathcal{L}|,m}$ (respectively, $NI_{\kappa,|\mathcal{L}|,m}$). In particular, κ is almost $(\mathcal{L}, m)$-ineffable (respectively, is $(\mathcal{L}, m)$-ineffable) if and only if it is almost $(|\mathcal{L}|, m)$-ineffable (respectively, is $(|\mathcal{L}|, m)$-ineffable).

If λ is (almost) ineffable, then κ is (almost) λ-ineffable if and only if κ is λ-Shelah. The following is essentially due to Di Prisco and Zwicker [11].

Proposition 2.2. *Suppose λ is ineffable (respectively, almost ineffable) and κ is μ-Shelah for every cardinal μ with $\kappa \leq \mu < \lambda$. Then κ is (λ, m)-ineffable (respectively, almost (λ, m)-ineffable) for $m = 1, 2, 3, \ldots$*

In [2] Abe showed that if $\lambda^{<\kappa} = 2^\lambda$, then $NAI_{\kappa,\lambda} = NI_{\kappa,\lambda}$. His result generalizes easily.

Lemma 2.3 (Baumgartner (see Theorem 2.3 in [12]), Matet-Péan-Shelah [23]).

(i) *Suppose J is a κ-complete ideal on $P_\kappa(\mathcal{L})$ such that $J \subsetneq NS_{\kappa,\mathcal{L}}$ and $\mathrm{cof}(J) \leq |\mathcal{L}|$. Then $J \mid A = I_{\kappa,\mathcal{L}} \mid A$ for some $A \in NS^*_{\kappa,\mathcal{L}}$.*

(ii) *Suppose κ is Mahlo and J is a κ-complete ideal on $P_\kappa(\mathcal{L})$ such that $J \subseteq NS^{[\mathcal{L}]^{<\kappa}}_{\kappa,\mathcal{L}}$ and $\mathrm{cof}(J) \leq |\mathcal{L}|^{<\kappa}$. Then $J|A = I_{\kappa,\mathcal{L}}|A$ for some $A \in (NS^{[\mathcal{L}]^{<\kappa}}_{\kappa,\mathcal{L}})^*$.*

The following is essentially due to Abe [2].

Proposition 2.4. *If $\mathrm{cof}(NS_{\kappa,\mathcal{L}}) = |\mathcal{L}|^{<\kappa}$, then $NAI_{\kappa,\mathcal{L},m} = NI_{\kappa,\mathcal{L},m}$ for $m = 1, 2, \ldots$*

Proof. Use Lemma 2.3. □

Lemma 2.5 (Matet-Péan-Shelah [23]). *If λ is a strong limit cardinal of cofinality less than κ, then $\mathrm{cof}(NS_{\kappa,\lambda}) = \lambda^{<\kappa}$.*

Thus if λ is a strong limit cardinal with $cf(\lambda) < \kappa$, then $NAI_{\kappa,\lambda,m} = NI_{\kappa,\lambda,m}$ for $m = 1, 2, \ldots$

3. $NAI^{[\mathcal{L}]^{<\kappa}}_{\kappa,\mathcal{L},m}$ and $NI^{[\mathcal{L}]^{<\kappa}}_{\kappa,\mathcal{L},m}$

Let $NI^{[\lambda]^{<\kappa}}_{\kappa,\lambda}$ denote the set of all $A \subseteq P_\kappa(\lambda)$ for which one can find $t_a \subseteq P_{|a\cap\kappa|}(a)$ for $a \in A$ so that $\{a \in A : T \cap P_{|a\cap\kappa|}(a) = t_a\} \in NS^{[\lambda]^{<\kappa}}_{\kappa,\lambda}$ for all $T \subseteq P_\kappa(\lambda)$. In [15] Kamo established that $NI^{[\lambda]^{<\kappa}}_{\kappa,\lambda}$ is the projection of $NI_{\kappa,\lambda^{<\kappa}}$. That is, a subset B of $P_\kappa(\lambda)$ lies in $NI^{[\lambda]^{<\kappa}}_{\kappa,\lambda}$ if and only if the set $\{x \in P_\kappa(\lambda^{<\kappa}) : x\cap\lambda \in B\}$ lies in $NI_{\kappa,\lambda^{<\kappa}}$. The main purpose of this section is to prove generalizations of Kamo's result.

Definition. Given $0 < m < \omega$, $NAI^{[\mathcal{L}]^{<\kappa}}_{\kappa,\mathcal{L},m}$ (respectively, $NI^{[\mathcal{L}]^{<\kappa}}_{\kappa,\mathcal{L},m}$) is the set of all $A \subseteq P_\kappa(\mathcal{L})$ with the property that one can find $t_{a_1\ldots a_m} \subseteq P_{|a_1\cap\kappa|}(a_1)$ for

$(a_1,\dots,a_m)\in[A]^m_<$ so that there does not exist $T\subseteq P_\kappa(\mathcal{L})$ and B in $I^+_{\kappa,\mathcal{L}}\cap P(A)$ (respectively, $(NS^{[\mathcal{L}]^{<\kappa}}_{\kappa,\mathcal{L}})^+\cap P(A)$) such that $T\cap P_{|a_1\cap\kappa|}(a_1)=t_{a_1\dots a_m}$ whenever $(a_1,\dots,a_m)\in[B]^m_<$.

We let $NAI^{[\mathcal{L}]^{<\kappa}}_{\kappa,\mathcal{L}}=NAI^{[\mathcal{L}]^{<\kappa}}_{\kappa,\mathcal{L},1}$ and $NI^{[\mathcal{L}]^{<\kappa}}_{\kappa,\mathcal{L}}=NI^{[\mathcal{L}]^{<\kappa}}_{\kappa,\mathcal{L},1}$.

Lemma 3.1.
(i) $NAI^{[\mathcal{L}]^{<\kappa}}_{\kappa,\mathcal{L}}\subseteq NAI^{[\mathcal{L}]^{<\kappa}}_{\kappa,\mathcal{L},2}\subseteq NAI^{[\mathcal{L}]^{<\kappa}}_{\kappa,\mathcal{L},3}\subseteq\cdots$.
(ii) $NI^{[\mathcal{L}]^{<\kappa}}_{\kappa,\mathcal{L}}\subseteq NI^{[\mathcal{L}]^{<\kappa}}_{\kappa,\mathcal{L},2}\subseteq NI^{[\mathcal{L}]^{<\kappa}}_{\kappa,\mathcal{L},3}\subseteq\cdots$.
(iii) *Let* $0<m<\omega$. *Then* $NAI_{\kappa,\mathcal{L},m}\subseteq NAI^{[\mathcal{L}]^{<\kappa}}_{\kappa,\mathcal{L},m}\subseteq NI^{[\mathcal{L}]^{<\kappa}}_{\kappa,\mathcal{L},m}$ *and* $NI_{\kappa,\mathcal{L},m}\subseteq NI^{[\mathcal{L}]^{<\kappa}}_{\kappa,\mathcal{L},m}$.
(iv) *Suppose* $0<m<\omega$ *and* $P_\kappa(\mathcal{L})$ *does not lie in* $NAI^{[\mathcal{L}]^{<\kappa}}_{\kappa,\mathcal{L},m}$ *(respectively,* $NI^{[\mathcal{L}]^{<\kappa}}_{\kappa,\mathcal{L},m}$*). Then* $NAI^{[\mathcal{L}]^{<\kappa}}_{\kappa,\mathcal{L},m}$ *(respectively,* $NI^{[\mathcal{L}]^{<\kappa}}_{\kappa,\mathcal{L},m}$*) is a strongly normal ideal on* $P_\kappa(\mathcal{L})$.

Proof. (i), (ii) and (iii) are straightforward. As for (iv), let us establish that if $P_\kappa(\lambda)\notin NAI^{[\mathcal{L}]^{<\kappa}}_{\kappa,\mathcal{L},m}$, then $NAI^{[\mathcal{L}]^{<\kappa}}_{\kappa,\mathcal{L},m}$ is a strongly normal ideal on $P_\kappa(\mathcal{L})$. The corresponding assertion for $NI^{[\mathcal{L}]^{<\kappa}}_{\kappa,\mathcal{L},m}$ can be proved in the same way. Thus assume $P_\kappa(\lambda)\notin NAI^{[\mathcal{L}]^{<\kappa}}_{\kappa,\mathcal{L},m}$. It is simple to see that $NAI^{[\mathcal{L}]^{<\kappa}}_{\kappa,\mathcal{L},m}$ is an ideal on $P_\kappa(\mathcal{L})$. To show strong normality, let $B_e\in NAI^{[\mathcal{L}]^{<\kappa}}_{\kappa,\mathcal{L},m}$ for $e\in P_\kappa(\mathcal{L})$. For each $e\in P_\kappa(\mathcal{L})$, select $t^e_{a_1\dots a_m}\subseteq P_{|a_1\cap\kappa|}(a_1)$ for $(a_1,\dots,a_m)\in[B_e]^m_<$ so that there does not exist $T\subseteq P_\kappa(\mathcal{L})$ and $Y\in I^+_{\kappa,\mathcal{L}}\cap P(B_e)$ such that $T\cap P_{|a_1\cap\kappa|}(a_1)=t^e_{a_1\dots a_m}$ whenever $(a_1,\dots,a_m)\in[Y]^m_<$. Suppose toward a contradiction that $A\notin NAI^{[\mathcal{L}]^{<\kappa}}_{\kappa,\mathcal{L},m}$, where $A=\{a\in P_\kappa(\mathcal{L}):\exists e<a\ \ (a\in B_e)\}$. Fix a bijection $j:P_\kappa(\mathcal{L})\times 2\to P_\kappa(\mathcal{L})$ and let C be the set of all $a\in E_{\kappa,\mathcal{L}}$ such that $j''(P_{|a\cap\kappa|}(a)\times 2)=P_{|a\cap\kappa|}(a)$. Note that $C\in(NS^{[\mathcal{L}]^{<\kappa}}_{\kappa,\mathcal{L}})^*$, so $A\cap C\notin NAI^{[\mathcal{L}]^{<\kappa}}_{\kappa,\mathcal{L},m}$. Define $f:A\cap C\to P_\kappa(\mathcal{L})$ so that for every $a\in A\cap C$, $f(a)<a$ and $a\in B_{f(a)}$. For $(a_1,\dots,a_m)\in[A\cap C]^m_<$, set

$$w_{a_1\dots a_m}=\{j(f(a_1),0)\}\cup\{j(d,1):d\in t^{f(a_1)}_{a_1\dots a_m}\}.$$

Note that $w_{a_1\dots a_m}\subseteq P_{|a_1\cap\kappa|}(a_1)$. Pick $W\subseteq P_\kappa(\mathcal{L})$ and $X\in I^+_{\kappa,\mathcal{L}}\cap P(A\cap C)$ so that $W\cap P_{|a_1\cap\kappa|}(a_1)=w_{a_1\dots a_m}$ whenever $(a_1,\dots,a_m)\in[X]^m_<$. For $i=0,1$, set $R_i=\{r\in P_\kappa(\mathcal{L}):j(r,i)\in W\}$. It is readily seen that $|R_0|=1$. Let e be the unique element of R_0. Then $X\subseteq B_e$. Moreover, $R_1\cap P_{|a_1\cap\kappa|}(a_1)=t^e_{a_1\dots a_m}$ for every $(a_1,\dots,a_m)\in[X]^m_<$, which brings the desired contradiction. □

Lemma 3.2. *Suppose* $0<m<\omega$, $A\subseteq P_\kappa(\mathcal{L})$ *and* $j:\mathcal{L}\to|\mathcal{L}|$ *is a bijection. Then* A *lies in* $NAI^{[\mathcal{L}]^{<\kappa}}_{\kappa,\mathcal{L},m}$ *(respectively,* $NI^{[\mathcal{L}]^{<\kappa}}_{\kappa,\mathcal{L},m}$*) if and only if* $\{j''a:a\in A\}$ *lies in* $NAI^{[|\mathcal{L}|]^{<\kappa}}_{\kappa,|\mathcal{L}|,m}$ *(respectively,* $NI^{[|\mathcal{L}|]^{<\kappa}}_{\kappa,|\mathcal{L}|,m}$*).*

Proof. Let i denote the inverse of j. To prove the result, use the fact that there is $C\in NS^*_{\kappa,\mathcal{L}}$ such that $j''a=|\mathcal{L}|\cap a$ for all $a\in C$, and $D\in NS^*_{\kappa,|\mathcal{L}|}$ such that

$b = \mathcal{L} \cap i''b$ for all $b \in D$: define C as the set of all $a \in P_\kappa(\mathcal{L})$ such that (a) $j(\alpha) \in a$ for every $\alpha \in a$, and (b) $i(\beta) \in a$ for every $\beta \in |\mathcal{L}| \cap a$, and D as the set of all $b \in P_\kappa(|\mathcal{L}|)$ such that for each $\gamma \in b, j(\gamma) \in b$ and $i(\gamma) \in b \cup (\mathcal{L} \setminus |\mathcal{L}|)$. □

Proposition 3.3. *Let $0 < m < \omega$. Then the following hold:*

(i) *Given $B \subseteq P_\kappa(\lambda), B \in NAI^{[\lambda]^{<\kappa}}_{\kappa,\lambda,m}$ if and only if $\{x \in P_\kappa(\lambda^{<\kappa}) : x \cap \lambda \in B\} \in NAI_{\kappa,\lambda^{<\kappa},m}$.*

(ii) *Given $B \subseteq P_\kappa(\lambda), B \in NI^{[\lambda]^{<\kappa}}_{\kappa,\lambda,m}$ if and only if $\{x \in P_\kappa(\lambda^{<\kappa}) : x \cap \lambda \in B\} \in NI_{\kappa,\lambda^{<\kappa},m}$.*

Proof. (i) Select a bijection $j : P_\kappa(\lambda) \to \lambda^{<\kappa}$. Let A be the set of all $a \in E_{\kappa,\lambda}$ such that $a = \lambda \cap \{j(e) : e < a\}$, and W be the set of all $y \in E_{\kappa,\lambda^{<\kappa}}$ such that $y = \{j(e) : e < y \cap \lambda\}$. Note that $A \in (NS^{[\lambda]^{<\kappa}}_{\kappa,\lambda})^*$ and $W \in (NS^{[\lambda^{<\kappa}]^{<\kappa}}_{\kappa,\lambda^{<\kappa}})^*$. We define a bijection $h : A \to W$ by $h(a) = \{j(e) : e < a\}$. Note that $h(a) \cap \lambda = a$ for all $a \in A$, and $h(y \cap \lambda) = y$ for all $y \in W$. Note further that given $a, b \in A, a < b$ if and only if $h(a) < h(b)$. Now fix $B \subseteq P_\kappa(\lambda)$ and set $X = \{x \in P_\kappa(\lambda^{<\kappa}) : x \cap \lambda \in B\}$. Note that $h''(A \cap B) = W \cap X$. To establish that $X \notin NAI_{\kappa,\lambda^{<\kappa},m}$ if and only if $B \notin NAI^{[\lambda]^{<\kappa}}_{\kappa,\lambda,m}$, it clearly suffices to establish that $W \cap X \notin NAI_{\kappa,\lambda^{<\kappa},m}$ if and only if $A \cap B \notin NAI^{[\lambda]^{<\kappa}}_{\kappa,\lambda,m}$. First, suppose that $W \cap X \in (NAI_{\kappa,\lambda^{<\kappa},m})^+$. Let $t_{a_1 \dots a_m} \subseteq P_{a_1 \cap \kappa}(a_1)$ for $(a_1, \dots, a_m) \in [A \cap B]^m_<$. For $(a_1, \dots, a_m) \in [A \cap B]^m_<$, put $s_{h(a_1) \dots h(a_m)} = j''t_{a_1 \dots a_m}$. Note that $(h(a_1), \dots, h(a_m)) \in [W \cap X]^m_<$ and $s_{h(a_1) \dots h(a_m)} \subseteq h(a_1)$. Now select $S \subseteq \lambda^{<\kappa}$ and $Y \in I^+_{\kappa,\lambda^{<\kappa}} \cap P(W \cap X)$ so that $S \cap y_1 = s_{y_1 \dots y_m}$ for every $(y_1, \dots, y_m) \in [Y]^m_<$. Set $C = \{y \cap \lambda : y \in Y\}$. It is immediate that $C \in I^+_{\kappa,\lambda} \cap P(A \cap B)$. Moreover, $h''C = Y$. So given $(a_1, \dots, a_m) \in [C]^m_<$, we have $j^{-1}(S) \cap P_{a_1 \cap \kappa}(a_1) = t_{a_1 \dots a_m}$, since for each $e \in P_{a_1 \cap \kappa}(a_1), e \in t_{a_1 \dots a_m}$ if and only if $j(e) \in s_{h(a_1) \dots h(a_m)}$ if and only if $j(e) \in S$. Conversely, suppose that $A \cap B \in (NAI^{[\lambda]^{<\kappa}}_{\kappa,\lambda,m})^+$. Let $s_{x_1 \dots x_m} \subseteq x_1$ for $(x_1, \dots, x_m) \in [W \cap X]^m_<$. For $(a_1, \dots, a_m) \in [A \cap B]^m_<$, set $t_{a_1 \dots a_m} = \{e < a_1 : j(e) \in s_{h(a_1) \dots h(a_m)}\}$. We can find $Z \subseteq P_\kappa(\lambda)$ and $D \in I^+_{\kappa,\lambda} \cap P(A \cap B)$ so that $Z \cap P_{a_1 \cap \kappa}(a_1) = t_{a_1 \dots a_m}$ whenever $(a_1, \dots, a_m) \in [D]^m_<$. It is simple to check that $h''D \in I^+_{\kappa,\lambda^{<\kappa}} \cap P(W \cap X)$. Moreover given $(x_1, \dots, x_m) \in [h''D]^m_<$, we get $s_{x_1 \dots x_m} = x_1 \cap j''Z$ since for each $e \in P_\kappa(\lambda)$ with $j(e) \in x_1, j(e) \in s_{x_1 \dots x_m}$ if and only if $e \in t_{x_1 \cap \lambda \cdots x_m \cap \lambda}$ if and only if $e \in Z$.

(ii) Argue as for (i). □

Corollary 3.4. *Suppose $\lambda^{<\kappa} = \lambda$, and let $0 < m < \omega$. Then $NAI^{[\lambda]^{<\kappa}}_{\kappa,\lambda,m} = NAI_{\kappa,\lambda,m}$ and $NI^{[\lambda]^{<\kappa}}_{\kappa,\lambda,m} = NI_{\kappa,\lambda,m}$.*

Corollary 3.5. *$P_\kappa(\mathcal{L})$ does not lie in $NAI^{[\mathcal{L}]^{<\kappa}}_{\kappa,\mathcal{L},m}$ (respectively, $NI^{[\mathcal{L}]^{<\kappa}}_{\kappa,\mathcal{L},m}$) if and only if κ is almost $(|\mathcal{L}|^{<\kappa}, m)$-ineffable (respectively, is $(|\mathcal{L}|^{<\kappa}, m)$-ineffable).*

Proof. By Lemma 3.2 and Proposition 3.3. □

4. Antireflection results

In [2] Abe proved that if $Z \in NI^+_{\kappa,\lambda}$, then $\{a \in E_{\kappa,\lambda} : Z \cap P_{a\cap\kappa}(a) \in NI_{a\cap\kappa,a}\} \in NI^+_{\kappa,\lambda}$. In this section we establish variants of Abe's result.

Proposition 4.1. *Suppose $0 < m < \omega$ and κ is almost $(\lambda^{<\kappa}, m)$-ineffable. Then*

$$\{a \in Z : Z \cap P_{a\cap\kappa}(a) \in NAI^{[a]^{<(a\cap\kappa)}}_{a\cap\kappa,a,m}\} \notin NAI^{[\lambda]^{<\kappa}}_{\kappa,\lambda,m}$$

for every $Z \in (NAI^{[\lambda]^{<\kappa}}_{\kappa,\lambda,m})^+$ with $Z \subseteq E_{\kappa,\lambda}$.

Proposition 4.1 will be established through a sequence of lemmas. But first, let us state the following corollary.

Corollary 4.2. *Suppose $0 < m < \omega$ and κ is almost $(\lambda^{<\kappa}, m)$-ineffable. Then*

$$\{a \in E_{\kappa,\lambda} : a \cap \kappa \textit{ is not almost } (|a|^{<(a\cap\kappa)}, m)\textit{-ineffable}\} \notin NAI^{[\lambda]^{<\kappa}}_{\kappa,\lambda,m}.$$

Definition. For $0 < m < \omega$, $\mathcal{A}_m$ is the set of all ordered pairs (τ, b) such that

(a) τ is a regular uncountable cardinal,
(b) $\tau \subseteq b$, and
(c) τ is almost $(|b|^{<\tau}, m)$-ineffable.

Lemma 4.3. *Suppose that $0 < m < \omega$, $(\tau, b) \in \mathcal{A}_m$ and $W \subseteq E_{\tau,b}$ is such that $\mathcal{C}_W \notin NAI^{[b]^{<\tau}}_{\tau,b}$, where $\mathcal{C}_W$ is the set of all $c \in E_{\tau,b}$ such that $W \cap P_{c\cap\tau}(c) \notin NAI^{[c]^{<(c\cap\tau)}}_{c\cap\tau,c,m}$. Then $W \notin NAI^{[b]^{<\tau}}_{\tau,b,m}$.*

Proof. Let $t_{a_1\ldots a_m} \subseteq P_{a_1\cap\kappa}(a_1)$ for $(a_1, \ldots, a_m) \in [W]^m_<$. For each $c \in \mathcal{C}_W$, select $T_c \subseteq P_{c\cap\tau}(c)$ and $D_c \in I^+_{c\cap\tau,c} \cap P(W)$ so that $T_c \cap P_{a_1\cap\tau}(a_1) = t_{a_1\ldots a_m}$ for every $(a_1, \ldots, a_m) \in [D_c]^m_<$. Select a bijection $j : 2 \times b \to b$ and let $A = \{c \in P_\tau(b) : j''(2 \times c) = c\}$. Now pick $B \in I^+_{\tau,b} \cap P(A \cap \mathcal{C}_W)$ and $Y \subseteq P_\tau(b)$ so that

$$Y \cap P_{c\cap\tau}(c) = \{j''(\{0\} \times e) : e \in T_c\} \cup \{j''(\{1\} \times d) : d \in D_c\}$$

for every $c \in B$. Put $T = \{e \in P_\tau(b) : j''(\{0\} \times e) \in Y\}$ and $D = \{d \in P_\tau(b) : j''(\{1\} \times d) \in Y\}$. Note that for every $c \in B, T \cap P_{c\cap\tau}(c) = T_c$ and $D \cap P_{c\cap\tau}(c) = D_c$. It follows that $D \in I^+_{\tau,b} \cap P(W)$. Given $(a_1, \ldots, a_m) \in [D]^m_<$, pick $c \in B$ so that $a_m \in P_{c\cap\tau}(c)$. Then $(a_1, \ldots, a_m) \in [D_c]^m_<$, so $t_{a_1\ldots a_m} = T_c \cap P_{a_1\cap\tau}(a_1) = T \cap P_{a_1\cap\tau}(a_1)$. □

Definition. Let $0 < m < \omega$. For $(\tau, b) \in \mathcal{A}_m$ and $Z \subseteq E_{\tau,b}$, let $W^{\tau,b}_{Z,m} = \{c \in Z : Z \cap P_{c\cap\tau}(c) \in NAI^{[c]^{<(c\cap\tau)}}_{c\cap\tau,c,m}\}$.

$\mathcal{B}_m$ is the set of all $(\tau, b) \in \mathcal{A}_m$ such that $W^{\tau,b}_{Z,m} \notin NAI^{[b]^{<\tau}}_{\tau,b,m}$ for every $Z \subseteq E_{\tau,b}$ with $Z \notin NAI^{[b]^{<\tau}}_{\tau,b,m}$.

Lemma 4.4. *Let $0 < m < \omega$, and let $(\tau, b) \in \mathcal{A}_m$ be such that $(c \cap \tau, c) \in \mathcal{B}_m$ for every $c \in E_{\tau,b}$ such that $(c \cap \tau, c) \in \mathcal{A}_m$. Then $(\tau, b) \in \mathcal{B}_m$.*

Proof. Fix $Z \subseteq E_{\tau,b}$ with $Z \notin NAI^{[b]^{<\tau}}_{\tau,b,m}$. Set $C = Z \setminus W^{\tau,b}_{Z,m}$. If $C \in NAI^{[b]^{<\tau}}_{\tau,b}$, then clearly $W^{\tau,b}_{Z,m} \notin NAI^{[b]^{<\tau}}_{\tau,b,m}$, so we are done. Now assume that $C \notin NAI^{[b]^{<\tau}}_{\tau,b}$. Given $c \in C$, we have $c \in E_{\tau,b}$ and $Z \cap P_{c\cap\tau}(c) \notin NAI^{[c]^{<(c\cap\tau)}}_{c\cap\tau,c,m}$, so $(c\cap\tau,c) \in \mathcal{A}_m$ and in fact $(c\cap\tau,c) \in \mathcal{B}_m$. Hence, $W^{c\cap\tau,c}_{Z\cap P_{c\cap\tau}(c),m} \notin NAI^{[c]^{<(c\cap\tau)}}_{c\cap\tau,c,m}$. Note that $W^{c\cap\tau,c}_{Z\cap P_{c\cap\tau}(c),m} = W^{\tau,b}_{Z,m} \cap P_{c\cap\kappa}(c)$. It now follows from Lemma 4.3 that $W^{\tau,b}_{Z,m} \notin NAI^{[b]^{<\tau}}_{\tau,b,m}$. □

Lemma 4.5. *Let $0 < m < \omega$. Then $\mathcal{B}_m = \mathcal{A}_m$.*

Proof. Suppose otherwise and select $(\tau,b) \in \mathcal{A}_m \setminus \mathcal{B}_m$. Using Lemma 4.4, define by induction c_k for $k \in \omega$ so that (a) $c_0 = b$, and (b) $c_{k+1} \in E_{c_k\cap\tau,c_k}$ and $(c_{k+1}\cap\tau, c_{k+1}) \in \mathcal{A}_m \setminus \mathcal{B}_m$. Now $c_0 \cap \tau > c_1\cap\tau > c_2\cap\tau > \dots$, which yields the desired contradiction. □

Proposition 4.1 immediately follows from Lemma 4.5.

The proof of the following is similar to that of Proposition 4.1.

Proposition 4.6. *Suppose $0 < m < \omega$ and κ is $(\lambda^{<\kappa}, m)$-ineffable. Then*

$$\{a \in Z : Z \cap P_{a\cap\kappa}(a) \in NI^{[a]^{<(a\cap\kappa)}}_{a\cap\kappa,a,m}\} \notin NI^{[\lambda]^{<\kappa}}_{\kappa,\lambda,m}$$

for every $Z \in (NI^{[\lambda]^{<\kappa}}_{\kappa,\lambda,m})^+$ with $Z \subseteq E_{\kappa,\lambda}$.

Corollary 4.7. *Suppose $0 < m < \omega$ and κ is $(\lambda^{<\kappa}, m)$-ineffable. Then*

$$\{a \in E_{\kappa,\lambda} : a\cap\kappa \textit{ is not } (|a|^{<(a\cap\kappa)}, m)\textit{-ineffable}\} \notin NI^{[\lambda]^{<\kappa}}_{\kappa,\lambda,m}.$$

5. Hierarchy results

In this section we generalize two results. The first one, due to Kamo [14], asserts that the set of all $a \in E_{\kappa,\lambda}$ such that $a\cap\kappa$ is not almost a-ineffable lies in $NI^{[\lambda]^{<\kappa}}_{\kappa,\lambda}$. The second result is due to Abe [2]. It asserts that if $cf(\lambda) \geq \kappa$, then the set of all $a \in E_{\kappa,\lambda}$ such that $a\cap\kappa$ is not a-Shelah lies in $NAI_{\kappa,\lambda}$.

Proposition 5.1. *Suppose that κ is $\lambda^{<\kappa}$-ineffable and almost $(\lambda^{<\kappa}, m)$-ineffable, where $0 < m < \omega$. Then*

$$\{b \in E_{\kappa,\lambda} : b\cap\kappa \textit{ is almost } (|b|^{<(b\cap\kappa)}, m)\textit{-ineffable}\} \in (NI^{[\lambda]^{<\kappa}}_{\kappa,\lambda})^*.$$

Proof. Let A be the set of all $b \in E_{\kappa,\lambda}$ such that $b\cap\kappa$ is not almost $(|b|^{<(b\cap\kappa)}, m)$-ineffable. Suppose toward a contradiction that $A \notin NI^{[\lambda]^{<\kappa}}_{\kappa,\lambda}$. Pick a bijection $j : \lambda\times(m+1) \to \lambda$, and let B the set of all $b \in P_\kappa(\lambda)$ such that $j''(b\times(m+1)) = b$. Note that $B \in NS^*_{\kappa,\lambda}$. Given $b \in A\cap B$, pick $t^b_{a_1\dots a_m} \subseteq P_{|a_1\cap\kappa|}(a_1)$ for $(a_1,\dots,a_m) \in [P_{b\cap\kappa}(b)]^m_<$ so that there does not exist $T \subseteq P_{b\cap\kappa}(b)$ and $W \in I^+_{b\cap\kappa,b}$

such that $t^b_{a_1\dots a_m} = T \cap P_{|a_1\cap\kappa|}(a_1)$ for all $(a_1,\dots,a_m) \in [W]^m_<$. For $(v_0,\dots,v_m) \in [P_\kappa(\lambda)]^{m+1}_<$, set $u_{v_0\dots v_m} = \{j(\alpha,k) : k < m+1 \text{ and } \alpha \in v_k\}$. Now put

$$r_b = \{u_{ea_1\dots a_m} : (a_1,\dots,a_m) \in [P_{b\cap\kappa}(b)]^m_< \text{ and } e \in t^b_{a_1\dots a_m}\}.$$

Note that $r_b \subseteq P_{b\cap\kappa}(b)$. Select $R \subseteq P_\kappa(\lambda)$ and $C \in (NS^{[\lambda]^{<\kappa}}_{\kappa,\lambda})^+ \cap P(A \cap B)$ so that $R \cap P_{b\cap\kappa}(b) = r_b$ for every $b \in C$. For $(a_1,\dots,a_m) \in [A \cap B]^m_<$, set $s_{a_1\dots a_m} = \{e \in P_{|a_1\cap\kappa|}(a_1) : u_{ea_1\dots a_m} \in R\}$. Since $A \notin NAI^{[\lambda]^{<\kappa}}_{\kappa,\lambda,m}$ by Corollary 4.2, one can find $S \subseteq P_\kappa(\lambda)$ and $D \in I^+_{\kappa,\lambda} \cap P(A \cap B)$ so that $S \cap P_{|a_1\cap\kappa|}(a_1) = s_{a_1\dots a_m}$ for every $(a_1,\dots,a_m) \in [D]^m_<$. Let X be the set of all $b \in E_{\kappa,\lambda}$ such that $D \cap P_{b\cap\kappa}(b) \in I^+_{b\cap\kappa,b}$. It is simple to see that $X \in (NS^{[\lambda]^{<\kappa}}_{\kappa,\lambda})^*$. Now pick $b \in C \cap X$. Clearly, $b \in A \cap B$ and $D \cap P_{b\cap\kappa}(b) \in I^+_{b\cap\kappa,b}$. Moreover for each $(a_1,\dots,a_m) \in [D \cap P_{b\cap\kappa}(b)]^m_<$,

$$\begin{aligned} S \cap P_{|a_1\cap\kappa|}(a_1) = s_{a_1\dots a_m} &= \{e \in P_{|a_1\cap\kappa|}(a_1) : u_{ea_1\dots a_m} \in R\} \\ &= \{e \in P_{|a_1\cap\kappa|}(a_1) : u_{ea_1\dots a_m} \in r_b\} = t^b_{a_1\dots a_m}. \end{aligned}$$

Contradiction! □

Definition. $NSu^{[\lambda]^{<\kappa}}_{\kappa,\lambda}$ is the set of all $A \subseteq P_\kappa(\lambda)$ such that one can find $D \in (NS^{[\lambda]^{<\kappa}}_{\kappa,\lambda})^*$ and $s_a \subseteq P_{|a\cap\kappa|}(a)$ for $a \in A$ so that $s_b \cap P_{|a\cap\kappa|}(a) \neq s_a$ whenever $(a,b) \in [A \cap D]^2_<$.

The following is readily checked.

Lemma 5.2. $NSu^{[\lambda]^{<\kappa}}_{\kappa,\lambda} \subseteq NAI^{[\lambda]^{<\kappa}}_{\kappa,\lambda}$.

Definition. $R_{\kappa,\lambda}$ is the set of all $a \in E_{\kappa,\lambda}$ such that $a \cap \kappa$ is not $|a|$-Shelah.

Lemma 5.3 (Abe [2]). $R_{\kappa,\lambda} \in NSu^{[\lambda]^{<\kappa}}_{\kappa,\lambda}$.

The following is immediate from Lemmas 5.2 and 5.3.

Proposition 5.4. $R_{\kappa,\lambda} \in NAI^{[\lambda]^{<\kappa}}_{\kappa,\lambda}$.

6. $(NAI^{[\lambda]^{<\kappa}}_{\kappa,\lambda,m}|A)^+ \to (I^+_{\kappa,\lambda})^{m+1}$

Definition. For $A \subseteq P_\kappa(\lambda)$ and $0 < r < \omega$, we let $[A]^r = \{(a_1,\dots,a_r) \in A \times \cdots \times A : a_1 \subset \cdots \subset a_r\}$.

Definition. Given $X, Y \subseteq P(P_\kappa(\lambda))$ and $0 < r < \omega, X \to (Y)^r$ means that for every $F : [P_\kappa(\lambda)]^r \to 2$ and every $A \in X$, there is $B \in Y \cap P(A)$ such that F is constant on $[Y]^r$.

In this section we prove that if κ is almost $(\lambda^{<\kappa}, m)$-ineffable, then there is a set $Q_{\kappa,\lambda}$ such that $(NAI^{[\lambda]^{<\kappa}}_{\kappa,\lambda,m}|Q_{\kappa,\lambda})^+ \to (I^+_{\kappa,\lambda})^{m+1}$. Our first concern is to define $Q_{\kappa,\lambda}$.

Definition. ρ_λ denotes the largest strong limit cardinal μ such that $\mu \leq \lambda$.

Definition. Let τ be an infinite cardinal. We define $\psi(\tau, \delta)$ by induction on $\delta \in \text{On}$ by

(i) $\psi(\tau, 0) = \tau$,
(ii) $\psi(\tau, \delta + 1) = 2^{\psi(\tau,\delta)}$, and
(iii) $\psi(\tau, \delta) = \bigcup_{\gamma<\delta} \psi(\tau, \gamma)$ if δ is an infinite limit ordinal.

Note that if δ is an infinite limit ordinal, then $\psi(\tau, \delta)$ is a strong limit cardinal of cofinality $cf(\delta)$.

Definition. n_λ denotes the unique $p \in \omega$ such that $\psi(\rho_\lambda, p) \leq \lambda < \psi(\rho_\lambda, p + 1)$.

Note that $\psi(\rho_\lambda, n_\lambda) = \psi(\kappa, \delta)$ for some $\delta \in \text{On}$.

Definition. Assuming κ is inaccessible, we define $h_{\kappa,\lambda} : E_{\kappa,\lambda} \to \kappa$ by $h_{\kappa,\lambda}(a) =$ the least strong limit cardinal τ such that $\tau > |a|$, i.e., $h_{\kappa,\lambda}(a) = \psi(|a|, \omega)$.

Definition. $Q_{\kappa,\lambda}$ is the set of all $a \in E_{\kappa,\lambda}$ such that $a \cap \kappa$ is not $\psi(h_{\kappa,\lambda}(a), n_\lambda)$-Shelah.

We need to show that if κ is almost $(\lambda^{<\kappa}, m)$-ineffable, then $Q_{\kappa,\lambda} \in (NAI^{[\lambda]^{<\kappa}}_{\kappa,\lambda,m})^+$.

Definition. κ is *λ-supercompact* if there exists a prime normal ideal on $P_\kappa(\lambda)$.
κ is *supercompact* if it is τ-supercompact for every cardinal $\tau \geq \kappa$.

The following is essentially due to Magidor [19].

Lemma 6.1. *Suppose κ is λ-supercompact, J is a prime normal ideal on $P_\kappa(\lambda)$ and $0 < m < \omega$. Then $NI^{[\lambda]^{<\kappa}}_{\kappa,\lambda,m} \subseteq J$. In fact, given $t_{a_1 \ldots a_m} \subseteq P_{|a_1 \cap \kappa|}(a_1)$ for $(a_1, \ldots, a_m) \in [P_\kappa(\lambda)]^m_<$, there is $T \subseteq P_\kappa(\lambda)$ and $B \in J^*$ such that $T \cap P_{|a_1 \cap \kappa|}(a_1) = t_{a_1 \ldots a_m}$ whenever $(a_1, \ldots, a_m) \in [B]^m_<$.*

Lemma 6.2 (Carr [7]). *If κ is $2^{\lambda^{<\kappa}}$-Shelah, then κ is λ-supercompact.*

Lemma 6.3. *Suppose $0 < m < \omega$ and κ is almost $(\lambda^{<\kappa}, m)$-ineffable. Then $Q_{\kappa,\lambda} \notin NAI^{[\lambda]^{<\kappa}}_{\kappa,\lambda,m}$.*

Proof. If $a \in E_{\kappa,\lambda}$ and $a \cap \kappa$ is $h_{\kappa,\lambda}(a)$-Shelah, then $a \cap \kappa$ is $2^{|a|^{<(a\cap\kappa)}}$-Shelah and hence $|a|$-supercompact by Lemma 6.2. Thus by Lemma 6.1,

$$\{a \in E_{\kappa,\lambda} : a \cap \kappa \text{ is not almost } (|a|^{<(a\cap\kappa)}, m)\text{-ineffable}\} \subseteq Q_{\kappa,\lambda}.$$

That $Q_{\kappa,\lambda} \notin NAI^{[\lambda]^{<\kappa}}_{\kappa,\lambda}$ now follows from Corollary 4.2. □

It remains to show that $(NAI^{[\lambda]^{<\kappa}}_{\kappa,\lambda,m} \mid Q_{\kappa,\lambda})^+ \to (I^+_{\kappa,\lambda})^{m+1}$. The proof will proceed through a sequence of lemmas. Let us first recall a few facts.

Lemma 6.4. *Let $\mu \leq \lambda$ be an infinite cardinal. Then the following hold:*
(i) $\{a \in P_\kappa(\lambda) : |a \cap 2^\mu] \leq 2^{|a\cap\mu|}\} \in NS^*_{\kappa,\lambda}$.
(ii) **(Abe [1])** *If* $2^\mu \leq \lambda$, *then* $\{a \in P_\kappa(\lambda) : |a \cap 2^\mu| < 2^{|a\cap\mu|}\} \in NSh_{\kappa,\lambda}$.

Lemma 6.5 (Abe [1]). *Let $\mu \leq \lambda$ be an infinite strong limit cardinal. Then*

$$\{a \in P_\kappa(\lambda) : |a \cap \mu| \textit{ is not a strong limit cardinal}\} \in NSh_{\kappa,\lambda}.$$

Lemma 6.6. *Suppose κ is λ-Shelah. Then one can find $A \in NSh^*_{\kappa,\lambda}$ and $f : \kappa \to \kappa$ so that $|a \cap \psi(\rho_\lambda, n_\lambda)| = f(a \cap \kappa)$ for all $a \in (A \cap Q_{\kappa,\lambda}) \setminus R_{\kappa,\lambda}$.*

Proof. Let A be the set of all $a \in E_{\kappa,\lambda}$ such that
(a) $|a| \leq 2^{|a\cap\psi(\rho_\lambda,n_\lambda)|}$,
(b) $|a \cap \psi(\rho_\lambda, q+1)| = 2^{|a\cap\psi(\rho_\lambda,q)|}$ for every $q < n_\lambda$, and
(c) $|a \cap \rho_\lambda|$ is a strong limit cardinal.

Then $A \in NSh^*_{\kappa,\lambda}$ by Lemmas 6.4 and 6.5. Now assume toward a contradiction that one can find $a, b \in (A\cap Q_{\kappa,\lambda})\setminus R_{\kappa,\lambda}$ so that $a \cap \kappa = b \cap \kappa$ and $|a\cap\psi(\rho_\lambda, n_\lambda)| \neq |b\cap\psi(\rho_\lambda, n_\lambda)|$. Then $|a\cap\rho_\lambda| < |b\cap\rho_\lambda|$, and so $|a| < |b\cap\rho_\lambda|$. It follows that $h_{\kappa,\lambda}(a) \leq |b \cap \rho_\lambda|$. But $b \cap \kappa$ is $|b \cap \psi(\rho_\lambda, n_\lambda)|$-Shelah and $|b\cap\psi(\rho_\lambda, n_\lambda)| = \psi(|b \cap \rho_\lambda|, n_\lambda)$, so $a \cap \kappa$ is $\psi(h_{\kappa,\lambda}(a), n_\lambda)$-Shelah. Contradiction! □

Lemma 6.7 (Johnson [13]). *Suppose κ is λ-Shelah, and let μ be a cardinal with $\kappa \leq \mu \leq \lambda$. Then there is $A \in NSh^*_{\kappa,\lambda}$ such that $|a \cap \mu| < |b \cap \mu|$ for all $a, b \in A$ with $a \cap \mu \subset b \cap \mu$.*

Lemma 6.8. *Suppose κ is λ-Shelah. Then there is $B \in NSh^*_{\kappa,\lambda}$ such that $a < b$ for all $a, b \in (B \cap Q_{\kappa,\lambda}) \setminus R_{\kappa,\lambda}$ with $a \cap \psi(\rho_\lambda, n_\lambda) \subset b \cap \psi(\rho_\lambda, n_\lambda)$.*

Proof. Let A and f be as in the statement of Lemma 6.6. By Lemma 6.7, there is $C \in NSh^*_{\kappa,\lambda}$ such that $|a \cap \psi(\rho_\lambda, n_\lambda)| < |b \cap \psi(\rho_\lambda, n_\lambda)|$ for all $a, b \in C$ with $a \cap \psi(\rho_\lambda, n_\lambda) \subset b \cap \psi(\rho_\lambda, n_\lambda)$. Let D be the set of all $a \in E_{\kappa,\lambda}$ such that $|a| \leq 2^{|a\cap\psi(\rho_\lambda,n_\lambda)|}$, and W be the set of all $b \in P_\kappa(\lambda)$ such that $2^{|f(\alpha)|} < |b \cap \kappa|$ for all $\alpha \in b \cap \kappa$. Then W lies in $NS^*_{\kappa,\lambda}$ and by Lemma 6.4, so does D. Now set $B = A\cap C\cap D\cap W$. Given $a, b \in (B\cap Q_{\kappa,\lambda})\setminus R_{\kappa,\lambda}$ with $a\cap\psi(\rho_\lambda, n_\lambda) \subset b\cap\psi(\rho_\lambda, n_\lambda)$, we must have $a \cap \kappa < b \cap \kappa$, and hence $|a| \leq 2^{f(a\cap\kappa)} < b \cap \kappa$. □

The following is essentially due to Kamo (see the proof of Theorem 5.1 in [16]).

Lemma 6.9. *Suppose $0 < m < \omega$ and κ is almost $(\lambda^{<\kappa}, m)$-ineffable. Let ν be a cardinal with $\nu < \lambda < 2^\nu$, and $A \in (NAI^{[\lambda]^{<\kappa}}_{\kappa,\lambda,m})^+$ be such that $|a \cap \nu| < |b \cap \kappa|$ for all $a, b \in A$ with $a \cap \nu \subset b \cap \nu$. Then $\{A\} \to (I^+_{\kappa,\lambda})^{m+1}$.*

Proof. By Lemma 6.7 there is $B \in NSh^*_{\kappa,\lambda}$ such that $|a| < |b|$ for every $(a, b) \in [B]^2$. Let C be the set of all $a \in A \cap B \cap E_{\kappa,\lambda}$ such that $|a| \leq 2^{|a\cap\nu|}$. Clearly, $C \in (NAI^{[\lambda]^{<\kappa}}_{\kappa,\lambda,m})^+$. Moreover, $a < b$ for all $a, b \in C$ with $a \cap \nu \subset b \cap \nu$. Select a bijection $j_\rho : 2^\rho + 1 \to P(\rho)$ for each infinite cardinal $\rho < \kappa$, and a bijection

$i_e : |e| \to e$ for each $e \in P_\kappa(\nu)$. For $a \in C$, set $s_a = \{i_{a\cap\nu}(\zeta) : \zeta \in j_{|a\cap\nu|}(|a|)\}$. Note that $s_a \neq s_b$ for every $(a, b) \in [C]^2$ with $a \cap \nu = b \cap \nu$.

Now let $F : [C]^{m+1} \to 2$. For $(a_1, \ldots, a_m) \in [C]^m_<$, set

$$t_{a_1 \ldots a_m} = \{c \in C \cap P_{a_1\cap\kappa}(a_1) : F(c, a_1, \ldots, a_m) = 0\} \cup \{\{\delta\} : \delta \in s_{a_1}\}.$$

Pick $D \in I^+_{\kappa,\lambda} \cap P(C)$ and $T \subseteq P_\kappa(\lambda)$ so that $T \cap P_{a_1\cap\kappa}(a_1) = t_{a_1,\ldots,a_m}$ whenever $(a_1, \ldots, a_m) \in [D]^m_<$. Put $X = \{x \in T : |x| = 1\}, Y = \cup X$ and $Z = T \setminus X$. Then $Y \subseteq \nu$. Moreover, $(*)$ $Y \cap a = s_a$ for every $a \in D$, and $(**)$ $Z \cap P_{a_1\cap\kappa}(a_1) = \{c \in C \cap P_{a_1\cap\kappa}(a_1) : F(c, a_1, \ldots, a_m) = 0\}$ whenever $(a_1, \ldots, a_m) \in [D]^m_<$. It follows that (i) $a < b$ for every $(a, b) \in [D]^2$, and (ii) there is $g : D \to 2$ such that $F(a_0, \ldots, a_m) = g(a_0)$ whenever $(a_0, \ldots, a_m) \in [D]^{m+1}$. Select $W \in I^+_{\kappa,\lambda} \cap P(D)$ so that g is constant on W. Then F is constant on $[W]^{m+1}$. □

The following is essentially due to Carr [8].

Lemma 6.10. *Suppose $0 < m < \omega$ and κ is almost $(\lambda^{<\kappa}, m)$-ineffable, and let $F : [P_\kappa(\lambda)]^{m+1}_< \to 2$ and $A \in (NAI^{[\lambda]^{<\kappa}}_{\kappa,\lambda,m})^+$. Then there is $B \in I^+_{\kappa,\lambda} \cap P(A)$ such that F is constant on $[B]^{m+1}_<$.*

Proposition 6.11. *Suppose $0 < m < \omega$ and κ is almost $(\lambda^{<\kappa}, m)$-ineffable. Then $(NAI^{[\lambda]^{<\kappa}}_{\kappa,\lambda,m}|Q_{\kappa,\lambda})^+ \to (I^+_{\kappa,\lambda})^{m+1}$.*

Proof. If $\psi(\rho_\lambda, n_\lambda) = \lambda$, the conclusion follows from Lemmas 6.8 and 6.10. Otherwise, use Lemmas 6.8 and 6.9. □

7. Supercompactness

The conclusion of Proposition 6.11 is not as weak as one might think. In fact, we will show that if for every cardinal $\tau \geq \kappa$, there is a normal ideal J on $P_\kappa(\tau)$ such that $J^+ \to (I^+_{\kappa,\tau})^2$, then κ is a supercompact cardinal.

Proposition 7.1 (Magidor [19], Menas [25]). *κ is supercompact if and only if* Part$^*(\kappa, \tau)$ *holds for every cardinal $\tau \geq \kappa$.*

Definition. Let δ be an ordinal with $\kappa \leq \delta < \lambda$. An ideal J on $P_\kappa(\lambda)$ is *δ-normal* if $\{a \in P_\kappa(\lambda) : \forall\alpha \in a \cap \delta \ \ (a \in A_\alpha)\} \in J^*$ whenever $A_\alpha \in J^*$ for $\alpha \in \delta$.

$NS^\delta_{\kappa,\lambda}$ denotes the smallest δ-normal ideal on $P_\kappa(\lambda)$.

Lemma 7.2 (Matet-Péan-Shelah [22]). *Let μ be a cardinal with $\kappa \leq \mu < \lambda$. Then* $\mathrm{cof}(NS^\mu_{\kappa,\lambda}) \leq \lambda^\mu$.

Lemma 7.3. *Suppose μ is a cardinal such that $\kappa \leq \mu < \lambda$ and $\{P_\kappa(\lambda)\} \to ((NS^\mu_{\kappa,\lambda})^+)^2$. Then κ is μ-Shelah.*

Proof. Let $g_c : c \to c$ for $c \in P_\kappa(\mu)$. Define $F : [P_\kappa(\lambda)]^2 \to 2$ by $F(a, b) = 0$ if and only if (a) there exists $\alpha \in a$ such that $g_{a\cap\mu}(\alpha) \neq g_{b\cap\mu}(\alpha)$, and (b) for the least such α, $g_{a\cap\mu}(\alpha) < g_{b\cap\mu}(\alpha)$. Then F is constant on $[A]^2$ for some $A \in (NS^\mu_{\kappa,\lambda})^+$. Using induction, we define $\xi_\beta < \mu$ for $\beta < \mu$ so that $\{a \in A : g_{a\cap\mu}(\beta) \neq \xi_\beta\} \in NS^\mu_{\kappa,\lambda}$. Suppose ξ_γ has already been defined for each $\gamma < \beta$. For $\zeta < \mu$, set $X^\beta_\zeta = \{a \in P_\kappa(\lambda) : g_{a\cap\mu}(\beta) = \zeta\}$. There is $\xi_\beta < \mu$ such that $A \cap X^\beta_{\xi_\beta} \notin NS^\mu_{\kappa,\lambda}$. Suppose to the contrary that $A \setminus X^\beta_{\xi_\beta} \notin NS^\mu_{\kappa,\lambda}$. Then there is $\zeta < \mu$ such that $(A \cap X^\beta_\zeta) \setminus X^\beta_{\xi_\beta} \notin NS^\mu_{\kappa,\lambda}$. Let Y be the set of all $a \in A$ such that $g_{a\cap\mu}(\gamma) \neq \xi_\gamma$ for some $\gamma \in a \cap \beta$. Note that $Y \in NS^\mu_{\kappa,\lambda}$. Now pick $a, b' \in (A \cap X^\beta_{\xi_\beta}) \setminus Y$ and $a', b \in (A \cap X^\beta_\zeta) \setminus (Y \cup X^\beta_{\xi_\beta})$ so that $\beta \in a, \beta \in a', a \subset b$ and $a' \subset b'$. Then $F(a, b) \neq F(a', b')$, which brings the desired contradiction.

Define $f : \mu \to \mu$ by $f(\beta) = \xi_\beta$. Given $d \in P_\kappa(\mu)$, select $a \in A$ so that $d \subseteq a$ and for every $\beta \in d$, $g_{a\cap\mu}(\beta) = \xi_\beta$. Then $d \subseteq a \cap \mu$ and $g_{a\cap\mu} \restriction d = f \restriction d$. □

Proposition 7.4. *The following are equivalent:*

(i) *κ is supercompact.*

(ii) *For every cardinal $\tau \geq \kappa$, $NS^*_{\kappa,\tau} \to (I^+_{\kappa,\tau})^2$.*

(iii) *For every cardinal $\mu \geq \kappa$, there is a cardinal $\tau > \mu$ such that $\{P_\kappa(\tau)\} \to ((NS^\mu_{\kappa,\tau})^+)^2$.*

Proof. (i) $\to$ (ii): By Proposition 7.1.

(ii) $\to$ (iii): Suppose (ii) holds, and fix a cardinal $\mu \geq \kappa$. Set $\tau = 2^\mu$. By Lemmas 2.3 and 7.2, there is $A \in NS^*_{\kappa,\tau}$ such that $NS^\mu_{\kappa,\tau} \mid A = I_{\kappa,\tau} \mid A$. Now $\{A\} \to (I^+_{\kappa,\tau})^2$, so $\{P_\kappa(\tau)\} \to ((NS^\mu_{\kappa,\tau})^+)^2$.

(iii) $\to$ (i): By Lemmas 6.2 and 7.3. □

8. $(NAI^{[\lambda]^{<\kappa}}_{\kappa,\lambda,m} \mid A)^+ \to (K^+)^{m+1}$

Under some assumptions we may strengthen the conclusion of Proposition 6.11. Let us first consider the possibility of replacing $I_{\kappa,\lambda}$ by a bigger ideal. We will use the following simple observation. Suppose (a) J is an ideal on $P_\kappa(\lambda)$ such that $J^+ \to (I^+_{\kappa,\lambda})^{m+1}$, where $0 < m < \omega$, and (b) K is an ideal on $P_\kappa(\lambda)$ such that $K \mid A = I_{\kappa,\lambda} \mid A$ for some $A \in J^*$. Then $J^+ \to (K^+)^{m+1}$. In particular, if J is a normal ideal on $P_\kappa(\lambda)$ such that $J^+ \to (I^+_{\kappa,\lambda})^{m+1}$, and μ is a cardinal such that $\kappa \leq \mu < \lambda$ and $\mathrm{cof}(NS^\mu_{\kappa,\lambda}) \leq \lambda$, then by Lemma 2.3, $J^+ \to ((NS^\mu_{\kappa,\lambda})^+)^{m+1}$.

Definition. An ideal J on $P_\kappa(\lambda)$ is *seminormal* if it is δ-normal for every ordinal δ with $\kappa \leq \delta < \lambda$.

$NSS_{\kappa,\lambda}$ denotes the smallest seminormal ideal on $P_\kappa(\lambda)$.

If $cf(\lambda) < \kappa$, then every seminormal ideal on $P_\kappa(\lambda)$ is normal, so $NSS_{\kappa,\lambda} = NS_{\kappa,\lambda}$. If $cf(\lambda) \geq \kappa$ and $\lambda > \kappa$, then by a result of [22], $NSS_{\kappa,\lambda} = \bigcup_{\kappa \leq \delta < \lambda} NS^{\delta}_{\kappa,\lambda}$. Note that if $\kappa \leq cf(\lambda) < \lambda$, then there exists $A \in NS^*_{\kappa,\lambda}$ such that $NS_{\kappa,\lambda} = NSS_{\kappa,\lambda} \mid A$.

By Lemma 2.3, if $\mathrm{cof}(NSS_{\kappa,\lambda}) \leq \lambda$ and J is a normal ideal on $P_\kappa(\lambda)$ such that $J^+ \longrightarrow (I^+_{\kappa,\lambda})^{m+1}$, then $J^+ \longrightarrow (NSS^+_{\kappa,\lambda})^{m+1}$.

Proposition 8.1.

(i) **(Johnson [12])** *If* $\lambda^{<\lambda} = \lambda$, *then* $\mathrm{cof}(NSS_{\kappa,\lambda}) = \lambda$.

(ii) **(Matet-Péan-Shelah [22])** *For each* $n \in \omega$, $\mathrm{cof}(NSS_{\kappa,\kappa^{+(n+1)}}) = \mathrm{cof}(NS_{\kappa,\kappa^{+n}})$.

By a result of Landver [18], $\mathrm{cof}(NS_{\kappa,\kappa}) = \mathfrak{d}_\kappa$, where $\mathfrak{d}_\kappa$ is the least cardinality of any family F of functions from κ to κ with the property that for every $g : \kappa \longrightarrow \kappa$, there is $f \in F$ such that $g(\alpha) \leq f(\alpha)$ for every $\alpha < \kappa$. Hence by Proposition 8.1, $\mathrm{cof}(NSS_{\kappa,\kappa^+}) = \mathfrak{d}_\kappa$.

Definition. Let $\theta < \kappa$ be an uncountable cardinal. An ideal J on $P_\kappa(\lambda)$ is $[\lambda]^{<\theta}$-*normal* if $\{a \in P_\kappa(\lambda) : \forall e \in P_\theta(a)\ (a \in A_e)\} \in J^*$ whenever $A_e \in J^*$ for $e \in P_\theta(\lambda)$.

If there exists a $[\lambda]^{<\theta}$-normal ideal on $P_\kappa(\lambda)$, then $NS^{[\lambda]^{<\theta}}_{\kappa,\lambda}$ denotes the smallest such ideal.

By Lemma 2.3, if $\mathrm{cof}(NS^{[\lambda]^{<\theta}}_{\kappa,\lambda}) \leq \lambda^{<\kappa}$ and J is a strongly normal ideal on $P_\kappa(\lambda)$ such that $J^+ \longrightarrow (I^+_{\kappa,\lambda})^{m+1}$, then $J^+ \longrightarrow \left((NS^{[\lambda]^{<\theta}}_{\kappa,\lambda})^+\right)^{m+1}$. Let us consider the case when $\theta = cf(\lambda)$.

Lemma 8.2 (Matet-Péan-Shelah [22]). *Suppose* κ *is Mahlo,* $cf(\lambda) < \kappa$ *and* $\tau^{<cf(\lambda)} < \lambda$ *for every cardinal* $\tau < \lambda$. *Then* $NS^{[\lambda]^{<cf(\lambda)}}_{\kappa,\lambda} = NS_{\kappa,\lambda}|A$ *for some* $A \in NS^+_{\kappa,\lambda}$.

Lemma 8.3 (Johnson [13]). *If* κ *is* λ*-Shelah and* $cf(\lambda) \geq \kappa$, *then* $\lambda^{<\kappa} = \lambda$.

Proposition 8.4. *Suppose* κ *is* λ*-Shelah,* $cf(\lambda) < \kappa$ *and* $\mathrm{cof}(NS_{\kappa,\lambda}) = \lambda^{<\kappa}$. *Then* $\mathrm{cof}(NS^{[\lambda]^{<cf(\lambda)}}_{\kappa,\lambda}) \leq \lambda^{<\kappa}$.

Proof. By Lemmas 8.2 and 8.3. □

In particular, by Lemma 2.5, if κ is λ-Shelah and λ is a strong limit cardinal of cofinality less than κ, then $\mathrm{cof}(NS^{[\lambda]^{<cf(\lambda)}}_{\kappa,\lambda}) \leq \lambda^{<\kappa}$.

9. $(NAI^{[\lambda]^{<\kappa}}_{\kappa,\lambda,m})^+ \longrightarrow (I^+_{\kappa,\lambda})^{m+1}$

Our goal in this section is to improve Proposition 6.11 by replacing $Q_{\kappa,\lambda}$ by a bigger set. Let us first try to modify the definition of $h_{\kappa,\lambda}$. This approach will work for some values of $cf(\rho_\lambda)$.

Lemma 9.1. *Let* $\mu \leq \lambda$ *be an infinite cardinal. Then the following hold:*

(i) $\{a \in P_\kappa(\lambda) : |a \cap \mu^+| \leq |a \cap \mu|^+\} \in NS^*_{\kappa,\lambda}$.

(ii) **(Johnson [13])** $\{a \in P_\kappa(\lambda) : |a \cap \mu^+| < |a \cap \mu|^+\} \in NSh_{\kappa,\lambda}$.

Proposition 9.2. *Suppose* κ *is* λ*-Shelah and* $\delta < \kappa^+$ *is such that* $\kappa^{+\delta} \leq \lambda$*. Then there is* $f : \kappa \to \kappa$ *such that* $\{a \in E_{\kappa,\lambda} : |a \cap \kappa^{+\delta}| = f(a \cap \kappa)\} \in NSh^*_{\kappa,\lambda}$.

Proof. We proceed by induction on δ. For $\delta = 0$ we can take for f the identity on κ. Next suppose that $\delta = \beta + 1$ and there is $g : \kappa \to \kappa$ such that

$$\{a \in E_{\kappa,\lambda} : |a \cap \kappa^{+\beta}| = g(a \cap \kappa)\} \in NSh^*_{\kappa,\lambda}.$$

Then by Lemma 9.1,

$$\{a \in E_{\kappa,\lambda} : |a \cap \kappa^{+\delta}| = (g(a \cap \kappa))^+\} \in NSh^*_{\kappa,\lambda}.$$

Finally, suppose δ is an infinite limit ordinal. Set $\nu = cf(\delta)$. Select a strictly increasing, continuous sequence $< \delta_i : i < \nu >$ of ordinals so that $\delta_0 = 0$ and $\bigcup_{i<\nu} \delta_i = \delta$. Assume that for $i < \nu$, there is $f_i : \kappa \to \kappa$ and $A_i \in NSh^*_{\kappa,\lambda} \cap P(E_{\kappa,\lambda})$ such that $|a \cap \kappa^{+\delta_i}| = f_i(a \cap \kappa)$ for all $a \in A_i$.

Case $\nu < \kappa$: Set $A = \{a \in \bigcap_{i\in\nu} A_i : \nu \subseteq a\}$. Then $A \in NSh^*_{\kappa,\lambda}$. Moreover, $|a \cap \kappa^{+\delta}| = |\bigcup_{i<\nu}(a \cap \kappa^{+\delta_i})| = \bigcup_{i<\nu} f_i(a \cap \kappa)$ for every $a \in A$.

Case $\nu = \kappa$: Let B be the set of all $a \in P_\kappa(\lambda)$ such that for every $\gamma \in (a \cap \kappa^{+\delta}) \setminus \kappa$, $i_\gamma \in a$, where i_γ is the unique i such that $\kappa^{+\delta_i} \leq \gamma < \kappa^{+\delta_{i+1}}$. Note that $B \in NS^*_{\kappa,\lambda}$. Let C be the set of all $a \in E_{\kappa,\lambda}$ such that $a \in A_{i+1}$ for every $i \in a \cap \kappa$. By normality of $NSh_{\kappa,\lambda}$, $C \in NSh^*_{\kappa,\lambda}$. Now for every $a \in B \cap C$,

$$|a \cap \kappa^{+\delta}| = |\bigcup_{i\in a\cap\kappa} (a \cap \kappa^{+\delta_{i+1}})| = \bigcup_{i\in a\cap\kappa} |a \cap \kappa^{+\delta_{i+1}}| = \bigcup_{i\in a\cap\kappa} f_{i+1}(a \cap \kappa). \qquad \square$$

Immediately from (the proof of) Proposition 9.2 we have :

Corollary 9.3.

(i) *Let* $\delta < \kappa$ *be such that* $\kappa^{+\delta} \leq \lambda$*. Then*

$$\{a \in E_{\kappa,\lambda} : |a \cap \kappa^{+\delta}| \neq (a \cap \kappa)^{+\delta}\} \in NSh_{\kappa,\lambda}.$$

(ii) *If* $\kappa^{+\kappa} \leq \lambda$*, then*

$$\{a \in E_{\kappa,\lambda} : |a \cap \kappa^{+\kappa}| \neq (a \cap \kappa)^{+(a\cap\kappa)}\} \in NSh_{\kappa,\lambda}.$$

Lemma 9.4 (Abe [1]). *Let* $\mu \leq \lambda$ *be an infinite cardinal. Then*

$$\{a \in E_{\kappa,\lambda} : cf(|a \cap \mu|) \neq |a \cap cf(\mu)|\} \in NSh_{\kappa,\lambda}.$$

It follows from Proposition 9.2 and Lemma 9.4 that if $cf(\rho_\lambda) < \kappa^{+\kappa^+}$ and we modify the definition of $h_{\kappa,\lambda}$ by setting $h_{\kappa,\lambda}(a) = \psi(|a|, |a \cap cf(\rho)|)$, then the results of Section 6 still hold. If ρ_λ is inaccessible, then by Lemma 9.4 we may take $h_{\kappa,\lambda}(a) =$ the least inaccessible cardinal τ such that $\tau > |a|$.

For small values of λ, $Q_{\kappa,\lambda}$ can be replaced by $P_\kappa(\lambda)$. We need a few lemmas.

Lemma 9.5 (Abe [1]). *If $\lambda \geq 2^\kappa$, then the set of all $a \in E_{\kappa,\lambda}$ such that $a \cap \kappa$ is not a measurable cardinal lies in $NSh_{\kappa,\lambda}$.*

The following is essentially due to Johnson [13].

Lemma 9.6. *Let $\mu \leq \lambda$ be an infinite cardinal. Then the set of all $a \in P_\kappa(\lambda)$ such that o.t.$(a \cap \mu)$ is not a cardinal lies in $NSh_{\kappa,\lambda}$.*

Definition. The *beth function* is defined by induction on $\alpha \in$ On by
(i) $\beth_0 = \aleph_0$,
(ii) $\beth_{\alpha+1} = 2^{\beth_\alpha}$, and
(iii) $\beth_\alpha = \bigcup_{\beta<\alpha} \beth_\beta$ if α is an infinite limit ordinal.

Notice that κ is inaccessible if and only if $\beth_\kappa = \kappa$.

Lemma 9.7. *Let μ be a cardinal such that $\kappa \leq \mu \leq \lambda$ and μ is not Ramsey. Then the set of all $a \in P_\kappa(\lambda)$ such that $|a \cap \mu|$ is a Ramsey cardinal lies in $NSh_{\kappa,\lambda}$.*

Proof. Suppose to the contrary that $A \in NSh^+_{\kappa,\lambda}$, where A is the set of all $a \in P_\kappa(\lambda)$ such that $|a \cap \mu|$ is Ramsey. Then by Lemma 9.6, $B \in NSh^+_{\kappa,\lambda}$, where B is the set of all $a \in A$ such that $o.t.(a \cap \mu) = |a \cap \mu|$. Now fix $F : P_\omega(\mu) \to 2$. For each $a \in B$, select a strictly increasing function $g_a : a \cap \mu \to a \cap \mu$ so that for every positive integer n, F is constant on $\{e \subset \text{ran}(g_a) : |e| = n\}$. Then there is $f : \mu \to \mu$ with the property that for every $c \in P_\kappa(\mu)$, one can find $a \in B$ such that $c \subseteq a$ and $f \restriction c = g_a \restriction c$. Clearly, f is a strictly increasing function. Moreover for each positive integer n, F is constant on $\{e \subset \text{ran}(f) : |e| = n\}$. Thus μ is a Ramsey cardinal. Contradiction! □

Lemma 9.8. *Suppose κ is λ-Shelah, and let $\alpha \in On$ be such that*
(i) $\beth_{\kappa+\alpha} \leq \lambda$, and
(ii) κ is the greatest Mahlo cardinal less than or equal to $\beth_{\kappa+\alpha}$.
Then there is $A \in NSh^*_{\kappa,\lambda}$ such that $|a \cap \beth_{\kappa+\alpha}| < |b \cap \kappa|$ for all $a, b \in A$ with $a \cap \beth_{\kappa+\alpha} \subset b \cap \beth_{\kappa+\alpha}$.

Proof. We proceed by induction on α. For $\alpha = 0$ we can take $A = E_{\kappa,\lambda}$. Next suppose that (a) $\alpha = \beta + 1$, (b) $\beth_{\kappa+\alpha} \leq \lambda$, and (c) there is $B \in NSh^*_{\kappa,\lambda}$ such that $|a \cap \beth_{\kappa+\beta}| < |b \cap \kappa|$ for all $a, b \in B$ with $a \cap \beth_{\kappa+\beta} \subset b \cap \beth_{\kappa+\beta}$. By Lemma 6.7 there exists $C \in NSh^*_{\kappa,\lambda}$ such that $|a \cap \beth_{\kappa+\alpha}| < |b \cap \beth_{\kappa+\alpha}|$ for all $a, b \in C$ with $a \cap \beth_{\kappa+\alpha} \subset b \cap \beth_{\kappa+\alpha}$. Set $D = \{a \in E_{\kappa,\lambda} : |a \cap \beth_{\kappa+\alpha}| = 2^{|a \cap \beth_{\kappa+\beta}|}\}$. Note that $D \in NSh^*_{\kappa,\lambda}$ by Lemma 6.4. Let us check that $A = B \cap C \cap D$ is as desired.

Thus let $a, b \in A$ be such that $a \cap \beth_{\kappa+\alpha} \subset b \cap \beth_{\kappa+\alpha}$. We have $| a \cap \beth_{\kappa+\alpha} |= 2^{|a \cap \beth_{\kappa+\beta}|}$ and $| b \cap \beth_{\kappa+\alpha} |= 2^{|b \cap \beth_{\kappa+\beta}|}$. Since $| a \cap \beth_{\kappa+\alpha} | \neq | b \cap \beth_{\kappa+\alpha} |$, it follows that $a \cap \beth_{\kappa+\beta} \subset b \cap \beth_{\kappa+\beta}$. Hence $| a \cap \beth_{\kappa+\beta} |< b \cap \kappa$ and therefore by inaccessibility of $b \cap \kappa$, $| a \cap \beth_{\kappa+\alpha} |< b \cap \kappa$.

Let us now deal with the case when α is an infinite limit ordinal such that $\beth_{\kappa+\alpha} \leq \lambda$ and $\beth_{\kappa+\alpha}$ is not an inaccessible cardinal. Setting $\nu = cf(\alpha)$, we get $\nu = cf(\beth_{\kappa+\alpha}) < \beth_{\kappa+\alpha}$. Put $\xi =$ the least $\beta < \alpha$ such that $\nu \leq \beth_{\kappa+\beta}$. Select a strictly increasing, continuous sequence $< \alpha_i : i < \nu >$ of ordinals so that $\alpha_0 = \xi$ and $\bigcup_{i<\nu} \alpha_i = \alpha$. Suppose that for $i < \nu$, there is $A_i \in NSh^*_{\kappa,\lambda}$ such that $| a \cap \beth_{\kappa+\alpha_i} |<| b \cap \kappa |$ for all $a, b \in A_i$ with $a \cap \beth_{\kappa+\alpha_i} \subset b \cap \beth_{\kappa+\alpha_i}$.

Case $\nu < \kappa$: Set $A = \{a \in E_{\kappa,\lambda} \cap (\bigcap_{i<\nu} A_i) : \nu \subseteq a\}$. Given $a, b \in A$ with $a \cap \beth_{\kappa+\alpha} \subset b \cap \beth_{\kappa+\alpha}$, put $j =$ the least $i < \nu$ such that $a \cap \beth_{\kappa+\alpha_i} \neq b \cap \beth_{\kappa+\alpha_i}$. Then $| a \cap \beth_{\kappa+\alpha_i} |< b \cap \kappa$ for $j \leq i < \nu$. Since $b \cap \kappa$ is regular, it follows that $| a \cap \beth_{\kappa+\alpha} |= \bigcup_{i<\nu} | a \cap \beth_{\kappa+\alpha_i} |< b \cap \kappa$.

Case $\nu \geq \kappa$: Define $h : \beth_{\kappa+\alpha} \to \nu$ by $h(\gamma) =$ the least $i < \nu$ such that $\gamma \in \beth_{\kappa+\alpha_i}$. Then $X \in NS^*_{\kappa,\lambda}$, where X is the set of all $a \in P_\kappa(\lambda)$ such that $h(\gamma) \in a$ for all $\gamma \in a \cap \beth_{\kappa+\alpha}$. Let Y be the set of all $a \in E_{\kappa,\lambda}$ such that (1) $a \in A_i$ for every $i \in a \cap \nu$, and (2) $cf(| a \cap \beth_{\kappa+\alpha} |) =| a \cap \nu |$. By Lemmas 1.4 and 9.4, $Y \in NSh^*_{\kappa,\lambda}$. Let us verify that $A = X \cap Y$ is as desired. Thus let $a, b \in A$ with $a \cap \beth_{\kappa+\alpha} \subset b \cap \beth_{\kappa+\alpha}$. Clearly, $a \cap \beth_{\kappa+\alpha} = \bigcup_{i \in a \cap \nu} (a \cap \beth_{\kappa+\alpha_i})$ and $b \cap \beth_{\kappa+\alpha} = \bigcup_{i \in b \cap \nu} (b \cap \beth_{\kappa+\alpha_i})$. So there must be $i \in a \cap \nu$ such that $a \cap \beth_{\kappa+\alpha_i} \neq b \cap \beth_{\kappa+\alpha_i}$. Let k be the least such i. Since $| a \cap \beth_{\kappa+\alpha_i} |< b \cap \kappa$ for every $i \in a \cap \nu$ such that $i \geq k$, we get $| a \cap \beth_{\kappa+\alpha} |= \bigcup_{i \in a \cap \nu} | a \cap \beth_{\kappa+\alpha_i} | \leq b \cap \kappa$. Note that $| a \cap \beth_{\kappa+\alpha} |$ is a singular cardinal and $b \cap \kappa$ a regular cardinal, so $| a \cap \beth_{\kappa+\alpha} | \neq b \cap \kappa$. Consequently, $| a \cap \beth_{\kappa+\alpha} |< b \cap \kappa$.

Finally, let α be an infinite limit ordinal such that $\beth_{\kappa+\alpha} \leq \lambda$ and $\beth_{\kappa+\alpha}$ is an inaccessible cardinal that is not Mahlo. Note that $\beth_{\kappa+\alpha} = \alpha$. Suppose that for each $\delta < \alpha$, there is $A_\delta \in NSh^*_{\kappa,\lambda}$ such that $| a \cap \beth_{\kappa+\delta} |<| b \cap \kappa |$ whenever $a, b \in A_\delta$ and $a \cap \beth_{\kappa+\delta} \subset b \cap \beth_{\kappa+\delta}$. Let C be a closed unbounded subset of α that consists of singular infinite cardinals, and let $< \gamma_\xi : \xi < \alpha >$ be the increasing enumeration of the elements of C. Let A be the set of all $a \in E_{\kappa,\lambda}$ such that (i) $\beth_{\kappa+\gamma_0} \in a$, (ii) $a \cap \kappa$ is a measurable cardinal, (iii) $o.t.(a \cap \beth_{\kappa+\alpha})$ is a regular cardinal that is not Ramsey, (iv) $a \in A_\delta$ for all $\delta \in a \cap \alpha$, (v) if $\xi < \alpha$ is such that $\{\zeta \in a : \beth_{\kappa+\gamma_\xi} \leq \zeta < \beth_{\kappa+\gamma_{\xi+1}}\} \neq \phi$, then $\{\gamma_{\xi+1}, \beth_{\kappa+\gamma_\xi}, \beth_{\kappa+\gamma_{\xi+1}}\} \subset a$, and (vi) if $\xi < \alpha$ is such that $\beth_{\kappa+\gamma_\xi} \in a$, then $cf(| a \cap \beth_{\kappa+\gamma_\xi} |) =| a \cap cf(\beth_{\kappa+\gamma_\xi}) |<| a \cap \beth_{\kappa+\gamma_\xi} |$. Then $A \in NSh^*_{\kappa,\lambda}$ by Lemmas 9.1, 9.4, 9.5, 9.6 and 9.7. Let us check that A is as desired. Thus let $a, b \in A$ with $a \cap \beth_{\kappa+\alpha} \subset b \cap \beth_{\kappa+\alpha}$. Let η be the least element of $b \setminus a$.

Case $\eta < \cup(a \cap \beth_{\kappa+\alpha})$: We claim that $o.t.(a \cap \beth_{\kappa+\alpha}) \leq b \cap \kappa$. Suppose otherwise. Then we can find $\zeta \in (a \cap \beth_{\kappa+\alpha}) \setminus (\eta \cup \beth_{\kappa+\gamma_0})$ so that $o.t.(a \cap \zeta) \geq b \cap \kappa$. Let ξ

be such that $\beth_{\kappa+\gamma_\xi} \leq \zeta < \beth_{\kappa+\gamma_{\xi+1}}$. Then $|a \cap \beth_{\kappa+\gamma_{\xi+1}}| < b \cap \kappa$, a contradiction. It remains to observe that $o.t.(a \cap \beth_{\kappa+\alpha}) \neq b \cap \kappa$ since $b \cap \kappa$ is a measurable cardinal and $o.t.(a \cap \beth_{\kappa+\alpha})$ is not a Ramsey cardinal.

Case $\eta \geq \cup(a \cap \beth_{\kappa+\alpha})$: Let σ be such that $\beth_{\kappa+\gamma_\sigma} \leq \eta < \beth_{\kappa+\gamma_{\sigma+1}}$. Then

$$a \cap \beth_{\kappa+\alpha} = a \cap \beth_{\kappa+\gamma_\sigma} = b \cap \beth_{\kappa+\gamma_\sigma},$$

which cannot be since $|a \cap \beth_{\kappa+\alpha}|$ is a regular cardinal, whereas $cf(|b \cap \beth_{\kappa+\gamma_\sigma}|) = |b \cap cf(\sigma)| < |b \cap \beth_{\kappa+\gamma_\sigma}|$. □

Proposition 9.9. *Suppose that* $0 < m < \omega, \kappa$ *is almost* $(\lambda^{<\kappa}, m)$*-ineffable and there is no Mahlo cardinal* τ *with* $\kappa < \tau \leq \lambda$*. Then* $(NAI_{\kappa,\lambda,m}^{[\lambda]^{<\kappa}})^+ \to (I_{\kappa,\lambda}^+)^{m+1}$.

Proof. We have $\beth_{\kappa+\alpha} \leq \lambda < \beth_{\kappa+\alpha+1}$ for some $\alpha \in$ On. If $\beth_{\kappa+\alpha} = \lambda$, the conclusion follows from Lemmas 6.10 and 9.8. Otherwise, use Lemmas 6.9 and 9.8. □

Lemma 9.8 is proved under the assumption that κ is the greatest Mahlo cardinal less than or equal to $\beth_{\kappa+\alpha}$. We will show that some condition of this kind is necessary.

Definition. NSu_κ is the set of all $A \subseteq \kappa$ such that one can find $s_\alpha \subseteq \alpha$ for $\alpha \in A$ with the following property : there is a closed unbounded subset C of κ such that $s_\beta \cap \alpha \neq s_\alpha$ for all $\alpha, \beta \in A \cap C$ with $\alpha < \beta$.

κ is *subtle* if $\kappa \notin NSu_\kappa$.

Lemma 9.10 (Baumgartner [4]). *Suppose* κ *is subtle. Then* NSu_κ *is an ideal on* κ*. Moreover, the set of all* $\mu < \kappa$ *such that* μ *is not an inaccessible cardinal lies in* NSu_κ.

Note that by Lemma 6.2, if $\sigma > \kappa$ is a strong limit cardinal, and κ is τ-Shelah for every cardinal τ with $\kappa \leq \tau < \sigma$, then κ is μ-ineffable for every cardinal μ with $\kappa \leq \mu < \sigma$. The following is inspired by an argument of Kunen (see Lemma 1 in [17]).

Proposition 9.11. *Suppose that* $\sigma > \kappa$ *is a subtle cardinal, and* κ *is* τ*-Shelah for every cardinal* τ *with* $\kappa \leq \tau < \sigma$*. Let* X *be the set of all cardinals* μ *such that (a)* $\kappa \leq \mu < \sigma$*, and (b) for every* $B \in NI_{\kappa,\mu}^*$*, there is* $(c,d) \in [B]^2$ *with* $c = d \cap (\cup c)$*. Then* $X \in NSu_\sigma^*$.

Proof. Assume otherwise. Then letting Y be the set of all inaccessible cardinals μ with $\kappa \leq \mu < \sigma$, one can find $Z \in NSu_\sigma^+ \cap P(Y)$ and $B_\mu \in NI_{\kappa,\mu}^*$ for $\mu \in Z$ so that (i) $\cup c \notin c$ for every $c \in B_\mu$, and (ii) $c \neq d \cap (\cup c)$ for every $(c,d) \in [B_\mu]^2$. Select a one-to-one function $j : P_\kappa(\sigma) \to \sigma$. There is a closed unbounded subset C of σ such that $j(e) \in \mu$ whenever $\mu \in Z \cap C$ and $e \in P_\kappa(\mu)$. Put $s_\mu = j''B_\mu$ for $\mu \in Z \cap C$. Pick $\rho, \mu \in Z \cap C$ so that $\rho < \mu$ and $s_\mu \cap \rho = s_\rho$. Then clearly, $B_\mu \cap P_\kappa(\rho) = B_\rho$. Now setting $Q = \{a \in P_\kappa(\mu) : a \cap \rho \in B_\rho\}$, it is simple to check that $Q \in NI_{\kappa,\mu}^*$. So there must be $d \in Q \cap B_\mu$ with $\rho \in d$. Putting $c = d \cap \rho$, we have $(c,d) \in [B_\mu]^2$, which yields the desired contradiction. □

10. $(NI^{[\lambda]^{<\kappa}}_{\kappa,\lambda,m}|A)^+ \to ((NS^{[\lambda]^{<\kappa}}_{\kappa,\lambda})^+)^{m+1}$

Turning from almost ineffability to ineffability, we present a variant of Proposition 6.11 which is proved in the same way. Details are left to the reader.

Lemma 10.1. *Suppose $0 < m < \omega$ and κ is $(\lambda^{<\kappa}, m)$-ineffable. Then $Q_{\kappa,\lambda} \notin NI^{[\lambda]^{<\kappa}}_{\kappa,\lambda,m}$.*

As Lemma 6.9, the following is essentially due to Kamo [16].

Lemma 10.2. *Suppose $0 < m < \omega$ and κ is $(\lambda^{<\kappa}, m)$-ineffable. Let ν be a cardinal with $\nu < \lambda < 2^\nu$, and $A \in (NI^{[\lambda]^{<\kappa}}_{\kappa,\lambda,m})^+$ be such that $|a \cap \nu| < |b \cap \kappa|$ for all $a, b \in A$ with $a \cap \nu \subset b \cap \nu$. Then $\{A\} \to ((NS^{[\lambda]^{<\kappa}}_{\kappa,\lambda})^+)^{m+1}$.*

The following is essentially due to Carr [8] (see also [15]).

Lemma 10.3. *Suppose $0 < m < \omega$ and κ is $(\lambda^{<\kappa}, m)$-ineffable, and let $F : [P_\kappa(\lambda)]^{m+1}_< \to 2$ and $A \in (NI^{[\lambda]^{<\kappa}}_{\kappa,\lambda,m})^+$. Then there is $B \in (NS^{[\lambda]^{<\kappa}}_{\kappa,\lambda})^+ \cap P(A)$ such that F is constant on $[B]^{m+1}_<$.*

Proposition 10.4. *Suppose $0 < m < \omega$ and κ is $(\lambda^{<\kappa}, m)$-ineffable. Then*

$$(NI^{[\lambda]^{<\kappa}}_{\kappa,\lambda,m}|Q_{\kappa,\lambda})^+ \to ((NS^{[\lambda]^{<\kappa}}_{\kappa,\lambda})^+)^{m+1}.$$

As in Section 9, there are cases when the result may be strengthened. For example, it is easy to see that if μ is a weakly compact cardinal with $\kappa \leq \mu \leq \lambda$, then the set of all $a \in P_\kappa(\lambda)$ such that $|a \cap \mu|$ is not a weakly compact cardinal lies in $NI_{\kappa,\lambda}$. So in the case when ρ_λ is a weakly compact cardinal, we may set $h_{\kappa,\lambda}(a) =$ the least weakly compact cardinal τ such that $\tau > |a|$.

Proposition 10.5. *Suppose that $0 < m < \omega$, κ is $(\lambda^{<\kappa}, m)$-ineffable and there is no Mahlo cardinal τ with $\kappa < \tau \leq \lambda$. Then $(NI^{[\lambda]^{<\kappa}}_{\kappa,\lambda,m})^+ \to ((NS^{[\lambda]^{<\kappa}}_{\kappa,\lambda})^+)^{m+1}$.*

Lemma 10.6 (Magidor [19]). *If* Part*(κ, λ) *holds, then κ is λ-ineffable.*

As mentioned in the introduction, Kamo [16] established that if κ is λ-ineffable and $cf(\lambda) \geq \kappa$, then Part*(κ, λ) holds. Hence by Lemmas 8.3 and 10.6, assuming $cf(\lambda) \geq \kappa$, κ is $\lambda^{<\kappa}$-ineffable if and only if Part*$(\kappa, \lambda^{<\kappa})$ holds. We will see that this equivalence remains valid in the case when $cf(\lambda) < \kappa$.

Lemma 10.7 (Kamo [15]). *Suppose κ is Mahlo, and let $B \subseteq P_\kappa(\lambda)$. Then $\{B\} \to ((NS^{[\lambda]^{<\kappa}}_{\kappa,\lambda})^+)^2$ if and only if $\{\{x \in P_\kappa(\lambda^{<\kappa}) : x \cap \lambda \in B\}\} \to (NS^+_{\kappa,\lambda^{<\kappa}})^2$.*

Lemma 10.8 (Matet-Péan-Shelah [22]). *Suppose κ is Mahlo, $cf(\lambda) < \kappa$ and $\tau^{<\kappa} < \lambda$ for every cardinal $\tau < \lambda$. Then $NS^{[\lambda]^{<\kappa}}_{\kappa,\lambda} = NS^{[\lambda]^{<(cf(\lambda))^+}}_{\kappa,\lambda} \mid A$ for some $A \in (NS^{[\lambda]^{<\kappa}}_{\kappa,\lambda})^*$.*

Proposition 10.9. *Assuming $cf(\lambda) < \kappa$, the following are equivalent:*

(i) *κ is $\lambda^{<\kappa}$-ineffable.*

(ii) $\{P_\kappa(\lambda)\} \to ((NS_{\kappa,\lambda}^{[\lambda]^{<\kappa}})^+)^2$.

(iii) $(NS_{\kappa,\lambda}^{[\lambda]^{<\kappa}})^* \to ((NS_{\kappa,\lambda}^{[\lambda]^{<(cf(\lambda))^+}})^+)^2$.

(iv) Part$^*(\kappa, \lambda^{<\kappa})$ *holds.*

Proof. (i) $\to$ (ii): By Proposition 10.4.

(ii) $\to$ (iii): Trivial.

(iii) $\to$ (ii): By Lemmas 8.3, 10.6 and 10.8.

(ii) $\to$ (iv): By Lemma 10.7.

(iv) $\to$ (i): By Lemma 10.6. □

Abe [2] proved that if κ is $\lambda^{<\kappa}$-ineffable and $\lambda^{<\kappa} = 2^\lambda$, then the set of all $x \in E_{\kappa,\lambda^{<\kappa}}$ such that $x \cap \kappa$ is almost $|x \cap \lambda|^{<(x\cap\kappa)}$-ineffable and $|x \cap \lambda|$-ineffable but not $|x \cap \lambda|^{<(x\cap\kappa)}$-ineffable lies in $NI^+_{\kappa,\lambda^{<\kappa}}$. A similar result can be obtained as follows. Suppose $0 < m < \omega$ and there is a regular uncountable cardinal μ such that μ is almost $((\psi(\mu, \omega))^{<\mu}, m)$-ineffable. Suppose further that κ is the least such μ, and $\lambda = \psi(\kappa, \omega)$. Then κ is (λ, m)-ineffable by Proposition 2.4 and Lemma 2.5. Setting $A = \{a \in E_{\kappa,\lambda} : |a| = \psi(a \cap \kappa, \omega)\}$, we have $A \in NSh^*_{\kappa,\lambda}$ by Lemma 6.4. Clearly, for every $a \in A$, $a \cap \kappa$ is not almost $((\psi(a \cap \kappa, \omega))^{<(a\cap\kappa)}, m)$-ineffable. Hence by Proposition 5.1, κ is not $\lambda^{<\kappa}$-ineffable. Note that by Proposition 6.11 and Lemmas 2.3 and 2.5, $(NAI_{\kappa,\lambda,m}^{[\lambda]^{<\kappa}})^+ \to (NS^+_{\kappa,\lambda})^{m+1}$.

We do not know whether $(NI_{\kappa,\lambda,m}^{[\lambda]^{<\kappa}})^+ \to (NS^+_{\kappa,\lambda})^{m+1}$ always holds. We will see that it does hold in the case when λ is a strong limit cardinal of cofinality less than κ. In the general case we are only able to show that $(NI_{\kappa,\lambda,m}^{[\lambda]^{<\kappa}})^+ \to (I^+_{\kappa,\lambda})^{m+1}$.

Lemma 10.10. *Suppose κ is Mahlo and $B \in (NS_{\kappa,\lambda}^{[\lambda]^{<\kappa}})^+ \cap P(E_{\kappa,\lambda})$. Then there is $C \in I^+_{\kappa,\lambda} \cap P(B)$ such that $[C]^2 = [C]^2_<$.*

Proof. Let $P_\kappa(\lambda) = \{d_\xi : \xi < \lambda^{<\kappa}\}$. Construct by induction a_ξ for $\xi < \lambda^{<\kappa}$ so that

(a) $a_\xi \in B$.

(b) $d_\xi < a_\xi$.

(c) $a_\xi \not\subseteq a_\zeta$ for every $\zeta < \xi$.

(d) If $\zeta < \xi$ and $d_\zeta < a_\xi$, then $|a_\zeta| < a_\xi \cap \kappa$.

Then $C = \{a_\xi : \xi < \lambda^{<\kappa}\}$ is as desired. □

Proposition 10.11. *Suppose κ is $(\lambda^{<\kappa}, m)$-ineffable, where $0 < m < \omega$, and let K be a κ-complete ideal on $P_\kappa(\lambda)$ such that $K \subseteq NS_{\kappa,\lambda}^{[\lambda]^{<\kappa}}$ and* cof$(K) = \lambda^{<\kappa}$. *Then* $(NI_{\kappa,\lambda,m}^{[\lambda]^{<\kappa}})^+ \to (K^+)^{m+1}$.

Proof. By Lemmas 2.3, 10.3 and 10.10. □

Corollary 10.12. *Suppose κ is $(\lambda^{<\kappa}, m)$-ineffable, where $0 < m < \omega$. Then the following hold:*

(i) $(NI^{[\lambda]^{<\kappa}}_{\kappa,\lambda,m})^+ \to (I^+_{\kappa,\lambda})^{m+1}$.

(ii) *If* $cf(\lambda) < \kappa$ *and* $\mathrm{cof}(NS_{\kappa,\lambda}) = \lambda^{<\kappa}$, *then* $(NI^{[\lambda]^{<\kappa}}_{\kappa,\lambda,m})^+ \to ((NS^{[\lambda]^{<cf(\lambda)}}_{\kappa,\lambda})^+)^{m+1}$.

Proof. Use Proposition 10.11 and Lemma 8.2. □

11. $\mathrm{Part}^*(\kappa, \kappa^+)$ at a nonmeasurable cardinal

Definition. For $2 \leq r < \omega$, $\mathrm{Part}^*(\kappa, \lambda)^r$ means that $\{P_\kappa(\lambda)\} \to (NS^+_{\kappa,\lambda})^r$.

By Lemmas 6.2, 9.5 and 10.6, if $\mathrm{Part}^*(\kappa, \lambda)^2$ holds and $\lambda \geq 2^\kappa$, then κ is a measurable limit of measurable cardinals. On the other hand, a result of Apter and Shelah can be used to establish the consistency of "κ is not measurable but $\mathrm{Part}^*(\kappa, \kappa^+)^r$ holds for $r = 2, 3, \ldots$".

Lemma 11.1 (Carr [9], Matet [21]). *Let $D \in NS^+_{\kappa,\lambda}$. Then*

$$\{b \in E_{\kappa,\lambda} : D \cap P_{b\cap\kappa}(b) \in NS_{b\cap\kappa,b}\} \in NSh_{\kappa,\lambda}.$$

Lemma 11.2. *Suppose κ is λ-supercompact and $0 < m < \omega$. Then*

$$\{b \in E_{\kappa,\lambda} : b \cap \kappa \text{ is } (|b|^{<(b\cap\kappa)}, m)\text{-ineffable}\} \in J^*$$

for every prime normal ideal J on $P_\kappa(\lambda)$.

Proof. Argue as for Proposition 5.1 and use Lemmas 6.1 and 11.1. □

Proposition 11.3. *It is consistent relative to the existence of a cardinal ρ that is ρ^+-supercompact that there is a nonmeasurable regular infinite cardinal χ such that $\mathrm{Part}^*(\chi, \chi^+)^r$ holds for every r with $2 \leq r < \omega$.*

Proof. Suppose that the GCH holds and κ is κ^+-supercompact. Then by a result of Apter and Shelah [3], there is a cardinal and cofinality preserving generic extension $V[G]$ of the universe in which the following hold : (a) κ is κ^+-supercompact, (b) κ is the least measurable cardinal, and (c) $2^\kappa = 2^{\kappa^+} = \kappa^{++}$. In $V[G]$, select a prime normal ideal J on $P_\kappa(\kappa^+)$, and let A be the set of all $a \in E_{\kappa,\kappa^+}$ such that $|a| = (a \cap \kappa)^+$, and B be the set of all $a \in E_{\kappa,\kappa^+}$ such that $a \cap \kappa$ is $(|a|^{<(a\cap\kappa)}, m)$-ineffable for every m with $0 < m < \omega$. Then by Lemmas 9.1 and 11.2, $A \cap B \in J^*$. For $a \in A \cap B$ and $0 < m < \omega$, we have

$$(NI_{a\cap\kappa,(a\cap\kappa)^+,m})^+ \to (NS^+_{a\cap\kappa,(a\cap\kappa)^+})^{m+1}$$

by Corollary 3.4 and Proposition 10.5. □

References

[1] Y. Abe. *Combinatorial characterization of Π^1_1-indescribability in $P_\kappa\lambda$;* Archive for Mathematical Logic 37 (1998), 261–272.

[2] Y. Abe. *Notes on subtlety and ineffability in $P_\kappa\lambda$;* Archive for Mathematical Logic 44 (2005), 619–631.

[3] A.W. Apter and S. Shelah. *Menas' result is best possible;* Transactions of the American Mathematical Society 349 (1997), 2007–2034.

[4] J.E. Baumgartner . *Ineffability properties of cardinals I;* in: Infinite and Finite Sets (A. Hajnal, R. Rado and V.T. Sós, eds.), Colloquia Mathematica Societatis János Bolyai vol. 10, North-Holland, Amsterdam, 1975, pp. 109–130.

[5] D.M. Carr. *The minimal normal filter on $P_\kappa\lambda$;* Proceedings of the American Mathematical Society 86 (1982), 316–320.

[6] D.M. Carr. *$P_\kappa\lambda$-Generalizations of weak compactness;* Zeitschrift für mathematische Logik und Grundlagen der Mathematik 31 (1985), 393–401.

[7] D.M. Carr. *The structure of ineffability properties of $P_\kappa\lambda$;* Acta Mathematica Hungarica 47 (1986), 325–332.

[8] D.M. Carr. *$P_\kappa\lambda$ Partition relations;* Fundamenta Mathematicae 128 (1987), 181–195.

[9] D.M. Carr. *A note on the λ-Shelah property;* Fundamenta Mathematicae 128 (1987), 197–198.

[10] D.M. Carr and D.H. Pelletier. *Towards a structure theory for ideals on $P_\kappa\lambda$;* in: Set Theory and its Applications (J. Steprāns and S. Watson, eds.), Lecture Notes in Mathematics 1401, Springer, Berlin, 1989, pp. 41–54.

[11] C.A. Di Prisco and W. Zwicker. *A remark on* Part*(κ, λ); Acta Cientifica Venezolana 29 (1978), 365–366.

[12] C.A. Johnson. *Seminormal λ-generated ideals on $P_\kappa\lambda$;* Journal of Symbolic Logic 53 (1988), 92–102.

[13] C.A. Johnson. *Some partition relations for ideals on $P_\kappa\lambda$;* Acta Mathematica Hungarica 56 (1990), 269–282.

[14] S. Kamo. *Remarks on $\mathcal{P}_\kappa\lambda$-combinatorics;* Fundamenta Mathematicae 145 (1994), 141–151.

[15] S. Kamo. *Ineffability and partition property on $\mathcal{P}_\kappa\lambda$;* Journal of the Mathematical Society of Japan 49 (1997), 125–143.

[16] S. Kamo. *Partition properties on $\mathcal{P}_\kappa\lambda$;* Journal of the Mathematical Society of Japan 54 (2002), 123–133.

[17] K. Kunen and D.H. Pelletier. *On a combinatorial property of Menas related to the partition property for measures on supercompact cardinals;* Journal of Symbolic Logic 48 (1983), 475–481.

[18] A. Landver. *Singular Baire numbers and related topics;* Ph. D. Thesis, University of Wisconsin-Madison, 1990.

[19] M. Magidor. *Combinatorial characterization of supercompact cardinals;* Proceedings of the American Mathematical Society 42 (1974), 279–285.

[20] P. Matet. *Un principe combinatoire en relation avec l'ultranormalité des idéaux;* Comptes Rendus de l'Académie des Sciences de Paris, Série I, 307 (1988), 61–62.

[21] P. MATET. *Concerning stationary subsets of* $[\lambda]^{<\kappa}$; in: Set Theory and its Applications (J. Steprāns and S. Watson, eds.), Lecture Notes in Mathematics 1401, Springer, Berlin, 1989, pp. 119–127.

[22] P. MATET, C. PÉAN AND S. SHELAH. *Cofinality of normal ideals on* $P_\kappa(\lambda)$ *I;* preprint.

[23] P. MATET, C. PÉAN AND S. SHELAH. *Cofinality of normal ideals on* $P_\kappa(\lambda)$ *II;* Israel Journal of Mathematics 150 (2005), 253–283.

[24] T.K. MENAS. *On strong compactness and supercompactness;* Annals of Mathematical Logic 7 (1974), 327–359.

[25] T.K. MENAS. *A combinatorial property of* $p_\kappa\lambda$; Journal of Symbolic Logic 41 (1976), 225–234.

Pierre Matet
Université de Caen – CNRS
Mathématiques
BP 5186
F-14032 CAEN CEDEX, France
e-mail: matet@math.unicaen.fr

Set Theory
Trends in Mathematics, 345–400

Local Connectedness and Distance Functions

Charles Morgan

Abstract. Local connectedness functions for $(\kappa, 1)$-simplified morasses, localisations of the coupling function c studied in [M96, §1], are defined and their elementary properties discussed. Several different useful canonical ways of arriving at the functions are examined. This analysis is then used to give explicit formulae for generalisations of the local distance functions introduced with a recursive definition in [K00], leading to simple proofs of the principal properties of those functions. It is then also extended to the properties of local connectedness functions in the context of κ-$\mathbb{M}$-proper forcing for successor κ. The functions are shown to enjoy substantial strengthenings of the properties (particularly the Δ-properties) hitherto proved for both the function c and for Todorcevic's ρ-functions in the special case $\kappa = \omega_1$. A couple of examples of the use of local connectedness functions in consort with κ-$\mathbb{M}$-proper forcing are then given.

Introduction

In *The two cultures of mathematics* ([G]) Gowers argues extremely persuasively that although combinatorics appears to have fewer layers of theory through which one has to wade before one can prove, or even appreciate, significant results, one should not imagine that it is more superficial than other areas of mathematics. He points out that while in algebraic number theory one deep theorem will use another, which in turn uses a third ..., in combinatorial parts of mathematics there would often be no hope of proving one result if one was unaware of the general principles introduced in the proof of another, which in turn would appear to be unprovable if one was unaware of the main ideas of the proof of a third ..., even if there is no logical dependence between results. So to make progress in combinatorial areas one must command a wide collection of typical themes of arguments. These important ideas can often, or perhaps even usually, be pithily expressed as general principles of extensive applicability. This paper can be seen as a contribution to clarifying some new insights about one such general principle, *stepping up.*

The idea behind stepping up is that when one can prove something about a collection of objects of a certain size and wants to prove an analogue for things of a larger size one can frequently try to use a similar proof technique, reducing problems that arise about larger size objects to ones about smaller size ones. That both the finite version of Ramsey's theorem and Erdős, Hajnal and Rado's result ([EHR65]) that the Erdős-Rado theorem is optimal are proved using stepping up arguments evidences that they have a long and illustrious history in both finite and infinite combinatorics.

While some stepping up arguments can be carried out 'by hand' it is often helpful, and sometimes necessary, to use *auxiliary apparatus*. A good, simple example of this is Erdős and Hajnal's proof ([EH]) that Chang's conjecture is equivalent to $\omega_2 \longrightarrow [\omega_1]^2_{\omega_1,<\omega_1}$. One uses a function $e : [\omega_2]^2 \longrightarrow \omega_1$ such that each $e(.,\alpha)$ is an injection of α into ω_1 to reduce questions about functions on a large set (the finite subsets of ω_2) to questions about functions on a smaller one (the finite subsets of ω_1) which one already knew could be dealt with successfully. Then, using the auxiliary apparatus once more, one steps the results back up again.

Another interesting example from the same period is Baumgartner's proof ([B]) that it is consistent that there is family of subsets of ω_1 of size ω_2 which is *strongly almost disjoint*, that is, the intersection of any two of which is finite. This *generic stepping up argument* appears to be somewhat different as one does not step up a smaller s.a.d. family *per se*, but rather steps up a ccc forcing notion for adding a family of ω_2-many (disjoint mod-finite) subsets of ω to one for adding the desired s.a.d. family of subsets of ω_1. Nevertheless at the broadest level the structure of the proof is as before. The crux of the argument is showing that the stepped up forcing is ccc. There a family of ω_2-many subsets of ω_1, the intersection of any two of which is *countable*, is used to reduce the problem of how to limit the size of various collections of finite subsets of ω_1 that one has to deal with (to being countable) to a fact that one already knows: that the number of finite subsets of any countable set is countable.

However stronger tools having properties beyond those that can be shown to exist in ZFC alone have also proven so widely useful as to constitute basic ideas, or perhaps basic subideas, in themselves. Most of these tools can crudely be fitted into one of two collections, collections for which Velleman's simplified morasses and Todorčević ρ-functions are limning emblems.

Tools of the first of these types are used in stepping up arguments in which something of cardinality greater than κ is built by an induction of length $\kappa + 1$. They are frameworks which enable one to amalgamate and fit together (through embeddings) pieces of size less than κ of some desired final object. At limit stages in the inductions, including the final one, limits are taken along some collection of functions more complicated than a simple chain in order to allow the final object to have size greater than κ. The power of these frameworks generally comes from their additional coherence properties. Examples of the tools from this circle of ideas include morasses, simplified morasses ([V84]), some uses of $\square_\kappa$, historicised forcing and pseudo- and semi-morasses ([SS], [K95], [M*3]), the uniform forcing axioms from [SHL], and, a little more distantly, the Abraham-Shore technique ([AS], M*1]).

The second type of tool is (something coded by) a function from $[\kappa^+]^2$ to κ, usually with some additional helpful properties. Families $\langle B_\alpha \mid \alpha < \omega_2 \rangle$ of almost disjoint sets (as in [B]) can be fitted under this rubric (by setting $b(\alpha, \beta) = \sup(B_\alpha \cap B_\beta)$), as can various types of κ^+-Aronszajn tree and κ-Kurepa tree (with or without restrictions on the sorts of subtrees that they have). A landmark in the development of this line of thought was Roitman's paper [R] in which such a function was used to help prove that there is a ccc forcing to add a superatomic Boolean algebra analogue of a Kurepa tree.

The crucial property of the function, which Roitman called the *new* Δ-*property*, is that for every uncountable disjoint family $\mathcal{A}$ of finite subsets of ω_2 and every $\delta < \omega_1$ there are some a, $b \in \mathcal{A}$ such that $\delta < e(\alpha, \beta)$ for every $\alpha \in a$ and $\beta \in b$. This property clearly implies the negation of Chang's conjecture since if $X \in [\omega_2]^{\omega_1}$ then $e``\{\{x\} \mid x \in X\}$ is uncountable, showing that e witnesses the negation of the Erdős-Hajnal partition calculus equivalent of Chang's conjecture discussed above. (In fact it turns out that a Skolemisation argument shows that the existence of a function with the new Δ-property is also an equivalent of the negation of Chang's conjecture.) But Chang's conjecture is consistent with ZFC assuming the consistency of relatively mild large cardinals (an ω_1-Erdős cardinal), so the assumption of the existence of a function with the new Δ-property goes beyond what can be harvested from working in ZFC alone. (Assuming, of course, as there seems every reason to, that ω_1-Erdős cardinals are consistent.) Roitman's paper was particularly significant because it was the first time such a function was used in generic stepping up.

There are several separate ways in which Roitman's precedent has been followed up in order to attack other problems involving generic stepping up for which it seemed that functions with even stronger 'Δ'-like properties would be helpful. Interestingly, each of these various ways involves a piggy-backing of stepping up arguments: stepping up is used to obtain a tool, and the tool is then used to help with a further stepping up argument.

The principal direction, probably first pursued in [Veličk] and later exploited for example in [KM] and [K98a], is to use functions which had already been used for non-generic stepping up and which are extracted directly from coherent frameworks. The best known of these are Todorčević's function ρ ([T87]) defined from $\square_\kappa$ and the coupling function c defined from a $(\kappa, 1)$-simplified morass and analysed in [M96]. Proofs that ρ and c have the new Δ-property when $\kappa = \omega_1$ are implicit in [Veličk] and [M96], respectively.

Another direction, followed in papers such as [BS], [Ra] and [Kom], is to use functions obtained by a historicised forcing as auxiliary apparatuses. However, as pointed out in [M96], the fact that the functions are obtained generically is not necessary in order to prove the properties these papers exploit: the historicised forcings have the side-effect of adding a simplified morass, the functions used are immediately definable from the c coming from this simplified morass as

$f(\nu,\tau) = \{\xi < \nu \mid c(\xi,\nu) \leq c(\nu,\tau)\}$, and the properties used can all be derived for f defined in this way from the coupling function c of *any* simplified morass. (The property used in [BS] can also be proved from the properties of ρ if ρ is substituted for c in the definition of f and $\square_{\omega_1}$ holds.)

Some people have gone beyond this and argue that, because it has been possible to go back to the definitions of ρ and c and prove further properties when necessary, ρ and c are inately superior to functions given as one-offs, for example through forcing. That argument seems tendentious unless backed by arguments that any stepping up apparatus that can be forced already exists in L and that ρ and/or c encapsulate all of the useful stepping up apparatus of L. While the study of ρ and c certainly has been very fruitful and while this paper itself focuses on related ideas, these two claims overstate the case.

There have been three primary sorts of development of the properties of ρ and c building on the new Δ-property, sometimes working in combination with one another. One of these is to improve the property to one that would facilitate proofs of the Knaster property for forcings rather than merely the countable chain condition. With this stronger unboundedness property one would be able to start with a family $\mathcal{A}$ of ω_1 many domains of conditions (finite subsets of ω_2) and some $\delta < \omega_1$ and thin the collection to some Δ-system $\mathcal{B}$ so that one still has ω_1 many domains of conditions and $\delta < e(\alpha,\beta)$ for all $\alpha \in a \setminus b$ and $\beta \in b \setminus a$ for any a, $b \in \mathcal{B}$. (cf. [M*7].) The point of this is that one can then thin $\mathcal{B}$ further to get a pair of conditions satisfying some other properties needed for compatibility, whereas the new Δ-property itself has to be the last thinning applied in any argument for the ccc. (See, for example, [K98a] for an argument of this type.)

Another evidently desirable generalisation is to get the analogue of the new Δ-property for arbitrary κ^+ instead of for ω_1. With this analogue one can aim to prove the κ^+-chain condition in place of the ccc in generic stepping up arguments when the domains of conditions have size $< \kappa$ rather than being finite. This was done for c in [KM] and used to prove the analogue of Roitman's theorem for arbitrary successor cardinals.

The third sort of development is to strengthen the new Δ-property, for example, to ensure that given a Δ-system like $\mathcal{A}$ from two paragraphs above one can find a pair of domains of conditions a, b such that $e(\alpha,\beta) \geq e(\alpha,\gamma)$ or $e(\beta,\gamma)$ for all triples $\alpha \in a \setminus b$, $\beta \in b \setminus a$ and $\gamma \in a \cap b$. As shown by a minute argument, given in [M*7], this is a strengthening of the new Δ-property under the mild assumption, certainly satisfied by all of the functions considered so far and a key to them being useful for stepping up, that $\{\xi \mid e(\xi,\zeta) \leq \delta\}$ is countable for all $\zeta < \omega_2$ and $\delta < \omega_1$. This variation has been used, for example, to give ccc forcings that add a Kurepa tree ([Veličk]) or a chain of length ω_2 in $\mathcal{P}(\omega_1)$ mod finite ([K98a]), and, implicitly, in adding superatomic Boolean algebras with Cantor-Bendixson rank $\omega_2 + 1$ and a countable collection of atoms at each stage of the Cantor-Bendixson process ([BS] and [Ra]). For more on this see [M*7].

As Chang's conjecture is not compatible with the existence of a superatomic Boolean algebra of the type added by Roitman, a Kurepa tree or a chain of length

ω_2 in $\mathcal{P}(\omega_1)$ mod finite ([K98a]), none of these objects can be added by a ccc forcing without assuming the negation of Chang's conjecture, or, equivalently, the existence of a function with the new Δ-property. So in these cases, at least, one really *had* to find auxiliary apparatuses beyond ZFC in order to solve the problems at hand.

Some problems are even more unassailable, so much so that it is no good looking for an argument similar to the generic steppings up considered so far but with an original twist on the properties of the stepping up function. For example, Koszmider ([K00][1]) showed that if the continuum hypothesis holds there is no ccc forcing *at all* which adds a chain of functions of length ω_2 in $\omega_1^{\omega_1}$ ordered by domination mod finite. In order to solve such problems in *two-cardinal combinatorics* Koszmider had the extremely original, if in retrospect logical, idea of trying to step up forcings with finite domains to ones which, even though not ccc, have some variant of properness and so will preserve ω_1. Koszmider was inspired by his understanding of Todorčević's method of 'forcing with chains of models as side conditions' ([T85], [T89], [K98b]) as a way of implementing the amalgamation arguments common in the proof of the countable chain condition in a more general setting.

Koszmider's insightful new method of $\mathbb{M}$-*proper forcing* is a melding of the two techniques of generic stepping up using auxiliary functions and forcing with chains or, more exactly, matrices of models as side conditions. Koszmider attaches finite sets of morass maps as 'side conditions' to conditions in a naive forcing, 'working parts.' Constraints by a single function, such as c, are replaced (implicitly) by constraints *synthetically* derived from these finite sets in the manner of proper forcing with matrices of models as side conditions. But properties of the simplified morass are used when carrying out amalgamation arguments, to prove $\mathbb{M}$-properness or the ω_2-chain condition, to give information about the relationship between constraints in the conditions which one is aiming to amalgamate. This information goes far beyond what could be gleaned merely from their coming from matrices (of $\in$-chains). As arguments using properness and matrices of models as side conditions seem to be limited to producing objects of size ω_1, Koszmider's method, which shows how to build objects of size ω_2, represents a considerable advance.

In §1 of this paper I analyse local versions of the coupling function c. These local versions are derived from a $(\kappa, 1)$-simplified morass in a similar way to c but locally, using collections of maps, rather than globally, using all of the morass maps. As the local functions are defined from chains of couplings perhaps it is reasonable to call them connectedness functions – and to imagine that this was how c got its name as well! As well as proving some basic facts about these functions I investigate a couple of canonical ways of defining them. These canonical definitions give one a more concrete grasp on the functions than their abstract definition.

In §2 I continue the analysis of these local connectedness functions. Two further canonical ways of defining them are given. In §3 the analysis of these canonical definitions of the local connectedness functions is used to give explicit

[1] I am grateful to Piotr Koszmider for making a preprint of [K00] available to me in early 2000.

formulae for local distance functions. Local distance functions are another very interesting new notion Koszmider introduced in [K00] giving a recursive definition. The explicit formulae again give a more concrete grasp on the distance functions and make analysis and use of them easier. For example, the triangle equality for local distance functions is an immediate corollary of the explicit formulae, as are some remarks on when different localisations give rise to the same distances.

It is perhaps not surprising, given what happens when constraints are given *a priori*, that one often needs analogues for local connectedness functions of, for example, the variant of the new Δ-property used by [Veličk] and [BS], in order to ensure that the amalgamations necessary to prove κ-$\mathbb{M}$-properness can be carried out. One also needs to understand how the different local connectedness functions derived from the sets of maps from pairs of conditions one is trying to amalgamate are related.

Thus, in §4 I define κ-$\mathbb{M}$-properness and build on the analysis of §1 to give generally applicable results of this sort, including analogues of the properties of c and ρ which have already proven so useful. In fact (the conclusions of) these analogues are stronger than the known results for c and ρ. The coherence properties of the simplified morass also ensure that the finite collections of maps from different conditions fit together in a compatible way and this gives one a reasonable chance of proving that κ^+ is preserved (often by showing that the κ^+-chain condition holds).

In §5, at the behest of the referee, I give two applications of the method, one topological, to scattered spaces of size κ^+, and one combinatorial. The reader is referred to that section for the details.

It may seem paradoxical to devote time to abstract analysis of these techniques after having started by highlighting a defence of the depth of combinatorial mathematics which emphasizes its accumulation of general principles for solving problems rather than on theory-building within it. However, that would be to confound this defence with the heroic fallacy that, although general heuristics may be useful, one can always solve combinatorial problems *ab initio*.

Actually, sometimes it is simply *useful* to have information about the abstract ways in which general principles have and/or can be implemented rather than reinventing the wheel on each occasion. Although Gowers ([G]) emphasizes that not all general principles can be precisely formulated, he also points out that some can be, and have then been "applied again and again."

In this case it proved possible and, moreover, convenient to give proofs of a variety of different results in §5 below, [M*4], [M*5], [M*6] and [M*8] in a more or less uniform way using the facts proven here in §§1–4. As these facts are independent of the specifics of the various forcings, it seems clearer to present them separately. But it would be an anathema to have aimed to set out the one true way of κ-$\mathbb{M}$-proper forcing (using local connectedness functions), and I hope that there remains plenty of scope for productive variations to be devised, especially regarding §4.

After extolling the fruitfulness of these generic stepping up techniques let me close this discussion on a downbeat note. Unfortunately κ-$\mathbb{M}$-proper forcing does not seem to open the way to analogous constructions of objects of size greater than κ^+. In particular, there seems to be no reasonable analogue of κ-$\mathbb{M}$-proper forcing for $(\kappa, 2)$-simplified morasses and there are technical reasons regarding $(\kappa, 2)$-simplified morasses why it seems unlikely such analogues will be found, even when $\kappa = \omega_1$. The general question of generically stepping up forcings with working parts of size less than κ^- (the cardinal predecessor of κ) to get objects of size at least κ^{++} remains an extremely important and radically open problem area.

For the most part the set theoretic notation used is standard. One piece of notation worth revising is that if $X \subseteq \mathrm{On}$ then $\mathrm{ssup}(X)$, the *strong supremum of* X, is the least ordinal α such that $X \subseteq \alpha$. Another is that $[\nu, \tau)$ is the interval of ordinals $\{\xi \mid \nu \leq \xi < \tau\}$, and similarly for (ν, τ) and $[\nu, \tau]$. A third is that if κ is a successor cardinal then κ^- is the cardinal predecessor of κ.

Definitions, Theorems, Lemmas and so on are numbered separately in each section. Thus, for example, Definition (8) of §1 is referred to as Definition (8) within §1 and as Definition (1.8) elsewhere.

I conclude this introduction by recalling some well-known definitions and facts about gap-one simplified morasses. I include give some sketches which may help comprehension. Similar sketches litter the remainder of the paper, but the reader may find it useful to supply more of their own to help their visualisation. The reader can consult [V84] and [M96] for more extensive expositions of the properties of gap-one simplified morasses, although this paper is, I hope, self-contained.

Definition 1. Let κ be a regular cardinal. $\mathbb{M} = \langle \langle \theta_\alpha \mid \alpha \leq \kappa \rangle, \langle \mathcal{F}_{\alpha\beta} \mid \alpha \leq \beta \leq \kappa \rangle \rangle$ *is a* $(\kappa, 1)$*-simplified morass* if $\langle \theta_\alpha \mid \alpha < \kappa \rangle$ is an increasing sequence of ordinals less than κ and $\theta_\kappa = \kappa^+$,

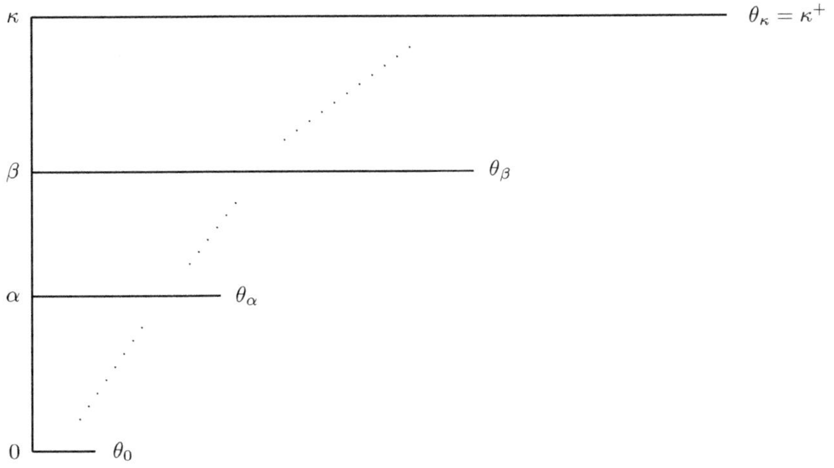

DIAGRAM 1. The ordinal part of a $(\kappa, 1)$-simplified morass. A vertical line represents $\kappa + 1$, the indices of the θ_α, and horizontal lines the θ_α.

and each $\mathcal{F}_{\alpha\beta}$ is a collection of maps from θ_α to θ_β

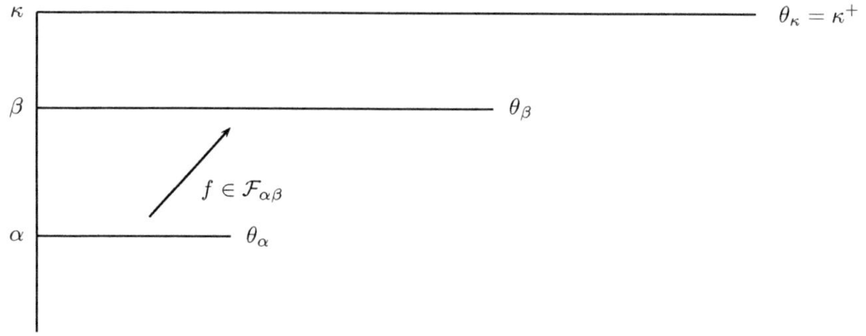

DIAGRAM 2. A map f in $\mathcal{F}_{\alpha\beta}$.

such that the following properties hold:

(i) $\forall\alpha \leq \kappa\ \mathcal{F}_{\alpha\alpha} = \{\mathrm{id}\}$

(ii) $\forall\alpha \leq \beta \leq \gamma \leq \kappa\ \mathcal{F}_{\alpha\gamma} = \{g \cdot f \mid f \in \mathcal{F}_{\alpha\beta}\ \&\ g \in \mathcal{F}_{\beta\gamma}\}$

(iii) $\forall\alpha < \kappa\ \big(\mathcal{F}_{\alpha\alpha+1} = \{\mathrm{id}\}\ \&\ \theta_{\alpha+1} = \theta_\alpha + 1$ or $\mathcal{F}_{\alpha\alpha+1} = \{\mathrm{id}, h\}\ \&\ \exists\sigma < \theta_\alpha\ \&$ $h \restriction \sigma = \mathrm{id}\ \&\ \forall\tau\ (\sigma + \tau < \theta_\alpha \longrightarrow h(\sigma+\tau) = \theta_\alpha + \tau)\big)$

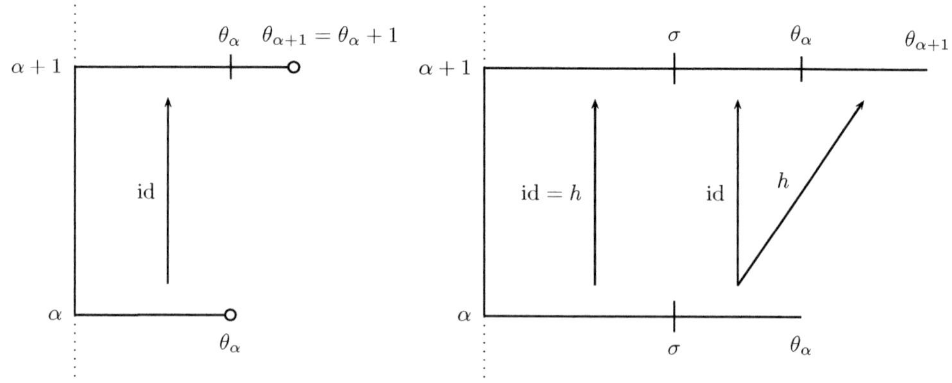

DIAGRAM 3. The two cases allowed in (iii).

(iv) $\forall \alpha \leq \kappa\ (\alpha$ is a limit ordinal $\longrightarrow \forall \beta_0, \beta_1 < \alpha\ \forall f_0 \in \mathcal{F}_{\beta_0 \alpha}\ \forall f_1 \in \mathcal{F}_{\beta_1 \alpha}$
$\exists \gamma \in [\beta_0 \cup \beta_1, \alpha)\ \exists h \in \mathcal{F}_{\gamma\alpha}\ \exists g_0 \in \mathcal{F}_{\beta_0\gamma}\ \exists g_1 \in \mathcal{F}_{\beta_1\gamma}\ (f_0 = h \cdot g_0\ \&\ f_1 = h \cdot g_1))$

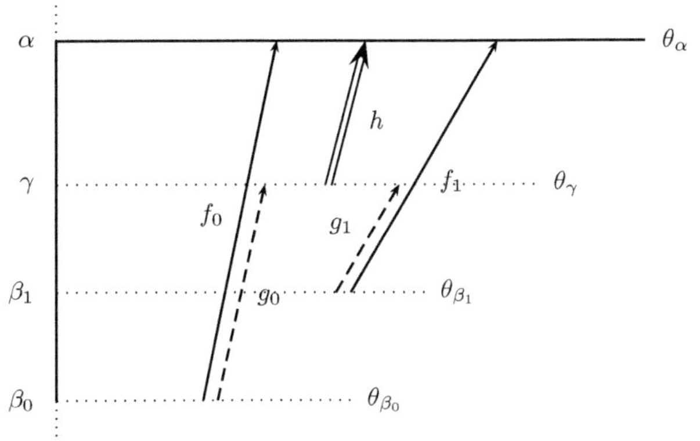

DIAGRAM 4. Illustration of (iv)

(v) $\bigcup\{ f``\theta_\alpha \mid \alpha < \kappa\ \&\ f \in \mathcal{F}_{\alpha\kappa} \} = \kappa^+$.

Definition 2. Let $\mathbb{M} = \langle \langle \theta_\alpha \mid \alpha \leq \kappa \rangle, \langle \mathcal{F}_{\alpha\beta} \mid \alpha \leq \beta \leq \kappa \rangle \rangle$ be a $(\kappa, 1)$-simplified morass. Then $\mathcal{F} = \bigcup\{ (\alpha, f) \mid \alpha < \kappa\ \&\ f \in \mathcal{F}_{\alpha\kappa} \}$.

The following well-known, fundamental fact due to Velleman always comes into play when dealing with simplified morasses.

Fact 3 (Velleman, [V84, Lemma 3.2]). For any $\alpha \leq \beta \leq \kappa$, f, $g \in \mathcal{F}_{\alpha\beta}$ with $\nu \in \mathrm{rge}(f) \cap \mathrm{rge}(g)$ there is some $\overline{\nu} < \theta_\alpha$ such that $f(\overline{\nu}) = \nu = g(\overline{\nu})$ and $f \restriction \overline{\nu} + 1 = g \restriction \overline{\nu} + 1$.

Proof. By induction on β for each α. □

Another two useful facts from [V84] are the following.

Fact 4 (Stanley, [V84, Theorem 3.9]). If $\alpha \leq \beta \leq \kappa$, $f \in \mathcal{F}_{\alpha\beta}$ and $\nu < \theta_\alpha$ then there is some $g \in \mathcal{F}_{\alpha\beta}$ such that $g \restriction \nu = f \restriction \nu$ and $g(\nu + \xi) = \mathrm{ssup}(g``\nu) + \xi$ for $\nu + \xi < \theta_\alpha$.

Proof. Again by induction on β for each α. □

Fact 5 (Velleman, [V84, Corollary 3.5]). If $\langle f_i \mid i < \chi \rangle$ is a collection of maps with each $f_i \in \mathcal{F}_{\alpha_i\beta}$ for $i < \chi$ and $\chi < \mathrm{cf}(\beta)$ then there is some $\alpha \in [\sup(\{ \alpha_i \mid i < \chi \}, \beta)$, some $f \in \mathcal{F}_{\alpha\beta}$ and maps $g_i \in \mathcal{F}_{\alpha_i\alpha}$ for all $i < \chi$ such that $f_i = f \cdot g_i$ for all $i < \chi$. In particular any collection of fewer than κ maps in $\mathcal{F}$ can be factored through some single common map $(\alpha, f) \in \mathcal{F}$. □

The following notation is now more or less standard for simplified morasses (see, for example, [M96]).

Definition 6. Let $\alpha \leq \beta \leq \kappa$, $f \in \mathcal{F}_{\alpha\beta}$ and $f(\overline{\nu}) = \nu$. Then $\psi_{\langle\alpha,\overline{\nu}\rangle,\langle\beta,\nu\rangle}$ is the function $f \restriction \overline{\nu}+1$. By Fact (3) this is a good definition. If $\beta = \kappa$ write ψ^{α}_{ν} for $\psi_{\langle\alpha,\overline{\nu}\rangle,\langle\beta,\nu\rangle}$. Write ν_α for the $\overline{\nu}$ such that $\psi^{\nu}_{\alpha}(\overline{\nu}) = \nu$.

Definition 7. If $\beta \leq \kappa$ and $\nu < \tau < \theta_\beta$ then $c_\beta(\nu,\tau) =$ the least $\alpha \leq \beta$ such that there is some $f \in \mathcal{F}_{\alpha\beta}$ with $\nu, \tau \in \mathrm{rge}(f)$. Write c for c_κ and note that $c(\nu,\tau) < \kappa$ for all $\nu, \tau < \kappa^+$ by properties (iii) and (v) of gap-one simplified morasses. c is the *coupling function* for $\mathbb{M}$, *see Diagram* 5.

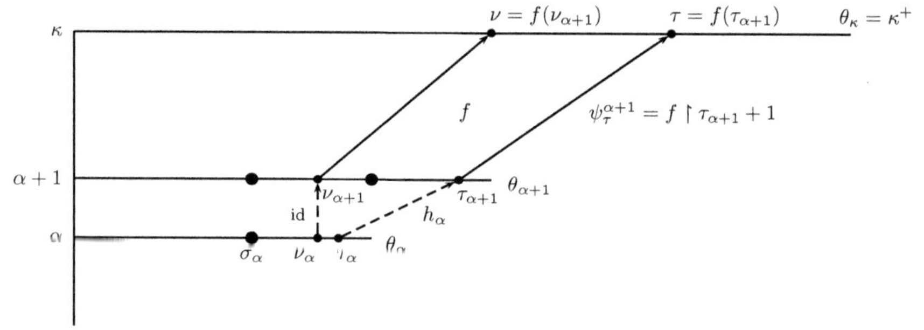

DIAGRAM 5. Illustration of $c(\nu,\tau) = \alpha+1$ being witnessed by $f \in \mathcal{F}_{\alpha\beta}$ as in Definition (7) and of the notation set up in Definition (6).

As motivation for some of the properties of functions defined in §1 I also recall the following fact.

Fact 8 ([M96]). c has the following properties.

If $\alpha < \kappa$ and $\tau < \kappa^+$ then $\overline{\overline{\{\xi \mid c(\xi,\tau) \leq \alpha\}}} < \kappa$.

c is *subadditive*: if $\mu < \nu < \tau < \kappa^+$ then

(i) $$c(\mu,\nu) \leq \max(\{c(\mu,\tau), c(\nu,\tau)\})$$

(ii) $$c(\mu,\tau) \leq \max(\{c(\mu,\nu), c(\nu,\tau)\})$$

If $\xi < \lambda = \bigcup\lambda < \mu$ then there is $\zeta \in [\xi,\lambda)$ such that for all $\nu \in [\zeta,\lambda)$ one has $c(\nu,\lambda) \leq c(\nu,\mu)$. □

Definition 9. Let $\nu < \tau < \kappa^+$ and suppose that $c_\kappa(\nu,\tau) \leq \alpha$. Then $D_\alpha(\nu,\tau) = \mathrm{otp}([\nu_\alpha,\tau_\alpha))$, *i.e.* $= \mathrm{otp}(\tau_\alpha \setminus \nu_\alpha)$. Let $D(\nu,\tau) = D_{c(\nu,\tau)}(\nu,\tau)$.

In §§4, 5 the local connectedness functions will be used in cardinal preservation arguments. In this context it will be necessary that the simplified morass is *stationary*, a notion now defined.

Definition 10. Let $\mathcal{L}$ be a first-order language. If $\mathcal{A} = (A,\dots)$ is an $\mathcal{L}$-structure then $|\mathcal{A}| = A$.

Definition 11. Let $\mathbb{M} = \langle\langle\theta_\alpha \mid \alpha \leq \kappa\rangle, \langle\mathcal{F}_{\alpha\beta} \mid \alpha \leq \beta \leq \kappa\rangle\rangle$ be a $(\kappa,1)$-simplified morass. $\mathbb{M}$ is *stationary* if $\{\mathrm{rge}(f) \mid \exists\alpha\,(\alpha,f) \in \mathcal{F}\}$ is stationary in $[\kappa^+]^{<\kappa}$.

Finally, in applications, *e.g.*, in §5, a couple more facts about simplified morasses are useful. The first is a simple, well-known extension of Fact (1.3).

Fact 12. Let $\alpha \leq \beta \leq \kappa$, and $f,\, g \in \mathcal{F}_{\alpha\beta}$. If $\nu,\, \tau \leq \theta_\alpha$ and $\mathrm{ssup}(f``\nu) = \mathrm{ssup}(g``\tau)$ then $\nu = \tau$ and $f \restriction \nu = g \restriction \nu$.

Proof. By induction on β for each α. (For $\nu,\, \tau < \theta_\alpha$ this is an immediate corollary of Facts (5) and (6), without the necessity of a separate inductive proof.) □

In contrast the following Δ-system type proposition does not seem to have been noted previously. The salient point is that no cardinal arithmetic assumption is necessary.

Proposition 13 (A Δ-system lemma for morass maps.). Let $\langle (\alpha_\eta, F_\eta) \mid \eta < \kappa^+ \rangle$ be a collection of distinct maps in $\mathcal{F}$. Then there is a (stationary) set $S \subseteq \kappa^+$ such that $\langle \mathrm{rge}(F_\eta) \mid \eta \in S \rangle$ forms a Δ-system.

Proof. Let $S_\kappa = \{\gamma < \kappa^+ \mid \mathrm{cf}(\gamma) = \kappa\}$. By Fodor's lemma one may as well assume that there is some $\alpha < \kappa$ and a stationary set $S \subseteq S_\kappa$ such that $\alpha_\eta = \alpha$ for all $\alpha \in S$.

Next set $G(\eta) = \mathrm{ssup}(\mathrm{rge}(F_\eta) \cap \eta)$ for $\eta \in S$. As $\mathrm{cf}(\eta) = \kappa$ and $\mathrm{otp}(\mathrm{rge}(F_\eta)) = \theta_\alpha < \kappa$ for all $\eta \in S$, one has that $G(\eta) < \eta$ for every $\eta < \kappa^+$. By Fodor's lemma again one has that there is some $\tau < \kappa^+$ and some stationary $T \subseteq S$ such that $G(\eta) = \tau$ for all $\eta \in T$. Hence, for each $\eta \in T$ one has that there is some $\tau_\eta \leq \theta_\alpha$ such that $\tau = \mathrm{ssup}(F_\eta ``\tau_\eta)$.

By Fact (12) one has that $\tau_\eta = \tau_\xi, = \overline{\tau}$, say, and that $F_\eta \restriction \overline{\tau} = F_\xi \restriction \overline{\tau}$ for all $\eta,\, \xi \in T$. Consequently $\mathrm{rge}(F_\eta) \cap \eta = \mathrm{rge}(F_\xi) \cap \xi$ $(\subseteq \tau)$ for all $\eta,\, \xi \in T$. (It is by the use of Fact (12) here that the necessity for cardinal arithmetic assumptions is avoided.)

Note also that $\tau < \theta_\alpha$ as otherwise all of the maps F_η for $\eta \in T$ are identical. Finally, choose by induction, $\{\eta_i \mid i < \kappa^+\} \subseteq T$ by letting η_ζ be the least $\eta \in T$ such that $\bigcup \{\mathrm{rge}(F_{\eta_i}) \mid i < \zeta\} \subseteq \eta$. □

1. On being 'connected in a set of maps' and the function c_A

Let κ be a regular. As will be the case throughout the paper, suppose that $(\kappa, 1)$-simplified morasses exist. Fix

$$\mathbb{M} = \langle \langle \theta_\alpha \mid \alpha \leq \kappa \rangle, \langle \mathcal{F}_{\alpha\beta} \mid \alpha \leq \beta \leq \kappa \rangle \rangle$$

a $(\kappa, 1)$-simplified morass, and let $\mathcal{F}$ be as in Definition (I.2).

Definition 1 ([K00, Definition 2.9] for $\kappa = \omega_1$ and A finite.). Let $A \subseteq \mathcal{F}$ and $\alpha < \kappa$. Then $\nu \leq \tau < \kappa^+$ are *α-connected in A* if and only if there are (α_0, f_0), $\ldots$, $(\alpha_{k-1}, f_{k-1}) \in A$, and $\xi_k = \nu \leq \xi_{k-1} \leq \cdots \leq \xi_0 = \tau$ such that

$$\forall i < k \; (\alpha_i \leq \alpha \quad \& \quad f_i \in \mathcal{F}_{\alpha_i \kappa} \quad \& \quad \xi_i,\, \xi_{i+1} \in \mathrm{rge}(f_i)).$$

(If $\nu \neq \tau$ one may as well assume strict inequality holds in the sequence $\xi_0, \ldots \xi_k$.) $\langle \xi \mid i \leq k \rangle$ and $\langle (\alpha_i, f_i) \mid i < k \rangle$ are *witnesses* that ν and τ are α-connected in A.

The key definition of this paper is the following definition of *local connectedness functions*.

Definition 2. Let $A \subseteq \mathcal{F}$. Define $c_A(\nu, \tau)$ = the least $\alpha < \kappa$ such that ν, τ are α-connected in A if ν, τ are α-connected in A for some $\alpha < \kappa$, and $= \kappa$ otherwise, (*see Diagram* 6).

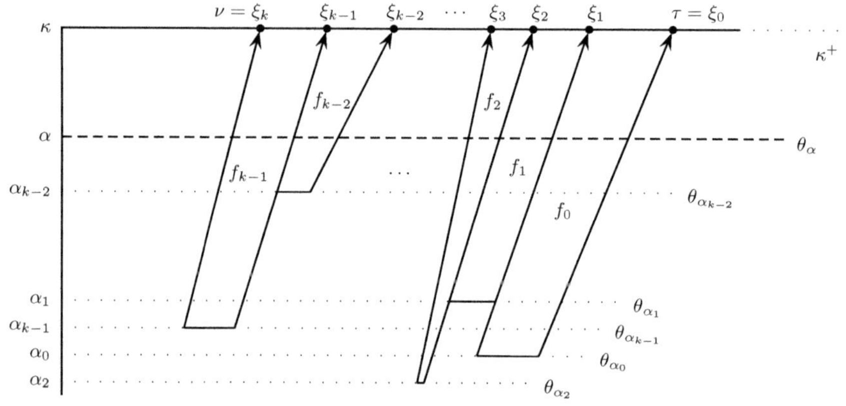

DIAGRAM 6. Illustrating $c_A(\nu, \tau) \leq \alpha$, *cf.* Definitions (1) and (2), where $f_i \in \mathcal{F}_{\alpha_i \kappa}$, $(\alpha_i, f_i) \in A$ and ξ_i, $\xi_{i+1} \in \mathrm{rge}(f_i)$ for all $i < k$.

Hence, for $\alpha < \kappa$, one has ν, τ are α-connected in A if and only if $c_A(\nu,\tau) \leq \alpha$. If $\langle \xi_i \mid i \leq k \rangle$ and $\langle (\alpha_i, f_i) \mid i < k \rangle$ witness that $c(\nu,\tau) = \alpha$ then $\max(\{\alpha_i \mid i \leq k\}) = \alpha$.

Lemma 3. $\forall A,\ B \subseteq \mathcal{F}\ \ \forall \nu < \tau < \kappa^+\ \ (A \subseteq B \longrightarrow c_B(\nu, \tau) \leq c_A(\nu, \tau))$.

Proof. Immediate from the definitions of c_A and c_B. □

The next lemma is fairly straightforward from the definition of the local connectedness functions and the basic facts about simplified morasses adumbrated in the introduction but will be used time and again.

Lemma 4. Suppose that $A \subseteq \mathcal{F}$ and $\nu < \tau < \kappa^+$. If $f \in \mathcal{F}_{\beta\kappa}$, $\tau \in \mathrm{rge}(f)$ and $c_A(\nu, \tau) \leq \beta$ then $\nu \in \mathrm{rge}(f)$.

Proof. Let $c_A(\nu, \tau) = \alpha$, say, with $\alpha \leq \beta$, and let $\nu = \xi_k < \cdots < \xi_0 = \tau$ and $\langle (\alpha_i, g_i) \mid i < k \rangle$ witness this. By induction on $i \leq k$ one has that $\xi_i \in \mathrm{rge}(f)$. For by the hypothesis $\xi_0 = \tau \in \mathrm{rge}(f)$. One also has $\alpha_i \leq \alpha \leq \beta$ so that that $g_i = h \cdot j$ for some $j \in \mathcal{F}_{\alpha_i \beta}$ and $h \in \mathcal{F}_{\beta\kappa}$ (by Definition (I.1.ii)). If one has that $\xi_i \in \mathrm{rge}(f)$ then this gives that $\xi_i \in \mathrm{rge}(f) \cap \mathrm{rge}(h)$, so, by Fact (I.3), there is some $\bar{\xi} < \theta_\beta$ such that $f(\bar{\xi}) = h(\bar{\xi}) = \xi_i$ and $f \restriction (\bar{\xi}+1) = h \restriction (\bar{\xi}+1)$. ($\bar{\xi} = (\xi_i)_\beta$ in the notation of Definition (I.5).) Then the fact that $\xi_{i+1} \in \mathrm{rge}(h)$ gives that $\xi_{i+1} \in \mathrm{rge}(f)$. Consequently $\nu = \xi_k \in \mathrm{rge}(f)$. □

The next three lemmas show that Lemma (4) can be exploited to give very useful results about how local connectedness functions vary when one varies the parameter A even without any further understanding of the witnesses to connectedness.

Lemma 5. Suppose $A, B \subseteq \mathcal{F}$, $A \subseteq B$, and that for all $(\alpha, f) \in A$ and $(\beta, g) \in B \backslash A$ one has that $\alpha \leq \beta$. Suppose $\nu < \tau < \zeta < \kappa^+$. Then if $c_A(\nu, \tau) < \kappa$ and $c_A(\nu, \tau) \leq c_A(\tau, \zeta)$ one has that $c_B(\nu, \tau) \leq c_B(\tau, \zeta)$.

Proof. Let $\beta^* = \min(\{\beta < \kappa \mid \exists(\beta, g) \in B \setminus A\})$. Clearly, as $c_A(\nu, \tau) < \kappa$ and since none of the maps in $B \backslash A$ can help reduce $c_B(\nu, \tau)$, one has that $c_B(\nu, \tau) = c_A(\nu, \tau)$. Similarly, if $c_A(\tau, \zeta) < \kappa$ then $c_B(\tau, \zeta) = c_A(\tau, \zeta)$. On the other hand, if $c_A(\tau, \zeta) = \kappa$ then $c_A(\nu, \tau) = c_B(\nu, \tau) \leq \beta^* \leq c_B(\tau, \zeta)$, since if $c_B(\tau, \zeta) < \kappa$ then one of any sequence of maps witnessing this is in $B \setminus A$. □

Lemma 6. Suppose $\forall A \subseteq \mathcal{F}$, $A = B \cup \{(\beta, g)\}$, and $\alpha \leq \beta$ for all $(\alpha, f) \in A$. Suppose $\nu < \tau < \zeta < \kappa^+$ and that $\nu \in \mathrm{rge}(g)$ if $\tau \in \mathrm{rge}(g)$. Then $c_A(\nu, \tau) \leq c_A(\tau, \zeta)$ implies that $c_B(\nu, \tau) \leq c_B(\tau, \zeta)$.

Proof. If $c_A(\nu, \tau) < \kappa$ the lemma is immediate from Lemma (5). If $c_A(\nu, \tau) = \kappa$ and $c_B(\tau, \zeta) < \kappa$ then (β, g) is in any witnessing sequence for the latter, whence $c_B(\tau, \zeta) = \beta$ and $\tau \in \mathrm{rge}(g)$. By the hypothesis of the lemma this gives that $\nu \in \mathrm{rge}(g)$, so that $c_A(\nu, \zeta) = \beta$. □

Remark 7. $c = c_{\mathcal{F}}$.

Proof. Let $\alpha = c_{\mathcal{F}}(\nu, \tau)$. If $\nu < \tau < \kappa^+$ then $c_{\mathcal{F}}(\nu, \tau) \leq c(\nu, \tau)$ by the definition of $c_{\mathcal{F}}$. On the other hand, if $f \in \mathcal{F}_{\alpha\kappa}$ and $\tau \in \mathrm{rge}(f)$ then $\nu \in \mathrm{rge}(f)$ by Lemma (4), so $c(\nu, \tau) \leq c_{\mathcal{F}}(\nu, \tau)$. But there is some such f since $\tau \in \mathrm{rge}(g)$ for any $g \in \mathcal{F}_{\beta\kappa}$ with $\beta < \kappa$ the first map of a witnessing sequence to the value of $c_{\mathcal{F}}(\nu, \tau)$ and g factors as some $f \cdot h$ with $f \in \mathcal{F}_{\alpha\kappa}$ and $h \in \mathcal{F}_{\beta\kappa}$ by (I.1.ii). (Or, more simply, by the neatness of $\mathbb{M}$.) □

Thus the functions c_A generalise the coupling function c and can be thought of as localisations of it. The remaining results of this section and §§2-4 will demonstrate that the c_A share some of c's most important properties and so are good generalisations of it.

The first of these results is a lemma showing that one can assume without loss of generality that the sequence of witnessing maps to the value of any $c_A(\nu, \tau)$ come from increasingly high up in the morass.

Definition 8. Let $A \subseteq \mathcal{F}$ and $\alpha < \kappa$. Suppose that $\nu < \tau < \kappa^+$ and that $c_A(\nu, \tau) \leq \alpha$. Then $\langle \xi_i \mid i \leq k \rangle$ and $\langle (\alpha_i, f_i) \mid i < k \rangle$ are *good witnesses* that $c_A(\nu, \tau) \leq \alpha$ or *good witnessing sequences* to $c_A(\nu, \tau) \leq \alpha$ if they witness $c_A(\nu, \tau) \leq \alpha$ as in Definitions (1) and (3) and if $\langle \alpha_i \mid i < k \rangle$ is strictly increasing. They are *good witnesses that* $c_A(\nu, \tau) = \beta$ or *to the value of* $c_A(\nu, \tau)$ if they are good witnesses that $c_A(\nu, \tau) \leq \beta$ and $c_A(\nu, \tau) = \alpha_{k-1} = \beta$, (*see Diagram* 7).

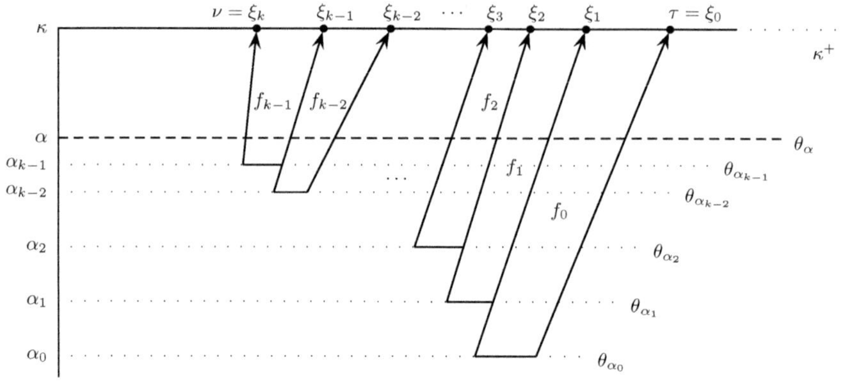

DIAGRAM 7. Illustrating good witnesses to $c_A(\nu,\tau) \leq \alpha$, *cf.* Definition (8) above, where $(\alpha_i, f_i) \in A$.

Lemma 9. Suppose that $A \subseteq \mathcal{F}$, $\alpha < \kappa$, $\nu < \tau < \kappa^+$ and $c_A(\nu,\tau) = \alpha$. Then there are good witnesses $\langle \xi_i \mid i \leq k \rangle$ and $\langle (\alpha_i, f_i) \mid i < k \rangle$ to the fact that $c_A(\nu,\tau) = \alpha$.

Proof. Let $\langle \xi_i \mid i \leq k \rangle$ and $\langle (\alpha_i, f_i) \mid i < k \rangle$ be any witnesses that $c_A(\nu,\tau) = \alpha$. Suppose $\alpha_{i+1} \leq \alpha_i$. One has that $c_A(\xi_{i+2}, \xi_{i+1}) \leq \alpha_i$, since $\xi_{i+2}, \xi_{i+1} \in \mathrm{rge}(f_{i+1}))$, and $\xi_{i+1} \in \mathrm{rge}(f_i)$, by the definition of a witnessing sequence. Hence, by Lemma (4), one has that $\xi_{i+2} \in \mathrm{rge}(f_i)$, (*see Diagram* 8). Thus the sequences generated by iteratively omitting every pair (α_{i+1}, f_{i+1}) and ξ_{i+1} such that $\alpha_{i+1} \leq \alpha_i$ are also witnesses that $c_A(\nu,\tau) = \alpha$. □

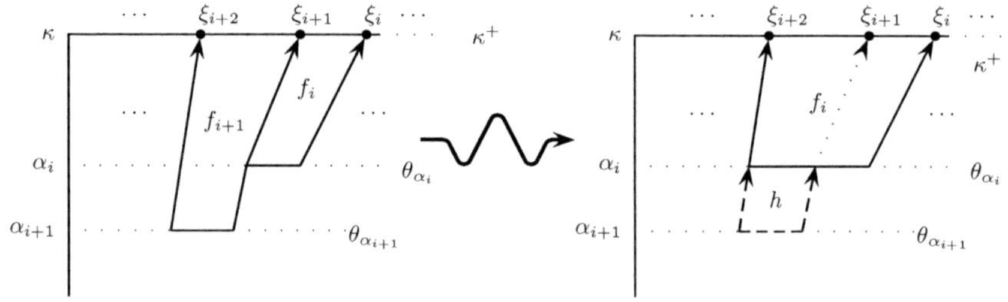

DIAGRAM 8. Illustrating the induction step towards obtaining good witnesses in Lemma (9). As $f_{i+1} \in \mathcal{F}_{\alpha_{i+1}\kappa}$ there is some (unique) $g \in \mathcal{F}_{\alpha_{i+1}\alpha_i}$ and some $h \in \mathcal{F}_{\alpha_i\kappa}$ such that $f_{i+1} = h \cdot g$. Since $\xi_{i+1} \in \mathrm{rge}(f_i) \cap \mathrm{rge}(f_{i+1})$ one has that $h \restriction (\xi_{i+1})_{\alpha_i}+1 = f_i \restriction (\xi_{i+1})_{\alpha_i}+1$, so $\xi_{i+2} \in \mathrm{rge}(f_i)$.

Note 10. Suppose $A \subseteq \mathcal{F}$, $\alpha < \kappa$, $\nu < \tau < \kappa^+$ and $c_A(\nu,\tau) = \alpha$. If $\langle \xi_i \mid i \leq k \rangle$ and $\langle (\alpha_i, f_i) \mid i < k \rangle$ are witnesses to the fact that $c_A(\nu,\tau) = \alpha$ such that $\alpha_i \leq \alpha_j$ for all $i < j < k$, then $c_A(\xi_j, \xi_i) \leq \alpha_{j-1}$ for all $i < j < k$.

With this information about the sort of witnessing sequences that are available in hand one can extract some more information about inequalities of the type considered in Lemmas (5) and (6). This inequality for decreasing parameter A is completely general (unlike Lemmas (5) and (6) which relied on increasing A only in a very controlled, albeit important, way).

Lemma 11.

$$\forall A,\, B \subseteq \mathcal{F}\, \forall \nu < \tau < \zeta < \kappa^+ \big(A \subseteq B \longrightarrow \\ (c_B(\nu,\tau) \leq c_B(\tau,\zeta) \longrightarrow c_A(\nu,\tau) \leq c_A(\tau,\zeta)\big).$$

Proof. If $c_A(\tau,\zeta) = \kappa$ there is nothing to be shown as $c_A(\nu,\tau) \leq \kappa$ by definition. If $c_A(\tau,\zeta) = \beta$, say, with $\beta < \kappa$, let $\xi_k = \tau < \xi_{k-1} < \ldots \xi_0 = \zeta$ and $\langle (\gamma_i, g_i) \mid i < k\rangle$ be a good witness showing this. Thus $\tau \in \mathrm{rge}(g_{k-1})$.

One has that $c_B(\nu,\tau) \leq c_B(\tau,\zeta) \leq c_A(\tau,\zeta)$ hypothesis and Lemma (11). So by Lemma (4) one has that $\nu \in \mathrm{rge}(g_{k-1})$. Hence $c_A(\tau,\zeta) \leq \beta$, as required, since $\nu < \xi_{k-1} < \ldots \xi_0 = \zeta$ and $\langle (\gamma_i, g_i) \mid i < k\rangle$ are good witnesses that $c_A(\nu,\tau) \leq c_A(\tau,\zeta)$. □

It is also now possible to prove one of the most important facts about the c_A, that they are subadditive.

Theorem 12. If $A \subseteq \mathcal{F}$ and $\nu < \tau < \zeta < \kappa^+$ then

(i) $$c_A(\nu,\tau) \leq \max(\{c_A(\nu,\zeta), c_A(\tau,\zeta)\}),$$

(ii) $$c_A(\nu,\zeta) \leq \max(\{c_A(\nu,\tau), c_A(\tau,\zeta)\}).$$

Proof. (i) Let $\langle \xi_i \mid i \leq k\rangle$ and $\langle (\alpha_i, f_i) \mid i < k\rangle$ be good witnesses to the value of $c_A(\tau,\zeta) = \alpha$, so that $\langle \alpha_i \mid i < k\rangle$ is strictly increasing, and let $\langle \xi'_i \mid i \leq l\rangle$ and $\langle (\beta_i, g_i) \mid i < l\rangle$ be good witnesses to the value of $c_A(\nu,\zeta) = \beta$, so that $\langle \beta_i \mid i < l\rangle$ is strictly increasing. Note that $\alpha = \alpha_{k-1}$ and $\beta = \beta_{k-1}$.

Suppose first of all that $c_A(\nu,\zeta) > c_A(\tau,\zeta)$. I claim that there is a witnessing sequence to the value of $c_A(\nu,\zeta)$ which features τ as one of the members of the sequence of ordinals. Let $i < l$ be greatest such that $\tau \leq \xi'_j$. There are two cases depending on whether $\beta_j \leq \alpha_{k-1}$ or not.

Suppose $\beta_j \leq \alpha_{k-1}$ and let $j' \in (j,l)$ be least such that $\alpha_{k-1} < \beta_{j'}$. Let $F \in \mathcal{F}_{\alpha\kappa}$ be such that $\zeta \in \mathrm{rge}(F)$. There is some such F because $\zeta \in \mathrm{rge}(f_0)$, by (I.i.ii) in the definition of $(\kappa,1)$-simplified morasses. (Note, I do not claim that $F \in A$ – in most cases $F \notin A$.) By Lemma (4) one has that τ, $\xi_{k-1} \in \mathrm{rge}(F)$ and so, letting $\bar{\xi}$ be $(\xi_{k-1})_\alpha$, whence $f_{k-1}(\bar{\xi}) = \xi_{k-1}$, one has that $F \restriction \bar{\xi}+1 = f_{k-1} \restriction \bar{\xi}+1$ and $\mathrm{rge}(F) \cap \xi_{k-1} = \mathrm{rge}(f_{k-1}) \cap \xi_{k-1}$. By Note (10) one has that $c_A(\xi'_{j'},\zeta) \leq \alpha$, so, by Lemma (4) again, one has that $\xi'_{j'} \in \mathrm{rge}(F)$. As $\xi'_{j'} < \tau$ one has that $\xi'_{j'} \in \mathrm{rge}(f_{k-1})$. Hence

$$\langle \xi_i \mid i \leq k\rangle^\frown\langle \xi'_i \mid i \in [j', l)\rangle$$

and

$$\langle (\alpha_i, f_i) \mid i < k\rangle^\frown(\alpha, f_{k-1})^\frown\langle (\beta_i, g_i \mid i \in [j', k)\rangle$$

are witnesses that $c_A(\nu,\zeta) = \beta$.

Now suppose that $\beta_j > \alpha_{k-1}$. Let $G \in \mathcal{F}_{\beta_j\kappa}$ be such that $\zeta \in \text{rge}(G)$. Note, again, I do not claim that $G \in A$, merely that there is some such G by (I.1.ii) since $\zeta \in \text{rge}(g_0)$. As in the previous argument, Lemma (4) ensures that both ξ'_j and $\tau \in \text{rge}(G)$. So $\tau \in \text{rge}(g_j)$. Consequently $\langle \xi_i \,|\, i \leq k\rangle^\frown\langle \xi'_i \,|\, i \in (j,l)\rangle$ and $\langle (\alpha_i, f_i) \,|\, i < k\rangle^\frown\langle (\beta_i, g_i \,|\, i \in [j,k)\rangle$ are witnesses that $c_A(\nu,\zeta) = \beta$.

Now it is clear that $c_A(\nu,\tau) = c_A(\nu,\zeta)$ since if $c_A(\nu,\tau) > c(\nu,\zeta)$ then truncating the initial part of the witnessing sequences just produced which include τ in the sequence of ordinals would give a contradiction by Note (10).

Finally, suppose that $c_A(\nu,\zeta) \leq c_A(\tau,\zeta)$. Again, there is some $F \in \mathcal{F}_{\alpha\kappa}$ such that $\tau \in \text{rge}(F)$. By Lemma (4), again, one has that $\tau \in \text{rge}(F)$ and that $\nu \in \text{rge}(F)$. Arguing as in the first case above, one has that $\text{rge}(F) \cap \xi_{k-1} = \text{rge}(f_{k-1}) \cap \xi_{k-1}$. So $\nu \in \text{rge}(f_{k-1})$, and $\langle \tau, \nu\rangle$ and $\langle (\alpha, f_{k-1})\rangle$ are witnesses that $c_A(\nu,\tau) \leq \alpha = c_A(\tau,\zeta)$.

(ii) It is clear that $c_A(\nu,\zeta) \leq \max(\{c_A(\nu,\tau), c_A(\tau,\zeta)\})$ since if one has witnessing sequences to the values of $c_A(\nu,\tau)$, $c_A(\tau,\zeta)$ then their concatentation is a witness to the inequality. □

Observation 13. Let $A \subseteq \mathcal{F}$. If $\alpha < \kappa$ and $\tau < \kappa^+$ then $\overline{\overline{\{\nu \,|\, c_A(\nu,\tau) \leq \alpha\}}} < \kappa$ (since $\bigcup\{\text{rge}(f) \,|\, (\alpha, f) \in A\}$ is of size less than κ). This gives another useful property shared by c_A and c.

The next lemma is interpolated here because the proof is elementary, although the lemma itself will be useful in arguments for the κ^+-chain condition for various κ-$\mathbb{M}$-proper forcings, (*cf.* §4, [M*4] and [M*5]).

Lemma 14. Suppose $\xi < \zeta < \kappa^+$, $\alpha < \kappa$, $D \subseteq \bigcup\{\mathcal{F}_{\gamma\alpha} \,|\, \gamma < \alpha\}$ and (α, F), $(\alpha, G) \in \mathcal{F}$. Set

$$\begin{aligned} A &= \{(\gamma, F\cdot g) \,|\, \gamma < \alpha \;\&\; g \in D \cap \mathcal{F}_{\gamma\alpha}\},\\ B &= \{(\gamma, G\cdot g) \,|\, \gamma < \alpha \;\&\; g \in D \cap \mathcal{F}_{\gamma\alpha}\}\\ \text{and} \qquad E &= A \cup B. \end{aligned}$$

Suppose $c_E(\xi,\zeta) \leq \alpha$. If $\zeta \in \text{rge}(F)$, then $c_E(\xi,\zeta) = c_A(\xi,\zeta)$, and if $\zeta \in \text{rge}(G)$ then $c_E(\xi,\zeta) = c_B(\xi,\zeta)$.

Proof. Suppose that $\zeta \in \text{rge}(F)$. Let $c_E(\xi,\zeta) = \beta$ and let $\langle \tau_i \,|\, i \leq k\rangle$ and $\langle (\gamma_i, g_i) \,|\, i < k\rangle$ be canonical short witnesses to this. As $\zeta \in \text{rge}(F)$ one has that $\tau_i \in \text{rge}(F)$ by Lemma (4) for each $i < k$. Suppose that $(\gamma_i, g_i) \in B$ for some $i < k$. Then $g_i = G \cdot h_i$ for some $h_i \in \mathcal{F}_{\gamma_i\alpha}$ with $(\alpha_i, h_i) \in A$. However then $\tau_i \in \text{rge}(F) \cap \text{rge}(G)$. So one has

$$F((\tau_i)_\alpha) = \tau_i = G((\tau_i)_\alpha) \quad \text{and} \quad F((\tau_i)_\alpha + 1) = G \restriction ((\tau_i)_\alpha + 1)$$

by Fact (I.3). Consequently one can replace the use of (γ_i, g_i) in the witnessing sequence by $(\gamma_i, F\cdot h_i)$, noting that $(\gamma_i, F\cdot h_i) \in A$ by the definition of A Hence $c_E(\xi,\zeta) \geq c_A(\xi,\zeta)$. So $c_E(\xi,\zeta) = c_A(\xi,\zeta)$ by Lemma (3).

Exactly the same argument with the rôles of A and B exchanged, gives that $c_E(\xi,\zeta) = c_B(\xi,\zeta)$ if $\zeta \in \text{rge}(G)$. □

The following definition and lemmas allow one to get a much more concrete grasp on the kind of witnessing sequences to the value of $c_A(\nu,\tau)$ that are available.

Definition 15. Let $A \subseteq \mathcal{F}$, $\tau < \kappa^+$ and $\alpha < \kappa$. Let $\xi_0 = \tau$ and define, by induction, $\xi_{i+1} = \min(\{\mathrm{rge}(g) \cap \xi_i \mid (\gamma, g) \in A \ \&\ \gamma \leq \alpha \ \&\ \xi_i \in \mathrm{rge}(g)\} \setminus \nu)$ and let γ_i be the least $\gamma \leq \alpha$ such that there is some $(\gamma, g) \in A$ with ξ_i, $\xi_{i+1} \in \mathrm{rge}(g)$. Let k be greatest such that ξ_k is defined. If $\nu < \tau$ and $c_A(\nu,\tau) = \alpha$ then $\langle \xi_i \mid i \leq k\rangle$ and $\langle (\gamma_i, g_i) \mid i < k\rangle$ are *canonical short witnesses* to the value of $c_A(\nu,\tau)$ if ξ_i, $\xi_{i+1} \in \mathrm{rge}(g_i)$ for all $i < k$, (*see Diagram* 9).

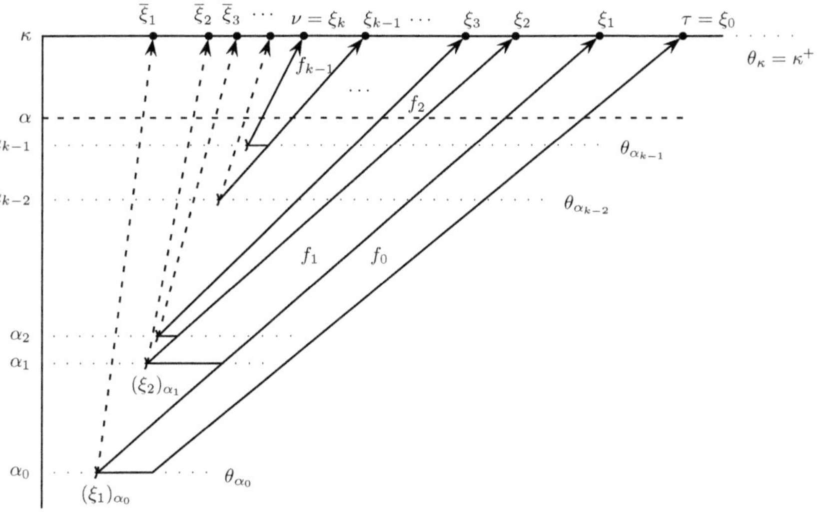

DIAGRAM 9. Illustrating *short* witnesses to $c_A(\nu,\tau) \leq \alpha$, *cf.* Definition (15), where $(\alpha_i, f_i) \in A$, $\xi_{i+1} = f_i((\xi_{i+1})_{\alpha_i})$ and $\bar\xi_{i+1} = \mathrm{ssup}(f_i``(\xi_{i+1})_{\alpha_i})$.

Clearly the ξ_i and γ_i are uniquely defined and the g_i are only defined uniquely up to agreeing on $(\xi_i)_{\alpha_i} + 1$.

Lemma 16. Let $A \subseteq \mathcal{F}$, $\nu < \tau < \kappa^+$ and $c_A(\nu,\tau) = \alpha < \kappa$. If $\langle \xi_i \mid i \leq k\rangle$ and $\langle (\gamma_i, g_i) \mid i < k\rangle$ are canonical short witnesses that $c_A(\nu,\tau) = \alpha$ then $\xi_k = \nu$ and the canonical short witnesses really are witnesses that $c_A(\nu,\tau) = \alpha$.

Proof. It is clear, by induction that $c_A(\xi_i,\tau) \leq \alpha$, by Note (10), and hence that $c_A(\nu,\xi_i) \leq \alpha$ by Theorem (12). As $\langle \xi_i \mid i \leq k\rangle$ is decreasing one has that the inductive definition of the ξ_{i+1} terminates only when one reaches some $\xi_k = \nu$. Thus canonical short witnesses are witnesses to the value of $c_A(\nu,\tau)$ and $\gamma_{k-1} = \alpha$ as otherwise one would have a contradiction to the minimality of α. $\square$

(Answering a question of the referee as to whether this Lemma is a triviality, notice that it is *not* assumed in Definition (15) that there will be some k with $\xi_k = \nu$, and so it has to be proven that there will indeed be some such k. This involves appealing to Theorem (12).)

The appellation ‘canonical short’ is justified by the following fact.

Proposition 17. If $A \subseteq \mathcal{F}$, $\nu < \tau < \kappa^+$, $c_A(\nu, \tau) = \alpha < \kappa$, and $\langle \xi_i \mid i \leq k \rangle$ and $\langle (\gamma_i, g_i) \mid i < k \rangle$ are canonical short witness to the value $c_A(\nu, \tau)$, then k is the least length of any witnessing sequence showing that $c_A(\nu, \tau) = \alpha$.

Proof. Let $\langle \xi_i \mid i \leq k \rangle$ and $\langle (\gamma_i, g_i) \mid i < k \rangle$ be canonical short witnesses to $c_A(\nu, \tau) = \alpha < \kappa$. Suppose that $\langle \zeta_i \mid i \leq l \rangle$ and $\langle (\beta_i, f_i) \mid i < l \rangle$ are witnesses to the value of $c_A(\nu, \tau)$ and of minimal length. Then $\langle \beta_i \mid i < l \rangle$ is strictly increasing or else one could decrease the length of the sequence by refining it as in Lemma (4), contradicting the minimality of l. Moreover suppose that the agreement between initial segments of $\langle \zeta_i \mid i \leq l \rangle$ and $\langle \xi_i \mid i \leq k \rangle$ is of maximal possible length. Suppose this agreement is not of length l. Let i be least such that $\xi_{i+1} \neq \zeta_{i+1}$. Then $\xi_{i+1} < \zeta_{i+1}$ since $\xi_i = \zeta_i$, $\zeta_{i+1} \in \mathrm{rge}(f_i)$ and $f_i \in A$, so $\xi_{i+1} \leq \min(\mathrm{rge}(f_i) \cap \nu) \leq \zeta_{i+1}$ by the definition of ξ_{i+1}. Consequently $\xi_{i+1} \neq \nu$ by the minimality of l. Let j be greatest such that $\xi_{i+1} < \zeta_j$.

If $\gamma_i \leq \beta_j$ then $\xi_{i+1} \in \mathrm{rge}(f_j)$ since there is some $F \in \mathcal{F}_{\beta_j \kappa}$ with $\xi_i \in \mathrm{rge}(F)$ by (I.1.ii). Now $c_A(\zeta_j, \zeta_i) \leq \beta_j$ by Note (10), so $\zeta_j \in \mathrm{rge}(F)$ by Lemma (4). Hence $\mathrm{rge}(F) \cap \zeta_j = \mathrm{rge}(f_j) \cap \zeta_j$ and one can apply Lemma (4) to get that $\xi_{i+1} \in \mathrm{rge}(f_j)$ since $c_A(\xi_{i+1}, \xi) \leq \beta_j$. Now the sequence $\langle \zeta_m \mid m < j \rangle^\frown \xi_{i+1}{}^\frown \langle \zeta_m \mid m \in (j+1, l) \rangle$ and $\langle (\beta_m, f_m) \mid m < l \rangle$ also witnesses the value of $c_A(\nu, \tau)$, has minimal length but has a longer agreement of initial segments with the canonical witness sequence than $\langle \zeta_m \mid m < l \rangle$ has, contradicting the maximality of that agreement.

So suppose that $\gamma_i > \beta_j$. Let j' be greatest such that $\gamma_i > \beta_{j'}$. Note that there is some such $j' < l - 1$ since $\xi_{i+1} \neq \nu$. Then $c_A(\zeta_{j'+1}, \zeta_i) < \gamma_i$ so $\zeta_{j'+1} \in \mathrm{rge}(g_i)$ by Lemma (4) (since $\zeta_i = \xi_i$). Then $\langle \zeta_m \mid m < i \rangle^\frown \xi_{i+1}{}^\frown \langle \zeta_m \mid m \in [j'+1, l) \rangle$ and $\langle (\gamma_m, g_m) \mid m \leq i \rangle^\frown \langle (\beta_m, f_m) \mid m \in [j'+1, l) \rangle$ also witnesses the value of $c_A(\nu, \tau) = \alpha$. This new witness either has length contradicting the minimality of l (if $i \neq j'$) or has minimal length but has a longer agreement of initial segments with the canonical witness sequence, contradicting the maximality of agreement that was assumed.

Hence one has that $k = l$ and the proposition is proven. □

2. Canonical long witnesses to connectedness

Notational reminder. In this section I will make heavy use of the notational conventions recalled in Definition (I.6). In particular, if $\nu < \kappa^+$, $(\beta, f) \in \mathcal{F}$ and $\nu \in \mathrm{rge}(f)$ then $f(\nu_\beta) = \nu$ and ψ_ν^β is the map $f \restriction \nu_\beta + 1$. Both ν_β and ψ_ν^β depend only on β and ν and not on f, by Fact (I.3). Sometimes I shall need to apply this notation to an ordinal signified by a Greek letter with a subscript, such as ξ_i. If $\xi_i \in \mathrm{rge}(f)$ for some $(\beta, f) \in \mathcal{F}$, I then write $(\xi_i)_\beta$ for the ordinal $\bar{\xi}$ such that $f(\bar{\xi}) = \xi_i$.

Definition 1. Let $A \subseteq \mathcal{F}$, $\alpha < \kappa$ and $\nu < \kappa^+$. Set

$$A_\alpha(\nu) = \{\xi < \nu \mid \exists(\gamma, g) \in A\ (\gamma \leq \alpha\ \ \&\ \ \nu,\, \xi \in \mathrm{rge}(g))\},$$
$$\Gamma(\nu) = \{\gamma \leq \alpha \mid \exists(\gamma, g) \in A\ \ \nu \in \mathrm{rge}(g)\},$$
$$\gamma^* = \gamma^*(\nu) = \sup(\Gamma(\nu)) \quad \text{and} \quad \psi_\nu = \psi_\nu^{\gamma^*(\nu)}.$$

So $A_\alpha(\nu)$ is the collection of ordinals ξ less than ν such that $c(\xi, \nu) \leq \alpha$ and this is demonstrated by some map in A. Obviously $\Gamma(\nu)$, γ^* and ψ_ν depend on A and α as well as ν, but as A and α will be fixed in the discussion in this section and the next the notation suppresses this dependence.

Lemma 2. Let $A \subseteq \mathcal{F}$, $\alpha < \kappa$ and $\nu < \kappa^+$. Then

$$A_\alpha(\nu) = \bigcup\{\psi_\nu^\gamma\text{“}\nu_\gamma \mid \exists(\gamma, g) \in A\ (\gamma \leq \alpha\ \ \&\ \ \nu,\, \xi \in \mathrm{rge}(g))\}.$$

Proof. Immediate from the definition of the ψ_ν^γ. □

The next proposition shows that each $A_\alpha(\nu)$ is the range of a single map derived from $\mathbb{M}$, albeit that this map need not be (a truncation of) a map in A.

Proposition 3. Let $A \subseteq \mathcal{F}$, $\alpha < \kappa$ and $\nu < \kappa^+$. Then $A_\alpha(\nu) = \psi_\nu\text{“}\nu_{\gamma^*}$.

Proof. It is clear that $A_\alpha(\nu) \subseteq \psi_\nu\text{“}\nu_{\gamma^*}$. For if $\nu \in \mathrm{rge}(g)$, $(\gamma, g) \in A$ and $\gamma \leq \alpha$ then $\gamma \leq \gamma^*$, by the definition of γ^*, and there are some $(\gamma^*, g^*) \in \mathcal{F}$ and $h \in \mathcal{F}_{\gamma\gamma^*}$ such that $g = g^* \cdot h$, by Definition (I.1.ii). But then $\nu = g(\nu_\gamma) = g^*(h(\nu_\gamma))$ and $h(\nu_\gamma) = \nu_{\gamma^*}$, and $h\text{“}\nu_\gamma \subseteq h(\nu_\gamma)$, so $g\text{“}\nu_\gamma \subseteq g^*\text{“}\nu_{\gamma^*}$.

The converse inequality is clear by Lemma (2) if γ^* is the maximal element of $\Gamma(\nu)$. On the other hand, if $\Gamma(\nu)$ has no maximal element then γ^* is a limit ordinal and if $\xi \in \psi_\nu\text{“}\nu_{\gamma^*}$ there is some $\gamma < \gamma^*$ and some map $h \in \mathcal{F}_{\gamma\gamma^*}$ such that ξ_{γ^*}, $\nu_{\gamma^*} \in \mathrm{rge}(h)$, by Definition (I.1.iv). Thus $\xi_{\gamma^*} \in \psi_{\langle\gamma,\nu_\gamma\rangle,\langle\gamma^*,\nu_{\gamma^*}\rangle}\text{“}\nu_\gamma$. So for any $\gamma' \in \Gamma(\nu) \setminus \gamma$ one has $\xi_{\gamma^*} \in \psi_{\langle\gamma',\nu_{\gamma'}\rangle,\langle\gamma^*,\nu_{\gamma^*}\rangle}\text{“}\nu_{\gamma'}$ as well (by the same argument as in the previous paragraph using (I.i.ii)). Hence $\xi \in \psi_\nu \cdot \psi_{\langle\gamma',\nu_{\gamma'}\rangle,\langle\gamma^*,\nu_{\gamma^*}\rangle}\text{“}\nu_{\gamma'} = \psi_\nu^{\gamma'}\text{“}\nu_{\gamma'}$, while $\psi_\nu^{\gamma'}\text{“}\nu_{\gamma'} \subseteq A_\alpha(\tau)$, by Lemma (2). □

Definition 4. Let $A \subseteq \mathcal{F}$, $\alpha < \kappa$ and $\nu < \kappa^+$. In view of Lemma (3), say that ψ_ν *instantiates* $A_\alpha(\nu)$. Occasionally I say that (γ, g), or ψ_τ^γ, *instantiates* $A_\alpha(\nu)$ if $\gamma = \gamma^*(\nu)$ and $g \restriction \nu_\gamma + 1 = \psi_\nu$, or $\psi_\tau^\gamma \restriction \nu_\gamma + 1 = \psi_\nu$, respectively,

Note it may be that there is no map $(\gamma^*, g) \in A$ such that $\nu \in \mathrm{rge}(g)$. However if there is some such (γ^*, g) then at times it will also be said that this map instantiates $A_\alpha(\nu)$ since $\psi_\nu \restriction \nu + 1 = g \restriction \nu + 1$.

Lemma 5. Let $A \subseteq \mathcal{F}$ and $\alpha < \kappa$. If $\nu < \kappa^+$, $(\beta, f) \in A$, $\beta \leq \alpha$ and $\nu \in \mathrm{rge}(f)$ then $\mathrm{rge}(f) \cap \nu \subseteq A_\alpha(\nu)$.

Proof. Immediate from the definition of $A_\alpha(\nu)$. □

Corollary 6. Let $A \subseteq \mathcal{F}$ and $\alpha < \kappa$. If $\nu < \tau < \kappa^+$ and $\nu \in A_\alpha(\tau)$ then $A_\alpha(\tau) \cap \nu \subseteq A_\alpha(\nu)$. Equality holds if $\gamma^*(\nu) = \gamma^*(\tau)$.

Proof. Let $(\gamma, g) \in A$ be such that $\gamma \leq \alpha$ and $\tau, \nu \in \mathrm{rge}(g)$. Then for all $\gamma' \in [\gamma, \gamma^*]$ one has that there are $f \in \mathcal{F}_{\gamma\gamma'}$ and $(\gamma', h) \in \mathcal{F}$ such that $g = h \cdot f$, and if $(\gamma', g') \in A$ and $\tau \in \mathrm{rge}(g)$ then $g' \restriction \tau_{\gamma'} = h \restriction \tau_{\gamma'}$, and hence $\nu \in \mathrm{rge}(g')$. Thus if $\xi \in A_\alpha(\tau) \cap \nu$ then there is some $(\gamma, g) \in A$ with ν, τ and $\xi \in \mathrm{rge}(g)$. So $A_\alpha(\tau) \cap \nu \subseteq A_\alpha(\nu)$.

Suppose $\gamma^*(\nu) = \gamma^*(\tau)$, $= \gamma^*$, say. One has $A_\alpha(\tau) \cap \nu = (\psi_\tau^{\gamma^*} \text{“}\tau_{\gamma^*}) \cap \nu$ and $A_\alpha(\nu) = \psi_\nu^{\gamma^*} \text{“}\nu_{\gamma^*}$. But $\nu \in \mathrm{rge}\, \psi_\tau^{\gamma^*}$, so $\psi_\tau^{\gamma^*} \restriction \nu_{\gamma^*} = \psi_\nu^{\gamma^*} \restriction \nu_{\gamma^*}$, by Fact (I.4), giving that $A_\alpha(\nu) = A_\alpha(\tau) \cap \nu$. □

The next proposition is similar in spirit to the definition of canonical short witnessing sequences, and one could define canonical short witnessing sequences through the inductive definition outlined in the proposition.

Proposition 7. Let $A \subseteq \mathcal{F}$, $\alpha < \kappa$, and $\nu < \tau < \kappa^+$. Define a sequence of ordinals ξ_i inductively. Set $\xi_0 = \tau$ and let $\xi_{i+1} = \min(A_\alpha(\xi_i) \setminus \nu)$ if $A_\alpha(\xi_i) \setminus \nu$ is non-empty, and stop the induction otherwise. As $\xi_{i+1} < \xi_i$ if both are defined the induction clearly stops after finitely many steps by the well-foundedness of the ordinals. Let $k-1$ be the greatest i such that $\nu < \xi_i$, so that either $\nu \in A_\alpha(\xi_{k-1})$ and $\xi_k = \nu$ or $A_\alpha(\xi_{k-1}) \setminus \nu = \emptyset$ and ξ_k is undefined. (Observe that if ξ_{i+1} is defined then there is some $(\gamma, g) \in A$ with $\gamma \leq \gamma^*(\xi_i)$ and ξ_{i+1}, $\xi_i \in \mathrm{rge}(g)$.)

If $c_A(\nu, \tau) \leq \alpha$, then (a) $c_A(\nu, \xi_i)$, $c_A(\xi_i, \tau) \leq \alpha$, and (b) $A_\alpha(\xi_i) \setminus \nu \neq \emptyset$, for all $i < k$. Consequently, $\xi_k = \nu$.

One also has that $\langle \gamma^*(\xi_i) \,|\, i < k \rangle$ is strictly increasing. And that if $\langle (\gamma_i, g_i) \,|\, i < k \rangle$ is such that $(\gamma_i, g_i) \in A$ and $\xi_{i+1}, \xi_i \in \mathrm{rge}(g_i)$ then $\gamma_i \leq \gamma^*(\xi_i) < \gamma_j$ for $i < j < k$ and $\langle \xi_i \,|\, i \leq k \rangle$ and $\langle (\gamma_i, g_i) \,|\, i < k \rangle$ are good witnesses to $c_A(\nu, \tau) \leq \alpha$. They are canonical short witnesses to the value of $c_A(\nu, \tau)$ if $\gamma^*(\xi_{k-1}) \leq c_A(\nu, \tau)$ and γ_i is the minimal γ such that there is some $(\gamma, g) \in A$ with ξ_{i+1}, $\xi_i \in \mathrm{rge}(g)$ for each $i < k$. Finally, k is the minimal length of a witnessing sequence to $c_A(\nu, \tau) \leq \alpha$.

Proof. The proof of (a) and (b) is by an interwoven induction on $k-1$. It is trivial that (a) holds for $i = 0$. Suppose one has that (a) for i. If (β, f) is the first element of A in any sequence showing that $c_A(\nu, \xi_i) \leq \alpha$ then $\mathrm{rge}(f) \cap [\nu, \xi_i) \neq \emptyset$ and so $A_\alpha(\xi_i) \setminus \nu \neq \emptyset$ as $\mathrm{rge}(f) \cap \nu \subseteq A_\alpha(\xi_i)$ by Lemma (5), so (b) holds for i. On the other hand, if (b) holds for i then ξ_{i+1} is defined and $\xi_{i+1} < \xi_i$. As $c_A(\xi_i, \tau) \leq \alpha$ and $c_A(\xi_{i+1}, \xi_i) \leq \alpha$ one has that (a) holds for $i+1$ by the subadditivity of c_A.

As noted in the definition of the ξ_i in the statement of the proposition, the consequence of (b) for $i = k-1$ is that $\xi_k = \nu$.

$\langle \gamma_i \,|\, i < k \rangle$ is a strictly increasing sequence by Lemma (6). If $\langle (\gamma_i, g_i) \,|\, i < k \rangle$ is as in the statement of the Lemma it is clearly a sequence witnessing that $c_A(\nu, \tau) \leq \alpha$, and such sequences do exist by the observation in the statement of the Lemma. Also $\gamma^*(\xi_i) < \gamma_{i+1}$ if $i+1 < k$ since otherwise one has a contradiction to the minimality of ξ_{i+1} in $A_\alpha(\xi_i) \setminus \nu$. Thus these witnesses are good. They are clearly canonical short witnesses if $\gamma^*(\xi_{k-1}) \leq c_A(\nu, \tau)$. Finally k is the minimal length of any witness sequence that $c_A(\nu, \tau) \leq \alpha$ by an argument almost identical to Proposition (1.10). I leave the details to the reader. □

Definition 8. Let $A \subseteq \mathcal{F}$, $\alpha < \kappa$, and $\nu < \tau < \kappa^+$. As in the statement of Proposition (7), let $\xi_0 = \tau$ and inductively set $\xi_{i+1} = \min(A_\alpha(\xi_i) \setminus \nu)$ if $A_\alpha(\xi_i) \setminus \nu$ is non-empty, stopping the induction otherwise. Suppose that $c_A(\nu, \tau) \leq \alpha$. Then $\langle \xi_i \mid i \leq k \rangle$ and $\langle \psi_{\xi_i} \mid i < k \rangle$ are *canonical short high witnesses* that $c_A(\nu, \tau) \leq \alpha$.

The idea behind canonical short high witnesses is much less complicated than the portmanteau Proposition (7) may make it appear and, in fact, is quite simple. One would like to start at τ and find a map in A which starts as far up the morass below $\alpha + 1$ as possible with τ in its range. One would then go as far along the range of this map towards ν as possible and make a note of the ordinal one arrives at. One would repeat the process, bounding along towards ν as fast as one could until one actually arrived at ν in this way, which one would do, with no possibility of overstepping it, provided that $c_A(\nu, \alpha) \leq \alpha$. However, if A is not finite, one cannot necessarily find maps in A with maximal heights below $\alpha + 1$ with particular ordinals in their range, so one has to do the next best thing and use the morass maps ψ_{ξ_i} derived from A instead. Making use of these maps instead, one repeatedly travels along their ranges until one finds ν. (Of course if A is finite then one actually proceeds in the fashion desired at the start of the paragraph.)

Lemma (7) says that this is the fastest way to get from τ to ν using maps from A which are from levels at or below α, in the sense of having the fewest intermediate ordinals to be noted down and ranges of maps to be traced along, *i.e.* the lemma shows canonical short high witnesses are the shortest possible sequences witnessing $c(\nu, \tau) \leq \alpha$.

But it turns out that canonical short (high) witnesses are by no means the only useful canonical (good) way of journeying from τ to(wards) ν. The next part of this section discusses canonical *long* witnesses. The idea here is essentially that one takes steps that are as short as is reasonable. Instead of going as far along each successive map towards ν as possible, one only goes as far towards ν as one can while it is still true that the map instantiates $A_\alpha(\eta)$ for the η that one is passing through.

Then one has to stumble about a little to find the next map to journey along. Fortunately the same map is fine for a final segment of the ordinals where one might chance to start the next step, so one really can journey along the length of the range of this map, until it too no longer instantiates the $A_\alpha(\eta)$ and one changes map again. Eventually one does finally reach ν if $c_A(\nu, \tau) \leq \alpha$. It is reasonably clear that this procedure takes (in general) many more steps than canonical short high witnesses, and as the definition can plausibly be described as canonical the name of canonical long witnesses is perhaps reasonable. In §3, when one sees the rôle played by the canonical long witnesses in the analysis of Koszmider's local distance function from [K00], justification for calling this sequence of maps and ordinals *the* canonical long witnesses will become more evident, because while one can increase the length of the sequence witnessing $c_A(\nu, \tau) \leq \alpha$ one cannot in this way increase the distance which will be derived from the witnessing sequence.

Definition 9. Let $A \subseteq \mathcal{F}$ and $\alpha < \kappa$. Let ψ_ν instantiate $A_\alpha(\nu)$. Let

$$C_\alpha(\nu) = \{\xi \in A_\alpha(\nu) \mid \psi_\nu \restriction \xi_{\gamma^*(\nu)} \text{ instantiates } A_\alpha(\xi)\}.$$

By Corollary (6) one has that $C_\alpha(\nu) = \{\xi \in A_\alpha(\nu) \mid \gamma^*(\nu) = \gamma^*(\xi)\}$.

Lemma 10. Let $A \subseteq \mathcal{F}$, $\alpha < \kappa$ and $\nu < \kappa^+$. If $\xi < \zeta$ and $\xi, \zeta \in A_\alpha(\nu)$ then there is some $(\beta, f) \in A$ such that ξ, ζ, and $\nu \in \text{rge}(f)$.

Proof. If $\Gamma(\nu)$ has a maximal element then any pair $(\gamma^*(\nu), h)$ with $\nu \in \text{rge}(h)$ also has $\xi, \zeta \in \text{rge}(h)$. If $\Gamma(\nu)$ has no maximal element then $\gamma^*(\nu)$ is a limit ordinal and there is some $(\varepsilon, h) \in A$ with $\xi, \zeta, \nu \in \text{rge}(h)$ by Definition (I.1.ii). □

Lemma 11. Let $A \subseteq \mathcal{F}$, $\alpha < \kappa$ and $\nu < \kappa^+$. If $\xi < \zeta$ and $\xi, \zeta \in A_\alpha(\nu)$ then $\gamma^*(\zeta) \leq \gamma^*(\xi)$.

Proof. By Lemma (10), fix some $(\beta, f) \in A$ such that ξ, ζ, and $\nu \in \text{rge}(f)$. If one had that $\gamma^*(\zeta) > \gamma^*(\xi)$ then there is some $(\gamma, g) \in A$ such that $\gamma^*(\xi) < \gamma \leq \gamma^*(\zeta)$ and $\zeta \in \text{rge}(g)$. Factor f as $f = f'' \cdot f'$ where $(\gamma, f'') \in \mathcal{F}$ and $f' \in \mathcal{F}_{\beta\gamma}$. Then $\zeta \in \text{rge}(f'') \cap \text{rge}(g)$, so $f'' \restriction \zeta_\gamma = g \restriction \zeta_\gamma$ and, hence, $\xi \in \text{rge}(g)$. This contradicts the supposition that $\gamma^*(\xi) < \gamma$. □

Lemma 12. Let $A \subseteq \mathcal{F}$ and $\alpha < \kappa$. Then $C_\alpha(\nu)$ is a final segment of $A_\alpha(\nu)$ for each $\nu < \kappa^+$.

Proof. If $C_\alpha(\nu) = \emptyset$ the lemma is trivially true. Otherwise, let $\xi \in C_\alpha(\nu)$ and let $\zeta \in A_\alpha(\nu) \setminus \xi + 1$. By Lemma (11) $\gamma^*(\xi) \geq \gamma^*(\zeta)$. As $\gamma^*(\nu) = \gamma^*(\xi)$ this gives that $\gamma^*(\nu) = \gamma^*(\zeta)$. □

Corollary 13. Let $A \subseteq \mathcal{F}$ and $\alpha < \kappa$. If $\xi, \zeta \in C_\alpha(\nu)$ and $\xi < \zeta$ then $\xi \in C_\alpha(\zeta)$ and $A_\alpha(\zeta) \setminus \xi = A_\alpha(\nu) \cap [\xi, \zeta)$. □

Note 14. Even if $\xi \in A_\alpha(\nu)$, $A_\alpha(\xi) = A_\alpha(\nu) \cap \xi$ and $\xi < \zeta \in A_\alpha(\nu)$ it may be the case that $A_\alpha(\zeta) \neq A_\alpha(\nu) \cap \zeta$. For it could be, for example, that (β, f) instantiates $A_\alpha(\nu)$, but that there is some $(\gamma, g) \in A$ with $\beta < \gamma$ and $\zeta \in \text{rge}(g)$ and, letting $f = k \cdot h$ where $k \in \mathcal{F}_{\gamma\kappa}$ and $h \in \mathcal{F}_{\beta\gamma}$, one has that $h \restriction (\xi_\beta + 1) = \text{id} \restriction (\xi_\beta + 1)$ and $h(\nu_\beta) \neq \nu_\beta$. Thus if $c_A(\nu, \tau) \leq \alpha$ and $\eta \in A_\alpha(\xi_i) \setminus \nu$, whence $c_A(\nu, \tau) \leq \alpha$, it may be that the canonical short high witnesses to the latter are not just a truncation of the canonical short high witnesses to the former.

Lemma 15. Suppose $A \subseteq \mathcal{F}$, $\alpha < \kappa$ and $\nu < \kappa^+$. There is some $\eta < \nu$ and some $\beta \leq \alpha$ such that ψ^β_τ instantiates $A_\alpha(\tau)$ for all $\tau \in A_\alpha(\nu) \setminus \eta$. (Note that for all $\tau \in A_\alpha(\nu) \setminus \eta$ one has that $\psi^\beta_\tau = \psi^\beta_\nu \restriction \tau + 1$.)

Proof. If $A_\alpha(\nu)$ has a maximal element then the lemma is trivial. If not, then, by Lemma (11), $\langle \gamma^*(\tau) \mid \tau \in A_\alpha(\nu) \rangle$ is a decreasing sequence of ordinals, so attains its minimum at some $\eta \in A_\alpha(\nu)$. By Lemma (11), ψ^β_τ instantiates $A_\alpha(\tau)$ for all $\tau \in A_\alpha(\nu) \setminus \eta$. □

Definition 16. Suppose that $A \subseteq \mathcal{F}$, $\alpha < \kappa$ and $\nu < \kappa^+$. Let $\eta < \nu$ and $\beta \leq \alpha$ be as given by Lemma (15), so that ψ_τ^β instantiates $A_\alpha(\tau)$ for all $\tau \in A_\alpha(\nu) \setminus \eta$. (Recall, again, that $\psi_\tau^\beta = \psi_\nu^\beta \restriction \tau_\beta$ for all $\tau \in A_\alpha(\nu)$.) Set

$$\psi_\nu^* = \psi_\nu^\beta \quad \text{and} \quad C'_\alpha(\nu) = \{\xi \in A_\alpha(\nu) \mid \psi_\nu^* \text{ instantiates } A_\alpha(\xi)\}.$$

Lemma 17. Suppose $A \subseteq \mathcal{F}$, $\alpha < \kappa$ and $\nu < \kappa^+$. If $\xi \in C'_\alpha(\nu)$ and $\rho \in A_\alpha(\nu) \setminus \xi$, then $\rho \in C'_\alpha(\nu)$.

Proof. Immediate from Lemma (11). □

Lemma 18. Suppose $A \subseteq \mathcal{F}$, $\alpha < \kappa$ and $\nu < \kappa^+$. If $C_\alpha(\nu) \neq \emptyset$ then $C'_\alpha(\nu) = C_\alpha(\nu)$.

Proof. Immediate from the definitions. □

Lemma 19. If $A \subseteq \mathcal{F}$, $\alpha < \kappa$, $\nu < \kappa^+$ and $\rho \in C'_\alpha(\nu)$, then $C_\alpha(\rho) \cap A_\alpha(\nu) = C'_\alpha(\nu) \cap \rho$.

Proof. Again immediate from the definitions. □

Definition 20. Let $A \subseteq \mathcal{F}$ and $\alpha < \kappa$. Suppose $\nu < \tau < \kappa^+$. Set $\zeta_0 = \tau$ and make the following inductive definition. If $\nu \in C'_\alpha(\zeta_i)$ then let $\zeta_{i+1} = \nu$. If $\nu \notin C'_\alpha(\zeta_i)$ let $\zeta_{i+1} = \min(C'_\alpha(\zeta_i))$. (So the induction continues as long as ν is not found in one of the sets mentioned in the definition and stops, without the next ζ being defined, when ν is found or one of the sets is empty.) Let $\psi_{\zeta_i}^*$ instantiate $A_\alpha(\eta)$ for $\eta \in C'_\alpha(\zeta_i)$ be as given in the definition of $C'_\alpha(\zeta_i)$. (Note that $\gamma^*(\zeta_i) < \gamma_i$, so that $\psi_{\zeta_i}^* \neq \psi_{\zeta_i}$, for $1 \leq i < l$.)

Theorem 21. The sequence of 'ζ_i's defined in Definition (20) is strictly decreasing and so has finite length, say l. If $c_A(\nu, \tau) \leq \alpha$ then $\nu \in C'_\alpha(\zeta_{l-1})$.

Proof. It is clear since each of the sets $C_\alpha(\zeta_i)$, $C'_\alpha(\zeta_i) \subseteq \zeta_i$ that $\zeta_{i+1} < \zeta_i$ whenever ζ_{i+1} is defined, so the inductive definition does stop.

If $\nu \in C_\alpha(\tau)$ then $\zeta_1 = \nu$ and there is nothing more to prove. So suppose that $\nu \notin C_\alpha(\tau)$. The remainder of the proof of the theorem is broken into a series of lemmas (Lemmas (22) to Corollary (27)).

Lemma 22. If $\eta \in C_\alpha(\tau) \cup \{\tau\}$ then $c_A(\eta, \tau) \leq \alpha$ and $c_A(\nu, \eta) \leq \alpha$.

Proof. As $c_A(\nu, \tau) \leq \alpha$, by taking the first map in any sequence witnessing this, it is clear that there is some $(\beta, f) \in A$ such that $\beta \leq \alpha$ and $\tau \in \text{rge}(f)$. So $\gamma^*(\tau)$ is defined, and is at most α. Now, by Lemma (10), for any $\rho \in A_\alpha(\tau)$ one has that there is some $(\gamma, g) \in A$ such that $\rho, \tau \in \text{rge}(g)$. Then $\gamma \leq \gamma^*(\tau) \leq \alpha$. So one has that $c_A(\eta, \tau) \leq \alpha$ and, by Theorem (1.12), the subadditivity of c_A, that $c_A(\nu, \eta) \leq \alpha$. □

Lemma 23. If $\eta \in C'_\alpha(\zeta_i) \setminus \nu$ then $c_A(\nu, \eta)$, $c_A(\eta, \tau) \leq \alpha$.

Proof. By induction on i. Suppose one has that $\nu \leq \zeta_i$ and that $c_A(\nu, \zeta_i) \leq \alpha$. (For $i = 0$ this is true by the hypothesis of the theorem.) If $\eta \in C'_\alpha(\zeta_i) \setminus \nu \subseteq A_\alpha(\zeta_i)$ then, by Lemma (10), clearly $c_A(\eta, \zeta_i) \leq \gamma^*(\zeta_i) \leq \alpha$. And, by the subadditivity of c_A, Theorem (1.12), again, $c_A(\nu, \eta)$, $c_A(\eta, \tau) \leq \alpha$. As $\zeta_{i+1} = \min(C'_\alpha(\zeta_i))$ if it is defined, the induction hypothesis is maintained if $\nu \leq \zeta_{i+1}$. □

Lemma 24. If ζ_i is defined and $\nu<\zeta_i$ one has that $A_\alpha(\zeta_i)\setminus\nu\neq\emptyset$ and so $C'_\alpha(\zeta_i)\setminus\nu\neq\emptyset$.

Proof. As $c_A(\nu,\zeta_i)\leq\alpha$ by Lemma (23), let (β,f) be the first map in any sequence witnessing this. One has that $\zeta_i\in\mathrm{rge}(f)$ and ζ_i is not minimal in $\mathrm{rge}(f)\setminus\nu$. Since $\mathrm{rge}(f)\cap[\nu,\zeta_i)\subseteq(\psi_{\zeta_i}``(\zeta_i)_{\gamma^*(\zeta_i)}\setminus\nu)=A_\alpha(\zeta_i)\setminus\nu$ one has that $A_\alpha(\zeta_i)\setminus\nu\neq\emptyset$. As, by definition, $C'_\alpha(\zeta_i)$ is a final segment of $A_\alpha(\zeta_i)$, the second half of the conclusion is immediate. □

Corollary 25. If $\nu<\zeta_i$ then ζ_{i+1} can only be undefined if

$$\nu\in C'_\alpha(\zeta_i)\quad\text{or}\quad\min(C'_\alpha(\zeta_i))<\nu.$$ □

Lemma 26. Suppose $\nu<\zeta_i$ and $\min(C'_\alpha(\zeta_i))\leq\nu$. Then $\nu\in C'_\alpha(\zeta_i)$.

Proof. If $\nu=\min(C'_\alpha(\zeta_i))$ there is nothing to prove. So suppose otherwise. One has that $\psi^*_{\zeta_i}\restriction\eta_{\gamma^*(\zeta_i)}+1$ instantiates $A_\alpha(\eta)$ for every $\eta\in C'_\alpha(\zeta_i)$. As $C'_\alpha(\zeta_i)$ is a final segment of $A_\alpha(\zeta_i)$ there is no $\eta\in A_\alpha(\zeta_i)\setminus\nu$ such that there is some $(\beta,f)\in A$ with $\eta\in\mathrm{rge}(f)$ and $\beta>\gamma_i$. But $c_A(\nu,\zeta_i)\leq\alpha$, so, by applying Lemma (1.4), it must be that $\nu\in\mathrm{rge}(\psi^*_{\zeta_i})$. Hence $\nu\in C'_\alpha(\zeta_i)$ by Lemma (11) □

Corollary 27. The definition of the ζ_i stops at some finite stage, say, l, and $\nu\in C'_{\zeta_{l-1}}$, whence $\zeta_l=\nu$, thus completing the proof of Theorem (21). □

Lemma 28. Let $A\subseteq\mathcal{F}$, $\alpha<\kappa$, and $\nu<\tau<\kappa^+$. Let $\langle\zeta_i\mid i<l\rangle$ be as defined in Definition (20). Suppose that $c_A(\nu,\tau)\leq\alpha$. Let $\eta\in C'_\alpha(\zeta_i)\setminus\nu$ (or $\eta=\tau$ if $C_\alpha(\tau)\neq\emptyset$). Then $A_\alpha(\eta)\cap C'_\alpha(\zeta_{i+1})\neq\emptyset$ if $i<l-1$.

Proof. Immediate as $A_\alpha(\eta)\cap\zeta_{i+1}=A_{\zeta_{i+1}}$ and $C'_\alpha(\zeta_{i+1})$ is a non-empty final segment of $A_{\zeta_{i+1}}$ if $i<l-1$. □

Definition 29. Suppose $c_A(\nu,\tau)\leq\alpha$. The sequences $\nu=\zeta_l<\cdots<\zeta_0=\tau$ and $\psi^*_{\zeta_i}$ for $i<l$ given by Definition (20) are *canonical long witnesses* that this is so. Note that Theorem (21) proves that one could re-write the definition of the ζ_i as $\zeta_0=\tau$ and $\zeta_{i+1}=\min(C'_\alpha(\zeta_i)\setminus\nu)$. (*cf.* Diagram (10) after Theorem (3.7).)

Canonical long witnesses are not witnesses in the sense that they directly show that $c_A(\nu,\tau)\leq\alpha$. As well as it being possible that the maps $\psi^*_{\zeta_i}$ are not restrictions of maps in A, one also has that although $\zeta_{i+1}\in\mathrm{rge}(\psi^*_{\zeta_i})$ nevertheless $\zeta_i\notin\mathrm{rge}(\psi^*_{\zeta_i})$ for $0<i<l$ (and for $i=0$ as well if $C_\alpha(\tau)=\emptyset$). Rather they are a sort of limit of the possible witnesses. By Lemma (28) one can pick ζ'_{i+1} for $i<l-1$ with $\zeta'_{i+1}\in A_\alpha(\zeta_i)$ and $A_\alpha(\zeta'_{i+1})=A_\alpha(\zeta_{i+1})$ which together with the maps $\psi_{\zeta'_i}(=\psi^*_{\zeta_i}\restriction(\zeta'_i)_{\gamma^*(\zeta_i)}+1)$ are witnesses in the sense that canonical short high witnesses are (making the convention of setting $\zeta'_0=\zeta_0=\tau$). That is one can replace each of the witnessing maps by maps which genuinely are in A which do witness that $c(\zeta'_{i+1},\zeta_i)\leq\gamma^*(\zeta_i)$. However one cannot find witnesses higher up the morass, or, in a sense discussed in the following section, slower.

Canonical long witnesses are different from canonical short high witnesses in an important way in that if $c_A(\nu,\tau)\leq\alpha$ and $\langle\zeta_i\mid i\leq l\rangle$ is the sequence of ordinals featured in canonical long witnesses and $\eta \in C'_\alpha(\zeta_i) \setminus \nu$ then a truncation (*i.e.* final segment) of a canonical long witness that $c_A(\nu, \tau) \leq \alpha$ is a canonical long witness that $c_A(\nu, \eta) \leq \alpha$. (By truncation, I mean truncation up to changing the initial ordinal if η is not one of the ζ_i.) Although this observation is immediate from Definition (20) it is sufficiently important to be glorified with the title of a lemma.

Lemma 30. Suppose that $c_A(\nu, \tau) \leq \alpha$, $\langle \zeta_i \mid i \leq l \rangle$ and $\psi^*_{\zeta_i}$ for $i < l$ are canonical long witnesses to this and $\eta \in C'_\alpha(\zeta_{i^*}) \setminus \nu$ for some $i^* < l$. Then $\langle \zeta_i \mid i \leq i^* \rangle^\frown \eta$ and $\psi^*_{\zeta_i}$ for $i \leq i^*$ are canonical long witnesses that $c_A(\eta, \tau) \leq \alpha$ and $\eta^\frown \langle \zeta_i \mid i^* < i \rangle$ and $\psi^*_{\zeta_i}$ for $i^* < i$ are canonical long witnesses that $c_A(\nu, \eta) \leq \alpha$.

Proof. Immediate from Definition (20). (This is the point of making Definition (20) the way is done there (and proving Theorem (21)) rather than using the equivalent as proved by Theorem (21) that $\zeta_{i+1} = \min(C'_\alpha(\zeta_i) \setminus \nu)$.) □

Lemma 31. Suppose $c_A(\nu, \tau) \leq \alpha$. Then $A_\alpha(\tau) \setminus \nu \subseteq \bigcup \{ C'_\alpha(\zeta_i) \setminus \nu \mid \nu < \zeta_i \}$.

Proof. Let $\eta \in A_\tau \setminus \nu$. Then there is some $i < l - 1$ such that $\zeta_{i+1} \leq \eta < \zeta_i$. For all $j < l$ one has that $\zeta_{j+1} \in A_\alpha(\zeta_j)$, so, by repeatedly applying Corollary (6), $\eta \in A_\alpha(\zeta_i)$. As $C'_\alpha(\zeta_i) \cap [\zeta_{i+1}, \zeta_i) = A_\alpha(\zeta_i) \cap [\zeta_{i+1}, \zeta_i)$, one has that $\eta \in C'_\alpha(\zeta_i)$ for some $i < l$. □

Corollary 32. Suppose $c_A(\nu, \tau) \leq \alpha$ and $\langle \zeta_i \mid i \leq l \rangle$ is the sequence of ordinals from the canonical long witness for this. Suppose $\eta \in C'_\alpha(\zeta_i)$ for $i < l$. Then

$$A_\alpha(\eta) \setminus \nu \subseteq \bigcup \{ C'_\alpha(\zeta_j) \setminus \nu \mid \nu < \zeta_j \ \ \& \ \ \zeta_j < \eta \}.$$

Proof. Immediate from Lemma (30) and Lemma (31). □

This concludes the current consideration of the canonical long witnesses to $c_A(\nu, \tau) \leq \alpha$ for some $\nu < \tau < \kappa^+$. There are other interesting canonical witnessing sequences which can be defined. One example, just to demonstrate the possibilities, is the following.

Definition 33. Let $A \subseteq \mathcal{F}$ and $\alpha < \kappa$. Let $B_\alpha(\nu) = \mathrm{rge}(f) \cap \nu$ where $(\beta, f) \in A$, $\nu \in \mathrm{rge}(f)$ and β is *minimal* in α showing this.

Definition 34. Let $A \subseteq \mathcal{F}$ and $\alpha < \kappa$. Let $\nu < \kappa^+$ and $(\beta, f) \in A$ be such that $\nu \in \mathrm{rge}(f)$ and β is *minimal* in α showing this. Let $(\gamma, g) \in A$ be such that γ is minimal greater than β such that $\nu \in \mathrm{rge}(g)$. Let

$$B'_\alpha(\nu) = \{ \xi \in A_\alpha(\nu) \mid (\gamma, g) \in A \text{ is minimal such that } \xi \in \mathrm{rge}(g) \}.$$

Definition 35. Suppose that $c_A(\nu, \tau) \leq \alpha$. Let $\xi_0 = \tau$, $\xi_1 = \min(B_\alpha(\tau))$, and $\xi_{i+1} = \min(B'_\alpha(\xi_i))$ for $i > 0$, and let $(\gamma_i, g_i) \in A$ be as in the definition of $B'_\alpha(\xi_i)$.

Corollary 36 (to proof of Proposition (7).). One could insist that $\xi_{i+1} \in B'_\alpha(\xi_i)$ for each $i < k$ in the conclusion of Proposition (7) instead of $\xi_{i+1} \in A_\alpha(\xi_i)$. □

The procedure of Definition (34) amounts to going as far down into the simplified morass as possible to find witnessing maps rather than staying as close to the surface as possible as is the case with canonical short high witnesses. Where canonical long witnesses give rise to 'long' distances (see the following section), these witnesses seem to give rise to 'shortest' distances.

3. Local distance functions

In this section I show how the analysis of the function c_A and the canonical witnesses in the previous section helps give a more explicit insight into the local distance function defined in [K00].

The natural (global) distance function associated with the coupling function c of Definition (I.7) is the function D given in Definition (I.9) by

$$D(\nu, \tau) = \mathrm{otp}(\tau_{c(\nu,\tau)} \setminus \nu_{c(\nu,\tau)}).$$

There are a number of ways of defining local distance functions analogous to D in the sense that c_A is an analogue of c. For example one can measure tracing along the ranges of the maps of a canonical short high witness or along the analogous sequences given by the $B'_\alpha(\eta)$. Koszmider took a seemingly different approach.

Definition 1 ([K00, Definition 2.11] for $\kappa = \omega_1$ and A finite.). Suppose $A \subseteq \mathcal{F}$ and $\alpha < \kappa$. For $\nu \leq \tau < \kappa^+$ set

$$d_{A,\alpha}(\nu, \tau) = \begin{cases} \sup(\{d_{A,\alpha}(\nu, \xi) + \mathrm{otp}(A_\alpha(\tau) \setminus \xi) \mid \xi \in A_\alpha(\tau) \setminus \nu\}), \\ \qquad\qquad \text{if } c_A(\nu, \tau) \leq \alpha \\ 0, \qquad\qquad \text{otherwise.} \end{cases}$$

Lemma 2. If $\nu \leq \tau < \kappa^+$ then

$$d_{A,\alpha}(\nu, \tau) = \sup(\{d_{A,\alpha}(\nu, \xi) + \mathrm{otp}(A_\alpha(\tau) \setminus \xi) \mid \xi \in A_\alpha(\tau) \setminus \nu \text{ and } c_A(\nu, \xi) \leq \alpha\}),$$

i.e. $d_{A,\alpha}(\nu, \tau) = 0$ if and only if $\alpha < c_A(\nu, \tau)$ – ν and τ are not α-connected in A – or $\nu = \tau$. □

However, as will become apparent, this definition is in fact equivalent to measuring along a canonical long witness. Fix $A \subseteq \mathcal{F}$ and $\alpha < \kappa$ for the remainder of this section.

Lemma 3. If $\nu < \kappa^+$ and $\xi \in C_\alpha(\nu)$ then $d_{A,\alpha}(\xi, \nu) = \mathrm{otp}(A_\alpha(\nu) \setminus \xi)$.

Proof. The proof is by induction on $\zeta \in A_\alpha(\nu) \setminus \xi$. If $\zeta = \xi$ then

$$d_{A,\alpha}(\xi, \zeta) = 0 = \mathrm{otp}(A_\alpha(\zeta) \setminus \xi).$$

Now suppose that ζ succeeds ζ' in $A_\alpha(\nu) \setminus \xi$. First of all note that by the hypothesis of the lemma and the case assumptions one has that $c_A(\xi, \nu)$, $c_A(\zeta, \nu)$ and $c_A(\zeta', \nu) \leq \alpha$, so one has that $c_A(\xi, \zeta) \leq \alpha$ by the subadditivity of c_A, Theorem (1.12). Also ζ, $\zeta' \in C_\alpha(\nu)$ as $\xi \in C_\alpha(\nu)$ and $C_\alpha(\nu)$ is a final segment of $A_\alpha(\nu)$ by Lemma (2.12). By Corollary (2.13) one has that $A_\alpha(\zeta) \setminus \xi = A_\alpha(\nu) \cap [\xi, \zeta)$ and

$A_\alpha(\zeta') \setminus \xi = A_\alpha(\nu) \cap [\xi, \zeta')$. Hence $\mathrm{otp}(A_\alpha(\zeta) \setminus \tau) = \mathrm{otp}(A_\alpha(\zeta') \setminus \tau) + 1$ for all $\tau \in (A_\alpha(\zeta) \setminus \xi) \cup \{\zeta'\} = A_\alpha(\zeta') \setminus \xi$. Thus

$$\begin{aligned} d_{A,\alpha}(\xi,\zeta) = \max\big(&d_{A,\alpha}(\xi,\zeta') + 1, \\ &\sup(\{d_{A,\alpha}(\xi,\tau) + \mathrm{otp}(A_\alpha(\zeta') \setminus \xi) + 1 \mid \tau \in A_\alpha(\zeta') \setminus \xi \ \ \& \ \ c_A(\xi,\tau) \le \alpha\})\big). \end{aligned}$$

But

$$\begin{aligned} &\sup(\{d_{A,\alpha}(\xi,\tau) + \mathrm{otp}(A_\alpha(\zeta') \setminus \xi) + 1 \mid \tau \in A_\alpha(\zeta') \setminus \xi \ \ \& \ \ c_A(\xi,\tau) \le \alpha\}) \\ &\quad \le \sup(\{d_{A,\alpha}(\xi,\tau) + \mathrm{otp}(A_\alpha(\zeta') \setminus \xi) \mid \tau \in A_\alpha(\zeta') \setminus \xi \ \ \& \ \ c_A(\xi,\tau) \le \alpha\}) + 1 \\ &\quad = d_{A,\alpha}(\xi,\zeta') + 1, \end{aligned}$$

so $d_{A,\alpha}(\xi,\zeta) = d_{A,\alpha}(\xi,\zeta') + 1$. By the induction hypothesis and Corollary (2.13) again this immediately gives that $d(\xi,\zeta) = \mathrm{otp}(A_\alpha(\zeta) \setminus \xi)$.

Finally, suppose that ζ is a limit point of $A_\alpha(\nu) \setminus \xi$. Then

$$\begin{aligned} d_{A,\alpha}(\xi,\zeta) &= \sup(\{d_{A,\alpha}(\xi,\tau) + \mathrm{otp}(A_\alpha(\zeta) \setminus \tau) \mid \tau \in A_\alpha(\zeta) \setminus \xi \ \ \& \ \ c_A(\xi,\tau) \le \alpha\}) \\ &= \sup(\{\mathrm{otp}(A_\alpha(\tau) \setminus \xi) + \mathrm{otp}(A_\alpha(\zeta) \setminus \tau) \mid \tau \in A_\alpha(\zeta) \setminus \xi\}) \end{aligned}$$

by the induction hypothesis, Lemma (2.12) and Corollary (2.13). Hence

$$d_{A,\alpha}(\xi,\zeta) = \mathrm{otp}(A_\alpha(\zeta) \setminus \xi)$$

by the definition of $C_\alpha(\zeta)$ and Corollary (2.13). □

Corollary 4. Lemma (3) says, in the notation of Definition (I.6), that if $\xi \in C_\alpha(\nu)$ then $d_{A,\alpha}(\xi,\nu) = \mathrm{otp}([\xi_\beta, \nu_\beta)) = D_\beta(\xi,\nu)$ if ψ_ν^β instantiates $A_\alpha(\nu)$. □

Lemma 5. If $\nu < \kappa^+$, $\xi \le \rho$ and ξ, $\rho \in C'_\alpha(\nu)$ then

$$d_{A,\alpha}(\xi,\rho) = \mathrm{otp}(A_\alpha(\rho) \setminus \xi).$$

Proof. $\xi \in C_\alpha(\rho)$ by Lemma (2.19), so the conclusion is immediate from Lemma (3). □

Lemma 6. If $\nu < \kappa^+$, $\xi \in C'_\alpha(\nu)$ and ψ_ν^* instantiates $A_\alpha(\eta)$ for all $\eta \in C'_\alpha(\nu)$ then

$$d_{A,\alpha}(\xi,\nu) = \mathrm{otp}\left(\mathrm{rge}(\psi_\nu^*) \cap \left[\xi, \mathrm{ssup}(A_\alpha(\nu))\right)\right).$$

Proof. If $C_\alpha(\nu) \ne \emptyset$ then $C'_\alpha(\nu) = C_\alpha(\nu)$, by Lemma (2.19). So ψ_ν^* instantiates $A_\alpha(\nu)$ and thus $\mathrm{rge}(\psi_\nu^*) \cap \left[\xi, \mathrm{ssup}(A_\alpha(\nu))\right) = A_\alpha(\nu) \setminus \xi$ and the conclusion is immediate from Lemma (3) again.

So suppose that $C_\alpha(\nu) = \emptyset$. As $\xi \in C'_\alpha(\nu)$ one has that $C'_\alpha(\nu) \setminus \xi = A_\alpha(\nu) \setminus \xi$ and that $c_A(\xi,\eta) \le \alpha$ if $\eta \in A_\alpha(\nu) \setminus \xi$, by Corollary (2.6) and the remark immediately after Definition (2.1), the definition of $A_\alpha(\eta)$. By Lemma (5) this gives that

$$d_{A,\alpha}(\xi,\nu) = \sup(\{\mathrm{otp}(A_\alpha(\eta) \setminus \xi) + \mathrm{otp}(A_\alpha(\nu) \setminus \eta) \mid \eta \in A_\alpha(\nu) \setminus \xi\}).$$

Recall that $A_\alpha(\nu) \cap [\xi,\eta) \subseteq A_\alpha(\eta) \setminus \xi$ if $\eta \in A_\alpha \setminus \xi$, by Corollary (2.6), so

$$\mathrm{otp}(A_\alpha(\eta') \setminus \xi) + \mathrm{otp}(A_\alpha(\nu) \setminus \eta') \le \mathrm{otp}(A_\alpha(\eta) \setminus \xi) + \mathrm{otp}(A_\alpha(\nu) \setminus \eta)$$

if η', $\eta \in A_\alpha(\nu) \setminus \xi$ and $\eta' < \eta$.

Thus if $A_\alpha \setminus \xi$ has a maximal element, say, ρ, then clearly

$$d_{A,\alpha}(\xi,\nu) = d_{A,\alpha}(\xi,\rho) + 1 = \mathrm{otp}\left(\mathrm{rge}(\psi^*_\nu) \cap \left[\xi, \mathrm{ssup}(A_\alpha(\nu))\right)\right).$$

Now it is also clear from the definition of $d_{A,\alpha}$ that $d_{A,\alpha}(\xi,\eta) \leq d_{A,\alpha}(\xi,\nu)$ for all $\eta \in A_\alpha(\nu) \setminus \xi$. Hence if $A_\alpha(\nu) \setminus \xi$ has no maximal element then

$$\begin{aligned}\sup(\{d_{A,\alpha}(\xi,\eta) \mid \eta \in A_\alpha(\nu) \setminus \xi\}) &= \sup(\{\mathrm{otp}(A_\alpha(\eta) \setminus \xi) \mid \eta \in A_\alpha(\nu) \setminus \xi\}) \\ &= \mathrm{otp}\left(\mathrm{rge}(\psi^*_\nu) \cap \left[\xi, \mathrm{ssup}(A_\alpha(\nu))\right)\right) \leq d_{A,\alpha}(\xi,\nu).\end{aligned}$$

On the other hand,

$$\mathrm{otp}(A_\alpha(\eta) \setminus \xi) + \mathrm{otp}(A_\alpha(\nu) \setminus \eta) \leq \mathrm{otp}\left(\mathrm{rge}(\psi^*_\nu) \cap \left[\xi, \mathrm{ssup}(A_\alpha(\nu))\right)\right)$$

for all $\eta \in A_\alpha(\nu) \setminus \xi$. Thus $d_{A,\alpha}(\xi,\nu) = \mathrm{otp}\left(\mathrm{rge}(g) \cap \left[\xi, \mathrm{ssup}(A_\alpha(\nu))\right)\right)$ as required. □

Theorem 7. Suppose $c_A(\nu,\tau) \leq \alpha$. Let $\langle \zeta_i \mid i \leq l\rangle$ and $\langle (\psi^*_{\zeta_i}) \mid i < l\rangle$ be canonical long witnesses to this. (Recall: $\nu = \zeta_l$, $\tau = \zeta_0$.) Write γ_i for $\gamma^*(\zeta_i)$, ψ_i for $\psi^*_{\zeta_i}$ and ζ^*_i for $\mathrm{ssup}(A_\alpha(\zeta_i))$. Then (*see Diagram* 10)

$$\begin{aligned}d_{A,\alpha}(\nu,\tau) &= \mathrm{otp}\left(\mathrm{rge}(\psi_{l-1}) \cap [\nu, \zeta^*_{l-1})\right) + \mathrm{otp}\left(\mathrm{rge}(\psi_{l-2}) \cap [\zeta_{l-1}, \zeta^*_{l-2})\right) \\ &\qquad + \ldots + \mathrm{otp}\left(\mathrm{rge}(\psi_1) \cap [\zeta_2, \zeta^*_1)\right) + \mathrm{otp}\left(\mathrm{rge}(\psi_0) \cap [\zeta_1, \tau^*)\right). \\ &= D_{\gamma_{l-1}}(\nu, \zeta^*_{l-1}) + \ldots + D_{\gamma_1}(\zeta_2, \zeta^*_1) + D_{\gamma_0}(\zeta_1, \tau^*).\end{aligned}$$

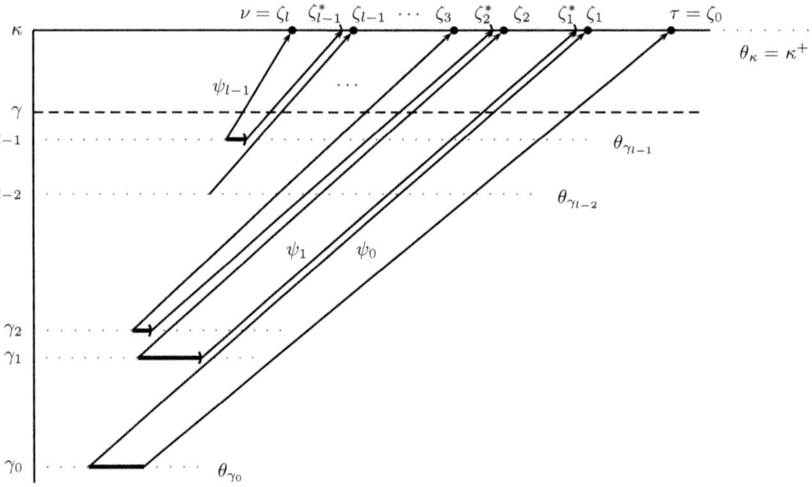

Diagram 10. Illustrating canonical *long* witnesses to $c_A(\nu,\tau) \leq \gamma$ and the local distance function $d_{A,\gamma}(\nu,\tau)$, *cf.* Theorem (7). $d_{A,\gamma}(\nu,\tau)$ is the sum, reading from top to bottom, of the ordertypes of the boldface lines.

Proof. The proof is by induction on l. (Recall that l is determined by ν and τ if $c_A(\nu,\tau)\le\alpha$, so this is a well defined proof strategy!) For $l=1$ the theorem is immediate from Lemmas (3) and (6). If $l=l'+1$ then by the theorem for l' applied to ν, ζ_1 and Lemma (2.31) one has that $d_{A,\alpha}(\nu,\xi)+\mathrm{otp}(A_\alpha(\tau)\cap[\xi,\zeta_1))\le d_{A,\alpha}(\nu,\zeta_1)$ for any $\xi\in A_\alpha(\tau)\cap[\nu,\zeta_1]$, so in order to calculate $d(\nu,\tau)$ one need only calculate the supremum of the $d_{A,\alpha}(\nu,\xi)+\mathrm{otp}(A_\alpha(\tau)\cap[\xi,\zeta_1))$ for $\xi\in A_\alpha(\tau)\cap[\zeta_1,\tau)$.

If $C_\alpha(\tau)\neq\emptyset$ the theorem now follows from calculating that

$$d_{A,\alpha}(\nu,\xi)=d_{A,\alpha}(\nu,\zeta_1)+\mathrm{otp}(A_\alpha(\tau)\setminus\zeta_1)$$

by induction along $C_\alpha(\tau)\neq\emptyset$ exactly as in the proof of Lemma (3), while if $C_\alpha(\tau)=\emptyset$ the theorem follows as in the proof of Lemma (6). □

Corollary 8 ([K00, Fact 2.13(3)] for $\kappa=\omega_1$ and A finite.). If $\nu<\eta<\tau<\kappa^+$ and $c_A(\nu,\eta),\ c_A(\eta,\tau)\le\alpha$ then

$$d_{A,\alpha}(\nu,\tau)=d_{A,\alpha}(\nu,\eta)+d_{A,\alpha}(\eta,\tau).$$

Proof. Immediate from Theorem (7) and Lemma (2.30). □

I draw a couple more simple corollaries of Theorem (7).

Corollary 9. Suppose $\nu<\tau<\kappa^+$ and $\alpha<\kappa$. Let $A\subseteq B\subseteq\mathcal{F}$ and suppose whenever $\langle\xi_i\mid i\le k\rangle$ and $\langle(\gamma_i,g_i)\mid i<k\rangle$ are sequences witnessing that $c_B(\nu,\tau)\le\alpha$ there are $\langle(\beta_i,f_i)\mid i<k\rangle$ with $\gamma_i\le\beta_i$ for all $i<k$ and such that $\langle\xi_i\mid i\le k\rangle$ and $\langle(\beta_i,f_i)\mid i<k\rangle$ witness that $c_A(\nu,\tau)\le\alpha$. (In words: whenever one has witnesses that $c_B(\nu,\tau)\le\alpha$ one can replace any maps used from $B\setminus A$ by maps in A which are start at least as far up in the simplified morass.) Then $d_{A,\alpha}(\nu,\tau)=d_{B,\alpha}(\nu,\tau)$.

Proof. Immediate from Theorem (7) and Definition (2.16), the definition of the $C'_\alpha(\nu)$. □

Corollary 10. Suppose $A\subseteq\mathcal{F}$, $\alpha<\kappa$ and $\nu<\tau<\kappa^+$. Suppose $c_A(\nu,\tau)=\beta<\alpha$ and

$$\{\gamma\mid\exists(\gamma,g)\in A\ \exists\text{ a witnessing sequence to } c_A(\nu,\tau)<\beta\text{ in which }(\gamma,g)\text{ is used}\}\cap[\beta,\alpha)=\emptyset.$$

Then $d_{A,\gamma}(\nu,\tau)=d_{A,\beta}(\nu,\tau)$ for all $\gamma\in[\beta,\alpha)$.

Proof. Immediate from Theorem (7) and Definition (2.16), the definition of the $C'_\alpha(\nu)$. □

Remark 11. It is trivial from Theorem (7) that for all $A\subseteq B\subseteq\mathcal{F}$, $\alpha<\kappa$ and $\nu<\tau<\kappa^+$ one has that $d_{A,\alpha}(\nu,\tau)\le d_{B,\alpha}(\nu,\tau)$.

While $d_{A,\alpha}(\nu,\tau)\le D_\alpha(\nu,\tau)$ there are reasons for considering another distance function.

Definition 12. Suppose $A\subseteq\mathcal{F}$ and $\alpha<\kappa$. For $\nu<\tau<\kappa^+$ set

$$\mathfrak{d}_{A,\alpha}(\nu,\tau)=\sup(\{d_{B,\alpha}(\nu,\tau)\mid B\in[A]^{<\omega}\}).$$

Note 13. By Remark (11) it is clear that for $A \subseteq \mathcal{F}$, $\alpha < \kappa$ and $\nu < \tau < \kappa^+$ one has $\eth_{A,\alpha}(\nu,\tau) \leq d_{A,\alpha}(\nu,\tau)$.

It is easy to see that $\eth_{A,\alpha}$ shares many useful properties with $d_{A,\alpha}$, in particular the conclusions of Corollaries (8)–(10) and Remark (11). (In order to prove these properties of $\eth$ one only needs Theorem (7) and Corollary (8) for finite A.)

Lemma 14. $A \subseteq B \subseteq \mathcal{F}$, $\alpha < \kappa$ and $\nu < \tau < \kappa^+$ one has that $\eth_{A,\alpha}(\nu,\tau) \leq \eth_{B,\alpha}(\nu,\tau)$.

Proof. Immediate from Remark (11). □

Lemma 15. If $A \subseteq \mathcal{F}$, $\alpha < \kappa$, $\nu < \eta < \tau < \kappa^+$ and $c_A(\nu,\eta)$, $c_A(\eta,\tau) \leq \alpha$ then

$$\eth_{A,\alpha}(\nu,\tau) = \eth_{A,\alpha}(\nu,\eta) + \eth_{A,\alpha}(\eta,\tau).$$

Proof. For each $B \in [A]^{<\omega}$ one has that $d_{B,\alpha}(\nu,\tau) = d_{B,\alpha}(\nu,\eta) + d_{B,\alpha}(\eta,\tau)$ by Corollary (8). Thus $\eth_{A,\alpha}(\nu,\tau) \leq \eth_{A,\alpha}(\nu,\eta) + \eth_{A,\alpha}(\eta,\tau)$.

On the other hand,

$$\eth_{A,\alpha}(\nu,\eta) + \eth_{A,\alpha}(\eta,\tau) = \sup(\{ d_{C,\alpha}(\nu,\eta) + d_{D,\alpha}(\eta,\tau) \mid C,\, D \in [A]^{<\omega} \}).$$

If C, $D \in [A]^{<\omega}$ and $B = C \cup D$ then $B \in [A]^{<\omega}$ and one has that

$$d_{C,\alpha}(\nu,\eta) \leq d_{B,\alpha}(\nu,\eta) \quad \text{and} \quad d_{D,\alpha}(\eta,\tau) \leq d_{B,\alpha}(\eta,\tau).$$

Thus $d_{C,\alpha}(\nu,\eta) + d_{D,\alpha}(\eta,\tau) \leq d_{B,\alpha}(\nu,\eta) + d_{B,\alpha}(\eta,\tau)$, $= d_{B,\alpha}(\nu,\tau)$, by Corollary (8). Hence $\eth_{A,\alpha}(\nu,\eta) + \eth_{A,\alpha}(\eta,\tau) \leq \eth_{A,\alpha}(\nu,\tau)$. □

Lemma 16. Suppose $\nu < \tau < \kappa^+$ and $\alpha < \kappa$. Let $A \subseteq B \subseteq \mathcal{F}$ and suppose whenever $\langle \xi_i \mid i \leq k \rangle$ and $\langle (\gamma_i, g_i) \mid i < k \rangle$ are sequences witnessing that $c_B(\nu,\tau) \leq \alpha$ there are $\langle (\beta_i, f_i) \mid i < k \rangle$ with $\gamma_i \leq \beta_i$ for all $i < k$ and such that $\langle \xi_i \mid i \leq k \rangle$ and $\langle (\beta_i, f_i) \mid i < k \rangle$ witness that $c_A(\nu,\tau) \leq \alpha$. (In words: whenever one has witnesses that $c_B(\nu,\tau) \leq \alpha$ one can replace any maps used from $B \setminus A$ by maps in A which are start at least as far up in the simplified morass.) Then $\eth_{A,\alpha}(\nu,\tau) = \eth_{B,\alpha}(\nu,\tau)$.

Proof. By the hypothesis, if $b \in [B]^{<\omega}$ there is some $a \in [A]^{<\omega}$ such that $d_{b,\alpha}(\nu,\tau) \leq d_{a,\alpha}(\nu,\tau)$. Hence $\eth_{B,\alpha}(\nu,\tau) \leq \eth_{A,\alpha}(\nu,\tau)$. Thus, by Lemma (14), $\eth_{A,\alpha}(\nu,\tau) = \eth_{B,\alpha}(\nu,\tau)$. □

Lemma 17. Suppose $A \subseteq \mathcal{F}$, $\alpha < \kappa$ and $\nu < \tau < \kappa^+$. Suppose $c_A(\nu,\tau) = \beta < \alpha$ and

$$\{\gamma \mid \exists (\gamma, g) \in A\ \exists \text{ a witnessing sequence to } c_A(\nu,\tau) < \beta \text{ in which } (\gamma, g) \text{ is used } \} \cap [\beta,\alpha) = \emptyset.$$

Then $\eth_{A,\gamma}(\nu,\tau) = \eth_{A,\beta}(\nu,\tau)$ for all $\gamma \in [\beta,\alpha)$.

Proof. Set $B = \{(\varepsilon, g) \in A \mid \varepsilon \leq \beta\}$. The conclusion is now immediate from Lemma (16) since $\eth_{A,\beta} = \eth_{B,\gamma}$ and $B \subseteq A$. □

One advantage of $\eth_{A,\alpha}$ over $d_{A,\alpha}$ is that it is better adapted for closure arguments.

Lemma 18. Let $\alpha < \kappa$ and $\nu < \tau < \kappa^+$. Suppose that $\langle A_i \mid i < \xi \rangle$ is a chain of subsets of $\mathcal{F}$, so that $A_i \subseteq A_j \subseteq \mathcal{F}$ for $i < j < \xi$. Let $A = \bigcup\{ A_i \mid i < \xi \}$. Then $\eth_{A,\alpha}(\nu, \tau) = \mathrm{ssup}(\{ \eth_{A_i,\alpha} \mid i < \xi \})$.

Proof. If the sequence is eventually constant the conclusion is trivial. If not then ξ is a limit ordinal and whenever $B \in [A]^{<\omega}$ there is some $i < \xi$ such that $B \in [A_i]^{<\omega}$. Hence $d_{B,\alpha}(\nu, \tau) \leq \eth_{A_i,\alpha}(\nu, \tau)$, and so $\eth_{A,\alpha}(\nu, \tau) \leq \sup(\{ \eth_{A_i,\alpha} \mid i < \xi \})$. Thus $\eth_{A,\alpha}(\nu, \tau) = \mathrm{ssup}(\{ \eth_{A_i,\alpha} \mid i < \xi \})$, by Lemma (14). □

4. Using c_A in forcing arguments

Throughout this section let κ be an arbitrary successor cardinal with cardinal predecessor μ, so that $\mu = \kappa^-$ and $\kappa = \mu^+$. Also suppose throughout this section that $\mathbb{M}$ is stationary (*cf.* Definition (1.11)).

In the section I examine the functions c_A in the context of abstract arguments about κ-$\mathbb{M}$-proper forcing when κ is an arbitrary successor cardinal, concentrating particularly on proving general properties of the functions c_A which help in proofs of κ-$\mathbb{M}$-properness. The results of §1 are used extensively, but not those of §§2,3. I start, however, with a general and well-known general fact about cardinal preservation.

Definition 1. Let $\mathbb{P}$ be forcing notion. Suppose that $\mathbb{P} \in H_\lambda$ for some regular cardinal λ. Let $(\mathcal{N}, \in) \prec (H_\lambda, \in)$ with $\overline{\overline{\mathcal{N}}} = \mu$, $\mu \subseteq \mathcal{N} \cap \kappa \in \kappa$ and $\mathbb{P} \in \mathcal{N}$. Then $p^* \in \mathbb{P}$ is $(\mathbb{P}, \mathcal{N})$*-generic* if whenever $\mathcal{D} \in \mathcal{N}$ is a dense and open subset of $\mathbb{P}$, $q \in \mathcal{D}$ and $q \leq p^*$ then there is some $s \in \mathcal{D} \cap \mathcal{N}$ compatible with q, *i.e.*, such that there is some $r \in \mathbb{P}$ such that $r \leq q$, s.

Fact 2 (Well-known folklore). If for every $p \in \mathbb{P}$ and every $\mathbb{P}$-name $\dot{f}$ with $p \Vdash \dot{f} : \mu \longrightarrow \kappa$ there is some $(\mathcal{N}, \epsilon) \prec (H_\lambda, \in)$ with $\overline{\overline{\mathcal{N}}} = \mu$, $\mu \subseteq \mathcal{N} \cap \kappa \in \kappa$, $\dot{f}$, p, $\mathbb{P} \in \mathcal{N}$, and some $p^* \leq p$ which is $(\mathbb{P}, \mathcal{N})$-generic then $\Vdash_{\mathbb{P}} \mu^+ = \kappa$.

Proof. This is just as the usual proof that properness preserves ω_1. Let $\dot{f}$, p, $\mathbb{P}$ be such that $p \Vdash \dot{f} : \mu \longrightarrow \kappa$. Let $\mathcal{N}$ be as in the statement of the fact. Let p^* be $(\mathbb{P}, \mathcal{N})$-generic. For each $\alpha < \mu$ let $\mathcal{D}_\alpha = \{ p' \leq p \mid \exists \gamma_{p'} \in \mathrm{On}\ p' \Vdash \text{“}\dot{f}(\alpha) = \hat{\gamma}_{p'}\text{”} \}$. Clearly each $\mathcal{D}_\alpha$ is a dense and open subset of $\mathbb{P}$. Let A_α be a maximal antichain in $\mathbb{P}$ for each $\alpha < \mu$. Suppose that $x \in A_\alpha$, so $x \leq p$, and suppose that x is compatible with p^*. Let $D'_\alpha = \{ y \in D_\alpha \mid \exists z \in A_\alpha\ y \leq z \}$. Clearly D'_α is dense and open. Let $q \in \mathcal{D}'_\alpha$ be such that $q \leq x$, p^*. Then there is some $s \in \mathcal{D}'_\alpha \cap \mathcal{N}$ and some $t \in \mathbb{P}(\cap \mathcal{D}'_\alpha)$ with $t \leq s$, q. As $t \leq x$ one has that $s \leq x$ (otherwise t would be below two (incompatible) elements of A_α). But $s \Vdash \text{“}\dot{f}(\alpha) \in \mathcal{N} \cap \kappa\text{”}$ since $s \in \mathcal{N}$. So $x \Vdash \text{“}\dot{f}(\alpha) \in \mathcal{N} \cap \kappa\text{”}$ as well. Thus one also has that $p^* \Vdash \text{“}\dot{f}(\alpha) \in \mathcal{N} \cap \kappa\text{”}$.

This shows that $\{ w \in \mathbb{P} \mid \exists \delta < \kappa\ w \Vdash \dot{f}(\alpha) < \delta \}$ is dense below any p such that $p \Vdash \dot{f} : \mu \longrightarrow \kappa$. Hence $\Vdash_{\mathbb{P}} \mu^+ = \kappa$. □

Fact (2) has been known for a considerable time – for example it is implicit in the brief remark after [AShe, Theorem 3] that (what is essentially) Baumgartner's forcing for adding a club subset of ω_1 with finite conditions "can be easily generalised to the case $\kappa = \mu^+ \ldots$ without adding sets of size $< \mu$ and without collapsing cardinals." Of course most attention has been paid to the case $\kappa = \omega_1$, because one can prove iteration theorems for proper forcing, while this is much harder in the general case (*cf.* [S*]). However, as one is primarily interested in using κ-$\mathbb{M}$-properness to help guarantee cardinal preservation for single-step forcings this problem is not important.

Definition 3. Let $\mathbb{P}$ be forcing notion with $\mathbb{P} \in H_\lambda$ for some regular cardinal λ. Call $\mathcal{N}$ a *good (sub) model (of H_λ)* if $(\mathcal{N},\epsilon) \prec (H_\lambda,\in)$, $\overline{\overline{\mathcal{N}}} = \mu$, $\{\mathbb{M},\ \mathcal{F},\ c,\ \kappa,\ \kappa^+\} \cup \mu \subseteq \mathcal{N}$, $\mathcal{N}^{<\mu} \subseteq \mathcal{N}$, $\delta = \mathcal{N} \cap \kappa \in \kappa$, there is some $F \in \mathcal{F}_{\delta\kappa}$ such that $\mathrm{rge}(F) = \mathcal{N} \cap \kappa^+$, and for each $(\alpha, f) \in \mathcal{F}$ with $\alpha < \delta$ if there is some $f' \in \mathcal{F}_{\alpha\delta}$ such that $f = F \cdot f'$ then $(\alpha, f) \in \mathcal{N}$. $\mathbb{P}$ is *κ-$\mathbb{M}$-proper* if there is some $x \in [H_\lambda]^\mu$, such that whenever $p \in \mathbb{P}$ and $\mathcal{N}$ is a good model with $\{p,\ \mathbb{P}\} \cup x \subseteq \mathcal{N}$ there is some $p^* \leq p$ which is $(\mathbb{P}, \mathcal{N})$-generic.

Note that if $\mathcal{N}$ is good and $(\alpha, f) \in \mathcal{N}$ then one has that there is some $f' \in \mathcal{F}_{\alpha\delta}$ such that $f = F \cdot f'$.

Lemma 4. If $\mathbb{P}$ is κ-$\mathbb{M}$-proper then $\Vdash_{\mathbb{P}} \mu^+ = \kappa$.

Proof. Given any $p \in \mathbb{P}$ and $\mathbb{P}$-name $\dot{f}$ such that $p \Vdash \dot{f} : \mu \longrightarrow \kappa$ let $\mathcal{N}$ be as in the definition of $\mathbb{M}$-κ-proper with $\dot{f} \in \mathcal{N}$. There are a closed unbounded in $[H_\lambda]^\mu$ set of such $\mathcal{N}$ without the requirement on F, and stationarily many with it by $\mathbb{M}$'s stationarity. The conclusion of the lemma follows from Fact (2) immediately. □

I concentrate on a particular type of forcings analogous to forcings with side conditions which are matrices of models. Conditions in these forcings have *working parts* with an associated *realm*, a subset of size less than μ of κ^+, which is the part of the domain (an unbounded subset of κ^+) which it will contribute to any generic object to which it belongs. They also have *side condition* parts which are some $A \in [\mathcal{F}]^{<\mu}$. So I start by considering pairs $p = (a_p, A_p)$ with $a_p \in [\kappa^+]^{<\mu}$ and $A_p \in [\mathcal{F}]^{<\mu}$. In analysing these forcings below it is convenient to fix some pieces of notation and working assumptions as one goes. These pieces of notation and assumptions are flagged with a $(\star)$.

Some, but by no means all, of the basic results below are extracted from Koszmider's proof in [K00] of $\mathbb{M}$-properness for a forcing to add a chain of length ω_2 in ${}^{\omega_1}\omega_1/fin$. More accurately, these results are abstracted from a construction to add a chain of length ω_2 in $\mathcal{P}(\omega_1)/fin$ using $\mathbb{M}$-proper forcing. Although in [K98a] this is done via a ccc forcing using a stepping up function, Koszmider revealed in personal communication that his original proof used (ω_1-)$\mathbb{M}$-proper forcing, and, indeed, such a construction can reasonably easily be read off from [K00] with the aid of [K98a]. The main technical difference between this construction and that in [K00] is that there is no need for the local distance functions $d_{A,\alpha}$ discussed in

§3 in order to add a chain of length ω_2 in $\mathcal{P}(\omega_1)/\mathit{fin}$. Conditions in Koszmider's forcings, for example, have a working part and a side condition part components as discussed above. The working parts have form $F_p : a_p \times b_p \longrightarrow 2$ or ω_1, where a_p is the realm of the condition and $b_p \in [\omega_1]^{<\omega}$.

Definition 5. Let $\mathbb{P}_0$ be $\{p \mid p = (a_p, A_p)$ where $a_p \in [\kappa^+]^{<\mu}$ and $A_p \in [\mathcal{F}]^{<\mu}\}$ ordered by reverse inclusion.

Of course, any two conditions in $\mathbb{P}_0$ are compatible since if p, $q \in \mathbb{P}_0$ then $(a_p \cup a_q, A_p \cup A_q) \leq p, q$, so forcing with $\mathbb{P}_0$ itself is uninteresting and preserves all cardinals. However this is not the case for more elaborate forcing notions whose conditions merely have retracts in $\mathbb{P}_0$.

($\star$) Suppose that $\mathbb{P}$ is a forcing and that conditions in $\mathbb{P}$ are of the form $p = (a_p, A_p, \dots)$ where $(a_p, A_p) \in \mathbb{P}_0$, a_p is the realm of p and A_p is the side condition part of p.

($\star$) Fix $p \in \mathbb{P}$. Suppose that $\mathcal{N} \prec H_{\kappa^{++}}$ is good with $(\delta, F) \in \mathcal{F}$ such that $\mathcal{N} \cap \kappa^+ = \mathrm{rge}(F)$.

In order to prove that $\mathbb{P}$ is κ-$\mathbb{M}$-proper the typical strategy, and the only one which will be dealt with in this paper, is to show that $(a_p, A_p \cup \{(\delta, F)\}, \dots)$ is $(P, \mathcal{N})$-generic. In practice one also has to *show* that this is a condition and is stronger than p, but in this section the aim is to analyse the situation in which this is possible, so simply assume that this is so.

($\star$) Assume that $(a_p, A_p \cup \{(\delta, F)\}, \dots) \in \mathbb{P}$ and that $(a_p, A_p \cup \{(\delta, F)\}, \dots) \leq p$.

Remark 6. One may as well assume that $\mathcal{N}$ satisfies the following

(i) If $(\alpha, f) \in \mathcal{N}$ then $\exists f' \in \mathcal{F}_{\alpha\delta}\ (f = F \cdot f')$.

(ii) If $\alpha < \delta$, $g \in \mathcal{F}_{\alpha\kappa}$, $g' \in \mathcal{F}_{\alpha\delta}$, $g'' \in \mathcal{F}_{\delta\kappa}$ and $g = g'' \cdot g'$ then $h(g) = F \cdot g' \in \mathcal{F}_{\alpha\kappa} \cap \mathcal{N}$ (and $(\alpha, h(g)) \in \mathcal{F} \cap \mathcal{N}$).

Proof. These remarks are standard and follow from the stationarity of $\mathbb{M}$. By the stationarity of $\mathbb{M}$ one may as well assume that one has an $\mathcal{N}$ for which properties such as these hold since, eg, $\mathcal{N} \cap \kappa^+$ is club in $\mathcal{P}_{\leq\mu}(\kappa^+)$. The point is that if $j : \overline{\mathcal{N}} \longrightarrow \mathcal{N}$ is the inverse of the transitive collapse of $\mathcal{N}$ then

$$j^{-1} \restriction \mathcal{N} \cap \kappa^+ = F^{-1} \restriction \mathcal{N} \cap \kappa^+$$

and if $j(\langle \langle \overline{\theta}_\alpha \mid \alpha \leq \delta \rangle, \langle \overline{\mathcal{F}}_{\alpha\beta} \mid \alpha \leq \beta \leq \delta \rangle \rangle) = \mathbb{M}$ then

$$\langle \langle \overline{\theta}_\alpha \mid \alpha \leq \delta \rangle, \langle \overline{\mathcal{F}}_{\alpha\beta} \mid \alpha \leq \beta \leq \delta \rangle \rangle = \langle \langle \theta_\alpha \mid \alpha \leq \delta \rangle, \langle \mathcal{F}_{\alpha\beta} \mid \alpha \leq \beta \leq \delta \rangle \rangle. \qquad \square$$

($\star$) Suppose that $\mathcal{D}$ is a dense, open subset of $\mathbb{P}$ with $\mathcal{D} \in \mathcal{N}$. Suppose also that $q \in \mathcal{D}$ and that $q \leq (a_p, A_p \cup \{(\delta, F)\}, \dots)$. Let $q \restriction \mathcal{N} = q \cap \mathcal{N} = (a_q \cap \mathcal{N}, A_q \cap \mathcal{N}, \dots)$.

In practice one now needs to *show* that $q \restriction \mathcal{N} \in \mathbb{P} \cap \mathcal{N}$ and that $q \leq q \restriction \mathcal{N}$, and this may be a challenging part of the proof of κ-$\mathbb{M}$-properness. Again, here the aim is to analyse the situation in which this is possible, so simply assume that this can be done.

($\star$) $q \restriction \mathcal{N} \in \mathbb{P} \cap \mathcal{N}$ and $q \leq q \restriction \mathcal{N}$.

If one wants to show that $(a_p, A_p \cup \{(\delta, F)\}, \ldots)$ is $(\mathbb{P}, \mathcal{N})$-generic it is typically helpful to explicitly reflect some of the properties of q and of the relationship between q and $q \restriction \mathcal{N}$ to $\mathcal{N}$ so that one can get an $s \in \mathcal{N} \cap \mathcal{D}$ which one can actually prove is amalgamable with q. In order to chrystalise these ideas a slew of notation and definitions, given in Notation (7) below, is unavoidable.

It would also be useful if one could without loss of generality smooth out A_q, by adding some further maps if necessary, in order to make its internal structure more uniform. This would be useful because the uniformity together with its reflection to s might also be helpful in performing the amalgamation of q and s. One can do this smoothing out for the partial order $\mathbb{P}_0$, since $\mathcal{D}$ is dense and open, and if $t \in \mathbb{P}_0$ and $A_t \subseteq A \subseteq \mathcal{F}$ then $(a_t, A) \leq t$, and so adding maps to A_q will give a stronger condition and also not take one outside $\mathcal{D}$. But $\mathbb{P}_0$ is deceptive because one of the essential points of κ-$\mathbb{M}$-proper forcing, as of proper forcing with chains of models as side conditions, is that typically one will give constraints on the working part of a condition which are dependent on the interaction between its realm and the side condition part of the condition. Thus one cannot simply throw extra maps into the side condition part of a condition willy-nilly and expect that conflicts with these constraints will not arise. It is likely that one will end up either with a condition that is not stronger than the one with which one started or with something that is not a condition at all.

However, fortunately, in the cases that have been looked at so far it has been possible to throw *some* additional maps into A_q and get a condition stronger than q.

The minimal amount of smoothing that seems to be necessary to get a satisfactory theory is to ensure that one has $c_{A_q}(\nu, \tau) < \kappa$ for all $\nu, \tau \in a_q$. Stronger conditions that may be useful are that there is some $(\alpha, f) \in A_p$ such that $a_p \subseteq \mathrm{rge}(f)$, or that

$$\exists(\alpha, f) \in A_p \, \big((\forall (\beta, g) \in A_p \setminus \{(\alpha, f)\} \\ (\beta < \alpha \;\&\; \exists h \in \mathcal{F}_{\beta\alpha} \; g = f \cdot h)) \;\&\; a_p \subseteq \mathrm{rge}(f) \big).$$

These conditions *have* been achievable in the cases considered so far. (The point is that one can factor all of the maps in A_q through some single map $(\alpha, f) \in \mathcal{F}$ with α arbitrary large below κ by Fact (I.4.b) and certainly above $c_{A_q}(\nu, \tau) < \alpha$ for all $\nu, \tau \in a_q$ with $c_{A_q}(\nu, \tau) < \kappa$, and add this map to A_q. This reduces $c_{A_q}(\nu, \tau)$ (to α) for $\nu, \tau \in a_q$ if (and only if) it previously took the value κ and this tends to be benign if α is chosen large enough.)

($\star$) Consequently I assume from now onwards that $c_{A_q}(\nu, \tau) < \kappa$ for all $\nu, \tau \in a_q$.

The following are three examples of the sort of uniformities in A_q that might conceivably be useful but for which it is harder to check that they can be obtained. I do not know of any occasions in which they have been checked and used (and the referee of this paper commented that to them they "seem bizarre"!).

(i) $\forall(\alpha, f), (\alpha, f'), (\beta, g) \in A_q\big((\beta < \alpha \;\&\; \exists h \in \mathcal{F}_{\beta\alpha}\; g = f \cdot h) \longrightarrow (\beta, f' \cdot h \in A_q)\big)$,

(ii) $\forall(\alpha, f), (\beta, g) \in A_q\; \big(\beta < \alpha \longrightarrow \exists f' \in \mathcal{F}_{\alpha\kappa}\; \exists h \in \mathcal{F}_{\beta\alpha}\; g = f' \cdot h \;\&\; (\alpha, f') \in A_q\big)$,

(iii) $\exists(\alpha, f) \in (A_q \cap \mathcal{N})\big(\alpha < \delta \;\&\; \exists h \in \mathcal{F}_{\alpha\delta}\; (f = F \cdot h \;\&\; a_q \cap \mathcal{N} \subseteq \mathrm{rge}(f) \;\&\; \forall(\beta, g) \in (A_q \cap \mathcal{N}) \setminus \{(\alpha, f)\}\; (\beta < \alpha \;\&\; \exists h \in \mathcal{F}_{\beta\alpha}\; g = f \cdot h))\big)$.

I now introduce the promised notation. (The definition of Y_i follows Koszmider in [K00] for $\kappa = \omega_1$ and a_q, A_q finite.) Recall, from Remark (6) that if $(\alpha, f) \in \mathcal{F}$, $\alpha < \delta$ and $f = f'' \cdot f'$ where $(\delta, f'') \in \mathcal{F}$ and $f' \in \mathcal{F}_{\alpha\delta}$ then $h(f) = F \cdot f'$.

Notation 7. Let $\langle(\beta_i, f_i) \mid i < \chi\rangle$ enumerate $\langle(\alpha, f) \in A_q \mid \alpha < \delta\rangle$ for some $\chi < \mu$.

Let $Y_i = \mathrm{rge}(f_i) \cap \mathrm{rge}(h(f_i))$ for $i < \chi$.

Note that $Y_i = \mathrm{rge}(f_i) \cap \mathrm{rge}(F)$ for all $i < \chi$ as well.

Let $\rho_i = \mathrm{ssup}(\{\rho < \theta_{\alpha_i} \mid f_i(\rho) = h(f_i)(\rho)\})$.

Then $Y_i = \mathrm{rge}(f_i \restriction \rho_i) = \mathrm{rge}(\psi_{(\alpha_i, \rho_i), (\kappa, f_i(\rho_i))} \restriction \rho_i)$.

Let $\beta^* = \mathrm{ssup}(\{\beta_i \mid i < \chi\})$.

Let $b_q = \{c_{A_q}(\zeta, \xi) \mid \zeta, \xi \in a_q\} = c_{A_q}``[a_q]^2$.

Let $e_q \in [\kappa]^{<\mu}$ be arbitrary, $e_{q\restriction\mathcal{N}} = e_q \cap \mathcal{N}$, and $\beta^\dagger = \max(\{\beta^*, \mathrm{ssup}(e_{q\restriction\mathcal{N}})\})$.

As $\overline{\overline{Y_i}}$ is an initial segment of $\mathrm{rge}(h(f_i)) \subseteq \mathrm{rge}(F) = \mathcal{N}$ for $i < \chi$ (by Fact (I.3)), one has that $Y_i \subseteq \mathcal{N}$ for each $i < \chi$. Note also that $Y_i \in \mathcal{N}$ and as $\mathcal{N}^{<\mu} \subseteq \mathcal{N}$ one has that $\langle Y_i \mid i < \chi\rangle \in \mathcal{N}$.

(Koszmider ([K00]), incidentally, also found the following notational definitions useful in connection with properties of $d_{A_q,\alpha}$ for various α. They will not be used in the remainder of the paper, so the reader may wish to skip over them.

Let $\{\xi_i \mid i < \lambda\}$ enumerate $(a_q \setminus \mathcal{N}) \cap \sup(\mathcal{N} \cap \kappa^+)$.

Let $\eta_i = \min(\mathcal{N} \setminus \xi_i)$ and $\nu_i = \sup(\{\sup(Y_j \cap \eta_i) \mid j < \chi\})$ for $i < \lambda$.)

Let $\phi(x)$ be the conjunction of the following.

(i) $x \in \mathcal{D}$,
(ii) $x \leq q \restriction \mathcal{N}$, and
(iii) $\exists(\alpha^*, h^*) \in A_x\;\; \beta^\dagger < \alpha^*$ such that
(1) $\forall(\beta, f) \in A_x\; \big(\beta < \beta^* \longrightarrow \exists i < \chi\; (\mathrm{rge}(f) \cap \mathrm{rge}(h^*) = Y_i)\big)$,
(2) $\forall i < \chi\; \exists(\beta_i, f) \in A_x\; \big(\mathrm{rge}(f) \cap \mathrm{rge}(h^*) = Y_i \;\&\; \exists f'' \in \mathcal{F}_{\alpha^*\kappa}\; \exists f' \in \mathcal{F}_{\beta\alpha^*}\; f = f'' \cdot f' \longrightarrow \mathrm{rge}(h^* \cdot f') \cap \mathrm{rge}(f) = Y_i\big)$,
(3) $a_{q\restriction\mathcal{N}} \subseteq \mathrm{rge}(h^*)$,
(4) $(a_x \setminus a_{q\restriction\mathcal{N}}) \cap \mathrm{rge}(h^*) = \emptyset$,
(5) $e_{q\restriction\mathcal{N}} \subseteq e_x \;\&\; (e_x \setminus \beta^\dagger) \cap \alpha^* = \emptyset$,
(6) $\forall\gamma \in b_x\; \gamma < \kappa$.

The following sentences would correspond to the stronger conditions on A_q mentioned above, but are *not* part of the definition of ϕ.

(α) $\exists(\alpha, f) \in A_x\ \forall(\beta, g) \in A_x \setminus \{(\alpha, f)\}$
$\big((\beta < \alpha\ \ \&\ \ \exists k \in \mathcal{F}_{\beta\alpha}\ g = f \cdot k)\ \ \&\ \ a_x \subseteq \mathrm{rge}(f)\big)$

(β) $\forall(\alpha, f),\ (\alpha, f'),\ (\beta, g) \in A_x$
$\big((\beta < \alpha\ \ \&\ \ \exists k \in \mathcal{F}_{\beta\alpha}\ g = f \cdot k) \longrightarrow (\beta, f' \cdot k) \in A_x\big)$

(γ) $\forall(\alpha, f),\ (\beta, g) \in A_x\ \big(\beta < \alpha \longrightarrow \exists f' \in \mathcal{F}_{\alpha\kappa}\ \exists k \in \mathcal{F}_{\beta\alpha}\ (g = f' \cdot k$
$\&\ \ (\alpha, f') \in A_x)\big)$

(δ) $\exists(\alpha, f) \in A_{q\restriction\mathcal{N}}\ \big(\alpha < \alpha^*\ \ \&\ \ \exists h \in \mathcal{F}_{\alpha\alpha^*}\ f = h^* \cdot k\ \ \&\ \ a_{q\restriction\mathcal{N}} \subseteq \mathrm{rge}(f)\ \ \&$
$\forall(\beta, g) \in A_{q\restriction\mathcal{N}} \setminus \{(\alpha, f)\}\ (\beta < \alpha\ \ \&\ \ \exists k \in \mathcal{F}_{\beta\alpha}\ g = f \cdot k)\big)$.

Finally let $\phi'(x)$ be the conjunction of $\phi(x)$ with

(7) $\forall i < \lambda\ \big(\nu_i,\ \eta_i \in \mathrm{rge}(h^*)\ \ \&\ \ \exists \zeta_i \in (\nu_i, \eta_i) \cap \mathrm{rge}(h^*)$
$\big(\text{there is a good model } \mathcal{M} \text{ such that } \mathrm{otp}([\nu_i, \zeta_i) \cap \mathrm{rge}(h^*)) \notin \mathcal{M}\big)\big)$.

(7) will not be used in the remainder of this section in dealing with the properties of c_{A_q}. It is included only as a gesture towards completeness as it proved useful in [K00] when dealing with local distance functions.

Lemma 8. $H_{\kappa^{++}} \models \phi(q)$ and $H_{\kappa^{++}} \models \phi'(q)$.

Proof. This is easy to see using (δ, F) as the witness (α^*, h^*) for (iii) in the definition of ϕ. □

If one could think of more things one could say (or assume and say) about F with parameters from $\mathcal{N}$ these could be added to the above list. (Perhaps the following examination of what can be deduced from the properties (1)–(6) will help in the search to formulate further useful properties.)

($\star$) Now suppose, by the elementarity of $\mathcal{N}$ in $H_{\kappa^{++}}$, that $s \in \mathcal{N}$ and $\mathcal{N} \models \phi(s)$. Let $(\alpha^*, h^*) \in \mathcal{N}$ be the witness to (iii) in the definition of ϕ for s in $\mathcal{N}$.

Of course in the skeletal case of forcing with $\mathbb{P}_0$ itself one does not need any further properties in order to be able to amalgamate such conditions s and q (and hence show that $\mathbb{P}_0$ is $\mathbb{M}$-proper), but it is useful when dealing with forcings which are more fleshed out to have considered properties relating the realms of s and q for which one can find conditions s by reflection.

($\star$) So, set $a_r = a_s \cup a_q$ and $A_r = A_s \cup A_q$.

The intended application of this is that in order to finish the proof of the compatability of q and s one will prove that some $r = (a_r, A_r, \dots)$ is an element of $\mathbb{P}$ and is stronger than each of q and s.

Key notational definition. Write $c^q(\zeta, \xi)$ in place of $c_{A_q}(\zeta, \xi)$, and similarly for other conditions and putative conditions in $\mathbb{P}$, in order to save on subscripts! So c^s is c_{A_s} and c^r is c_{A_r}. Note that c_r depends only on A_r and not, for example, on having successfully shown that $r \in \mathbb{P}$.

I derive a series of extremely useful general facts concerning a_q, a_s, a_r, c^q, c^s and c^r. In order to complete the proof for specific κ-$\mathbb{M}$-proper forcings one needs to use these facts in conjunction with the specifics concealed in the '...' part of the working part of conditions.

In particular, Corollary (24) below is a considerable strengthening of (the conclusion of) the Δ-properties of ρ and c. See [M*7] for more on these properties.

Lemma 9. $\forall \zeta_0 < \zeta_1 < \kappa^+ \big(\zeta_0,\, \zeta_1 \in a_s \longrightarrow c^s(\zeta_0,\zeta_1) < \delta\big)$.

Proof. By Remark (6.i) and clause (7.iii.6) in the definition of ϕ. □

Lemma 10. $\forall \xi < \kappa^+ \, \forall (\gamma, g) \in A_q \, \big((\gamma < \delta \;\&\; \xi \in \mathrm{rge}(g) \;\&\; \xi \in \mathcal{N}) \longrightarrow \big(\mathrm{rge}(g) \cap \xi \subseteq \mathcal{N} \;\&\; \xi \in \mathrm{rge}(h^*)\big)\big)$.

Proof. Suppose that ξ and (γ, g) are as in the antecedent of the lemma. Thus $\xi \in \mathcal{N} = \mathrm{rge}(F)$. As $\gamma < \delta$ there are some $(\delta, g'') \in \mathcal{F}$ and $g' \in \mathcal{F}_{\gamma\delta}$ such that $g = g'' \cdot g'$. By Fact (I.3), $g'' \restriction \xi_\delta = F \restriction \xi_\delta$, so $\mathrm{rge}(g) \cap \xi \subseteq \mathrm{rge}(F) = \mathcal{N} \cap \kappa^+$. Also as $\gamma < \delta$ there is some $i < \chi$ such that $(\gamma, g) = (\beta_i, f_i)$. As $\mathcal{N} \cap \kappa^+ = \mathrm{rge}(F)$, by the definition of Y_i one has that $\xi \in Y_i$. By (7.iii.2) for s there is some $(\beta_i, f) \in A_s$ such that $Y_i = \mathrm{rge}(f) \cap \mathrm{rge}(h^*)$, so $\xi \in \mathrm{rge}(f) \cap \mathrm{rge}(h^*)$ and $f \restriction \xi_\gamma = g \restriction \xi_\gamma$. □

Lemma 11. $\forall \zeta_0 < \zeta_1 < \kappa^+ \big((\zeta_1 \in \mathcal{N} \;\&\; c^r(\zeta_0,\zeta_1) < \delta) \longrightarrow \big(\zeta_0 \in \mathcal{N} \;\&\; c^s(\zeta_0,\zeta_1) = c^r(\zeta_0,\zeta_1)\big)\big)$.

Proof. Suppose $\zeta_0 < \zeta_1 < \kappa^+$, $\beta < \delta$, and $\zeta_1 \in \mathcal{N}$ and that $\langle \xi_i \,|\, i \le k\rangle$ and $\langle (\gamma_i, g_i) \,|\, i < k\rangle$ are witnesses that $c^r(\zeta_0,\zeta_1) = \alpha \le \beta$. Now, $\xi_0 (= \zeta_1) \in \mathcal{N}$, and if $\xi_i \in \mathcal{N}$ then either $(\gamma_i, g_i) \in A_s$, when $\xi_{i+1} \in \mathrm{rge}(g_i) \subseteq \mathcal{N}$, or $(\gamma_i, g_i) \in A_q \setminus A_s$, when by the definition of a witnessing sequence of maps and Lemma (10) one has that $\xi_{i+1} \in \mathrm{rge}(g_i) \subseteq \mathcal{N}$ and $g_i \restriction (\xi_i)_{\gamma_i} = f \restriction (\xi_i)_{\gamma_i}$ for some $(\beta_i, f) \in A_s$. So one has that each $\xi_i \in \mathrm{rge}(F) = \mathcal{N} \cap \kappa^+$ and may replace each $(\gamma_i, g_i) \in A_q \setminus A_s$ by some $(\gamma_i, f) \in A_s$. Thus $\zeta_0 \in \mathcal{N}$ and $c^r(\zeta_0,\zeta_1) = c^s(\zeta_0,\zeta_1) \le \beta$. □

Corollary 12. $\forall \zeta_0 < \zeta_1 < \kappa^+ \big(\zeta_0,\, \zeta_1 \in a_s \longrightarrow c^s(\zeta_0,\zeta_1) = c^r(\zeta_0,\zeta_1) < \delta\big)$. □

Corollary 13. $\forall \xi < \zeta < \kappa^+ \big((\xi \notin \mathcal{N} \;\&\; \zeta \in a_s) \longrightarrow \delta \le c^r(\xi,\zeta)\big)$. □

Lemma 14. $e_s \cap e_q = e_q \cap \mathcal{N} \subseteq \beta^\dagger < \alpha^*$.

Proof. $e_s \subseteq \delta$, $e_q \backslash (e_q \cap \mathcal{N}) \cap \delta = \emptyset$, as $\mathcal{N} \cap \kappa = \delta$, and $e_q \cap \mathcal{N} \subseteq e_s$, so $e_s \cap b_q = b_q \cap \mathcal{N}$. For the subset inclusion use (7.iii.5). □

Lemma 15. $\forall \nu < \xi < \kappa^+ \big(\xi \in a_q \;\&\; \beta < \alpha^* \longrightarrow (c^r(\nu,\xi) = \beta \longleftrightarrow c^q(\nu,\xi) = \beta)\big)$.

Proof. (Similar to the proof of Lemma (11).) Let ν and ξ be as in the antecedent of the lemma. Suppose $\langle \xi_i \,|\, i \le k\rangle$ and $\langle (\gamma_i, g_i) \,|\, i < k\rangle$ are good witnesses that $c^r(\nu,\xi) = \beta < \alpha^*$. If there is no i such that $\xi_i \in \mathcal{N}$ then, by Remark (6.i), there is no i such that $(\gamma_i, g_i) \in A_s$, so $c^q(\nu,\xi) = \beta$. If there is some i such that $\xi_i \in \mathcal{N}$ then let i^* be the least such. As $c^r(\xi_i, \xi_{i^*}) \le \gamma_{i-1} < \beta < \alpha^*$ for all $i \in (i^*, k]$, Lemma (11) shows that one then has that $\xi_i \in \mathcal{N}$ for all $i \in [i^*, k]$ and $c^r(\nu, \xi_{i^*}) = c^s(\nu, \xi_{i^*})$. By (7.iii.5) one has that $c^s(\nu, \xi_{i^*}) < \beta^*$. But then, by (7.iii.2), each $(\gamma_i, g_i) \in A_s$, for $i \in [i^*, k)$, can be replaced by some $(\gamma_i, g_i') \in A_q$. Thus, again, $c^q(\nu,\xi) = \alpha$. □

Lemma 16. $\forall \xi_0 < \xi_1 < \kappa^+,\ \forall \beta \in [\delta, \kappa)\ \big(c^q(\xi_0, \xi_1) \leq \beta \longleftrightarrow c^r(\xi_0, \xi_1) \leq \beta\big)$.

Proof. Let $\xi_0 < \xi_1 < \kappa^+$, $\beta \in [\delta, \kappa)$ and suppose that $c^r(\xi_0, \xi_1) \leq \beta$. Simply replace any $(\gamma, f) \in A_s$ in a witnessing sequence for this by (δ, F) $(\in A_q)$. The converse is immediate from Lemma (1.3). □

Corollary 17. $\forall \xi < \zeta < \kappa^+\ \big(\xi \in a_q \setminus a_s\ \&\ \zeta \in a_s \longrightarrow \delta \leq c^r(\xi, \zeta) = c^q(\xi, \zeta)\big)$.

Proof. Immediate from Corollary (13) and Lemmas (16). □

Lemma 18. Suppose $\zeta \in a_s \setminus a_q$, $\xi \in a_q$, $\zeta < \xi$ and $c^r(\zeta, \xi) < \delta$. Suppose also that $\langle \xi_i \mid i \leq k \rangle$ and $\langle (\gamma_i, g_i) \mid i < k \rangle$ are witnesses that $c^r(\zeta, \xi) = \alpha < \delta$. Then there is some $i < k$ such that $\xi_i \in \mathrm{rge}(h^*)$.

Proof. If $\xi \in \mathrm{rge}(h^*)$ there is nothing more to prove. Otherwise let $i < k$ be least such that $\xi_{i+1} \in \mathcal{N}$. As $\xi_i \notin \mathcal{N}$ one has that $(\gamma_i, g_i) \notin \mathcal{N}$ and hence $(\gamma_i, g_i) \in A_q$. Thus $\xi_{i+1} \in \mathrm{rge}(g_i) \cap \mathcal{N}$, and so $\xi_{i+1} \in \mathrm{rge}(h^*)$ by Lemma (10). Finally, $i < k$ since $\xi_k = \zeta \notin \mathrm{rge}(h^*)$, by (7.iii.4). □

Lemma 19. $\forall \zeta < \xi < \kappa^+\ \big((\zeta \in a_s \setminus a_q\ \&\ \xi \in a_q) \longrightarrow \alpha^* < c^r(\zeta, \xi)\big)$.

Proof. Let ζ and ξ be as in the antecedent of the lemma. Suppose that $\langle \xi_i \mid i \leq k \rangle$ and $\langle (\gamma_i, g_i) \mid i < k \rangle$ are good witnesses that $c^r(\zeta, \xi) = \alpha$ and that $\alpha \leq \alpha^*$. As $(a_s \setminus a_q) \cap \mathrm{rge}(h^*) = \emptyset$, by (7.iii.4) for s, it suffices to show that $\zeta \in \mathrm{rge}(h^*)$ to obtain a contradiction.

If $\alpha < \alpha^*$ then $c^r(\zeta, \xi) = \alpha = c^q(\zeta, \xi)$ by Lemma (15), so one may as well assume that $(\gamma_i, g_i) \in A_q$ for all $i < k$. But $\zeta \in \mathrm{rge}(g_{k-1})$ and $\zeta \in \mathcal{N}$, so $\zeta \in \mathrm{rge}(h^*)$ by Lemma (10).

If $\alpha = \alpha^*$, applying Lemma (18), let i^* be minimal such that $\xi_{i^*} \in \mathrm{rge}(h^*)$. As $c^r(\zeta, \xi_{i^*}) \leq \alpha$, Lemma (1.4) gives that $\zeta \in \mathrm{rge}(h^*)$. □

Lemma 20. $\forall \xi_0 < \xi_1 < \kappa^+\ \big(\xi_0, \xi_1 \in a_q \setminus a_s \longrightarrow (c^r(\xi_0, \xi_1) < \delta \longleftrightarrow c^r(\xi_0, \xi_1) < \beta^*)\big)$.

Proof. Let ξ_0 and ξ_1 be as in the antecedent of the lemma. Suppose that $c^r(\xi_0, \xi_1) = \alpha < \delta$ and let $\langle \mu_i \mid i < k \rangle$ and $\langle (\gamma_i, g_i \mid i < k \rangle$ be witnesses to this. By Lemma (11), if one of the $\mu_i \in \mathcal{N}$ then $\xi_0 \in a_q \cap \mathcal{N} = a_q \cap a_s$, a contradiction. So, by Remark (6.i) and the fact that $A_s \subseteq \mathcal{N}$, each $(\gamma_i, g_i) \in A_q \setminus A_s$. Hence $c^q(\xi_0, \xi_1) = \alpha$. But $c^q(\xi_0, \xi_1) < \delta$ if and only if $c^q(\xi_0, \xi_1) < \beta^*$, by the definition of β^*. □

Lemma 21. $\forall \tau < \xi < \kappa^+\ \big(\tau \in a_s \cap a_q\ \&\ \xi \in a_q \setminus a_s \longrightarrow (c^r(\tau, \xi) \leq \alpha^* \text{ or } c^r(\tau, \xi) \geq \delta)\big)$.

Proof. Let τ and κ be as in the antecedent of the lemma. Suppose $\langle \mu_i \mid i \leq k \rangle$ and $\langle (\gamma_i, g_i) \mid i < k \rangle$ are good witnesses that $c^r(\tau, \xi) = \beta < \delta$. Then there is a least $i < k$ such that $\mu_{i+1} \in \mathcal{N}$ since $\mu_k \in \mathcal{N}$. As $\mu_j \notin \mathcal{N}$ for $j \leq i$ one has that $(\gamma_j, g_j) \notin \mathcal{N}$, and $\gamma_j < \beta^* < \alpha^*$ since $\gamma_j < \delta$, so $c^r(\mu_{i+1}, \tau) \leq \gamma_i < \alpha^*$. If $k = i + 1$ the proof is finished. Otherwise one has that $\mu_{i+1} \in \mathrm{rge}(g_i) \cap \mathcal{N}$, and thus $\mu_{i+1} \in \mathrm{rge}(h^*)$ by Lemma (10). But then $(\alpha^*, h^*) \in A_s$ and $\tau \in a_{q \restriction \mathcal{N}} \subseteq \mathrm{rge}(h^*)$, so τ, $\mu_{i+1} \in \mathrm{rge}(h^*)$. Hence $c^r(\tau, \mu_{i+1}) \leq \alpha^*$, so one has that $c^r(\tau, \xi) \leq \alpha^*$ by the subadditivity of c^r, Theorem (1.12). □

Proposition 22. $\forall \zeta < \tau,\ \xi < \kappa^+ \ \big(\zeta \in a_s \setminus a_q \ \ \&\ \ \tau \in a_s \cap a_q \ \ \&\ \ \xi \in a_q \setminus a_s \longrightarrow c^r(\zeta,\tau) \leq c^r(\zeta,\xi)\big).$

Proof. Suppose that $c^r(\zeta,\xi) \leq c^r(\zeta,\tau)$. This gives that $\alpha^* < c^r(\zeta,\xi) \leq c^r(\zeta,\tau) < \delta \leq c^r\{\tau,\xi\}$, using Lemma (19), Lemma (9), and Corollary (13) or Lemma (21) for the first, third and fourth inequalities, respectively. Let $\nu = \min(\mathrm{rge}(h^*) \setminus \zeta)$. One has that $\zeta < \nu \leq \min(\{\tau,\xi\})$, the former since $\zeta \notin \mathrm{rge}(h^*)$ as $\mathrm{rge}(h^*) \cap a_s \setminus a_q = \emptyset$ by (7.iii.4) for s, and the latter because $\tau \in \mathrm{rge}(h^*)$, if $\tau < \xi$, and by Lemma (18), which shows that there is some element of $\mathrm{rge}(h^*) \cap [\zeta,\xi]$, if $\xi < \tau$.

As $\nu,\tau \in \mathrm{rge}(h^*)$ one has $c^r(\nu,\tau) \leq \alpha^*$. Since $\alpha^* < c^r(\zeta,\tau)$, the subadditivity of c^r gives that $c^r(\zeta,\nu) = c^r(\zeta,\tau)$.

However, if $\langle \xi_i \mid i \leq k\rangle$ and $\langle (\gamma_i, g_i) \mid i < k\rangle$ are good witnesses to the value of $c^r(\zeta,\xi)$ then as noted above, by Lemma (18), there is some $i^* < k-1$ such that $\xi_{i^*} \in \mathrm{rge}(h^*)$. Now, $\langle \gamma_i \mid i < k\rangle$ is strictly increasing, as the witnesses are good, and $c^r(\xi_{i^*},\xi) \leq \gamma_{i^*-1} < \gamma_{i^*} \leq c^r(\zeta,\xi)$, so $c^r(\zeta,\xi_{i^*}) = c^r(\zeta,\xi)$, by the subadditivity of c^r, again. One also has that $\nu,\ \xi_{i^*} \in \mathrm{rge}(h^*)$, so $c^r(\nu,\xi_{i^*}) \leq \alpha^*$. Thus, using the subadditivity of c^r once more, one has that $c^r(\zeta,\nu) = c^r(\zeta,\xi_{i^*}) = c^r(\zeta,\xi)$.

Taken together these deductions show that $c^r(\zeta,\tau) = c^r(\zeta,\nu) = c^r(\zeta,\xi)$. □

Notation 23. If $\nu,\ \mu < \kappa^+$ and $A \in [\mathcal{F}]^{<\omega}$ let $c_A\{\nu,\mu\}$ be $c_A(\nu,\mu)$ if $\nu < \mu$ and be $c_A(\mu,\nu)$ if $\mu < \nu$.

Corollary 24. $\forall \zeta,\ \tau,\ \xi < \kappa^+ \ \big(\zeta \in a_s \setminus a_q \ \ \&\ \ \tau \in a_s \cap a_q \ \ \&\ \ \xi \in a_q \setminus a_s \longrightarrow c^r\{\zeta,\tau\} \leq c^r\{\zeta,\xi\}\big).$

Proof. If $\xi < \zeta$ then $c^r\{\tau,\zeta\} < \delta < c^r(\xi,\zeta)$ by Lemma (9) and Corollary (13). If $\zeta < \tau,\ \xi$ then $c^r(\zeta,\tau) \leq c^r(\zeta,\xi)$ by Proposition (22). If $\tau < \zeta < \xi$ and $c^r(\tau,\xi) < \delta$ then $c^r(\tau,\xi) \leq \alpha^*$ by Lemma (21) and $\alpha^* < c^r(\zeta,\xi)$ by Lemma (19), so $c^r(\tau,\zeta) \leq c^r(\zeta,\xi)$ by the subadditivity of c^r. Finally, if $\tau < \zeta < \xi$ and $\delta \leq c^r(\tau,\xi)$ then $c(\tau,\zeta) < \delta \leq c^r(\tau,\xi) \leq c(\zeta,\xi)$ by the subadditivity of c^r, again. □

The reader should compare this result with the strengthening of the new Δ-property that was used in [Veličk] and [BS] discussed in the introduction. Corollary (24) gives one a property for c_r which is substantially more powerful again.

Corollary 25. If $\zeta,\ \tau,\ \xi < \kappa^+$, $\zeta \in a_s \setminus a_q$, $\tau \in a_s \cap a_q$ and $\xi \in a_q \setminus a_s$ then

$$\tau,\ \zeta < \xi \longrightarrow \{\gamma \mid c^r\{\gamma,\zeta\} \leq c^r\{\zeta,\tau\}\} \subseteq \{\gamma \mid c^r\{\gamma,\zeta\} \leq c^r\{\zeta,\xi\}\}$$

Corollary 26. Summarising, perhaps repetiously, if $\zeta,\ \tau,\ \xi < \kappa^+$, $\zeta \in a_s \setminus a_q$, $\tau \in a_s \cap a_q$ and $\xi \in a_q \setminus a_s$ then

$$\begin{aligned}
\tau < \zeta < \xi &\longrightarrow c^r(\tau,\zeta),\ c^r(\tau,\xi) \leq c^r(\zeta,\xi),\\
\tau < \xi < \zeta &\longrightarrow c^r(\tau,\zeta),\ c^r(\tau,\xi) \leq c^r(\xi,\zeta),\\
\zeta < \tau < \xi &\longrightarrow c^r(\zeta,\tau) \leq c^r(\zeta,\xi) \leq c(\tau,\xi),\\
\xi < \tau < \zeta &\longrightarrow c^r(\tau,\zeta) < c^r(\xi,\tau) \leq c^r(\xi,\zeta),\\
\zeta < \xi < \tau &\longrightarrow c^r(\zeta,\tau) \leq c^r(\zeta,\xi) \leq c(\xi,\tau),\\
\xi < \zeta < \tau &\longrightarrow c^r(\zeta,\tau) < c^r(\xi,\zeta) = c^r(\xi,\tau).
\end{aligned}$$

Lemma 27. $\forall \xi_0 < \xi_1 < \kappa^+ \ (\xi_0,\ \xi_1 \in a_q \setminus a_s \longrightarrow c^r(\xi_0,\xi_1) = c^q(\xi_0,\xi_1))$.

Proof. Let ξ_0 and ξ_1 be as in the antecedent of the lemma. By Lemma (16), if $c^r(\xi_0,\xi_1) \geq \delta$ then $c^r(\xi_0,\xi_1) = c^q(\xi_0,\xi_1)$. By Lemma (20), $c^r(\xi_0,\xi_1) < \delta$ implies $c^r(\xi_0,\xi_1) < \alpha^*$. But Lemma (15) gives that $c^r(\xi_0,\xi_1) < \alpha^*$ implies $c^r(\xi_0,\xi_1) = c^q(\xi_0,\xi_1)$. □

Lemma 28. $\forall \tau_0 < \tau_1 < \kappa^+ \ (\tau_0,\ \tau_1 \in a_q \cap a_s \longrightarrow c^r(\tau_0,\tau_1) \leq \alpha^*)$.

Proof. This is immediate from the fact, already used several times, that $a_q \cap \mathcal{N} \subseteq \mathrm{rge}(h^*)$, which is given by clause (7.iii.3) of the definition of ϕ. □

Lemma 29. $\forall \tau_0 < \tau_1 < \tau_2 < \kappa^+ \ (\tau_0,\ \tau_1,\ \tau_2 \in a_q \cap a_s \longrightarrow$
$(c^q(\tau_0,\tau_1) \leq c^q(\tau_1,\tau_2) \longrightarrow c^{q\restriction\mathcal{N}}(\tau_0,\tau_1) \leq c^{q\restriction\mathcal{N}}(\tau_1,\tau_2)))$.

Proof. This is immediate from Lemma (1.11). □

Lemma 30. If $\xi \in a_q$, $\tau \in a_q \cap a_s$, $\xi < \tau$ and $c^r(\xi,\tau) \leq \delta$ then $\xi \in \mathrm{rge}(F)$, so $\xi \in a_q \cap a_s$ and $c^r(\xi,\tau) \leq \alpha^*$.

Proof. It is immediate that $\xi \in \mathrm{rge}(F)$ by Lemma (1.4). The remainder of the lemma follows from the fact that $a_s \cap a_q = a_q \cap \mathcal{N}$ and Lemma (28). □

I give a couple of final corollaries of the work of this section showing some circumstances when the hypotheses of Corollaries (3.9) and (3.10) are satisfied.

Corollary 31. Suppose that $\nu < \tau < \kappa^+$. If $\tau \in \mathcal{N}$, $\alpha < \delta$, $A = A_s$ and $B = A_r$, or if $\tau \in a_q$ (or even merely $\tau \in (\kappa^+ \setminus \mathcal{N}) \cup \mathrm{rge}(h^*)$), $\alpha < \alpha^*$, $A = A_q$ and $B = A_r$, or if $\alpha \in [\delta,\kappa)$, $A = A_q$ and $B = A_r$, then the hypotheses of Corollary (3.9) are satisfied.

Proof. This is shown in the proofs of Lemmas (11), (15) and (16), respectively. □

Corollary 32. If ν, $\tau \in \kappa^+ \setminus \mathcal{N}$, $\beta^* \leq \beta$ and $\alpha < \delta$ then the hypotheses of Corollary (3.10) are satisfied.

Proof. This is immediate from Lemma (10). □

This concludes the discussion of the relationship between c^q, c^s and c^r based on the fact that $\mathcal{N} \models \phi(s)$ and using the analysis of c_A given in §1.

5. Two applications

5.1. Scattered spaces of size μ^{++}.

Recall the following topological definitions.

Definition 1. A topological space X is *μ-Lindelöf* if every open cover has a subcover of size *less than* μ. Thus ω-Lindeöf means compact and ω_1-Lindelöf means Lindelöf. X is locally μ-Lindelöf if every point has an open neighborhood whose closure is

μ-Lindelof. A space is 0-dimensional if its topology has a basis of sets which are both closed and open. A space X is *scattered* or *right-separated* if there is a well ordering $<_X$ of it such that for every $x \in X$ there is an open neighborhood $\mathcal{O}$ of x such that $\mathcal{O} \cap \{y \in X \mid x <_X y\} = \emptyset$. A space X is μ-tight if for every $Y \subseteq X$ and x in the closure of Y there is some $Z \in [Y]^{\leq\mu}$ such that x is in the closure of Z. The *Cantor-Bendixson derivative* of a space is the space obtained by discarding all isolated points with the relative topology. The *height* of a scattered space is its Cantor-Bendixson rank – the ordinal number of times one can iterate taking the Cantor-Bendixson derivative of a space (taking intersections at limit stages of the iteration). The *width* of a scattered space is the supremum of the size of the sets discarded at each stage when iterating taking the Cantor-Bendixson derivative.

Theorem 2. If there is a stationary $(\kappa, 1)$-simplified morass and κ is the successor of a regular cardinal μ with ${\kappa^+}^{<\mu} = \kappa^+$ then there is a cardinal preserving forcing which adds a 0-dimensional, locally μ-Lindelöf, μ-tight, scattered space of separability degree μ, size and height κ^+ and width μ. The one-point μ-Lindelöfization of the space has the same properties, but is μ-Lindelöf rather than locally μ-Lindelöf. $\square$

A space as described in Theorem (2) was added for $\mu = \omega$ by [BS] – although they worked in terms of superatomic Boolean algebras, Stone duality gives an immediate translation. It would be nice to be able to prove that one can force to add a 0-dimensional, locally compact, countably tight, scattered space of size μ^{++}, however it seems unlikely that one will be able to do this with conditions whose working parts are not finite. I sketch why this is so at the end of the proof of Theorem (2).

Definition 3. Define a partial order $\mathbb{P}^-$ with elements of the form $p = (a_p, \leq_p, i_p)$, where $a_p \in [\kappa^+]^{<\mu}$, $\leq_p$ is a partial order on a_p such that $\alpha \leq_p \beta$ implies $\alpha \leq \beta$, and $i_p : [a_p]^2 \longrightarrow [a_p]^{<\mu}$ is defined by

$$i_p(\alpha, \beta) = \{\gamma \in a_p \mid \gamma \leq \alpha,\ \beta\ \&\ \forall \varepsilon \in a_p\ (\neg(\gamma <_p \varepsilon \leq_p \alpha,\ \beta))\}$$

and is included in the condition p purely for notational purposes. (Here $\gamma <_p \varepsilon$ abbreviates $\gamma \leq_p \varepsilon\ \&\ \gamma \neq \varepsilon$, as usual.) The order on $\mathbb{P}^-$ is $p' \leq p$ if $a_p \subseteq a_{p'}$, $\leq_p = \leq_{p'} \restriction a_p \times a_p$ and $i_p = i_{p'} \restriction [a_p]^2$.

Of course one may equally well write $p = (a_p, \leq_p, i_p)$ as $p = (a_p, h_p, i_p)$ where $h_p : a_p \times a_p \longrightarrow 2$ is a function such that

$$\forall \alpha \in a_p\ \ h_p(\alpha, \alpha) = 1; \qquad \forall \alpha,\ \beta \in a_p\ \ (\alpha < \beta \longrightarrow h_p(\alpha, \beta) = 0;$$

and

$$\forall \alpha,\ \beta,\ \ \gamma \in a_p\ (h_p(\alpha, \beta) = 1\ \&\ h_p(\beta, \gamma) = 1 \longrightarrow h_p(\alpha, \gamma) = 1),$$

that is, that h_p satisfies the requirements to be the characteristic function of a partial order on a_p.

Note that $\mathbb{P}^-$ does not have the κ-chain condition: if one takes κ many conditions and applies the Δ-system lemma to their 'a's then it may still be the case that no two of the 'i's agree on their restriction to the root of the Δ-system.

Definition 4. Define a forcing notion $\mathbb{P}$ with conditions $p = (a_p, \leq_p, i_p, A_p)$ where $(a_p, \leq_p, i_p) \in \mathbb{P}^-$, $A_p \in \mathcal{F}$ and, writing c^p for c_{A_p},

$$\forall\alpha,\ \beta \in a_p\ (\alpha < \beta \longrightarrow i_p(\alpha,\beta) \subseteq \{\gamma \leq \alpha \mid c^p(\gamma,\alpha) \leq c^p(\alpha,\beta)\}).$$

The order is the product of the order on $\mathbb{P}^-$ and reverse inclusion, *i.e.* $p' \leq p$ if $a_p \subseteq a_{p'}$, $\leq_p = \leq_{p'} \restriction a_p \times a_p$, $i_p = i_{p'} \restriction [a_p]^2$ and $A_p \subseteq A_{p'}$. a_p is the *realm* of p and A_p is the *side condition part* of p.

The intention is that if G is $\mathbb{P}$-generic one will take

$$B_\alpha = \{\beta \leq \alpha \mid \exists p \in G\ (\beta,\ \alpha \in a_p\ \ \&\ \ \beta \leq_p \alpha)\}.$$

Taking $\mathcal{B} = \{B_\alpha \setminus \bigcup\{B_\beta \mid \beta \in b\} \mid \alpha < \kappa^+\ \ \&\ \ b \in [\alpha]^{<\mu}\}$ will give a basis for a topology on κ^+ which satisfies the theorem provided that cardinals are preserved. I prove the topological properties after showing that cardinals are indeed preserved by the forcing.

Proposition 5. $\mathbb{P}$ is κ-$\mathbb{M}$-proper.

Proof. So suppose that $p \in \mathbb{P}$ and that $\mathcal{N}$ is as in the general framework of §4 but for this specific $\mathbb{P}$.

Step A. Checking that $p' = (a_p, \leq_p, i_p, A_p \cup \{(\delta, F)\})$ is a condition and is stronger than p.

Proof. Observe that $(\varepsilon, f) \in \mathcal{N}$ for each $(\varepsilon, f) \in A_p$, since $p \subseteq \mathcal{N}$. So for each $(\varepsilon, f) \in A_p$ one has, by (i) of the properties of $\mathcal{N}$ given in §4, that there is some $f' \in \mathcal{F}_{\varepsilon\delta}$ such that $f = F \cdot f'$. Suppose that α, β, $\gamma \in a_p$, $\gamma < \alpha < \beta$ and $c^p(\gamma,\alpha) \leq c^p(\alpha,\beta)$. As $a_p \subseteq \mathcal{N} \cap \kappa^+ = \mathrm{rge}(F)$ one has that $c^{p'}(\gamma,\alpha) \leq c^{p'}(\alpha,\beta)$ by Lemma (1.6). This shows that $i_{p'}(\alpha,\beta) \subseteq \{\gamma < \alpha \mid c^{p'}(\gamma,\alpha) \leq c^{p'}(\alpha,\beta)\}$. Hence $p' \in \mathbb{P}$.

Suppose that $\mathcal{D}$ is a dense, open subset of $\mathbb{P}$, $\mathcal{D} \in \mathcal{N}$ and $q \in \mathcal{D}$ where $q \leq p'$.

Step B. Show that without loss of generality one may assume that $c^q(\nu,\tau) < \kappa$ for all ν, $\tau \in a_q$.

Proof. For if q does not satisfy this one may simply add to A_q a map (β, f) such that $a_q \subseteq \mathrm{rge}(f)$ and $\alpha \leq \beta$ for all $(\alpha, g) \in A_q$ by applying Fact (I.5). By Lemma (1.6), this augmentation of q is still a condition, it is stronger than q by the definition of $\leq_{\mathbb{P}}$, and is in $\mathcal{D}$ because $\mathcal{D}$ is open.

Let $q \restriction \mathcal{N} = q \cap \mathcal{N}$.

Step C. Show that $q \restriction \mathcal{N} \in \mathbb{P} \cap \mathcal{N}$ and $q \leq q \restriction \mathcal{N}$.

Lemma 6. If τ_0, $\tau_1 \in a_q \cap \mathcal{N}$ then $i_q(\tau_0,\tau_1) \subseteq \mathcal{N}$. And if $\gamma < \tau_0 < \tau_1$ and $c^q(\gamma,\tau_0) \leq c^q(\tau_0,\tau_1)$ then $c^{q\restriction\mathcal{N}}(\gamma,\tau_0) \leq c^{q\restriction\mathcal{N}}(\tau_0,\tau_1)$.

Proof. As τ_0, $\tau_1 \in \mathcal{N} = \mathrm{rge}(F)$, one has that $c^q(\tau_0, \tau_1) \leq \delta$. If $\gamma \in i_q(\tau_0, \tau_1)$ then $c^q(\gamma, \tau_0) \leq \delta$ by the constraint on i_q in the definition of what it is to be a condition in $\mathbb{P}$. So by Lemma (1.4) one has that $\gamma \in \mathrm{rge}(F)$. By Lemma (1.11), for every $\gamma < \tau_0$, if $c^q(\gamma, \tau_0) \leq c^q(\tau_0, \tau_1)$ one has that $c^{q \restriction \mathcal{N}}(\gamma, \tau_0) \leq c^{q \restriction \mathcal{N}}(\tau_0, \tau_1)$. □

Corollary 7. $q \leq q \restriction \mathcal{N} \in \mathbb{P} \cap \mathcal{N}$.

Proof. If τ_0, $\tau_1 \in a_q \cap \mathcal{N}$ and $\gamma \in i_q(\tau_0, \tau_1)$ then, by the first half of Lemma (6), $\gamma \in a_{q \restriction N}$ and so $i_{q \restriction \mathcal{N}}(\tau_0, \tau_1) = i_q(\tau_0, \tau_1)$. So $i_{q \restriction \mathcal{N}}(\tau_0, \tau_1) = i_q(\tau_0, \tau_1) \subseteq \{\gamma \leq \tau_0 \mid c^q(\gamma, \tau_0) \leq c^q(\tau_0, \tau_1)\} \subseteq \{\gamma \leq \tau_0 \mid c^{q \restriction \mathcal{N}}(\tau_0, \tau_1) \leq c^{q \restriction \mathcal{N}}(\gamma, \tau_0)\}$, by the second half of Lemma (6). Hence as far as the pair τ_0, τ_1 are concerned, $q \restriction \mathcal{N}$ satisfies the requirement for being a condition, and q and $q \restriction \mathcal{N}$ satisfy the requirement for q to be a stronger condition than $q \restriction \mathcal{N}$. □

Now obtain s as in §2 using b_q as the arbitrary set e_q in the section of Notation in §4.

Step D. Define some r with realm $a_s \cup a_q$ and side condition part $A_s \cup A_q$. Show that $r \in \mathbb{P}$ and that $r \leq s$, q.

Proof. I define r as follows. Let $a_r = a_s \cup a_q$ and $A_r = A_s \cup A_q$. Let $\alpha \leq_r \beta$ if $\alpha \leq_q \beta$ for $\alpha \leq_s \beta$ or $\exists \gamma\ (\alpha \leq_s \gamma \leq_q \beta)$ or $\exists \gamma\ (\alpha \leq_q \gamma \leq_s \beta)$.

Lemma 8. If $\{\alpha, \gamma\} \in [a_q]^2$, $\beta \in a_s$ and ε, $\nu \in a_s \cap a_q$ are such that $\alpha \leq_q \varepsilon \leq_s \beta \leq_s \nu \leq_q \gamma$ then $\alpha \leq_q \gamma$. The same conclusion also holds with s and q exchanged.

Proof. One has that $\alpha \leq_q \varepsilon \leq_s \nu \leq_q \gamma$ by the transitivity of $\leq_s$. Moreover, $\varepsilon \leq_q \nu$ since $\leq_q \restriction a_{q \restriction \mathcal{N}} = \leq_s \restriction a_{q \restriction \mathcal{N}}$, so $\alpha \leq_q \gamma$ by the transitivity of $\leq_q$. The proof with q and s exchanged is identical. □

Corollary 9. $\leq_r \restriction a_q \times a_q = \leq_q$ and $\leq_r \restriction a_s \times a_s = \leq_s$. □

Lemma 10. $i_r \restriction [a_q]^2 = i_q$ and $i_r \restriction [a_s]^2 = i_s$.

Proof. Suppose $\xi_0, \xi_1 \in a_q$ and $\nu \leq_r \xi_0, \xi_1$. If $\nu \in a_q$ then $\nu \leq_q \xi_0, \xi_1$, by Corollary (9), so there is some $\xi \in i_q(\xi_0, \xi_1)$ such that $\nu \leq_q \xi$. If $\nu \in a_s \setminus a_q$ then there are τ_0, $\tau_1 \in a_s \cap a_q$ such that $\nu \leq_s \tau_0 \leq_q \xi_0$ and $\nu \leq_s \tau_1 \leq_q \xi_1$. Let $\tau \in i_s(\tau_0, \tau_1)$ be such that $\nu \leq_s \tau$. Now, as q, $s \leq q \restriction \mathcal{N}$ one has that $i_q(\tau_0, \tau_1) = i_{q \restriction \mathcal{N}}(\tau_0, \tau_1) = i_s(\tau_0, \tau_1)$. So $\tau \in i_q(\tau_0, \tau_1)$. Let $\xi \in i_q(\xi_0, \xi_1)$ be such that $\tau \leq_q \xi$. Then $\nu \leq_r \tau \leq_r \xi \leq_r \xi_0$, ξ_1, so $\nu \leq_r \xi \in i_q(\xi_0, \xi_1)$. This shows that $i_r \restriction [a_q]^2 = i_q$. The argument that $i_r \restriction [a_s]^2 = i_s$ is identical with the rôles of s and q swapped. □

Lemma 11. $\leq_r$ is a partial order on $a_r \times a_r$.

Proof. $\leq_r$ is clearly reflexive and anti-symmetric, so only transitivity needs to be shown. Suppose $\alpha \leq_r \beta \leq_r \gamma$. If $\{\alpha, \beta, \gamma\} \in [a_q]^3 \cup [a_s]^3$ then $\alpha \leq_r \gamma$. If $\{\alpha, \gamma\} \in [a_q]^2$ and $\beta \in a_s$ then $\alpha \leq_q \gamma$ by Lemma (8). If $\{\alpha, \beta\} \in [a_q]^2$ and $\gamma \in a_s$ let $\varepsilon \in a_s \cap a_q$ be such that $\alpha \leq_q \beta \leq_q \varepsilon \leq_s \gamma$. Then $\alpha \leq_q \varepsilon \leq_s \gamma$ by the transitivity of $\leq_q$. The argument in each of the other cases is identical to one of these latter two. □

The arguments for Lemmas (8), (10) and (11) are essentially the same as analogous arguments used in [BS, §8] in the proof that the subcollection of $\mathbb{P}^-$ (when $\kappa = \omega_1$) consisting of those conditions p with $i_p(\alpha,\beta) \subseteq \{\gamma \leq \alpha \mid c(\gamma,\alpha) \leq c(\alpha,\beta)\}$ for all α, $\beta \in a_p$, under the suborder of $\leq_{\mathbb{P}^-}$ such that that i remains fixed between a condition and a stronger one, has the ccc. Similarly to that proof, as a consequence of Lemmas (8), (10) and (11) it suffices to show that $r \in \mathbb{P}$ in order to show that q and s are compatible in $\mathbb{P}$. Here this will show that $\mathbb{P}$ is $\mathbb{M}$-proper. What needs to be done in order to complete this is to show that if α, $\beta \in a_r$ with $\alpha < \beta$ and $\gamma \in i_r(\alpha,\beta)$ then $c^r(\gamma,\alpha) \leq c^r(\alpha,\beta)$. But the proof of this is markedly different from (and harder than) the analogous proof in [BS, §8]. In conformity with the usage in §4, I try to use τ, possibly adorned with subscripts, for variables in $a_s \cap a_q$, ζ for variables in a_s and ξ for variables in a_q.

Proposition 12. $r \in \mathbb{P}$.

Proof. **Case A.** ξ_0, $\xi_1 \in a_q$ and $\xi_0 < \xi_1$. Lemma (10) shows that $i_r(\xi_0,\xi_1) = i_q(\xi_0,\xi_1)$. So what needs to be shown is that if $\xi \in i_q(\xi_0,\xi_1)$ then $c^r(\xi,\xi_0) \leq c^r(\xi_0,\xi_1)$. Suppose that $\xi \in i_q(\xi_0,\xi_1)$.

Subcase A.i. ξ_0, $\xi_1 \in a_q \cap a_s = a_{q\restriction\mathcal{N}}$. If $\xi \in i_q(\xi_0,\xi_1)$ then $\xi \in a_q \cap \mathcal{N} = a_q \cap a_s$, by Lemma (6) and clauses (iii.3) and (iii.4) of the definition of ϕ for q, respectively. Thus $\xi \in i_{q\restriction\mathcal{N}}(\xi_0,\xi_1)$. As $s \leq q \restriction \mathcal{N}$ one has that $\xi \in i_s(\xi_0,\xi_1)$, and hence that $c^s(\xi,\xi_0) \leq c^s(\xi_0,\xi_1)$. By (4.12) one has that $c^s(\xi,\xi_0) = c^r(\xi,\xi_0)$ and $c^s(\xi_0,\xi_1) = c^r(\xi_0,\xi_1)$. Thus $c^r(\xi,\xi_0) \leq c^r(\xi_0,\xi_1)$.

Subcase A.ii. ξ_0, $\xi_1 \in a_q \setminus a_s$. By Lemma (1.3) one has that $c^r(\xi,\xi_0) \leq c^q(\xi,\xi_0)$, and one also has that $c^q(\xi,\xi_0) \leq c^q(\xi_0,\xi_1)$ since $q \in \mathbb{P}$ and $\xi \in i_q(\xi_0,\xi_1)$. Moreover, $c^q(\xi_0,\xi_1) = c^r(\xi_0,\xi_1)$ by (4.27). Hence $c^r(\xi,\xi_0) \leq c^r(\xi_0,\xi_1)$.

Subcase A.iii. $\xi_0 \in a_q \setminus a_s$ and $\xi_1 \in a_q \cap a_s = a_{q\restriction\mathcal{N}}$. One has $c^r(\xi,\xi_0) \leq c^q(\xi,\xi_0)$ by Lemma (1.3), again, and $c^q(\xi,\xi_0) \leq c^q(\xi_0,\xi_1)$ since $q \in \mathbb{P}$ and $\xi \in i_q(\xi_0,\xi_1)$. One also has that $\delta \leq c^r(\xi_0,\xi_1) = c^q(\xi_0,\xi_1)$ by (4.17). Hence $c^r(\xi,\xi_0) \leq c^r(\xi_0,\xi_1)$.

Subcase A.iv. $\xi_0 \in a_q \cap a_s$ and $\xi_1 \in a_q \setminus a_s$. If $\delta \leq c^r(\xi_0,\xi_1)$ then $c^r(\xi_0,\xi_1) = c^q(\xi_0,\xi_1)$ by (4.16), when $c^r(\xi,\xi_0) \leq c^r(\xi_0,\xi_1)$ as in Cases (A.ii) and (A.iii) again. Otherwise, $c^r(\xi_0,\xi_1) \leq \alpha^*$ by (4.21). If $c^r(\xi_0,\xi_1) < \alpha^*$ then $c^r(\xi_0,\xi_1) = c^q(\xi_0,\xi_1)$ by (4.15), and $c^r(\xi,\xi_0) \leq c^r(\xi_0,\xi_1)$ as in Cases (A.ii) and (A.iii) once more. The remaining possibility is that $c^r(\xi_0,\xi_1) = \alpha^*$ and $c^q(\xi_0,\xi_1) = \delta$. As $\xi \in a_q$ one has that $c^q(\xi,\xi_0) \leq c^q(\xi,\xi_1) = \delta$. By Lemma (1.4) one has that $\xi \in \mathrm{rge}(F) = \mathcal{N} \cap \kappa^+$, so that $\xi \in a_q \cap a_s$. By clause (iii.3) of the definition of ϕ for s one then has that ξ, $\xi_0 \in \mathrm{rge}(h^*)$, so that $c^r(\xi,\xi_0) \leq \alpha^*$, as required.

Case B. ζ_0, $\zeta_1 \in a_s$ and $\zeta_0 < \zeta_1$. Lemma (10) shows that $i_r(\zeta_0,\zeta_1) = i_s(\zeta_0,\zeta_1)$. So what needs to be shown is that if $\zeta \in i_s(\zeta_0,\zeta_1)$ then $c^r(\zeta,\zeta_0) \leq c^r(\zeta_0,\zeta_1)$. So suppose that $\zeta \in i_q(\zeta_0,\zeta_1)$. As previously, one has that $c^r(\zeta,\zeta_0) \leq c^s(\zeta,\zeta_0)$ by Lemma (1.3), and that $c^s(\zeta,\zeta_0) \leq c^s(\zeta_0,\zeta_1)$ since $s \in \mathbb{P}$ and $\zeta \in i_s(\zeta_0,\zeta_1)$. But (4.12) shows that $c^r(\zeta_0,\zeta_1) = c^s(\zeta_0,\zeta_1)$. So $c^r(\zeta,\zeta_0) \leq c^r(\zeta_0,\zeta_1)$ as required.

Case C. $\zeta \in a_s \setminus a_q$ and $\xi \in a_q \setminus a_s$. If $\zeta \leq_r \xi$ then $i_r(\zeta,\xi) = \{\zeta\}$ and, clearly, $c^r(\zeta,\zeta) \leq c^r(\zeta,\xi)$. Similarly if $\xi \leq_r \zeta$ then $i_r(\xi,\zeta) = \{\xi\}$ and $c^r(\xi,\xi) \leq c^r(\xi,\zeta)$. So suppose that $\zeta \not\leq_r \xi$ and $\xi \not\leq_r \zeta$.

Let $\nu \in i_r\{\xi,\zeta\}$. In order to complete the proof one has to show that $c^r(\nu,\zeta) \leq c^r\{\zeta,\xi\}$.

SUBCASE C.I. $\nu \in a_s \cap a_q$. By (4.24) one immediately has $c^r(\nu,\zeta) \leq c^r\{\zeta,\xi\}$.

SUBCASE C.II. $\nu \in a_s \setminus a^q$.

By the definition of $\leq_r$, let $\tau \in a_s \cap a_q$ be such that $\nu \leq_s \tau \leq_q \xi$. Clearly $\nu \in i_r\{\tau,\zeta\}$, and so by CASE (B) above one has that $c^r(\nu,\tau)$, $c^r(\nu,\zeta) \leq c^r\{\tau,\zeta\}$. However, by (4.24) again, one has that $c^r\{\tau,\zeta\} \leq c^r\{\zeta,\xi\}$. So $c^r(\nu,\zeta) \leq c^r\{\zeta,\xi\}$, and by the subadditivity of c^r one has that $c^r(\nu,\zeta)$, $c^r(\nu,\xi) \leq c^r\{\zeta,\xi\}$.

SUBCASE C.III. $\nu \in a_q \setminus a_s$. Exactly as SUBCASE (C.II) but switching the rôles of s, ζ and q, ξ, and using CASE (A) in place of CASE (B).

Hence Cases (A), (B) and (C) taken together show that if α, $\beta \in a_r$ with $\alpha < \beta$ and $\gamma \in i_r(\alpha,\beta)$ then $c^r(\gamma,\alpha) \leq c^r(\alpha,\beta)$, and so that $r \in \mathbb{P}$, thus concluding the proof Proposition (12). □

This shows that q and s are compatible and hence $\mathbb{P}$ is κ-$\mathbb{M}$-proper, concluding the proof of Proposition (5). Hence forcing with $\mathbb{P}$ preserves κ. □

One great advantage of κ-$\mathbb{M}$-proper forcing over proper forcing is that one has a reasonable chance of preserving κ^+ (and all greater cardinals) as well as κ!

Proposition 13. $\mathbb{P}$ has the κ^+-cc.

Proof. Let $\{p'_\nu \mid \nu < \kappa^+\}$ be a subset of $\mathbb{P}$ of size κ^+. Extend each p'_ν in the original collection one is given to a stronger condition p_ν by adding a single map to $A_{p'_\nu}$ if necessary in order to ensure that there is a map $(\alpha_\nu, F_\nu) \in A_{p_\nu}$ such that for all $(\beta, f) \in A_{p_\nu}$ one has $\beta \leq \alpha$ and there is some $g \in \mathcal{F}_{\beta\alpha}$ such that $f = F \cdot g$ and such that $a_p \subseteq \mathrm{rge}(F)$. This is possible by Fact (I.5), as in the argument in the paragraph two prior to Lemma (6) above where q is introduced and then assumed to have this property. If one can find η, $\nu < \delta$ such that p_η and p_ν are compatible then, clearly, one will have that p'_η and p'_ν are compatible as well.

By thinning using the Δ-system lemma if necessary one may as well assume that there is some $a \in [\kappa^+]^{<\mu}$ such that a is an initial segment of each a_{p_ν}. One may as well also assume that if $\eta < \nu < \kappa^+$ then $\max(a_{p_\eta}) < \min(a_{p_\nu} \setminus a)$, and there is an order preserving isomorphism from $(a_{p_\eta}, \leq_{p_\eta})$ to $(a_{p_\nu}, \leq_{p_\nu})$. Moreover, one may as well assume that there is some $\alpha < \kappa$, some $A \in [\bigcup\{\{\beta\} \times \mathcal{F}_{\beta\alpha} \mid \beta < \alpha\}]^{<\mu}$ and for each $\nu < \kappa^+$ there is a map $F_\nu \in \mathcal{F}_{\alpha\kappa}$ such that

$$A_{p_\nu} = \{(\beta, F_\nu \cdot h) \mid (\beta, h) \in A\} \cup \{(\alpha, F_\nu)\}$$

and such that $F_\nu \cdot F_\eta^{-1} \restriction a_{p_\eta}$ is the order preserving isomorphism from $(a_{p_\eta}, \leq_{p_\eta})$ to $(a_{p_\nu}, \leq_{p_\nu})$. This is possible using Proposition (I.13).

Clearly for each $\nu < \kappa^+$ and $\xi < \zeta < \kappa^+$ one has that $c^{p_\nu}(\xi,\zeta) \leq \alpha$ if and only if $c^{p_\nu}(\xi,\zeta) < \kappa$. Let $\eta < \nu < \kappa^+$ and set $a_r = a_{p_\eta} \cup a_{p_\nu}$, $\leq_r = \leq_{p_\eta} \cup \leq_{p_\nu}$ and

$A_r = A_{p_\eta} \cup A_{p_\nu}$. Note that $i_r(\zeta_0, \zeta_1) = i_{p_\eta}$ if ζ_0, $\zeta_1 \in a_{p_\eta}$ and $i_r(\xi_0, \xi_1) = i_{p_\nu}$ if ξ_0, $\xi_1 \in a_{p_\nu}$. It is also clear that if $\xi < \zeta < \kappa^+$ then $c^r(\xi, \zeta) \leq \alpha$ if and only if $c^r(\xi, \zeta) < \kappa$. By Lemma (1.14), one has that if $\zeta \in \mathrm{rge}(F_\eta)$ then $c^r(\xi, \zeta) = c^{p_\eta}(\xi, \zeta)$ and if $\zeta \in \mathrm{rge}(F_\nu)$ then $c^r(\xi, \zeta) = c^{p_\nu}(\xi, \zeta)$.

Claim 14. If $\xi \in a_{p_\eta} \setminus a_{p_\nu}$ then $\xi \notin \mathrm{rge}(F_\nu)$.

Proof. If $\xi \in \mathrm{rge}(F_\nu)$ then $\xi \in \mathrm{rge}(F_\nu) \cap \mathrm{rge}(F_\eta)$ since $F_\nu \cdot F_\eta^{-1} \restriction a_{p_\eta}$ is the order preserving isomorphism from a_{p_η} to a_{p_ν}. But then one has that $F_\nu(\xi_\alpha) = \xi = F_\eta(\xi_\alpha)$ by Fact (I.3), so $F_\nu \cdot F_\eta^{-1}(\xi) = \xi$ and hence $\xi \in a_{p_\nu} \cap a_{p_\eta}$, a contradiction. □

It is now immediate that if $\xi \in a_{p_\eta} \setminus a_{p_\nu}$ and $\zeta \in a_{p_\nu} \setminus a_{p_\eta}$ then $\xi \notin \mathrm{rge}(F_\nu)$, by Claim (14), so that $c^{p_\nu}(\xi, \zeta) = \kappa$, and thus $c^r(\xi, \zeta) = \kappa$ from Claim (14). Hence for such ξ, ζ one has that $i_r(\xi, \zeta) \subseteq \{\tau \leq \xi \,|\, c^r(\tau, \xi) \leq c^r(\xi, \zeta)\}$.

By Claim (14), as well, if ξ_0, $\xi_1 \in a_{p_\eta}$ then $c^r(\xi_0, \xi_1) = c^{p_\eta}(\xi_0, \xi_1)$; and so if $\xi_0 < \xi_1$ one has that

$$i_r(\zeta_0, \zeta_1) \subseteq \{\tau \leq \xi_0 \,|\, e^{p_\eta}(\tau, \xi_0) \leq e^{p_\eta}(\xi_0, \xi_1)\} = \{\tau \leq \xi_0 \,|\, e^r(\tau, \xi_0) \leq e^r(\xi_0, \xi_1)\}.$$

(One gets equality rather than just the necessary inclusion by applying Claim (14) again for $\tau < \xi_0 \in \mathrm{rge}(F_\eta)$.) Similarly, if ζ_0, $\zeta_1 \in a_{p_\nu}$ then $c^r(\zeta_0, \zeta_1) = c^{p_\nu}(\zeta_0, \zeta_1)$. So $i_r(\zeta_0, \zeta_1) \subseteq \{\tau \leq \zeta_0 \,|\, e^r(\tau, \zeta_0) \leq e^r(\zeta_0, \zeta_1)\}$ if $\zeta_0 < \zeta_1$.

Hence $r \in \mathbb{P}$, and thus $r \leq_{\mathbb{P}} p_\eta$, p_ν. So $\mathbb{P}$ does have the κ^+-cc. □

Thus cardinals above μ are preserved by forcing with $\mathbb{P}$. The next proposition shows that cardinals up to μ are also preserved.

Proposition 15. $\mathbb{P}$ is μ-closed.

Proof. Let $\langle p_\varepsilon \,|\, \varepsilon < \chi \rangle$ be a descending sequence of conditions for some limit ordinal $\chi < \mu$. Define p by setting $a_p = \bigcup \{a_{p_\varepsilon} \,|\, \varepsilon < \chi\}$, $\nu \leq_p \tau$ if there is some $\varepsilon < \chi$ such that $\nu \leq_{p_\varepsilon} \tau$ and $A_p = \bigcup \{A_{p_\varepsilon} \,|\, \varepsilon < \chi\}$.

First of all, I show that $p \in \mathbb{P}$. Suppose that η, ν, $\tau \in a_p$ and $\eta \in i_p(\nu, \tau)$. Let $\langle \xi_i \,|\, i \leq k \rangle$ and $\langle (\gamma_i, g_i) \,|\, i < k \rangle$ and $\langle \zeta_j \,|\, j \leq l \rangle$ and $\langle (\beta_j, f_j) \,|\, j < l \rangle$ be witnesses to the values of $c^p(\eta, \nu)$ and $c^p(\nu, \tau)$ respectively. Then there is some $\varepsilon < \chi$ such that $\{(\gamma_i, g_i) \,|\, i < k\} \cup \{(\beta_j, f_j) \,|\, j < l\} \subseteq A_{p_\varepsilon}$ and η, ν, $\tau \in a_{p_\varepsilon}$. First of all note, by the definition of i_{p_ε} and the fact that $a_{p_\varepsilon} \subseteq a_p$, that $\eta \in i_{p_\varepsilon}(\nu, \tau)$. As p_ε is a condition this implies that $c^{p_\varepsilon}(\eta, \nu) \leq c^{p_\varepsilon}(\nu, \tau)$. But as $c^p(\eta, \nu) \leq c^{p_\varepsilon}(\eta, \nu)$ and $c^p(\nu, \tau) \leq c^{p_\varepsilon}(\nu, \tau)$, by Lemma (1.3), one has that $\langle \xi_i \,|\, i \leq k \rangle$ and $\langle (\gamma_i, g_i) \,|\, i < k \rangle$ and $\langle \zeta_j \,|\, j \leq l \rangle$ and $\langle (\beta_j, f_j) \,|\, j < l \rangle$ are witnesses to the values of $c^{p_\varepsilon}(\eta, \nu)$ and $c^{p_\varepsilon}(\nu, \tau)$ respectively. Thus $c^p(\eta, \nu) = c^{p_\varepsilon}(\eta, \nu)$ and $c^p(\nu, \tau) = c^{p_\varepsilon}(\nu, \tau)$ and one has $c^p(\eta, \nu) \leq c^p(\nu, \tau)$ as required. Hence $p \in \mathbb{P}$.

In order to show that $p \leq p_\varepsilon$ for $\varepsilon < \chi$ one needs only to show that $i_p(\nu, \tau) = i_{p_\varepsilon}(\nu, \tau)$ for ν, $\tau \in a_{p_\varepsilon}$.

If $\eta \in i_p(\nu, \tau)$ there is some $\varepsilon' \in [\varepsilon, \chi)$ such that $\eta \in a_{p_{\varepsilon'}}$, and hence $\eta \in i_{p_{\varepsilon'}}(\nu, \tau)$ by the definition of $i_{p_{\varepsilon'}}$ (again). But $p_{\varepsilon'} \leq p_\varepsilon$, so $\eta \in a_{p_\varepsilon}(\nu, \tau)$.

Conversely, suppose towards a contradiction that $\eta \in i_{p_\varepsilon}(\nu, \tau)$ and there is some $\xi \in a_p$ with $\eta <_p \xi \leq_p \nu$, τ. Then there is some $\varepsilon' \in [\varepsilon, \chi)$ such that $\xi \in a_{p_{\varepsilon'}}$.

Hence $\eta <_{p_{\varepsilon'}} \xi \leq_{p_{\varepsilon'}} \nu, \tau$. But this contradicts the fact that $i_{p_{\varepsilon'}}(\nu,\tau) = i_{p_\varepsilon}(\nu,\tau)$, which is true because $p_{\varepsilon'} \leq p_\varepsilon$. □

Lemma 16. For each $\xi < \kappa^+$ the collection $\{p \in \mathbb{P} \mid \xi \in a_p\}$ is a dense (and open) subset of $\mathbb{P}$.

Proof. Let $\xi < \kappa^+$ and $p \in \mathbb{P}$. If $\xi \in a_p$ there is nothing to prove. Otherwise define q by $a_q = a_p \cup \{\xi\}$, $\leq_q = \leq_p$, $A_q = A_p$. Then $q \in \mathbb{P}$ and $\xi \in a_q$. □

Definition 17. Let G be $\mathbb{P}$-generic over V and define $B_\alpha = \{\beta \leq \alpha \mid \exists p \in G\ \beta \leq_p \alpha\}$ for each $\alpha < \kappa^+$. Let $i(\alpha,\beta) = \{\gamma \mid \exists p \in G\ \gamma \in i_p(\alpha,\beta)\}$ for $\alpha < \beta < \kappa^+$. Note that $\overline{\overline{i(\alpha,\beta)}} < \mu$ for all $\alpha < \beta < \kappa^+$.

Let $\mathcal{B} = \{B_\alpha \setminus \bigcup\{B_\beta \mid \beta \in b\} \mid \alpha < \kappa^+ \ \&\ b \in [\alpha]^{<\mu}\}$. Let τ be the topology on κ^+ generated by taking $\mathcal{B}$ as a sub-basis.

Lemma 18. $\mathcal{B}$ is a basis for a topology on κ^+, not just a sub-basis.

Proof. Let $\mathcal{O}_0 = B_\alpha \setminus \bigcup\{B_\gamma \mid \gamma \in a\}$ and $\mathcal{O}_1 = B_\beta \setminus \bigcup\{B_\varepsilon \mid \varepsilon \in b\}$ be sets in $\mathcal{B}$. Then $\mathcal{O}_0 \cap \mathcal{O}_1 = B_\alpha \cap B_\beta \cap (\kappa^+ \setminus \bigcup\{B\gamma \mid \gamma \in a \cup b\})$.

Now $B_\alpha \cap B_\beta = \bigcup\{B_\varepsilon \mid \varepsilon \in i(\alpha,\beta)\}$, so

$$\mathcal{O}_0 \cap \mathcal{O}_1 = \bigcup\{B_\varepsilon \setminus \bigcup\{B\gamma \mid \gamma \in a \cup b\} \mid \varepsilon \in i(\alpha,\beta)\}.$$

So in order to show that $\mathcal{O}_0 \cap \mathcal{O}_1$ is a union of sets in $\mathcal{B}$ it suffices to show that each $B_\varepsilon \setminus \bigcup\{B_\gamma \mid \gamma \in a \cup b\}$ is in $\mathcal{B}$.

But $B_\varepsilon \setminus \bigcup\{B_\gamma \mid \gamma \in a \cup b\} = B_\varepsilon \setminus \bigcup\{B_\varepsilon \cap B_\gamma \mid \gamma \in a \cup b\}$ and each $B_\varepsilon \cap B_\gamma = \bigcup\{B_\nu \mid \nu \in i\{\gamma,\varepsilon\}\}$ (recalling that $i\{\gamma,\varepsilon\}$ denotes $i(\gamma,\varepsilon)$ if $\gamma \leq \varepsilon$, and $i(\varepsilon,\gamma)$ if instead $\varepsilon \leq \gamma$). Since each $i(\gamma,\varepsilon)$ has size less than μ, one has that $\{\nu \mid \exists\gamma \in a \cup b\ \nu \in i\{\gamma,\varepsilon\}\}$ is a union of fewer than μ sets each of size less than μ and so, as μ is regular, itself has size less than μ. Thus $B_\varepsilon \setminus \bigcup\{B_\gamma \mid \gamma \in a \cup b\} \in \mathcal{B}$ as required and so $\mathcal{O}_1 \cap \mathcal{O}_2$ is a union of sets in $\mathcal{B}$. □

Lemma 19. $\mathcal{B}$ is a clopen basis and so τ is 0-dimensional.

Proof. If $\mathcal{O} = B_\alpha \setminus \bigcup\{B_\gamma \mid \gamma \in a\}$ then $\kappa^+ \setminus \mathcal{O} = (\kappa^+ \setminus B_\alpha) \cup \{B_\gamma \mid \gamma \in a\}$. Clearly $B_\gamma \in \mathcal{B}$ for $\gamma \in a$. Also $\kappa^+ \setminus B_\alpha = \bigcup\{B_\beta \setminus B_\alpha \mid \beta < \kappa^+\}$. But

$$B_\beta \setminus B_\alpha = B_\beta \setminus (B_\beta \cap B_\alpha) = B_\beta \setminus \bigcup\{B_\nu \mid \nu \in i\{\alpha,\beta\}\},$$

as in the proof of Lemma (18). Thus each $B_\beta \setminus B_\alpha$ is a union of sets in $\mathcal{B}$, and hence $\kappa^+ \setminus \mathcal{O}$ is also a union of sets in $\mathcal{B}$. □

Lemma 20. τ is right-separated.

Proof. If $\alpha < \kappa^+$ then $\alpha \in B_\alpha$, but $B_\alpha \subseteq \alpha + 1$, so $\beta \notin B_\alpha$ for $\alpha < \beta < \kappa^+$. Of course, each $B_\alpha \in \mathcal{B}$. □

Lemma 21. (κ^+, τ) is locally μ-Lindelöf.

Proof. Let $\alpha < \kappa^+$. $\alpha \in B_\alpha$, and B_α is closed (by Lemma (19)). So it suffices to show that each B_α is μ-Lindelöf. This is done by induction on κ^+.

Suppose that B_γ is μ-Lindelöf for each $\gamma < \alpha$. Let $\langle \mathcal{O}_i \mid i \in I \rangle$ be an open cover of B_α. So there is some $i \in I$ and some $B_\beta \setminus \bigcup\{B_{\beta_j} \mid \beta_j \in b\} \subseteq \mathcal{O}_i$ such that $\alpha \in B_\beta \setminus \bigcup\{B_{\beta_j} \mid \beta_j \in b\} \in \mathcal{B}$. So B_α has a subcover of size less than μ if and only if $B_\alpha \cap B_{\beta_j}$ has a subcover of size less than μ for each $j \in b$ (using the facts that $\overline{\overline{b}} < \mu$ and that μ is regular). But $B_\alpha \cap B_{\beta_j} = \bigcup\{B_\gamma \mid \gamma \in i\{\alpha, \beta_j\}\}$ for each $j \in b$, and thus is the union of fewer than μ many sets of the form B_ε with $\varepsilon < \alpha$ since each $i\{\alpha, \beta_j\}$ has size less than μ.

By the inductive hypothesis each B_ε, $\subseteq B_\alpha \subseteq \bigcup\{\mathcal{O}_i \mid i \in I\}$, has a subcover of size less than μ, and hence $B_\alpha \cap B_{\beta_j}$ has the union of these subcovers as a subcover of size less than μ. Hence B_α is μ-Lindelöf in τ. □

Lemma 22. Suppose $\mathcal{O} \in \mathcal{B}$. Then $\mathcal{O} \cap \mu \neq \emptyset$.

Proof. Let $p \in \mathbb{P}$ be such that $p \Vdash$ "$\mathcal{O} = B_\alpha \setminus \bigcup\{B_\gamma \mid \gamma \in a\} \in \mathcal{B}$" and suppose, without loss of generality, by the μ-closure of $\mathbb{P}$, that $\{\alpha\} \cup a \subseteq a_p$. Let $\beta \in (\mu \cap \alpha) \setminus a_p$. Define q by $a_q = a_p \cup \{\beta\}$, $A_q = A_p$ and $\leq_q \restriction a_p \times a_p = \leq_p$ and for each $\varepsilon \in a_p$ set $\beta \leq_q \varepsilon$ if and only if $\alpha \leq_p \varepsilon$. It is clear that $q \in \mathbb{P}$ and $q \leq p$. And it is also clear that $q \Vdash \beta \in \mathcal{O} \cap \mu$. □

Corollary 23. The separability degree of τ is μ.

Proof. Lemma (22) gives μ as an upper bound, but it is clear from the definition of $\mathbb{P}$ that given any smaller sized set and some condition there is a stronger condition that forces some set in $\mathcal{B}$ to miss it. □

Proposition 24. τ has tightness μ.

Proof. Write $cl(Z)$ for the closure of Z in the topology generated by $\mathcal{B}$. I prove by induction that if $\gamma \in \mathrm{cl}(Z)$ for some $Z \subseteq \kappa^+$ then there is some $Y \in [Z]^\mu$ such that $\gamma \in \mathrm{cl}(Y)$. (*cf.* [M*2, Theorem 4.25]).

So fix $Z \subseteq \kappa^+$, and $\gamma \in \mathrm{cl}(Z)$. First of all $Z \setminus B_\gamma \subseteq \kappa^+ \setminus B_\gamma$, so $\mathrm{cl}(Z \setminus B_\gamma) \subseteq \mathrm{cl}(\kappa^+ \setminus B_\gamma) = \kappa^+ \setminus B_\gamma$, since B_γ is open in τ. Thus $\mathrm{cl}(Z \setminus B_\gamma) \cap B_\gamma = \emptyset$.

This gives us that $\gamma \in \mathrm{cl}(Z \cap B_\gamma)$, because $\mathrm{cl}(Z) = \mathrm{cl}(Z \cap B_\gamma) \cup \mathrm{cl}(Z \setminus B_\gamma)$ since $\mathrm{cl}(.)$ is a closure operator.

Now B_γ is closed in τ by Lemma (19), so $\mathrm{cl}(Z \cap B_\gamma) \subseteq B_\gamma$. I now proceed by a series of claims.

Claim 25. There is some $Y \subseteq Z$ such that $\gamma \in \mathrm{cl}(Y)$ and $\mathrm{cl}(Y) \cap \gamma$ has no maximal element.

Proof. Let $X_0 = Z \cap B_\gamma$ and inductively set δ_i to be maximal in $\mathrm{cl}(X_i) \cap \gamma$ and $X_{i+1} = X_i \setminus B_{\delta_i}$ for as long as possible. By the above observations applied to X_i in place of Z, $\mathrm{cl}(X_{i+1}) \subseteq (\kappa^+ \setminus B_{\delta_i})$ and $\mathrm{cl}(X_i \cap B_{\delta_i}) \subseteq B_{\delta_i}$, so that $\gamma \in \mathrm{cl}(X_{i+1}) \setminus \mathrm{cl}(X_i \cap B_{\delta_i})$ and $\delta_i \in \mathrm{cl}(X_i \cap B_{\delta_i}) \setminus \mathrm{cl}(X_{i+1})$ for each i for which δ_i is defined. As $X_{i+1} \subseteq X_i$ one has that $\delta_{i+1} \leq \delta_i$ for all i for which δ_{i+1} is defined. But $\delta_{i+1} \in \kappa^+ \setminus B_{\delta_i}$, so $\delta_{i+1} \neq \delta_i$ and, hence, $\delta_{i+1} < \delta_i$ for each i for which δ_{i+1} is defined. So let $k < \omega$ be least such that $\mathrm{cl}(X_k) \cap \gamma$ has no maximal element. Set $Y = X_k$. Then $Y \subseteq X_0 = Z$, $\gamma \in \mathrm{cl}(Y)$ and $\mathrm{cl}(Y) \cap \gamma$ has no maximal element. □

Claim 26. There is some $Y' \in [\mathrm{cl}(Y) \cap \gamma]^{\mu}$ such that $\gamma \in \mathrm{cl}(Y')$.

Proof. Write W for $\mathrm{cl}(Y) \cap \gamma$, and set $W^* = \mu \cap \bigcup\{B_\delta \mid \delta \in W\}$. Note that $Y \subseteq W$ since $\mathrm{cl}(Y) \subseteq B_\gamma$ and $\gamma \notin Y$, and that $\mathrm{cl}(W) = \mathrm{cl}(Y)$. Choose $Y' \in [W]^{\mu}$ so that $W^* \subseteq \bigcup\{B_\delta \mid \delta \in Y'\}$. □

Subclaim 27. $\gamma \in \mathrm{cl}(W^*)$.

Proof. Suppose that $\mathcal{O}$ is an open set and that $\mathcal{O} \cap W^* = \emptyset$. Then for all $\delta \in W$ one has that $\mathcal{O} \cap B_\delta \cap \mu = \emptyset$. By Lemma (22) this implies that $\mathcal{O} \cap B_\delta = \emptyset$ for all $\delta \in W$. Hence $\mathcal{O} \cap \bigcup\{B_\delta \mid \delta \in W\} = \emptyset$. But $\gamma \in \mathrm{cl}(Y) = \mathrm{cl}(W) \subseteq \mathrm{cl}(\bigcup\{B_\delta \mid \delta \in W\}$, so one must have $\gamma \notin \mathcal{O}$. □

Since $\mathrm{cl}(Y') \subseteq \mathrm{cl}(Y) \subseteq B_\gamma$, one has that $\mathrm{cl}(Y')$ is μ-Lindelöf in τ. Suppose that $\gamma \notin \mathrm{cl}(Y')$.

Then $\{B_\delta \mid \delta \in W\}$ is a cover of $\mathrm{cl}(Y')$ and so there is some $\chi < \mu$ and some $\{\varepsilon_i \mid i < \chi\}$ such that $\mathrm{cl}(Y') \subseteq \bigcup\{B_{\varepsilon_i} \mid i < \chi\}$. Hence if $\delta \in \mathrm{cl}(Y')$ one has that there is some $i < \chi$ such that $\delta \in B_{\varepsilon_i}$, and so $B_\delta \subseteq B_{\varepsilon_i}$.

Thus $\bigcup\{B_\delta \mid \delta \in W\} \subseteq \bigcup\{B_{\varepsilon_i} \mid i < \chi\}$. Consequently $W^* \subseteq \bigcup\{B_{\varepsilon_i} \mid i < \chi\}$ and so, trivially, $\mathrm{cl}(W^*) \subseteq \mathrm{cl}(\bigcup\{B_{\varepsilon_i} \mid i < \chi\})$.

However, $B_\gamma \setminus \bigcup\{B_{\varepsilon_i} \mid i < \chi\}$ is an open set in τ which contains γ and has empty intersection with $\bigcup\{B_{\varepsilon_i} \mid i < \chi\}$, and thus shows that $\gamma \notin \mathrm{cl}(\bigcup\{B_{\varepsilon_i} \mid i < \chi\})$, and hence that $\gamma \notin \mathrm{cl}(W^*)$, contradicting Subclaim (27). Thus $\gamma \in \mathrm{cl}(Y')$ as required. □

Conclusion of proof of Proposition (24). If $\delta \in Y'$ then by the inductive hypothesis there is some $W_\delta \in [Z]^{\mu}$ such that $\delta \in \mathrm{cl}(W_\delta)$. Set $Y'' = \bigcup\{W_\delta \mid \delta \in Y'\}$. Then $\overline{\overline{Y''}} \leq \mu$ and $Y' \subseteq \mathrm{cl}(Y'')$, and so $\gamma \in \mathrm{cl}(Y') \subseteq \mathrm{cl}(\mathrm{cl}(Y'')) = \mathrm{cl}(Y'')$. □

By Lemmas (18) to (23) and Proposition (24), (κ^+, τ) has the topological properties described in Theorem (2) and cardinals are preserved in the generic extension by Propositions (5), (13) and (15).

Comment. As remarked after the statement of Theorem (2) it seems unlikely that one can force to add a 0-dimensional, locally compact, countably tight, scattered space of size κ^+ by κ-$\mathbb{M}$-proper forcing when μ is uncountable. It is hard to see how to amalgamate two conditions whose working parts are infinite, albeit of size less than μ, and keep the size of the ranges of an analogue of i_r finite. Of course the consistency of there being such spaces is a major, and radically open, problem.

5.2. Chains in $\mathcal{P}(\kappa)$ mod $<\mu$.

Theorem 28. If there is a stationary $(\kappa,1)$-simplified morass and κ is the successor of a regular cardinal μ with $\kappa^{<\mu} = \kappa$ then there is a cardinal preserving forcing which adds a chain of length κ^+ in $\mathcal{P}(\kappa)$ mod $<\mu$. $\square$

In this subsection I show how to use κ-$\mathbb{M}$-proper forcing to add a chain of length κ^+ in $\mathcal{P}(\kappa)$ mod $<\mu$, that is a sequence $\langle X_\nu \mid \nu < \kappa^+\rangle$ of subsets of κ such that $\overline{\overline{X_\nu \setminus X_\tau}} < \mu$ and $\overline{\overline{X_\tau \setminus X_\nu}} = \kappa$ for each $\nu < \tau < \kappa^+$. One has that if the Chang conjecture $(\kappa^{++},\kappa^+) \longrightarrow (\kappa^+,\kappa)$ holds then there is no such chain – Koszmider's proof ([K98, Fact 1]) that this is so when $\kappa = \omega_1$ goes through for arbitrary successor κ. The forcing is the obvious generalisation of Koszmider's forcing from [K98] with sets of maps from the simplified morass of size less than μ as side conditions.

Definition 29. Let the forcing $\mathbb{P}$ consist of conditions $p = (a_p, b_p, f_p, A_p)$ where $a_p \in [\kappa^+]^{<\mu}$, $b_p \in [\kappa]^{<\mu}$, $f_p : a_p \times b_p \longrightarrow 2$ and $A_p \in [\mathcal{F}]^{<\mu}$ and, writing c^p for c_{A_p},

$$(\bullet) \qquad \forall \nu, \tau \in a_p\ \forall \beta \in b_p\ \big((\nu < \tau\ \&\ c^p(\nu,\tau) \leq \beta) \longrightarrow \\ (f_p(\nu,\beta) = 1 \longrightarrow f_p(\tau,\beta) = 1)\big).$$

$q \leq p$ if $a_p \subseteq a_q$, $b_p \subseteq b_q$, $f_p = f_q \restriction a_p \times b_p$, $A_p \subseteq A_q$ and

$$(\circ) \qquad \forall \nu, \tau \in a_p\ \forall \beta \in b_q \setminus b_p\ \big(\nu < \tau \longrightarrow (f_q(\nu,\beta) = 1 \longrightarrow f_q(\tau,\beta) = 1)\big).$$

The idea behind the forcing is that if G is $\mathbb{P}$-generic over the ground model then one gets the desired chain in $\mathcal{P}(\kappa)$ mod $<\mu$ by setting

$$X_\nu = \{\beta < \kappa \mid \exists p \in G\ f_p(\nu,\beta) = 1\} \quad \text{for} \quad \nu < \kappa^+.$$

The following two lemmas say that this will happen.

Lemma 30. $\{p \in \mathbb{P} \mid \beta \in b_p\}$ is dense and open for each $\beta < \kappa$.

Proof. If $\beta \notin b_p$ then define q by setting $a_q = a_p$, $A_q = A_p$, $b_q = b_p \cup \{\beta\}$, $f_q \restriction a_p \times b_p = f_p$ and $f_q(\nu,\beta) = 0$ for all $\nu \in a_p$. Then clearly $q \in \mathbb{P}$, $q \leq p$ and $q \in \{p \in \mathbb{P} \mid \beta \in b_p\}$. $\square$

Lemma 31. $\{p \in \mathbb{P} \mid \tau \in a_p\}$ is dense and open for each $\tau < \kappa^+$.

Proof. If $\tau \notin a_p$ then define q by setting $a_q = a_p \cup \{\tau\}$, $b_q = b_p$, $A_q = A_p$, $f_q \restriction a_p \times b_p = f_p$, and by setting, for each $\beta \in b_p$, $f_q(\tau,\beta) = 1$ if there is some $\nu \in a_p \cap \tau$ such that $c^q(\nu,\tau) \leq \beta$ and $f_p(\nu,\beta) = 1$, and $f_q(\tau,\beta) = 0$ otherwise. If $\beta \in b_p$, and $f_q(\tau,\beta) = 1$ then by the definition just given of f_q there is some $\nu < \tau$ such $c^q(\nu,\tau) \leq \beta$ and $f_p(\nu,\beta) = 1$. If one also has $\xi \in a_p \setminus \tau$ and $c^q(\tau,\xi) \leq \beta$ then $c^q(\nu,\xi) \leq \beta$ by the subadditivity of c^q. As $c^q = c^p$, the condition $(\bullet)$ for p applied to ν and ξ ensures that $f_p(\xi,\beta) = 1$. This shows that the condition $(\bullet)$ also holds for q, so that $q \in \mathbb{P}$. It is clear that $q \leq p$ (the condition $(\circ)$ holds vacuously since $b_q \setminus b_p = \emptyset$). $\square$

It remains to show that $\mathbb{P}$ preserves cardinals.

Lemma 32. $\mathbb{P}$ is μ-closed.

Proof. Let $\langle p_i \mid i < \chi\rangle$ be a descending sequence of conditions for some limit ordinal $\chi < \mu$. Define p by setting $a_p = \bigcup\{a_{p_i} \mid i < \chi\}$, $b_p = \bigcup\{b_{p_i} \mid i < \chi\}$, $f_p = \bigcup\{f_{p_i} \mid i < \chi\}$, and $A_p = \bigcup\{A_{p_i} \mid i < \chi\}$.

First of all I show that ($\bullet$) holds for p and hence that $p \in \mathbb{P}$. Suppose that ν, $\tau \in a_p$, $\beta \in b_p$, $c^p(\nu,\tau) \leq \beta$ and $f_p(\nu,\beta) = 1$. Let $c^p(\nu,\tau) = \alpha$ and let $\langle \xi_j \mid j \leq k\rangle$ and $\langle(\gamma_j, g_j) \mid j < k\rangle$ witness this. Then there is some $i < \chi$ such that ν, $\tau \in a_{p_i}$, $\beta \in b_{p_i}$ and $(\gamma_j, g_j) \in A_{p_i}$ for each $j < k$. So $f_{p_i}(\nu,\beta) = 1$ and $\langle \xi_j \mid j \leq k\rangle$ and $\langle(\gamma_j, g_j) \mid j < k\rangle$ witness that $c^{p_i}(\nu,\tau) \leq \alpha$. (As $A_{p_i} \subseteq A_p$ this in fact gives that $c^{p_i}(\nu,\tau) = c^p(\nu,\tau)$ by Lemma (1.3).) By ($\bullet$) for p_i one has that $f_{p_i}(\tau,\beta) = 1$, and hence $f_p(\tau,\beta) = 1$.

Next I show that $p \leq p_i$ for all $i < \chi$ by checking that ($\circ$) holds between p and each p_i. So let $i < \chi$ with ν, $\tau \in a_{p_i}$, $\nu < \tau$, $\beta \in b_p \setminus b_{p_i}$ and $f_p(\nu,\beta) = 1$. Then there is some $i' \in [i,\chi)$ such that $\beta \in p_{i'}$, whence $f_{p_{i'}}(\nu,\beta) = 1$ since $f_p \restriction a_{p_{i'}} \times b_{p_{i'}} = f_{p_{i'}}$. By ($\circ$) for $p_{i'}$ and p_i one has that $f_{p_{i'}}(\tau,\beta) = 1$ and hence that $f_p(\tau,\beta) = 1$, as required. □

Lemma 33. $\mathbb{P}$ has the κ^+-chain condition.

Proof. Let $\{p'_i \mid i < \kappa^+\}$ be a subset of $\mathbb{P}$ of size κ^+. For each $p' \in \{p'_i \mid i < \kappa^+\}$, by Fact (I.5), let $(\gamma, g) \in \mathcal{F}$ be such that $a_{p'} \subseteq \mathrm{rge}(g)$, $\mathrm{ssup}(b_{p'}) < \gamma$ and for all $(\beta, f) \in A_{p'}$ there is some $f' \in \mathcal{F}_{\beta\gamma}$ with $f = g \cdot f'$. Define a condition p by letting $a_p = a_{p'}$, $b_p = b_{p'}$, $f_p = f_{p'}$ and $A_p = A_{p'} \cup \{(\gamma, g)\}$. Now it is clear that $c^{p'}(\nu,\tau) \leq \beta$ if and only if $c^p(\nu,\tau) \leq \beta$ for any pair ν, $\tau \in a_p$ and $\beta \in b_p$ since any sequence of maps witnessing the latter cannot contain (γ, g) (as $\beta < \gamma$) and, hence, witnesses the former. Hence p satisfies ($\bullet$) and so is a condition. As ($\circ$) holds vacuously and all of the other conditions for being an extension hold by definition one also has that $p \leq p'$. If one can find i, $j < \kappa^+$ such that p_i and p_j are compatible then, clearly, one will have that p'_i and p'_j are compatible as well.

By thinning using the Δ-system lemma if necessary one may as well assume that there is some $b \in [\kappa]^{<\mu}$ such that $b = b_{p_i}$ for all $i < \kappa^+$ since $\kappa^{<\mu} = \kappa$. One may as well also assume there is some $a \in [\kappa^+]^{<\mu}$ such that a is an initial segment of each a_{p_i}, that if $i < j < \kappa^+$ then $\max(a_{p_i}) < \min(a_{p_j} \setminus a)$, and there is an order preserving isomorphism from a_{p_i} to a_{p_j} which extends to an isomorphism from f_{p_i} to f_{p_j}. Moreover, one may as well assume that there is some $\alpha < \kappa$, some $A \in [\bigcup\{\{\beta\} \times \mathcal{F}_{\beta\alpha} \mid \beta < \alpha\}]^{<\mu}$ and a map $F_i \in \mathcal{F}_{\alpha\kappa}$ for each $\nu < \kappa^+$ such that $A_{p_i} = \{(\beta, F_\nu \cdot h) \mid (\beta, h) \in A\} \cup \{(\alpha, F_i)\}$ and such that $F_i \cdot F_j^{-1} \restriction a_{p_j}$ is the order preserving isomorphism from (a_{p_j}, f_{p_j}) to (a_{p_i}, f_{p_j}). This is possible using the Δ-system lemma for morass maps, Proposition (I.13). Clearly for each $i < \kappa^+$ and $\xi < \zeta < \kappa^+$ one has that $c^{p_i}(\xi,\zeta) \leq \alpha$ if and only if $c^{p_i}(\xi,\zeta) < \kappa$.

Let $i < j < \kappa^+$ and set $a_r = a_{p_i} \cup a_{p_j}$, $b_r = b$, $f_r \restriction a_{p_i} \times b = f_{p_i}$ and $f_r \restriction a_{p_j} \times b = f_{p_j}$, and $A_r = A_{p_i} \cup A_{p_j}$.

It is also clear that if $\xi < \zeta < \kappa^+$ then $c^r(\xi, \zeta) \leq \alpha$ if and only if $c^r(\xi, \zeta) < \kappa$. By Lemma (1.14), one has that if $\zeta \in \mathrm{rge}(F_i)$ then $c^r(\xi, \zeta) = c^{p_i}(\xi, \zeta)$ and if $\zeta \in \mathrm{rge}(F_j)$ then $c^r(\xi, \zeta) = c^{p_j}(\xi, \zeta)$. □

Claim 34. If $\xi \in a_{p_i} \setminus a_{p_j}$ then $\xi \notin \mathrm{rge}(F_j)$.

Proof. If $\xi \in \mathrm{rge}(F_j)$ then $\xi \in \mathrm{rge}(F_j) \cap \mathrm{rge}(F_i)$ since $F_j \cdot F_i^{-1} \restriction a_{p_i}$ is the order preserving isomorphism from a_{p_i} to a_{p_j}. But then one has that $F_j(\xi_\alpha) = \xi = F_i(\xi_\alpha)$ by Fact (I.3), so $F_j \cdot F_i^{-1}(\xi) = \xi$ and hence $\xi \in a_{p_i} \cap a_{p_j}$, a contradiction. □

It is now immediate that if $\xi \in a_{p_i} \setminus a_{p_j}$ and $\zeta \in a_{p_j} \setminus a_{p_i}$ then $\xi \notin \mathrm{rge}(F_j)$, by Claim (34), so that $c^{p_i}(\xi, \zeta) = \kappa$, and thus $c^r(\xi, \zeta) = \kappa$ from Lemma (1.14). Hence for such ξ, ζ one has that $\beta < c^r(\xi, \zeta)$ for all $\beta \in b$. Hence ($\bullet$) is vacuously true for all such pairs ξ, ζ.

By Claim (34) as well, if ξ_0, $\xi_1 \in a_{p_i}$ then $c^r(\xi_0, \xi_1) = c^{p_i}(\xi_0, \xi_1)$. And so if $\xi_0 < \xi_1$, $c(\xi_0, \xi_1) \leq \beta$ and $f_r(\xi_0, \beta) = 1$ for some $\beta \in b$ then (by ($\bullet$) for p_i) one also has that $f_r(\xi_1, \beta) = 1$. Similarly, if ζ_0, $\zeta_1 \in a_{p_j}$, $c(\xi_0, \xi_1) \leq \beta$ and $f_r(\zeta_0, \beta) = 1$ for some $\beta \in b$ then (by ($\bullet$) for p_j) one also has that $f_r(\zeta_1, \beta) = 1$.

Hence $r \in \mathbb{P}$, and thus $r \leq_{\mathbb{P}} p_i$, p_j (since ($\circ$) holds vacuously). So $\mathbb{P}$ does have the κ^+-cc. □

The previous two lemmas show that cardinals below κ and above κ, respectively, are preserved. So in order to complete the proof it suffices to show that $\mathbb{P}$ is κ-$\mathbb{M}$-proper and hence that κ itself is preserved.

Proposition 35. $\mathbb{P}$ is κ-$\mathbb{M}$-proper.

Proof. The proof strategy will be to work through the framework of §4 checking that one can actually perform each of the four steps enumerated there for $\mathbb{P}$.

So suppose that $p \in \mathbb{P}$ and that $\mathcal{N}$ is as in the general framework but for this specific $\mathbb{P}$. Obviously, a_p is the realm of p and A_p is the side condition part.

Step A. Checking $(a_p, b_p, f_p, A_p \cup \{(\delta, F)\})$ is a condition and is stronger than p.

Proof. This is just as in the argument of the first paragraph of the proof of the κ^+-chain condition (Lemma (33)). Let $p' = (a_p, b_p, f_p, A_p \cup \{(\delta, F)\})$. Clearly p' is a condition if it satisfies ($\bullet$) and is stronger than p if it is a condition because ($\circ$) will hold vacuously. But if ν, $\tau \in a_p$, $\beta \in b_p$ and $c^{p'}(\nu, \tau) \leq \beta$ then any maps in a witnessing sequence to this cannot include (δ, F) because $p \in \mathcal{N}$ and has size less than μ, so $b_p \subseteq \mathcal{N} \cap \kappa = \delta$. Hence any maps from such a witnessing sequence witness that $c^p(\nu, \tau) \leq \beta$ as well. Thus if $p(\nu, \beta) = 1$ one also has that $p(\tau, \beta) = 1$ by ($\bullet$) for p. □

Now let $\mathcal{D}$ be a dense, open subset of $\mathbb{P}$ and $q \in \mathcal{D}$ with $q \leq p'$.

Step B. One may assume without loss of generality that $c^q(\nu, \tau) < \kappa$ for all ν, $\tau \in a_q$.

Proof. The argument is almost exactly the same argument as in the first paragraph of the proof of the κ^+-chain condition (Lemma (33)) and the argument used for

Step (A). Explicitly, suppose one has a q without this property. Then for each $\nu \in a_q$ choose a map $(\gamma_\nu, g_\nu) \in \mathcal{F}$ such that $\nu \in \mathrm{rge}(g_\nu)$. Now apply Fact (I.5) to this collection of maps and select some $(\gamma, g) \in \mathcal{F}$ through which they all factor and with γ greater than $\mathrm{ssup}(\{\alpha \mid \exists(\alpha, f) \in A_q\} \cup b_p)$. Add this single map (γ, g) to A^q. As in the argument for Step (A) the resulting object is still a condition since the new map cannot play a rôle in witnessing that $c(\nu, \tau) \leq \beta$ for any $\beta \in b_q$ and ν, $\tau \in a_q$. □

Next let $q \restriction \mathcal{N} = (a_q \cap \mathcal{N}, b_q \cap \mathcal{N}, f_q \cap \mathcal{N}, A_q \cap \mathcal{N})$. Note that $f_q \cap \mathcal{N} = f_q \restriction (a_q \cap \mathcal{N} \times b_q \cap \mathcal{N})$, since the range of f_q is (a subset of) $\{0, 1\}$. This makes the proof that $q \restriction \mathcal{N} \in \mathbb{P}$ particularly simple.

Step C. Show that $q \restriction \mathcal{N} \in \mathbb{P} \cap \mathcal{N}$ and that $q \leq q \restriction \mathcal{N}$.

Proof. Each of the four components is a set of size less than μ which is an element of $\mathcal{N}$ and so is itself an element of $\mathcal{N}$ since the latter is closed under sequences of length less than μ. Thus $q \restriction \mathcal{N} \in \mathcal{N}$.

Now I prove that $q \restriction \mathcal{N} \in \mathbb{P}$ by showing that $(\bullet)$ holds. Let ν, $\tau \in a_q \cap \mathcal{N}$, $\beta \in b_q \cap \mathcal{N} (= b_q \cap \delta)$, $\nu < \tau$, $c^{q\restriction\mathcal{N}}(\nu, \tau) \leq \beta$ and $f_q(\nu, \beta) = 1$. By Lemma (1.3) one has that $c^q(\nu, \tau) \leq c^{q\restriction\mathcal{N}}(\nu, \tau) \leq \beta$, so one has that $f_q(\tau, \beta) = 1$ by $(\bullet)$ for q. Thus $q \restriction \mathcal{N}$ does satisfy $(\bullet)$ and so $q \restriction \mathcal{N} \in \mathbb{P}$.

Finally I prove that $q \leq q \restriction \mathcal{N}$ by demonstrating that $(\circ)$ holds. Let ν, $\tau \in a_q \cap \mathcal{N}$, again, with $\nu < \tau$, let $\beta \in b_q \setminus \mathcal{N}$ and suppose that $f_q(\nu, \beta) = 1$. As ν, $\tau \in \mathcal{N} = \mathrm{rge}(F)$, $(\delta, F) \in A_q$ and $\beta \notin \mathcal{N}$ one has that $c^q(\nu, \tau) \leq \delta \leq \beta$. Hence, by $(\bullet)$ for q, one has that $f_q(\tau, \beta) = 1$. Thus $(\circ)$ does hold between q and $q \restriction \mathcal{N}$ and one does have that $q \leq q \restriction \mathcal{N}$. □

Now obtain s as in §4 using b_q as the arbitrary set e_q in Notation (7) in §4.

Step D. Define some r with realm $a_s \cup a_q$ and side condition part $A_s \cup A_q$. Show that $r \in \mathbb{P}$ and that $r \leq s$, q.

Proof. Write a for $a_s \cap a_q$, a_0 for $a_s \setminus a_q$, a_1 for $a_q \setminus a_s$, b for $b_s \cap b_q$, b_0 for $b_s \setminus b_q$, b_1 for $b_q \setminus b_s$.

Define r as follows. Let $a_r = a_s \cup a_q$, $b_r = b_s \cup b_q$ and $A_r = A_s \cup A_q$. If $(\tau, \beta) \in a_s \times b_s$ let $f_r(\tau, \beta) = f_s(\tau, \beta)$ and if $(\tau, \beta) \in a_q \times b_q$ let $f_r(\tau, \beta) = f_q(\tau, \beta)$. It remains to define $f_r \restriction a_0 \times b_1 \cup a_1 \times b_0$. This is done separately for each β and by induction on τ.

If $\beta \in b_1$ and $\tau \in a_0$ let $f_r(\tau, \beta) = 1$ if there is some $\nu \in a_r \cap \tau$ such that $c^r(\nu, \tau) \leq \beta$ and $f_r(\nu, \beta) = 1$ or if there is some $\nu \in a_0 \cap \tau$ such that $f_r(\nu, \beta) = 1$. Otherwise set $f_r(\nu, \tau) = 0$.

Similarly, if $\beta \in b_0$ and $\tau \in a_1$ let $f_r(\tau, \beta) = 1$ if there is some $\nu \in a_r \cap \tau$ such that $c^r(\nu, \tau) \leq \beta$ and $f_r(\nu, \beta) = 1$ or there is some $\nu \in a_1 \cap \tau$ such that $f_r(\nu, \beta) = 1$. Otherwise set $f_r(\nu, \tau) = 0$.

Observe that if $\tau \in a_1$ and $f_r(\tau, \beta) = 1$ because there is some $\nu \in a_1$ such that the inductive process already set $f_r(\nu, \beta) = 1$ then there is some such ν (obtainable by taking the least such ν) for which there is some $\zeta \in a_s$ with $c^r(\zeta, \nu) < \beta$ and

$f_s(\zeta,\beta) = 1$. A similar observation is true for a_0. These observations are used repeatedly below, without explicit comment.

I show, first of all, that r is a condition by proving that it satisfies ($\bullet$). Suppose that ν, $\tau \in a_r$ with $\nu < \tau$, $\beta \in b_r$, $c^r(\nu,\tau) \leq \beta$ and $f_r(\nu,\beta) = 1$. If $\tau \in a_1$ and $\beta \in b_0$ or $\tau \in a_0$ and $\beta \in b_1$ then $f_r(\tau,\beta) = 1$ by the definition of f_r just made.

Case A. $\tau \in a_q$ and $\beta \in b_q$. Then there is some $\xi \in a_q \cap \tau$ such that $f_r(\xi,\beta) = 1$. For if $\nu \in a_q$ then take $\xi = \nu$. Otherwise, if $\nu \in a_0$ then there is some $\zeta \leq \nu$ such that $\zeta \in a_0$ and some $\xi \in a_q \cap \zeta$ such that $c^r(\xi,\zeta) \leq \beta$ and $f_r(\xi,\beta) = 1$. As ν, $\zeta \in a_s$ one has that $c^r(\zeta,\nu) < \delta$ (by (4.12)), so $c^r(\xi,\tau) \leq \beta$ by the subadditivity of c^r (Theorem (1.12)).

SUBCASE A.I. $\beta \in b$. Then $\beta < \alpha^*$ (by clause (iii) of the definition of ϕ for s) and one has that $c^r(\xi,\tau) = c^q(\xi,\tau)$ by (4.15). Thus $f_q(\tau,\beta) = 1$ by ($\bullet$) for q and so $f_r(\tau,\beta) = 1$.

SUBCASE A.II. $\beta \in b_1$. One has that $\delta \leq \beta$ (by clause (iii.5) of the definition of ϕ from Notation (7) of §4). Hence, by (4.16), one has that $c^q(\xi,\tau) \leq \beta$ So one has that $f_q(\tau,\beta) = 1$ by ($\bullet$) for q and hence $f_r(\xi,\tau) = 1$.

Case B. $\tau \in a_s$ and $\beta \in b_s$. One cannot have that $\nu \in a_1$ because then one would have $\beta < \delta \leq c^r(\xi,\tau)$, by (4.17). So suppose that $\nu \in a_s$. Then $c^s(\nu,\tau) = c^r(\nu,\tau) \leq \beta < \delta$ by (4.12), and so $f_s(\tau,\beta) = 1$ by ($\bullet$) for s, whence $f_r(\tau,\beta) = 1$ again.

Now I show that $r \leq q$. Let ν, $\tau \in a_q$, $\nu < \tau$, $\beta \in b_0$ and $f_r(\nu,\beta) = 1$. If $\tau \in a_0$ then $f_r(\tau,\beta) = 1$ by the definition of $f_r(.\,,\beta)$. So suppose that $\tau \in a_s \cap a_q$.

Case A. $\nu \in a_s \cap a_q$. Then ν, $\tau \in \text{rge}(h^*)$. This shows that $c^r(\nu,\tau) \leq \beta < \delta$ for otherwise one has $\beta < c^r(\nu,\tau) \leq \alpha^*$, but $b_s \setminus b_q \cap \alpha^* = \emptyset$, a contradiction. Now, $c^s(\nu,\tau) = c^r(\nu,\tau) \leq \beta$, by (4.12), so $f_s(\tau,\beta) = 1$ by ($\bullet$) for s, and hence $f_r(\tau,\beta) = 1$.

Case B. $\nu \in a_1$. Let $\xi \in a_1 \cap \nu$ and $\zeta \in a_s \cap \xi$ be such that $c^r(\zeta,\xi) \leq \beta$ and $f_r(\zeta,\beta) = 1$. If $\zeta \in a_s \cap a_q$ then Case (A) deals with the situation. If $\zeta \in a_0$ then $c^s(\zeta,\tau) = c^r(\zeta,\tau) \leq c^r(\zeta,\xi)$, by (4.12) and (4.22) respectively. So $c^s(\zeta,\tau) \leq \beta$, and $f_s(\tau,\beta) = 1$ by ($\bullet$) for s. Hence $f_r(\tau,\beta) = 1$.

Finally, I show that $r \leq s$. Let ν, $\tau \in a_s$, $\nu < \tau$, $\beta \in b_1$ and $f_r(\nu,\beta) = 1$. If $\tau \in a_1$ then $f_r(\tau,\beta) = 1$ by the definition of $f_r(.\,,\beta)$. So suppose that $\tau \in a_s \cap a_q$.

Case A. $\nu \in a_s \cap a_q$. Then $c^q(\nu,\tau) \leq \delta \leq \beta$ since ν, $\tau \in \text{rge}(F)$ so one has that has that $f_q(\tau,\beta) = 1$ by ($\bullet$) for q, whence $f_r(\tau,\beta) = 1$.

Case B. $\nu \in a_0$. Let $\zeta \in a_0 \cap \nu$ and $\xi \in a_s \cap \zeta$ be such that $c^r(\xi,\zeta) \leq \beta$ and $f_r(\xi,1) = 1$. If $\xi \in a_s \cap a_q$ then one is back in Case (A) again and so finished. If $\xi \in a_1$ then $\delta \leq c^r(\xi,\zeta) = c^r(\xi,\nu) = c^q(\xi,\nu)$ by (4.17), the subadditivity of c^r and the fact that $c^r(\zeta,\nu) < \delta$ by (4.12). But, again by (4.12), $c^r(\nu,\tau) < \delta$. So $\delta \leq c^r(\xi,\nu) = c^r(\xi,\tau) = c^q(\xi,\tau)$, by the subadditivity of c^r once more. Thus $c^q(\xi,\tau) \leq \beta$, and $f_r(\tau,\beta) = 1$ by ($\bullet$) for q. Hence $f_r(\xi,\tau) = 1$.

This concludes the proof that $\mathbb{P}$ is κ-$\mathbb{M}$-proper and so the proof of Theorem (28). $\square$

Question. Is it consistent that there are (κ^+, κ^+)-gaps in $\mathcal{P}(\kappa)$ mod $< \mu$? It is believed that this is unknown even in the case $\kappa = \omega_1$.

Acknowledgments
I should like to thank the Departmento de Matemática of the Universidade Federal de Bahia, Salvador, Brazil, where most of research which makes up this paper was carried out, for providing a very enlivening working environment, and CAPES, Brazil, for funding my visit to UFBa in late 2000. I also wish to thank my extremely kind hosts Regina and Davidson Fernandes, without whose generosity my visit to Salvador, and consequently this research, would not have been possible, let alone so enjoyable. While this paper was being written I was employed at the University of East Anglia via a grant from the Leverhulme foundation and was a research fellow at University College London.

Parts of this work was presented at the CRM during the Set Theory research programme in 2003–2004. I would like to thank Joan Bagaria and Juan Carlos Martinez their help in facilitating my spending a very productive 2003 at the CRM, Joan and Stevo Todorcevic for organising a fascinating research programme and Prof. Manuel Castellet, Neus Portet, Consol Roca and Maria Julia for their warm hospitality at the CRM as well as their unflagging help in making my visit run smoothly. I am also grateful to the Spanish Ministry of Education for their funding of this visit to Barcelona.

References

[AShe] U. Abraham and S. Shelah, *Forcing closed unbounded sets*, Journal of Symbolic Logic, **48**, (1983), pp. 643–657.

[AS] U. Abraham and R. Shore, *Initial segements of the degrees of size* $\aleph_1$, Israel Journal of Mathematics, **53**, (1986), pp. 1–51.

[B] J. Buamgartner, *Almost-disjoint sets, the dense set problem and the partition calculus*, Annals of Mathematical Logic, **10**, (1976), pp. 401–439.

[BS] J. Baumgartner and S. Shelah, *Remarks on superatomic Boolean algebras*, Annals of Pure and Applied Logic, **33**, (1987), pp. 109–129.

[EH] P. Erdős and A. Hajnal, *Unsolved and solved problems in set theory*, Proceedings of the Symposia in Pure Mathematics, **25**, (1974), pp. 267–287.

[G] W. Gowers, *The two cultures of mathematics*, in *Mathematics: frontiers and perspectives*, ed. V.I. Arnold, M.A. Atiyah, P. Lax and B. Mazur, American Mathematical Society, Providence, Rhode Island, U.S.A., 2000, pp. 65–78.

[KM] P. Koepke and J.C. Martinez, *Superatomic Boolean algebras constructed from morasses*, Journal of Symbolic Logic, **60**, (1995), pp. 940–951.

[Kom] P. Komjath, *A consistency results concerning set mappings*, Acta Mathematica Hungarica, **64**, (1994), pp. 93–99.

[K95] P. Koszmider, *Semimorasses and nonreflection a singular cardinals*, Annals of Pure and Applied Logic, **72**, (1995), pp. 1–23.

[K98a] P. Koszmider, *On the existence of strong chains in* $\mathcal{P}(\omega_1)/Fin$, Journal of Symbolic Logic, **63**, (1998), pp. 1055–1062.

[K98b] P. Koszmider, *Models as side conditions*, in *Set theory*, ed. C.A. Di Prisco, J. Larson, J. Bagaria and A.R.D. Mathias, Kluwer, Amsterdam, Holland, (1998), pp. 99–107.

[K00] P. Koszmider, *On strong chains of uncountable functions*, Israel Journal of Mathematics, **118**, (2000), pp. 289–315.

[M96] C.J.G. Morgan, *Morasses, square and forcing axioms*, Annals of Pure and Applied Logic, **80**, (1996), pp. 139–163.

[M*1] C.J.G. Morgan, *A gap cohomology group, II: the Abraham-Shore technique*, Journal of Symbolic Logic, to appear.

[M*2] C.J.G. Morgan, *pcf-Structures, I: characterisation*, preprint, (2001).

[M*3] C.J.G. Morgan, *The uses and abuses of historicised forcing*, handwritten notes, (1992), revised preprint, (2001).

[M*4] C.J.G. Morgan, *pcf-Structures, II: the consistency of pcf structures on* ω_2, preprint, (2001).

[M*5] C.J.G. Morgan, *A few gentle stretching exercises*, CRM preprint. bf 554, (2003).

[M*6] C.J.G. Morgan, *Adding club subsets of* ω_2 *using conditions with finite working parts*, CRM preprint, (2004).

[M*7] C.J.G. Morgan, *On the Velickovic* Δ*-property for the stepping up functions* c *and* ρ, submitted.

[M*8] C.J.G. Morgan, *pcf-structures, III: structures below* ω_3, in preparation.

[Ra] M. Rabus, *An* ω_2*-minimal Boolean algebra*, Proceedings of The American Mathematical Society, **348**, (1996), pp. 3235–3244.

[R] J. Roitman, *Height and width of superatomic Boolean algebras*, Proceedings of the American Mathematical Society, **94**, (1985), pp. 9–14.

[S*] S. Shelah, *Not collapsing cardinals* $\leq \kappa$ *in* $(< \kappa)$*-support iterations*, preprint.

[SHL] S. Shelah, C. Laflamme and B. Hart, *Models with second-order properties, V, a general principle*, Annals of Pure and Applied Logic, **64**, (1993), pp. 169–194.

[SS] S. Shelah, L. Stanley, *A theorem and some consistency results in partition calculus*, Annals of Pure and Applied Logic, **36**, (1987), pp. 119–152.

[T85] S. Todorčević, *Directed sets and cofinal types*, Transactions of the American Mathematical Society, **290**, (1985), pp. 711–723.

[T87] S. Todorčević, *Partitioning pairs of countable ordinals*, Acta Mathematica, **159**, (1987), pp. 261–294.

[T89] S. Todorčević, *Partition problems in topology*, Contemporary Mathematics, **84**, American Mathematical Society, Providence, Rhode Island, U.S.A., 1989.

[V84] D. Velleman, *Simplified morasses*, Journal of Symbolic Logic, **49**, (1984), pp. 257–271.

[Veličk] B. Veličković, *Forcing axioms and stationary sets*, Advances in Mathematics, **94**, (1992), pp. 256–284

Charles Morgan
Department of Mathematics, University College London
Gower Street, London, WC1E 6BT, Great Britain
e-mail: `charles.morgan@ucl.ac.uk`

Set Theory
Trends in Mathematics, 401–406

Bounded Martin's Maximum and Strong Cardinals

Ralf Schindler

Abstract. We show that if Bounded Martin's Maximum (BMM) holds then for every $X \in V$ there is an inner model with a strong cardinal containing X. We also discuss various open questions which are related to BMM.

1. Introduction and statement of the result

This paper strengthens one of the results of [6]. Bounded Martin's Maximum (BMM, for short) is the statement that whenever $\mathbb{P} \in V$ is a stationary set preserving forcing notion then

$$H_{\omega_2}^V \prec_{\Sigma_1} H_{\omega_2}^{V^{\mathbb{P}}}.$$

Bounded forcing axioms were introduced in [2] (as weakenings of the "unbounded" forcing axioms PFA and MM). Todorcevic showed (cf. [8]) that BMM implies that $2^{\aleph_0} = \aleph_2$. (This was later improved by J. Moore who showed that already the Bounded Proper Forcing Axiom implies $2^{\aleph_0} = \aleph_2$; cf. [4].) We refer the reader to [9, Section 10.3] for a discussion of BMM. We proved in [6, Theorem 1.3] that if BMM holds then for every $X \in V$, $X^{\#}$ exists. The purpose of the present note is to prove the following.

Theorem 1.1. *Suppose that* BMM *holds. Then for every $X \in V$ there is an inner model with a strong cardinal containing X.*

We do not know if Theorem 1.1 gives the optimal lower bound for the consistency strength of BMM. Woodin has shown in unpublished work (cf. [10]) that $\omega + 1$ many Woodin cardinals are an upper bound. We refer the reader to [6] for further background information.

The main result of this note was proven in February 2004 while the author was a guest at the CRM in Barcelona. He would like to thank Neus Portet and Joan Bagaria for their warm hospitality.

Whereas [6] constructs, assuming $\mathsf{BMM} + V$ is not closed under the $\#$ operator, a strictly decreasing sequence of functions from ω_1 to ω_1, the key new idea here is to construct, assuming $\mathsf{BMM}+$ there is some set X which is not in an inner model with a strong cardinal, a strictly decreasing sequence of functions from ω_1 to the set of all countable mice, where "decreasing" means "decreasing in the mouse order."

2. The proof

Let us assume that BMM holds throughout this section. We shall prove that there is an inner model with a strong cardinal. The reader will gladly verify that the argument to follow "relativizes" to any $X \in V$, yielding a proof of Theorem 1.1.

Let us suppose that there is no inner model with a strong cardinal. We may thus let K denote the core model (cf. [3]). If $\mathcal{M}$ is a premouse and $\alpha \leq \mathcal{M} \cap \mathrm{OR}$ then we say that α is overlapped in $\mathcal{M}$ just in case there is some extender $E_\nu^{\mathcal{M}}$ such that $\mathrm{crit}(E_\nu^{\mathcal{M}}) < \alpha$ and $\nu \geq \alpha$. As there is no inner model with a strong cardinal, in K there is no E_ν^K such that $\mathrm{crit}(E_\nu^K)$ is overlapped in K.

We shall use the following notation. Let $\mathcal{M} = J_\alpha[E]$ be a premouse, and let $A \subset \bar{\alpha}$ for some $\bar{\alpha} < \alpha$. Then by $\mathcal{M}[A]$ we denote the transitive set (structure) $J_\alpha[E, A]$. Of course, in general $\mathcal{M}[A]$ will not be a premouse (or not even be a model of a reasonable fragment of ZFC). Nevertheless, models of the form $\mathcal{M}[A]$ will play a key rôle in what follows.

By $\leq^*$ we shall denote the pre-well-ordering of mice (cf. for instance [7]). I.e., if $\mathcal{M}$ and $\mathcal{N}$ are mice and $\mathcal{T}, \mathcal{U}$ is the coiteration of $\mathcal{M}$, $\mathcal{N}$ then $\mathcal{M} \leq^* \mathcal{N}$ if and only if $\mathcal{M}_\infty^{\mathcal{T}} \trianglelefteq \mathcal{M}_\infty^{\mathcal{U}}$ (in which case $[0, \infty)_{\mathcal{T}}$ contains no drop). We shall write $\mathcal{M} <^* \mathcal{N}$ iff $\mathcal{M} \leq^* \mathcal{N}$ and $\neg(\mathcal{N} \leq^* \mathcal{M})$.

Using the Dodd-Jensen Lemma and the fact that there are no degenerate iterations of mice, one can show that $\leq^*$ is indeed a pre-well-ordering. The following Lemma will thus give the desired contradiction.

Lemma 2.1. ($\mathsf{BMM}+$ *there is no inner model with a strong cardinal.*) *There is a sequence* $\mathcal{S} = (A_n, C_n, (\mathcal{N}_{n,\alpha} \colon \alpha \in C_n) : n < \omega)$ *such that for every* $n < \omega$, $A_n \subset \omega_1$, C_n *is a club subset of* ω_1, $C_n \subset C_{n-1}$ *if* $n > 0$, *and for every* $\alpha \in C_n$, $\mathcal{N}_{n,\alpha}$ *is a sound mouse with* $\rho_\omega(\mathcal{N}_{n,\alpha}) = \kappa \geq \alpha$, *where* κ *is the largest cardinal of* $\mathcal{N}_{n,\alpha}$, *and* κ *is not overlapped in* $\mathcal{N}_{n,\alpha}$, $(A_n \cap \alpha)_{\mathrm{odd}}$ *codes the mouse* $\mathcal{N}_{n,\alpha}||\kappa$,[1] *and* $\mathcal{N}_{n,\alpha}[A_n \cap \alpha] \models$ *"α is countable"; moreover,* $\mathcal{N}_{n,\alpha} <^* \mathcal{N}_{n-1,\alpha}$ *if* $n > 0$ *and* $\alpha \in C_n$.

The proof of Lemma 2.1 exploits a version of the "faster reshaping forcing" which we had introduced in [6].

Let $n \in \omega$, and let us assume that $\mathcal{S} \restriction n$ has been constructed. We aim to construct A_n, C_n, and $(\mathcal{N}_{n,\alpha} \colon \alpha \in C_n)$.

---

[1]If A is a set of ordinals then by A_{odd} we mean the set $\{\alpha \colon 2\alpha + 1 \in A\}$. The coding here and in what follows is to be understood as being according to some standard soft coding device. For instance, if M is transitive and of size ζ then there is some $E \subset \zeta^2$ and some isomorphism $\sigma \colon (\zeta; E) \cong (M; \in)$; via Gödel's pairing function, E may be construed as a subset of ζ which codes M.

By BMM, it suffices to force the existence of these objects with the desired properties by a stationary set preserving forcing notion.

By [5], there is a σ-closed forcing notion $\mathbb{P}$ such that in $V^{\mathbb{P}}$, CH holds, there is some $A^+ \subset \omega_1$ with $H_{\omega_2} = K||\omega_2[A^+]$, and $K||\omega_2$ has a largest cardinal κ which is not overlapped in $K||\omega_2$. We may assume that A^+ is chosen such that $H_{\omega_1} = K||\omega_1[A^+]$.

Let us for the rest of this argument assume that $n > 0$. The case $n = 0$ is an easier variant of what is to come. We thus may and shall assume that A^+_{odd} is the join of A_{n-1} and a code of $K||\kappa$. Let $C \subset C_{n-1}$ be club and such that there is a sequence $(\mathcal{M}_\alpha \colon \alpha \in C)$ of mice such that for every $\alpha \in C$, $(A^+ \cap \alpha)_{\text{odd}}$ codes the join of $A_{n-1} \cap \alpha$ and a code of $\mathcal{M}_\alpha$. Let κ_α denote the height of $\mathcal{M}_\alpha$ for $\alpha \in C$. The following will be crucial.

Claim 1. Let $\pi \colon \bar{K}[A^+ \cap \alpha] \cong X \prec (K||\omega_2[A^+]; \in, A^+)$, where X is countable and $\alpha = X \cap \omega_1 \in C$. Then $\mathcal{M}_\alpha \lhd \bar{K}$, $\kappa_\alpha = \pi^{-1}(\kappa)$ is the largest cardinal of $\bar{K}$, and $\bar{K} \leq^* \mathcal{N}_{n-1,\alpha}$.

Proof of Claim 1. Everything except for $\bar{K} \leq^* \mathcal{N}_{n-1,\alpha}$ easily follows by the elementarity of π. Let us write $\mathcal{N} = \mathcal{N}_{n-1,\alpha}$. Suppose Claim 1 to be false; i.e., $\mathcal{N} <^* \bar{K}$. We shall argue that $\bar{K}[A^+ \cap \alpha] \models$ "α is countable," which is of course nonsense.

Because $\bar{K}$ does not have any active extenders with indices between $\pi^{-1}(\kappa)$ and its height, it is easy to see that there must be now some $\xi < \bar{K} \cap \mathrm{OR}$ such that $\mathcal{N} <^* \bar{K}||\xi$. Let $\mathcal{T}$, $\mathcal{U}$ denote the coiteration of $\mathcal{N}$, $\bar{K}||\xi$. We know that $[0, \infty)_{\mathcal{T}}$ contains no drops, and that $\mathcal{M}^{\mathcal{T}}_\infty \trianglelefteq \mathcal{M}^{\mathcal{U}}_\infty$.

Let $\rho_\omega(\mathcal{N}) = \eta$. Because $(A_{n-1} \cap \alpha)_{\text{odd}}$ codes $\mathcal{N}||\eta$, $\mathcal{N}||\eta \in L[A_{n-1} \cap \alpha] \subset L[A^+ \cap \alpha]$.

Let $\bar{\mathcal{T}}$, $\bar{\mathcal{U}}$ denote the coiteration of $\mathcal{N}||\eta$, $\bar{K}||\xi$. Clearly, $\bar{\mathcal{T}}$ is an "initial segment" of $\mathcal{T}$, and $\bar{\mathcal{U}}$ is an "initial segment" of $\mathcal{U}$. Moreover, $\mathcal{N}||\eta <^* \bar{K}||\xi$, so that $[0, \infty)_{\bar{\mathcal{T}}}$ contains no drops and $\mathcal{M}^{\bar{\mathcal{T}}}_\infty \trianglelefteq \mathcal{M}^{\bar{\mathcal{U}}}_\infty$. If we had $\mathrm{lh}(\mathcal{T}) > \mathrm{lh}(\bar{\mathcal{T}})$ then, as η is not overlapped in $\mathcal{N}$ and $\rho_\omega(\mathcal{N}) = \eta$, $\mathcal{M}^{\mathcal{T}}_\infty$ would be non-sound, although $\mathcal{N} <^* \bar{K}$. Therefore, $\mathrm{lh}(\mathcal{T}) = \mathrm{lh}(\bar{\mathcal{T}})$. But then we must also have that $\mathrm{lh}(\mathcal{U}) = \mathrm{lh}(\bar{\mathcal{U}})$, as η is the largest cardinal of $\mathcal{N}$.

Because $\mathcal{N}||\eta \in L[A^+ \cap \alpha]$ and $\bar{K}||\xi \in L[\bar{K}||\xi]$, we know that $\pi^{\bar{\mathcal{T}}}_{0\infty}$ as well as $\mathcal{M}^{\bar{\mathcal{U}}}_\infty = \mathcal{M}^{\mathcal{U}}_\infty$ are elements of $L[A^+ \cap \alpha, \bar{K}||\xi]$.

As $\mathcal{M}^{\mathcal{T}}_\infty \trianglelefteq \mathcal{M}^{\mathcal{U}}_\infty$, we thus have that $\mathcal{M}^{\mathcal{T}}_\infty \in L[A^+ \cap \alpha, \bar{K}||\xi]$. But

$$\mathcal{N} \cong h^{\mathcal{M}^{\mathcal{T}}_\infty}(\mathrm{ran}(\pi^{\bar{\mathcal{T}}}_{0\infty}) \cup \{\pi^{\mathcal{T}}_{0\infty}(p_{\mathcal{N}})\}),$$

where $p_{\mathcal{N}}$ is the standard parameter of $\mathcal{N}$ and $h^{\mathcal{M}^{\mathcal{T}}_\infty}$ is an appropriate Skolem hull operator. This yields that in fact $\mathcal{N} \in L[A^+ \cap \alpha, \bar{K}||\xi]$.

Now let $a \in {}^\omega\omega \cap \mathcal{N}[A_n \cap \alpha]$ code a well-order of ω of order-type α. We have shown that $a \in L[A^+ \cap \alpha, \bar{K}||\xi]$. By [6], $V^{\mathbb{P}}$ is closed under the # operator. We therefore have that $(A^+ \cap \alpha, \bar{K}||\xi)^\#$ exists and $a \in (A^+ \cap \alpha, \bar{K}||\xi)^\#$. But $A^+ \cap \alpha$ and $\bar{K}||\xi$ are both elements of $\bar{K}[A^+ \cap \alpha]$. Due to π, $\bar{K}[A^+ \cap \alpha]$ is closed under the # operator, so that in fact $a \in \bar{K}[A^+ \cap \alpha]$. But then α is countable in $\bar{K}[A^+ \cap \alpha]$, a contradiction. □ (Claim 1)

In $V^{\mathbb{P}}$, we shall now consider the following forcing notion, denoted by $\mathbb{Q}$. We let $(c,p) \in \mathbb{Q}$ if and only if c is a countable closed subset of C, $p\colon \max(c) \to 2$, and for every $\alpha \in c$ there is some sound mouse $\mathcal{N} \trianglerighteq \mathcal{M}_\alpha$ such that κ_α is the largest cardinal of $\mathcal{N}$, κ_α is not overlapped in $\mathcal{N}$, $\rho_\omega(\mathcal{N}) = \kappa_\alpha$, $\mathcal{N}[A^+ \cap \alpha, p \restriction \alpha] \models$ "α is countable," and $\mathcal{N} <^* \mathcal{N}_{n-1,\alpha}$. A condition (c',q) is stronger than (c,p) iff $\max(c') \geq \max(c)$, $c' \cap (\max(c)+1) = c$, and $q \restriction \max(c) = p$.

Claim 2. ("Extendability Lemma") If $(c,p) \in \mathbb{Q}$ and $\alpha < \omega_1$ then there is some $(c',q) \leq (c,p)$ such that $\max(c') \geq \alpha$.

Proof of Claim 2. Given (c,p) and α, let us assume w.l.o.g. that $\alpha \geq \max(c) + \omega$, $\alpha \in C$, and $\alpha = X \cap \omega_1$ for some $X \prec K||\omega_2[A^+]$, so that $X \cong \bar{K}$ for some $\bar{K} \triangleright \mathcal{M}_\alpha$ and $\bar{K} \leq^* \mathcal{N}_{n-1,\alpha}$ by Claim 1. Pick a code $x \in {}^\omega\omega$ for the ordinal α, and let $\mathrm{dom}(q) = \alpha$, where $q \restriction \mathrm{dom}(p) = p$, $q(\max(c)+n) = x(n)$ for all $n < \omega$, and $q(\gamma) = 0$ for all $\gamma \geq \mathrm{dom}(c) + \omega$. There will be some $\mathcal{P}$ with $\bar{K}||\pi^{-1}(\kappa) \triangleleft \mathcal{P} \triangleleft \bar{K}$ such that $\rho_\omega(\mathcal{P}) = \pi^{-1}(\kappa)$, $\pi^{-1}(\kappa)$ is the largest cardinal of $\mathcal{P}$, and $\mathcal{P}[A^+ \cap \alpha, q] \models$ "α is countable." Therefore, $(c \cup \{\alpha\}, q)$ is as desired. □ (Claim 2)

In order to finish the proof of Lemma 2.1 (and hence of Theorem 1.1) it now obviously suffices to verify the following.

Claim 3. $\mathbb{Q}$ is stationary set preserving.

Proof of Claim 3. Let $S \subset \omega_1$ be stationary, and let $(c,p) \Vdash$ "$\dot{C}$ is a club subset of $\check{\omega}_1$." We aim to construct some $(c',q) \leq (c,p)$ such that $(c',q) \Vdash$ "$\dot{C} \cap \check{S} \neq \emptyset$."

Let $X \prec (H_{\omega_2}; \in, A^+, \mathbb{Q}, (c,p))$ be transitive and of size $\aleph_1$, and let $(X_i\colon i \leq \omega_1)$ be a continuous chain of countable elementary submodels of X approaching X (i.e., $X = X_{\omega_1}$). Recall that $H_{\omega_2} = K||\omega_2[A^+]$ (in $V^{\mathbb{P}}$). Let

$$\pi\colon \bar{K}[A^+ \cap \alpha] \cong Y \prec (H_{\omega_2}; \in, A^+, \mathbb{Q}, (c,p), (X_i\colon i \leq \omega_1)),$$

where Y is countable and $\alpha = Y \cap \omega_1 = \omega_1^{\bar{H}[A^+ \cap \alpha]} < \omega_1$. We also may and shall assume that $\alpha \in S$. Let us write $\alpha_i = X_i \cap \omega_1 < \omega_1$ for $i < \omega_1$. Of course $(\alpha_i\colon i < \alpha) \in \bar{K}[A^+ \cap \alpha]$, where $(\alpha_i\colon i < \alpha)$ is cofinal in α. Let us pick (externally, i.e., in $V^{\mathbb{P}}$) a sequence $(i_n\colon n < \omega)$ which is cofinal in α; hence $(\alpha_{i_n}\colon n < \omega)$ is cofinal in α as well. As $H_{\omega_1} = K||\omega_1[A^+]$, $(H_{\omega_1})^{\bar{K}[A^+ \cap \alpha]} \subset \bigcup_{i<\alpha} X_i$, so that we may assume that in fact $(c,p) \in X_{i_0}$.

We shall now recursively construct sequences $((c_n,p_n)\colon n<\omega)$ and $((c'_n,p'_n)\colon n<\omega)$ of conditions in $\mathbb{Q}$ with the following properties.

(1) $(c_0,p_0) = (c,p)$,
(2) $(c'_n,p'_n) \leq (c_n,p_n)$, and $(c'_n,p'_n) \in X_{i_{n+1}}$,
(3) for all $\xi \in (\mathrm{dom}(p'_n) \setminus \mathrm{dom}(p_n)) \cap \{\alpha_i\colon i < \alpha\}$, $p'_n(\xi) = 1$ if and only if $\xi = \alpha_{i_n}$,
(4) $(c_{n+1},p_{n+1}) \leq (c'_n,p'_n)$, and $(c_{n+1},p_{n+1}) \in X_{i_{n+1}}$,
(5) there is some $\beta > \alpha_{i_n}$ such that $(c_{n+1},p_{n+1}) \Vdash$ "$\check{\beta} \in \dot{C}$, and
(6) $\max(c_{n+1}) > \alpha_{i_n}$.

In the light of (the proof of) Claim 2, there is no problem with this recursion.

Let us now set $c^* = \bigcup_{n<\omega} c_n \cup \{\alpha\}$ and $p^* = \bigcup_{n<\omega} p_n$. We're done if we can show that (c^*, p^*) is a condition, because then $(c^*, p^*) \Vdash$ "$\check{\alpha} \in \check{S} \cap \dot{C}$."

Well, we have that $\bar{K}[A^+ \cap \alpha, p^*] \models$ "α is countable," because for $\xi \in \{\alpha_i : i_0 \leq i < \alpha\}$, $p^*(\xi) = 1$ if and only if $\xi = \alpha_{i_n}$ for some $n < \omega$, so that $(\alpha_{i_n} : n < \omega) \in \bar{K}[A^+ \cap \alpha, p^*]$.

By Claim 1, $\bar{K} \rhd \mathcal{M}_\alpha = \bar{K}||\pi^{-1}(\kappa)$. Moreover, there will again certainly be some $\mathcal{P}$ with $\bar{K}||\pi^{-1}(\kappa) \lhd \mathcal{P} \lhd \bar{K}$ such that $\rho_\omega(\mathcal{P}) = \pi^{-1}(\kappa)$, $\pi^{-1}(\kappa)$ is the largest cardinal of $\mathcal{P}$, and $\mathcal{P}[A^+ \cap \alpha, p^*] \models$ "α is countable." By Claim 1, $\mathcal{P} <^* \mathcal{N}_{n-1,\alpha}$, so that (c^*, p^*) is really a condition. □ (Claim 3)

□ (Lemma 2.1, Theorem 1.1)

3. Some problems

Let $(f_\alpha : \alpha < \omega_2)$ denote "the" sequence of canonical functions from ω_1 to ω_1. I.e., $f_\alpha(\nu) = \text{otp}\ g_\alpha{}''\nu$, where $g_\alpha : \omega_1 \to \alpha$ is bijective. The Club Bounding Principle, CBP for short, says that for every $f : \omega_1 \to \omega_1$ there is some $\alpha < \omega_2$ such that $f < f_\alpha$ on a club, i.e., such that $\{\nu : f(\nu) < f_\alpha(\nu)\}$ contains a club subset of ω_1.

P. Larson has shown that CBP implies $P_2(\omega_1)^+$ (cf. [1, Fact 3.4]; cf. [1, Definition 3.1 (b)] on the definition of $P_2(\omega_1)^+$). On the other hand one has:

Lemma 3.1. *If* BMM *and* $P_2(\omega_1)^+$ *hold then* CBP *holds.*

Proof sketch. Let $f : \omega_1 \to \omega_1$. If $P_2(\omega_1)^+$ holds then the natural forcing $\mathbb{P}_f$ for adding a canonical function above f is easily seen to be stationary set preserving. $\mathbb{P}_f$ is just the collection of all (c, p) such that c is a closed countable subset of ω_1 and $p : \max(c) \to \omega_2$ is such that for all $\nu \in c$, $f(\nu) < \text{otp}\ p''\nu$. □ (Lemma 3.1)

We do not know if BMM alone implies the Club Bounding Principle. In order to answer this question we need a better understanding of how to construct models of BMM. The same remark applies to the following questions.

Woodin introduced the principle ψ_{AC} and showed that ψ_{AC} implies that $2^{\aleph_0} = \aleph_2$. (Cf. [9, Definition 5.12 and Lemma 5.15].) We do not know if BMM implies that ψ_{AC} holds (we know that BMM implies that $2^{\aleph_0} = \aleph_2$, though, cf. [8]). But we have:

Lemma 3.2. *If* BMM *and* $P_1(\omega_1)^+$ *hold then* ψ_{AC} *holds.*

Proof sketch. Cf. [1, Definition 3.1 (d)] on the definition of $P_1(\omega_1)^+$. Let $S, T \subset \omega_1$ both be stationary and costationary. By $P_1(\omega_1)^+$, the canonical forcing $\mathbb{P}_{S,T}$ for adding a witness to ψ_{AC} is easily seen to be stationary set preserving. $\mathbb{P}_{S,T}$ consists of all (c, p) such that c is a closed countable subset of ω_1, $p : \max(c) \to \omega_2$, and for all $\nu \in c$, $\text{otp}\ p''\nu \in T \longleftrightarrow \nu \in S$. □ (Lemma 3.2)

It is not known if ψ_{AC} implies $P_1(\omega_1)^+$. By [1, Fact 3.1], ψ_{AC} implies CBP.

Another interesting question is whether BMM implies that $u_2 = \omega_2$. This question makes sense by [6]. If BMM holds then every real has a #, and hence we may ask if u_2, the second uniform indiscernible, is equal to ω_2.

References

[1] Aspero, D., and Welch, Ph., *Bounded Martin's Maximum, weak Erdös cardinals, and* ψ_{AC}, Journal Symb. Logic **67** (2002), pp. 1141–1152.

[2] Goldstern, M., and Shelah, S., *The bounded proper forcing axiom*, Journal Symb. Logic **60** (1995), pp. 58–73.

[3] Jensen, R., *The core model for non-overlapping extender sequences*, handwritten notes, Oxford.

[4] Moore, J., *Set mapping reflection*, preprint.

[5] Schindler, R., *Coding into K by reasonable forcing*, Trans. Amer. Math. Soc. **353** (2000), pp. 479–489.

[6] Schindler, R., *Semi-proper forcing, remarkable cardinals, and Bounded Martin's Maximum*, Mathematical Logic Quarterly, submitted.

[7] Steel, J., and Welch, Ph., Σ^1_3 *absoluteness and the second uniform indiscernible*, Israel Journal of Mathematics **104** (1998), pp. 157–190.

[8] Todorcevic, S., *Generic absoluteness and the continuum*, Mathematical Research Letters **9** (2002), pp. 1–7.

[9] Woodin, H., *The axiom of determinacy, forcing axioms, and the nonstationary ideal*, de Gruyter 1999.

[10] Woodin, H., private communication, Aug. 03.

Ralf Schindler
Institut für Mathematische Logik und Grundlagenforschung
Westfälische Wilhelms-Universität Münster
Einsteinstr. 62
D-48149 Münster, Germany
e-mail: `rds@math.uni-muenster.de`
URL: `http://wwwmath.uni-muenster.de/math/inst/logik/org/staff/rds/`

GPSR Compliance

The European Union's (EU) General Product Safety Regulation (GPSR) is a set of rules that requires consumer products to be safe and our obligations to ensure this.

If you have any concerns about our products, you can contact us on ProductSafety@springernature.com

In case Publisher is established outside the EU, the EU authorized representative is:

Springer Nature Customer Service Center GmbH
Europaplatz 3
69115 Heidelberg, Germany

Batch number: 10732470

Printed by Printforce, the Netherlands